T0325825

Mathematics for Physics

A Guided Tour for Graduate Students

An engagingly written account of mathematical tools and ideas, this book provides a graduate-level introduction to the mathematics used in research in physics.

The first half of the book focuses on the traditional mathematical methods of physics: differential and integral equations, Fourier series and the calculus of variations. The second half contains an introduction to more advanced subjects, including differential geometry, topology and complex variables.

The authors' exposition avoids excess rigour whilst explaining subtle but important points often glossed over in more elementary texts. The topics are illustrated at every stage by carefully chosen examples, exercises and problems drawn from realistic physics settings. These make it useful both as a textbook in advanced courses and for self-study. Password-protected solutions to the exercises are available to instructors at www.cambridge.org/9780521854030.

MICHAEL STONE is a Professor in the Department of Physics at the University of Illinois at Urbana-Champaign. He has worked on quantum field theory, superconductivity, the quantum Hall effect and quantum computing.

PAUL GOLDBART is a Professor in the Department of Physics at the University of Illinois at Urbana-Champaign, where he directs the Institute for Condensed Matter Theory. His research ranges widely over the field of condensed matter physics, including soft matter, disordered systems, nanoscience and superconductivity.

MATHEMATICS FOR PHYSICS

A Guided Tour for Graduate Students

MICHAEL STONE
University of Illinois at Urbana-Champaign

and

PAUL GOLDBART
University of Illinois at Urbana-Champaign

CAMBRIDGE
UNIVERSITY PRESS

University Printing House, Cambridge CB2 8BS, United Kingdom

One Liberty Plaza, 20th Floor, New York, NY 10006, USA

477 Williamstown Road, Port Melbourne, VIC 3207, Australia

314-321, 3rd Floor, Plot 3, Splendor Forum, Jasola District Centre, New Delhi-110025, India

79 Anson Road, #06-04/06, Singapore 079906

Cambridge University Press is part of the University of Cambridge.

It furthers the University's mission by disseminating knowledge in the pursuit of education, learning and research at the highest international levels of excellence.

www.cambridge.org
Information on this title: www.cambridge.org/9780521854030

First published 2009
Reprinted 2010

A catalogue record for this publication is available from the British Library

ISBN 978-0-521-85403-0 Hardback

To the memory of Mike's mother, Aileen Stone: $9 \times 9 = 81$.

To Paul's mother and father, Carole and Colin Goldbart.

Contents

Preface

This book is based on a two-semester sequence of courses taught to incoming graduate students at the University of Illinois at Urbana-Champaign, primarily physics students but also some from other branches of the physical sciences. The courses aim to introduce students to some of the mathematical methods and concepts that they will find useful in their research. We have sought to enliven the material by integrating the mathematics with its applications. We therefore provide illustrative examples and problems drawn from physics. Some of these illustrations are classical but many are small parts of contemporary research papers. In the text and at the end of each chapter we provide a collection of exercises and problems suitable for homework assignments. The former are straightforward applications of material presented in the text; the latter are intended to be interesting, and take rather more thought and time.

We devote the first, and longest, part (Chapters 1–9, and the first semester in the classroom) to traditional mathematical methods. We explore the analogy between linear operators acting on function spaces and matrices acting on finite-dimensional spaces, and use the operator language to provide a unified framework for working with ordinary differential equations, partial differential equations and integral equations. The mathematical prerequisites are a sound grasp of undergraduate calculus (including the vector calculus needed for electricity and magnetism courses), elementary linear algebra and competence at complex arithmetic. Fourier sums and integrals, as well as basic ordinary differential equation theory, receive a quick review, but it would help if the reader had some prior experience to build on. Contour integration is not required for this part of the book.

The second part (Chapters 10–14) focuses on modern differential geometry and topology, with an eye to its application to physics. The tools of calculus on manifolds, especially the exterior calculus, are introduced, and used to investigate classical mechanics, electromagnetism and non-abelian gauge fields. The language of homology and cohomology is introduced and is used to investigate the influence of the global topology of a manifold on the fields that live in it and on the solutions of differential equations that constrain these fields.

Chapters 15 and 16 introduce the theory of group representations and their applications to quantum mechanics. Both finite groups and Lie groups are explored.

The last part (Chapters 17–19) explores the theory of complex variables and its applications. Although much of the material is standard, we make use of the exterior calculus, and discuss rather more of the topological aspects of analytic functions than is customary.

A cursory reading of the Contents of the book will show that there is more material here than can be comfortably covered in two semesters. When using the book as the basis for lectures in the classroom, we have found it useful to tailor the presented material to the interests of our students.

Acknowledgments

A great many people have encouraged us along the way:

- Our teachers at the University of Cambridge, the University of California-Los Angeles, and Imperial College London.
- Our students – your questions and enthusiasm have helped shape our understanding and our exposition.
- Our colleagues – faculty and staff – at the University of Illinois at Urbana-Champaign – how fortunate we are to have a community so rich in both accomplishment and collegiality.
- Our friends and family: Kyre and Steve and Ginna; and Jenny, Ollie and Greta – we hope to be more attentive now that this book is done.
- Our editor Simon Capelin at Cambridge University Press – your patience is appreciated.
- The staff of the US National Science Foundation and the US Department of Energy, who have supported our research over the years.

Our sincere thanks to you all.

1
Calculus of variations

We begin our tour of useful mathematics with what is called the *calculus of variations*. Many physics problems can be formulated in the language of this calculus, and once they are there are useful tools to hand. In the text and associated exercises we will meet some of the equations whose solution will occupy us for much of our journey.

1.1 What is it good for?

The classical problems that motivated the creators of the calculus of variations include:

(i) *Dido's problem*: In Virgil's *Aeneid*, Queen Dido of Carthage must find the largest area that can be enclosed by a curve (a strip of bull's hide) of fixed length.
(ii) *Plateau's problem*: Find the surface of minimum area for a given set of bounding curves. A soap film on a wire frame will adopt this minimal-area configuration.
(iii) *Johann Bernoulli's brachistochrone*: A bead slides down a curve with fixed ends. Assuming that the total energy $\frac{1}{2}mv^2 + V(x)$ is constant, find the curve that gives the most rapid descent.
(iv) *Catenary*: Find the form of a hanging heavy chain of fixed length by minimizing its potential energy.

These problems all involve finding maxima or minima, and hence equating some sort of derivative to zero. In the next section we define this derivative, and show how to compute it.

1.2 Functionals

In variational problems we are provided with an expression $J[y]$ that "eats" whole functions $y(x)$ and returns a single number. Such objects are called *functionals* to distinguish them from ordinary functions. An ordinary function is a map $f : \mathbb{R} \to \mathbb{R}$. A functional J is a map $J : C^\infty(\mathbb{R}) \to \mathbb{R}$ where $C^\infty(\mathbb{R})$ is the space of smooth (having derivatives of all orders) functions. To find the function $y(x)$ that maximizes or minimizes a given functional $J[y]$ we need to define, and evaluate, its *functional derivative*.

1.2.1 The functional derivative

We restrict ourselves to expressions of the form

$$J[y] = \int_{x_1}^{x_2} f(x, y, y', y'', \cdots y^{(n)})\, dx, \tag{1.1}$$

where f depends on the value of $y(x)$ and only finitely many of its derivatives. Such functionals are said to be *local* in x.

Consider first a functional $J = \int f dx$ in which f depends only x, y and y'. Make a change $y(x) \to y(x) + \varepsilon\eta(x)$, where ε is a (small) x-independent constant. The resultant change in J is

$$\begin{aligned}
J[y + \varepsilon\eta] - J[y] &= \int_{x_1}^{x_2} \{f(x, y + \varepsilon\eta, y' + \varepsilon\eta') - f(x, y, y')\}\, dx \\
&= \int_{x_1}^{x_2} \left\{\varepsilon\eta\frac{\partial f}{\partial y} + \varepsilon\frac{d\eta}{dx}\frac{\partial f}{\partial y'} + O(\varepsilon^2)\right\} dx \\
&= \left[\varepsilon\eta\frac{\partial f}{\partial y'}\right]_{x_1}^{x_2} + \int_{x_1}^{x_2} (\varepsilon\eta(x)) \left\{\frac{\partial f}{\partial y} - \frac{d}{dx}\left(\frac{\partial f}{\partial y'}\right)\right\} dx \\
&\quad + O(\varepsilon^2).
\end{aligned}$$

If $\eta(x_1) = \eta(x_2) = 0$, the variation $\delta y(x) \equiv \varepsilon\eta(x)$ in $y(x)$ is said to have "fixed endpoints". For such variations the integrated-out part $[\ldots]_{x_1}^{x_2}$ vanishes. Defining δJ to be the $O(\varepsilon)$ part of $J[y + \varepsilon\eta] - J[y]$, we have

$$\begin{aligned}
\delta J &= \int_{x_1}^{x_2} (\varepsilon\eta(x)) \left\{\frac{\partial f}{\partial y} - \frac{d}{dx}\left(\frac{\partial f}{\partial y'}\right)\right\} dx \\
&= \int_{x_1}^{x_2} \delta y(x) \left(\frac{\delta J}{\delta y(x)}\right) dx.
\end{aligned} \tag{1.2}$$

The function

$$\frac{\delta J}{\delta y(x)} \equiv \frac{\partial f}{\partial y} - \frac{d}{dx}\left(\frac{\partial f}{\partial y'}\right) \tag{1.3}$$

is called the *functional* (or *Fréchet*) derivative of J with respect to $y(x)$. We can think of it as a generalization of the partial derivative $\partial J/\partial y_i$, where the discrete subscript "i" on y is replaced by a continuous label "x", and sums over i are replaced by integrals over x:

$$\delta J = \sum_i \frac{\partial J}{\partial y_i}\delta y_i \to \int_{x_1}^{x_2} dx \left(\frac{\delta J}{\delta y(x)}\right) \delta y(x). \tag{1.4}$$

1.2.2 The Euler–Lagrange equation

Suppose that we have a differentiable function $J(y_1,y_2,\ldots,y_n)$ of n variables and seek its *stationary points* – these being the locations at which J has its maxima, minima and saddle points. At a stationary point $(y_1,y_2,\ldots,y_n)$ the variation

$$\delta J = \sum_{i=1}^{n} \frac{\partial J}{\partial y_i}\delta y_i \tag{1.5}$$

must be zero for all possible δy_i. The necessary and sufficient condition for this is that all partial derivatives $\partial J/\partial y_i$, $i = 1,\ldots,n$ be zero. By analogy, we expect that a functional $J[y]$ will be stationary under fixed-endpoint variations $y(x) \to y(x) + \delta y(x)$, when the functional derivative $\delta J/\delta y(x)$ vanishes for all x. In other words, when

$$\boxed{\frac{\partial f}{\partial y(x)} - \frac{d}{dx}\left(\frac{\partial f}{\partial y'(x)}\right) = 0, \quad x_1 < x < x_2.} \tag{1.6}$$

The condition (1.6) for $y(x)$ to be a stationary point is usually called the *Euler–Lagrange* equation.

That $\delta J/\delta y(x) \equiv 0$ is a *sufficient* condition for δJ to be zero is clear from its definition in (1.2). To see that it is a *necessary* condition we must appeal to the assumed smoothness of $y(x)$. Consider a function $y(x)$ at which $J[y]$ is stationary but where $\delta J/\delta y(x)$ is non-zero at some $x_0 \in [x_1,x_2]$. Because $f(y,y',x)$ is smooth, the functional derivative $\delta J/\delta y(x)$ is also a smooth function of x. Therefore, by continuity, it will have the same sign throughout some open interval containing x_0. By taking $\delta y(x) = \varepsilon\eta(x)$ to be zero outside this interval, and of one sign within it, we obtain a non-zero δJ – in contradiction to stationarity. In making this argument, we see why it was essential to integrate by parts so as to take the derivative off δy: when y is fixed at the endpoints, we have $\int \delta y'\, dx = 0$, and so we cannot find a $\delta y'$ that is zero everywhere outside an interval and of one sign within it.

When the functional depends on more than one function y, then stationarity under all possible variations requires one equation

$$\frac{\delta J}{\delta y_i(x)} = \frac{\partial f}{\partial y_i} - \frac{d}{dx}\left(\frac{\partial f}{\partial y_i'}\right) = 0 \tag{1.7}$$

for each function $y_i(x)$.

If the function f depends on higher derivatives, y'', $y^{(3)}$, etc., then we have to integrate by parts more times, and we end up with

$$0 = \frac{\delta J}{\delta y(x)} = \frac{\partial f}{\partial y} - \frac{d}{dx}\left(\frac{\partial f}{\partial y'}\right) + \frac{d^2}{dx^2}\left(\frac{\partial f}{\partial y''}\right) - \frac{d^3}{dx^3}\left(\frac{\partial f}{\partial y^{(3)}}\right) + \cdots. \tag{1.8}$$

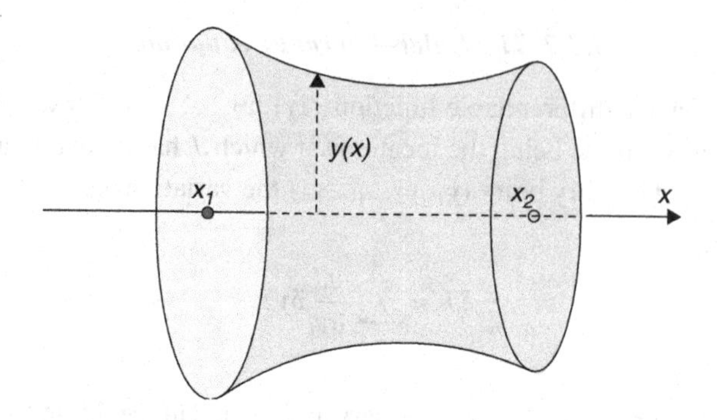

Figure 1.1 Soap film between two rings.

1.2.3 Some applications

Now we use our new functional derivative to address some of the classic problems mentioned in the introduction.

Example: *Soap film supported by a pair of coaxial rings* (Figure 1.1). This is a simple case of Plateau's problem. The free energy of the soap film is equal to twice (once for each liquid–air interface) the surface tension σ of the soap solution times the area of the film. The film can therefore minimize its free energy by minimizing its area, and the axial symmetry suggests that the minimal surface will be a surface of revolution about the x-axis. We therefore seek the profile $y(x)$ that makes the area

$$J[y] = 2\pi \int_{x_1}^{x_2} y\sqrt{1+y'^2}\,dx \tag{1.9}$$

of the surface of revolution the least among all such surfaces bounded by the circles of radii $y(x_1) = y_1$ and $y(x_2) = y_2$. Because a minimum is a stationary point, we seek candidates for the minimizing profile $y(x)$ by setting the functional derivative $\delta J/\delta y(x)$ to zero.

We begin by forming the partial derivatives

$$\frac{\partial f}{\partial y} = 4\pi\sqrt{1+y'^2}, \quad \frac{\partial f}{\partial y'} = \frac{4\pi y y'}{\sqrt{1+y'^2}} \tag{1.10}$$

and use them to write down the Euler–Lagrange equation

$$\sqrt{1+y'^2} - \frac{d}{dx}\left(\frac{yy'}{\sqrt{1+y'^2}}\right) = 0. \tag{1.11}$$

Performing the indicated derivative with respect to x gives

$$\sqrt{1+y'^2} - \frac{(y')^2}{\sqrt{1+y'^2}} - \frac{yy''}{\sqrt{1+y'^2}} + \frac{y(y')^2y''}{(1+y'^2)^{3/2}} = 0. \tag{1.12}$$

After collecting terms, this simplifies to

$$\frac{1}{\sqrt{1+y'^2}} - \frac{yy''}{(1+y'^2)^{3/2}} = 0. \tag{1.13}$$

The differential equation (1.13) still looks a trifle intimidating. To simplify further, we multiply by y' to get

$$\begin{aligned} 0 &= \frac{y'}{\sqrt{1+y'^2}} - \frac{yy'y''}{(1+y'^2)^{3/2}} \\ &= \frac{d}{dx}\left(\frac{y}{\sqrt{1+y'^2}}\right). \end{aligned} \tag{1.14}$$

The solution to the minimization problem therefore reduces to solving

$$\frac{y}{\sqrt{1+y'^2}} = \kappa, \tag{1.15}$$

where κ is an as yet undetermined integration constant. Fortunately this nonlinear, first-order, differential equation is elementary. We recast it as

$$\frac{dy}{dx} = \sqrt{\frac{y^2}{\kappa^2} - 1} \tag{1.16}$$

and separate variables

$$\int dx = \int \frac{dy}{\sqrt{\frac{y^2}{\kappa^2} - 1}}. \tag{1.17}$$

We now make the natural substitution $y = \kappa \cosh t$, whence

$$\int dx = \kappa \int dt. \tag{1.18}$$

Thus we find that $x + a = \kappa t$, leading to

$$y = \kappa \cosh \frac{x+a}{\kappa}. \tag{1.19}$$

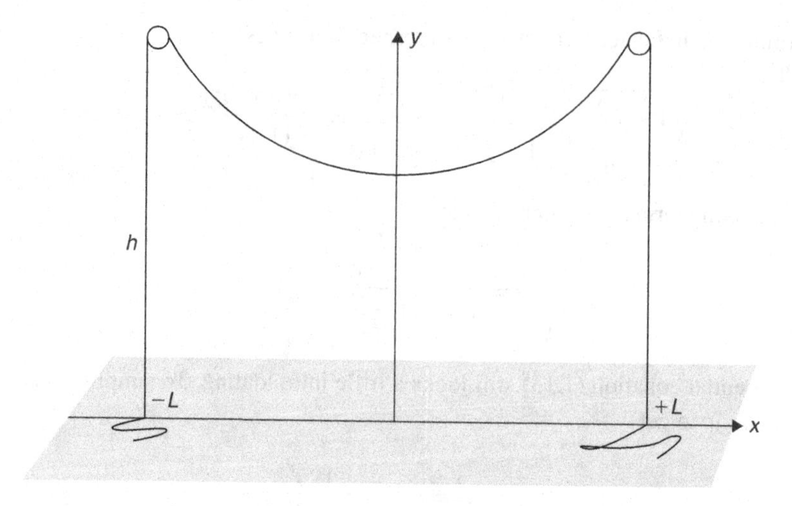

Figure 1.2 Hanging chain.

We select the constants κ and a to fit the endpoints $y(x_1) = y_1$ and $y(x_2) = y_2$.

Example: *Heavy chain over pulleys*. We cannot yet consider the form of the catenary, a hanging chain of fixed length, but we can solve a simpler problem of a heavy flexible cable draped over a pair of pulleys located at $x = \pm L$, $y = h$, and with the excess cable resting on a horizontal surface as illustrated in Figure 1.2.

The potential energy of the system is

$$P.E. = \sum mgy = \rho g \int_{-L}^{L} y\sqrt{1 + (y')^2}dx + \text{const.} \tag{1.20}$$

Here the constant refers to the unchanging potential energy

$$2 \times \int_0^h mgy\, dy = mgh^2 \tag{1.21}$$

of the vertically hanging cable. The potential energy of the cable lying on the horizontal surface is zero because y is zero there. Notice that the tension in the suspended cable is being tacitly determined by the weight of the vertical segments.

The Euler–Lagrange equations coincide with those of the soap film, so

$$y = \kappa \cosh \frac{(x + a)}{\kappa} \tag{1.22}$$

where we have to find κ and a. We have

$$\begin{aligned} h &= \kappa \cosh(-L + a)/\kappa, \\ &= \kappa \cosh(L + a)/\kappa, \end{aligned} \tag{1.23}$$

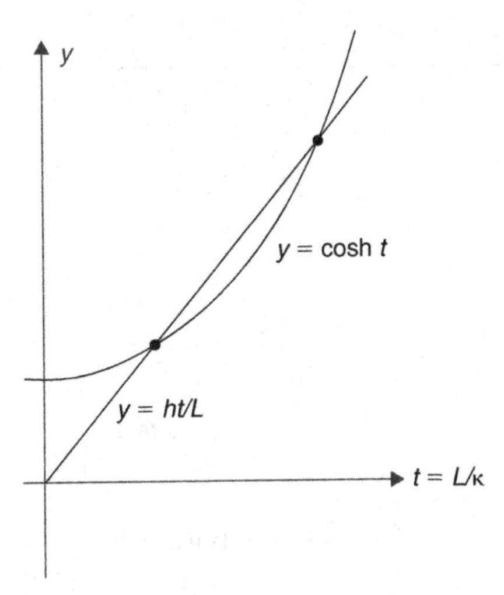

Figure 1.3 Intersection of $y = ht/L$ with $y = \cosh t$.

so $a = 0$ and $h = \kappa \cosh L/\kappa$. Setting $t = L/\kappa$ this reduces to

$$\left(\frac{h}{L}\right) t = \cosh t. \tag{1.24}$$

By considering the intersection of the line $y = ht/L$ with $y = \cosh t$ (Figure 1.3) we see that if h/L is too small there is no solution (the weight of the suspended cable is too big for the tension supplied by the dangling ends) and once h/L is large enough there will be two possible solutions. Further investigation will show that the solution with the larger value of κ is a point of stable equilibrium, while the solution with the smaller κ is unstable.

Example: *The brachistochrone*. This problem was posed as a challenge by Johann Bernoulli in 1696. He asked what shape should a wire with endpoints $(0, 0)$ and (a, b) take in order that a frictionless bead will slide from rest down the wire in the shortest possible time (Figure 1.4). The problem's name comes from Greek: βραχιστος means shortest and χρονος means time.

When presented with an ostensibly anonymous solution, Johann made his famous remark: "*Tanquam ex unguem leonem*" (I recognize the lion by his clawmark) meaning that he recognized that the author was Isaac Newton.

Johann gave a solution himself, but that of his brother Jacob Bernoulli was superior and Johann tried to pass it off as his. This was not atypical. Johann later misrepresented the publication date of his book on hydraulics to make it seem that he had priority in this field over his own son, Daniel Bernoulli.

We begin *our* solution of the problem by observing that the total energy

$$E = \frac{1}{2}m(\dot{x}^2 + \dot{y}^2) - mgy = \frac{1}{2}m\dot{x}^2(1 + y'^2) - mgy, \tag{1.25}$$

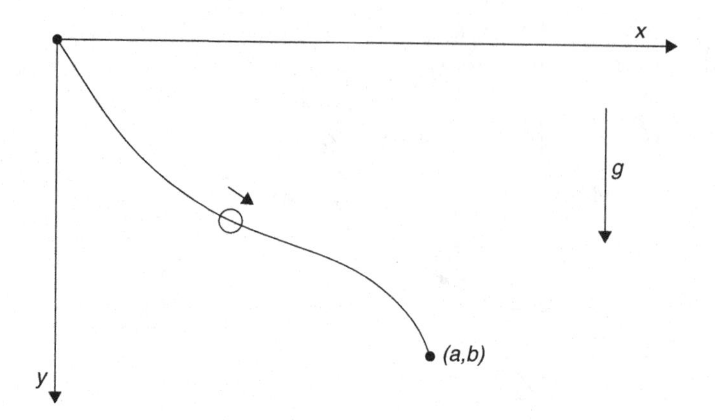

Figure 1.4 Bead on a wire.

of the bead is constant. From the initial condition we see that this constant is zero. We therefore wish to minimize

$$T = \int_0^T dt = \int_0^a \frac{1}{\dot{x}} dx = \int_0^a \sqrt{\frac{1+y'^2}{2gy}} dx \tag{1.26}$$

so as to find $y(x)$, given that $y(0) = 0$ and $y(a) = b$. The Euler–Lagrange equation is

$$yy'' + \frac{1}{2}(1+y'^2) = 0. \tag{1.27}$$

Again this looks intimidating, but we can use the same trick of multiplying through by y' to get

$$y'\left(yy'' + \frac{1}{2}(1+y'^2)\right) = \frac{1}{2}\frac{d}{dx}\left\{y(1+y'^2)\right\} = 0. \tag{1.28}$$

Thus

$$2c = y(1+y'^2). \tag{1.29}$$

This differential equation has a parametric solution

$$x = c(\theta - \sin\theta),$$
$$y = c(1 - \cos\theta), \tag{1.30}$$

(as you should verify) and the solution is the cycloid shown in Figure 1.5. The parameter c is determined by requiring that the curve does in fact pass through the point (a, b).

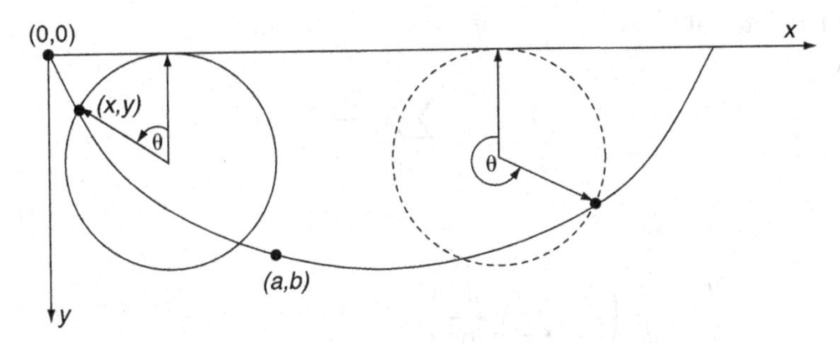

Figure 1.5 A wheel rolls on the x-axis. The dot, which is fixed to the rim of the wheel, traces out a cycloid.

1.2.4 First integral

How did we know that we could simplify both the soap-film problem and the brachistochrone by multiplying the Euler equation by y'? The answer is that there is a general principle, closely related to energy conservation in mechanics, that tells us when and how we can make such a simplification. The y' trick works when the f in $\int f\,dx$ is of the form $f(y, y')$, i.e. has no *explicit* dependence on x. In this case the last term in

$$\frac{df}{dx} = y'\frac{\partial f}{\partial y} + y''\frac{\partial f}{\partial y'} + \frac{\partial f}{\partial x} \tag{1.31}$$

is absent. We then have

$$\begin{aligned}\frac{d}{dx}\left(f - y'\frac{\partial f}{\partial y'}\right) &= y'\frac{\partial f}{\partial y} + y''\frac{\partial f}{\partial y'} - y''\frac{\partial f}{\partial y'} - y'\frac{d}{dx}\left(\frac{\partial f}{\partial y'}\right)\\ &= y'\left(\frac{\partial f}{\partial y} - \frac{d}{dx}\left(\frac{\partial f}{\partial y'}\right)\right),\end{aligned} \tag{1.32}$$

and this is zero if the Euler–Lagrange equation is satisfied.

The quantity

$$I = f - y'\frac{\partial f}{\partial y'} \tag{1.33}$$

is called a *first integral* of the Euler–Lagrange equation. In the soap-film case

$$f - y'\frac{\partial f}{\partial y'} = y\sqrt{1+(y')^2} - \frac{y(y')^2}{\sqrt{1+(y')^2}} = \frac{y}{\sqrt{1+(y')^2}}. \tag{1.34}$$

When there are a number of dependent variables y_i, so that we have

$$J[y_1, y_2, \dots y_n] = \int f(y_1, y_2, \dots y_n; y_1', y_2', \dots y_n')\,dx \tag{1.35}$$

then the first integral becomes

$$I = f - \sum_i y_i' \frac{\partial f}{\partial y_i'}. \tag{1.36}$$

Again

$$\begin{aligned}
\frac{dI}{dx} &= \frac{d}{dx}\left(f - \sum_i y_i' \frac{\partial f}{\partial y_i'}\right) \\
&= \sum_i \left(y_i' \frac{\partial f}{\partial y_i} + y_i'' \frac{\partial f}{\partial y_i'} - y_i'' \frac{\partial f}{\partial y_i'} - y_i' \frac{d}{dx}\left(\frac{\partial f}{\partial y_i'}\right)\right) \\
&= \sum_i y_i' \left(\frac{\partial f}{\partial y_i} - \frac{d}{dx}\left(\frac{\partial f}{\partial y_i'}\right)\right),
\end{aligned} \tag{1.37}$$

and this is zero if the Euler–Lagrange equation is satisfied for each y_i.

Note that there is only *one* first integral, no matter how many y_i's there are.

1.3 Lagrangian mechanics

In his *Mécanique Analytique* (1788) Joseph-Louis de La Grange, following Jean d'Alembert (1742) and Pierre de Maupertuis (1744), showed that most of classical mechanics can be recast as a variational condition: the *principle of least action*. The idea is to introduce the *Lagrangian* function $L = T - V$ where T is the kinetic energy of the system and V the potential energy, both expressed in terms of *generalized coordinates* q^i and their time derivatives $\dot{q}^i$. Then, Lagrange showed, the multitude of Newton's $\mathbf{F} = m\mathbf{a}$ equations, one for each particle in the system, can be reduced to

$$\frac{d}{dt}\left(\frac{\partial L}{\partial \dot{q}^i}\right) - \frac{\partial L}{\partial q^i} = 0, \tag{1.38}$$

one equation for each generalized coordinate q. Quite remarkably – given that Lagrange's derivation contains no mention of maxima or minima – we recognize that this is precisely the condition that the *action functional*

$$S[q] = \int_{t_{initial}}^{t_{final}} L(t, q^i; q'^i)\, dt \tag{1.39}$$

be stationary with respect to variations of the trajectory $q^i(t)$ that leave the initial and final points fixed. This fact so impressed its discoverers that they believed they had uncovered the unifying principle of the universe. Maupertuis, for one, tried to base a proof of the existence of God on it. Today the action integral, through its starring role in

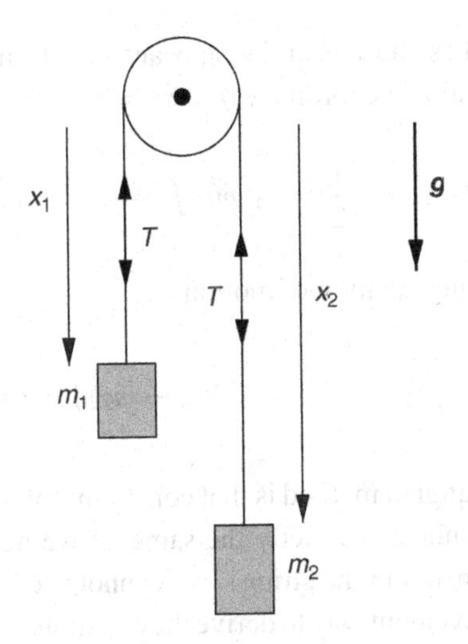

Figure 1.6 Atwood's machine.

the Feynman path-integral formulation of quantum mechanics, remains at the heart of theoretical physics.

1.3.1 One degree of freedom

We shall not attempt to derive Lagrange's equations from d'Alembert's extension of the principle of virtual work – leaving this task to a mechanics course – but instead satisfy ourselves with some examples which illustrate the computational advantages of Lagrange's approach, as well as a subtle pitfall.

Consider, for example, *Atwood's machine* (Figure 1.6). This device, invented in 1784 but still a familiar sight in teaching laboratories, is used to demonstrate Newton's laws of motion and to measure g. It consists of two weights connected by a light string of length l which passes over a light and frictionless pulley.

The elementary approach is to write an equation of motion for each of the two weights

$$\begin{aligned} m_1\ddot{x}_1 &= m_1 g - T, \\ m_2\ddot{x}_2 &= m_2 g - T. \end{aligned} \tag{1.40}$$

We then take into account the constraint $\dot{x}_1 = -\dot{x}_2$ and eliminate $\ddot{x}_2$ in favour of $\ddot{x}_1$:

$$\begin{aligned} m_1\ddot{x}_1 &= m_1 g - T, \\ -m_2\ddot{x}_1 &= m_2 g - T. \end{aligned} \tag{1.41}$$

Finally we eliminate the constraint force and the tension T, and obtain the acceleration

$$(m_1 + m_2)\ddot{x}_1 = (m_1 - m_2)g. \tag{1.42}$$

Lagrange's solution takes the constraint into account from the very beginning by introducing a single generalized coordinate $q = x_1 = l - x_2$, and writing

$$L = T - V = \frac{1}{2}(m_1 + m_2)\dot{q}^2 - (m_2 - m_1)gq. \tag{1.43}$$

From this we obtain a single equation of motion

$$\frac{d}{dt}\left(\frac{\partial L}{\partial \dot{q}^i}\right) - \frac{\partial L}{\partial q^i} = 0 \quad \Rightarrow \quad (m_1 + m_2)\ddot{q} = (m_1 - m_2)g. \tag{1.44}$$

The advantage of the Lagrangian method is that constraint forces, which do no net work, never appear. The disadvantage is exactly the same: if we need to find the constraint forces – in this case the tension in the string – we cannot use Lagrange alone.

Lagrange provides a convenient way to derive the equations of motion in non-cartesian coordinate systems, such as plane polar coordinates.

Consider the central force problem with $F_r = -\partial_r V(r)$. Newton's method begins by computing the acceleration in polar coordinates. This is most easily done by setting $z = re^{i\theta}$ and differentiating twice:

$$\begin{aligned} \dot{z} &= (\dot{r} + ir\dot{\theta})e^{i\theta}, \\ \ddot{z} &= (\ddot{r} - r\dot{\theta}^2)e^{i\theta} + i(2\dot{r}\dot{\theta} + r\ddot{\theta})e^{i\theta}. \end{aligned} \tag{1.45}$$

Reading off the components parallel and perpendicular to $e^{i\theta}$ gives the radial and angular acceleration (Figure 1.7)

$$\begin{aligned} a_r &= \ddot{r} - r\dot{\theta}^2, \\ a_\theta &= r\ddot{\theta} + 2\dot{r}\dot{\theta}. \end{aligned} \tag{1.46}$$

Newton's equations therefore become

$$\begin{aligned} m(\ddot{r} - r\dot{\theta}^2) &= -\frac{\partial V}{\partial r} \\ m(r\ddot{\theta} + 2\dot{r}\dot{\theta}) &= 0, \quad \Rightarrow \quad \frac{d}{dt}(mr^2\dot{\theta}) = 0. \end{aligned} \tag{1.47}$$

Setting $l = mr^2\dot{\theta}$, the conserved angular momentum, and eliminating $\dot{\theta}$ gives

$$m\ddot{r} - \frac{l^2}{mr^3} = -\frac{\partial V}{\partial r}. \tag{1.48}$$

(If this were Kepler's problem, where $V = GmM/r$, we would now proceed to simplify this equation by substituting $r = 1/u$, but that is another story.)

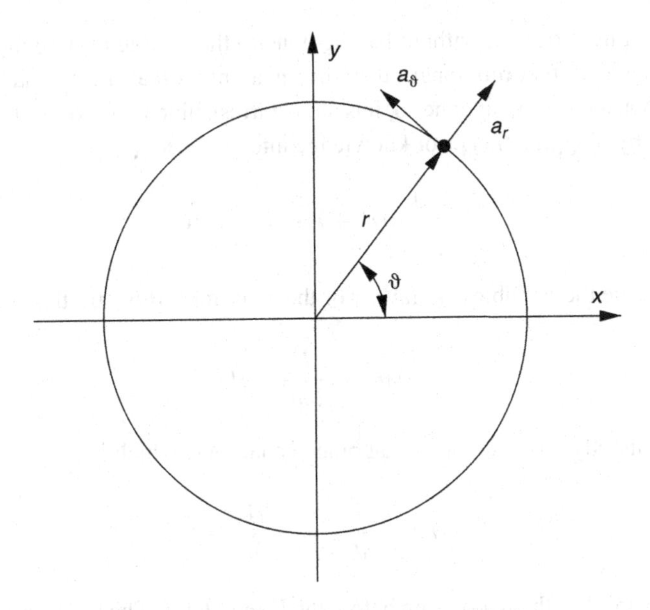

Figure 1.7 Polar components of acceleration.

Following Lagrange we first compute the kinetic energy in polar coordinates (this requires less thought than computing the acceleration) and set

$$L = T - V = \frac{1}{2}m(\dot{r}^2 + r^2\dot{\theta}^2) - V(r). \tag{1.49}$$

The Euler–Lagrange equations are now

$$\begin{aligned} \frac{d}{dt}\left(\frac{\partial L}{\partial \dot{r}}\right) - \frac{\partial L}{\partial r} &= 0, \Rightarrow m\ddot{r} - mr\dot{\theta}^2 + \frac{\partial V}{\partial r} = 0, \\ \frac{d}{dt}\left(\frac{\partial L}{\partial \dot{\theta}}\right) - \frac{\partial L}{\partial \theta} &= 0, \Rightarrow \frac{d}{dt}(mr^2\dot{\theta}) = 0, \end{aligned} \tag{1.50}$$

and coincide with Newton's.

The first integral is

$$\begin{aligned} E &= \dot{r}\frac{\partial L}{\partial \dot{r}} + \dot{\theta}\frac{\partial L}{\partial \dot{\theta}} - L \\ &= \frac{1}{2}m(\dot{r}^2 + r^2\dot{\theta}^2) + V(r). \end{aligned} \tag{1.51}$$

which is the total energy. Thus the constancy of the first integral states that

$$\frac{dE}{dt} = 0, \tag{1.52}$$

or that energy is conserved.

Warning: we might realize, without having gone to the trouble of deriving it from the Lagrange equations, that rotational invariance guarantees that the angular momentum $l = mr^2\dot{\theta}$ is constant. Having done so, it is almost irresistible to try to short-circuit some of the labour by plugging this prior knowledge into

$$L = \frac{1}{2}m(\dot{r}^2 + r^2\dot{\theta}^2) - V(r) \tag{1.53}$$

so as to eliminate the variable $\dot{\theta}$ in favour of the constant l. If we try this we get

$$L \overset{?}{\rightarrow} \frac{1}{2}m\dot{r}^2 + \frac{l^2}{2mr^2} - V(r). \tag{1.54}$$

We can now directly write down the Lagrange equation r, which is

$$m\ddot{r} + \frac{l^2}{mr^3} \overset{?}{=} -\frac{\partial V}{\partial r}. \tag{1.55}$$

Unfortunately this has the wrong sign before the l^2/mr^3 term! The lesson is that we must be very careful in using consequences of a variational principle to modify the principle. It can be done, and in mechanics it leads to the *Routhian* or, in more modern language, to *Hamiltonian reduction*, but it requires using a Legendre transform. The reader should consult a book on mechanics for details.

1.3.2 Noether's theorem

The time-independence of the first integral

$$\frac{d}{dt}\left\{\dot{q}\frac{\partial L}{\partial \dot{q}} - L\right\} = 0, \tag{1.56}$$

and of angular momentum

$$\frac{d}{dt}\{mr^2\dot{\theta}\} = 0, \tag{1.57}$$

are examples of *conservation laws*. We obtained them both by manipulating the Euler–Lagrange equations of motion, but also indicated that they were in some way connected with symmetries. One of the chief advantages of a variational formulation of a physical problem is that this connection

Symmetry ⇔ Conservation law

can be made explicit by exploiting a strategy due to Emmy Noether. She showed how to proceed directly from the action integral to the conserved quantity without having to fiddle about with the individual equations of motion. We begin by illustrating her technique in the case of angular momentum, whose conservation is a consequence of the rotational symmetry of the central force problem. The action integral for the central force problem is

$$S = \int_0^T \left\{ \frac{1}{2} m(\dot{r}^2 + r^2\dot{\theta}^2) - V(r) \right\} dt. \tag{1.58}$$

Noether observes that the integrand is left unchanged if we make the variation

$$\theta(t) \to \theta(t) + \varepsilon\alpha \tag{1.59}$$

where α is a fixed angle and ε is a small, time-independent, parameter. This invariance is the symmetry we shall exploit. It is a mathematical identity: it does not require that r and θ obey the equations of motion. She next observes that since the *equations of motion* are equivalent to the statement that S is left stationary under *any* infinitesimal variations in r and θ, they necessarily imply that S is stationary under the specific variation

$$\theta(t) \to \theta(t) + \varepsilon(t)\alpha \tag{1.60}$$

where now ε is allowed to be time-dependent. This stationarity of the action is no longer a mathematical identity, but, because it requires r, θ, to obey the equations of motion, has physical content. Inserting $\delta\theta = \varepsilon(t)\alpha$ into our expression for S gives

$$\delta S = \alpha \int_0^T \left\{ mr^2\dot{\theta} \right\} \dot{\varepsilon}\, dt. \tag{1.61}$$

Note that this variation depends only on the time derivative of ε, and not ε itself. This is because of the invariance of S under time-independent rotations. We now assume that $\varepsilon(t) = 0$ at $t = 0$ and $t = T$, and integrate by parts to take the time derivative off ε and put it on the rest of the integrand:

$$\delta S = -\alpha \int \left\{ \frac{d}{dt}(mr^2\dot{\theta}) \right\} \varepsilon(t)\, dt. \tag{1.62}$$

Since the equations of motion say that $\delta S = 0$ under all infinitesimal variations, and in particular those due to any time-dependent rotation $\varepsilon(t)\alpha$, we deduce that the equations of motion imply that the coefficient of $\varepsilon(t)$ must be zero, and so, provided $r(t)$, $\theta(t)$, obey the equations of motion, we have

$$0 = \frac{d}{dt}(mr^2\dot{\theta}). \tag{1.63}$$

As a second illustration we derive energy (first integral) conservation for the case that the system is invariant under time translations – meaning that L does not depend explicitly on time. In this case the action integral is invariant under constant time shifts $t \to t + \varepsilon$ in the argument of the dynamical variable:

$$q(t) \to q(t+\varepsilon) \approx q(t) + \varepsilon \dot{q}. \tag{1.64}$$

The equations of motion tell us that the action will be stationary under the variation

$$\delta q(t) = \varepsilon(t)\dot{q}, \tag{1.65}$$

where again we now permit the parameter ε to depend on t. We insert this variation into

$$S = \int_0^T L\, dt \tag{1.66}$$

and find

$$\delta S = \int_0^T \left\{ \frac{\partial L}{\partial q}\dot{q}\varepsilon + \frac{\partial L}{\partial \dot{q}}(\ddot{q}\varepsilon + \dot{q}\dot{\varepsilon}) \right\} dt. \tag{1.67}$$

This expression contains undotted ε's. Because of this the change in S is not obviously zero when ε is time independent, but the absence of any explicit t dependence in L tells us that

$$\frac{dL}{dt} = \frac{\partial L}{\partial q}\dot{q} + \frac{\partial L}{\partial \dot{q}}\ddot{q}. \tag{1.68}$$

As a consequence, for time-independent ε, we have

$$\delta S = \int_0^T \left\{ \varepsilon \frac{dL}{dt} \right\} dt = \varepsilon [L]_0^T, \tag{1.69}$$

showing that the change in S comes entirely from the endpoints of the time interval. These fixed endpoints explicitly break time-translation invariance, but in a trivial manner. For general $\varepsilon(t)$ we have

$$\delta S = \int_0^T \left\{ \varepsilon(t) \frac{dL}{dt} + \frac{\partial L}{\partial \dot{q}}\dot{q}\dot{\varepsilon} \right\} dt. \tag{1.70}$$

This equation is an identity. It does not rely on q obeying the equation of motion. After an integration by parts, taking $\varepsilon(t)$ to be zero at $t = 0, T$, it is equivalent to

$$\delta S = \int_0^T \varepsilon(t) \frac{d}{dt} \left\{ L - \frac{\partial L}{\partial \dot{q}}\dot{q} \right\} dt. \tag{1.71}$$

Now we assume that $q(t)$ does obey the equations of motion. The variation principle then says that $\delta S = 0$ for any $\varepsilon(t)$, and we deduce that for $q(t)$ satisfying the equations of motion we have

$$\frac{d}{dt}\left\{L - \frac{\partial L}{\partial \dot{q}}\dot{q}\right\} = 0. \tag{1.72}$$

The general strategy that constitutes "Noether's theorem" must now be obvious: we look for an invariance of the action under a symmetry transformation with a time-independent parameter. We then observe that if the dynamical variables obey the equations of motion, then the action principle tells us that the action will remain stationary under such a variation of the dynamical variables even after the parameter is promoted to being time dependent. The resultant variation of S can only depend on time derivatives of the parameter. We integrate by parts so as to take all the time derivatives off it, and on to the rest of the integrand. Because the parameter is arbitrary, we deduce that the equations of motion tell us that that its coefficient in the integral must be zero. This coefficient is the time derivative of something, so this something is conserved.

1.3.3 Many degrees of freedom

The extension of the action principle to many degrees of freedom is straightforward. As an example consider the small oscillations about equilibrium of a system with N degrees of freedom. We parametrize the system in terms of deviations from the equilibrium position and expand out to quadratic order. We obtain a Lagrangian

$$L = \sum_{i,j=1}^{N}\left\{\frac{1}{2}M_{ij}\dot{q}^i\dot{q}^j - \frac{1}{2}V_{ij}q^iq^j\right\}, \tag{1.73}$$

where M_{ij} and V_{ij} are $N \times N$ symmetric matrices encoding the inertial and potential energy properties of the system. Now we have one equation

$$0 = \frac{d}{dt}\left(\frac{\partial L}{\partial \dot{q}^i}\right) - \frac{\partial L}{\partial q^i} = \sum_{j=1}^{N}\left(M_{ij}\ddot{q}^j + V_{ij}q^j\right) \tag{1.74}$$

for each i.

1.3.4 Continuous systems

The action principle can be extended to field theories and to continuum mechanics. Here one has a continuous infinity of dynamical degrees of freedom, either one for each point in space and time or one for each point in the material, but the extension of the variational derivative to functions of more than one variable should possess no conceptual difficulties.

Suppose we are given an action functional $S[\varphi]$ depending on a field $\varphi(x^\mu)$ and its first derivatives

$$\varphi_\mu \equiv \frac{\partial \varphi}{\partial x^\mu}. \tag{1.75}$$

Here x^μ, $\mu = 0, 1, \ldots, d$, are the coordinates of $(d+1)$-dimensional space-time. It is traditional to take $x^0 \equiv t$ and the other coordinates space-like. Suppose further that

$$S[\varphi] = \int L\, dt = \int \mathcal{L}(x^\mu, \varphi, \varphi_\mu)\, d^{d+1}x, \tag{1.76}$$

where $\mathcal{L}$ is the *Lagrangian density*, in terms of which

$$L = \int \mathcal{L}\, d^d x, \tag{1.77}$$

and the integral is over the space coordinates. Now

$$\begin{aligned} \delta S &= \int \left\{ \delta\varphi(x) \frac{\partial \mathcal{L}}{\partial \varphi(x)} + \delta(\varphi_\mu(x)) \frac{\partial \mathcal{L}}{\partial \varphi_\mu(x)} \right\} d^{d+1}x \\ &= \int \delta\varphi(x) \left\{ \frac{\partial \mathcal{L}}{\partial \varphi(x)} - \frac{\partial}{\partial x^\mu}\left(\frac{\partial \mathcal{L}}{\partial \varphi_\mu(x)} \right) \right\} d^{d+1}x. \end{aligned} \tag{1.78}$$

In going from the first line to the second, we have observed that

$$\delta(\varphi_\mu(x)) = \frac{\partial}{\partial x^\mu} \delta\varphi(x) \tag{1.79}$$

and used the divergence theorem,

$$\int_\Omega \left(\frac{\partial A^\mu}{\partial x^\mu} \right) d^{n+1}x = \int_{\partial\Omega} A^\mu n_\mu dS, \tag{1.80}$$

where Ω is some space-time region and $\partial\Omega$ its boundary, to integrate by parts. Here dS is the element of area on the boundary, and n_μ the outward normal. As before, we take $\delta\varphi$ to vanish on the boundary, and hence there is no boundary contribution to variation of S. The result is that

$$\frac{\delta S}{\delta\varphi(x)} = \frac{\partial \mathcal{L}}{\partial \varphi(x)} - \frac{\partial}{\partial x^\mu}\left(\frac{\partial \mathcal{L}}{\partial \varphi_\mu(x)} \right), \tag{1.81}$$

and the equation of motion comes from setting this to zero. Note that a sum over the repeated coordinate index μ is implied. In practice it is easier not to use this formula. Instead, make the variation by hand – as in the following examples.

Example: *The vibrating string*. The simplest continuous dynamical system is the transversely vibrating string (Figure 1.8). We describe the string displacement by $y(x, t)$.

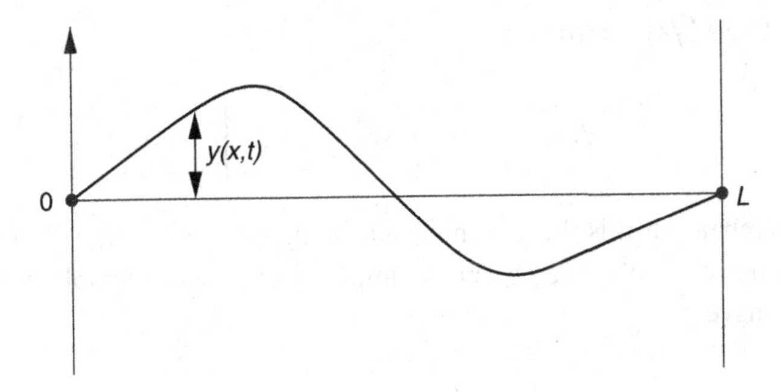

Figure 1.8 Transversely vibrating string.

Let us suppose that the string has fixed ends, a mass per unit length of ρ and is under tension T. If we assume only small displacements from equilibrium, the Lagrangian is

$$L = \int_0^L dx \left\{ \frac{1}{2}\rho \dot{y}^2 - \frac{1}{2} T y'^2 \right\}. \tag{1.82}$$

The dot denotes a partial derivative with respect to t, and the prime a partial derivative with respect to x. The variation of the action is

$$\begin{aligned} \delta S &= \iint_0^L dt dx \left\{ \rho \dot{y}\, \delta \dot{y} - T y' \delta y' \right\} \\ &= \iint_0^L dt dx \left\{ \delta y(x,t) \left(-\rho \ddot{y} + T y'' \right) \right\}. \end{aligned} \tag{1.83}$$

To reach the second line we have integrated by parts, and, because the ends are fixed, and therefore $\delta y = 0$ at $x = 0$ and L, there is no boundary term. Requiring that $\delta S = 0$ for all allowed variations δy then gives the equation of motion

$$\rho \ddot{y} - T y'' = 0. \tag{1.84}$$

This is the wave equation describing transverse waves propagating with speed $c = \sqrt{T/\rho}$. Observe that from (1.83) we can read off the functional derivative of S with respect to the variable $y(x,t)$ as being

$$\frac{\delta S}{\delta y(x,t)} = -\rho \ddot{y}(x,t) + T y''(x,t). \tag{1.85}$$

In writing down the first integral for this continuous system, we must replace the sum over discrete indices by an integral:

$$E = \sum_i \dot{q}_i \frac{\partial L}{\partial \dot{q}_i} - L \to \int dx \left\{ \dot{y}(x) \frac{\delta L}{\delta \dot{y}(x)} \right\} - L. \tag{1.86}$$

When computing $\delta L/\delta\dot{y}(x)$ from

$$L = \int_0^L dx \left\{ \frac{1}{2}\rho\dot{y}^2 - \frac{1}{2}Ty'^2 \right\},$$

we must remember that it is the continuous analogue of $\partial L/\partial\dot{q}_i$, and so, in contrast to what we do when computing $\delta S/\delta y(x)$, we must treat $\dot{y}(x)$ as a variable independent of $y(x)$. We then have

$$\frac{\delta L}{\delta\dot{y}(x)} = \rho\dot{y}(x), \tag{1.87}$$

leading to

$$E = \int_0^L dx \left\{ \frac{1}{2}\rho\dot{y}^2 + \frac{1}{2}Ty'^2 \right\}. \tag{1.88}$$

This, as expected, is the total energy, kinetic plus potential, of the string.

The energy–momentum tensor

If we consider an action of the form

$$S = \int \mathcal{L}(\varphi, \varphi_\mu)\, d^{d+1}x, \tag{1.89}$$

in which $\mathcal{L}$ does not depend *explicitly* on any of the coordinates x^μ, we may refine Noether's derivation of the law of conservation of total energy and obtain accounting information about the position-dependent energy *density*. To do this we make a variation of the form

$$\varphi(x) \to \varphi(x^\mu + \varepsilon^\mu(x)) = \varphi(x^\mu) + \varepsilon^\mu(x)\partial_\mu\varphi + O(|\varepsilon|^2), \tag{1.90}$$

where ε depends on $x \equiv (x^0, \ldots, x^d)$. The resulting variation in S is

$$\begin{aligned}
\delta S &= \int \left\{ \frac{\partial\mathcal{L}}{\partial\varphi}\varepsilon^\mu\partial_\mu\varphi + \frac{\partial\mathcal{L}}{\partial\varphi_\nu}\partial_\nu(\varepsilon^\mu\partial_\mu\varphi) \right\} d^{d+1}x \\
&= \int \varepsilon^\mu(x)\frac{\partial}{\partial x^\nu}\left\{ \mathcal{L}\delta^\nu_\mu - \frac{\partial\mathcal{L}}{\partial\varphi_\nu}\partial_\mu\varphi \right\} d^{d+1}x.
\end{aligned} \tag{1.91}$$

When φ satisfies the equations of motion, this δS will be zero for arbitrary $\varepsilon^\mu(x)$. We conclude that

$$\frac{\partial}{\partial x^\nu}\left\{ \mathcal{L}\delta^\nu_\mu - \frac{\partial\mathcal{L}}{\partial\varphi_\nu}\partial_\mu\varphi \right\} = 0. \tag{1.92}$$

The $(d+1)$-by-$(d+1)$ array of functions

$$T^{\nu}{}_{\mu} \equiv \frac{\partial \mathcal{L}}{\partial \varphi_{\nu}} \partial_{\mu}\varphi - \delta^{\nu}_{\mu}\mathcal{L} \tag{1.93}$$

is known as the *canonical energy–momentum tensor* because the statement

$$\partial_{\nu} T^{\nu}{}_{\mu} = 0 \tag{1.94}$$

often provides bookkeeping for the flow of energy and momentum.

In the case of the vibrating string, the $\mu = 0, 1$ components of $\partial_{\nu} T^{\nu}{}_{\mu} = 0$ become the two following *local* conservation equations:

$$\frac{\partial}{\partial t}\left\{\frac{\rho}{2}\dot{y}^2 + \frac{T}{2}y'^2\right\} + \frac{\partial}{\partial x}\left\{-T\dot{y}y'\right\} = 0, \tag{1.95}$$

and

$$\frac{\partial}{\partial t}\left\{-\rho\dot{y}y'\right\} + \frac{\partial}{\partial x}\left\{\frac{\rho}{2}\dot{y}^2 + \frac{T}{2}y'^2\right\} = 0. \tag{1.96}$$

It is easy to verify that these are indeed consequences of the wave equation. They are "local" conservation laws because they are of the form

$$\frac{\partial q}{\partial t} + \operatorname{div} \mathbf{J} = 0, \tag{1.97}$$

where q is the local density, and $\mathbf{J}$ the flux, of the globally conserved quantity $Q = \int q\, d^d x$. In the first case, the local density q is

$$T^{0}{}_{0} = \frac{\rho}{2}\dot{y}^2 + \frac{T}{2}y'^2, \tag{1.98}$$

which is the energy density. The energy flux is given by $T^{1}{}_{0} \equiv -T\dot{y}y'$, which is the rate that a segment of string is doing work on its neighbour to the right. Integrating over x, and observing that the fixed-end boundary conditions are such that

$$\int_0^L \frac{\partial}{\partial x}\left\{-T\dot{y}y'\right\} dx = \left[-T\dot{y}y'\right]_0^L = 0, \tag{1.99}$$

gives us

$$\frac{d}{dt}\int_0^L \left\{\frac{\rho}{2}\dot{y}^2 + \frac{T}{2}y'^2\right\} dx = 0, \tag{1.100}$$

which is the global energy conservation law we obtained earlier.

The physical interpretation of ${T^0}_1 = -\rho\dot{y}y'$, the locally conserved quantity appearing in (1.96), is less obvious. If this were a relativistic system, we would immediately identify $\int {T^0}_1 \, dx$ as the x-component of the energy–momentum 4-vector, and therefore ${T^0}_1$ as the density of x-momentum. Now any real string will have some motion in the x-direction, but the magnitude of this motion will depend on the string's elastic constants and other quantities unknown to our Lagrangian. Because of this, the ${T^0}_1$ derived from L cannot be the string's x-momentum density. Instead, it is the density of something called *pseudo-momentum*. The distinction between true and pseudo-momentum is best appreciated by considering the corresponding Noether symmetry. The symmetry associated with Newtonian momentum is the invariance of the action integral under an x-translation of the entire apparatus: the string, and any wave on it. The symmetry associated with pseudo-momentum is the invariance of the action under a shift $y(x) \to y(x-a)$ of the location of the wave on the string – the string itself not being translated. Newtonian momentum is conserved if the *ambient space* is translationally invariant. Pseudo-momentum is conserved only if the *string* is translationally invariant – i.e. if ρ and T are position-independent. A failure to realize that the presence of a medium (here the string) requires us to distinguish between these two symmetries is the origin of much confusion involving "wave momentum".

Maxwell's equations

Michael Faraday and James Clerk Maxwell's description of electromagnetism in terms of dynamical vector fields gave us the first modern field theory. D'Alembert and Maupertuis would have been delighted to discover that the famous equations of Maxwell's *A Treatise on Electricity and Magnetism* (1873) follow from an action principle. There is a slight complication stemming from gauge invariance but, as long as we are not interested in exhibiting the covariance of Maxwell under Lorentz transformations, we can sweep this under the rug by working in the *axial gauge*, where the scalar electric potential does not appear.

We will start from Maxwell's equations

$$\begin{aligned}
\operatorname{div} \mathbf{B} &= 0, \\
\operatorname{curl} \mathbf{E} &= -\frac{\partial \mathbf{B}}{\partial t}, \\
\operatorname{curl} \mathbf{H} &= \mathbf{J} + \frac{\partial \mathbf{D}}{\partial t}, \\
\operatorname{div} \mathbf{D} &= \rho,
\end{aligned} \tag{1.101}$$

and show that they can be obtained from an action principle. For convenience we shall use *natural* units in which $\mu_0 = \varepsilon_0 = 1$, and so $c = 1$ and $\mathbf{D} \equiv \mathbf{E}$ and $\mathbf{B} \equiv \mathbf{H}$.

The first equation $\operatorname{div} \mathbf{B} = 0$ contains no time derivatives. It is a constraint which we satisfy by introducing a vector potential $\mathbf{A}$ such that $\mathbf{B} = \operatorname{curl} \mathbf{A}$. If we set

$$\mathbf{E} = -\frac{\partial \mathbf{A}}{\partial t}, \tag{1.102}$$

then this automatically implies Faraday's law of induction

$$\operatorname{curl}\mathbf{E} = -\frac{\partial \mathbf{B}}{\partial t}. \tag{1.103}$$

We now guess that the Lagrangian is

$$L = \int d^3x \left[\frac{1}{2}\left\{ \mathbf{E}^2 - \mathbf{B}^2 \right\} + \mathbf{J}\cdot\mathbf{A} \right]. \tag{1.104}$$

The motivation is that L looks very like $T - V$ if we regard $\frac{1}{2}\mathbf{E}^2 \equiv \frac{1}{2}\dot{\mathbf{A}}^2$ as being the kinetic energy and $\frac{1}{2}\mathbf{B}^2 = \frac{1}{2}(\operatorname{curl}\mathbf{A})^2$ as being the potential energy. The term in $\mathbf{J}$ represents the interaction of the fields with an external current source. In the axial gauge the electric charge density ρ does not appear in the Lagrangian. The corresponding action is therefore

$$S = \int L\,dt = \iint d^3x \left[\frac{1}{2}\dot{\mathbf{A}}^2 - \frac{1}{2}(\operatorname{curl}\mathbf{A})^2 + \mathbf{J}\cdot\mathbf{A} \right] dt. \tag{1.105}$$

Now vary $\mathbf{A}$ to $\mathbf{A} + \delta\mathbf{A}$, whence

$$\delta S = \iint d^3x \left[-\ddot{\mathbf{A}}\cdot\delta\mathbf{A} - (\operatorname{curl}\mathbf{A})\cdot(\operatorname{curl}\delta\mathbf{A}) + \mathbf{J}\cdot\delta\mathbf{A} \right] dt. \tag{1.106}$$

Here, we have already removed the time derivative from $\delta\mathbf{A}$ by integrating by parts in the time direction. Now we do the integration by parts in the space directions by using the identity

$$\operatorname{div}(\delta\mathbf{A} \times (\operatorname{curl}\mathbf{A})) = (\operatorname{curl}\mathbf{A})\cdot(\operatorname{curl}\delta\mathbf{A}) - \delta\mathbf{A}\cdot(\operatorname{curl}(\operatorname{curl}\mathbf{A})) \tag{1.107}$$

and taking $\delta\mathbf{A}$ to vanish at spatial infinity, so the surface term, which would come from the integral of the total divergence, is zero. We end up with

$$\delta S = \iint d^3x \left\{ \delta\mathbf{A}\cdot\left[-\ddot{\mathbf{A}} - \operatorname{curl}(\operatorname{curl}\mathbf{A}) + \mathbf{J} \right] \right\} dt. \tag{1.108}$$

Demanding that the variation of S be zero thus requires

$$\frac{\partial^2 \mathbf{A}}{\partial t^2} = -\operatorname{curl}(\operatorname{curl}\mathbf{A}) + \mathbf{J}, \tag{1.109}$$

or, in terms of the physical fields,

$$\operatorname{curl}\mathbf{B} = \mathbf{J} + \frac{\partial \mathbf{E}}{\partial t}. \tag{1.110}$$

This is Ampère's law, as modified by Maxwell so as to include the displacement current.

How do we deal with the last Maxwell equation, Gauss' law, which asserts that $\operatorname{div}\mathbf{E} = \rho$? If ρ were equal to zero, this equation would hold if $\operatorname{div}\mathbf{A} = 0$, i.e. if $\mathbf{A}$ were solenoidal. In this case we might be tempted to impose the constraint $\operatorname{div}\mathbf{A} = 0$ on the vector potential, but doing so would undo all our good work, as we have been assuming that we can vary $\mathbf{A}$ freely.

We notice, however, that the three Maxwell equations we already possess tell us that

$$\frac{\partial}{\partial t}(\operatorname{div}\mathbf{E} - \rho) = \operatorname{div}(\operatorname{curl}\mathbf{B}) - \left(\operatorname{div}\mathbf{J} + \frac{\partial \rho}{\partial t}\right). \tag{1.111}$$

Now $\operatorname{div}(\operatorname{curl}\mathbf{B}) = 0$, so the left-hand side is zero provided charge is conserved, i.e. provided

$$\dot{\rho} + \operatorname{div}\mathbf{J} = 0. \tag{1.112}$$

We assume that this is so. Thus, if Gauss' law holds initially, it holds eternally. We arrange for it to hold at $t = 0$ by imposing initial conditions on $\mathbf{A}$. We first choose $\mathbf{A}|_{t=0}$ by requiring it to satisfy

$$\mathbf{B}|_{t=0} = \operatorname{curl}\left(\mathbf{A}|_{t=0}\right). \tag{1.113}$$

The solution is not unique, because may we add any $\nabla\phi$ to $\mathbf{A}|_{t=0}$, but this does not affect the physical $\mathbf{E}$ and $\mathbf{B}$ fields. The initial "velocities" $\dot{\mathbf{A}}|_{t=0}$ are then fixed uniquely by $\dot{\mathbf{A}}|_{t=0} = -\mathbf{E}|_{t=0}$, where the initial $\mathbf{E}$ satisfies Gauss' law. The subsequent evolution of $\mathbf{A}$ is then uniquely determined by integrating the second-order equation (1.109).

The first integral for Maxwell is

$$\begin{aligned} E &= \sum_{i=1}^{3} \int d^3x \left\{ \dot{A}_i \frac{\delta L}{\delta \dot{A}_i} \right\} - L \\ &= \int d^3x \left[\frac{1}{2} \left\{ \mathbf{E}^2 + \mathbf{B}^2 \right\} - \mathbf{J}\cdot\mathbf{A} \right]. \end{aligned} \tag{1.114}$$

This will be conserved if $\mathbf{J}$ is time-independent. If $\mathbf{J} = 0$, it is the total field energy.

Suppose $\mathbf{J}$ is neither zero nor time-independent. Then, looking back at the derivation of the time-independence of the first integral, we see that if L *does* depend on time, we instead have

$$\frac{dE}{dt} = -\frac{\partial L}{\partial t}. \tag{1.115}$$

In the present case we have

$$-\frac{\partial L}{\partial t} = -\int \dot{\mathbf{J}}\cdot\mathbf{A}\, d^3x, \tag{1.116}$$

so that

$$-\int \dot{\mathbf{J}}\cdot\mathbf{A}\, d^3x = \frac{dE}{dt} = \frac{d}{dt}(\text{Field Energy}) - \int \left\{ \mathbf{J}\cdot\dot{\mathbf{A}} + \dot{\mathbf{J}}\cdot\mathbf{A} \right\} d^3x. \tag{1.117}$$

Thus, cancelling the duplicated term and using $\mathbf{E} = -\dot{\mathbf{A}}$, we find

$$\frac{d}{dt}(\text{Field Energy}) = -\int \mathbf{J} \cdot \mathbf{E}\, d^3x. \tag{1.118}$$

Now $\int \mathbf{J} \cdot (-\mathbf{E})\, d^3x$ is the rate at which the power source driving the current is doing work against the field. The result is therefore physically sensible.

Continuum mechanics

Because the mechanics of discrete objects can be derived from an action principle, it seems obvious that so must the mechanics of continua. This is certainly true if we use the *Lagrangian* description where we follow the history of each particle composing the continuous material as it moves through space. In fluid mechanics it is more natural to describe the motion by using the *Eulerian* description in which we focus on what is going on at a particular point in space by introducing a velocity field $\mathbf{v}(\mathbf{r}, t)$. Eulerian action principles can still be found, but they seem to be logically distinct from the Lagrangian mechanics action principle, and mostly were not discovered until the twentieth century.

We begin by showing that Euler's equation for the irrotational motion of an inviscid compressible fluid can be obtained by applying the action principle to a functional

$$S[\phi, \rho] = \int dt\, d^3x \left\{ \rho \frac{\partial \phi}{\partial t} + \frac{1}{2}\rho(\nabla\phi)^2 + u(\rho) \right\}, \tag{1.119}$$

where ρ is the mass density and the flow velocity is determined from the velocity potential ϕ by $\mathbf{v} = \nabla\phi$. The function $u(\rho)$ is the internal energy density.

Varying $S[\phi, \rho]$ with respect to ρ is straightforward, and gives a time-dependent generalization of (Daniel) Bernoulli's equation

$$\frac{\partial \phi}{\partial t} + \frac{1}{2}\mathbf{v}^2 + h(\rho) = 0. \tag{1.120}$$

Here $h(\rho) \equiv du/d\rho$ is the specific enthalpy.[1] Varying with respect to ϕ requires an integration by parts, based on

$$\text{div}\,(\rho\, \delta\phi\, \nabla\phi) = \rho(\nabla\delta\phi) \cdot (\nabla\phi) + \delta\phi\, \text{div}\,(\rho\nabla\phi), \tag{1.121}$$

and gives the equation of mass conservation

$$\frac{\partial \rho}{\partial t} + \text{div}\,(\rho\mathbf{v}) = 0. \tag{1.122}$$

[1] The enthalpy $H = U + PV$ per unit mass. In general u and h will be functions of both the density and the specific entropy. By taking u to depend only on ρ we are tacitly assuming that specific entropy is constant. This makes the resultant flow *barotropic*, meaning that the pressure is a function of the density only.

Taking the gradient of Bernoulli's equation, and using the fact that for potential flow the *vorticity* $\boldsymbol{\omega} \equiv \operatorname{curl} \mathbf{v}$ is zero and so $\partial_i v_j = \partial_j v_i$, we find that

$$\frac{\partial \mathbf{v}}{\partial t} + (\mathbf{v} \cdot \nabla)\mathbf{v} = -\nabla h. \tag{1.123}$$

We now introduce the pressure P, which is related to h by

$$h(P) = \int_0^P \frac{dP}{\rho(P)}. \tag{1.124}$$

We see that $\rho \nabla h = \nabla P$, and so obtain Euler's equation

$$\rho \left(\frac{\partial \mathbf{v}}{\partial t} + (\mathbf{v} \cdot \nabla)\mathbf{v} \right) = -\nabla P. \tag{1.125}$$

For future reference, we observe that combining the mass-conservation equation

$$\partial_t \rho + \partial_j \left\{ \rho v_j \right\} = 0 \tag{1.126}$$

with Euler's equation

$$\rho(\partial_t v_i + v_j \partial_j v_i) = -\partial_i P \tag{1.127}$$

yields

$$\partial_t \left\{ \rho v_i \right\} + \partial_j \left\{ \rho v_i v_j + \delta_{ij} P \right\} = 0, \tag{1.128}$$

which expresses the local conservation of momentum. The quantity

$$\Pi_{ij} = \rho v_i v_j + \delta_{ij} P \tag{1.129}$$

is the *momentum-flux tensor*, and is the j-th component of the flux of the i-th component $p_i = \rho v_i$ of momentum density.

The relations $h = du/d\rho$ and $\rho = dP/dh$ show that P and u are related by a Legendre transformation: $P = \rho h - u(\rho)$. From this, and the Bernoulli equation, we see that the integrand in the action (1.119) is equal to minus the pressure:

$$-P = \rho \frac{\partial \phi}{\partial t} + \frac{1}{2} \rho (\nabla \phi)^2 + u(\rho). \tag{1.130}$$

This Eulerian formulation cannot be a "follow the particle" action principle in a clever disguise. The mass conservation law is only a consequence of the equation of motion, and is not built in from the beginning as a constraint. Our variations in ϕ are therefore conjuring up new matter rather than merely moving it around.

1.4 Variable endpoints

We now relax our previous assumption that all boundary or surface terms arising from integrations by parts may be ignored. We will find that variation principles can be very useful for working out what boundary conditions we should impose on our differential equations.

Consider the problem of building a railway across a parallel sided isthmus (Figure 1.9). Suppose that the cost of construction is proportional to the length of the track, but the cost of sea transport being negligible, we may locate the terminal seaports wherever we like. We therefore wish to minimize the length

$$L[y] = \int_{x_1}^{x_2} \sqrt{1+(y')^2}dx, \tag{1.131}$$

by allowing both the path $y(x)$ and the endpoints $y(x_1)$ and $y(x_2)$ to vary. Then

$$\begin{aligned} L[y+\delta y] - L[y] &= \int_{x_1}^{x_2} (\delta y') \frac{y'}{\sqrt{1+(y')^2}} dx \\ &= \int_{x_1}^{x_2} \left\{ \frac{d}{dx}\left(\delta y \frac{y'}{\sqrt{1+(y')^2}} \right) - \delta y \frac{d}{dx}\left(\frac{y'}{\sqrt{1+(y')^2}} \right) \right\} dx \\ &= \delta y(x_2) \frac{y'(x_2)}{\sqrt{1+(y')^2}} - \delta y(x_1) \frac{y'(x_1)}{\sqrt{1+(y')^2}} \\ &\quad - \int_{x_1}^{x_2} \delta y \frac{d}{dx}\left(\frac{y'}{\sqrt{1+(y')^2}} \right) dx. \end{aligned} \tag{1.132}$$

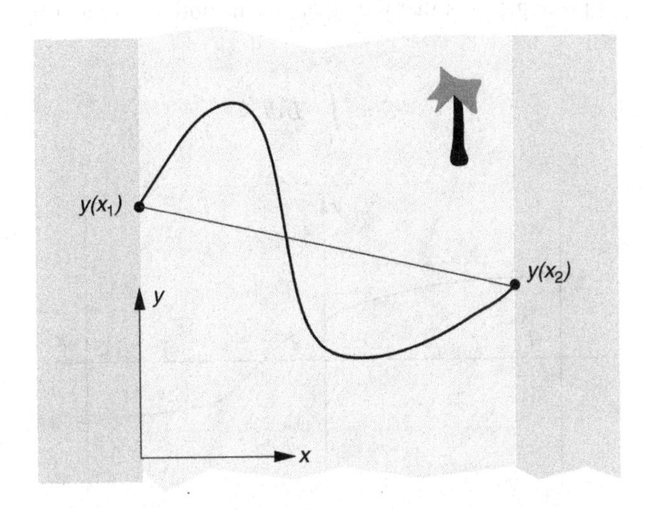

Figure 1.9 Railway across an isthmus.

We have stationarity when both

(i) The coefficient of $\delta y(x)$ in the integral,

$$-\frac{d}{dx}\left(\frac{y'}{\sqrt{1+(y')^2}}\right), \tag{1.133}$$

is zero. This requires that $y' = \text{const.}$, i.e. the track should be straight.

(ii) The coefficients of $\delta y(x_1)$ and $\delta y(x_2)$ vanish. For this we need

$$0 = \frac{y'(x_1)}{\sqrt{1+(y')^2}} = \frac{y'(x_2)}{\sqrt{1+(y')^2}}. \tag{1.134}$$

This in turn requires that $y'(x_1) = y'(x_2) = 0$.

The integrated-out bits have determined the boundary conditions that are to be imposed on the solution of the differential equation. In the present case they require us to build perpendicular to the coastline, and so we go straight across the isthmus. When boundary conditions are obtained from endpoint variations in this way, they are called *natural boundary conditions*.

Example: *Sliding string*. A massive string of linear density ρ is stretched between two smooth posts separated by distance $2L$ (Figure 1.10). The string is under tension T, and is free to slide up and down the posts. We consider only small deviations of the string from the horizontal.

As we saw earlier, the Lagrangian for a stretched string is

$$L = \int_{-L}^{L} \left\{ \frac{1}{2}\rho\dot{y}^2 - \frac{1}{2}T(y')^2 \right\} dx. \tag{1.135}$$

Now, Lagrange's principle says that the equation of motion is found by requiring the action

$$S = \int_{t_i}^{t_f} L\, dt \tag{1.136}$$

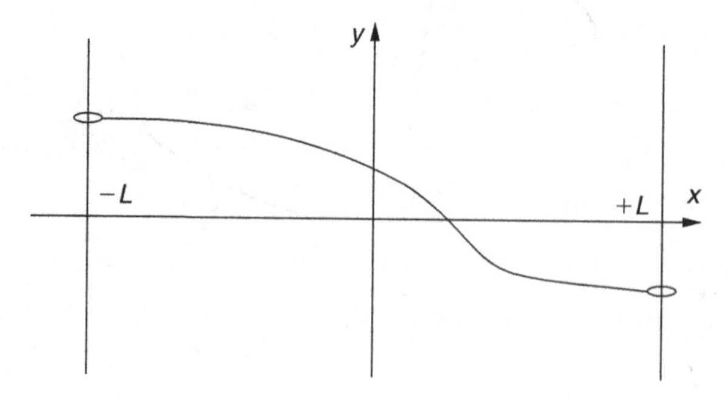

Figure 1.10 Sliding string.

to be stationary under variations of $y(x,t)$ that vanish at the initial and final times, t_i and t_f. It does *not* demand that δy vanish at the ends of the string, $x = \pm L$. So, when we make the variation, we must not assume this. Taking care not to discard the results of the integration by parts in the x-direction, we find

$$\delta S = \int_{t_i}^{t_f}\int_{-L}^{L} \delta y(x,t)\left\{-\rho\ddot{y} + Ty''\right\} dxdt - \int_{t_i}^{t_f} \delta y(L,t)Ty'(L)\,dt$$
$$+ \int_{t_i}^{t_f} \delta y(-L,t)Ty'(-L)\,dt. \tag{1.137}$$

The equation of motion, which arises from the variation within the interval, is therefore the wave equation

$$\rho\ddot{y} - Ty'' = 0. \tag{1.138}$$

The boundary conditions, which come from the variations at the endpoints, are

$$y'(L,t) = y'(-L,t) = 0, \tag{1.139}$$

at all times t. These are the physically correct boundary conditions, because any up-or-down component of the tension would provide a finite force on an infinitesimal mass. The string must therefore be horizontal at its endpoints.

Example: *Bead and string*. Suppose now that a bead of mass M is free to slide up and down the y axis, and is attached to the $x = 0$ end of our string (Figure 1.11). The Lagrangian for the string–bead contraption is

$$L = \frac{1}{2}M[\dot{y}(0)]^2 + \int_0^L \left\{\frac{1}{2}\rho\dot{y}^2 - \frac{1}{2}Ty'^2\right\} dx. \tag{1.140}$$

Here, as before, ρ is the mass per unit length of the string and T is its tension. The end of the string at $x = L$ is fixed. By varying the action $S = \int L dt$, and taking care not to

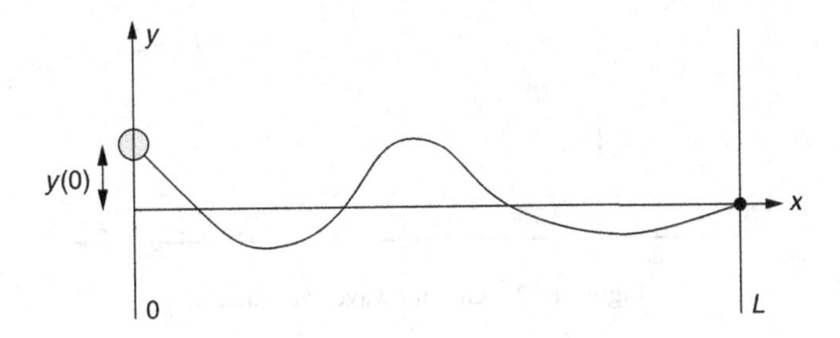

Figure 1.11 A bead connected to a string.

throw away the boundary part at $x = 0$ we find that

$$\delta S = \int_{t_i}^{t_f} \left[Ty' - M\ddot{y}\right]_{x=0} \delta y(0,t)\,dt + \int_{t_i}^{t_f}\int_0^L \{Ty'' - \rho\ddot{y}\}\,\delta y(x,t)\,dxdt. \tag{1.141}$$

The Euler–Lagrange equations are therefore

$$\begin{aligned} \rho\ddot{y}(x) - Ty''(x) &= 0, \quad 0 < x < L, \\ M\ddot{y}(0) - Ty'(0) &= 0, \quad y(L) = 0. \end{aligned} \tag{1.142}$$

The boundary condition at $x = 0$ is the equation of motion for the bead. It is clearly correct, because $Ty'(0)$ is the vertical component of the force that the string tension exerts on the bead.

These examples led to boundary conditions that we could easily have figured out for ourselves without the variational principle. The next example shows that a variational formulation can be exploited to obtain a set of boundary conditions that might be difficult to write down by purely "physical" reasoning.

Harder example: Gravity waves on the surface of water (Figure 1.12). An action suitable for describing water waves is given by[2] $S[\phi, h] = \int L\,dt$, where

$$L = \int dx \int_0^{h(x,t)} \rho_0 \left\{ \frac{\partial\phi}{\partial t} + \frac{1}{2}(\nabla\phi)^2 + gy \right\} dy. \tag{1.143}$$

Here ϕ is the velocity potential and ρ_0 is the density of the water. The density will not be varied because the water is being treated as incompressible. As before, the flow velocity is given by $\mathbf{v} = \nabla\phi$. By varying $\phi(x, y, t)$ and the depth $h(x, t)$, and taking care not

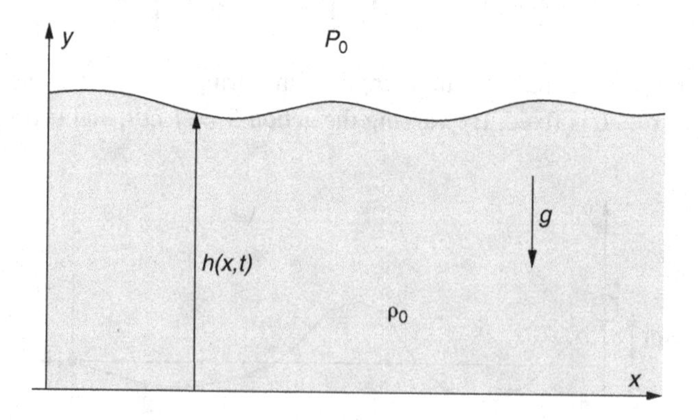

Figure 1.12 Gravity waves on water.

[2] J. C. Luke, *J. Fluid Dynamics,* **27** (1967) 395.

to throw away any integrated-out parts of the variation at the physical boundaries, we obtain:

$$\begin{aligned}
\nabla^2\phi &= 0, \quad \text{within the fluid}\\
\frac{\partial\phi}{\partial t} + \frac{1}{2}(\nabla\phi)^2 + gy &= 0, \quad \text{on the free surface}\\
\frac{\partial\phi}{\partial y} &= 0, \quad \text{on} \quad y = 0\\
\frac{\partial h}{\partial t} - \frac{\partial\phi}{\partial y} + \frac{\partial h}{\partial x}\frac{\partial\phi}{\partial x} &= 0, \quad \text{on the free surface}
\end{aligned} \tag{1.144}$$

The first equation comes from varying ϕ within the fluid, and it simply confirms that the flow is incompressible, i.e. obeys $\operatorname{div}\mathbf{v} = 0$. The second comes from varying h, and is the Bernoulli equation stating that we have $P = P_0$ (atmospheric pressure) everywhere on the free surface. The third, from the variation of ϕ at $y = 0$, states that no fluid escapes through the lower boundary.

Obtaining and interpreting the last equation, involving $\partial h/\partial t$, is somewhat trickier. It comes from the variation of ϕ on the upper boundary. The variation of S due to $\delta\phi$ is

$$\delta S = \int \rho_0 \left\{ \frac{\partial}{\partial t}\delta\phi + \frac{\partial}{\partial x}\left(\delta\phi\frac{\partial\phi}{\partial x}\right) + \frac{\partial}{\partial y}\left(\delta\phi\frac{\partial\phi}{\partial y}\right) - \delta\phi\,\nabla^2\phi \right\} dtdxdy. \tag{1.145}$$

The first three terms in the integrand constitute the three-dimensional divergence $\operatorname{div}(\delta\phi\,\mathbf{\Phi})$, where, listing components in the order t, x, y,

$$\mathbf{\Phi} = \left[1, \frac{\partial\phi}{\partial x}, \frac{\partial\phi}{\partial y}\right]. \tag{1.146}$$

The integrated-out part on the upper surface is therefore $\int(\mathbf{\Phi}\cdot\mathbf{n})\delta\phi\, d|S|$. Here, the outward normal is

$$\mathbf{n} = \left(1 + \left(\frac{\partial h}{\partial t}\right)^2 + \left(\frac{\partial h}{\partial x}\right)^2\right)^{-1/2} \left[-\frac{\partial h}{\partial t}, -\frac{\partial h}{\partial x}, 1\right], \tag{1.147}$$

and the element of area

$$d|S| = \left(1 + \left(\frac{\partial h}{\partial t}\right)^2 + \left(\frac{\partial h}{\partial x}\right)^2\right)^{1/2} dtdx. \tag{1.148}$$

The boundary variation is thus

$$\delta S|_{y=h} = -\int \left\{ \frac{\partial h}{\partial t} - \frac{\partial\phi}{\partial y} + \frac{\partial h}{\partial x}\frac{\partial\phi}{\partial x} \right\} \delta\phi\Big(x, h(x,t), t\Big)\, dxdt. \tag{1.149}$$

Requiring this variation to be zero for arbitrary $\delta\phi\big(x, h(x,t), t\big)$ leads to

$$\frac{\partial h}{\partial t} - \frac{\partial \phi}{\partial y} + \frac{\partial h}{\partial x}\frac{\partial \phi}{\partial x} = 0. \tag{1.150}$$

This last boundary condition expresses the geometrical constraint that the surface moves with the fluid it bounds, or, in other words, that a fluid particle initially on the surface stays on the surface. To see that this is so, define $f(x,y,t) = h(x,t) - y$. The free surface is then determined by $f(x,y,t) = 0$. Because the surface particles are carried with the flow, the convective derivative of f,

$$\frac{df}{dt} \equiv \frac{\partial f}{\partial t} + (\mathbf{v}\cdot\nabla)f, \tag{1.151}$$

must vanish on the free surface. Using $\mathbf{v} = \nabla\phi$ and the definition of f, this reduces to

$$\frac{\partial h}{\partial t} + \frac{\partial \phi}{\partial x}\frac{\partial h}{\partial x} - \frac{\partial \phi}{\partial y} = 0, \tag{1.152}$$

which is indeed the last boundary condition.

1.5 Lagrange multipliers

Figure 1.13 shows the contour map of a hill of height $h = f(x,y)$. The hill is traversed by a road whose points satisfy the equation $g(x,y) = 0$. Our challenge is to use the data $f(x,y)$ and $g(x,y)$ to find the highest point on the road.

When $\mathbf{r}$ changes by $d\mathbf{r} = (dx, dy)$, the height f changes by

$$df = \nabla f \cdot d\mathbf{r}, \tag{1.153}$$

where $\nabla f = (\partial_x f, \partial_y f)$. The highest point, being a stationary point, will have $df = 0$ for all displacements $d\mathbf{r}$ that stay on the road – that is for all $d\mathbf{r}$ such that $dg = 0$. Thus

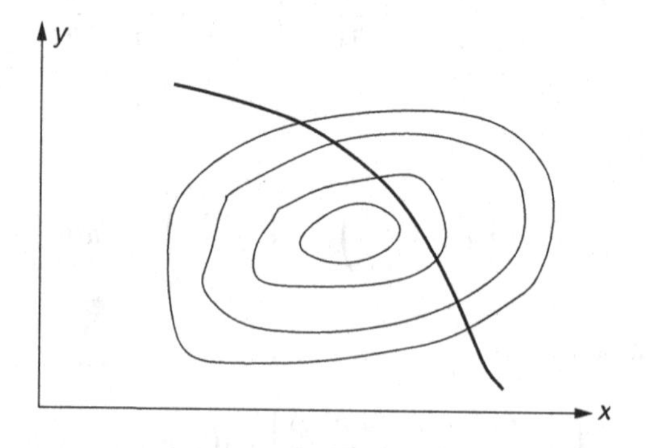

Figure 1.13 Road on hill.

$\nabla f \cdot d\mathbf{r}$ must be zero for those $d\mathbf{r}$ such that $0 = \nabla g \cdot d\mathbf{r}$. In other words, at the highest point ∇f will be orthogonal to all vectors that are orthogonal to ∇g. This is possible only if the vectors ∇f and ∇g are parallel, and so $\nabla f = \lambda \nabla g$ for some λ.

To find the stationary point, therefore, we solve the equations

$$\begin{aligned} \nabla f - \lambda \nabla g &= 0, \\ g(x,y) &= 0, \end{aligned} \tag{1.154}$$

simultaneously.

Example: Let $f = x^2 + y^2$ and $g = x + y - 1$. Then $\nabla f = 2(x,y)$ and $\nabla g = (1,1)$. So

$$2(x,y) - \lambda(1,1) = 0 \quad \Rightarrow \quad (x,y) = \frac{\lambda}{2}(1,1)$$

$$x + y = 1 \quad \Rightarrow \quad \lambda = 1 \quad \Longrightarrow \quad (x,y) = \left(\frac{1}{2}, \frac{1}{2}\right).$$

When there are n constraints, $g_1 = g_2 = \cdots = g_n = 0$, we want ∇f to lie in

$$(\langle \nabla g_i \rangle^{\perp})^{\perp} = \langle \nabla g_i \rangle, \tag{1.155}$$

where $\langle \mathbf{e}_i \rangle$ denotes the space spanned by the vectors $\mathbf{e}_i$ and $\langle \mathbf{e}_i \rangle^{\perp}$ is its orthogonal complement. Thus ∇f lies in the space spanned by the vectors ∇g_i, so there must exist n numbers λ_i such that

$$\nabla f = \sum_{i=1}^{n} \lambda_i \nabla g_i. \tag{1.156}$$

The numbers λ_i are called *Lagrange multipliers*. We can therefore regard our problem as one of finding the stationary points of an auxiliary function

$$F = f - \sum_i \lambda_i g_i, \tag{1.157}$$

with the n undetermined multipliers $\lambda_i, i = 1, \ldots, n$, subsequently being fixed by imposing the n requirements that $g_i = 0, i = 1, \ldots, n$.

Example: Find the stationary points of

$$F(\mathbf{x}) = \frac{1}{2}\mathbf{x} \cdot \mathbf{A}\mathbf{x} = \frac{1}{2} x_i A_{ij} x_j \tag{1.158}$$

on the surface $\mathbf{x} \cdot \mathbf{x} = 1$. Here A_{ij} is a symmetric matrix.

Solution: We look for stationary points of

$$G(\mathbf{x}) = F(\mathbf{x}) - \frac{1}{2}\lambda |\mathbf{x}|^2. \tag{1.159}$$

The derivatives we need are

$$\begin{aligned}\frac{\partial F}{\partial x^k} &= \frac{1}{2}\delta_{ki}A_{ij}x_j + \frac{1}{2}x_i A_{ij}\delta_{jk} \\ &= A_{kj}x_j,\end{aligned} \tag{1.160}$$

and

$$\frac{\partial}{\partial x_k}\left(\frac{\lambda}{2}x_j x_j\right) = \lambda x_k. \tag{1.161}$$

Thus, the stationary points must satisfy

$$\begin{aligned}A_{kj}x_j &= \lambda x_k, \\ x^i x^i &= 1,\end{aligned} \tag{1.162}$$

and so are the normalized eigenvectors of the matrix **A**. The Lagrange multiplier at each stationary point is the corresponding eigenvalue.

Example: *Statistical mechanics.* Let Γ denote the classical phase space of a mechanical system of n particles governed by a Hamiltonian $H(p,q)$. Let $d\Gamma$ be the Liouville measure $d^{3n}p\, d^{3n}q$. In statistical mechanics we work with a probability density $\rho(p,q)$ such that $\rho(p,q)d\,\Gamma$ is the probability of the system being in a state in the small region $d\Gamma$. The *entropy* associated with the probability distribution is the functional

$$S[\rho] = -\int_\Gamma \rho \ln \rho \, d\Gamma. \tag{1.163}$$

We wish to find the $\rho(p,q)$ that maximizes the entropy for a given energy

$$\langle E\rangle = \int_\Gamma \rho H \, d\Gamma. \tag{1.164}$$

We cannot vary ρ freely as we should preserve both the energy and the normalization condition

$$\int_\Gamma \rho \, d\Gamma = 1 \tag{1.165}$$

that is required of any probability distribution. We therefore introduce two Lagrange multipliers, $1+\alpha$ and β, to enforce the normalization and energy conditions, and look for stationary points of

$$F[\rho] = \int_\Gamma \{-\rho \ln \rho + (\alpha+1)\rho - \beta\rho H\}\, d\Gamma. \tag{1.166}$$

Now we can vary ρ freely, and hence find that

$$\delta F = \int_\Gamma \{-\ln \rho + \alpha - \beta H\}\, \delta\rho\, d\Gamma. \tag{1.167}$$

Requiring this to be zero gives us

$$\rho(p,q) = e^{\alpha - \beta H(p,q)}, \tag{1.168}$$

where α, β are determined by imposing the normalization and energy constraints. This probability density is known as the *canonical distribution*, and the parameter β is the inverse temperature $\beta = 1/T$.

Example: *The catenary.* At last we have the tools to solve the problem of the hanging chain of fixed length. We wish to minimize the potential energy

$$E[y] = \int_{-L}^{L} y\sqrt{1 + (y')^2}dx, \tag{1.169}$$

subject to the constraint

$$l[y] = \int_{-L}^{L} \sqrt{1 + (y')^2}dx = \text{const.}, \tag{1.170}$$

where the constant is the length of the chain. We introduce a Lagrange multiplier λ and find the stationary points of

$$F[y] = \int_{-L}^{L} (y - \lambda)\sqrt{1 + (y')^2}dx, \tag{1.171}$$

so, following our earlier methods, we find

$$y = \lambda + \kappa \cosh\frac{(x + a)}{\kappa}. \tag{1.172}$$

We choose κ, λ, a to fix the two endpoints (two conditions) and the length (one condition).

Example: *Sturm–Liouville problem.* We wish to find the stationary points of the quadratic functional

$$J[y] = \int_{x_1}^{x_2} \frac{1}{2}\left\{p(x)(y')^2 + q(x)y^2\right\} dx, \tag{1.173}$$

subject to the boundary conditions $y(x) = 0$ at the endpoints x_1, x_2 and the normalization

$$K[y] = \int_{x_1}^{x_2} y^2\, dx = 1. \tag{1.174}$$

Taking the variation of $J - (\lambda/2)K$, we find

$$\delta J = \int_{x_1}^{x_2} \{-(py')' + qy - \lambda y\}\, \delta y\, dx. \tag{1.175}$$

Stationarity therefore requires

$$-(py')' + qy = \lambda y, \quad y(x_1) = y(x_2) = 0. \tag{1.176}$$

This is the *Sturm–Liouville eigenvalue problem*. It is an infinite-dimensional analogue of the $F(\mathbf{x}) = \frac{1}{2}\mathbf{x} \cdot \mathbf{A}\mathbf{x}$ problem.

Example: *Irrotational flow again.* Consider the action functional

$$S[\mathbf{v}, \phi, \rho] = \int \left\{ \frac{1}{2}\rho \mathbf{v}^2 - u(\rho) + \phi \left(\frac{\partial \rho}{\partial t} + \operatorname{div} \rho \mathbf{v} \right) \right\} dt d^3x. \tag{1.177}$$

This is similar to our previous action for the irrotational barotropic flow of an inviscid fluid, but here $\mathbf{v}$ is an independent variable and we have introduced infinitely many Lagrange multipliers $\phi(x, t)$, one for each point of space-time, so as to enforce the equation of mass conservation $\dot{\rho} + \operatorname{div} \rho\mathbf{v} = 0$ everywhere, and at all times. Equating $\delta S/\delta \mathbf{v}$ to zero gives $\mathbf{v} = \nabla \phi$, and so these Lagrange multipliers become the velocity potential as a consequence of the equations of motion. The Bernoulli and Euler equations now follow almost as before. Because the equation $\mathbf{v} = \nabla\phi$ does not involve time derivatives, this is one of the cases where it is legitimate to substitute a consequence of the action principle back into the action. If we do this, we recover our previous formulation.

1.6 Maximum or minimum?

We have provided many examples of stationary points in function space. We have said almost nothing about whether these stationary points are maxima or minima. There is a reason for this: investigating the character of the stationary point requires the computation of the second functional derivative

$$\frac{\delta^2 J}{\delta y(x_1) \delta y(x_2)}$$

and the use of the functional version of Taylor's theorem to expand about the stationary point $y(x)$:

$$J[y + \varepsilon\eta] = J[y] + \varepsilon \int \eta(x) \left.\frac{\delta J}{\delta y(x)}\right|_y dx + \frac{\varepsilon^2}{2} \int \eta(x_1)\eta(x_2) \left.\frac{\delta^2 J}{\delta y(x_1)\delta y(x_2)}\right|_y dx_1 dx_2 + \cdots. \tag{1.178}$$

Since $y(x)$ is a stationary point, the term with $\delta J/\delta y(x)|_y$ vanishes. Whether $y(x)$ is a maximum, a minimum, or a saddle therefore depends on the number of positive and negative eigenvalues of $\delta^2 J/\delta(y(x_1))\delta(y(x_2))$, a matrix with a continuous infinity of rows and columns, these being labelled by x_1 and x_2, respectively. It is not easy to diagonalize a continuously infinite matrix! Consider, for example, the functional

$$J[y] = \int_a^b \frac{1}{2}\left\{p(x)(y')^2 + q(x)y^2\right\} dx, \tag{1.179}$$

with $y(a) = y(b) = 0$. Here, as we already know,

$$\frac{\delta J}{\delta y(x)} = Ly \equiv -\frac{d}{dx}\left(p(x)\frac{d}{dx}y(x)\right) + q(x)y(x), \tag{1.180}$$

and, except in special cases, this will be zero only if $y(x) \equiv 0$. We might reasonably expect the second derivative to be

$$\frac{\delta}{\delta y}(Ly) \stackrel{?}{=} L, \tag{1.181}$$

where L is the Sturm–Liouville differential operator

$$L = -\frac{d}{dx}\left(p(x)\frac{d}{dx}\right) + q(x). \tag{1.182}$$

How can a differential operator be a matrix like $\delta^2 J/\delta(y(x_1))\delta(y(x_2))$?

We can formally compute the second derivative by exploiting the Dirac delta "function" $\delta(x)$ which has the property that

$$y(x_2) = \int \delta(x_2 - x_1)y(x_1)\, dx_1. \tag{1.183}$$

Thus

$$\delta y(x_2) = \int \delta(x_2 - x_1)\delta y(x_1)\, dx_1, \tag{1.184}$$

from which we read off that

$$\frac{\delta y(x_2)}{\delta y(x_1)} = \delta(x_2 - x_1). \tag{1.185}$$

Using (1.185), we find that

$$\frac{\delta}{\delta y(x_1)}\left(\frac{\delta J}{\delta y(x_2)}\right) = -\frac{d}{dx_2}\left(p(x_2)\frac{d}{dx_2}\delta(x_2 - x_1)\right) + q(x_2)\delta(x_2 - x_1). \tag{1.186}$$

How are we to make sense of this expression? We begin in the next chapter where we explain what it means to differentiate $\delta(x)$, and show that (1.186) does indeed correspond to the differential operator L. In subsequent chapters we explore the manner in which differential operators and matrices are related. We will learn that just as some matrices can be diagonalized so can some differential operators, and that the class of diagonalizable operators includes (1.182).

If all the eigenvalues of L are positive, our stationary point was a minimum. For each negative eigenvalue, there is direction in function space in which $J[y]$ decreases as we move away from the stationary point.

1.7 Further exercises and problems

Here is a collection of problems relating to the calculus of variations. Some date back to the sixteenth century, others are quite recent in origin.

Exercise 1.1: A smooth path in the xy-plane is given by $\mathbf{r}(t) = (x(t), y(t))$ with $\mathbf{r}(0) = \mathbf{a}$, and $\mathbf{r}(1) = \mathbf{b}$. The length of the path from $\mathbf{a}$ to $\mathbf{b}$ is therefore

$$S[\mathbf{r}] = \int_0^1 \sqrt{\dot{x}^2 + \dot{y}^2}\, dt,$$

where $\dot{x} \equiv dx/dt$, $\dot{y} \equiv dy/dt$. Write down the Euler–Lagrange conditions for $S[\mathbf{r}]$ to be stationary under small variations of the path that keep the endpoints fixed, and hence show that the shortest path between two points is a straight line.

Exercise 1.2: *Fermat's principle*. A medium is characterized optically by its *refractive index* n, such that the speed of light in the medium is c/n. According to Fermat (1657), the path taken by a ray of light between any two points makes the travel time stationary between those points. Assume that the ray propagates in the xy-plane in a layered medium with refractive index $n(x)$. Use Fermat's principle to establish Snell's law in its general form $n(x) \sin\psi = \text{constant}$, by finding the equation giving the stationary paths $y(x)$ for

$$F_1[y] = \int n(x)\sqrt{1 + y'^2}dx.$$

(Here the prime denotes differentiation with respect to x.) Repeat this exercise for the case that n depends only on y and find a similar equation for the stationary paths of

$$F_2[y] = \int n(y)\sqrt{1 + y'^2}dx.$$

By using suitable definitions of the angle of incidence ψ in each case, show that the two formulations of the problem give physically equivalent answers. In the second formulation you will find it easiest to use the first integral of Euler's equation.

Problem 1.3: *Hyperbolic geometry*. This problem introduces a version of the Poincaré model for the non-Euclidean geometry of Lobachevski.

(a) Show that the stationary paths for the functional

$$F_3[y] = \int \frac{1}{y}\sqrt{1+y'^2}\,dx,$$

with $y(x)$ restricted to lying in the upper half-plane, are semicircles of arbitrary radius and with centres on the x-axis. These paths are the *geodesics*, or minimum length paths, in a space with Riemann metric

$$ds^2 = \frac{1}{y^2}(dx^2 + dy^2), \qquad y > 0.$$

(b) Show that if we call these geodesics "lines", then one and only one line can be drawn though two given points.
(c) Two lines are said to be *parallel* if, and only if, they meet at "infinity", i.e. on the x-axis. (Verify that the x-axis is indeed infinitely far from any point with $y > 0$.) Show that given a line q and a point A not lying on that line, there are *two* lines passing through A that are parallel to q, and that between these two lines lies a pencil of lines passing through A that never meet q.

Problem 1.4: *Elastic rods*. The elastic energy per unit length of a bent steel rod is given by $\frac{1}{2}YI/R^2$. Here R is the radius of curvature due to the bending, Y is the Young's modulus of the steel and $I = \iint y^2 dxdy$ is the moment of inertia of the rod's cross-section about an axis through its centroid and perpendicular to the plane in which the rod is bent. If the rod is only slightly bent into the yz-plane and lies close to the z-axis, show that this elastic energy can be approximated as

$$U[y] = \int_0^L \frac{1}{2} YI\,(y'')^2\,dz,$$

where the prime denotes differentiation with respect to z and L is the length of the rod. We will use this approximate energy functional to discuss two practical problems.

(a) *Euler's problem: The buckling of a slender column*. The rod is used as a column which supports a compressive load Mg directed along the z-axis (which is vertical; see Figure (1.14a)). Show that when the rod buckles slightly (i.e. deforms with both ends remaining on the z-axis) the total energy, including the gravitational potential energy of the loading mass M, can be approximated by

$$U[y] = \int_0^L \left\{ \frac{YI}{2}(y'')^2 - \frac{Mg}{2}(y')^2 \right\} dz.$$

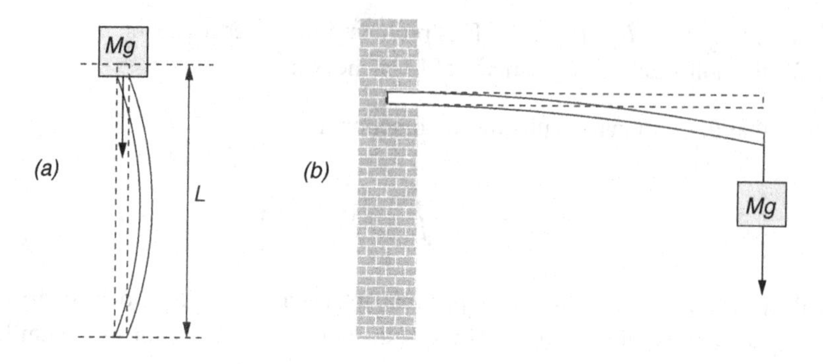

Figure 1.14 A rod used as: (a) a column, (b) a cantilever.

By considering small deformations of the form

$$y(z) = \sum_{n=1}^{\infty} a_n \sin \frac{n\pi z}{L}$$

show that the column is unstable to buckling and collapse if $Mg \geq \pi^2 YI/L^2$.

(b) *Leonardo da Vinci's problem: The light cantilever.* Here we take the z-axis as horizontal and the y-axis as being vertical (Figure 1.14b). The rod is used as a beam or cantilever and is fixed into a wall so that $y(0) = 0 = y'(0)$. A weight Mg is hung from the end $z = L$ and the beam sags in the $(-y)$-direction. We wish to find $y(z)$ for $0 < z < L$. We will ignore the weight of the beam itself.

- Write down the complete expression for the energy, including the gravitational potential energy of the weight.
- Find the differential equation and boundary conditions at $z = 0, L$ that arise from minimizing the total energy. In doing this take care not to throw away any term arising from the integration by parts. You may find the following identity to be of use:

$$\frac{d}{dz}(f'g'' - fg''') = f''g'' - fg''''.$$

- Solve the equation. You should find that the displacement of the end of the beam is $y(L) = -\frac{1}{3}MgL^3/YI$.

Exercise 1.5: Suppose that an elastic body Ω of density ρ is slightly deformed so that the point that was at cartesian coordinate x_i is moved to $x_i + \eta_i(x)$. We define the resulting *strain tensor* e_{ij} by

$$e_{ij} = \frac{1}{2}\left(\frac{\partial \eta_j}{\partial x_i} + \frac{\partial \eta_i}{\partial x_j}\right).$$

It is automatically symmetric in its indices. The Lagrangian for small-amplitude elastic motion of the body is

$$L[\eta] = \int_\Omega \left\{ \frac{1}{2}\rho \dot{\eta}_i^2 - \frac{1}{2} e_{ij} c_{ijkl} e_{kl} \right\} d^3x.$$

Here, c_{ijkl} is the tensor of *elastic constants*, which has the symmetries

$$c_{ijkl} = c_{klij} = c_{jikl} = c_{ijlk}.$$

By varying the η_i, show that the equation of motion for the body is

$$\rho \frac{\partial^2 \eta_i}{\partial t^2} - \frac{\partial}{\partial x_j} \sigma_{ji} = 0,$$

where

$$\sigma_{ij} = c_{ijkl} e_{kl}$$

is the *stress tensor*. Show that variations of η_i on the boundary $\partial\Omega$ give as boundary conditions

$$\sigma_{ij} n_j = 0,$$

where n_i are the components of the outward normal on $\partial\Omega$.

Problem 1.6: *The catenary revisited.* We can describe a catenary curve in parametric form as $x(s)$, $y(s)$, where s is the arc-length. The potential energy is then simply $\int_0^L \rho g y(s) ds$ where ρ is the mass per unit length of the hanging chain. The x, y are not independent functions of s, however, because $\dot{x}^2 + \dot{y}^2 = 1$ at every point on the curve. Here a dot denotes a derivative with respect to s.

(a) Introduce infinitely many Lagrange multipliers $\lambda(s)$ to enforce the $\dot{x}^2 + \dot{y}^2$ constraint, one for each point s on the curve. From the resulting functional derive two coupled equations describing the catenary, one for $x(s)$ and one for $y(s)$. By thinking about the forces acting on a small section of the cable, and perhaps by introducing the angle ψ where $\dot{x} = \cos\psi$ and $\dot{y} = \sin\psi$, so that s and ψ are *intrinsic coordinates* for the curve, interpret these equations and show that $\lambda(s)$ is proportional to the position-dependent tension $T(s)$ in the chain.

(b) You are provided with a lightweight line of length $\pi a/2$ and some lead shot of total mass M. By using equations from the previous part (suitably modified to take into account the position dependent $\rho(s)$) or otherwise, determine how the lead should be distributed along the line if the loaded line is to hang in an arc of a circle of radius a (see Figure 1.15) when its ends are attached to two points at the same height.

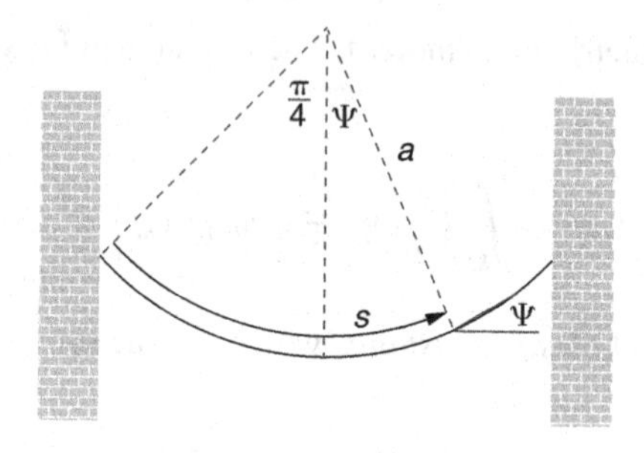

Figure 1.15 Weighted line.

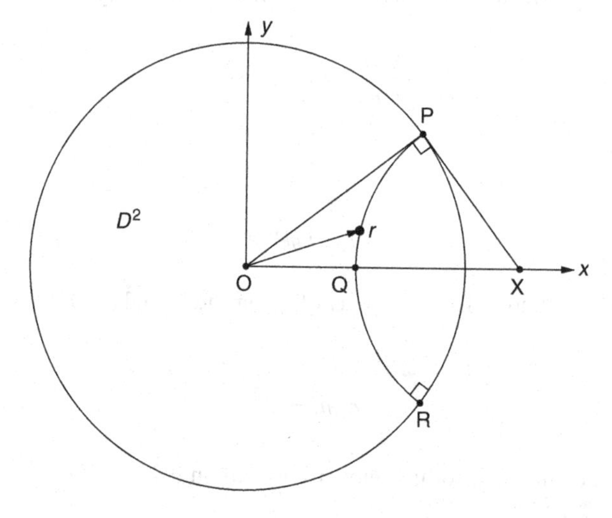

Figure 1.16 The Poincaré disc of Exercise 1.7. The radius OP of the Poincaré disc is unity, while the radius of the geodesic arc PQR is PX = QX = RX = R. The distance between the centres of the disc and arc is OX = x_0. Your task in part (c) is to show that ∠OPX = ∠ORX = 90°.

Problem 1.7: Another model for Lobachevski geometry (see Exercise 1.3) is the *Poincaré disc* (Figure 1.16). This space consists of the interior of the unit disc $D^2 = \{(x,y) \in \mathbb{R}^2 : x^2 + y^2 \leq 1\}$ equipped with the Riemann metric

$$ds^2 = \frac{dx^2 + dy^2}{(1 - x^2 - y^2)^2}.$$

The geodesic paths are found by minimizing the arc-length functional

$$s[\mathbf{r}] \equiv \int ds = \int \left\{ \frac{1}{1 - x^2 - y^2} \sqrt{\dot{x}^2 + \dot{y}^2} \right\} dt,$$

where $\mathbf{r}(t) = (x(t), y(t))$ and a dot indicates a derivative with respect to the parameter t.

(a) Either by manipulating the two Euler–Lagrange equations that give the conditions for $s[\mathbf{r}]$ to be stationary under variations in $\mathbf{r}(t)$, or, more efficiently, by observing that $s[\mathbf{r}]$ is invariant under the infinitesimal rotation

$$\begin{aligned}\delta x &= \varepsilon y\\ \delta y &= -\varepsilon x\end{aligned}$$

and applying Noether's theorem, show that the parametrized geodesics obey

$$\frac{d}{dt}\left(\frac{1}{1-x^2-y^2}\frac{x\dot{y}-y\dot{x}}{\sqrt{\dot{x}^2+\dot{y}^2}}\right)=0.$$

(b) Given a point (a,b) within D^2, and a direction through it, show that the equation you derived in part (a) determines a unique geodesic curve passing through (a,b) in the given direction, but does not determine the parametrization of the curve.

(c) Show that there exists a solution to the equation in part (a) in the form

$$\begin{aligned}x(t) &= R\cos t + x_0\\ y(t) &= R\sin t.\end{aligned}$$

Find a relation between x_0 and R, and from it deduce that the geodesics are circular arcs that cut the bounding unit circle (which plays the role of the line at infinity in the Lobachevski plane) at right angles.

Exercise 1.8: The Lagrangian for a particle of charge q is

$$L[\mathbf{x},\dot{\mathbf{x}}]=\frac{1}{2}m\dot{\mathbf{x}}^2-q\phi(\mathbf{x})+q\dot{\mathbf{x}}\cdot\mathbf{A}(\mathbf{x}).$$

Show that Lagrange's equation leads to

$$m\ddot{\mathbf{x}}=q(\mathbf{E}+\dot{\mathbf{x}}\times\mathbf{B}),$$

where

$$\mathbf{E}=-\nabla\phi-\frac{\partial\mathbf{A}}{\partial t},\qquad \mathbf{B}=\operatorname{curl}\mathbf{A}.$$

Exercise 1.9: Consider the action functional

$$S[\boldsymbol{\omega},\mathbf{p},\mathbf{r}]=\int\left(\frac{1}{2}I_1\omega_1^2+\frac{1}{2}I_2\omega_2^2+\frac{1}{2}I_3\omega_3^2+\mathbf{p}\cdot(\dot{\mathbf{r}}+\boldsymbol{\omega}\times\mathbf{r})\right\}dt,$$

where $\mathbf{r}$ and $\mathbf{p}$ are time-dependent 3-vectors, as is $\boldsymbol{\omega}=(\omega_1,\omega_2,\omega_3)$. Apply the action principle to obtain the equations of motion for $\mathbf{r},\mathbf{p},\boldsymbol{\omega}$ and show that they lead to Euler's

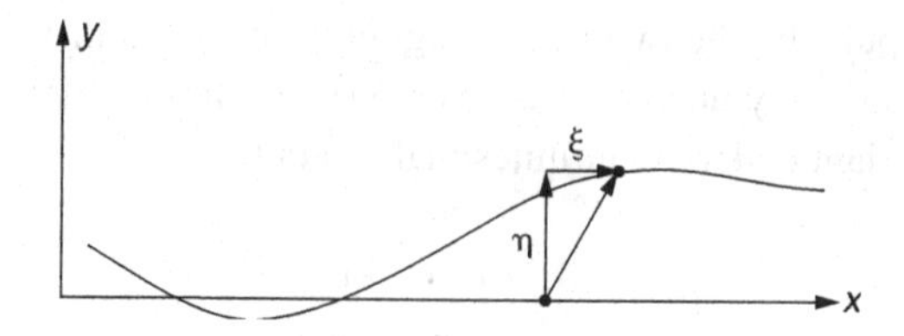

Figure 1.17 Vibrating piano string.

equations

$$
\begin{aligned}
I_1\dot{\omega}_1 - (I_2 - I_3)\omega_2\omega_3 &= 0, \\
I_2\dot{\omega}_2 - (I_3 - I_1)\omega_3\omega_1 &= 0, \\
I_3\dot{\omega}_3 - (I_1 - I_2)\omega_1\omega_2 &= 0,
\end{aligned}
$$

governing the angular velocity of a freely rotating rigid body.

Problem 1.10: *Piano string*. An elastic piano string can vibrate both transversely and longitudinally, and the two vibrations influence one another (Figure 1.17). A Lagrangian that takes into account the lowest-order effect of stretching on the local string tension, and can therefore model this coupled motion, is

$$
L[\xi,\eta] = \int dx \left\{ \frac{1}{2}\rho_0 \left[\left(\frac{\partial \xi}{\partial t}\right)^2 + \left(\frac{\partial \eta}{\partial t}\right)^2 \right] - \frac{\lambda}{2}\left[\frac{\tau_0}{\lambda} + \frac{\partial \xi}{\partial x} + \frac{1}{2}\left(\frac{\partial \eta}{\partial x}\right)^2 \right]^2 \right\}.
$$

Here $\xi(x,t)$ is the longitudinal displacement and $\eta(x,t)$ the transverse displacement of the string. Thus, the point that in the undisturbed string had coordinates $[x, 0]$ is moved to the point with coordinates $[x+\xi(x,t), \eta(x,t)]$. The parameter τ_0 represents the tension in the undisturbed string, λ is the product of Young's modulus and the cross-sectional area of the string and ρ_0 is the mass per unit length.

(a) Use the action principle to derive the two coupled equations of motion, one involving $\dfrac{\partial^2 \xi}{\partial t^2}$ and one involving $\dfrac{\partial^2 \eta}{\partial t^2}$.
(b) Show that when we linearize these two equations of motion, the longitudinal and transverse motions decouple. Find expressions for the longitudinal (c_{L}) and transverse (c_{T}) wave velocities in terms of τ_0, ρ_0 and λ.
(c) Assume that a given transverse pulse $\eta(x,t) = \eta_0(x - c_{\mathrm{T}}t)$ propagates along the string. Show that this induces a concurrent longitudinal pulse of the form $\xi(x - c_{\mathrm{T}}t)$. Show further that the longitudinal Newtonian momentum density in this concurrent pulse is given by

$$
\rho_0 \frac{\partial \xi}{\partial t} = \frac{1}{2}\frac{c_{\mathrm{L}}^2}{c_{\mathrm{L}}^2 - c_{\mathrm{T}}^2} T^0{}_1
$$

where

$$T^0{}_1 \equiv -\rho_0 \frac{\partial \eta}{\partial x} \frac{\partial \eta}{\partial t}$$

is the associated pseudo-momentum density.

The forces that created the transverse pulse will also have created other longitudinal waves that travel at c_L. Consequently the Newtonian x-momentum moving at c_T is not the only x-momentum on the string, and the total "true" longitudinal momentum density is not simply proportional to the pseudo-momentum density.

Exercise 1.11: Obtain the canonical energy–momentum tensor $T^\nu{}_\mu$ for the barotropic fluid described by (1.119). Show that its conservation leads to both the momentum conservation equation (1.128), and the energy conservation equation

$$\partial_t \mathcal{E} + \partial_i \{v_i(\mathcal{E} + P)\},$$

where the energy density is

$$\mathcal{E} = \frac{1}{2}\rho(\nabla\phi)^2 + u(\rho).$$

Interpret the energy flux as being the sum of the convective transport of energy together with the rate of working by an element of fluid on its neighbours.

Problem 1.12: Consider the action functional[3]

$$S[\mathbf{v}, \rho, \phi, \beta, \gamma] = \int d^4x \left\{ -\frac{1}{2}\rho \mathbf{v}^2 - \phi\left(\frac{\partial \rho}{\partial t} + \operatorname{div}(\rho \mathbf{v})\right) \right. \\ \left. + \rho\beta\left(\frac{\partial \gamma}{\partial t} + (\mathbf{v}\cdot\nabla)\gamma\right) + u(\rho)\right\},$$

which is a generalization of (1.177) to include two new scalar fields β and γ. Show that varying $\mathbf{v}$ leads to

$$\mathbf{v} = \nabla\phi + \beta\nabla\gamma.$$

This is the *Clebsch representation* of the velocity field. It allows for flows with non-zero vorticity

$$\boldsymbol{\omega} \equiv \operatorname{curl} \mathbf{v} = \nabla\beta \times \nabla\gamma.$$

[3] H. Bateman, *Proc. Roy. Soc. Lond.* A, **125** (1929) 598; C. C. Lin, *Liquid Helium* in *Proc. Int. Sch. Phys. "Enrico Fermi", Course XXI* (Academic Press, 1965).

Show that the equations that arise from varying the remaining fields ρ, ϕ, β, γ together imply the mass conservation equation

$$\frac{\partial \rho}{\partial t} + \operatorname{div}(\rho \mathbf{v}) = 0,$$

and Bernoulli's equation in the form

$$\frac{\partial \mathbf{v}}{\partial t} + \boldsymbol{\omega} \times \mathbf{v} = -\nabla\left(\frac{1}{2}\mathbf{v}^2 + h\right).$$

(Recall that $h = du/d\rho$.) Show that this form of Bernoulli's equation is equivalent to Euler's equation

$$\frac{\partial \mathbf{v}}{\partial t} + (\mathbf{v} \cdot \nabla)\mathbf{v} = -\nabla h.$$

Consequently S provides an action principle for a general inviscid barotropic flow.

Exercise 1.13: *Drums and membranes*. The shape of a distorted drumskin is described by the function $h(x, y)$, which gives the height to which the point (x, y) of the flat undistorted drumskin is displaced.

(a) Show that the area of the distorted drumskin is equal to

$$\text{Area}[h] = \int dx\, dy \sqrt{1 + \left(\frac{\partial h}{\partial x}\right)^2 + \left(\frac{\partial h}{\partial y}\right)^2},$$

where the integral is taken over the area of the flat drumskin.

(b) Show that for small distortions, the area reduces to

$$\mathcal{A}[h] = \text{const.} + \frac{1}{2}\int dx\, dy\, |\nabla h|^2.$$

(c) Show that if h satisfies the two-dimensional Laplace equation then $\mathcal{A}$ is stationary with respect to variations that vanish at the boundary.

(d) Suppose the drumskin has mass ρ_0 per unit area, and surface tension T. Write down the Lagrangian controlling the motion of the drumskin, and derive the equation of motion that follows from it.

Problem 1.14: *The Wulff construction.* The surface-area functional of the previous exercise can be generalized so as to find the equilibrium shape of a crystal. We describe the crystal surface by giving its height $z(x, y)$ above the xy-plane, and introduce the direction-dependent surface tension (the surface free-energy per unit area) $\alpha(p, q)$, where

$$p = \frac{\partial z}{\partial x}, \quad q = \frac{\partial z}{\partial y}. \qquad (\star)$$

We seek to minimize the total surface free energy

$$F[z] = \int dxdy \left\{ \alpha(p,q)\sqrt{1+p^2+q^2} \right\},$$

subject to the constraint that the volume of the crystal

$$V[z] = \int z\, dxdy$$

remains constant.

(a) Enforce the volume constraint by introducing a Lagrange multiplier $2\lambda^{-1}$, and so obtain the Euler–Lagrange equation

$$\frac{\partial}{\partial x}\left(\frac{\partial f}{\partial p}\right) + \frac{\partial}{\partial y}\left(\frac{\partial f}{\partial q}\right) = 2\lambda^{-1}.$$

Here

$$f(p,q) = \alpha(p.q)\sqrt{1+p^2+q^2}.$$

(b) Show in the isotropic case, where α is constant, that

$$z(x,y) = \sqrt{(\alpha\lambda)^2 - (x-a)^2 - (y-b)^2} + \text{const.}$$

is a solution of the Euler–Lagrange equation. In this case, therefore, the equilibrium shape is a sphere.

An obvious way to satisfy the Euler–Lagrange equation in the general anisotropic case would be to arrange things so that

$$x = \lambda\frac{\partial f}{\partial p}, \quad y = \lambda\frac{\partial f}{\partial q}. \tag{$\star\star$}$$

(c) Show that ($\star\star$) is exactly the relationship we would have if $z(x,y)$ and $\lambda f(p,q)$ were Legendre transforms of each other, i.e. if

$$\lambda f(p,q) = px + qy - z(x,y),$$

where the x and y on the right-hand side are functions of p, q obtained by solving ($\star$). Do this by showing that the inverse relation is

$$z(x,y) = px + qy - \lambda f(p,q)$$

where now the p, q on the right-hand side become functions of x and y, and are obtained by solving ($\star\star$).

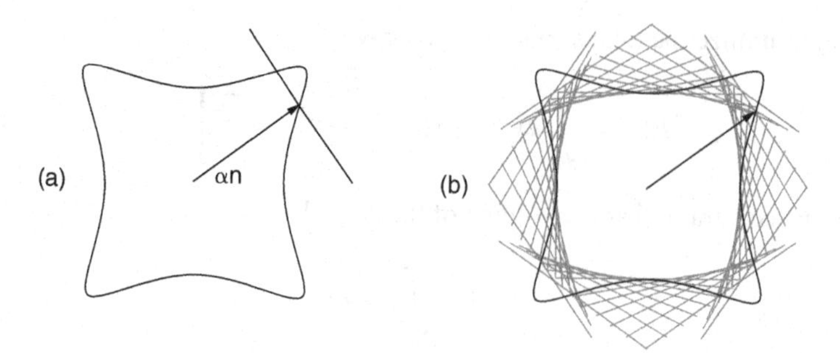

Figure 1.18 Two-dimensional Wulff crystal. (a) Polar plot of surface tension α as a function of the normal $\mathbf{n}$ to a crystal face, together with a line perpendicular to $\mathbf{n}$ at distance α from the origin. (b) Wulff's construction of the corresponding crystal surface as the envelope of the family of perpendicular lines. In this case, the minimum-energy crystal has curved faces, but sharp corners. The envelope continues beyond the corners, but these parts are unphysical.

For real crystals, $\alpha(p,q)$ can have the property of being a continuous-but-nowhere-differentiable function, and so the differential calculus used in deriving the Euler–Lagrange equation is inapplicable. The Legendre transformation, however, has a geometric interpretation that is more robust than its calculus-based derivation.

Recall that if we have a two-parameter family of surfaces in $\mathbb{R}^3$ given by $F(x,y,z;p,q) = 0$, then the equation of the *envelope* of the surfaces is found by solving the equations

$$0 = F = \frac{\partial F}{\partial p} = \frac{\partial F}{\partial q}$$

so as to eliminate the parameters p, q.

(d) Show that the equation

$$F(x,y,z;p,q) \equiv px + qy - z - \lambda\alpha(p,q)\sqrt{1+p^2+q^2} = 0$$

describes a family of planes perpendicular to the unit vectors

$$\mathbf{n} = \frac{(p,q,-1)}{\sqrt{1+p^2+q^2}}$$

and at a distance $\lambda\alpha(p,q)$ away from the origin.

(e) Show that the equations to be solved for the envelope of this family of planes are exactly those that determine $z(x,y)$. Deduce that, for smooth $\alpha(p,q)$, the profile $z(x,y)$ is this envelope.

Wulff conjectured[4] that, even for non-smooth $\alpha(p,q)$, the minimum-energy shape is given by an equivalent geometric construction: erect the planes from part (d) and, for

[4] G. Wulff, *Zeitschrift für Kristallografie,* **34** (1901) 449.

each plane, discard the half-space of $\mathbb{R}^3$ that lies on the far side of the plane from the origin. The convex region consisting of the intersection of the retained half-spaces is the crystal. When $\alpha(p, q)$ is smooth this "Wulff body" is bounded by part of the envelope of the planes. (The parts of the envelope not bounding the convex body – the "swallowtails" visible in Figure 1.18 – are unphysical.) When $\alpha(p, q)$ has cusps, these singularities can give rise to flat facets which are often joined by rounded edges. A proof of Wulff's claim had to wait 43 years until 1944, when it was established by use of the Brunn–Minkowski inequality.[5]

[5] A. Dinghas, *Zeitshrift für Kristallografie,* **105** (1944) 304. For a readable modern account see: R. Gardner, *Bulletin Amer. Math. Soc.* **39** (2002) 355.

2
Function spaces

Many differential equations of physics are relations involving *linear differential operators*. These operators, like matrices, are linear maps acting on vector spaces. The new feature is that the elements of the vector spaces are functions, and the spaces are infinite dimensional. We can try to survive in these vast regions by relying on our experience in finite dimensions, but sometimes this fails, and more sophistication is required.

2.1 Motivation

In the previous chapter we considered two variational problems:

(1) Find the stationary points of

$$F(\mathbf{x}) = \frac{1}{2}\mathbf{x} \cdot \mathbf{A}\mathbf{x} = \frac{1}{2}x_i A_{ij} x_j \tag{2.1}$$

on the surface $\mathbf{x} \cdot \mathbf{x} = 1$. This led to the matrix eigenvalue equation

$$\mathbf{A}\mathbf{x} = \lambda\mathbf{x}. \tag{2.2}$$

(2) Find the stationary points of

$$J[y] = \int_a^b \frac{1}{2}\left\{p(x)(y')^2 + q(x)y^2\right\} dx, \tag{2.3}$$

subject to the conditions $y(a) = y(b) = 0$ and

$$K[y] = \int_a^b y^2\, dx = 1. \tag{2.4}$$

This led to the differential equation

$$-(py')' + qy = \lambda y, \quad y(a) = y(b) = 0. \tag{2.5}$$

There will be a solution that satisfies the boundary conditions only for a discrete set of values of λ.

The stationary points of both function and functional are therefore determined by *linear eigenvalue problems*. The only difference is that the finite matrix in the first is replaced in the second by a linear differential operator. The theme of the next few chapters is an exploration of the similarities and differences between finite matrices and linear differential operators. In this chapter we will focus on how the functions on which the derivatives act can be thought of as vectors.

2.1.1 Functions as vectors

Consider $F[a,b]$, the set of all real (or complex) valued functions $f(x)$ on the interval $[a,b]$. This is a *vector space* over the field of the real (or complex) numbers: given two functions $f_1(x)$ and $f_2(x)$, and two numbers λ_1 and λ_2, we can form the sum $\lambda_1 f_1(x) + \lambda_2 f_2(x)$ and the result is still a function on the same interval. Examination of the axioms listed in Appendix A will show that $F[a,b]$ possesses all the other attributes of a vector space as well. We may think of the array of numbers $(f(x))$ for $x \in [a,b]$ as being the components of the vector. Since there is an infinity of independent components – one for each point x – the space of functions is infinite dimensional.

The set of *all* functions is usually too large for us. We will restrict ourselves to subspaces of functions with nice properties, such as being continuous or differentiable. There is some fairly standard notation for these spaces: the space of C^n functions (those which have n continuous derivatives) is called $C^n[a,b]$. For smooth functions (those with derivatives of all orders) we write $C^\infty[a,b]$. For the space of analytic functions (those whose Taylor expansion actually converges to the function) we write $C^\omega[a,b]$. For C^∞ functions defined on the whole real line we write $C^\infty(\mathbb{R})$. For the subset of functions with compact support (those that vanish outside some finite interval) we write $C_0^\infty(\mathbb{R})$. There are no non-zero analytic functions with compact support: $C_0^\omega(\mathbb{R}) = \{0\}$.

2.2 Norms and inner products

We are often interested in "how large" a function is. This leads to the idea of *normed* function spaces. There are many measures of function size. Suppose $R(t)$ is the number of inches per hour of rainfall. If you are a farmer you are probably most concerned with the total amount of rain that falls. A big rain has big $\int |R(t)|\, dt$. If you are the Urbana city engineer worrying about the capacity of the sewer system to cope with a downpour, you are primarily concerned with the maximum value of $R(t)$. For you a big rain has a big "sup $|R(t)|$".[1]

[1] Here "sup", short for *supremum*, is synonymous with the "least upper bound" of a set of numbers, i.e. the smallest number that is exceeded by no number in the set. This concept is more useful than "maximum" because the supremum need not be an element of the set. It is an axiom of the real number system that any bounded set of real numbers has a least upper bound. The "greatest lower bound" is denoted "inf", for *infimum*.

2.2.1 Norms and convergence

We can seldom write down an exact solution function to a real-world problem. We are usually forced to use numerical methods, or to expand as a power series in some small parameter. The result is a sequence of approximate solutions $f_n(x)$, which we hope will converge to the desired exact solution $f(x)$ as we make the numerical grid smaller, or take more terms in the power series.

Because there is more than one way to measure of the "size" of a function, the convergence of a sequence of functions f_n to a limit function f is not as simple a concept as the convergence of a sequence of numbers x_n to a limit x. Convergence means that the distance between the f_n and the limit function f gets smaller and smaller as n increases, so each different measure of this distance provides a new notion of what it means to converge. We are not going to make much use of formal "ε, δ" analysis, but you must realize that this distinction between different forms of convergence is not merely academic: real-world engineers must be precise about the kind of errors they are prepared to tolerate, or else a bridge they design might collapse. Graduate-level *engineering* courses in mathematical methods therefore devote much time to these issues. While physicists do not normally face the same legal liabilities as engineers, we should at least have it clear in our own minds what we mean when we write that $f_n \to f$.
Here are some common forms of convergence:

(i) If, for each x in its domain of definition $\mathcal{D}$, the set of numbers $f_n(x)$ converges to $f(x)$, then we say the sequence converges *pointwise*.

(ii) If the maximum separation

$$\sup_{x\in\mathcal{D}} |f_n(x) - f(x)| \tag{2.6}$$

goes to zero as $n \to \infty$, then we say that f_n converges to f *uniformly* on $\mathcal{D}$.

(iii) If

$$\int_{\mathcal{D}} |f_n(x) - f(x)|\, dx \tag{2.7}$$

goes to zero as $n \to \infty$, then we say that f_n converges *in the mean* to f on $\mathcal{D}$.

Uniform convergence implies pointwise convergence, but not *vice versa*. If $\mathcal{D}$ is a finite interval, then uniform convergence implies convergence in the mean, but convergence in the mean implies neither uniform nor pointwise convergence.

Example: Consider the sequence $f_n = x^n$ ($n = 1, 2, \ldots$) and $\mathcal{D} = [0, 1)$. Here, the round and square bracket notation means that the point $x = 0$ (Figure 2.1) is included in the interval, but the point 1 is excluded.

As n becomes large we have $x^n \to 0$ pointwise in $\mathcal{D}$, but the convergence is *not* uniform because

$$\sup_{x\in\mathcal{D}} |x^n - 0| = 1 \tag{2.8}$$

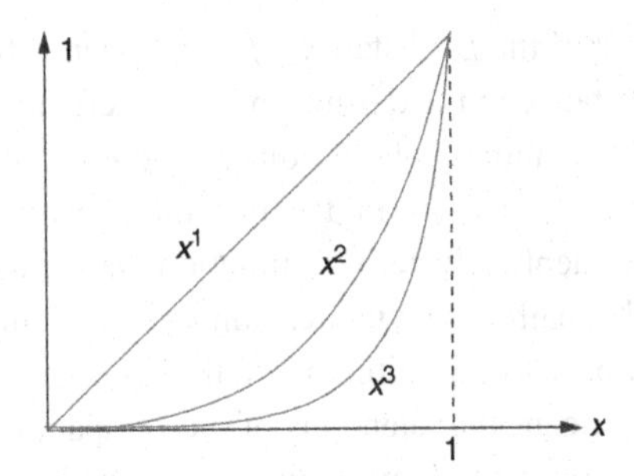

Figure 2.1 $x^n \to 0$ on $[0, 1)$, but not uniformly.

for all n.

Example: Let $f_n = x^n$ with $\mathcal{D} = [0, 1]$. Now the two square brackets mean that both $x = 0$ and $x = 1$ are to be included in the interval. In this case we have neither uniform nor pointwise convergence of the x^n to zero, but $x^n \to 0$ in the mean.

We can describe uniform convergence by means of a *norm* – a generalization of the usual measure of the length of a vector. A norm, denoted by $\|f\|$, of a vector f (a function, in our case) is a real number that obeys

(i) positivity: $\|f\| \geq 0$, and $\|f\| = 0 \Leftrightarrow f = 0$;
(ii) the *triangle inequality*: $\|f + g\| \leq \|f\| + \|g\|$;
(iii) linear homogeneity: $\|\lambda f\| = |\lambda| \|f\|$.

One example is the "sup" norm, which is defined by

$$\|f\|_\infty = \sup_{x \in \mathcal{D}} |f(x)|. \tag{2.9}$$

This number is guaranteed to be finite if f is continuous and $\mathcal{D}$ is compact. In terms of the sup norm, uniform convergence is the statement that

$$\lim_{n \to \infty} \|f_n - f\|_\infty = 0. \tag{2.10}$$

2.2.2 Norms from integrals

The space $L^p[a, b]$, for any $1 \leq p < \infty$, is defined to be our $F[a, b]$ equipped with

$$\|f\|_p = \left(\int_a^b |f(x)|^p \, dx \right)^{1/p}, \tag{2.11}$$

as the measure of length, and with a restriction to functions for which $\|f\|_p$ is finite.

We say that $f_n \to f$ in L^p if the L^p distance $\|f - f_n\|_p$ tends to zero. We have already seen the L^1 measure of distance in the definition of convergence in the mean. As in that case, convergence in L^p says nothing about pointwise convergence.

We would like to regard $\|f\|_p$ as a norm. It is possible, however, for a function to have $\|f\|_p = 0$ without f being identically zero – a function that vanishes at all but a finite set of points, for example. This pathology violates number (i) in our list of requirements for something to be called a norm, but we circumvent the problem by simply declaring such functions to be zero. This means that elements of the L^p spaces are not really functions, but only *equivalence classes* of functions – two functions being regarded as the same if they differ by a function of zero length. Clearly these spaces are not for use when anything significant depends on the value of the function at any precise point. They *are* useful in physics, however, because we can never measure a quantity at an exact position in space or time. We usually measure some sort of local average.

The L^p norms satisfy the triangle inequality for all $1 \le p \le \infty$, although this is not exactly trivial to prove.

An important property for any space to have is that of being *complete*. Roughly speaking, a space is complete if when some sequence of elements of the space look as if they are converging, then they are indeed converging and their limit is an element of the space. To make this concept precise, we need to say what we mean by the phrase "look as if they are converging". This we do by introducing the idea of a *Cauchy sequence*.

Definition: *A sequence f_n in a normed vector space is Cauchy if for any $\varepsilon > 0$ we can find an N such that $n, m > N$ implies that $\|f_m - f_n\| < \varepsilon$.*

This definition can be loosely paraphrased to say that the elements of a Cauchy sequence get arbitrarily close to each other as $n \to \infty$.

A normed vector space is *complete* with respect to its norm if every Cauchy sequence actually converges to some element in the space. Consider. for example, the normed vector space $\mathbb{Q}$ of rational numbers with distance measured in the usual way as $\|q_1 - q_2\| \equiv |q_1 - q_2|$. The sequence

$$
\begin{aligned}
q_0 &= 1.0, \\
q_1 &= 1.4, \\
q_2 &= 1.41, \\
q_3 &= 1.414, \\
&\vdots
\end{aligned}
$$

consisting of successive decimal approximations to $\sqrt{2}$, obeys

$$
|q_n - q_m| < \frac{1}{10^{\min(n,m)}} \tag{2.12}
$$

and so is Cauchy. Pythagoras famously showed that $\sqrt{2}$ is irrational, however, and so this sequence of rational numbers has no limit in $\mathbb{Q}$. Thus $\mathbb{Q}$ is *not* complete. The space $\mathbb{R}$ of real numbers is constructed by filling in the gaps between the rationals, and so *completing* $\mathbb{Q}$. A real number such as $\sqrt{2}$ is defined as a Cauchy sequence of rational numbers (by giving a rule, for example, that determines its infinite decimal expansion), with two rational sequences q_n and q'_n defining the same real number if $q_n - q'_n$ converges to zero.

A complete normed vector space is called a *Banach space*. If we interpret the norms as Lebesgue integrals[2] then the $L^p[a,b]$ are complete, and therefore Banach spaces. The theory of Lebesgue integration is rather complicated, however, and is not really necessary. One way of avoiding it is explained in Exercise 2.2.

Exercise 2.1: Show that any convergent sequence is Cauchy.

2.2.3 Hilbert space

The Banach space $L^2[a,b]$ is special in that it is also a *Hilbert space*. This means that its norm is derived from an inner product. If we define the inner product

$$\langle f,g\rangle = \int_a^b f^* g\,dx \tag{2.13}$$

then the $L^2[a,b]$ norm can be written

$$\|f\|_2 = \sqrt{\langle f,f\rangle}. \tag{2.14}$$

When we omit the subscript on a norm, we mean it to be this one. You are probably familiar with this Hilbert space from your quantum mechanics classes.

Being positive definite, the inner product satisfies the *Cauchy–Schwarz–Bunyakovsky inequality*

$$|\langle f,g\rangle| \le \|f\|\,\|g\|. \tag{2.15}$$

That this is so can be seen by observing that

$$\langle \lambda f + \mu g, \lambda f + \mu g\rangle = \begin{pmatrix}\lambda^*, & \mu^*\end{pmatrix}\begin{pmatrix}\|f\|^2 & \langle f,g\rangle \\ \langle f,g\rangle^* & \|g\|^2\end{pmatrix}\begin{pmatrix}\lambda \\ \mu\end{pmatrix} \tag{2.16}$$

must be non-negative for any choice of λ and μ. We therefore select $\lambda = \|g\|$, $\mu = -\langle f,g\rangle^*\|g\|^{-1}$, in which case the non-negativity of (2.16) becomes the statement that

$$\|f\|^2\|g\|^2 - |\langle f,g\rangle|^2 \ge 0. \tag{2.17}$$

[2] The "L" in L^p honours Henri Lebesgue. Banach spaces are named after Stefan Banach, who was one of the founders of functional analysis, a subject largely developed by him and other habitués of the Scottish Café in Lvóv, Poland.

From Cauchy–Schwarz–Bunyakovsky we can establish the triangle inequality:

$$\begin{aligned}\|f+g\|^2 &= \|f\|^2 + \|g\|^2 + 2\mathrm{Re}\langle f,g\rangle \\ &\le \|f\|^2 + \|g\|^2 + 2|\langle f,g\rangle|, \\ &\le \|f\|^2 + \|g\|^2 + 2\|f\|\|g\|, \\ &= (\|f\| + \|g\|)^2, \end{aligned} \tag{2.18}$$

so

$$\|f+g\| \le \|f\| + \|g\|. \tag{2.19}$$

A second important consequence of Cauchy–Schwarz–Bunyakovsky is that if $f_n \to f$ in the sense that $\|f_n - f\| \to 0$, then

$$\begin{aligned}|\langle f_n, g\rangle - \langle f, g\rangle| &= |\langle (f_n - f), g\rangle| \\ &\le \|f_n - f\|\,\|g\| \end{aligned} \tag{2.20}$$

tends to zero, and so

$$\langle f_n, g\rangle \to \langle f, g\rangle. \tag{2.21}$$

This means that the inner product $\langle f,g\rangle$ is a *continuous* functional of f and g. Take care to note that this continuity hinges on $\|g\|$ being finite. It is for this reason that we do not permit $\|g\| = \infty$ functions to be elements of our Hilbert space.

Orthonormal sets

Once we are in possession of an inner product, we can introduce the notion of an *orthonormal set*. A set of functions $\{u_n\}$ is orthonormal if

$$\langle u_n, u_m\rangle = \delta_{nm}. \tag{2.22}$$

For example,

$$2\int_0^1 \sin(n\pi x)\sin(m\pi x)\,dx = \delta_{nm}, \quad n, m = 1, 2, \ldots \tag{2.23}$$

so the set of functions $u_n = \sqrt{2}\sin n\pi x$ is orthonormal on [0, 1]. This set of functions is also *complete* – in a different sense, however, from our earlier use of this word. An orthonormal set of functions is said to be complete if any function f for which $\|f\|^2$ is finite, and hence f an element of the Hilbert space, has a convergent expansion

$$f(x) = \sum_{n=0}^{\infty} a_n u_n(x).$$

If we assume that such an expansion exists, and that we can freely interchange the order of the sum and integral, we can multiply both sides of this expansion by $u_m^*(x)$, integrate over x, and use the orthonormality of the u_n's to read off the expansion coefficients as $a_n = \langle u_n, f \rangle$. When

$$\|f\|^2 = \int_0^1 |f(x)|^2 \, dx \tag{2.24}$$

and $u_n = \sqrt{2}\sin(n\pi x)$, the result is the half-range sine Fourier series.

Example: *Expanding unity*. Suppose $f(x) = 1$. Since $\int_0^1 |f|^2 dx = 1$ is finite, the function $f(x) = 1$ can be represented as a convergent sum of the $u_n = \sqrt{2}\sin(n\pi x)$.

The inner product of f with the u_n's is

$$\langle u_n, f \rangle = \int_0^1 \sqrt{2}\sin(n\pi x)\, dx = \begin{cases} 0, & \text{n even,} \\ \frac{2\sqrt{2}}{n\pi}, & \text{n odd.} \end{cases}$$

Thus,

$$1 = \sum_{n=0}^{\infty} \frac{4}{(2n+1)\pi} \sin\Big((2n+1)\pi x\Big), \quad \text{in} \quad L^2[0,1]. \tag{2.25}$$

It is important to understand that the sum converges to the left-hand side in the closed interval $[0, 1]$ only in the L^2 sense. The series does not converge pointwise to unity at $x = 0$ or $x = 1$ – every term is zero at these points.

Figure 2.2 shows the sum of the series up to and including the term with $n = 30$. The $L^2[0, 1]$ measure of the distance between $f(x) = 1$ and this sum is

$$\int_0^1 \left| 1 - \sum_{n=0}^{30} \frac{4}{(2n+1)\pi} \sin\Big((2n+1)\pi x\Big) \right|^2 dx = 0.00654. \tag{2.26}$$

We can make this number as small as we desire by taking sufficiently many terms.

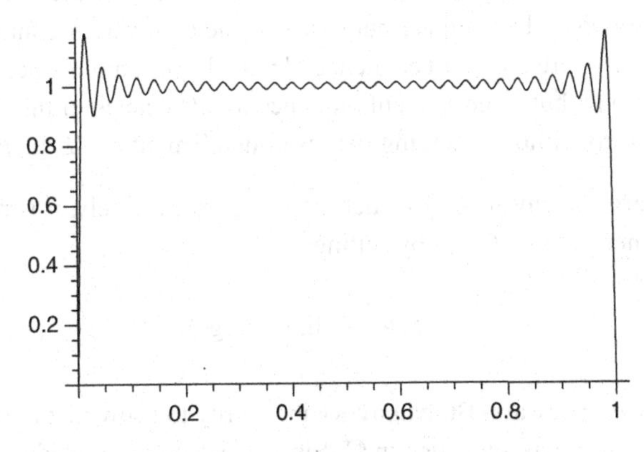

Figure 2.2 The sum of the first 31 terms in the sine expansion of $f(x) = 1$.

It is perhaps surprising that a set of functions that vanish at the endpoints of the interval can be used to expand a function that does not vanish at the ends. This exposes an important technical point: any finite sum of continuous functions vanishing at the endpoints is also a continuous function vanishing at the endpoints. It is therefore tempting to talk about the "subspace" of such functions. This set is indeed a vector space, and a subset of the Hilbert space, but it is not itself a Hilbert space. As the example shows, a Cauchy sequence of continuous functions vanishing at the endpoints of an interval can converge to a continuous function that does *not* vanish there. The "subspace" is therefore not complete in our original meaning of the term. The set of continuous functions vanishing at the endpoints fits into the whole Hilbert space much as the rational numbers fit into the real numbers: a finite sum of rationals is a rational number, but an infinite sum of rationals is not in general a rational number and we can obtain any real number as the limit of a sequence of rational numbers. The rationals $\mathbb{Q}$ are therefore a *dense* subset of the reals, and, as explained earlier, the reals are obtained by completing the set of rationals by adding to this set its limit points. In the same sense, the set of continuous functions vanishing at the endpoints is a dense subset of the whole Hilbert space and the whole Hilbert space is its completion.

Exercise 2.2: In this technical exercise we will explain in more detail how we "complete" a Hilbert space. The idea is to mirror the construction to the real numbers and *define* the elements of the Hilbert space to be Cauchy sequences of continuous functions. To specify a general element of $L^2[a,b]$ we must therefore exhibit a Cauchy sequence $f_n \in C[a,b]$. The choice is not unique: two Cauchy sequences $f_n^{(1)}(x)$ and $f_n^{(2)}(x)$ will specify the same element if

$$\lim_{n\to\infty} \|f_n^{(1)} - f_n^{(2)}\| = 0.$$

Such sequences are said to be *equivalent*. For convenience, we will write "$\lim_{n\to\infty} f_n = f$" but bear in mind that, in this exercise, this means that the sequence f_n *defines* the symbol f, and not that f is the limit of the sequence, as this limit need have no prior existence. We have deliberately written "f", and not "$f(x)$", for the "limit function" to warn us that f is assigned no unique numerical value at any x. A continuous function $f(x)$ can still be considered to be an element of $L^2[a,b]$ – take a sequence in which every $f_n(x)$ is equal to $f(x)$ – but an equivalent sequence of $f_n(x)$ can alter the limiting $f(x)$ on a set of measure zero without changing the resulting element $f \in L^2[a,b]$.

(i) If f_n and g_n are Cauchy sequences defining f, g, respectively, it is natural to try to define the inner product $\langle f,g\rangle$ by setting

$$\langle f,g\rangle \equiv \lim_{n\to\infty} \langle f_n, g_n\rangle.$$

Use the Cauchy–Schwarz–Bunyakovsky inequality to show that the numbers $F_n = \langle f_n, g_n\rangle$ form a Cauchy sequence in $\mathbb{C}$. Since $\mathbb{C}$ is complete, deduce that this limit exists. Next show that the limit is unaltered if either f_n or g_n is replaced by an equivalent sequence. Conclude that our tentative inner product is well defined.

(ii) The next, and harder, task is to show that the "completed" space is indeed complete. The problem is to show that any given Cauchy sequence $f_k \in L^2[a,b]$, where the f_k are *not* necessarily in $C[a,b]$, has a limit in $L^2[a,b]$. Begin by taking Cauchy sequences $f_{ki} \in C[a,b]$ such that $\lim_{i\to\infty} f_{ki} = f_k$. Use the triangle inequality to show that we can select a subsequence $f_{k,i(k)}$ that is Cauchy and so defines the desired limit.

Later we will show that the elements of $L^2[a,b]$ can be given a concrete meaning as *distributions*.

Best approximation

Let $u_n(x)$ be an orthonormal set of functions. The sum of the first N terms of the Fourier expansion of $f(x)$ in the u_n is the closest – measuring distance with the L^2 norm – that one can get to f whilst remaining in the space spanned by $u_1, u_2, \ldots, u_N$.

To see this, consider the square of the error-distance:

$$\begin{aligned}\Delta \stackrel{\text{def}}{=} \|f - \sum_1^N a_n u_n\|^2 &= \left\langle f - \sum_{m=1}^N a_m u_m, f - \sum_{n=1}^N a_n u_n \right\rangle \\ &= \|f\|^2 - \sum_{n=1}^N a_n \langle f, u_n\rangle - \sum_{m=1}^N a_m^* \langle u_m, f\rangle + \sum_{n,m=1}^N a_m^* a_n \langle u_m, u_n\rangle \\ &= \|f\|^2 - \sum_{n=1}^N a_n \langle f, u_n\rangle - \sum_{m=1}^N a_m^* \langle u_m, f\rangle + \sum_{n=1}^N |a_n|^2. \end{aligned} \tag{2.27}$$

In the last line we have used the orthonormality of the u_n. We can complete the squares, and rewrite Δ as

$$\Delta = \|f\|^2 - \sum_{n=1}^N |\langle u_n, f\rangle|^2 + \sum_{n=1}^N |a_n - \langle u_n, f\rangle|^2. \tag{2.28}$$

We seek to minimize Δ by a suitable choice of coefficients a_n. The smallest we can make it is

$$\Delta_{\min} = \|f\|^2 - \sum_{n=1}^N |\langle u_n, f\rangle|^2, \tag{2.29}$$

and we attain this bound by setting each of the $|a_n - \langle u_n, f\rangle|$ equal to zero. That is, by taking

$$a_n = \langle u_n, f\rangle. \tag{2.30}$$

Thus the Fourier coefficients $\langle u_n, f\rangle$ are the optimal choice for the a_n.

Suppose we have some *non-orthogonal* collection of functions g_n, $n = 1, \ldots, N$, and we have found the best approximation $\sum_{n=1}^{N} a_n g_n(x)$ to $f(x)$. Now suppose we are given a g_{N+1} to add to our collection. We may then seek an improved approximation $\sum_{n=1}^{N+1} a'_n g_n(x)$ by including this new function – but finding this better fit will generally involve tweaking *all* the a_n, not just trying different values of a_{N+1}. The great advantage of approximating by orthogonal functions is that, given another member of an orthonormal family, we can improve the precision of the fit by adjusting only the coefficient of the new term. We do not have to perturb the previously obtained coefficients.

Parseval's theorem

The "best approximation" result from the previous section allows us to give an alternative definition of a "complete orthonormal set", and to obtain the formula $a_n = \langle u_n, f\rangle$ for the expansion coefficients without having to assume that we can integrate the infinite series $\sum a_n u_n$ term-by-term. Recall that we said that a set of points S is a *dense* subset of a space T if any given point $x \in T$ is the limit of a sequence of points in S, i.e. there are elements of S lying arbitrarily close to x. For example, the set of rational numbers $\mathbb{Q}$ is a dense subset of $\mathbb{R}$. Using this language, we say that a set of orthonormal functions $\{u_n(x)\}$ is complete if the set of all finite linear combinations of the u_n is a dense subset of the entire Hilbert space. This guarantees that, by taking N sufficently large, our best approximation will approach arbitrarily close to our target function $f(x)$. Since the best approximation containing all the u_n up to u_N is the N-th partial sum of the Fourier series, this shows that the Fourier series actually converges to f.

We have therefore proved that if we are given $u_n(x)$, $n = 1, 2, \ldots$, a complete orthonormal set of functions on $[a, b]$, then any function for which $\|f\|^2$ is finite can be expanded as a convergent Fourier series

$$f(x) = \sum_{n=1}^{\infty} a_n u_n(x), \tag{2.31}$$

where

$$a_n = \langle u_n, f\rangle = \int_a^b u_n^*(x) f(x)\, dx. \tag{2.32}$$

The convergence is guaranteed only in the L^2 sense that

$$\lim_{N\to\infty} \int_a^b \left| f(x) - \sum_{n=1}^{N} a_n u_n(x) \right|^2 dx = 0. \tag{2.33}$$

Equivalently

$$\Delta_N = \|f - \sum_{n=1}^{N} a_n u_n\|^2 \to 0 \tag{2.34}$$

as $N \to \infty$. Now, we showed in the previous section that

$$\Delta_N = \|f\|^2 - \sum_{n=1}^{N} |\langle u_n, f\rangle|^2$$
$$= \|f\|^2 - \sum_{n=1}^{N} |a_n|^2, \tag{2.35}$$

and so the L^2 convergence is equivalent to the statement that

$$\|f\|^2 = \sum_{n=1}^{\infty} |a_n|^2. \tag{2.36}$$

This last result is called *Parseval's theorem*.

Example: In the expansion (2.25), we have $\|f^2\| = 1$ and

$$|a_n|^2 = \begin{cases} 8/(n^2\pi^2), & \text{n odd,} \\ 0, & \text{n even.} \end{cases} \tag{2.37}$$

Parseval therefore tells us that

$$\sum_{n=0}^{\infty} \frac{1}{(2n+1)^2} = 1 + \frac{1}{3^2} + \frac{1}{5^2} + \cdots = \frac{\pi^2}{8}. \tag{2.38}$$

Example: The functions $u_n(x) = \frac{1}{\sqrt{2\pi}} e^{inx}$, $n \in \mathbb{Z}$, form a complete orthonormal set on the interval $[-\pi, \pi]$. Let $f(x) = \frac{1}{\sqrt{2\pi}} e^{i\zeta x}$. Then its Fourier expansion is

$$\frac{1}{\sqrt{2\pi}} e^{i\zeta x} = \sum_{n=-\infty}^{\infty} c_n \frac{1}{\sqrt{2\pi}} e^{inx}, \quad -\pi < x < \pi, \tag{2.39}$$

where

$$c_n = \frac{1}{2\pi} \int_{-\pi}^{\pi} e^{i\zeta x} e^{-inx}\, dx = \frac{\sin(\pi(\zeta - n))}{\pi(\zeta - n)}. \tag{2.40}$$

We also have that

$$\|f\|^2 = \int_{-\pi}^{\pi} \frac{1}{2\pi} dx = 1. \tag{2.41}$$

Now Parseval tells us that

$$\|f\|^2 = \sum_{n=-\infty}^{\infty} \frac{\sin^2(\pi(\zeta - n))}{\pi^2(\zeta - n)^2}, \tag{2.42}$$

the left-hand side being unity.

Finally, as $\sin^2(\pi(\zeta - n)) = \sin^2(\pi\zeta)$, we have

$$\operatorname{cosec}^2(\pi\zeta) \equiv \frac{1}{\sin^2(\pi\zeta)} = \sum_{n=-\infty}^{\infty} \frac{1}{\pi^2(\zeta - n)^2}. \tag{2.43}$$

The end result is a quite non-trivial expansion for the square of the cosecant.

2.2.4 Orthogonal polynomials

A useful class of orthonormal functions are the sets of *orthogonal polynomials* associated with an interval $[a, b]$ and a positive weight function $w(x)$ such that $\int_a^b w(x)\,dx$ is finite. We introduce the Hilbert space $L^2_w[a, b]$ with the real inner product

$$\langle u, v\rangle_w = \int_a^b w(x)u(x)v(x)\,dx, \tag{2.44}$$

and apply the *Gram–Schmidt procedure* to the monomial powers $1, x, x^2, x^3, \ldots$ so as to produce an orthonomal set. We begin with

$$P_0(x) \equiv 1/\|1\|_w, \tag{2.45}$$

where $\|1\|_w = \sqrt{\int_a^b w(x)\,dx}$, and define recursively

$$P_{n+1}(x) = \frac{xP_n(x) - \sum_0^n P_i(x)\langle P_i, xP_n\rangle_w}{\|xP_n - \sum_0^n P_i\langle P_i, xP_n\rangle\|_w}. \tag{2.46}$$

Clearly $P_n(x)$ is an n-th order polynomial, and by construction

$$\langle P_n, P_m\rangle_w = \delta_{nm}. \tag{2.47}$$

All such sets of polynomials obey a three-term recurrence relation

$$xP_n(x) = b_nP_{n+1}(x) + a_nP_n(x) + b_{n-1}P_{n-1}(x). \tag{2.48}$$

That there are only three terms, and that the coefficients of P_{n+1} and P_{n-1} are related, is due to the identity

$$\langle P_n, xP_m\rangle_w = \langle xP_n, P_m\rangle_w. \tag{2.49}$$

This means that the matrix (in the P_n basis) representing the operation of multiplication by x is symmetric. Since multiplication by x takes us from P_n only to P_{n+1}, the matrix has just one non-zero entry above the main diagonal, and hence, by symmetry, only one below.

The completeness of a family of polynomials orthogonal on a finite interval is guaranteed by the *Weierstrass approximation theorem*, which asserts that for any continuous real function $f(x)$ on $[a, b]$, and for any $\varepsilon > 0$, there exists a polynomial $p(x)$ such that $|f(x) - p(x)| < \varepsilon$ for all $x \in [a, b]$. This means that polynomials are dense in the space of continuous functions equipped with the $\| \dots \|_\infty$ norm. Because $|f(x) - p(x)| < \varepsilon$ implies that

$$\int_a^b |f(x) - p(x)|^2 w(x)\, dx \le \varepsilon^2 \int_a^b w(x)\, dx, \tag{2.50}$$

they are also a dense subset of the continuous functions in the sense of $L^2_w[a, b]$ convergence. Because the Hilbert space $L^2_w[a, b]$ is defined to be the completion of the space of continuous functions, the continuous functions are automatically dense in $L^2_w[a, b]$. Now the triangle inequality tells us that a dense subset of a dense set is dense in the larger set, so the polynomials are dense in $L^2_w[a, b]$ itself. The normalized orthogonal polynomials therefore constitute a complete orthonormal set.

For later use, we here summarize the properties of the families of polynomials named after Legendre, Hermite and Tchebychef.

Legendre polynomials

Legendre polynomials have $a = -1$, $b = 1$ and $w = 1$. The standard Legendre polynomials are not normalized by the scalar product, but instead by setting $P_n(1) = 1$. They are given by *Rodriguez'* formula

$$P_n(x) = \frac{1}{2^n n!} \frac{d^n}{dx^n} (x^2 - 1)^n. \tag{2.51}$$

The first few are

$$\begin{aligned} P_0(x) &= 1, \\ P_1(x) &= x, \\ P_2(x) &= \frac{1}{2}(3x^2 - 1), \\ P_3(x) &= \frac{1}{2}(5x^3 - 3x), \\ P_4(x) &= \frac{1}{8}(35x^4 - 30x^2 + 3). \end{aligned}$$

Their inner product is

$$\int_{-1}^1 P_n(x) P_m(x)\, dx = \frac{2}{2n+1} \delta_{nm}. \tag{2.52}$$

The three-term recurrence relation is

$$(2n+1)xP_n(x) = (n+1)P_{n+1}(x) + nP_{n-1}(x). \tag{2.53}$$

The P_n form a complete set for expanding functions on $[-1, 1]$.

Hermite polynomials

The Hermite polynomials have $a = -\infty$, $b = +\infty$ and $w(x) = e^{-x^2}$, and are defined by the generating function

$$e^{2tx-t^2} = \sum_{n=0}^{\infty} \frac{1}{n!} H_n(x) t^n. \tag{2.54}$$

If we write

$$e^{2tx-t^2} = e^{x^2-(x-t)^2}, \tag{2.55}$$

we may use Taylor's theorem to find

$$H_n(x) = \left.\frac{d^n}{dt^n} e^{x^2-(x-t)^2}\right|_{t=0} = (-1)^n e^{x^2} \frac{d^n}{dx^n} e^{-x^2}, \tag{2.56}$$

which is a useful alternative definition. The first few Hermite polynomials are

$$\begin{aligned}
H_0(x) &= 1, \\
H_1(x) &= 2x, \\
H_2(x) &= 4x^2 - 2 \\
H_3(x) &= 8x^3 - 12x \\
H_4(x) &= 16x^4 - 48x^2 + 12.
\end{aligned}$$

The normalization is such that

$$\int_{-\infty}^{\infty} H_n(x) H_m(x) e^{-x^2}\, dx = 2^n n! \sqrt{\pi}\, \delta_{nm}, \tag{2.57}$$

as may be proved by using the generating function. The three-term recurrence relation is

$$2xH_n(x) = H_{n+1}(x) + 2nH_{n-1}(x). \tag{2.58}$$

Exercise 2.3: Evaluate the integral

$$F(s,t) = \int_{-\infty}^{\infty} e^{-x^2} e^{2sx-s^2} e^{2tx-t^2}\, dx$$

and expand the result as a double power series in s and t. By examining the coefficient of $s^n t^m$, show that

$$\int_{-\infty}^{\infty} H_n(x)H_m(x)e^{-x^2}\,dx = 2^n n!\sqrt{\pi}\,\delta_{nm}.$$

Problem 2.4: Let

$$\varphi_n(x) = \frac{1}{\sqrt{2^n n!\sqrt{\pi}}} H_n(x)e^{-x^2/2}$$

be the *normalized Hermite functions*. They form a complete orthonormal set in $L^2(\mathbb{R})$. Show that

$$\sum_{n=0}^{\infty} t^n \varphi_n(x)\varphi_n(y) = \frac{1}{\sqrt{\pi(1-t^2)}} \exp\left\{\frac{4xyt - (x^2+y^2)(1+t^2)}{2(1-t^2)}\right\}, \quad 0 \le t < 1.$$

This is *Mehler's formula*. (Hint: expand the right-hand side as $\sum_{n=0}^{\infty} a_n(x,t)\varphi_n(y)$. To find $a_n(x,t)$, multiply by $e^{2sy-s^2-y^2/2}$ and integrate over y.)

Exercise 2.5: Let $\varphi_n(x)$ be the same functions as in the preceding problem. Define a Fourier-transform operator $F : L^2(\mathbb{R}) \to L^2(\mathbb{R})$ by

$$F(f) = \frac{1}{\sqrt{2\pi}} \int_{-\infty}^{\infty} e^{ixs} f(s)\,ds.$$

With this normalization of the Fourier transform, F^4 is the identity map. The possible eigenvalues of F are therefore ± 1, $\pm i$. Starting from (2.56), show that the $\varphi_n(x)$ are eigenfunctions of F, and that

$$F(\varphi_n) = i^n \varphi_n(x).$$

Tchebychef polynomials

Tchebychef polynomials are defined by taking $a = -1, b = +1$ and $w(x) = (1-x^2)^{\pm 1/2}$. The *Tchebychef polynomials of the first kind* are

$$T_n(x) = \cos(n \cos^{-1} x). \tag{2.59}$$

The first few are

$$\begin{aligned} T_0(x) &= 1, \\ T_1(x) &= x, \\ T_2(x) &= 2x^2 - 1, \\ T_3(x) &= 4x^3 - 3x. \end{aligned}$$

The *Tchebychef polynomials of the second kind* are

$$U_{n-1}(x) = \frac{\sin(n\cos^{-1}x)}{\sin(\cos^{-1}x)} = \frac{1}{n}T_n'(x) \tag{2.60}$$

and the first few are

$$\begin{aligned} U_{-1}(x) &= 0, \\ U_0(x) &= 1, \\ U_1(x) &= 2x, \\ U_2(x) &= 4x^2 - 1, \\ U_3(x) &= 8x^3 - 4x. \end{aligned}$$

T_n and U_n obey the same recurrence relation

$$\begin{aligned} 2xT_n &= T_{n+1} + T_{n-1}, \\ 2xU_n &= U_{n+1} + U_{n-1}, \end{aligned}$$

which are disguised forms of elementary trigonometric identities. The orthogonality is also a disguised form of the orthogonality of the functions $\cos n\theta$ and $\sin n\theta$. After setting $x = \cos\theta$ we have

$$\int_0^\pi \cos n\theta \cos m\theta \, d\theta = \int_{-1}^1 \frac{1}{\sqrt{1-x^2}} T_n(x)T_m(x)\, dx = h_n\delta_{nm}, \quad n, m, \geq 0, \tag{2.61}$$

where $h_0 = \pi$, $h_n = \pi/2$, $n > 0$, and

$$\int_0^\pi \sin n\theta \sin m\theta \, d\theta = \int_{-1}^1 \sqrt{1-x^2}\, U_{n-1}(x)U_{m-1}(x)\, dx = \frac{\pi}{2}\delta_{nm}, \quad n, m > 0. \tag{2.62}$$

The set $\{T_n(x)\}$ is therefore orthogonal and complete in $L^2_{(1-x^2)^{-1/2}}[-1, 1]$, and the set $\{U_n(x)\}$ is orthogonal and complete in $L^2_{(1-x^2)^{1/2}}[-1, 1]$. Any function continuous on the closed interval $[-1, 1]$ lies in both of these spaces, and can therefore be expanded in terms of either set.

2.3 Linear operators and distributions

Our theme is the analogy between linear differential operators and matrices. It is therefore useful to understand how we can think of a differential operator as a continuously indexed "matrix".

2.3.1 Linear operators

The action of a matrix on a vector $\mathbf{y} = \mathbf{A}\mathbf{x}$ is given in components by

$$y_i = A_{ij}x_j. \tag{2.63}$$

The function-space analogue of this, $g = Af$, is naturally to be thought of as

$$g(x) = \int_a^b A(x,y)f(y)\,dy, \tag{2.64}$$

where the summation over adjacent indices has been replaced by an integration over the dummy variable y. If $A(x,y)$ is an ordinary function then $A(x,y)$ is called an *integral kernel*. We will study such linear operators in the chapter on integral equations.

The identity operation is

$$f(x) = \int_a^b \delta(x-y)f(y)\,dy, \tag{2.65}$$

and so the Dirac delta function, which is **not** an ordinary function, plays the role of the identity matrix. Once we admit *distributions* such as $\delta(x)$, we can think of differential operators as continuously indexed matrices by using the distribution

$$\delta'(x) = \text{“}\frac{d}{dx}\delta(x)\text{”}. \tag{2.66}$$

The quotes are to warn us that we are not really taking the derivative of the highly singular delta function. The symbol $\delta'(x)$ is properly defined by its behaviour in an integral

$$\begin{aligned}
\int_a^b \delta'(x-y)f(y)\,dy &= \int_a^b \frac{d}{dx}\delta(x-y)f(y)\,dy \\
&= -\int_a^b f(y)\frac{d}{dy}\delta(x-y)\,dy \\
&= \int_a^b f'(y)\delta(x-y)\,dy \quad \text{(integration by parts)} \\
&= f'(x).
\end{aligned}$$

The manipulations here are purely formal, and serve only to motivate the defining property

$$\int_a^b \delta'(x-y)f(y)\,dy = f'(x). \tag{2.67}$$

It is, however, sometimes useful to think of a smooth approximation to $\delta'(x-a)$ being the genuine derivative of a smooth approximation to $\delta(x-a)$, as illustrated in Figure 2.3.

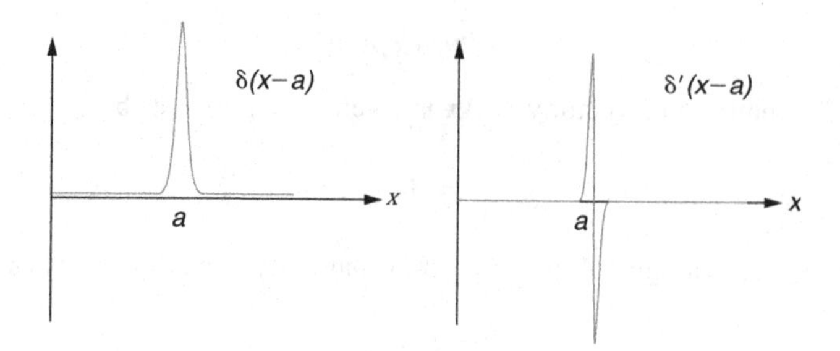

Figure 2.3 Smooth approximations to $\delta(x-a)$ and $\delta'(x-a)$.

We can now define higher "derivatives" of $\delta(x)$ by

$$\int_a^b \delta^{(n)}(x) f(x) dx = (-1)^n f^{(n)}(0), \tag{2.68}$$

and use them to represent any linear differential operator as a formal integral kernel.

Example: In Chapter 1 we formally evaluated a functional second derivative and ended up with the distributional kernel (1.186), which we here write as

$$\begin{aligned} k(x,y) &= -\frac{d}{dy}\left(p(y)\frac{d}{dy}\delta(y-x)\right) + q(y)\delta(y-x) \\ &= -p(y)\delta''(y-x) - p'(y)\delta'(y-x) + q(y)\delta(y-x). \end{aligned} \tag{2.69}$$

When k acts on a function u, it gives

$$\begin{aligned} \int k(x,y)u(y)\,dy &= \int \{-p(y)\delta''(y-x) - p'(y)\delta'(y-x) \\ &\qquad + q(y)\delta(y-x)\}\,u(y)\,dy \\ &= \int \delta(y-x)\left\{-[p(y)u(y)]'' + [p'(y)u(y)]' + q(y)u(y)\right\}\,dy \\ &= \int \delta(y-x)\left\{-p(y)u''(y) - p'(y)u'(y) + q(y)u(y)\right\}\,dy \\ &= -\frac{d}{dx}\left(p(x)\frac{du}{dx}\right) + q(x)u(x). \end{aligned} \tag{2.70}$$

The continuous matrix (1.186) therefore does, as indicated in Chapter 1, represent the Sturm–Liouville operator L defined in (1.182).

Exercise 2.6: Consider the distributional kernel

$$k(x,y) = a_2(y)\delta''(x-y) + a_1(y)\delta'(x-y) + a_0(y)\delta(x-y).$$

Show that

$$\int k(x,y)u(y)\,dy = (a_2(x)u(x))'' + (a_1(x)u(x))' + a_0(x)u(x).$$

Similarly show that

$$k(x,y) = a_2(x)\delta''(x-y) + a_1(x)\delta'(x-y) + a_0(x)\delta(x-y),$$

leads to

$$\int k(x,y)u(y)\,dy = a_2(x)u''(x) + a_1(x)u'(x) + a_0(x)u(x).$$

Exercise 2.7: The distributional kernel (2.69) was originally obtained as a functional second derivative

$$\begin{aligned} k(x_1,x_2) &= \frac{\delta}{\delta y(x_1)}\left(\frac{\delta J[y]}{\delta y(x_2)}\right) \\ &= -\frac{d}{dx_2}\left(p(x_2)\frac{d}{dx_2}\delta(x_2-x_1)\right) + q(x_2)\delta(x_2-x_1). \end{aligned}$$

By analogy with conventional partial derivatives, we would expect that

$$\frac{\delta}{\delta y(x_1)}\left(\frac{\delta J[y]}{\delta y(x_2)}\right) = \frac{\delta}{\delta y(x_2)}\left(\frac{\delta J[y]}{\delta y(x_1)}\right),$$

but x_1 and x_2 appear asymmetrically in $k(x_1,x_2)$. Define

$$k^T(x_1,x_2) = k(x_2,x_1),$$

and show that

$$\int k^T(x_1,x_2)u(x_2)\,dx_2 = \int k(x_1,x_2)u(x_2)\,dx_2.$$

Conclude that, superficial appearance notwithstanding, we do have $k(x_1,x_2) = k(x_2,x_1)$.

The example and exercises show that linear differential operators correspond to continuously infinite matrices having entries only infinitesimally close to their main diagonal.

2.3.2 Distributions and test-functions

It is possible to work most of the problems in this book with no deeper understanding of what a delta-function is than that presented in Section 2.3.1. At some point, however, the more careful reader will wonder about the logical structure of what we are doing, and

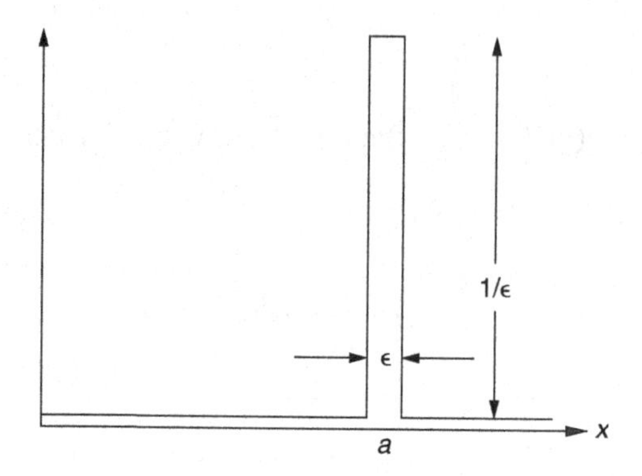

Figure 2.4 Approximation $\delta_\varepsilon(x-a)$ to $\delta(x-a)$.

will soon discover that too free a use of $\delta(x)$ and its derivatives can lead to paradoxes. How do such creatures fit into the function-space picture, and what sort of manipulations with them are valid?

We often think of $\delta(x)$ as being a "limit" of a sequence of functions whose graphs are getting narrower and narrower while their height grows to keep the area under the curve fixed. An example would be the spike function $\delta_\varepsilon(x-a)$ appearing in Figure 2.4.

The L^2 norm of δ_ε,

$$\|\delta_\varepsilon\|^2 = \int |\delta_\varepsilon(x)|^2\,dx = \frac{1}{\varepsilon}, \tag{2.71}$$

tends to infinity as $\varepsilon \to 0$, so δ_ε cannot be tending to any function in L^2. This delta function has infinite "length", and so is *not* an element of our Hilbert space.

The simple spike is not the only way to construct a delta function. In Fourier theory we meet

$$\delta_\Lambda(x) = \int_{-\Lambda}^{\Lambda} e^{ikx}\,\frac{dk}{2\pi} = \frac{1}{\pi}\frac{\sin \Lambda x}{x}, \tag{2.72}$$

which becomes a delta function when Λ becomes large. In this case

$$\|\delta_\Lambda\|^2 = \int_{-\infty}^{\infty} \frac{\sin^2 \Lambda x}{\pi^2 x^2}\,dx = \Lambda/\pi. \tag{2.73}$$

Again the "limit" has infinite length and cannot be accommodated in Hilbert space. This $\delta_\Lambda(x)$ is even more pathological than δ_ε. It provides a salutary counter-example to the often asserted "fact" that $\delta(x) = 0$ for $x \neq 0$. As Λ becomes large $\delta_\Lambda(0)$ diverges to infinity. At any fixed non-zero x, however, $\delta_\Lambda(x)$ oscillates between $\pm 1/x$ as Λ grows. Consequently the limit $\lim_{\Lambda\to\infty} \delta_\Lambda(x)$ exists nowhere. It therefore makes no sense to assign a numerical value to $\delta(x)$ at any x.

Given its wild behaviour, it is not surprising that mathematicians looked askance at Dirac's $\delta(x)$. It was only in 1944, long after its effectiveness in solving physics and engineering problems had become an embarrassment, that Laurent Schwartz was able to tame $\delta(x)$ by creating his *theory of distributions*. Using the language of distributions we can state precisely the conditions under which a manoeuvre involving singular objects such as $\delta'(x)$ is legitimate.

Schwartz' theory is built on a concept from linear algebra. Recall that the *dual space* V^* of a vector space V is the vector space of linear functions from the original vector space V to the field over which it is defined. We consider $\delta(x)$ to be an element of the dual space of a vector space $\mathcal{T}$ of *test functions*. When a test function $\varphi(x)$ is plugged in, the δ-machine returns the number $\varphi(0)$. This operation is a linear map because the action of δ on $\lambda\varphi(x) + \mu\chi(x)$ is to return $\lambda\varphi(0) + \mu\chi(0)$. Test functions are smooth (infinitely differentiable) functions that tend rapidly to zero at infinity. Exactly what class of function we chose for $\mathcal{T}$ depends on the problem at hand. If we are going to make extensive use of Fourier transforms, for example, we might select the *Schwartz space*, $\mathcal{S}(\mathbb{R})$. This is the space of infinitely differentiable functions $\varphi(x)$ such that the *seminorms*[3]

$$|\varphi|_{m,n} = \sup_{x\in\mathbb{R}} \left\{ |x|^n \left| \frac{d^m\varphi}{dx^m} \right| \right\} \tag{2.74}$$

are finite for all positive integers m and n. The Schwartz space has the advantage that if φ is in $\mathcal{S}(\mathbb{R})$, then so is its Fourier transform. Another popular space of test functions is $\mathcal{D}$ consisting of C^∞ functions of *compact support* – meaning that each function is identically zero outside some finite interval. Only if we want to prove theorems is a precise specification of $\mathcal{T}$ essential. For most physics calculations infinite differentiability and a rapid enough decrease at infinity for us to be able to ignore boundary terms is all that we need.

The "nice" behaviour of the test functions compensates for the "nasty" behaviour of $\delta(x)$ and its relatives. The objects, such as $\delta(x)$, composing the dual space of $\mathcal{T}$ are called *generalized functions*, or *distributions*. Actually, not every linear map $\mathcal{T} \to \mathbb{R}$ is to be included in the dual space because, for technical reasons, we must require the maps to be *continuous*. In other words, if $\varphi_n \to \varphi$, we want our distributions u to obey $u(\varphi_n) \to u(\varphi)$. Making precise what we mean by $\varphi_n \to \varphi$ is part of the task of specifying $\mathcal{T}$. In the Schwartz space, for example, we declare that $\varphi_n \to \varphi$ if $|\varphi_n - \varphi|_{n,m} \to 0$, for all positive m, n. When we restrict a dual space to continuous functionals, we usually denote it by V' rather than V^*. The space of distributions is therefore $\mathcal{T}'$.

When they wish to stress the dual-space aspect of distribution theory, mathematically minded authors use the notation

$$\delta(\varphi) = \varphi(0), \tag{2.75}$$

or

$$(\delta, \varphi) = \varphi(0), \tag{2.76}$$

[3] A seminorm $|\cdots|$ has all the properties of a norm except that $|\varphi| = 0$ does not imply that $\varphi = 0$.

in place of the common, but purely formal,

$$\int \delta(x)\varphi(x)\,dx = \varphi(0). \tag{2.77}$$

The expression (δ, φ) here represents the *pairing* of the element φ of the vector space $\mathcal{T}$ with the element δ of its dual space $\mathcal{T}'$. It should not be thought of as an inner product as the distribution and the test function lie in different spaces. The "integral" in the common notation is purely symbolic, of course, but the common notation should not be despised even by those in quest of rigour. It suggests correct results, such as

$$\int \delta(ax-b)\varphi(x)\,dx = \frac{1}{|a|}\varphi(b/a), \tag{2.78}$$

which would look quite unmotivated in the dual-space notation.

The distribution $\delta'(x)$ is now defined by the pairing

$$(\delta', \varphi) = -\varphi'(0), \tag{2.79}$$

where the minus sign comes from imagining an integration by parts that takes the "derivative" off $\delta(x)$ and puts it on to the smooth function $\varphi(x)$:

$$\text{“}\int \delta'(x)\varphi(x)\,dx\text{”} = -\int \delta(x)\varphi'(x)\,dx. \tag{2.80}$$

Similarly $\delta^{(n)}(x)$ is now defined by the pairing

$$(\delta^{(n)}, \varphi) = (-1)^n\varphi^{(n)}(0). \tag{2.81}$$

The "nicer" the class of test function we take, the "nastier" the class of distributions we can handle. For example, the Hilbert space L^2 is its own dual: the *Riesz–Fréchet theorem* (see Exercise 2.10) asserts that any continuous linear map $F : L^2 \to \mathbb{R}$ can be written as $F[f] = \langle l, f\rangle$ for some $l \in L^2$. The delta-function map is not continuous when considered as a map from $L^2 \to \mathbb{R}$, however. An arbitrarily small change, $f \to f + \delta f$, in a function (small in the L^2 sense of $\|\delta f\|$ being small) can produce an arbitrarily large change in $f(0)$. Thus L^2 functions are not "nice" enough for their dual space to be able to accommodate the delta function. Another way of understanding this is to remember that we regard two L^2 functions as being the same whenever $\|f_1 - f_2\| = 0$. This distance will be zero even if f_1 and f_2 differ from one another on a countable set of points. As we have remarked earlier, this means that elements of L^2 are not really functions at all – they do not have an assigned value at each point. They are, instead, only *equivalence classes* of functions. Since $f(0)$ is undefined, any attempt to interpret the statement $\int \delta(x)f(x)\,dx = f(0)$ for f an arbitrary element L^2 is necessarily doomed to failure. Continuous functions, however, do have well-defined values at every point. If we take the space of test functions $\mathcal{T}$ to consist of all continuous functions, but not demand that they be differentiable, then $\mathcal{T}'$ will include the delta function, but not its "derivative" $\delta'(x)$, as this requires us to evaluate $f'(0)$. If we require the test functions to be once-differentiable, then $\mathcal{T}'$ will include $\delta'(x)$ but not $\delta''(x)$, and so on.

When we add suitable spaces $\mathcal{T}$ and $\mathcal{T}'$ to our toolkit, we are constructing what is called a *rigged*[4] Hilbert space. In such a rigged space we have the inclusion

$$\mathcal{T} \subset L^2 \equiv [L^2]' \subset \mathcal{T}'. \tag{2.82}$$

The idea is to take the space $\mathcal{T}'$ big enough to contain objects such as the limit of our sequence of "approximate" delta functions δ_ε, which does not converge to anything in L^2.

Ordinary functions can also be regarded as distributions, and this helps illuminate the different senses in which a sequence u_n can converge. For example, we can consider the functions

$$u_n = \sin n\pi x, \quad 0 < x < 1, \tag{2.83}$$

as being either elements of $L^2[0,1]$ or as distributions. As distributions we evaluate them on a smooth function φ as

$$(u_n, \varphi) = \int_0^1 \varphi(x) u_n(x)\, dx. \tag{2.84}$$

Now

$$\lim_{n\to\infty} (u_n, \varphi) = 0, \tag{2.85}$$

since the high-frequency Fourier coefficients of any smooth function tend to zero. We deduce that *as a distribution* we have $\lim_{n\to\infty} u_n = 0$, the convergence being pointwise on the space of test functions. Considered as elements of $L^2[0,1]$, however, the u_n do not tend to zero. Their norm is $\|u_n\| = 1/2$ and so all the u_n remain at the same fixed distance from 0.

Exercise 2.8: Here we show that the elements of $L^2[a,b]$, which we defined in Exercise 2.2 to be the formal limits of Cauchy sequences of continuous functions, may be thought of as distributions.

(i) Let $\varphi(x)$ be a test function and $f_n(x)$ a Cauchy sequence of continuous functions defining $f \in L^2$. Use the Cauchy–Schwarz–Bunyakovsky inequality to show that the sequence of numbers $\langle \varphi, f_n \rangle$ is Cauchy and so deduce that $\lim_{n\to\infty} \langle \varphi, f_n \rangle$ exists.

(ii) Let $\varphi(x)$ be a test function and $f_n^{(1)}(x)$ and $f_n^{(2)}(x)$ be a pair of equivalent sequences defining the same element $f \in L^2$. Use Cauchy–Schwarz–Bunyakovsky to show that

$$\lim_{n\to\infty} \left\langle \varphi, f_n^{(1)} - f_n^{(2)} \right\rangle = 0.$$

[4] "Rigged" as in a sailing ship ready for sea, not "rigged" as in a corrupt election.

Combine this result with that of the preceding exercise to deduce that we can set

$$(\varphi, f) \equiv \lim_{n\to\infty} \langle \varphi^*, f_n \rangle,$$

and so define $f \equiv \lim_{n\to\infty} f_n$ as a distribution.

The interpretation of elements of L^2 as distributions is simultaneously simpler and more physical than the classical interpretation *via* the Lebesgue integral.

Weak derivatives

By exploiting the infinite differentiability of our test functions, we were able to make mathematical sense of the "derivative" of the highly singular delta function. The same idea of a formal integration by parts can be used to define the "derivative" for any distribution, and also for ordinary functions that would not usually be regarded as being differentiable.

We therefore define the *weak* or *distributional* derivative $v(x)$ of a distribution $u(x)$ by requiring its evaluation on a test function $\varphi \in \mathcal{T}$ to be

$$\int v(x)\varphi(x)\,dx \stackrel{\text{def}}{=} -\int u(x)\varphi'(x)\,dx. \tag{2.86}$$

In the more formal pairing notation we write

$$(v, \varphi) \stackrel{\text{def}}{=} -(u, \varphi'). \tag{2.87}$$

The right-hand side of (2.87) is a continuous linear function of φ, and so, therefore, is the left-hand side. Thus the weak derivative $u' \equiv v$ is a well-defined distribution for any u.

When $u(x)$ is an ordinary function that is differentiable in the conventional sense, its weak derivative coincides with the usual derivative. When the function is not conventionally differentiable the weak derivative still exists, but does not assign a numerical value to the derivative at each point. It is therefore a distribution and not a function.

The elements of L^2 are not quite functions – having no well-defined value at a point – but are particularly mild-mannered distributions, and their weak derivatives may themselves be elements of L^2. It is in this weak sense that we will, in later chapters, allow differential operators to act on L^2 "functions".

Example: In the weak sense

$$\frac{d}{dx}|x| = \operatorname{sgn}(x), \tag{2.88}$$

$$\frac{d}{dx}\operatorname{sgn}(x) = 2\delta(x). \tag{2.89}$$

The object $|x|$ is an ordinary function, but $\operatorname{sgn}(x)$ has no definite value at $x = 0$, whilst $\delta(x)$ has no definite value at any x.

Example: As a more subtle illustration, consider the weak derivative of the function $\ln|x|$. With $\varphi(x)$ a test function, the improper integral

$$I = -\int_{-\infty}^{\infty} \varphi'(x) \ln|x|\, dx \equiv -\lim_{\varepsilon,\varepsilon'\to 0}\left(\int_{-\infty}^{-\varepsilon} + \int_{\varepsilon'}^{\infty}\right) \varphi'(x) \ln|x|\, dx \tag{2.90}$$

is convergent and defines the pairing $(-\ln|x|, \varphi')$. We wish to integrate by parts and interpret the result as $([\ln|x|]', \varphi)$. The logarithm is differentiable in the conventional sense away from $x = 0$, and

$$[\ln|x|\varphi(x)]' = \frac{1}{x}\varphi(x) + \ln|x|\varphi'(x), \quad x \neq 0. \tag{2.91}$$

From this we find that

$$\begin{aligned} -(\ln|x|, \varphi') = \lim_{\varepsilon,\varepsilon'\to 0} \Bigg\{ &\left(\int_{-\infty}^{-\varepsilon} + \int_{\varepsilon'}^{\infty}\right) \frac{1}{x}\varphi(x)\, dx \\ &+ \Big(\varphi(\varepsilon')\ln|\varepsilon'| - \varphi(-\varepsilon)\ln|\varepsilon|\Big)\Bigg\}. \end{aligned} \tag{2.92}$$

So far ε and ε' are unrelated except in that they are both being sent to zero. If, however, we choose to make them equal, $\varepsilon = \varepsilon'$, then the integrated-out part becomes

$$\Big(\varphi(\varepsilon) - \varphi(-\varepsilon)\Big)\ln|\varepsilon| \sim 2\varphi'(0)\varepsilon\ln|\varepsilon|, \tag{2.93}$$

and this tends to zero as ε becomes small. In this case

$$-([\ln|x|], \varphi') = \lim_{\varepsilon\to 0}\left\{\left(\int_{-\infty}^{-\varepsilon} + \int_{\varepsilon}^{\infty}\right)\frac{1}{x}\varphi(x)\, dx\right\}. \tag{2.94}$$

By the definition of the weak derivative, the left-hand side of (2.94) is the pairing $([\ln|x|]', \varphi)$. We conclude that

$$\frac{d}{dx}\ln|x| = P\left(\frac{1}{x}\right), \tag{2.95}$$

where $P(1/x)$, the *principal-part* distribution, is defined by the right-hand side of (2.94). It is evaluated on the test function $\varphi(x)$ by forming $\int \varphi(x)/x\, dx$, but with an infinitesimal interval from $-\varepsilon$ to $+\varepsilon$, omitted from the range of integration. It is essential that this omitted interval lie symmetrically about the dangerous point $x = 0$. Otherwise the integrated-out part will not vanish in the $\varepsilon \to 0$ limit. The resulting *principal-part integral*, written $P\int \varphi(x)/x\, dx$, is then convergent and $P(1/x)$ is a well-defined distribution despite the singularity in the integrand. Principal-part integrals are common in physics. We will next meet them when we study Green functions.

For further reading on distributions and their applications we recommend M. J. Lighthill *Fourier Analysis and Generalised Functions*, or F. G. Friedlander *Introduction to the Theory of Distributions*. Both books are published by Cambridge University Press.

2.4 Further exercises and problems

The first two exercises lead the reader through a proof of the Riesz–Fréchet theorem. Although not an essential part of our story, they demonstrate how "completeness" is used in Hilbert space theory, and provide some practice with "ϵ, δ" arguments for those who desire it.

Exercise 2.9: Show that if a norm $\| \ \|$ is derived from an inner product, then it obeys the *parallelogram law*

$$\|f + g\|^2 + \|f - g\|^2 = 2(\|f\|^2 + \|g\|^2).$$

Let N be a complete linear subspace of a Hilbert space H. Let $g \notin N$, and let

$$\inf_{f \in N} \|g - f\| = d.$$

Show that there exists a sequence $f_n \in N$ such that $\lim_{n\to\infty} \|f_n - g\| = d$. Use the parallelogram law to show that the sequence f_n is Cauchy, and hence deduce that there is a unique $f \in N$ such that $\|g - f\| = d$. From this, conclude that $d > 0$. Now show that $\langle (g - f), h\rangle = 0$ for all $h \in N$.

Exercise 2.10: *Riesz–Fréchet theorem.* Let $L[h]$ be a continuous linear functional on a Hilbert space H. Here *continuous* means that

$$\|h_n - h\| \to 0 \Rightarrow L[h_n] \to L[h].$$

Show that the set $N = \{f \in H : L[f] = 0\}$ is a complete linear subspace of H.

Suppose now that there is a $g \in H$ such that $L(g) \neq 0$, and let $l \in H$ be the vector "$g - f$" from the previous problem. Show that

$$L[h] = \langle \alpha l, h\rangle, \quad \text{where } \alpha^* = L[g]/\langle l, g\rangle = L[g]/\|l\|^2.$$

A continuous linear functional can therefore be expressed as an inner product.

Next we have some problems on orthogonal polynomials and three-term recurrence relations. They provide an excuse for reviewing linear algebra, and also serve to introduce the theory behind some practical numerical methods.

Exercise 2.11: Let $\{P_n(x)\}$ be a family of polynomials orthonormal on $[a, b]$ with respect to a positive weight function $w(x)$, and with $\deg[P_n(x)] = n$. Let us also scale $w(x)$ so that $\int_a^b w(x)\,dx = 1$, and $P_0(x) = 1$.

(a) Suppose that the $P_n(x)$ obey the three-term recurrence relation

$$xP_n(x) = b_nP_{n+1}(x) + a_nP_n(x) + b_{n-1}P_{n-1}(x);$$
$$P_{-1}(x) = 0, \ P_0(x) = 1.$$

Define

$$p_n(x) = P_n(x)(b_{n-1}b_{n-2}\cdots b_0),$$

and show that

$$xp_n(x) = p_{n+1}(x) + a_n p_n(x) + b_{n-1}^2 p_{n-1}(x); \quad p_{-1}(x) = 0,\ p_0(x) = 1.$$

Conclude that the $p_n(x)$ are *monic* – i.e. the coefficient of their leading power of x is unity.

(b) Show also that the functions

$$q_n(x) = \int_a^b \frac{p_n(x) - p_n(\xi)}{x - \xi} w(\xi)\, d\xi$$

are degree $n-1$ monic polynomials that obey the same recurrence relation as the $p_n(x)$, but with initial conditions $q_0(x) = 0$, $q_1(x) \equiv \int_a^b w\, dx = 1$.

Warning: while the $q_n(x)$ polynomials defined in part (b) turn out to be very useful, they are *not* mutually orthogonal with respect to $\langle\ ,\ \rangle_w$.

Exercise 2.12: *Gaussian quadrature*. Orthogonal polynomials have application to numerical integration. Let the polynomials $\{P_n(x)\}$ be orthonormal on $[a,b]$ with respect to the positive weight function $w(x)$, and let x_ν, $\nu = 1,\ldots,N$, be the zeros of $P_N(x)$. You will show that if we define the weights

$$w_\nu = \int_a^b \frac{P_N(x)}{P_N'(x_\nu)(x - x_\nu)} w(x)\, dx$$

then the approximate integration scheme

$$\int_a^b f(x)w(x)\, dx \approx w_1 f(x_1) + w_2 f(x_2) + \cdots w_N f(x_N),$$

known as *Gauss' quadrature rule*, is *exact* for $f(x)$ any polynomial of degree less than or equal to $2N-1$.

(a) Let $\pi(x) = (x - \xi_1)(x - \xi_2)\cdots(x - \xi_N)$ be a polynomial of degree N. Given a function $F(x)$, show that

$$F_L(x) \stackrel{\text{def}}{=} \sum_{\nu=1}^{N} F(\xi_\nu) \frac{\pi(x)}{\pi'(\xi_\nu)(x - \xi_\nu)}$$

is a polynomial of degree $N-1$ that coincides with $F(x)$ at $x = \xi_\nu$, $\nu = 1,\ldots,N$. (This is *Lagrange's interpolation formula.*)

(b) Show that if $F(x)$ is a polynomial of degree $N-1$ or less then $F_L(x) = F(x)$.
(c) Let $f(x)$ be a polynomial of degree $2N-1$ or less. Cite the polynomial division algorithm to show that there exist polynomials $Q(x)$ and $R(x)$, each of degree $N-1$ or less, such that

$$f(x) = P_N(x)Q(x) + R(x).$$

(d) Show that $f(x_\nu) = R(x_\nu)$, and that

$$\int_a^b f(x)w(x)\,dx = \int_a^b R(x)w(x)\,dx.$$

(e) Combine parts (a), (b) and (d) to establish Gauss' result.
(f) Show that if we normalize $w(x)$ so that $\int w\,dx = 1$ then the weights w_ν can be expressed as $w_\nu = q_N(x_\nu)/p'_N(x_\nu)$, where $p_n(x)$, $q_n(x)$ are the monic polynomials defined in the preceding problem.

The ultimate large-N exactness of Gaussian quadrature can be expressed as

$$w(x) = \lim_{N\to\infty} \left\{ \sum_\nu \delta(x - x_\nu) w_\nu \right\}.$$

Of course, a sum of Dirac delta functions can never become a continuous function in any ordinary sense. The equality holds only after both sides are integrated against a smooth test function, i.e. when it is considered as a statement about distributions.

Exercise 2.13: The completeness of a set of polynomials $\{P_n(x)\}$, orthonormal with respect to the positive weight function $w(x)$, is equivalent to the statement that

$$\sum_{n=0}^{\infty} P_n(x)P_n(y) = \frac{1}{w(x)}\delta(x-y).$$

It is useful to have a formula for the partial sums of this infinite series.

Suppose that the polynomials $P_n(x)$ obey the three-term recurrence relation

$$xP_n(x) = b_nP_{n+1}(x) + a_nP_n(x) + b_{n-1}P_{n-1}(x); \quad P_{-1}(x) = 0, \; P_0(x) = 1.$$

Use this recurrence relation, together with its initial conditions, to obtain the *Christoffel–Darboux* formula

$$\sum_{n=0}^{N-1} P_n(x)P_n(y) = \frac{b_{N-1}[P_N(x)P_{N-1}(y) - P_{N-1}(x)P_N(y)]}{x-y}.$$

Exercise 2.14: Again suppose that the polynomials $P_n(x)$ obey the three-term recurrence relation

$$xP_n(x) = b_nP_{n+1}(x) + a_nP_n(x) + b_{n-1}P_{n-1}(x); \quad P_{-1}(x) = 0, \; P_0(x) = 1.$$

Consider the N-by-N tridiagonal matrix eigenvalue problem

$$\begin{bmatrix} a_{N-1} & b_{N-2} & 0 & 0 & \dots & 0 \\ b_{N-2} & a_{N-2} & b_{N-3} & 0 & \dots & 0 \\ 0 & b_{N-3} & a_{N-3} & b_{N-4} & \dots & 0 \\ \vdots & \ddots & \ddots & \ddots & \ddots & \vdots \\ 0 & \dots & b_2 & a_2 & b_1 & 0 \\ 0 & \dots & 0 & b_1 & a_1 & b_0 \\ 0 & \dots & 0 & 0 & b_0 & a_0 \end{bmatrix} \begin{bmatrix} u_{N-1} \\ u_{N-2} \\ u_{N-3} \\ \vdots \\ u_2 \\ u_1 \\ u_0 \end{bmatrix} = x \begin{bmatrix} u_{N-1} \\ u_{N-2} \\ u_{N-3} \\ \vdots \\ u_2 \\ u_1 \\ u_0 \end{bmatrix}.$$

(a) Show that the eigenvalues x are given by the zeros x_ν, $\nu = 1, \dots, N$, of $P_N(x)$, and that the corresponding eigenvectors have components $u_n = P_n(x_\nu)$, $n = 0, \dots, N-1$.

(b) Take the $x \to y$ limit of the Christoffel–Darboux formula from the preceding problem, and use it to show that the orthogonality and completeness relations for the eigenvectors can be written as

$$\sum_{n=0}^{N-1} P_n(x_\nu)P_n(x_\mu) = w_\nu^{-1}\delta_{\nu\mu},$$

$$\sum_{\nu=1}^{N} w_\nu P_n(x_\nu)P_m(x_\nu) = \delta_{nm}, \qquad n.m \le N-1,$$

where $w_\nu^{-1} = b_{N-1}P'_N(x_\nu)P_{N-1}(x_\nu)$.

(c) Use the original Christoffel–Darboux formula to show that, when the $P_n(x)$ are orthonormal with respect to the positive weight function $w(x)$, the normalization constants w_ν of this present problem coincide with the weights w_ν occurring in the Gauss quadrature rule. Conclude from this equality that the Gauss quadrature weights are *positive*.

Exercise 2.15: Write the N-by-N tridiagonal matrix eigenvalue problem from the preceding exercise as $\mathbf{H}\mathbf{u} = x\mathbf{u}$, and set $d_N(x) = \det(x\mathbf{I} - \mathbf{H})$. Similarly define $d_n(x)$ to be the determinant of the n-by-n tridiagonal submatrix with $x - a_{n-1}, \dots, x - a_0$ along its principal diagonal. Laplace-develop the determinant $d_n(x)$ about its first row, and hence obtain the recurrence

$$d_{n+1}(x) = (x - a_n)d_n(x) - b_{n-1}^2 d_{n-1}(x).$$

Conclude that

$$\det(x\mathbf{I} - \mathbf{H}) = p_N(x),$$

where $p_n(x)$ is the monic orthogonal polynomial obeying

$$xp_n(x) = p_{n+1}(x) + a_n p_n(x) + b_{n-1}^2 p_{n-1}(x); \quad p_{-1}(x) = 0,\ p_0(x) = 1.$$

Exercise 2.16: Again write the N-by-N tridiagonal matrix eigenvalue problem from the preceding exercises as $\mathbf{Hu} = x\mathbf{u}$.

(a) Show that the lowest and rightmost matrix element

$$\langle 0|(x\mathbf{I} - \mathbf{H})^{-1}|0\rangle \equiv (x\mathbf{I} - \mathbf{H})^{-1}_{00}$$

of the *resolvent matrix* $(x\mathbf{I} - \mathbf{H})^{-1}$ is given by a continued fraction $G_{N-1,0}(x)$ where, for example,

$$G_{3,z}(x) = \cfrac{1}{x - a_0 - \cfrac{b_0^2}{x - a_1 - \cfrac{b_1^2}{x - a_2 - \cfrac{b_2^2}{x - a_3 + z}}}}.$$

(b) Use induction on n to show that

$$G_{n,z}(x) = \frac{q_n(x)z + q_{n+1}(x)}{p_n(x)z + p_{n+1}(x)},$$

where $p_n(x), q_n(x)$ are the monic polynomial functions of x defined by the recurrence relations

$$xp_n(x) = p_{n+1}(x) + a_n p_n(x) + b_{n-1}^2 p_{n-1}(x), \quad p_{-1}(x) = 0,\ p_0(x) = 1,$$
$$xq_n(x) = q_{n+1}(x) + a_n q_n(x) + b_{n-1}^2 q_{n-1}(x), \quad q_0(x) = 0,\ q_1(x) = 1.$$

(c) Conclude that

$$\langle 0|(x\mathbf{I} - \mathbf{H})^{-1}|0\rangle = \frac{q_N(x)}{p_N(x)},$$

has a *pole* singularity when x approaches an eigenvalue x_ν. Show that the *residue* of the pole (the coefficient of $1/(x - x_n)$) is equal to the Gauss quadrature weight w_ν for $w(x)$, the weight function (normalized so that $\int w\,dx = 1$) from which the coefficients a_n, b_n were derived.

Continued fractions were introduced by John Wallis in his *Arithmetica Infinitorum* (1656), as was the recursion formula for their evaluation. Today, when combined with the output of the next exercise, they provide the mathematical underpinning of the *Haydock recursion method* in the band theory of solids. Haydock's method computes $w(x) = \lim_{N\to\infty} \left\{\sum_\nu \delta(x - x_\nu) w_\nu\right\}$, and interprets it as the local density of states that is measured in scanning tunnelling microscopy.

Exercise 2.17: *The Lanczos tridiagonalization algorithm*. Let V be an N-dimensional complex vector space equipped with an inner product $\langle\ ,\ \rangle$ and let $H : V \to V$ be a hermitian linear operator. Starting from a unit vector $\mathbf{u}_0$, and taking $\mathbf{u}_{-1} = \mathbf{0}$, recursively generate the unit vectors $\mathbf{u}_n$ and the numbers a_n, b_n and c_n by

$$H\mathbf{u}_n = b_n\mathbf{u}_{n+1} + a_n\mathbf{u}_n + c_{n-1}\mathbf{u}_{n-1},$$

where the coefficients

$$a_n \equiv \langle \mathbf{u}_n, H\mathbf{u}_n \rangle,$$
$$c_{n-1} \equiv \langle \mathbf{u}_{n-1}, H\mathbf{u}_n \rangle,$$

ensure that $\mathbf{u}_{n+1}$ is perpendicular to both $\mathbf{u}_n$ and $\mathbf{u}_{n-1}$, and

$$b_n = \|H\mathbf{u}_n - a_n\mathbf{u}_n - c_{n-1}\mathbf{u}_{n-1}\|,$$

a positive real number, makes $\|\mathbf{u}_{n+1}\| = 1$.

(a) Use induction on n to show that $\mathbf{u}_{n+1}$, although only constructed to be perpendicular to the previous *two* vectors, is in fact (and in the absence of numerical rounding errors) perpendicular to *all* $\mathbf{u}_m$ with $m \le n$.
(b) Show that a_n, c_n are *real*, and that $c_{n-1} = b_{n-1}$.
(c) Conclude that $b_{N-1} = 0$, and (provided that no earlier b_n happens to vanish) that the $\mathbf{u}_n$, $n = 0, \ldots, N-1$, constitute an orthonormal basis for V, in terms of which H is represented by the N-by-N real-symmetric tridiagonal matrix $\mathbf{H}$ of the preceding exercises.

Because the eigenvalues of a tridiagonal matrix are given by the numerically easy-to-find zeros of the associated monic polynomial $p_N(x)$, the Lanczos algorithm provides a computationally efficient way of extracting the eigenvalues from a large sparse matrix. In theory, the entries in the tridiagonal $\mathbf{H}$ can be computed while retaining only $\mathbf{u}_n$, $\mathbf{u}_{n-1}$ and $H\mathbf{u}_n$ in memory at any one time. In practice, with finite precision computer arithmetic, orthogonality with the earlier $\mathbf{u}_m$ is eventually lost, and spurious or duplicated eigenvalues appear. There exist, however, stratagems for identifying and eliminating these fake eigenvalues.

The following two problems are "toy" versions of the *Lax pair* and *tau function* constructions that arise in the general theory of soliton equations. They provide useful practice in manipulating matrices and determinants.

Problem 2.18: The monic orthogonal polynomials $p_i(x)$ have inner products

$$\langle p_i, p_j \rangle_w \equiv \int p_i(x)p_j(x)w(x)\,dx = h_i\delta_{ij},$$

and obey the recursion relation

$$xp_i(x) = p_{i+1}(x) + a_ip_i(x) + b_{i-1}^2p_{i-1}(x); \quad p_{-1}(x) = 0,\, p_0(x) = 1.$$

Write the recursion relation as

$$\mathbf{L}\mathbf{p} = x\mathbf{p},$$

where

$$\mathbf{L} \equiv \begin{bmatrix} \ddots & \ddots & \ddots & \ddots & \vdots \\ \cdots & 1 & a_2 & b_1^2 & 0 \\ \cdots & 0 & 1 & a_1 & b_0^2 \\ \cdots & 0 & 0 & 1 & a_0 \end{bmatrix}, \quad \mathbf{p} \equiv \begin{bmatrix} \vdots \\ p_2 \\ p_1 \\ p_0 \end{bmatrix}.$$

Suppose that

$$w(x) = \exp\left\{ -\sum_{n=1}^{\infty} t_n x^n \right\},$$

and consider how the $p_i(x)$ and the coefficients a_i and b_i^2 vary with the parameters t_n.

(a) Show that

$$\frac{\partial \mathbf{p}}{\partial t_n} = \mathbf{M}^{(n)}\mathbf{p},$$

where $\mathbf{M}^{(n)}$ is some strictly upper triangular matrix – i.e. all entries on and below its principal diagonal are zero.

(b) By differentiating $\mathbf{L}\mathbf{p} = x\mathbf{p}$ with respect to t_n show that

$$\frac{\partial \mathbf{L}}{\partial t_n} = [\mathbf{M}^{(n)}, \mathbf{L}].$$

(c) Compute the matrix elements

$$\langle i|\mathbf{M}^{(n)}|j\rangle \equiv M_{ij}^{(n)} = h_j^{-1}\left\langle p_j, \frac{\partial p_i}{\partial t_n}\right\rangle_w$$

(note the interchange of the order of i and j in the $\langle\ ,\ \rangle_w$ product!) by differentiating the orthogonality condition $\langle p_i, p_j\rangle_w = h_i\delta_{ij}$. Hence show that

$$\mathbf{M}^{(n)} = \left(\mathbf{L}^n\right)_+$$

where $(\mathbf{L}^n)_+$ denotes the strictly upper triangular projection of the n-th power of $\mathbf{L}$ – i.e. the matrix $\mathbf{L}^n$, but with its diagonal and lower triangular entries replaced by zero.

Thus

$$\frac{\partial \mathbf{L}}{\partial t_n} = \left[\left(\mathbf{L}^n\right)_+, \mathbf{L}\right]$$

describes a family of deformations of the semi-infinite matrix **L** that, in some formal sense, preserve its eigenvalues x.

Problem 2.19: Let the monic polynomials $p_n(x)$ be orthogonal with respect to the weight function

$$w(x) = \exp\left\{-\sum_{n=1}^{\infty} t_n x^n\right\}.$$

Define the "tau-function" $\tau_n(t_1, t_2, t_3 \ldots)$ of the parameters t_i to be the n-fold integral

$$\tau_n(t_1, t_2, \ldots) = \iint \cdots \int dx_x dx_2 \ldots dx_n \Delta^2(x) \exp\left\{-\sum_{\nu=1}^{n}\sum_{m=1}^{\infty} t_m x_\nu^m\right\}$$

where

$$\Delta(x) = \begin{vmatrix} x_1^{n-1} & x_1^{n-2} & \ldots & x_1 & 1 \\ x_2^{n-1} & x_2^{n-2} & \ldots & x_2 & 1 \\ \vdots & \vdots & \ddots & \vdots & \vdots \\ x_n^{n-1} & x_n^{n-2} & \ldots & x_n & 1 \end{vmatrix} = \prod_{\nu<\mu} (x_\nu - x_\mu)$$

is the n-by-n Vandermonde determinant.

(a) Show that

$$\begin{vmatrix} x_1^{n-1} & x_1^{n-2} & \ldots & x_1 & 1 \\ x_2^{n-1} & x_2^{n-2} & \ldots & x_2 & 1 \\ \vdots & \vdots & \ddots & \vdots & \vdots \\ x_n^{n-1} & x_n^{n-2} & \ldots & x_n & 1 \end{vmatrix} = \begin{vmatrix} p_{n-1}(x_1) & p_{n-2}(x_1) & \ldots & p_1(x_1) & p_0(x_1) \\ p_{n-1}(x_2) & p_{n-2}(x_2) & \ldots & p_1(x_2) & p_0(x_2) \\ \vdots & \vdots & \ddots & \vdots & \vdots \\ P_{n-1}(x_n) & p_{n-2}(x_n) & \ldots & p_1(x_n) & p_0(x_n) \end{vmatrix}.$$

(b) Combine the identity from part (a) with the orthogonality property of the $p_n(x)$ to show that

$$\begin{aligned} p_n(x) &= \frac{1}{\tau_n} \int dx_1 dx_2 \ldots dx_n \Delta^2(x) \prod_{\mu=1}^{n} (x - x_\mu) \exp\left\{-\sum_{\nu=1}^{n}\sum_{m=1}^{\infty} t_m x_\nu^m\right\} \\ &= x^n \frac{\tau_n(t_1', t_2', t_3', \ldots)}{\tau_n(t_1, t_2, t_3, \ldots)} \end{aligned}$$

where

$$t_m' = t_m + \frac{1}{m x^m}.$$

Here are some exercises on distributions:

Exercise 2.20: Let $f(x)$ be a continuous function. Observe that $f(x)\delta(x) = f(0)\delta(x)$. Deduce that

$$\frac{d}{dx}[f(x)\delta(x)] = f(0)\delta'(x).$$

If $f(x)$ were differentiable we might also have used the product rule to conclude that

$$\frac{d}{dx}[f(x)\delta(x)] = f'(x)\delta(x) + f(x)\delta'(x).$$

Show, by evaluating $f(0)\delta'(x)$ and $f'(x)\delta(x) + f(x)\delta'(x)$ on a test function $\varphi(x)$, that these two expressions for the derivative of $f(x)\delta(x)$ are equivalent.

Exercise 2.21: Let $\varphi(x)$ be a test function. Show that

$$\frac{d}{dt}\left\{P\int_{-\infty}^{\infty}\frac{\varphi(x)}{(x-t)}\,dx\right\} = P\int_{-\infty}^{\infty}\frac{\varphi(x)-\varphi(t)}{(x-t)^2}\,dx.$$

Show further that the right-hand side of this equation is equal to

$$-\left(\frac{d}{dx}P\left(\frac{1}{x-t}\right),\varphi\right) \equiv P\int_{-\infty}^{\infty}\frac{\varphi'(x)}{(x-t)}\,dx.$$

Exercise 2.22: Let $\theta(x)$ be the *step function* or *Heaviside distribution*

$$\theta(x) = \begin{cases} 1, & x > 0, \\ \text{undefined}, & x = 0, \\ 0, & x < 0. \end{cases}$$

By forming the weak derivative of both sides of the equation

$$\lim_{\varepsilon\to 0_+} \ln(x+i\varepsilon) = \ln|x| + i\pi\theta(-x),$$

conclude that

$$\lim_{\varepsilon\to 0_+}\left(\frac{1}{x+i\varepsilon}\right) = P\left(\frac{1}{x}\right) - i\pi\delta(x).$$

Exercise 2.23: Use induction on n to generalize Exercise 2.21 and show that

$$\begin{aligned}
&\frac{d^n}{dt^n}\left\{P\int_{-\infty}^{\infty}\frac{\varphi(x)}{(x-t)}\,dx\right\} \\
&\quad = P\int_{-\infty}^{\infty}\frac{n!}{(x-t)^{n+1}}\left[\varphi(x) - \sum_{m=0}^{n-1}\frac{1}{m!}(x-t)^m\varphi^{(m)}(t)\right]dx, \\
&\quad = P\int_{-\infty}^{\infty}\frac{\varphi^{(n)}}{x-t}\,dx.
\end{aligned}$$

Exercise 2.24: Let the non-local functional $S[f]$ be defined by

$$S[f] = \frac{1}{4\pi} \int_{-\infty}^{\infty} \int_{-\infty}^{\infty} \left\{ \frac{f(x) - f(x')}{x - x'} \right\}^2 dx dx'.$$

Compute the functional derivative of $S[f]$ and verify that it is given by

$$\frac{\delta S}{\delta f(x)} = \frac{1}{\pi} \frac{d}{dx} \left\{ P \int_{-\infty}^{\infty} \frac{f(x')}{x - x'} dx' \right\}.$$

See Exercise 6.10 for an occurence of this functional.

3

Linear ordinary differential equations

In this chapter we will discuss linear ordinary differential equations. We will not describe tricks for solving any particular equation, but instead focus on those aspects of the general theory that we will need later.

We will consider either *homogeneous equations*, $Ly = 0$ with

$$Ly \equiv p_0(x)y^{(n)} + p_1(x)y^{(n-1)} + \cdots + p_n(x)y, \tag{3.1}$$

or *inhomogeneous equations* $Ly = f$. In full,

$$p_0(x)y^{(n)} + p_1(x)y^{(n-1)} + \cdots + p_n(x)y = f(x). \tag{3.2}$$

We will begin with homogeneous equations.

3.1 Existence and uniqueness of solutions

The fundamental result in the theory of differential equations is the existence and uniqueness theorem for systems of first-order equations.

3.1.1 Flows for first-order equations

Let $x^1, \dots, x^n$ be a system of coordinates in $\mathbb{R}^n$, and let $X^i(x^1, x^2, \dots, x^n, t)$, $i = 1, \dots, n$, be the components of a t-dependent vector field. Consider the system of first-order differential equations

$$\begin{aligned}
\frac{dx^1}{dt} &= X^1(x^1, x^2, \dots, x^n, t),\\
\frac{dx^2}{dt} &= X^2(x^1, x^2, \dots, x^n, t),\\
&\vdots\\
\frac{dx^n}{dt} &= X^n(x^1, x^2, \dots, x^n, t).
\end{aligned} \tag{3.3}$$

For a sufficiently smooth vector field $(X^1, X^2, \dots, X^n)$ there is a unique solution $x^i(t)$ for any initial condition $x^i(0) = x^i_0$. Rigorous proofs of this claim, including a statement of exactly what "sufficiently smooth" means, can be found in any standard book on differential equations. Here, we will simply assume the result. It is of course "physically" plausible. Regard the X^i as being the components of the velocity field in a fluid flow, and the solution $x^i(t)$ as the trajectory of a particle carried by the flow. A particle initially at $x^i(0) = x^i_0$ certainly goes somewhere, and unless something seriously pathological is happening, that "somewhere" will be unique.

Now introduce a single function $y(t)$, and set

$$\begin{aligned} x^1 &= y, \\ x^2 &= \dot{y}, \\ x^3 &= \ddot{y}, \\ &\vdots \\ x^n &= y^{(n-1)}, \end{aligned} \tag{3.4}$$

and, given smooth functions $p_0(t), \ldots, p_n(t)$, with $p_0(t)$ nowhere vanishing, look at the particular system of equations

$$\begin{aligned} \frac{dx^1}{dt} &= x^2, \\ \frac{dx^2}{dt} &= x^3, \\ &\vdots \\ \frac{dx^{n-1}}{dt} &= x^n, \\ \frac{dx^n}{dt} &= -\frac{1}{p_0}\left(p_1 x^n + p_2 x^{n-1} + \cdots + p_n x^1\right). \end{aligned} \tag{3.5}$$

This system is equivalent to the single equation

$$p_0(t)\frac{d^n y}{dt^n} + p_1(t)\frac{d^{n-1}y}{dt^{n-1}} + \cdots + p_{n-1}(t)\frac{dy}{dt} + p_n(t)y(t) = 0. \tag{3.6}$$

Thus an n-th order ordinary differential equation (ODE) can be written as a first-order equation in n dimensions, and we can exploit the uniqueness result cited above. We conclude, provided p_0 never vanishes, that the differential equation $Ly = 0$ has a unique solution, $y(t)$, for each set of initial data $(y(0), \dot{y}(0), \ddot{y}(0), \ldots, y^{(n-1)}(0))$. Thus,

(i) If $Ly = 0$ and $y(0) = 0$, $\dot{y}(0) = 0$, $\ddot{y}(0) = 0$, …, $y^{(n-1)}(0) = 0$, we deduce that $y \equiv 0$.

(ii) If $y_1(t)$ and $y_2(t)$ obey the same equation $Ly = 0$, and have the same initial data, then $y_1(t) = y_2(t)$.

3.1.2 Linear independence

In this section we will assume that p_0 does not vanish in the region of x we are interested in, and that all the p_i remain finite and differentiable sufficiently many times for our formulæ to make sense.

Consider an n-th order linear differential equation

$$p_0(x)y^{(n)} + p_1(x)y^{(n-1)} + \cdots + p_n(x)y = 0. \tag{3.7}$$

The set of solutions of this equation constitutes a *vector space* because if $y_1(x)$ and $y_2(x)$ are solutions, then so is any linear combination $\lambda y_1(x) + \mu y_2(x)$. We will show that the dimension of this vector space is n. To see that this is so, let $y_1(x)$ be a solution with initial data

$$\begin{aligned} y_1(0) &= 1, \\ y_1'(0) &= 0, \\ &\vdots \\ y_1^{(n-1)} &= 0, \end{aligned} \tag{3.8}$$

let $y_2(x)$ be a solution with

$$\begin{aligned} y_2(0) &= 0, \\ y_2'(0) &= 1, \\ &\vdots \\ y_2^{(n-1)} &= 0, \end{aligned} \tag{3.9}$$

and so on, up to $y_n(x)$, which has

$$\begin{aligned} y_n(0) &= 0, \\ y_n'(0) &= 0, \\ &\vdots \\ y_n^{(n-1)} &= 1. \end{aligned} \tag{3.10}$$

We claim that the functions $y_i(x)$ are *linearly independent*. Suppose, to the contrary, that there are constants $\lambda_1, \ldots, \lambda_n$ such that

$$0 = \lambda_1 y_1(x) + \lambda_2 y_2(x) + \cdots + \lambda_n y_n(x). \tag{3.11}$$

Then

$$0 = \lambda_1 y_1(0) + \lambda_2 y_2(0) + \cdots + \lambda_n y_n(0) \quad \Rightarrow \lambda_1 = 0. \tag{3.12}$$

Differentiating once and setting $x = 0$ gives

$$0 = \lambda_1 y_1'(0) + \lambda_2 y_2'(0) + \cdots + \lambda_n y_n'(0) \quad \Rightarrow \lambda_2 = 0. \tag{3.13}$$

We continue in this manner all the way to

$$0 = \lambda_1 y_1^{(n-1)}(0) + \lambda_2 y_2^{(n-1)}(0) + \cdots + \lambda_n y_n^{(n-1)}(0) \quad \Rightarrow \lambda_n = 0. \tag{3.14}$$

All the λ_i are zero! There is therefore no non-trivial linear relation between the $y_i(x)$, and they are indeed linearly independent.

The solutions $y_i(x)$ also *span* the solution space, because the unique solution with initial data $y(0) = a_1, y'(0) = a_2, \ldots, y^{(n-1)}(0) = a_n$ can be written in terms of them as

$$y(x) = a_1 y_1(x) + a_2 y_2(x) + \cdots + a_n y_n(x). \tag{3.15}$$

Our chosen set of n solutions is therefore a *basis* for the solution space of the differential equation. The dimension of the solution space is therefore n, as claimed.

3.1.3 The Wronskian

If we manage to find a different set of n solutions, how will we know whether they are also linearly independent? The essential tool is the *Wronskian*:

$$W(y_1, \ldots, y_n; x) \stackrel{\text{def}}{=} \begin{vmatrix} y_1 & y_2 & \ldots & y_n \\ y_1' & y_2' & \ldots & y_n' \\ \vdots & \vdots & \ddots & \vdots \\ y_1^{(n-1)} & y_2^{(n-1)} & \ldots & y_n^{(n-1)} \end{vmatrix}. \tag{3.16}$$

Recall that the derivative of a determinant

$$D = \begin{vmatrix} a_{11} & a_{12} & \ldots & a_{1n} \\ a_{21} & a_{22} & \ldots & a_{2n} \\ \vdots & \vdots & \ddots & \vdots \\ a_{n1} & a_{n2} & \ldots & a_{nn} \end{vmatrix} \tag{3.17}$$

may be evaluated by differentiating row-by-row:

$$\frac{dD}{dx} = \begin{vmatrix} a'_{11} & a'_{12} & \dots & a'_{1n} \\ a_{21} & a_{22} & \dots & a_{2n} \\ \vdots & \vdots & \ddots & \vdots \\ a_{n1} & a_{n2} & \dots & a_{nn} \end{vmatrix} + \begin{vmatrix} a_{11} & a_{12} & \dots & a_{1n} \\ a'_{21} & a'_{22} & \dots & a'_{2n} \\ \vdots & \vdots & \ddots & \vdots \\ a_{n1} & a_{n2} & \dots & a_{nn} \end{vmatrix} + \cdots$$

$$\cdots + \begin{vmatrix} a_{11} & a_{12} & \dots & a_{1n} \\ a_{21} & a_{22} & \dots & a_{2n} \\ \vdots & \vdots & \ddots & \vdots \\ a'_{n1} & a'_{n2} & \dots & a'_{nn} \end{vmatrix}.$$

Applying this to the derivative of the Wronskian, we find

$$\frac{dW}{dx} = \begin{vmatrix} y_1 & y_2 & \dots & y_n \\ y'_1 & y'_2 & \dots & y'_n \\ \vdots & \vdots & \ddots & \vdots \\ y_1^{(n)} & y_2^{(n)} & \dots & y_n^{(n)} \end{vmatrix}. \tag{3.18}$$

Only the term where the very last row is being differentiated survives. All the other row derivatives give zero because they lead to a determinant with two identical rows. Now, if the y_i are all solutions of

$$p_0 y^{(n)} + p_1 y^{(n-1)} + \cdots + p_n y = 0, \tag{3.19}$$

we can substitute

$$y_i^{(n)} = -\frac{1}{p_0}\left(p_1 y_i^{(n-1)} + p_2 y_i^{(n-2)} + \cdots + p_n y_i\right), \tag{3.20}$$

use the row-by-row linearity of determinants,

$$\begin{vmatrix} \lambda a_{11} + \mu b_{11} & \lambda a_{12} + \mu b_{12} & \dots & \lambda a_{1n} + \mu b_{1n} \\ c_{21} & c_{22} & \dots & c_{2n} \\ \vdots & \vdots & \ddots & \vdots \\ c_{n1} & c_{n2} & \dots & c_{nn} \end{vmatrix}$$

$$= \lambda \begin{vmatrix} a_{11} & a_{12} & \dots & a_{1n} \\ c_{21} & c_{22} & \dots & c_{2n} \\ \vdots & \vdots & \ddots & \vdots \\ c_{n1} & c_{n2} & \dots & c_{nn} \end{vmatrix} + \mu \begin{vmatrix} b_{11} & b_{12} & \dots & b_{1n} \\ c_{21} & c_{22} & \dots & c_{2n} \\ \vdots & \vdots & \ddots & \vdots \\ c_{n1} & c_{n2} & \dots & c_{nn} \end{vmatrix}, \tag{3.21}$$

and find, again because most terms have two identical rows, that only the terms with p_1 survive. The end result is

$$\frac{dW}{dx} = -\left(\frac{p_1}{p_0}\right) W. \tag{3.22}$$

Solving this first-order equation gives

$$W(y_i;x) = W(y_i;x_0)\exp\left\{-\int_{x_0}^{x}\left(\frac{p_1(\xi)}{p_0(\xi)}\right)d\xi\right\}. \tag{3.23}$$

Since the exponential function itself never vanishes, $W(x)$ either vanishes at all x, or never. This is *Liouville's theorem*, and (3.23) is called *Liouville's formula*.

Now suppose that $y_1,\ldots,y_n$ are a set of C^n functions of x, not necessarily solutions of an ODE. Suppose further that there are constants λ_i, not all zero, such that

$$\lambda_1 y_1(x) + \lambda_2 y_2(x) + \cdots + \lambda_n y_n(x) \equiv 0 \tag{3.24}$$

(i.e. the functions are linearly dependent). Then the set of equations

$$\begin{aligned}
\lambda_1 y_1(x) + \lambda_2 y_2(x) + \cdots + \lambda_n y_n(x) &= 0,\\
\lambda_1 y_1'(x) + \lambda_2 y_2'(x) + \cdots + \lambda_n y_n'(x) &= 0,\\
&\vdots\\
\lambda_1 y_1^{(n-1)}(x) + \lambda_2 y_2^{(n-1)}(x) + \cdots + \lambda_n y_n^{(n-1)}(x) &= 0
\end{aligned} \tag{3.25}$$

has a non-trivial solution $\lambda_1,\lambda_2,\ldots,\lambda_n$, and so the determinant of the coefficients,

$$W = \begin{vmatrix} y_1 & y_2 & \cdots & y_n \\ y_1' & y_2' & \cdots & y_n' \\ \vdots & \vdots & \ddots & \vdots \\ y_1^{(n-1)} & y_2^{(n-1)} & \cdots & y_n^{(n-1)} \end{vmatrix}, \tag{3.26}$$

must vanish. Thus

$$\boxed{\text{Linear dependence} \Rightarrow W \equiv 0.}$$

There is a partial converse of this result: suppose that $y_1,\ldots,y_n$ are solutions to an n-th order ODE and $W(y_i;x) = 0$ at $x = x_0$. Then there must exist a set of λ_i, not all zero, such that

$$Y(x) = \lambda_1 y_1(x) + \lambda_2 y_2(x) + \cdots + \lambda_n y_n(x) \tag{3.27}$$

has $0 = Y(x_0) = Y'(x_0) = \cdots = Y^{(n-1)}(x_0)$. This is because the system of linear equations determining the λ_i has the Wronskian as its determinant. Now the function $Y(x)$ is a solution of the ODE and has vanishing initial data. It is therefore identically zero. We conclude that

$$\boxed{\text{ODE and } W = 0 \Rightarrow \text{Linear dependence.}}$$

If there is no ODE, the Wronskian may vanish without the functions being linearly dependent. As an example, consider

$$\begin{aligned} y_1(x) &= \begin{cases} 0, & x \le 0, \\ \exp\{-1/x^2\}, & x > 0, \end{cases} \\ y_2(x) &= \begin{cases} \exp\{-1/x^2\}, & x \le 0, \\ 0, & x > 0. \end{cases} \end{aligned} \tag{3.28}$$

We have $W(y_1, y_2; x) \equiv 0$, but y_1, y_2 are not proportional to one another, and so not linearly dependent. (Note that $y_{1,2}$ are smooth functions. In particular they have derivatives of all orders at $x = 0$.)

Given n linearly independent smooth functions y_i, can we always find an n-th order differential equation that has them as its solutions? The answer had better be "no", or there would be a contradiction between the preceding theorem and the counter-example to its extension. If the functions *do* satisfy a common equation, however, we can use a Wronskian to construct it: let

$$Ly = p_0(x)y^{(n)} + p_1(x)y^{(n-1)} + \cdots + p_n(x)y \tag{3.29}$$

be the differential polynomial in $y(x)$ that results from expanding

$$D(y) = \begin{vmatrix} y^{(n)} & y^{(n-1)} & \dots & y \\ y_1^{(n)} & y_1^{(n-1)} & \dots & y_1 \\ \vdots & \vdots & \ddots & \vdots \\ y_n^{(n)} & y_n^{(n-1)} & \dots & y_n \end{vmatrix}. \tag{3.30}$$

Whenever y coincides with any of the y_i, the determinant will have two identical rows, and so $Ly = 0$. The y_i are indeed n solutions of $Ly = 0$. As we have noted, this construction cannot always work. To see what can go wrong, observe that it gives

$$p_0(x) = \begin{vmatrix} y_1^{(n-1)} & y_1^{(n-2)} & \dots & y_1 \\ y_2^{(n-1)} & y_2^{(n-2)} & \dots & y_2 \\ \vdots & \vdots & \ddots & \vdots \\ y_n^{(n-1)} & y_n^{(n-2)} & \dots & y_n \end{vmatrix} = W(y; x). \tag{3.31}$$

If this Wronskian is zero, then our construction fails to deliver an n-th order equation. Indeed, taking y_1 and y_2 to be the functions in the example above yields an equation in which all three coeffecients p_0, p_1, p_2 are identically zero.

3.2 Normal form

In elementary algebra a polynomial equation

$$a_0 x^n + a_1 x^{n-1} + \cdots a_n = 0, \tag{3.32}$$

with $a_0 \neq 0$, is said to be in *normal form* if $a_1 = 0$. We can always put such an equation in normal form by defining a new variable $\tilde{x}$ with $x = \tilde{x} - a_1(na_0)^{-1}$.

By analogy, an n-th order linear ODE with no $y^{(n-1)}$ term is also said to be in normal form. We can put an ODE in normal form by the substitution $y = w\tilde{y}$, for a suitable function $w(x)$. Let

$$p_0 y^{(n)} + p_1 y^{(n-1)} + \cdots + p_n y = 0. \tag{3.33}$$

Set $y = w\tilde{y}$. Using Leibniz' rule, we expand out

$$(w\tilde{y})^{(n)} = w\tilde{y}^{(n)} + nw'\tilde{y}^{(n-1)} + \frac{n(n-1)}{2!} w''\tilde{y}^{(n-2)} + \cdots + w^{(n)}\tilde{y}. \tag{3.34}$$

The differential equation becomes, therefore,

$$(wp_0)\tilde{y}^{(n)} + (p_1 w + p_0 n w')\tilde{y}^{(n-1)} + \cdots = 0. \tag{3.35}$$

We see that if we chose w to be a solution of

$$p_1 w + p_0 n w' = 0, \tag{3.36}$$

for example

$$w(x) = \exp\left\{-\frac{1}{n}\int_0^x \left(\frac{p_1(\xi)}{p_0(\xi)}\right) d\xi\right\}, \tag{3.37}$$

then $\tilde{y}$ obeys the equation

$$(wp_0)\tilde{y}^{(n)} + \tilde{p}_2 \tilde{y}^{(n-2)} + \cdots = 0, \tag{3.38}$$

with no second-highest derivative.

Example: For a second-order equation,

$$y'' + p_1 y' + p_2 y = 0, \tag{3.39}$$

we set $y(x) = v(x)\exp\{-\frac{1}{2}\int_0^x p_1(\xi)d\xi\}$ and find that v obeys

$$v'' + \Omega v = 0, \tag{3.40}$$

where

$$\Omega = p_2 - \frac{1}{2}p_1' - \frac{1}{4}p_1^2. \tag{3.41}$$

Reducing an equation to normal form gives us the best chance of solving it by inspection. For physicists, another advantage is that a second-order equation in normal form can be thought of as a Schrödinger equation,

$$-\frac{d^2\psi}{dx^2} + (V(x) - E)\psi = 0, \tag{3.42}$$

and we can gain insight into the properties of the solution by bringing our physics intuition and experience to bear.

3.3 Inhomogeneous equations

A linear *inhomogeneous* equation is one with a source term:

$$p_0(x)y^{(n)} + p_1(x)y^{(n-1)} + \cdots + p_n(x)y = f(x). \tag{3.43}$$

It is called "inhomogeneous" because the source term $f(x)$ does not contain y, and so is different from the rest. We will devote an entire chapter to the solution of such equations by the method of Green functions. Here, we simply review some elementary material.

3.3.1 Particular integral and complementary function

One method of dealing with inhomogeneous problems, one that is especially effective when the equation has constant coefficients, is simply to try and guess a solution to (3.43). If you are successful, the guessed solution y_{PI} is then called a *particular integral*. We may add any solution y_{CF} of the homogeneous equation

$$p_0(x)y^{(n)} + p_1(x)y^{(n-1)} + \cdots + p_n(x)y = 0 \tag{3.44}$$

to y_{PI} and it will still be a solution of the inhomogeneous problem. We use this freedom to satisfy the boundary or initial conditions. The added solution, y_{CF}, is called the *complementary function*.

Example: Charging capacitor. The capacitor in the circuit in Figure 3.1 is initially uncharged. The switch is closed at $t = 0$.

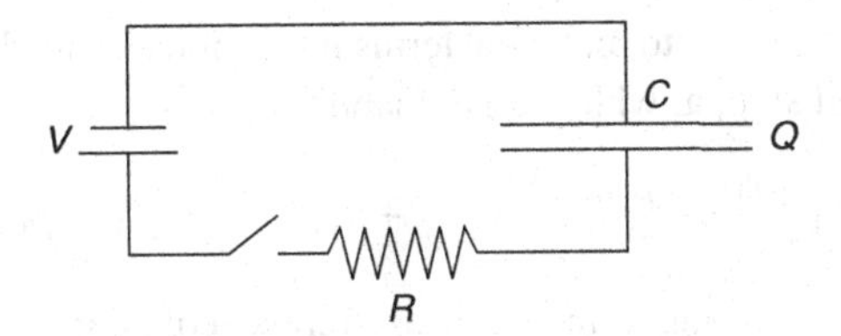

Figure 3.1 Capacitor circuit.

The charge on the capacitor, Q, obeys

$$R\frac{dQ}{dt} + \frac{Q}{C} = V, \tag{3.45}$$

where R, C, V are constants. A particular integral is given by $Q(t) = CV$. The complementary-function solution of the homogeneous problem is

$$Q(t) = Q_0 e^{-t/RC}, \tag{3.46}$$

where Q_0 is constant. The solution satisfying the initial conditions is

$$Q(t) = CV\left(1 - e^{-t/RC}\right). \tag{3.47}$$

3.3.2 Variation of parameters

We now follow Lagrange, and solve

$$p_0(x)y^{(n)} + p_1(x)y^{(n-1)} + \cdots + p_n(x)y = f(x) \tag{3.48}$$

by writing

$$y = v_1 y_1 + v_2 y_2 + \cdots + v_n y_n \tag{3.49}$$

where the y_i are the n linearly independent solutions of the homogeneous equation and the v_i are functions of x that we have to determine. This method is called *variation of parameters*.

Now, differentiating gives

$$y' = v_1 y_1' + v_2 y_2' + \cdots + v_n y_n' + \left\{v_1' y_1 + v_2' y_2 + \cdots + v_n' y_n\right\}. \tag{3.50}$$

We will choose the v's so as to make the terms in the braces vanish. Differentiate again:

$$y'' = v_1 y_1'' + v_2 y_2'' + \cdots + v_n y_n'' + \left\{v_1' y_1' + v_2' y_2' + \cdots + v_n' y_n'\right\}. \tag{3.51}$$

Again, we will choose the v's to make the terms in the braces vanish. We proceed in this way until the very last step, at which we demand

$$\left\{v_1'y_1^{(n-1)} + v_2'y_2^{(n-1)} + \cdots + v_n'y_n^{n-1}\right\} = f(x)/p_0(x). \tag{3.52}$$

If you substitute the resulting y into the differential equation, you will see that the equation is satisfied.

We have imposed the following conditions on v_i':

$$\begin{aligned} v_1'y_1 + v_2'y_2 + \cdots + v_n'y_n &= 0, \\ v_1'y_1' + v_2'y_2' + \cdots + v_n'y_n' &= 0, \\ &\vdots \\ v_1'y_1^{(n-1)} + v_2'y_2^{(n-1)} + \cdots + v_n'y_n^{n-1} &= f(x)/p_0(x). \end{aligned} \tag{3.53}$$

This system of linear equations will have a solution for $v_1', \ldots, v_n'$, provided the Wronskian of the y_i is non-zero. This, however, is guaranteed by the assumed linear independence of the y_i. Having found the $v_1', \ldots, v_n'$, we obtain the $v_1, \ldots, v_n$ themselves by a single integration.

Example: *First-order linear equation.* A simple and useful application of this method solves

$$\frac{dy}{dx} + P(x)y = f(x). \tag{3.54}$$

The solution to the homogeneous equation is

$$y_1 = e^{-\int_a^x P(s)\,ds}. \tag{3.55}$$

We therefore set

$$y = v(x)e^{-\int_a^x P(s)\,ds}, \tag{3.56}$$

and find that

$$v'(x)e^{-\int_a^x P(s)\,ds} = f(x). \tag{3.57}$$

We integrate once to find

$$v(x) = \int_b^x f(\xi)e^{\int_a^\xi P(s)\,ds}d\xi, \tag{3.58}$$

and so

$$y(x) = \int_b^x f(\xi)\left\{e^{-\int_\xi^x P(s)\,ds}\right\}d\xi. \tag{3.59}$$

We select b to satisfy the initial condition.

3.4 Singular points

So far in this chapter, we have been assuming, either explicitly or tacitly, that our coefficients $p_i(x)$ are smooth, and that $p_0(x)$ never vanishes. If $p_0(x)$ does become zero (or, more precisely, if one or more of the p_i/p_0 becomes singular) then dramatic things happen, and the location of the zero of p_0 is called a *singular point* of the differential equation. All other points are called *ordinary points*.

In physics application we often find singular points at the ends of the interval in which we wish to solve our differential equation. For example, the origin $r = 0$ is often a singular point when r is the radial coordinate in plane or spherical polar coordinates. The existence and uniqueness theorems that we have relied upon throughout this chapter may fail at singular endpoints. Consider, for example, the equation

$$xy'' + y' = 0, \tag{3.60}$$

which is singular at $x = 0$. The two linearly independent solutions for $x > 0$ are $y_1(x) = 1$ and $y_2(x) = \ln x$. The general solution is therefore $A + B \ln x$, but no choice of A and B can satisfy the initial conditions $y(0) = a$, $y'(0) = b$ when b is non-zero. Because of these complications, we will delay a systematic study of singular endpoints until Chapter 8.

3.4.1 Regular singular points

If, in the differential equation

$$p_0 y'' + p_1 y' + p_2 y = 0, \tag{3.61}$$

we have a point $x = a$ such that

$$p_0(x) = (x-a)^2 P(x), \quad p_1(x) = (x-a)Q(x), \quad p_2(x) = R(x), \tag{3.62}$$

where P and Q and R are analytic[1] and P and Q non-zero in a neighbourhood of a then the point $x = a$ is called a *regular singular point* of the equation. All other singular points are said to be *irregular*. Close to a regular singular point a the equation looks like

$$P(a)(x-a)^2 y'' + Q(a)(x-a)y' + R(a)y = 0. \tag{3.63}$$

The solutions of this reduced equation are

$$y_1 = (x-a)^{\lambda_1}, \quad y_2 = (x-a)^{\lambda_2}, \tag{3.64}$$

[1] A function is *analytic* at a point if it has a power-series expansion that converges to the function in a neighbourhood of the point.

where $\lambda_{1,2}$ are the roots of the *indicial equation*

$$\lambda(\lambda-1)P(a)+\lambda Q(a)+R(a)=0. \tag{3.65}$$

The solutions of the full equation are then

$$y_1=(x-a)^{\lambda_1}f_1(x), \quad y_2=(x-a)^{\lambda_2}f_2(x), \tag{3.66}$$

where $f_{1,2}$ have power series solutions convergent in a neighbourhood of a. An exception occurs when λ_1 and λ_2 coincide or differ by an integer, in which case the second solution is of the form

$$y_2=(x-a)^{\lambda_1}\Big(\ln(x-a)f_1(x)+f_2(x)\Big), \tag{3.67}$$

where f_1 is the same power series that occurs in the first solution, and f_2 is a new power series. You will probably have seen these statements proved by the tedious procedure of setting

$$f_1(x)=(x-a)^{\lambda}(b_0+b_1(x-a)+b_2(x-a)^2+\cdots, \tag{3.68}$$

and obtaining a recurrence relation determining the b_i. Far more insight is obtained, however, by extending the equation and its solution to the complex plane, where the structure of the solution is related to its *monodromy properties*. If you are familiar with complex analytic methods, you might like to look ahead to the discussion of monodromy in Section 19.2.

3.5 Further exercises and problems

Exercise 3.1: *Reduction of order.* Sometimes additional information about the solutions of a differential equation enables us to reduce the order of the equation, and so solve it.

(a) Suppose that we know that $y_1=u(x)$ is one solution to the equation

$$y''+V(x)y=0.$$

By trying $y=u(x)v(x)$ show that

$$y_2=u(x)\int^x \frac{d\xi}{u^2(\xi)}$$

is also a solution of the differential equation. Is this new solution ever merely a constant multiple of the old solution, or must it be linearly independent? (Hint: Evaluate the Wronskian $W(y_2,y_1)$.)

(b) Suppose that we are told that the product, y_1y_2, of the two solutions to the equation $y'' + p_1y' + p_2y = 0$ is a constant. Show that this requires $2p_1p_2 + p_2' = 0$.

(c) By using ideas from part (b) or otherwise, find the general solution of the equation

$$(x+1)x^2y'' + xy' - (x+1)^3y = 0.$$

Exercise 3.2: Show that the general solution of the differential equation

$$\frac{d^2y}{dx^2} - 2\frac{dy}{dx} + y = \frac{e^x}{1+x^2}$$

is

$$y(x) = Ae^x + Bxe^x - \tfrac{1}{2}e^x \ln(1+x^2) + xe^x \tan^{-1}x.$$

Exercise 3.3: Use the method of variation of parameters to show that if $y_1(x)$ and $y_2(x)$ are linearly independent solutions to the equation

$$p_0(x)\frac{d^2y}{dx^2} + p_1(x)\frac{dy}{dx} + p_2(x)y = 0,$$

then the general solution of the equation

$$p_0(x)\frac{d^2y}{dx^2} + p_1(x)\frac{dy}{dx} + p_2(x)y = f(x)$$

is

$$y(x) = Ay_1(x) + By_2(x) - y_1(x)\int^x \frac{y_2(\xi)f(\xi)}{p_0W(y_1,y_2)}d\xi + y_2(x)\int^x \frac{y_1(\xi)f(\xi)}{p_0W(y_1,y_2)}d\xi.$$

Problem 3.4: *One-dimensional scattering theory.* Consider the one-dimensional Schrödinger equation

$$-\frac{d^2\psi}{dx^2} + V(x)\psi = E\psi,$$

where $V(x)$ is zero except in a finite interval $[-a, a]$ near the origin (Figure 3.2).

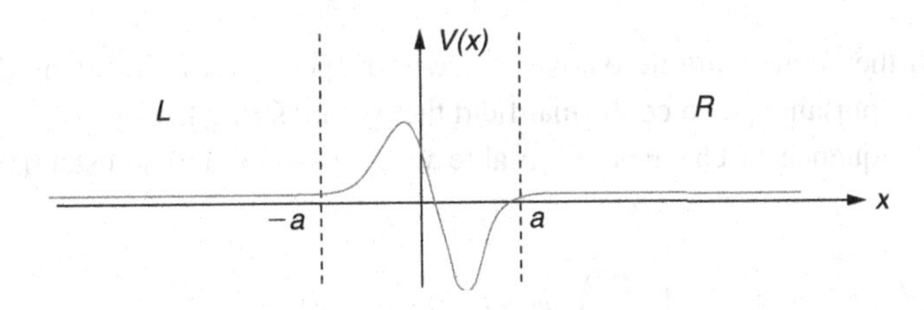

Figure 3.2 A typical potential V for Problem 3.4.

Let L denote the left asymptotic region, $-\infty < x < -a$, and similarly let R denote $a < x < \infty$. For $E = k^2$ there will be scattering solutions of the form

$$\psi_k(x) = \begin{cases} e^{ikx} + r_L(k)e^{-ikx}, & x \in L, \\ t_L(k)e^{ikx}, & x \in R, \end{cases}$$

which for $k > 0$ describe waves incident on the potential $V(x)$ from the left. There will be solutions with

$$\psi_k(x) = \begin{cases} t_R(k)e^{ikx}, & x \in L, \\ e^{ikx} + r_R(k)e^{-ikx}, & x \in R, \end{cases}$$

which for $k < 0$ describe waves incident from the right. The wavefunctions in $[-a, a]$ will naturally be more complicated. Observe that $[\psi_k(x)]^*$ is also a solution of the Schrödinger equation.

By using properties of the Wronskian, show that:

(a) $|r_{L,R}|^2 + |t_{L,R}|^2 = 1$.
(b) $t_L(k) = t_R(-k)$.
(c) Deduce from parts (a) and (b) that $|r_L(k)| = |r_R(-k)|$.
(d) Take the specific example of $V(x) = \lambda\delta(x - b)$ with $|b| < a$. Compute the transmission and reflection coefficients and hence show that $r_L(k)$ and $r_R(-k)$ may differ in phase.

Exercise 3.5: Suppose $\psi(x)$ obeys a Schrödinger equation

$$\left(-\frac{1}{2}\frac{d^2}{dx^2} + [V(x) - E]\right)\psi = 0.$$

(a) Make a smooth and invertible change of independent variable by setting $x = x(z)$ and find the second-order differential equation in z obeyed by $\psi(z) \equiv \psi(x(z))$. Reduce this equation to normal form, and show that the resulting equation is

$$\left(-\frac{1}{2}\frac{d^2}{dz^2} + (x')^2[V(x(z)) - E] - \frac{1}{4}\{x, z\}\right)\widetilde{\psi}(z) = 0,$$

where the primes denote differentiation with respect to z, and

$$\{x, z\} \stackrel{\text{def}}{=} \frac{x'''}{x'} - \frac{3}{2}\left(\frac{x''}{x'}\right)^2$$

is called the *Schwarzian* derivative of x with respect to z. Schwarzian derivatives play an important role in conformal field theory and string theory.

(b) Make a sequence of changes of variable $x \to z \to w$, and so establish *Cayley's identity*

$$\left(\frac{dz}{dw}\right)^2 \{x, z\} + \{z, w\} = \{x, w\}.$$

(Hint: if your proof takes more than one line, you are missing the point.)

4
Linear differential operators

In this chapter we will begin to take a more sophisticated approach to differential equations. We will define, with some care, the notion of a linear differential operator, and explore the analogy between such operators and matrices. In particular, we will investigate what is required for a linear differential operator to have a complete set of eigenfunctions.

4.1 Formal *vs.* concrete operators

We will call the object

$$L = p_0(x)\frac{d^n}{dx^n} + p_1(x)\frac{d^{n-1}}{dx^{n-1}} + \cdots + p_n(x), \tag{4.1}$$

which we also write as

$$p_0(x)\partial_x^n + p_1(x)\partial_x^{n-1} + \cdots + p_n(x), \tag{4.2}$$

a *formal linear differential operator*. The word "formal" refers to the fact that we are not yet worrying about what sort of functions the operator is applied to.

4.1.1 The algebra of formal operators

Even though they are not acting on anything in particular, we can still form products of operators. For example if v and w are smooth functions of x we can define the operators $\partial_x + v(x)$ and $\partial_x + w(x)$ and find

$$(\partial_x + v)(\partial_x + w) = \partial_x^2 + w' + (w + v)\partial_x + vw, \tag{4.3}$$

or

$$(\partial_x + w)(\partial_x + v) = \partial_x^2 + v' + (w + v)\partial_x + vw. \tag{4.4}$$

We see from this example that the operator algebra is not usually commutative.

The algebra of formal operators has some deep applications. Consider, for example, the operators

$$L = -\partial_x^2 + q(x) \tag{4.5}$$

and

$$P = \partial_x^3 + a(x)\partial_x + \partial_x a(x). \tag{4.6}$$

In the last expression, the combination $\partial_x a(x)$ means "first multiply by $a(x)$, and then differentiate the result", so we could also write

$$\partial_x a = a\partial_x + a'. \tag{4.7}$$

We can now form the commutator $[P, L] \equiv PL - LP$. After a little effort, we find

$$[P, L] = (3q' + 4a')\partial_x^2 + (3q'' + 4a'')\partial_x + q''' + 2aq' + a'''. \tag{4.8}$$

If we choose $a = -\frac{3}{4}q$, the commutator becomes a pure multiplication operator, with no differential part:

$$[P, L] = \frac{1}{4}q''' - \frac{3}{2}qq'. \tag{4.9}$$

The equation

$$\frac{dL}{dt} = [P, L], \tag{4.10}$$

or, equivalently,

$$\dot{q} = \frac{1}{4}q''' - \frac{3}{2}qq', \tag{4.11}$$

has a formal solution

$$L(t) = e^{tP}L(0)e^{-tP}, \tag{4.12}$$

showing that the time evolution of L is given by a similarity transformation, which (again formally) does not change its eigenvalues. The partial differential equation (4.11) is the famous Korteweg–de Vries (KdV) equation, which has "soliton" solutions whose existence is intimately connected with the fact that it can be written as (4.10). The operators P and L are called a *Lax pair*, after Peter Lax who uncovered much of the structure.

4.1.2 Concrete operators

We want to explore the analogies between linear differential operators and matrices acting on a finite-dimensional vector space. Because the theory of matrix operators makes much use of inner products and orthogonality, the analogy is closest if we work with a function space equipped with these same notions. We therefore let our differential operators act on $L^2[a, b]$, the Hilbert space of square-integrable functions on $[a, b]$. Now a differential operator cannot act on *every* function in the Hilbert space because not all of them are differentiable. Even though we will relax our notion of differentiability and permit weak derivatives, we must at least demand that the *domain* $\mathcal{D}$, the subset of functions on which we allow the operator to act, contains only functions that are sufficiently differentiable that the function resulting from applying the operator remains an element of $L^2[a, b]$. We will usually restrict the set of functions even further, by imposing boundary conditions at the endpoints of the interval. A *linear differential operator* is now defined as a formal linear differential operator, together with a specification of its domain $\mathcal{D}$.

The boundary conditions that we will impose will always be *linear* and *homogeneous*. This is so that the domain of definition is a vector space. In other words, if y_1 and y_2 obey the boundary conditions then so should $\lambda y_1 + \mu y_2$. Thus, for a second-order operator

$$L = p_0 \partial_x^2 + p_1 \partial_x + p_2 \tag{4.13}$$

on the interval $[a, b]$, we might impose

$$\begin{aligned} B_1[y] &= \alpha_{11} y(a) + \alpha_{12} y'(a) + \beta_{11} y(b) + \beta_{12} y'(b) = 0, \\ B_2[y] &= \alpha_{21} y(a) + \alpha_{22} y'(a) + \beta_{21} y(b) + \beta_{22} y'(b) = 0, \end{aligned} \tag{4.14}$$

but we will not, in defining the differential *operator*, impose *inhomogeneous* conditions, such as

$$\begin{aligned} B_1[y] &= \alpha_{11} y(a) + \alpha_{12} y'(a) + \beta_{11} y(b) + \beta_{12} y'(b) = A, \\ B_2[y] &= \alpha_{21} y(a) + \alpha_{22} y'(a) + \beta_{21} y(b) + \beta_{22} y'(b) = B, \end{aligned} \tag{4.15}$$

with non-zero A, B – even though we will solve differential *equations* with such boundary conditions.

Also, for an n-th order operator, we will not constrain derivatives of order higher than $n - 1$. This is reasonable:[1] if we seek solutions of $Ly = f$ with L a second-order operator, for example, then the values of y'' at the endpoints are already determined in terms of y' and y by the differential equation. We cannot choose to impose some other value. By differentiating the equation enough times, we can similarly determine all higher endpoint derivatives in terms of y and y'. These two derivatives, therefore, are all we can fix by fiat.

The boundary and differentiability conditions that we impose make $\mathcal{D}$ a subset of the entire Hilbert space. This subset will always be *dense*: any element of the Hilbert space can be obtained as an L^2 limit of functions in $\mathcal{D}$. In particular, there will never be a function in $L^2[a, b]$ that is orthogonal to all functions in $\mathcal{D}$.

[1] There is a deeper reason which we will explain in Section 9.7.2.

4.2 The adjoint operator

One of the important properties of matrices, established in Appendix A, is that a matrix that is *self-adjoint*, or *hermitian*, may be *diagonalized*. In other words, the matrix has sufficiently many eigenvectors for them to form a basis for the space on which it acts. A similar property holds for self-adjoint differential operators – but we must be careful in our definition of self-adjointness.

Before reading this section, we suggest you review the material on adjoint operators on *finite*-dimensional spaces that appears in Appendix A.

4.2.1 The formal adjoint

Given a formal differential operator

$$L = p_0(x)\frac{d^n}{dx^n} + p_1(x)\frac{d^{n-1}}{dx^{n-1}} + \cdots + p_n(x), \tag{4.16}$$

and a *weight function* $w(x)$, real and positive on the interval (a, b), we can find another such operator $L^\dagger$, such that, for any sufficiently differentiable $u(x)$ and $v(x)$, we have

$$w\left(u^*Lv - v(L^\dagger u)^*\right) = \frac{d}{dx}Q[u, v], \tag{4.17}$$

for some function Q, which depends bilinearly on u and v and their first $n-1$ derivatives. We call $L^\dagger$ the *formal adjoint* of L with respect to the weight w. The equation (4.17) is called *Lagrange's identity*. The reason for the name "adjoint" is that if we define an inner product

$$\langle u, v\rangle_w = \int_a^b wu^*v\,dx, \tag{4.18}$$

and if the functions u and v have boundary conditions that make $Q[u, v]|_a^b = 0$, then

$$\langle u, Lv\rangle_w = \langle L^\dagger u, v\rangle_w, \tag{4.19}$$

which is the defining property of the adjoint operator on a vector space. The word "formal" means, as before, that we are not yet specifying the domain of the operator.

The method for finding the formal adjoint is straightforward: integrate by parts enough times to get all the derivatives off v and on to u.

Example: If

$$L = -i\frac{d}{dx} \tag{4.20}$$

then let us find the adjoint $L^\dagger$ with respect to the weight $w \equiv 1$. We start from

$$u^*(Lv) = u^*\left(-i\frac{d}{dx}v\right),$$

and use the integration-by-parts technique once to get the derivative off v and onto u^*:

$$\begin{aligned} u^*\left(-i\frac{d}{dx}v\right) &= \left(i\frac{d}{dx}u^*\right)v - i\frac{d}{dx}(u^*v) \\ &= \left(-i\frac{d}{dx}u\right)^* v - i\frac{d}{dx}(u^*v) \\ &\equiv v(L^\dagger u)^* + \frac{d}{dx}Q[u,v]. \end{aligned} \tag{4.21}$$

We have ended up with the Lagrange identity

$$u^*\left(-i\frac{d}{dx}v\right) - v\left(-i\frac{d}{dx}u\right)^* = \frac{d}{dx}(-iu^*v), \tag{4.22}$$

and found that

$$L^\dagger = -i\frac{d}{dx}, \quad Q[u,v] = -iu^*v. \tag{4.23}$$

The operator $-id/dx$ (which you should recognize as the "momentum" operator from quantum mechanics) obeys $L = L^\dagger$, and is therefore, *formally self-adjoint*, or *hermitian*.

Example: Let

$$L = p_0\frac{d^2}{dx^2} + p_1\frac{d}{dx} + p_2, \tag{4.24}$$

with the p_i all real. Again let us find the adjoint $L^\dagger$ with respect to the inner product with $w \equiv 1$. Now, proceeding as above, but integrating by parts *twice*, we find

$$\begin{aligned} &u^*\left[p_0v'' + p_1v' + p_2v\right] - v\left[(p_0u)'' - (p_1u)' + p_2u\right]^* \\ &= \frac{d}{dx}\left[p_0(u^*v' - vu^{*\prime}) + (p_1 - p_0')u^*v\right]. \end{aligned} \tag{4.25}$$

From this we read off that

$$\begin{aligned} L^\dagger &= \frac{d^2}{dx^2}p_0 - \frac{d}{dx}p_1 + p_2 \\ &= p_0\frac{d^2}{dx^2} + (2p_0' - p_1)\frac{d}{dx} + (p_0'' - p_1' + p_2). \end{aligned} \tag{4.26}$$

What conditions do we need to impose on $p_{0,1,2}$ for this L to be formally self-adjoint with respect to the inner product with $w \equiv 1$? For $L = L^\dagger$ we need

$$\begin{aligned} p_0 &= p_0 \\ 2p_0' - p_1 &= p_1 \quad \Rightarrow \quad p_0' = p_1 \\ p_0'' - p_1' + p_2 &= p_2 \quad \Rightarrow \quad p_0'' = p_1'. \end{aligned} \tag{4.27}$$

We therefore require that $p_1 = p_0'$, and so

$$L = \frac{d}{dx}\left(p_0 \frac{d}{dx}\right) + p_2, \tag{4.28}$$

which we recognize as a *Sturm–Liouville* operator.

Example: *Reduction to Sturm–Liouville form.* Another way to make the operator

$$L = p_0 \frac{d^2}{dx^2} + p_1 \frac{d}{dx} + p_2 \tag{4.29}$$

self-adjoint is by a suitable choice of weight function w. Suppose that p_0 is positive on the interval (a, b), and that p_0, p_1, p_2 are all real. Then we may define

$$w = \frac{1}{p_0} \exp\left\{ \int_a^x \left(\frac{p_1}{p_0}\right) dx' \right\} \tag{4.30}$$

and observe that it is positive on (a, b), and that

$$Ly = \frac{1}{w}(w p_0 y')' + p_2 y. \tag{4.31}$$

Now

$$\langle u, Lv \rangle_w - \langle Lu, v \rangle_w = [w p_0 (u^* v' - u^{*\prime} v)]_a^b, \tag{4.32}$$

where

$$\langle u, v \rangle_w = \int_a^b w u^* v \, dx. \tag{4.33}$$

Thus, provided p_0 does not vanish, there is always *some* inner product with respect to which a real second-order differential operator is formally self-adjoint.

Note that with

$$Ly = \frac{1}{w}(w p_0 y')' + p_2 y, \tag{4.34}$$

the eigenvalue equation

$$Ly = \lambda y \tag{4.35}$$

can be written

$$(wp_0 y')' + p_2 wy = \lambda wy. \tag{4.36}$$

When you come across a differential equation where, in the term containing the eigenvalue λ, the eigenfunction is being multiplied by some other function, you should immediately suspect that the operator will turn out to be self-adjoint with respect to the inner product having this other function as its weight.

Illustration (Bargmann–Fock space): This is a more exotic example of a formal adjoint. You may have met it in quantum mechanics. Consider the space of polynomials $P(z)$ in the complex variable $z = x + iy$. Define an inner product by

$$\langle P, Q \rangle = \frac{1}{\pi} \int d^2 z \, e^{-z^* z} \, [P(z)]^* \, Q(z),$$

where $d^2 z \equiv dx\, dy$ and the integration is over the entire xy-plane. With this inner product, we have

$$\langle z^n, z^m \rangle = n! \delta_{nm}.$$

If we define

$$\hat{a} = \frac{d}{dz},$$

then

$$\begin{aligned}
\langle P, \hat{a}\, Q \rangle &= \frac{1}{\pi} \int d^2 z \, e^{-z^* z} \, [P(z)]^* \, \frac{d}{dz} Q(z) \\
&= -\frac{1}{\pi} \int d^2 z \left(\frac{d}{dz} e^{-z^* z} \, [P(z)]^* \right) Q(z) \\
&= \frac{1}{\pi} \int d^2 z \, e^{-z^* z} z^* \, [P(z)]^* \, Q(z) \\
&= \frac{1}{\pi} \int d^2 z \, e^{-z^* z} \, [zP(z)]^* \, Q(z) \\
&= \langle \hat{a}^\dagger P, \hat{Q} \rangle
\end{aligned}$$

where $\hat{a}^\dagger = z$, i.e. the operation of multiplication by z. In this case, the adjoint is not even a differential operator.[2]

[2] In deriving this result we have used the *Wirtinger calculus* where z and z^* are treated as independent variables so that

$$\frac{d}{dz} e^{-z^* z} = -z^* e^{-z^* z},$$

Exercise 4.1: Consider the differential operator $\hat{L} = id/dx$. Find the formal adjoint of L with respect to the inner product $\langle u, v\rangle_w = \int wu^* v\, dx$, and find the corresponding surface term $Q[u, v]$.

Exercise 4.2: *Sturm–Liouville forms.* By constructing appropriate weight functions $w(x)$ convert the following common operators into Sturm–Liouville form:

(a) $\hat{L} = (1 - x^2)\, d^2/dx^2 + [(\mu - \nu) - (\mu + \nu + 2)x]\, d/dx$;
(b) $\hat{L} = (1 - x^2)\, d^2/dx^2 - 3x\, d/dx$;
(c) $\hat{L} = d^2/dx^2 - 2x(1 - x^2)^{-1}\, d/dx - m^2\, (1 - x^2)^{-1}$.

4.2.2 A simple eigenvalue problem

A finite hermitian matrix has a complete set of orthonormal eigenvectors. Does the same property hold for a hermitian differential operator?

Consider the differential operator

$$T = -\partial_x^2, \quad \mathcal{D}(T) = \{y, Ty \in L^2[0, 1] \,:\, y(0) = y(1) = 0\}. \tag{4.37}$$

With the inner product

$$\langle y_1, y_2\rangle = \int_0^1 y_1^* y_2\, dx \tag{4.38}$$

we have

$$\langle y_1, Ty_2\rangle - \langle Ty_1, y_2\rangle = [y_1'^* y_2 - y_1^* y_2']_0^1 = 0. \tag{4.39}$$

The integrated-out part is zero because both y_1 and y_2 satisfy the boundary conditions. We see that

$$\langle y_1, Ty_2\rangle = \langle Ty_1, y_2\rangle \tag{4.40}$$

and so T is *hermitian* or *symmetric*.

The eigenfunctions and eigenvalues of T are

$$\left.\begin{aligned} y_n(x) &= \sin n\pi x \\ \lambda_n &= n^2\pi^2 \end{aligned}\right\} \quad n = 1, 2, \dots. \tag{4.41}$$

---

and observed that, because $[P(z)]^*$ is a function of z^* only,

$$\frac{d}{dz}[P(z)]^* = 0.$$

If you are uneasy at regarding z, z^*, as independent, you should confirm these formulae by expressing z and z^* in terms of x and y, and using

$$\frac{d}{dz} \equiv \frac{1}{2}\left(\frac{\partial}{\partial x} - i\frac{\partial}{\partial y}\right), \quad \frac{d}{dz^*} \equiv \frac{1}{2}\left(\frac{\partial}{\partial x} + i\frac{\partial}{\partial y}\right).$$

We see that:

(i) the eigenvalues are *real*;
(ii) the eigenfunctions for different λ_n are *orthogonal*,

$$2\int_0^1 \sin n\pi x \sin m\pi x\, dx = \delta_{nm}, \quad n = 1, 2, \ldots; \tag{4.42}$$

(iii) the normalized eigenfunctions $\varphi_n(x) = \sqrt{2}\sin n\pi x$ are *complete*: any function in $L^2[0, 1]$ has an (L^2) convergent expansion as

$$y(x) = \sum_{n=1}^{\infty} a_n\sqrt{2}\sin n\pi x \tag{4.43}$$

where

$$a_n = \int_0^1 y(x)\sqrt{2}\sin n\pi x\, dx. \tag{4.44}$$

This all looks very good – exactly the properties we expect for finite hermitian matrices. Can we carry over all the results of finite matrix theory to these hermitian operators? The answer sadly is *no*! Here is a counter-example:

Let

$$T = -i\partial_x, \quad \mathcal{D}(T) = \{y, Ty \in L^2[0, 1] : y(0) = y(1) = 0\}. \tag{4.45}$$

Again

$$\begin{aligned}\langle y_1, Ty_2\rangle - \langle Ty_1, y_2\rangle &= \int_0^1 dx\, \{y_1^*(-i\partial_x y_2) - (-i\partial_x y_1)^* y_2\}\\ &= -i[y_1^* y_2]_0^1 = 0.\end{aligned} \tag{4.46}$$

Once more, the integrated out part vanishes due to the boundary conditions satisfied by y_1 and y_2, so T is nicely hermitian. Unfortunately, T with these boundary conditions has *no* eigenfunctions at all, never mind a complete set! Any function satisfying $Ty = \lambda y$ will be proportional to $e^{i\lambda x}$, but an exponential function is never zero, and cannot satisfy the boundary conditions.

It seems clear that the boundary conditions are the problem. We need a better definition of "adjoint" than the formal one – one that pays more attention to boundary conditions. We will then be forced to distinguish between mere hermiticity, or *symmetry*, and true self-adjointness.

Exercise 4.3: *Another disconcerting example.* Let $p = -i\partial_x$. Show that the following operator on the infinite real line is formally self-adjoint:

$$H = x^3 p + px^3. \tag{4.47}$$

Now let

$$\psi_\lambda(x) = |x|^{-3/2} \exp\left\{-\frac{\lambda}{4x^2}\right\}, \tag{4.48}$$

where λ is real and positive. Show that

$$H\psi_\lambda = -i\lambda\psi_\lambda, \tag{4.49}$$

so ψ_λ is an eigenfunction with a purely imaginary eigenvalue. Examine the proof that hermitian operators have real eigenvalues, and identify at which point it fails. (Hint: H is formally self-adjoint because it is of the form $T + T^\dagger$. Now ψ_λ is square-integrable, and so an element of $L^2(\mathbb{R})$. Is $T\psi_\lambda$ an element of $L^2(\mathbb{R})$?)

4.2.3 Adjoint boundary conditions

The usual definition of the adjoint operator in linear algebra is as follows: given the operator $T : V \to V$ and an inner product $\langle\ ,\ \rangle$, we look at $\langle u, Tv\rangle$, and ask if there is a w such that $\langle w, v\rangle = \langle u, Tv\rangle$ for all v. If there is, then u is in the domain of $T^\dagger$, and we set $T^\dagger u = w$.

For finite-dimensional vector spaces V there always is such a w, and so the domain of $T^\dagger$ is the entire space. In an infinite-dimensional Hilbert space, however, not all $\langle u, Tv\rangle$ can be written as $\langle w, v\rangle$ with w a finite-length element of L^2. In particular delta functions are not allowed – but these are exactly what we would need if we were to express the boundary values appearing in the integrated out part, $Q(u, v)$, as an inner-product integral. We must therefore ensure that u is such that $Q(u, v)$ vanishes, but then accept *any* u with this property into the domain of $T^\dagger$. What this means in practice is that we look at the integrated out term $Q(u, v)$ and see what is required of u to make $Q(u, v)$ zero for any v satisfying the boundary conditions appearing in $\mathcal{D}(T)$. These conditions on u are the *adjoint boundary conditions*, and define the domain of $T^\dagger$.

Example: Consider

$$T = -i\partial_x, \quad \mathcal{D}(T) = \{y, Ty \in L^2[0,1] : y(1) = 0\}. \tag{4.50}$$

Now,

$$\begin{aligned}\int_0^1 dx\, u^*(-i\partial_x v) &= -i[u^*(1)v(1) - u^*(0)v(0)] + \int_0^1 dx(-i\partial_x u)^* v \\ &= -i[u^*(1)v(1) - u^*(0)v(0)] + \langle w, v\rangle, \end{aligned} \tag{4.51}$$

where $w = -i\partial_x u$. Since $v(x)$ is in the domain of T, we have $v(1) = 0$, and so the first term in the integrated out bit vanishes whatever value we take for $u(1)$. On the other hand,

$v(0)$ could be anything, so to be sure that the second term vanishes we must demand that $u(0) = 0$. This, then, is the adjoint boundary condition. It defines the domain of $T^\dagger$:

$$T^\dagger = -i\partial_x, \quad \mathcal{D}(T^\dagger) = \{y, Ty \in L^2[0,1] : y(0) = 0\}. \tag{4.52}$$

For our problematic operator

$$T = -i\partial_x, \quad \mathcal{D}(T) = \{y, Ty \in L^2[0,1] : y(0) = y(1) = 0\}, \tag{4.53}$$

we have

$$\begin{aligned}\int_0^1 dx\, u^*(-i\partial_x v) &= -i[u^* v]_0^1 + \int_0^1 dx(-i\partial_x u)^* v \\ &= 0 + \langle w, v\rangle,\end{aligned} \tag{4.54}$$

where again $w = -i\partial_x u$. This time *no* boundary conditions need be imposed on u to make the integrated out part vanish. Thus

$$T^\dagger = -i\partial_x, \quad \mathcal{D}(T^\dagger) = \{y, Ty \in L^2[0,1]\}. \tag{4.55}$$

Although any of these operators "$T = -i\partial_x$" is *formally* self-adjoint we have,

$$\mathcal{D}(T) \neq \mathcal{D}(T^\dagger), \tag{4.56}$$

so T and $T^\dagger$ are not the same operator and none of them is *truly* self-adjoint.

Exercise 4.4: Consider the differential operator $M = d^4/dx^4$. Find the formal adjoint of M with respect to the inner product $\langle u, v\rangle = \int u^* v\, dx$, and find the corresponding surface term $Q[u, v]$. Find the adjoint boundary conditions defining the domain of $M^\dagger$ for the case

$$\mathcal{D}(M) = \{y, y^{(4)} \in L^2[0,1] : y(0) = y'''(0) = y(1) = y'''(1) = 0\}.$$

4.2.4 Self-adjoint boundary conditions

A *formally* self-adjoint operator T is *truly* self-adjoint only if the domains of $T^\dagger$ and T coincide. From now on, the unqualified phrase "self-adjoint" will always mean "truly self-adjoint".

Self-adjointness is usually desirable in physics problems. It is therefore useful to investigate what boundary conditions lead to self-adjoint operators. For example, what are the most general boundary conditions we can impose on $T = -i\partial_x$ if we require the resultant operator to be self-adjoint? Now,

$$\int_0^1 dx\, u^*(-i\partial_x v) - \int_0^1 dx(-i\partial_x u)^* v = -i\Big(u^*(1)v(1) - u^*(0)v(0)\Big). \tag{4.57}$$

Demanding that the right-hand side be zero gives us, after division by $u^*(0)v(1)$,

$$\frac{u^*(1)}{u^*(0)} = \frac{v(0)}{v(1)}. \tag{4.58}$$

We require this to be true for any u and v obeying the same boundary conditions. Since u and v are unrelated, both sides must equal a constant κ, and furthermore this constant must obey $\kappa^* = \kappa^{-1}$ in order that $u(1)/u(0)$ be equal to $v(1)/v(0)$. Thus, the boundary condition is

$$\frac{u(1)}{u(0)} = \frac{v(1)}{v(0)} = e^{i\theta} \tag{4.59}$$

for some real angle θ. The domain is therefore

$$\mathcal{D}(T) = \{y, Ty \in L^2[0,1] \ : y(1) = e^{i\theta}y(0)\}. \tag{4.60}$$

These are *twisted periodic* boundary conditions.

With these generalized periodic boundary conditions, everything we expect of a self-adjoint operator actually works:

(i) The functions $u_n = e^{i(2\pi n+\theta)x}$, with $n = \ldots, -2, -1, 0, 1, 2\ldots$ are eigenfunctions of T with eigenvalues $k_n \equiv 2\pi n + \theta$.
(ii) The eigenvalues are real.
(iii) The eigenfunctions form a complete orthonormal set.

Because self-adjoint operators possess a complete set of mutually orthogonal eigenfunctions, they are compatible with the interpretational postulates of quantum mechanics, where the square of the inner product of a state vector with an eigenstate gives the probability of measuring the associated eigenvalue. In quantum mechanics, self-adjoint operators are therefore called *observables*.

Example: *The Sturm–Liouville equation*. With

$$L = \frac{d}{dx}p(x)\frac{d}{dx} + q(x), \quad x \in [a,b], \tag{4.61}$$

we have

$$\langle u, Lv\rangle - \langle Lu, v\rangle = [p(u^*v' - u'^*v)]_a^b. \tag{4.62}$$

Let us seek to impose boundary conditions separately at the two ends. Thus, at $x = a$ we want

$$(u^*v' - u'^*v)|_a = 0, \tag{4.63}$$

or

$$\frac{u'^*(a)}{u^*(a)} = \frac{v'(a)}{v(a)}, \tag{4.64}$$

and similarly at b. If we want the boundary conditions imposed on v (which define the domain of L) to coincide with those for u (which define the domain of $L^\dagger$) then we must have

$$\frac{v'(a)}{v(a)} = \frac{u'(a)}{u(a)} = \tan\theta_a \tag{4.65}$$

for some real angle θ_a, and similar boundary conditions with a θ_b at b. We can also write these boundary conditions as

$$\begin{aligned} \alpha_a y(a) + \beta_a y'(a) &= 0, \\ \alpha_b y(b) + \beta_b y'(b) &= 0. \end{aligned} \tag{4.66}$$

Deficiency indices and self-adjoint extensions

There is a general theory of self-adjoint boundary conditions, due to Hermann Weyl and John von Neumann. We will not describe this theory in any detail, but simply give their recipe for counting the number of parameters in the most general self-adjoint boundary condition: to find this number we define an initial domain $\mathcal{D}_0(L)$ for the operator L by imposing the strictest possible boundary conditions. This we do by setting to zero the boundary values of all the $y^{(n)}$ with n less than the order of the equation. Next count the number of square-integrable eigenfunctions of the resulting adjoint operator $T^\dagger$ corresponding to eigenvalue $\pm i$. The numbers, n_+ and n_-, of these eigenfunctions are called the *deficiency indices*. If they are not equal then there is no possible way to make the operator self-adjoint. If they are equal, $n_+ = n_- = n$, then there is an n^2 real-parameter family of *self-adjoint extensions* $\mathcal{D}(L) \supset \mathcal{D}_0(L)$ of the initial tightly restricted domain.

Example: *The sad case of the "radial momentum operator".* We wish to define the operator $P_r = -i\partial_r$ on the half-line $0 < r < \infty$. We start with the restrictive domain

$$P_r = -i\partial_r, \quad \mathcal{D}_0(T) = \{y, P_r y \in L^2[0,\infty] : y(0) = 0\}. \tag{4.67}$$

We then have

$$P_r^\dagger = -i\partial_r, \quad \mathcal{D}(P_r^\dagger) = \{y, P_r^\dagger y \in L^2[0,\infty]\} \tag{4.68}$$

with no boundary conditions. The equation $P_r^\dagger y = iy$ has a normalizable solution $y = e^{-r}$. The equation $P_r^\dagger y = -iy$ has no normalizable solution. The deficiency indices are therefore $n_+ = 1$, $n_- = 0$, and this operator cannot be rescued and made self-adjoint.

Example: *The Schrödinger operator.* We now consider $-\partial_x^2$ on the half-line. Set

$$T = -\partial_x^2, \quad \mathcal{D}_0(T) = \{y, Ty \in L^2[0,\infty] : y(0) = y'(0) = 0\}. \tag{4.69}$$

We then have

$$T^\dagger = -\partial_x^2, \quad \mathcal{D}(T^\dagger) = \{y, T^\dagger y \in L^2[0,\infty]\}. \tag{4.70}$$

Again $T^\dagger$ comes with *no* boundary conditions. The eigenvalue equation $T^\dagger y = iy$ has one normalizable solution $y(x) = e^{(i-1)x/\sqrt{2}}$, and the equation $T^\dagger y = -iy$ also has one normalizable solution $y(x) = e^{-(i+1)x/\sqrt{2}}$. The deficiency indices are therefore $n_+ = n_- = 1$. The Weyl–von Neumann theory now says that, by relaxing the restrictive conditions $y(0) = y'(0) = 0$, we can extend the domain of definition of the operator to find a one-parameter family of self-adjoint boundary conditions. These will be the conditions $y'(0)/y(0) = \tan\theta$ that we found above.

If we consider the operator $-\partial_x^2$ on the finite interval $[a, b]$, then both solutions of $(T^\dagger \pm i)y = 0$ are normalizable, and the deficiency indices will be $n_+ = n_- = 2$. There should therefore be $2^2 = 4$ real parameters in the self-adjoint boundary conditions. This is a larger class than those we found in (4.66), because it includes generalized boundary conditions of the form

$$B_1[y] = \alpha_{11}y(a) + \alpha_{12}y'(a) + \beta_{11}y(b) + \beta_{12}y'(b) = 0,$$
$$B_2[y] = \alpha_{21}y(a) + \alpha_{22}y'(a) + \beta_{21}y(b) + \beta_{22}y'(b) = 0.$$

Physics application: Semiconductor heterojunction

We now demonstrate why we have spent so much time on identifying self-adjoint boundary conditions: the technique is important in practical physics problems.

A *heterojunction* is an atomically smooth interface between two related semiconductors, such as GaAs and $Al_xGa_{1-x}As$, which typically possess different band masses. We wish to describe the conduction electrons by an effective Schrödinger equation containing these band masses (see Figure 4.1). What matching condition should we impose on the wavefunction $\psi(x)$ at the interface between the two materials? A first guess is that

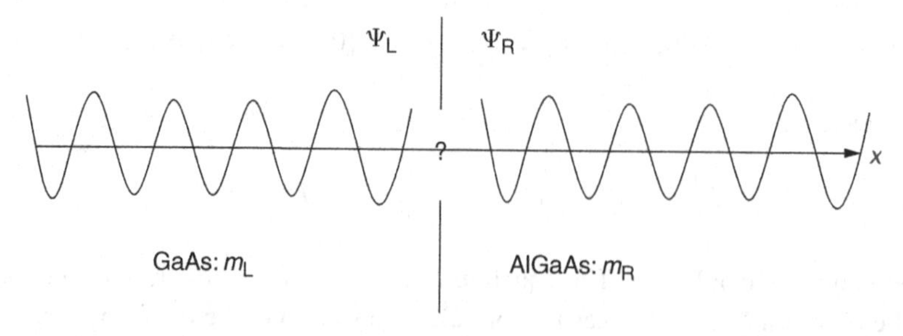

Figure 4.1 Heterojunction and wavefunctions.

the wavefunction must be continuous, but this is not correct because the "wavefunction" in an effective-mass band-theory Hamiltonian is not the actual wavefunction (which *is* continuous) but instead a slowly varying envelope function multiplying a Bloch wavefunction. The Bloch function is rapidly varying, fluctuating strongly on the scale of a single atom. Because the Bloch form of the solution is no longer valid at a discontinuity, the envelope function is not even defined in the neighbourhood of the interface, and certainly has no reason to be continuous. There must still be some linear relation between the ψ's in the two materials, but finding it will involve a detailed calculation on the atomic scale. In the absence of these calculations, we must use general principles to constrain the form of the relation. What are these principles?

We know that, were we to do the atomic-scale calculation, the resulting connection between the right and left wavefunctions would:

- be linear;
- involve no more than $\psi(x)$ and its first derivative $\psi'(x)$;
- make the Hamiltonian into a self-adjoint operator.

We want to find the most general connection formula compatible with these principles. The first two are easy to satisfy. We therefore investigate what matching conditions are compatible with self-adjointness.

Suppose that the band masses are m_L and m_R, so that

$$\begin{aligned} H &= -\frac{1}{2m_L}\frac{d^2}{dx^2} + V_L(x), \quad x < 0, \\ &= -\frac{1}{2m_R}\frac{d^2}{dx^2} + V_R(x), \quad x > 0. \end{aligned} \tag{4.71}$$

Integrating by parts, and keeping the terms at the interface, gives us

$$\begin{aligned} \langle\psi_1, H\psi_2\rangle - \langle H\psi_1, \psi_2\rangle &= \frac{1}{2m_L}\left\{\psi_{1L}^*\psi_{2L}' - \psi'^*_{1L}\psi_{2L}\right\} \\ &\quad - \frac{1}{2m_R}\left\{\psi_{1R}^*\psi_{2R}' - \psi'^*_{1R}\psi_{2R}\right\}. \end{aligned} \tag{4.72}$$

Here, $\psi_{L,R}$ refers to the boundary values of ψ immediately to the left or right of the junction, respectively. Now we impose general linear homogeneous boundary conditions on ψ_2:

$$\begin{pmatrix} \psi_{2L} \\ \psi_{2L}' \end{pmatrix} = \begin{pmatrix} a & b \\ c & d \end{pmatrix} \begin{pmatrix} \psi_{2R} \\ \psi_{2R}' \end{pmatrix}. \tag{4.73}$$

This relation involves four complex, and therefore eight real, parameters. Demanding that

$$\langle\psi_1, H\psi_2\rangle = \langle H\psi_1, \psi_2\rangle, \tag{4.74}$$

we find

$$\frac{1}{2m_L}\{\psi_{1L}^*(c\psi_{2R}+d\psi_{2R}')-\psi'^*_{1L}(a\psi_{2R}+b\psi_{2R}')\}=\frac{1}{2m_R}\{\psi_{1R}^*\psi_{2R}'-\psi'^*_{1R}\psi_{2R}\}, \tag{4.75}$$

and this must hold for arbitrary ψ_{2R}, ψ_{2R}', so, picking off the coefficients of these expressions and complex conjugating, we find

$$\begin{pmatrix}\psi_{1R}\\ \psi_{1R}'\end{pmatrix}=\left(\frac{m_R}{m_L}\right)\begin{pmatrix}d^* & -b^*\\ -c^* & a^*\end{pmatrix}\begin{pmatrix}\psi_{1L}\\ \psi_{1L}'\end{pmatrix}. \tag{4.76}$$

Because we wish the domain of $H^\dagger$ to coincide with that of H, these must be the same conditions that we imposed on ψ_2. Thus we must have

$$\begin{pmatrix}a & b\\ c & d\end{pmatrix}^{-1}=\left(\frac{m_R}{m_L}\right)\begin{pmatrix}d^* & -b^*\\ -c^* & a^*\end{pmatrix}. \tag{4.77}$$

Since

$$\begin{pmatrix}a & b\\ c & d\end{pmatrix}^{-1}=\frac{1}{ad-bc}\begin{pmatrix}d & -b\\ -c & a\end{pmatrix}, \tag{4.78}$$

we see that this requires

$$\begin{pmatrix}a & b\\ c & d\end{pmatrix}=e^{i\phi}\sqrt{\frac{m_L}{m_R}}\begin{pmatrix}A & B\\ C & D\end{pmatrix}, \tag{4.79}$$

where ϕ, A, B, C, D are real, and $AD-BC=1$. Demanding self-adjointness has therefore cut the original eight real parameters down to four. These can be determined either by experiment or by performing the microscopic calculation.[3] Note that $4=2^2$, a perfect square, as required by the Weyl–Von Neumann theory.

Exercise 4.5: Consider the Schrödinger operator $\hat{H}=-\partial_x^2$ on the interval [0, 1]. Show that the most general self-adjoint boundary condition applicable to $\hat{H}$ can be written as

$$\begin{bmatrix}\varphi(0)\\ \varphi'(0)\end{bmatrix}=e^{i\phi}\begin{bmatrix}a & b\\ c & d\end{bmatrix}\begin{bmatrix}\varphi(1)\\ \varphi'(1)\end{bmatrix},$$

where ϕ, a, b, c, d are real and $ad-bc=1$. Consider $\hat{H}$ as the quantum Hamiltonian of a particle on a ring constructed by attaching $x=0$ to $x=1$. Show that the self-adjoint boundary condition found above leads to unitary scattering at the point of join. Does the most general unitary point-scattering matrix correspond to the most general self-adjoint boundary condition?

[3] For example, see T. Ando, S. Mori, *Surf. Sci.*, **113** (1982) 124.

4.3 Completeness of eigenfunctions

Now that we have a clear understanding of what it means to be self-adjoint, we can reiterate the basic claim: an operator T that is self-adjoint with respect to an $L^2[a,b]$ inner product possesses a complete set of mutually orthogonal eigenfunctions. The proof that the eigenfunctions are orthogonal is identical to that for finite matrices. We will sketch a proof of the completeness of the eigenfunctions of the Sturm–Liouville operator in the next section.

The set of eigenvalues is, with some mathematical cavils, called the *spectrum* of T. It is usually denoted by $\sigma(T)$. An eigenvalue is said to belong to the *point* spectrum when its associated eigenfunction is normalizable, i.e. is a *bona fide* member of $L^2[a,b]$ having a finite length. Usually (but not always) the eigenvalues of the point spectrum form a discrete set, and so the point spectrum is also known as the *discrete spectrum*. When the operator acts on functions on an infinite interval, the eigenfunctions may fail to be normalizable. The associated eigenvalues are then said to belong to the *continuous spectrum*. Sometimes, e.g. the hydrogen atom, the spectrum is partly discrete and partly continuous. There is also something called the *residual spectrum*, but this does not occur for self-adjoint operators.

4.3.1 Discrete spectrum

The simplest problems have a purely discrete spectrum. We have eigenfunctions $\phi_n(x)$ such that

$$T\phi_n(x) = \lambda_n \phi_n(x), \tag{4.80}$$

where n is an integer. After multiplication by suitable constants, the ϕ_n are orthonormal,

$$\int \phi_n^*(x)\phi_m(x)\,dx = \delta_{nm}, \tag{4.81}$$

and complete. We can express the *completeness condition* as the statement that

$$\sum_n \phi_n(x)\phi_n^*(x') = \delta(x-x'). \tag{4.82}$$

If we take this representation of the delta function and multiply it by $f(x')$ and integrate over x', we find

$$f(x) = \sum_n \phi_n(x) \int \phi_n^*(x') f(x')\,dx'. \tag{4.83}$$

So,

$$f(x) = \sum_n a_n \phi_n(x) \tag{4.84}$$

with

$$a_n = \int \phi_n^*(x')f(x')\,dx'. \tag{4.85}$$

This means that if we can expand a delta function in terms of the $\phi_n(x)$, we can expand any (square integrable) function.

Warning: the convergence of the series $\sum_n \phi_n(x)\phi_n^*(x')$ to $\delta(x-x')$ is neither pointwise nor in the L^2 sense. The sum tends to a limit only in the sense of a distribution – meaning that we must multiply the partial sums by a smooth test function and integrate over x before we have something that actually converges in any meaningful manner. As an illustration consider our favourite orthonormal set: $\phi_n(x) = \sqrt{2}\sin(n\pi x)$ on the interval $[0,1]$. A plot of the first 70 terms in the sum

$$\sum_{n=1}^{\infty} \sqrt{2}\sin(n\pi x)\sqrt{2}\sin(n\pi x') = \delta(x-x')$$

is shown in Figure 4.2. The "wiggles" on both sides of the spike at $x = x'$ do not decrease in amplitude as the number of terms grows. They do, however, become of higher and higher frequency. When multiplied by a smooth function and integrated, the contributions from adjacent positive and negative wiggle regions tend to cancel, and it is only after this integration that the sum tends to zero away from the spike at $x = x'$.

Rayleigh–Ritz and completeness

For the Schrödinger eigenvalue problem

$$Ly = -y'' + q(x)y = \lambda y, \quad x \in [a,b], \tag{4.86}$$

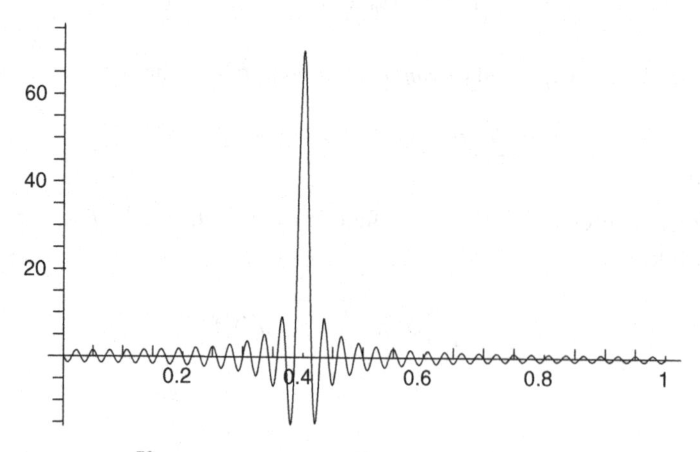

Figure 4.2 The sum $\sum_{n=1}^{70} 2\sin(n\pi x)\sin(n\pi x')$ for $x' = 0.4$. Take note of the very disparate scales on the horizontal and vertical axes.

the large eigenvalues are $\lambda_n \approx n^2\pi^2/(a-b)^2$. This is because the term qy eventually becomes negligible compared to λy, and we can then solve the equation with sines and cosines. We see that there is no upper limit to the magnitude of the eigenvalues. The eigenvalues of the Sturm–Liouville problem

$$Ly = -(py')' + qy = \lambda y, \quad x \in [a, b], \tag{4.87}$$

are similarly unbounded. We will use this unboundedness of the spectrum to make an estimate of the rate of convergence of the eigenfunction expansion for functions in the domain of L, and extend this result to prove that the eigenfunctions form a complete set.

We know from Chapter 1 that the Sturm–Liouville eigenvalues are the stationary values of $\langle y, Ly\rangle$ when the function y is constrained to have unit length, $\langle y, y\rangle = 1$. The lowest eigenvalue, λ_0, is therefore given by

$$\lambda_0 = \inf_{y\in\mathcal{D}(L)} \frac{\langle y, Ly\rangle}{\langle y, y\rangle}. \tag{4.88}$$

As the *variational principle*, this formula provides a well-known method of obtaining approximate ground state energies in quantum mechanics. Part of its effectiveness comes from the stationary nature of $\langle y, Ly\rangle$ at the minimum: a crude approximation to y often gives a tolerably good approximation to λ_0. In the wider world of eigenvalue problems, the variational principle is named after Rayleigh and Ritz.[4]

Suppose we have already found the first n normalized eigenfunctions $y_0, y_1, \ldots, y_{n-1}$. Let the space spanned by these functions be V_n. Then an obvious extension of the variational principle gives

$$\lambda_n = \inf_{y\in V_n^\perp} \frac{\langle y, Ly\rangle}{\langle y, y\rangle}. \tag{4.89}$$

We now exploit this variational estimate to show that if we expand an arbitrary y in the domain of L in terms of the full set of eigenfunctions y_m,

$$y = \sum_{m=0}^{\infty} a_m y_m, \tag{4.90}$$

where

$$a_m = \langle y_m, y\rangle, \tag{4.91}$$

then the sum does indeed converge to y.

Let

$$h_n = y - \sum_{m=0}^{n-1} a_m y_m \tag{4.92}$$

[4] J. W. Strutt (later Lord Rayleigh), *Phil. Trans.*, **161** (1870) 77; W. Ritz, *J. reine angew. Math.*, **135** (1908).

be the residual error after the first n terms. By definition, $h_n \in V_n^{\perp}$. Let us assume that we have adjusted, by adding a constant to q if necessary, L so that all the λ_m are positive. This adjustment will not affect the y_m. We expand out

$$\langle h_n, Lh_n \rangle = \langle y, Ly \rangle - \sum_{m=0}^{n-1} \lambda_m |a_m|^2, \tag{4.93}$$

where we have made use of the orthonormality of the y_m. The subtracted sum is guaranteed positive, so

$$\langle h_n, Lh_n \rangle \leq \langle y, Ly \rangle. \tag{4.94}$$

Combining this inequality with Rayleigh–Ritz tells us that

$$\frac{\langle y, Ly \rangle}{\langle h_n, h_n \rangle} \geq \frac{\langle h_n, Lh_n \rangle}{\langle h_n, h_n \rangle} \geq \lambda_n. \tag{4.95}$$

In other words

$$\frac{\langle y, Ly \rangle}{\lambda_n} \geq \|y - \sum_{m=0}^{n-1} a_m y_m\|^2. \tag{4.96}$$

Since $\langle y, Ly \rangle$ is independent of n, and $\lambda_n \to \infty$, we have $\|y - \sum_0^{n-1} a_m y_m\|^2 \to 0$. Thus the eigenfunction expansion indeed converges to y, and does so faster than λ_n^{-1} goes to zero.

Our estimate of the rate of convergence applies only to the expansion of functions y for which $\langle y, Ly \rangle$ is defined, i.e. to functions $y \in \mathcal{D}(L)$. The domain $\mathcal{D}(L)$ is always a dense subset of the entire Hilbert space $L^2[a,b]$, however, and, since a dense subset of a dense subset is also dense in the larger space, we have shown that the linear span of the eigenfunctions is a dense subset of $L^2[a,b]$. Combining this observation with the alternative definition of completeness in Section 2.2.3, we see that the eigenfunctions do indeed form a complete orthonormal set. Any square-integrable function therefore has a convergent expansion in terms of the y_m, but the rate of convergence may well be slower than that for functions $y \in \mathcal{D}(L)$.

Operator methods

Sometimes there are tricks for solving the eigenvalue problem.

Example: *Quantum harmonic oscillator.* Consider the operator

$$H = (-\partial_x + x)(\partial_x + x) + 1 = -\partial_x^2 + x^2. \tag{4.97}$$

This is in the form $Q^\dagger Q + 1$, where $Q = (\partial_x + x)$, and $Q^\dagger = (-\partial_x + x)$ is its formal adjoint. If we write these operators in the opposite order we have

$$QQ^\dagger = (\partial_x + x)(-\partial_x + x) = -\partial_x^2 + x^2 + 1 = H + 1. \tag{4.98}$$

Now, if ψ is an eigenfunction of $Q^\dagger Q$ with non-zero eigenvalue λ then $Q\psi$ is an eigenfunction of $QQ^\dagger$ with the same eigenvalue. This is because

$$Q^\dagger Q\psi = \lambda\psi \tag{4.99}$$

implies that

$$Q(Q^\dagger Q\psi) = \lambda Q\psi, \tag{4.100}$$

or

$$QQ^\dagger(Q\psi) = \lambda(Q\psi). \tag{4.101}$$

The only way that $Q\psi$ can fail to be an eigenfunction of $QQ^\dagger$ is if it happens that $Q\psi = 0$, but this implies that $Q^\dagger Q\psi = 0$ and so the eigenvalue was zero. Conversely, if the eigenvalue *is* zero then

$$0 = \langle\psi, Q^\dagger Q\psi\rangle = \langle Q\psi, Q\psi\rangle, \tag{4.102}$$

and so $Q\psi = 0$. In this way, we see that $Q^\dagger Q$ and $QQ^\dagger$ have exactly the same spectrum, with the possible exception of any zero eigenvalue.

Now notice that $Q^\dagger Q$ does have a zero eigenvalue because

$$\psi_0 = e^{-\frac{1}{2}x^2} \tag{4.103}$$

obeys $Q\psi_0 = 0$ and is normalizable. The operator $QQ^\dagger$, considered as an operator on $L^2[-\infty,\infty]$, does not have a zero eigenvalue because this would require $Q^\dagger\psi = 0$, and so

$$\psi = e^{+\frac{1}{2}x^2}, \tag{4.104}$$

which is not normalizable, and so not an element of $L^2[-\infty,\infty]$.

Since

$$H = Q^\dagger Q + 1 = QQ^\dagger - 1, \tag{4.105}$$

we see that ψ_0 is an eigenfunction of H with eigenvalue 1, and so an eigenfunction of $QQ^\dagger$ with eigenvalue 2. Hence $Q^\dagger\psi_0$ is an eigenfunction of $Q^\dagger Q$ with eigenvalue 2 and so an eigenfunction H with eigenvalue 3. Proceeding in this way we find that

$$\psi_n = (Q^\dagger)^n\psi_0 \tag{4.106}$$

is an eigenfunction of H with eigenvalue $2n+1$.

Since $Q^\dagger = -e^{\frac{1}{2}x^2}\partial_x e^{-\frac{1}{2}x^2}$, we can write

$$\psi_n(x) = H_n(x)e^{-\frac{1}{2}x^2}, \tag{4.107}$$

where

$$H_n(x) = (-1)^n e^{x^2}\frac{d^n}{dx^n}e^{-x^2} \tag{4.108}$$

are the *Hermite polynomials*.

This is a useful technique for any second-order operator that can be factorized – and a surprising number of the equations for "special functions" can be. You will see it later, both in the exercises and in connection with Bessel functions.

Exercise 4.6: Show that we have found all the eigenfunctions and eigenvalues of $H = -\partial_x^2 + x^2$. Hint: show that Q lowers the eigenvalue by 2 and use the fact that $Q^\dagger Q$ cannot have negative eigenvalues.

Problem 4.7: Schrödinger equations of the form

$$-\frac{d^2\psi}{dx^2} - l(l+1)\text{sech}^2 x\,\psi = E\psi$$

are known as *Pöschel–Teller equations*. By setting $u = l\tanh x$ and following the strategy of this problem one may relate solutions for l to those for $l-1$ and so find all bound states and scattering eigenfunctions for any integer l.

(a) Suppose that we know that $\psi = \exp\left\{-\int^x u(x')dx'\right\}$ is a solution of

$$L\psi \equiv \left(-\frac{d^2}{dx^2} + W(x)\right)\psi = 0.$$

Show that L can be written as $L = M^\dagger M$ where

$$M = \left(\frac{d}{dx} + u(x)\right), \quad M^\dagger = \left(-\frac{d}{dx} + u(x)\right),$$

the adjoint being taken with respect to the product $\langle u, v\rangle = \int u^* v\,dx$.

(b) Now assume L is acting on functions on $[-\infty, \infty]$ and that we do not have to worry about boundary conditions. Show that given an eigenfunction ψ_- obeying $M^\dagger M\psi_- = \lambda\psi_-$ we can multiply this equation on the left by M and so find an eigenfunction ψ_+ with the same eigenvalue for the differential operator

$$L' = MM^\dagger = \left(\frac{d}{dx} + u(x)\right)\left(-\frac{d}{dx} + u(x)\right)$$

and *vice versa*. Show that this correspondence $\psi_- \leftrightarrow \psi_+$ will fail if, *and only if*, $\lambda = 0$.

(c) Apply the strategy from part (b) in the case $u(x) = \tanh x$ and one of the two differential operators $M^\dagger M, MM^\dagger$ is (up to an additive constant)

$$H = -\frac{d}{dx}^2 - 2\,\mathrm{sech}^2 x.$$

Show that H has eigenfunctions of the form $\psi_k = e^{ikx} P(\tanh x)$ and eigenvalue $E = k^2$ for any k in the range $-\infty < k < \infty$. The function $P(\tanh x)$ is a polynomial in $\tanh x$ which you should be able to find explicitly. By thinking about the exceptional case $\lambda = 0$, show that H has an eigenfunction $\psi_0(x)$, with eigenvalue $E = -1$, that tends rapidly to zero as $x \to \pm\infty$. Observe that there is no corresponding eigenfunction for the other operator of the pair.

4.3.2 Continuous spectrum

Rather than give a formal discussion, we will illustrate this subject with some examples drawn from quantum mechanics.

The simplest example is the free particle on the real line. We have

$$H = -\partial_x^2. \tag{4.109}$$

We eventually want to apply this to functions on the entire real line, but we will begin with the interval $[-L/2, L/2]$, and then take the limit $L \to \infty$.

The operator H has formal eigenfunctions

$$\varphi_k(x) = e^{ikx}, \tag{4.110}$$

corresponding to eigenvalues $\lambda = k^2$. Suppose we impose periodic boundary conditions at $x = \pm L/2$:

$$\varphi_k(-L/2) = \varphi_k(+L/2). \tag{4.111}$$

This selects $k_n = 2\pi n/L$, where n is any positive, negative or zero integer, and allows us to find the normalized eigenfunctions

$$\chi_n(x) = \frac{1}{\sqrt{L}} e^{ik_n x}. \tag{4.112}$$

The completeness condition is

$$\sum_{n=-\infty}^{\infty} \frac{1}{L} e^{ik_n x} e^{-ik_n x'} = \delta(x - x'), \quad x, x' \in [-L/2, L/2]. \tag{4.113}$$

As L becomes large, the eigenvalues become so close that they can hardly be distinguished; hence the name *continuous spectrum*,[5] and the spectrum $\sigma(H)$ becomes the entire positive real line. In this limit, the sum on n becomes an integral

$$\sum_{n=-\infty}^{\infty}\left\{\ldots\right\} \to \int dn\left\{\ldots\right\} = \int dk\left(\frac{dn}{dk}\right)\left\{\ldots\right\}, \tag{4.114}$$

where

$$\frac{dn}{dk} = \frac{L}{2\pi} \tag{4.115}$$

is called the (momentum) density of states. If we divide this by L to get a density of states per unit length, we get an L independent "finite" quantity, the *local density of states*. We will often write

$$\frac{dn}{dk} = \rho(k). \tag{4.116}$$

If we express the density of states in terms of the eigenvalue λ then, by an abuse of notation, we have

$$\rho(\lambda) \equiv \frac{dn}{d\lambda} = \frac{L}{2\pi\sqrt{\lambda}}. \tag{4.117}$$

Note that

$$\frac{dn}{d\lambda} = 2\frac{dn}{dk}\frac{dk}{d\lambda}, \tag{4.118}$$

which looks a bit weird, but remember that *two* states, $\pm k_n$, correspond to the same λ and that the symbols

$$\frac{dn}{dk}, \quad \frac{dn}{d\lambda} \tag{4.119}$$

are ratios of measures, i.e. *Radon–Nikodym derivatives*, not ordinary derivatives.

In the limit $L \to \infty$, the completeness condition becomes

$$\int_{-\infty}^{\infty} \frac{dk}{2\pi} e^{ik(x-x')} = \delta(x - x'), \tag{4.120}$$

and the length L has disappeared.

[5] When L is strictly infinite, $\varphi_k(x)$ is no longer normalizable. Mathematicians do not allow such unnormalizable functions to be considered as true eigenfunctions, and so a point in the continuous spectrum is not, to them, actually an eigenvalue. Instead, they say that a point λ lies in the continuous spectrum if for any $\epsilon > 0$ there exists an *approximate eigenfunction* φ_ϵ such that $\|\varphi_\epsilon\| = 1$, but $\|L\varphi_\epsilon - \lambda\varphi_\epsilon\| < \epsilon$. This is not a profitable definition for us. We prefer to regard non-normalizable wavefunctions as being distributions in our rigged Hilbert space.

Suppose that we now apply boundary conditions $y = 0$ on $x = \pm L/2$. The normalized eigenfunctions are then

$$\chi_n = \sqrt{\frac{2}{L}} \sin k_n(x + L/2), \tag{4.121}$$

where $k_n = n\pi/L$. We see that the allowed k's are twice as close together as they were with periodic boundary conditions, but now n is restricted to being a positive non-zero integer. The momentum density of states is therefore

$$\rho(k) = \frac{dn}{dk} = \frac{L}{\pi}, \tag{4.122}$$

which is twice as large as in the periodic case, but the eigenvalue density of states is

$$\rho(\lambda) = \frac{L}{2\pi\sqrt{\lambda}}, \tag{4.123}$$

which is exactly the same as before.

That the number of states per unit energy per unit volume does not depend on the boundary conditions at infinity makes physical sense: no local property of the sublunary realm should depend on what happens in the sphere of fixed stars. This point was not fully grasped by physicists, however, until Rudolph Peierls[6] explained that the quantum particle had to actually travel to the distant boundary and back before the precise nature of the boundary could be felt. This journey takes time T (depending on the particle's energy) and from the energy–time uncertainty principle, we can distinguish one boundary condition from another only by examining the spectrum with an energy resolution finer than $\hbar/T$. Neither the distance nor the nature of the boundary can affect the coarse details, such as the local density of states.

The dependence of the spectrum of a general differential operator on boundary conditions was investigated by Hermann Weyl. Weyl distinguished two classes of singular boundary points: *limit-circle*, where the spectrum depends on the choice of boundary conditions, and *limit-point*, where it does not. For the Schrödinger operator, the point at infinity, which is "singular" simply because it is at infinity, is in the limit-point class. We will discuss Weyl's theory of singular endpoints in Chapter 8.

Phase shifts

Consider the eigenvalue problem

$$\left(-\frac{d^2}{dr^2} + V(r)\right)\psi = E\psi \tag{4.124}$$

[6] Peierls proved that the phonon contribution to the specific heat of a crystal could be correctly calculated by using periodic boundary conditions. Some sceptics had thought that such "unphysical" boundary conditions would give a result wrong by factors of two.

on the interval $[0,R]$, and with boundary conditions $\psi(0) = 0 = \psi(R)$. This problem arises when we solve the Schrödinger equation for a central potential in spherical polar coordinates, and assume that the wavefunction is a function of r only (i.e. S-wave, or $l = 0$). Again, we want the boundary at R to be infinitely far away, but we will start with R at a large but finite distance, and then take the $R \to \infty$ limit. Let us first deal with the simple case that $V(r) \equiv 0$; then the solutions are

$$\psi_k(r) \propto \sin kr, \tag{4.125}$$

with eigenvalue $E = k^2$, and with the allowed values being given by $k_n R = n\pi$. Since

$$\int_0^R \sin^2(k_n r)\, dr = \frac{R}{2}, \tag{4.126}$$

the normalized wavefunctions are

$$\psi_k = \sqrt{\frac{2}{R}} \sin kr, \tag{4.127}$$

and completeness reads

$$\sum_{n=1}^{\infty} \left(\frac{2}{R}\right) \sin(k_n r) \sin(k_n r') = \delta(r - r'). \tag{4.128}$$

As R becomes large, this sum goes over to an integral:

$$\begin{aligned} \sum_{n=1}^{\infty} \left(\frac{2}{R}\right) \sin(k_n r) \sin(k_n r') &\to \int_0^{\infty} dn \left(\frac{2}{R}\right) \sin(kr)\, \sin(kr'), \\ &= \int_0^{\infty} \frac{R dk}{\pi} \left(\frac{2}{R}\right) \sin(kr)\, \sin(kr'). \end{aligned} \tag{4.129}$$

Thus,

$$\left(\frac{2}{\pi}\right) \int_0^{\infty} dk\, \sin(kr) \sin(kr') = \delta(r - r'). \tag{4.130}$$

As before, the large distance, here R, no longer appears.

Now consider the more interesting problem which has the potential $V(r)$ included. We will assume, for simplicity, that there is an R_0 such that $V(r)$ is zero for $r > R_0$. In this case, we know that the solution for $r > R_0$ is of the form

$$\psi_k(r) = \sin\left(kr + \eta(k)\right), \tag{4.131}$$

where the *phase shift* $\eta(k)$ is a functional of the potential V. The eigenvalue is still $E = k^2$.

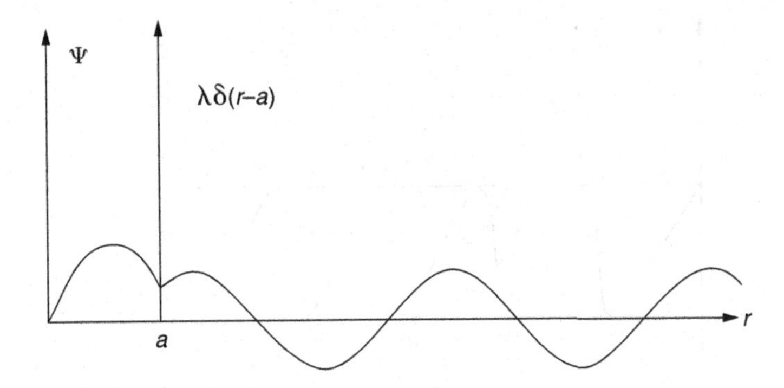

Figure 4.3 Delta-function shell potential.

Example: *A delta-function shell.* We take $V(r) = \lambda\delta(r-a)$. See Figure 4.3.

A solution with eigenvalue $E = k^2$ and satisfying the boundary condition at $r = 0$ is

$$\psi(r) = \begin{cases} A\sin(kr), & r < a, \\ \sin(kr+\eta), & r > a. \end{cases} \tag{4.132}$$

The conditions to be satisfied at $r = a$ are:

(i) continuity, $\psi(a-\epsilon) = \psi(a+\epsilon) \equiv \psi(a)$; and
(ii) jump in slope, $-\psi'(a+\epsilon) + \psi'(a-\epsilon) + \lambda\psi(a) = 0$.

Therefore,

$$\frac{\psi'(a+\epsilon)}{\psi(a)} - \frac{\psi'(a-\epsilon)}{\psi(a)} = \lambda, \tag{4.133}$$

or

$$\frac{k\cos(ka+\eta)}{\sin(ka+\eta)} - \frac{k\cos(ka)}{\sin(ka)} = \lambda. \tag{4.134}$$

Thus,

$$\cot(ka+\eta) - \cot(ka) = \frac{\lambda}{k}, \tag{4.135}$$

and

$$\eta(k) = -ka + \cot^{-1}\left(\frac{\lambda}{k} + \cot ka\right). \tag{4.136}$$

A sketch of $\eta(k)$ is shown in Figure 4.4. The allowed values of k are required by the boundary condition

$$\sin(kR + \eta(k)) = 0 \tag{4.137}$$

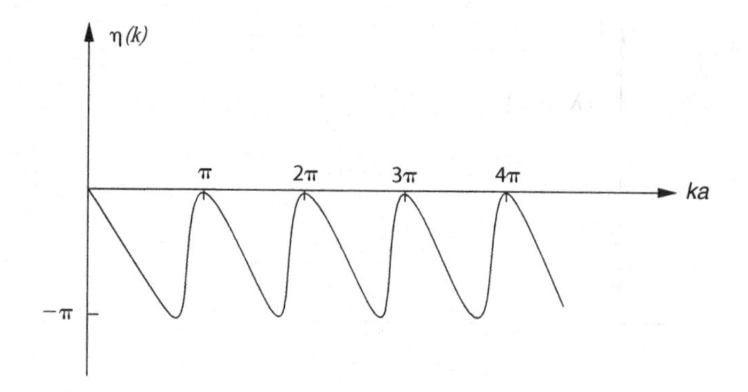

Figure 4.4 The phase shift $\eta(k)$ of Equation (4.136) plotted against ka.

to satisfy

$$kR + \eta(k) = n\pi. \tag{4.138}$$

This is a transcendental equation for k, and so finding the individual solutions k_n is not simple. We can, however, write

$$n = \frac{1}{\pi}\Big(kR + \eta(k)\Big) \tag{4.139}$$

and observe that, when R becomes large, only an infinitesimal change in k is required to make n increment by unity. We may therefore regard n as a "continuous" variable which we can differentiate with respect to k to find

$$\frac{dn}{dk} = \frac{1}{\pi}\left\{R + \frac{\partial \eta}{\partial k}\right\}. \tag{4.140}$$

The density of allowed k values is therefore

$$\rho(k) = \frac{1}{\pi}\left\{R + \frac{\partial \eta}{\partial k}\right\}. \tag{4.141}$$

For our delta-shell example, a plot of $\rho(k)$ appears in Figure 4.5. This figure shows a sequence of resonant bound states at $ka = n\pi$ superposed on the background continuum density of states appropriate to a large box of length $(R - a)$. Each "spike" contains one extra state, so the average density of states is that of a box of length R. We see that changing the potential does not create or destroy eigenstates, it just moves them around.

The spike is not exactly a delta function because of level repulsion between nearly degenerate eigenstates. The interloper elbows the nearby levels out of the way, and all the neighbours have to make do with a bit less room. The stronger the coupling between the states on either side of the delta shell, the stronger is the inter-level repulsion, and the broader the resonance spike.

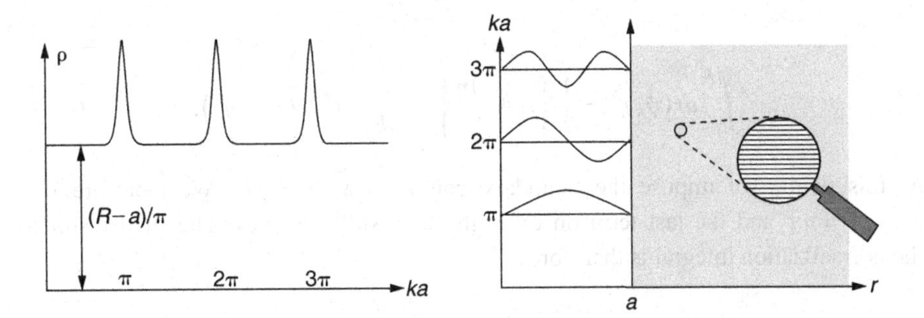

Figure 4.5 The density of states for the delta shell potential. The extended states are so close in energy that we need an optical aid to resolve individual levels. The almost-bound resonance levels have to squeeze in between them.

Normalization factor

We now evaluate

$$\int_0^R dr|\psi_k|^2 = N_k^{-2}, \tag{4.142}$$

so as to find the the normalized wavefunctions

$$\chi_k = N_k\psi_k. \tag{4.143}$$

Let $\psi_k(r)$ be a solution of

$$H\psi = \left(-\frac{d^2}{dr^2} + V(r)\right)\psi = k^2\psi \tag{4.144}$$

satisfying the boundary condition $\psi_k(0) = 0$, but not necessarily the boundary condition at $r = R$. Such a solution exists for any k. We scale ψ_k by requiring that $\psi_k(r) = \sin(kr + \eta)$ for $r > R_0$. We now use Lagrange's identity to write

$$\begin{aligned}(k^2 - k'^2)\int_0^R dr\, \psi_k\, \psi_{k'} &= \int_0^R dr\, \{(H\psi_k)\psi_{k'} - \psi_k(H\psi_{k'})\} \\ &= \left[\psi_k\psi'_{k'} - \psi'_k\psi_{k'}\right]_0^R \\ &= \sin(kR+\eta)k'\cos(k'R+\eta) \\ &\quad - k\cos(kR+\eta)\sin(k'R+\eta).\end{aligned} \tag{4.145}$$

Here, we have used $\psi_{k,k'}(0) = 0$, so the integrated out part vanishes at the lower limit, and have used the explicit form of $\psi_{k,k'}$ at the upper limit.

Now differentiate with respect to k, and then set $k = k'$. We find

$$2k\int_0^R dr(\psi_k)^2 = -\frac{1}{2}\sin\Big(2(kR+\eta)\Big) + k\left\{R + \frac{\partial\eta}{\partial k}\right\}. \tag{4.146}$$

In other words,

$$\int_0^R dr(\psi_k)^2 = \frac{1}{2}\left\{R + \frac{\partial \eta}{\partial k}\right\} - \frac{1}{4k}\sin\Big(2(kR+\eta)\Big). \tag{4.147}$$

At this point, we impose the boundary condition at $r = R$. We therefore have $kR + \eta = n\pi$ and the last term on the right-hand side vanishes. The final result for the normalization integral is therefore

$$\int_0^R dr|\psi_k|^2 = \frac{1}{2}\left\{R + \frac{\partial \eta}{\partial k}\right\}. \tag{4.148}$$

Observe that the same expression occurs in both the density of states and the normalization integral. When we use these quantities to write down the contribution of the normalized states in the continuous spectrum to the completeness relation we find that

$$\int_0^\infty dk \left(\frac{dn}{dk}\right) N_k^2 \psi_k(r)\psi_k(r') = \left(\frac{2}{\pi}\right)\int_0^\infty dk\, \psi_k(r)\psi_k(r'), \tag{4.149}$$

the density of states and normalization factor having cancelled and disappeared from the end result. This is a general feature of scattering problems: the completeness relation must give a delta function when evaluated far from the scatterer where the wavefunctions look like those of a free-particle. So, provided we normalize ψ_k so that it reduces to a free-particle wavefunction at large distance, the measure in the integral over k must also be the same as for the free particle.

Including any bound states in the discrete spectrum, the full statement of completeness is therefore

$$\sum_{\text{bound states}} \psi_n(r)\psi_n(r') + \left(\frac{2}{\pi}\right)\int_0^\infty dk\, \psi_k(r)\, \psi_k(r') = \delta(r - r'). \tag{4.150}$$

Example: We will exhibit a completeness relation for a problem on the entire real line. We have already met the Pöschel–Teller equation,

$$H\psi = \left(-\frac{d^2}{dx^2} - l(l+1)\,\mathrm{sech}^2 x\right)\psi = E\psi \tag{4.151}$$

in Exercise 4.7. When l is an integer, the potential in this Schrödinger equation has the special property that it is reflectionless.

The simplest non-trivial example is $l = 1$. In this case, H has a single discrete bound state at $E_0 = -1$. The normalized eigenfunction is

$$\psi_0(x) = \frac{1}{\sqrt{2}}\mathrm{sech}\, x. \tag{4.152}$$

The rest of the spectrum consists of a continuum of unbound states with eigenvalues $E(k) = k^2$ and eigenfunctions

$$\psi_k(x) = \frac{1}{\sqrt{1+k^2}} e^{ikx}(-ik + \tanh x). \tag{4.153}$$

Here, k is any real number. The normalization of $\psi_k(x)$ has been chosen so that, at large $|x|$, where $\tanh x \to \pm 1$, we have

$$\psi_k^*(x)\psi_k(x') \to e^{-ik(x-x')}. \tag{4.154}$$

The measure in the completeness integral must therefore be $dk/2\pi$, the same as that for a free particle.

Let us compute the difference

$$\begin{aligned}
I &= \delta(x-x') - \int_{-\infty}^{\infty} \frac{dk}{2\pi} \psi_k^*(x)\psi_k(x') \\
&= \int_{-\infty}^{\infty} \frac{dk}{2\pi} \left(e^{-ik(x-x)} - \psi_k^*(x)\psi_k(x') \right) \\
&= \int_{-\infty}^{\infty} \frac{dk}{2\pi} e^{-ik(x-x')} \frac{1 + ik(\tanh x - \tanh x') - \tanh x \tanh x'}{1+k^2}.
\end{aligned} \tag{4.155}$$

We use the standard integral,

$$\int_{-\infty}^{\infty} \frac{dk}{2\pi} e^{-ik(x-x')} \frac{1}{1+k^2} = \frac{1}{2} e^{-|x-x'|}, \tag{4.156}$$

together with its x' derivative,

$$\int_{-\infty}^{\infty} \frac{dk}{2\pi} e^{-ik(x-x')} \frac{ik}{1+k^2} = \operatorname{sgn}(x-x') \frac{1}{2} e^{-|x-x'|}, \tag{4.157}$$

to find

$$I = \frac{1}{2} \left\{ 1 + \operatorname{sgn}(x-x')(\tanh x - \tanh x') - \tanh x \tanh x' \right\} e^{-|x-x'|}. \tag{4.158}$$

Assume, without loss of generality, that $x > x'$; then this reduces to

$$\begin{aligned}
\frac{1}{2}(1+\tanh x)(1-\tanh x') e^{-(x-x')} &= \frac{1}{2} \operatorname{sech} x \operatorname{sech} x' \\
&= \psi_0(x)\psi_0(x').
\end{aligned} \tag{4.159}$$

Thus, the expected completeness condition,

$$\psi_0(x)\psi_0(x') + \int_{-\infty}^{\infty} \frac{dk}{2\pi} \psi_k^*(x)\psi_k(x') = \delta(x-x'), \tag{4.160}$$

is confirmed.

4.4 Further exercises and problems

We begin with a practical engineering eigenvalue problem.

Exercise 4.8: *Whirling drive shaft.* A thin flexible drive shaft is supported by two bearings that impose the conditions $x' = y' = x = y = 0$ at $z = \pm L$ (see Figure 4.6). Here $x(z), y(z)$ denote the transverse displacements of the shaft, and the primes denote derivatives with respect to z.

The shaft is driven at angular velocity ω. Experience shows that at certain critical frequencies ω_n the motion becomes unstable to *whirling* – a spontaneous vibration and deformation of the normally straight shaft. If the rotation frequency is raised above ω_n, the shaft becomes quiescent and straight again until we reach a frequency ω_{n+1}, at which the pattern is repeated. Our task is to understand why this happens.

The kinetic energy of the whirling shaft is

$$T = \frac{1}{2}\int_{-L}^{L} \rho\{\dot{x}^2 + \dot{y}^2\}\, dz,$$

and the strain energy due to bending is

$$V[x,y] = \frac{1}{2}\int_{-L}^{L} \gamma\{(x'')^2 + (y'')^2\}\, dz.$$

(a) Write down the Lagrangian, and from it obtain the equations of motion for the shaft.
(b) Seek whirling-mode solutions of the equations of motion in the form

$$x(z,t) = \psi(z)\cos\omega t,$$
$$y(z,t) = \psi(z)\sin\omega t.$$

Show that this quest requires the solution of the eigenvalue problem

$$\frac{\gamma}{\rho}\frac{d^4\psi}{dz^4} = \omega_n^2\psi, \quad \psi'(-L) = \psi(-L) = \psi'(L) = \psi(L) = 0.$$

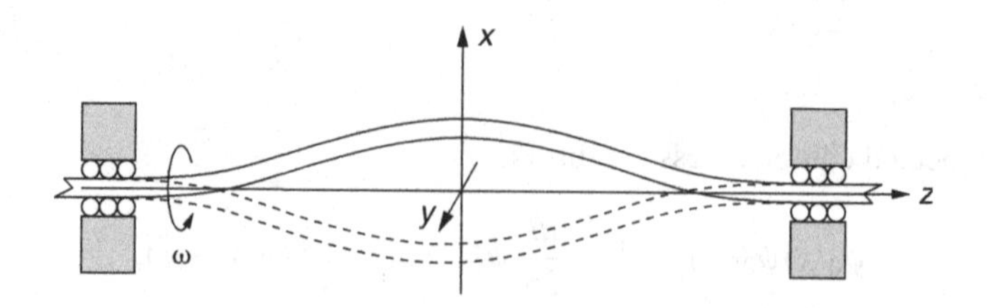

Figure 4.6 The $n = 1$ even-parity mode of a whirling shaft.

(c) Show that the critical frequencies are given in terms of the solutions ξ_n to the transcendental equation

$$\tanh \xi_n = \pm \tan \xi_n, \qquad (\star)$$

as

$$\omega_n = \sqrt{\frac{\gamma}{\rho}} \left(\frac{\xi_n}{L}\right)^2.$$

Show that the plus sign in ($\star$) applies to *odd* parity modes, where $\psi(z) = -\psi(-z)$, and the minus sign to *even* parity modes where $\psi(z) = \psi(-z)$.

Whirling, we conclude, occurs at the frequencies of the natural transverse vibration modes of the elastic shaft. These modes are excited by slight imbalances that have negligible effect except when the shaft is being rotated at the resonant frequency.

Insight into adjoint boundary conditions for an ODE can be obtained by thinking about how we would impose these boundary conditions in a numerical solution. The next problem illustrates this.

Problem 4.9: *Discrete approximations and self-adjointness.* Consider the second-order inhomogeneous equation $Lu \equiv u'' = g(x)$ on the interval $0 \leq x \leq 1$. Here $g(x)$ is known and $u(x)$ is to be found. We wish to solve the problem on a computer, and so set up a discrete approximation to the ODE in the following way:

- Replace the continuum of independent variables $0 \leq x \leq 1$ by the discrete lattice of points $0 \leq x_n \equiv \left(n - \frac{1}{2}\right)/N \leq 1$. Here N is a positive integer and $n = 1, 2, \ldots, N$.
- Replace the functions $u(x)$ and $g(x)$ by the arrays of real variables $u_n \equiv u(x_n)$ and $g_n \equiv g(x_n)$.
- Replace the continuum differential operator d^2/dx^2 by the difference operator $\mathcal{D}^2$, defined by $\mathcal{D}^2 u_n \equiv u_{n+1} - 2u_n + u_{n-1}$.

Now do the following problems:

(a) Impose continuum Dirichlet boundary conditions $u(0) = u(1) = 0$. Decide what these correspond to in the discrete approximation, and write the resulting set of algebraic equations in matrix form. Show that the corresponding matrix is real and symmetric.

(b) Impose the periodic boundary conditions $u(0) = u(1)$ and $u'(0) = u'(1)$, and show that these require us to set $u_0 \equiv u_N$ and $u_{N+1} \equiv u_1$. Again write the system of algebraic equations in matrix form and show that the resulting matrix is real and symmetric.

(c) Consider the non-symmetric N-by-N matrix operator

$$D^2u = \begin{pmatrix} 0 & 0 & 0 & 0 & 0 & \dots & 0 \\ 1 & -2 & 1 & 0 & 0 & \dots & 0 \\ 0 & 1 & -2 & 1 & 0 & \dots & 0 \\ \vdots & \vdots & \vdots & \ddots & \vdots & \vdots & \vdots \\ 0 & \dots & 0 & 1 & -2 & 1 & 0 \\ 0 & \dots & 0 & 0 & 1 & -2 & 1 \\ 0 & \dots & 0 & 0 & 0 & 0 & 0 \end{pmatrix} \begin{pmatrix} u_N \\ u_{N-1} \\ u_{N-2} \\ \vdots \\ u_3 \\ u_2 \\ u_1 \end{pmatrix}.$$

(i) What vectors span the null space of D^2?

(ii) To what continuum boundary conditions for d^2/dx^2 does this matrix correspond?

(iii) Consider the matrix $(D^2)^\dagger$. To what continuum boundary conditions does this matrix correspond? Are they the adjoint boundary conditions for the differential operator in part (ii)?

Exercise 4.10: Let

$$\begin{aligned} \widehat{H} &= \begin{pmatrix} -i\partial_x & m_1 - im_2 \\ m_1 + im_2 & i\partial_x \end{pmatrix} \\ &= -i\widehat{\sigma}_3\partial_x + m_1\widehat{\sigma}_1 + m_2\widehat{\sigma}_2 \end{aligned}$$

be a one-dimensional Dirac Hamiltonian. Here $m_1(x)$ and $m_2(x)$ are real functions and the $\widehat{\sigma}_i$ are the Pauli matrices. The matrix differential operator $\widehat{H}$ acts on the two-component "spinor"

$$\Psi(x) = \begin{pmatrix} \psi_1(x) \\ \psi_2(x) \end{pmatrix}.$$

(a) Consider the eigenvalue problem $\widehat{H}\Psi = E\Psi$ on the interval $[a, b]$. Show that the boundary conditions

$$\frac{\psi_1(a)}{\psi_2(a)} = \exp\{i\theta_a\}, \qquad \frac{\psi_1(b)}{\psi_2(b)} = \exp\{i\theta_b\},$$

where θ_a, θ_b are real angles, make $\widehat{H}$ into an operator that is self-adjoint with respect to the inner product

$$\langle \Psi_1, \Psi_2 \rangle = \int_a^b \Psi_1^\dagger(x)\Psi_2(x)\, dx.$$

(b) Find the eigenfunctions Ψ_n and eigenvalues E_n in the case that $m_1 = m_2 = 0$ and the $\theta_{a,b}$ are arbitrary real angles.

Here are three further problems involving the completeness of operators with a continuous spectrum:

Problem 4.11: *Missing state.* In Problem 4.4.7 you will have found that the Schrödinger equation

$$\left(-\frac{d^2}{dx^2} - 2\,\mathrm{sech}^2 x\right)\psi = E\,\psi$$

has eigensolutions

$$\psi_k(x) = e^{ikx}(-ik + \tanh x)$$

with eigenvalue $E = k^2$.

- For x large and positive $\psi_k(x) \approx A\, e^{ikx} e^{i\eta(k)}$, while for x large and negative $\psi_k(x) \approx A\, e^{ikx} e^{-i\eta(k)}$, the (complex) constant A being the same in both cases. Express the phase shift $\eta(k)$ as the inverse tangent of an algebraic expression in k.
- Impose periodic boundary conditions $\psi(-L/2) = \psi(+L/2)$ where $L \gg 1$. Find the allowed values of k and hence an explicit expression for the k-space density, $\rho(k) = \frac{dn}{dk}$, of the eigenstates.
- Compare your formula for $\rho(k)$ with the corresponding expression, $\rho_0(k) = L/2\pi$, for the eigenstate density of the zero-potential equation and compute the integral

$$\Delta N = \int_{-\infty}^{\infty} \{\rho(k) - \rho_0(k)\} dk.$$

- Deduce that one eigenfunction has gone missing from the continuum and becomes the localized bound state $\psi_0(x) = \frac{1}{\sqrt{2}}\mathrm{sech}\, x$.

Problem 4.12: *Continuum completeness.* Consider the differential operator

$$\hat{L} = -\frac{d^2}{dx^2}, \qquad 0 \le x < \infty$$

with self-adjoint boundary conditions $\psi(0)/\psi'(0) = \tan\theta$ for some fixed angle θ.

- Show that when $\tan\theta < 0$ there is a single normalizable negative-eigenvalue eigenfunction localized near the origin, but none when $\tan\theta > 0$.
- Show that there is a continuum of positive-eigenvalue eigenfunctions of the form $\psi_k(x) = \sin(kx + \eta(k))$ where the phase shift η is found from

$$e^{i\eta(k)} = \frac{1 + ik\tan\theta}{\sqrt{1 + k^2\tan^2\theta}}.$$

- Write down (no justification required) the appropriate completeness relation

$$\delta(x - x') = \int \frac{dn}{dk} N_k^2 \psi_k(x)\psi_k(x')\, dk + \sum_{\text{bound}} \psi_n(x)\psi_n(x')$$

with an explicit expression for the product (not the separate factors) of the density of states and the normalization constant N_k^2, and with the correct limits on the integral over k.

- Confirm that the ψ_k continuum on its own, or together with the bound state when it exists, form a complete set. You will do this by evaluating the integral

$$I(x,x') = \frac{2}{\pi}\int_0^\infty \sin(kx+\eta(k))\sin(kx'+\eta(k))\,dk$$

and interpreting the result. You will need the following standard integral

$$\int_{-\infty}^{\infty}\frac{dk}{2\pi}e^{ikx}\frac{1}{1+k^2t^2} = \frac{1}{2|t|}e^{-|x|/|t|}.$$

Take care! You should monitor how the bound state contribution switches on and off as θ is varied. Keeping track of the modulus signs $|\ldots|$ in the standard integral is essential for this.

Problem 4.13: *One-dimensional scattering redux.* Consider again the one-dimensional Schrödinger equation from Chapter 3, Problem 3.4:

$$-\frac{d^2\psi}{dx^2} + V(x)\psi = E\psi,$$

where $V(x)$ is zero except in a finite interval $[-a,a]$ near the origin (Figure 4.7).
For $k > 0$, consider solutions of the form

$$\psi(x) = \begin{cases} a_L^{\text{in}} e^{ikx} + a_L^{\text{out}} e^{-ikx}, & x \in L, \\ a_R^{\text{in}} e^{-ikx} + a_R^{\text{out}} e^{ikx}, & x \in R. \end{cases}$$

(a) Show that, in the notation of Problem 3.4, we have

$$\begin{bmatrix} a_L^{\text{out}} \\ a_R^{\text{out}} \end{bmatrix} = \begin{bmatrix} r_L(k) & t_R(-k) \\ t_L(k) & r_R(-k) \end{bmatrix}\begin{bmatrix} a_L^{\text{in}} \\ a_R^{\text{in}} \end{bmatrix},$$

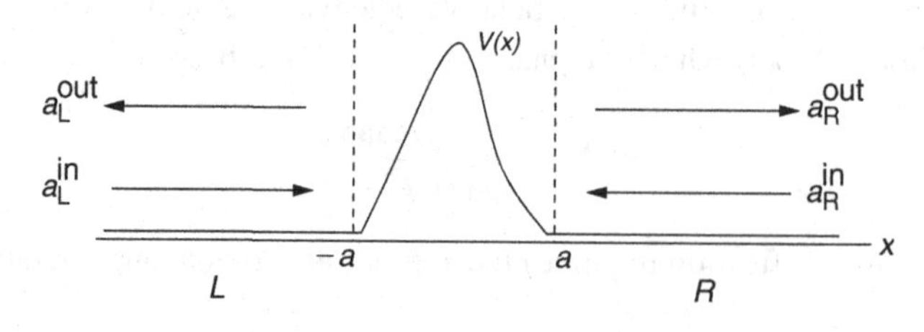

Figure 4.7 Incoming and outgoing waves in Problem 4.13. The asymptotic regions L and R are defined by $L = \{x < -a\}$ and $R = \{x > a\}$.

and show that the S-matrix

$$S(k) \equiv \begin{bmatrix} r_L(k) & t_R(-k) \\ t_L(k) & r_R(-k) \end{bmatrix}$$

is unitary.

(b) By observing that complex conjugation interchanges the "in" and "out" waves, show that it is natural to extend the definition of the transmission and reflection coefficients to all real k by setting $r_{L,R}(k) = r^*_{L,R}(-k)$, $t_{L,R}(k) = t^*_{L,R}(-k)$.

(c) In Problem 3.4 we introduced the particular solutions

$$\psi_k(x) = \begin{cases} e^{ikx} + r_L(k)e^{-ikx}, & x \in L, \\ t_L(k)e^{ikx}, & x \in R, \end{cases} \quad k > 0,$$

$$= \begin{cases} t_R(k)e^{ikx}, & x \in L, \\ e^{ikx} + r_R(k)e^{-ikx}, & x \in R. \end{cases} \quad k < 0.$$

Show that, together with any bound states $\psi_n(x)$, these $\psi_k(x)$ satisfy the completeness relation

$$\sum_{\text{bound}} \psi_n^*(x)\psi_n(x') + \int_{-\infty}^{\infty} \frac{dk}{2\pi} \psi_k^*(x)\psi_k(x') = \delta(x - x')$$

provided that

$$-\sum_{\text{bound}} \psi_n^*(x)\psi_n(x') = \int_{-\infty}^{\infty} \frac{dk}{2\pi} r_L(k)e^{-ik(x+x')}, \qquad x, x' \in L,$$

$$= \int_{-\infty}^{\infty} \frac{dk}{2\pi} t_L(k)e^{-ik(x-x')}, \qquad x \in L,\ x' \in R,$$

$$= \int_{-\infty}^{\infty} \frac{dk}{2\pi} t_R(k)e^{-ik(x-x')}, \qquad x \in R,\ x' \in L,$$

$$= \int_{-\infty}^{\infty} \frac{dk}{2\pi} r_R(k)e^{-ik(x+x')}, \qquad x, x' \in R.$$

(d) Compute $r_{L,R}(k)$ and $t_{L,R}(k)$ for the potential $V(x) = \lambda\delta(x - b)$, and verify that the conditions in part (c) are satisfied.

If you are familiar with complex variable methods, look ahead to Chapter 18 where Problem 18.22 shows you how to use complex variable methods to evaluate the Fourier transforms in part (c), and so confirm that the bound state $\psi_n(x)$ and the $\psi_k(x)$ together constitute a complete set of eigenfunctions.

Problem 4.14: Levinson's theorem and the Friedel sum rule. The interaction between an attractive impurity and (S-wave, and ignoring spin) electrons in a metal can be modelled by a one-dimensional Schrödinger equation

$$-\frac{d^2\chi}{dr^2}+V(r)\chi=k^2\chi.$$

Here r is the distance away from the impurity, $V(r)$ is the (spherically symmetric) impurity potential and $\chi(r)=\sqrt{4\pi}r\psi(r)$ where $\psi(r)$ is the three-dimensional wavefunction. The impurity attracts electrons to its vicinity. Let $\chi_k^0(r)=\sin(kr)$ denote the unperturbed wavefunction, and $\chi_k(r)$ denote the perturbed wavefunction that beyond the range of impurity potential becomes $\sin(kr+\eta(k))$. We fix the $2n\pi$ ambiguity in the definition of $\eta(k)$ by taking $\eta(\infty)$ to be zero, and requiring $\eta(k)$ to be a continuous function of k.

- Show that the continuous-spectrum contribution to the change in the number of electrons within a sphere of radius R surrounding the impurity is given by

$$\frac{2}{\pi}\int_0^{k_{\mathrm{f}}}\left(\int_0^R\left\{|\chi_k(x)|^2-|\chi_k^0(x)|^2\right\}dr\right)dk=\frac{1}{\pi}\left[\eta(k_{\mathrm{f}})-\eta(0)\right]+\text{oscillations}.$$

  Here k_{f} is the Fermi momentum, and "oscillations" refers to *Friedel oscillations* $\approx\cos(2(k_{\mathrm{f}}R+\eta))$. You should write down an explicit expression for the Friedel oscillation term, and recognize it as the Fourier transform of a function $\propto k^{-1}\sin\eta(k)$.
- Appeal to the Riemann–Lebesgue lemma to argue that the Friedel density oscillations make no contribution to the accumulated electron number in the limit $R\to\infty$. (Hint: you may want to look ahead to the next part of the problem in order to show that $k^{-1}\sin\eta(k)$ remains finite as $k\to 0$.)

The impurity-induced change in the number of unbound electrons in the interval $[0,R]$ is generically some fraction of an electron, and, in the case of an attractive potential, can be *negative* – the phase shift being positive and decreasing steadily to zero as k increases to infinity. This should not be surprising. Each electron in the Fermi sea speeds up as it enters an attractive potential well, spends less time there, and so makes a smaller contribution to the average local density than it would in the absence of the potential. We would, however, surely expect an attractive potential to accumulate a net *positive* number of electrons.

- Show that a negative continuous-spectrum contribution to the accumulated electron number is more than compensated for by a positive number

$$N_{\text{bound}}=\int_0^\infty(\rho_0(k)-\rho(k))dk=-\int_0^\infty\frac{1}{\pi}\frac{\partial\eta}{\partial k}dk=\frac{1}{\pi}\eta(0)$$

of electrons bound to the potential. After accounting for these bound electrons, show that the total number of electrons accumulated near the impurity is

$$Q_{\text{tot}} = \frac{1}{\pi}\eta(k_{\text{f}}).$$

This formula (together with its higher angular momentum versions) is known as the *Friedel sum rule*. The relation between $\eta(0)$ and the number of bound states is called *Levinson's theorem*. A more rigorous derivation of this theorem would show that $\eta(0)$ may take the value $(n + 1/2)\pi$ when there is a non-normalizable zero-energy "half-bound" state. In this exceptional case the accumulated charge will depend on R.

5

Green functions

In this chapter we will study strategies for solving the inhomogeneous linear differential equation $Ly = f$. The tool we use is the *Green function*, which is an integral kernel representing the inverse operator L^{-1}. Apart from their use in solving inhomogeneous equations, Green functions play an important role in many areas of physics.

5.1 Inhomogeneous linear equations

We wish to solve $Ly = f$ for y. Before we set about doing this, we should ask ourselves whether a solution *exists*, and, if it does, whether it is *unique*. The answers to these questions are summarized by the *Fredholm alternative*.

5.1.1 Fredholm alternative

The Fredholm alternative for operators on a finite-dimensional vector space is discussed in detail in the Appendix on linear algebra. You will want to make sure that you have read and understood this material. Here, we merely restate the results.

Let V be finite-dimensional vector space equipped with an inner product, and let A be a linear operator $A : V \to V$ on this space. Then

I. **Either**

   (i) $Ax = b$ has a *unique* solution,

   or

   (ii) $Ax = 0$ has a non-trivial solution.

II. If $Ax = 0$ has n linearly independent solutions, then so does $A^{\dagger}x = 0$.

III. If alternative (ii) holds, then $Ax = b$ has *no* solution unless b is perpendicular to all solutions of $A^{\dagger}x = 0$.

What is important for us in the present chapter is that this result continues to hold for linear differential operators L on a finite interval – provided that we define $L^{\dagger}$ as in the previous chapter, *and provided the number of boundary conditions is equal to the order of the equation.*

If the number of boundary conditions is *not* equal to the order of the equation then the number of solutions to $Ly = 0$ and $L^{\dagger}y = 0$ will differ in general. It is still true, however, that $Ly = f$ has *no* solution unless f is perpendicular to all solutions of $L^{\dagger}y = 0$.

Example: As an illustration of what happens when an equation possesses too many boundary conditions, consider

$$Ly = \frac{dy}{dx}, \quad y(0) = y(1) = 0. \tag{5.1}$$

Clearly $Ly = 0$ has only the trivial solution $y \equiv 0$. If a solution to $Ly = f$ exists, therefore, it will be unique.

We know that $L^\dagger = -d/dx$, with *no* boundary conditions on the functions in its domain. The equation $L^\dagger y = 0$ therefore has the non-trivial solution $y = 1$. This means that there should be no solution to $Ly = f$ unless

$$\langle 1, f\rangle = \int_0^1 f\, dx = 0. \tag{5.2}$$

If this condition is satisfied then

$$y(x) = \int_0^x f(x)\, dx \tag{5.3}$$

satisfies both the differential equation and the boundary conditions at $x = 0, 1$. If the condition is not satisfied, $y(x)$ is not a solution, because $y(1) \neq 0$.

Initially we only solve $Ly = f$ for homogeneous boundary conditions. After we have understood how to do this, we will extend our methods to deal with differential equations with inhomogeneous boundary conditions.

5.2 Constructing Green functions

We will solve $Ly = f$, a differential equation with homogeneous boundary conditions, by finding an inverse operator L^{-1}, so that $y = L^{-1}f$. This inverse operator L^{-1} will be represented by an integral kernel

$$(L^{-1})_{x,\xi} = G(x, \xi), \tag{5.4}$$

with the property

$$L_x G(x, \xi) = \delta(x - \xi). \tag{5.5}$$

Here, the subscript x on L indicates that L acts on the first argument, x, of G. Then

$$y(x) = \int G(x, \xi) f(\xi)\, d\xi \tag{5.6}$$

will obey

$$L_x y = \int L_x G(x, \xi) f(\xi)\, d\xi = \int \delta(x - \xi) f(\xi)\, d\xi = f(x). \tag{5.7}$$

The problem is how to construct $G(x,\xi)$. There are three necessary ingredients:

- the function $\chi(x) \equiv G(x,\xi)$ must have some discontinuous behaviour at $x = \xi$ in order to generate the delta function;
- away from $x = \xi$, the function $\chi(x)$ must obey $L\chi = 0$;
- the function $\chi(x)$ must obey the homogeneous boundary conditions required of y at the ends of the interval.

The last ingredient ensures that the resulting solution, $y(x)$, obeys the boundary conditions. It also ensures that the range of the integral operator G lies within the domain of L, a prerequisite if the product $LG = I$ is to make sense. The manner in which these ingredients are assembled to construct $G(x,\xi)$ is best explained through examples.

5.2.1 Sturm–Liouville equation

We begin by constructing the solution to the equation

$$(p(x)y')' + q(x)y(x) = f(x) \tag{5.8}$$

on the finite interval $[a,b]$ with homogeneous self-adjoint boundary conditions

$$\frac{y'(a)}{y(a)} = \tan\theta_L, \quad \frac{y'(b)}{y(b)} = \tan\theta_R. \tag{5.9}$$

We therefore seek a function $G(x,\xi)$ such that $\chi(x) = G(x,\xi)$ obeys

$$L\chi = (p\chi')' + q\chi = \delta(x-\xi). \tag{5.10}$$

The function $\chi(x)$ must also obey the homogeneous boundary conditions we require of $y(x)$.

Now (5.10) tells us that $\chi(x)$ must be continuous at $x = \xi$. For if not, the two differentiations applied to a jump function would give us the derivative of a delta function, and we want only a plain $\delta(x-\xi)$. If we write

$$G(x,\xi) = \chi(x) = \begin{cases} Ay_L(x)y_R(\xi), & x < \xi, \\ Ay_L(\xi)y_R(x), & x > \xi, \end{cases} \tag{5.11}$$

then $\chi(x)$ is automatically continuous at $x = \xi$. We take $y_L(x)$ to be a solution of $Ly = 0$, chosen to satisfy the boundary condition at the left-hand end of the interval. Similarly $y_R(x)$ should solve $Ly = 0$ and satisfy the boundary condition at the right-hand end. With these choices we satisfy (5.10) at all points away from $x = \xi$.

To work out how to satisfy the equation exactly *at* the location of the delta function, we integrate (5.10) from $\xi - \varepsilon$ to $\xi + \varepsilon$ and find that

$$p(\xi)[\chi'(\xi+\varepsilon) - \chi'(\xi-\varepsilon)] = 1. \tag{5.12}$$

With our product form for $\chi(x)$, this *jump condition* becomes

$$Ap(\xi)\Big(y_L(\xi)y_R'(\xi) - y_L'(\xi)y_R(\xi)\Big) = 1 \tag{5.13}$$

and determines the constant A. We recognize the Wronskian $W(y_L, y_R; \xi)$ on the left-hand side of this equation. We therefore have $A = 1/(pW)$ and

$$G(x,\xi) = \begin{cases} \frac{1}{pW} y_L(x) y_R(\xi), & x < \xi, \\ \frac{1}{pW} y_L(\xi) y_R(x), & x > \xi. \end{cases} \tag{5.14}$$

For the Sturm–Liouville equation the product pW is constant. This fact follows from Liouville's formula,

$$W(x) = W(0) \exp\left\{ -\int_0^x \left(\frac{p_1}{p_0}\right) d\xi \right\}, \tag{5.15}$$

and from $p_1 = p_0' = p'$ in the Sturm–Liouville equation. Thus

$$W(x) = W(0)\exp\Big(-\ln[p(x)/p(0)]\Big) = W(0)\frac{p(0)}{p(x)}. \tag{5.16}$$

The constancy of pW means that $G(x,\xi)$ is symmetric:

$$G(x,\xi) = G(\xi, x). \tag{5.17}$$

This is as it should be. The inverse of a symmetric matrix (and the real, self-adjoint, Sturm–Liouville operator is the function-space analogue of a real symmetric matrix) is itself symmetric.

The solution to

$$Ly = (p_0 y')' + qy = f(x) \tag{5.18}$$

is therefore

$$y(x) = \frac{1}{Wp}\left\{ y_L(x)\int_x^b y_R(\xi) f(\xi)\, d\xi + y_R(x)\int_a^x y_L(\xi) f(\xi)\, d\xi \right\}. \tag{5.19}$$

Take care to understand the ranges of integration in this formula. In the first integral $\xi > x$ and we use $G(x,\xi) \propto y_L(x) y_R(\xi)$. In the second integral $\xi < x$ and we use $G(x,\xi) \propto y_L(\xi) y_R(x)$. It is easy to get these the wrong way round.

Because we must divide by it in constructing $G(x,\xi)$, it is necessary that the Wronskian $W(y_L, y_R)$ not be zero. This is reasonable. If W were zero then $y_L \propto y_R$, and the single function y_R satisfies both $Ly_R = 0$ and the boundary conditions. This means that the differential operator L has y_R as a zero-mode, so there can be no unique solution to $Ly = f$.

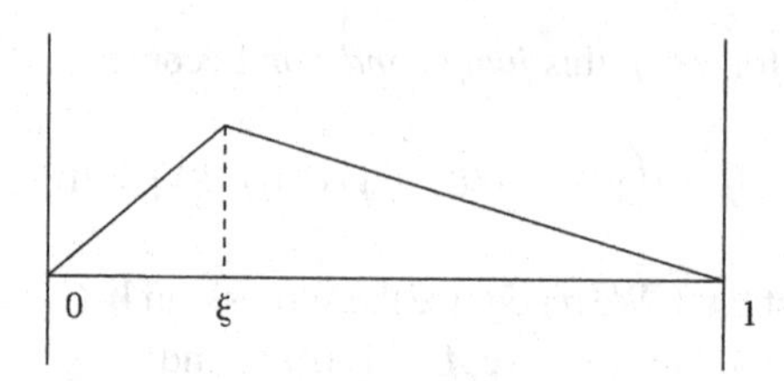

Figure 5.1 The function $\chi(x) = G(x, \xi)$.

Example: Solve

$$-\partial_x^2 y = f(x), \quad y(0) = y(1) = 0. \tag{5.20}$$

We have

$$\left.\begin{aligned} y_L &= x \\ y_R &= 1 - x \end{aligned}\right\} \quad \Rightarrow y'_L y_R - y_L y'_R \equiv 1. \tag{5.21}$$

We find that (Figure 5.1)

$$G(x, \xi) = \begin{cases} x(1-\xi), & x < \xi, \\ \xi(1-x), & x > \xi, \end{cases} \tag{5.22}$$

and

$$y(x) = (1-x)\int_0^x \xi f(\xi)\, d\xi + x\int_x^1 (1-\xi) f(\xi)\, d\xi. \tag{5.23}$$

5.2.2 Initial value problems

Initial value problems are those boundary value problems where all boundary conditions are imposed at one end of the interval, instead of some conditions at one end and some at the other. The same ingredients go into constructing the Green function, though.

Consider the problem

$$\frac{dy}{dt} - Q(t)y = F(t), \quad y(0) = 0. \tag{5.24}$$

We seek a Green function such that

$$L_t G(t, t') \equiv \left(\frac{d}{dt} - Q(t)\right) G(t, t') = \delta(t - t') \tag{5.25}$$

and $G(0, t') = 0$.

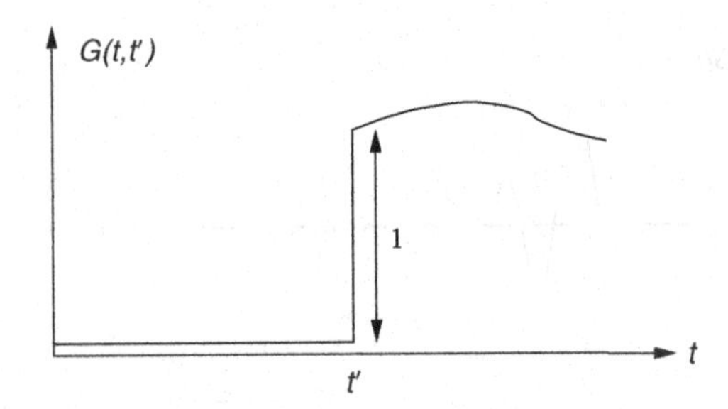

Figure 5.2 The Green function $G(t,t')$ for the first-order initial value problem.

We need $\chi(t) = G(t,t')$ to satisfy $L_t\chi = 0$, except at $t = t'$, and need $\chi(0) = 0$. The unique solution of $L_t\chi = 0$ with $\chi(0) = 0$ is $\chi(t) \equiv 0$. This means that $G(t,0) = 0$ for all $t < t'$. Near $t = t'$ we have the jump condition

$$G(t' + \varepsilon, t') - G(t' - \varepsilon, t') = 1. \tag{5.26}$$

The unique solution is (Figure 5.2)

$$G(t,t') = \theta(t - t') \exp\left\{\int_{t'}^{t} Q(s)ds\right\}, \tag{5.27}$$

where $\theta(t - t')$ is the Heaviside step distribution

$$\theta(t) = \begin{cases} 0, & t < 0, \\ 1, & t > 0. \end{cases} \tag{5.28}$$

Therefore

$$\begin{aligned} y(t) &= \int_0^\infty G(t,t')F(t')dt', \\ &= \int_0^t \exp\left\{\int_{t'}^t Q(s)\,ds\right\} F(t')\,dt' \\ &= \exp\left\{\int_0^t Q(s)\,ds\right\} \int_0^t \exp\left\{-\int_0^{t'} Q(s)\,ds\right\} F(t')\,dt'. \end{aligned} \tag{5.29}$$

We earlier obtained this solution *via* variation of parameters.

Example: *Forced, damped, harmonic oscillator*. An oscillator obeys the equation

$$\ddot{x} + 2\gamma\dot{x} + (\Omega^2 + \gamma^2)x = F(t). \tag{5.30}$$

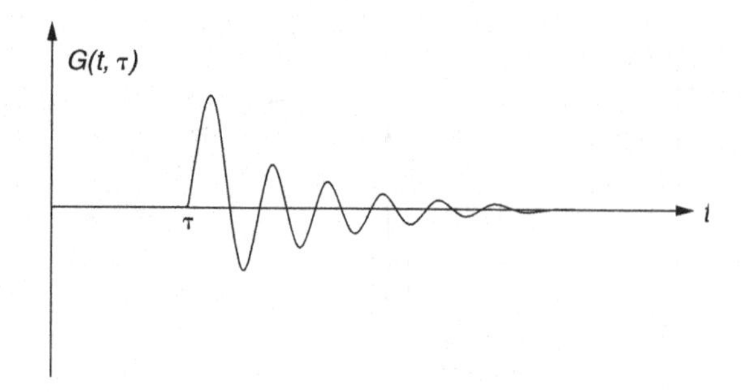

Figure 5.3 The Green function $G(t, \tau)$ for the damped oscillator problem.

Here $\gamma > 0$ is the friction coefficient. Assuming that the oscillator is at rest at the origin at $t = 0$, we will show that

$$x(t) = \left(\frac{1}{\Omega}\right) \int_0^t e^{-\gamma(t-\tau)} \sin \Omega(t-\tau) F(\tau) d\tau. \tag{5.31}$$

We seek a Green function $G(t, \tau)$ such that $\chi(t) = G(t, \tau)$ obeys $\chi(0) = \chi'(0) = 0$. Again, the unique solution of the differential equation with this initial data is $\chi(t) \equiv 0$. The Green function must be continuous at $t = \tau$, but its derivative must be discontinuous there, jumping from zero to unity to provide the delta function. Thereafter, it must satisfy the homogeneous equation. The unique function satisfying all these requirements is (see Figure 5.3)

$$G(t, \tau) = \theta(t-\tau) \frac{1}{\Omega} e^{-\gamma(t-\tau)} \sin \Omega(t-\tau). \tag{5.32}$$

Both these initial-value Green functions $G(t, t')$ are identically zero when $t < t'$. This is because the Green function is the response of the system to a kick at time $t = t'$, and in physical problems no effect comes before its cause. Such Green functions are said to be *causal*.

Physics application: Friction without friction – the Caldeira–Leggett model in real time

We now describe an application of the initial value problem Green function we found in the preceding example.

When studying the quantum mechanics of systems with friction, such as the viscously damped oscillator, we need a tractable model of the dissipative process. Such a model was introduced by Caldeira and Leggett.[1] They consider the Lagrangian

$$L = \frac{1}{2}\left(\dot{Q}^2 - (\Omega^2 - \Delta\Omega^2)Q^2\right) - Q\sum_i f_i q_i + \sum_i \frac{1}{2}\left(\dot{q}_i^2 - \omega_i^2 q_i^2\right), \tag{5.33}$$

[1] A. Caldeira, A. J. Leggett, *Phys. Rev. Lett.*, **46** (1981) 211.

which describes a macroscopic variable $Q(t)$, linearly coupled to an *oscillator bath* of very many simple systems q_i representing the environment. The quantity

$$\Delta\Omega^2 \stackrel{\text{def}}{=} -\sum_i \left(\frac{f_i^2}{\omega_i^2}\right) \tag{5.34}$$

is a counter-term that is inserted to cancel the frequency shift

$$\Omega^2 \to \Omega^2 - \sum_i \left(\frac{f_i^2}{\omega_i^2}\right), \tag{5.35}$$

caused by the coupling to the bath.[2]

The equations of motion are

$$\begin{aligned} \ddot{Q} + (\Omega^2 - \Delta\Omega^2)Q + \sum_i f_i q_i &= 0, \\ \ddot{q}_i + \omega_i^2 q_i + f_i Q &= 0. \end{aligned} \tag{5.36}$$

Using our initial value Green function, we solve for the q_i in terms of $Q(t)$:

$$f_i q_i = -\int_{-\infty}^{t} \left(\frac{f_i^2}{\omega_i}\right) \sin \omega_i (t-\tau) Q(\tau) d\tau. \tag{5.37}$$

The resulting motion of the q_i feeds back into the equation for Q to give

$$\ddot{Q} + (\Omega^2 - \Delta\Omega^2)Q + \int_{-\infty}^{t} F(t-\tau)Q(\tau)\, d\tau = 0, \tag{5.38}$$

where

$$F(t) \stackrel{\text{def}}{=} -\sum_i \left(\frac{f_i^2}{\omega_i}\right) \sin(\omega_i t) \tag{5.39}$$

is a *memory function*.

It is now convenient to introduce a *spectral function*

$$J(\omega) \stackrel{\text{def}}{=} \frac{\pi}{2} \sum_i \left(\frac{f_i^2}{\omega_i}\right) \delta(\omega - \omega_i), \tag{5.40}$$

[2] The shift arises because a static Q displaces the bath oscillators so that $f_i q_i = -(f_i^2/\omega_i^2)Q$. Substituting these values for the $f_i q_i$ into the potential terms shows that, in the absence of $\Delta\Omega^2 Q^2$, the effective potential seen by Q would be

$$\frac{1}{2}\Omega^2 Q^2 + Q\sum_i f_i q_i + \sum_i \frac{1}{2}\omega_i^2 q_i^2 = \frac{1}{2}\left(\Omega^2 - \sum_i \left(\frac{f_i^2}{\omega_i^2}\right)\right) Q^2.$$

which characterizes the spectrum of couplings and frequencies associated with the oscillator bath. In terms of $J(\omega)$ we can write

$$F(t) = -\frac{2}{\pi}\int_0^\infty J(\omega)\sin(\omega t)\,d\omega. \tag{5.41}$$

Although $J(\omega)$ is defined as a sum of delta function "spikes", the oscillator bath contains a very large number of systems and this makes $J(\omega)$ effectively a smooth function. This is just as the density of a gas (a sum of delta functions at the location of the atoms) is macroscopically smooth. By taking different forms for $J(\omega)$ we can represent a wide range of environments. Caldeira and Leggett show that to obtain a friction force proportional to $\dot{Q}$ we should make $J(\omega)$ proportional to the frequency ω. To see how this works, consider the choice

$$J(\omega) = \eta\omega\left[\frac{\Lambda^2}{\Lambda^2+\omega^2}\right], \tag{5.42}$$

which is equal to $\eta\omega$ for small ω, but tends to zero when $\omega \gg \Lambda$. The high-frequency cutoff Λ is introduced to make the integrals over ω converge. With this cutoff

$$\frac{2}{\pi}\int_0^\infty J(\omega)\sin(\omega t)\,d\omega = \frac{2}{2\pi i}\int_{-\infty}^\infty \frac{\eta\,\omega\Lambda^2 e^{i\omega t}}{\Lambda^2+\omega^2}\,d\omega = \mathrm{sgn}\,(t)\eta\,\Lambda^2 e^{-\Lambda|t|}. \tag{5.43}$$

Therefore,

$$\begin{aligned}\int_{-\infty}^t F(t-\tau)Q(\tau)\,d\tau &= -\int_{-\infty}^t \eta\Lambda^2 e^{-\Lambda|t-\tau|}Q(\tau)\,d\tau \\ &= -\eta\Lambda Q(t) + \eta\dot{Q}(t) - \frac{\eta}{2\Lambda}\ddot{Q}(t) + \cdots,\end{aligned} \tag{5.44}$$

where the second line results from expanding $Q(\tau)$ as a Taylor series

$$Q(\tau) = Q(t) + (\tau - t)\dot{Q}(t) + \cdots, \tag{5.45}$$

and integrating term-by-term. Now,

$$-\Delta\Omega^2 \equiv \sum_i\left(\frac{f_i^2}{\omega_i^2}\right) = \frac{2}{\pi}\int_0^\infty \frac{J(\omega)}{\omega}d\omega = \frac{2}{\pi}\int_0^\infty \frac{\eta\Lambda^2}{\Lambda^2+\omega^2}d\omega = \eta\Lambda. \tag{5.46}$$

The $-\Delta\Omega^2 Q$ counter-term thus cancels the leading term $-\eta\Lambda Q(t)$ in (5.44), which would otherwise represent a Λ-dependent frequency shift. After this cancellation we can safely let $\Lambda \to \infty$, and so ignore terms with negative powers of the cutoff. The only surviving term in (5.44) is then $\eta\dot{Q}$. This we substitute into (5.38), which becomes the equation for viscously damped motion:

$$\ddot{Q} + \eta\dot{Q} + \Omega^2 Q = 0. \tag{5.47}$$

The oscillators in the bath absorb energy but, unlike a pair of coupled oscillators which trade energy rhythmically back-and-forth, the incommensurate motion of the many q_i prevents them from cooperating for long enough to return any energy to $Q(t)$.

5.2.3 Modified Green function

When the equation $Ly = 0$ has a non-trivial solution, there can be no unique solution to $Ly = f$, but there will still be solutions provided f is orthogonal to all solutions of $L^\dagger y = 0$.

Example: Consider

$$Ly \equiv -\partial_x^2 y = f(x), \quad y'(0) = y'(1) = 0. \tag{5.48}$$

The equation $Ly = 0$ has one non-trivial solution, $y(x) = 1$. The operator L is self-adjoint, $L^\dagger = L$, and so there will be solutions to $Ly = f$ provided $\langle 1, f\rangle = \int_0^1 f\, dx = 0$.

We cannot define the Green function as a solution to

$$-\partial_x^2 G(x, \xi) = \delta(x - \xi), \tag{5.49}$$

because $\int_0^1 \delta(x - \xi)\, dx = 1 \neq 0$, but we can seek a solution to

$$-\partial_x^2 G(x, \xi) = \delta(x - \xi) - 1 \tag{5.50}$$

as the right-hand side integrates to zero.

A general solution to $-\partial_x^2 y = -1$ is

$$y = A + Bx + \frac{1}{2}x^2, \tag{5.51}$$

and the functions

$$\begin{aligned} y_L &= A + \frac{1}{2}x^2, \\ y_R &= C - x + \frac{1}{2}x^2, \end{aligned} \tag{5.52}$$

obey the boundary conditions at the left and right ends of the interval, respectively. Continuity at $x = \xi$ demands that $A = C - \xi$, and we are left with

$$G(x, \xi) = \begin{cases} C - \xi + \frac{1}{2}x^2, & 0 < x < \xi \\ C - x + \frac{1}{2}x^2, & \xi < x < 1. \end{cases} \tag{5.53}$$

There is no freedom left to impose the condition

$$G'(\xi - \varepsilon, \xi) - G'(\xi + \varepsilon, \xi) = 1, \tag{5.54}$$

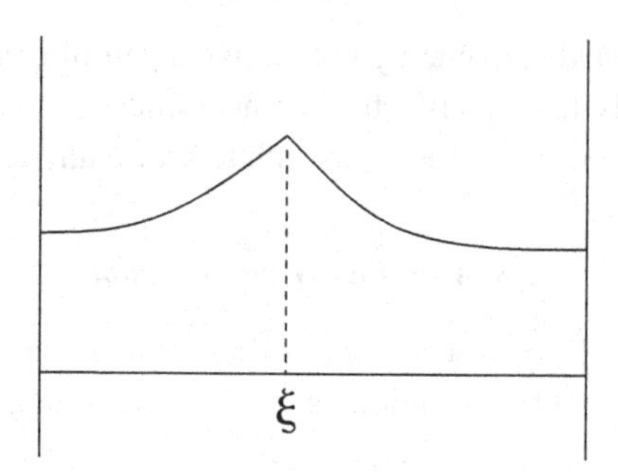

Figure 5.4 The modified Green function.

but it is *automatically satisfied*! Indeed,

$$G'(\xi - \varepsilon, \xi) = \xi$$
$$G'(\xi + \varepsilon, \xi) = -1 + \xi. \tag{5.55}$$

We may select a different value of C for each ξ, and a convenient choice is

$$C = \frac{1}{2}\xi^2 + \frac{1}{3} \tag{5.56}$$

which makes G symmetric (Figure 5.4):

$$G(x, \xi) = \begin{cases} \frac{1}{3} - \xi + \frac{x^2+\xi^2}{2}, & 0 < x < \xi, \\ \frac{1}{3} - x + \frac{x^2+\xi^2}{2}, & \xi < x < 1. \end{cases} \tag{5.57}$$

It also makes $\int_0^1 G(x, \xi)\, dx = 0$.

The solution to $Ly = f$ is

$$y(x) = \int_0^1 G(x, \xi) f(\xi)\, d\xi + A, \tag{5.58}$$

where A is arbitrary.

5.3 Applications of Lagrange's identity

5.3.1 Hermiticity of Green functions

Earlier we noted the symmetry of the Green function for the Sturm–Liouville equation. We will now establish the corresponding result for general differential operators.

Let $G(x, \xi)$ obey $L_x G(x, \xi) = \delta(x - \xi)$ with homogeneous boundary conditions B, and let $G^\dagger(x, \xi)$ obey $L_x^\dagger G^\dagger(x, \xi) = \delta(x - \xi)$ with adjoint boundary conditions $B^\dagger$. Then,

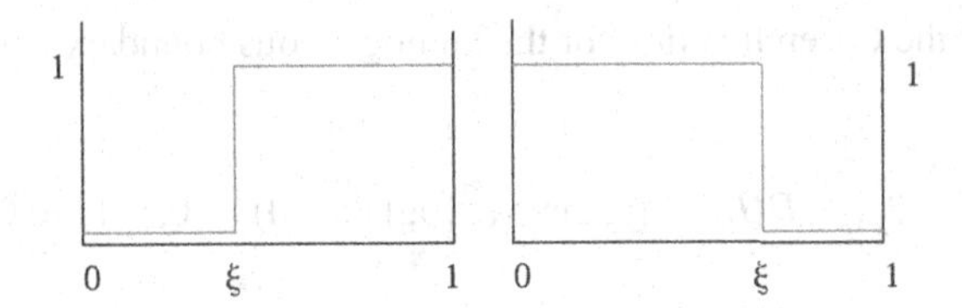

Figure 5.5 $G(x,\xi) = \theta(x-\xi)$, and $G^\dagger(x,\xi) = \theta(\xi - x)$.

from Lagrange's identity, we have

$$
\begin{aligned}
[Q(G,G^\dagger)]_a^b &= \int_a^b dx \left\{ \left(L_x^\dagger G^\dagger(x,\xi) \right)^* G(x,\xi') - (G^\dagger(x,\xi))^* L G(x,\xi') \right\} \\
&= \int_a^b dx \left\{ \delta(x-\xi) G(x,\xi') - \left(G^\dagger(x,\xi) \right)^* \delta(x-\xi') \right\} \\
&= G(\xi,\xi') - \left(G^\dagger(\xi',\xi) \right)^*.
\end{aligned} \tag{5.59}
$$

Thus, provided $[Q(G,G^\dagger)]_a^b = 0$, which is indeed the case because the boundary conditions for L, $L^\dagger$ are mutually adjoint, we have

$$
G^\dagger(\xi, x) = \left(G(x,\xi) \right)^*, \tag{5.60}
$$

and the Green functions, regarded as matrices with continuous rows and columns, are hermitian conjugates of one another.

Example: Let

$$
L = \frac{d}{dx}, \qquad \mathcal{D}(L) = \{y, Ly \in L^2[0,1] : y(0) = 0\}. \tag{5.61}
$$

In this case $G(x,\xi) = \theta(x-\xi)$; see Figure 5.5.

Now, we have

$$
L^\dagger = -\frac{d}{dx}, \qquad \mathcal{D}(L) = \{y, Ly \in L^2[0,1] : y(1) = 0\} \tag{5.62}
$$

and $G^\dagger(x,\xi) = \theta(\xi - x)$; see Figure 5.5.

5.3.2 Inhomogeneous boundary conditions

Our differential operators have been defined with linear *homogeneous* boundary conditions. We can, however, use them, and their Green-function inverses, to solve differential equations with *inhomogeneous* boundary conditions.

Suppose, for example, we wish to solve

$$
-\partial_x^2 y = f(x), \qquad y(0) = a, \quad y(1) = b. \tag{5.63}
$$

We already know the Green function for the homogeneous boundary condition problem with operator

$$L = -\partial_x^2, \quad \mathcal{D}(L) = \{y, Ly \in L^2[0,1] : y(0) = 0, y(1) = 0\}. \tag{5.64}$$

It is

$$G(x,\xi) = \begin{cases} x(1-\xi), & x < \xi, \\ \xi(1-x), & x > \xi. \end{cases} \tag{5.65}$$

Now we apply Lagrange's identity to $\chi(x) = G(x,\xi)$ and $y(x)$ to get

$$\begin{aligned} &\int_0^1 dx \left\{ G(x,\xi)\left(-\partial_x^2 y(x)\right) - y(x)\left(-\partial_x^2 G(x,\xi)\right) \right\} \\ &\quad = [G'(x,\xi)y(x) - G(x,\xi)y'(x)]_0^1. \end{aligned} \tag{5.66}$$

Here, as usual, $G'(x,\xi) = \partial_x G(x,\xi)$. The integral is equal to

$$\int dx\, \{G(x,\xi)f(x) - y(x)\delta(x-\xi)\} = \int G(x,\xi)f(x)\, dx - y(\xi), \tag{5.67}$$

whilst the integrated-out bit is

$$-(1-\xi)y(0) - 0\,y'(0) - \xi y(1) + 0\,y'(1). \tag{5.68}$$

Therefore, we have

$$y(\xi) = \int G(x,\xi)f(x)\, dx + (1-\xi)y(0) + \xi y(1). \tag{5.69}$$

Here the term with $f(x)$ is the particular integral, whilst the remaining terms constitute the complementary function (obeying the differential equation without the source term) which serves to satisfy the boundary conditions. Observe that the arguments in $G(x,\xi)$ are not in the usual order, but, in the present example, this does not matter because G is symmetric.

When the operator L is not self-adjoint, we need to distinguish between L and $L^\dagger$, and G and $G^\dagger$. We then apply Lagrange's identity to the unknown function $u(x)$ and $\chi(x) = G^\dagger(x,\xi)$.

Example: We will use the Green-function method to solve the differential equation

$$\frac{du}{dx} = f(x), \quad x \in [0,1], \qquad u(0) = a. \tag{5.70}$$

We can, of course, write down the answer to this problem directly, but it is interesting to see how the general strategy produces the solution. We first find the Green function

$G(x, \xi)$ for the operator with the corresponding homogeneous boundary conditions. In the present case, this operator is

$$L = \partial_x, \quad \mathcal{D}(L) = \{u, Lu \in L^2[0,1] : u(0) = 0\}, \tag{5.71}$$

and the appropriate Green function is $G(x, \xi) = \theta(x - \xi)$. From G we then read off the adjoint Green function as $G^\dagger(x, \xi) = \left(G(\xi, x)\right)^*$. In the present example, we have $G^\dagger(x,' x) = \theta(\xi - x)$. We now use Lagrange's identity in the form

$$\int_0^1 dx \left\{ \left(L_x^\dagger G^\dagger(x, \xi)\right)^* u(x) - \left(G^\dagger(x, \xi)\right)^* L_x u(x) \right\} = \left[Q\left(G^\dagger, u\right) \right]_0^1. \tag{5.72}$$

In all cases, the left-hand side is equal to

$$\int_0^1 dx \left\{ \delta(x - \xi) u(x) - G^T(x, \xi) f(x) \right\}, \tag{5.73}$$

where T denotes transpose, $G^T(x, \xi) = G(\xi, x)$. The left-hand side is therefore equal to

$$u(\xi) - \int_0^1 dx\, G(\xi, x) f(x). \tag{5.74}$$

The right-hand side depends on the details of the problem. In the present case, the integrated out part is

$$\left[Q(G^\dagger, u) \right]_0^1 = -\left[G^T(x, \xi) u(x) \right]_0^1 = u(0). \tag{5.75}$$

At the last step we have used the specific form $G^T(x, \xi) = \theta(\xi - x)$ to find that only the lower limit contributes. The end result is therefore the expected one:

$$u(y) = u(0) + \int_0^y f(x)\, dx. \tag{5.76}$$

Variations of this strategy enable us to solve any inhomogeneous boundary value problem in terms of the Green function for the corresponding homogeneous boundary value problem.

5.4 Eigenfunction expansions

Self-adjoint operators possess a complete set of eigenfunctions, and we can expand the Green function in terms of these. Let

$$L\varphi_n = \lambda_n \varphi_n. \tag{5.77}$$

Let us further suppose that none of the λ_n are zero. Then the Green function has the eigenfunction expansion

$$G(x,\xi) = \sum_n \frac{\varphi_n(x)\varphi_n^*(\xi)}{\lambda_n}. \tag{5.78}$$

That this is so follows from

$$\begin{aligned} L_x\left(\sum_n \frac{\varphi_n(x)\varphi_n^*(\xi)}{\lambda_n}\right) &= \sum_n \frac{\Big(L_x\varphi_n(x)\Big)\varphi_n^*(\xi)}{\lambda_n} \\ &= \sum_n \frac{\lambda_n\varphi_n(x)\varphi_n^*(\xi)}{\lambda_n} \\ &= \sum_n \varphi_n(x)\varphi_n^*(\xi) \\ &= \delta(x-\xi). \end{aligned} \tag{5.79}$$

Example: Consider our familiar exemplar

$$L = -\partial_x^2, \quad \mathcal{D}(L) = \{y, Ly \in L^2[0,1] : y(0) = y(1) = 0\}, \tag{5.80}$$

for which

$$G(x,\xi) = \begin{cases} x(1-\xi), & x < \xi, \\ \xi(1-x), & x > \xi. \end{cases} \tag{5.81}$$

Computing the Fourier series shows that

$$G(x,\xi) = \sum_{n=1}^{\infty}\left(\frac{2}{n^2\pi^2}\right)\sin(n\pi x)\sin(n\pi\xi). \tag{5.82}$$

Modified Green function

When one or more of the eigenvalues is zero, a modified Green function is obtained by simply omitting the corresponding terms from the series.

$$G_{\text{mod}}(x,\xi) = \sum_{\lambda_n \neq 0} \frac{\varphi_n(x)\varphi_n^*(\xi)}{\lambda_n}. \tag{5.83}$$

Then

$$L_x G_{\text{mod}}(x,\xi) = \delta(x-\xi) - \sum_{\lambda_n = 0} \varphi_n(x)\varphi_n^*(\xi). \tag{5.84}$$

We see that this G_{mod} is still hermitian, and, as a function of x, is orthogonal to the zero modes. These are the properties we elected when constructing the modified Green function in Equation (5.57).

5.5 Analytic properties of Green functions

In this section we study the properties of Green functions considered as functions of a complex variable. Some of the formulæ are slightly easier to derive using contour integral methods, but these are not necessary and we will not use them here. The only complex-variable prerequisite is a familiarity with complex arithmetic and, in particular, knowledge of how to take the logarithm and the square root of a complex number.

5.5.1 Causality implies analyticity

Consider a Green function of the form $G(t-\tau)$ and possessing the causal property that $G(t-\tau)=0$, for $t<\tau$. If the improper integral defining its Fourier transform,

$$\widetilde{G}(\omega)=\int_0^\infty e^{i\omega t}G(t)\,dt \stackrel{\text{def}}{=} \lim_{T\to\infty}\left\{\int_0^T e^{i\omega t}G(t)\,dt\right\}, \tag{5.85}$$

converges for real ω, it will converge even better when ω has a positive imaginary part. Consequently $\widetilde{G}(\omega)$ will be a well-behaved function of the complex variable ω everywhere in the upper half of the complex ω plane. Indeed, it will be *analytic* there, meaning that its Taylor series expansion about any point actually converges to the function. For example, the Green function for the damped harmonic oscillator

$$G(t)=\begin{cases}\frac{1}{\Omega}e^{-\gamma t}\sin(\Omega t), & t>0,\\ 0, & t<0,\end{cases} \tag{5.86}$$

has Fourier transform

$$\widetilde{G}(\omega)=\frac{1}{\Omega^2-(\omega+i\gamma)^2}, \tag{5.87}$$

which is always finite in the upper half-plane, although it has *pole* singularities at $\omega=-i\gamma\pm\Omega$ in the lower half-plane.

The only way that the Fourier transform $\widetilde{G}$ of a causal Green function can have a pole singularity in the upper half-plane is if G contains an exponential factor growing in time, in which case the system is *unstable* to perturbations (and the real-frequency Fourier transform does not exist). This observation is at the heart of the *Nyquist criterion* for the stability of linear electronic devices.

Inverting the Fourier transform, we have

$$G(t)=\int_{-\infty}^{\infty}\frac{1}{\Omega^2-(\omega+i\gamma)^2}e^{-i\omega t}\frac{d\omega}{2\pi}=\theta(t)\frac{1}{\Omega}e^{-\gamma t}\sin(\Omega t). \tag{5.88}$$

It is perhaps surprising that this integral is identically zero if $t<0$, and non-zero if $t>0$. This is one of the places where contour integral methods might cast some light, but because we have confidence in the Fourier inversion formula, we know that it must be correct.

Remember that in deriving (5.88) we have explicitly assumed that the damping coefficient γ is positive. It is important to realize that reversing the sign of γ on the left-hand side of (5.88) does more than just change $e^{-\gamma t} \to e^{\gamma t}$ on the right-hand side. Naïvely setting $\gamma \to -\gamma$ on both sides of (5.88) gives an equation that cannot possibly be true. The left-hand side would be the Fourier transform of a smooth function, and the Riemann–Lebesgue lemma tells us that such a Fourier transform must become zero when $|t| \to \infty$. The right-hand side, on the contrary, would be a function whose oscillations grow without bound as t becomes large and positive.

To find the correct equation, observe that we can legitimately effect the sign-change $\gamma \to -\gamma$ by first complex-conjugating the integral and then changing t to $-t$. Performing these two operations on both sides of (5.88) leads to

$$\int_{-\infty}^{\infty} \frac{1}{\Omega^2 - (\omega - i\gamma)^2} e^{-i\omega t} \frac{d\omega}{2\pi} = -\theta(-t) \frac{1}{\Omega} e^{\gamma t} \sin(\Omega t). \tag{5.89}$$

The new right-hand side represents an exponentially growing oscillation that is suddenly silenced by the kick at $t = 0$.

The effect of taking the damping parameter γ from an infinitesimally small positive value ε to an infinitesimally small negative value $-\varepsilon$ is therefore to turn the *causal* Green function (no motion before it is started by the delta-function kick) of the *undamped* oscillator into an *anti-causal* Green function (no motion after it is stopped by the kick); see Figure 5.6. Ultimately, this is because the differential operator corresponding to a harmonic oscillator with *initial* value data is not self-adjoint, and its adjoint operator corresponds to a harmonic oscillator with *final* value data.

This discontinuous dependence on an infinitesimal damping parameter is the subject of the next few sections.

Physics application: Caldeira–Leggett in frequency space

If we write the Caldeira–Leggett equations of motion (5.36) in Fourier frequency space by setting

$$Q(t) = \int_{-\infty}^{\infty} \frac{d\omega}{2\pi} Q(\omega) e^{-i\omega t}, \tag{5.90}$$

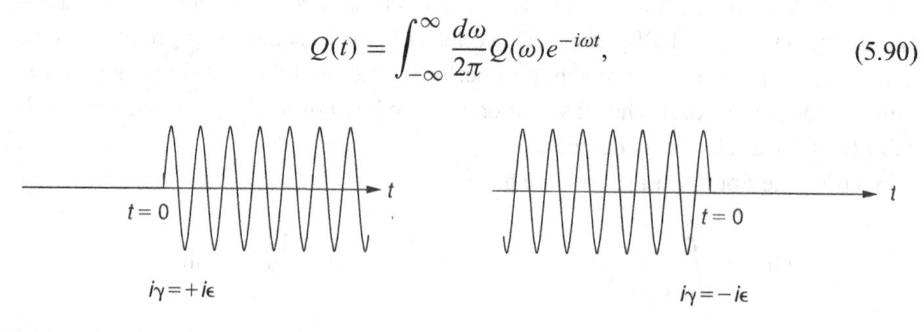

Figure 5.6 The effect on $G(t)$, the Green function of an undamped oscillator, of changing $i\gamma$ from $+i\varepsilon$ to $-i\varepsilon$.

and

$$q_i(t) = \int_{-\infty}^{\infty} \frac{d\omega}{2\pi} q_i(\omega) e^{-i\omega t}, \tag{5.91}$$

we have (after including an external force F_{ext} to drive the system)

$$\begin{gathered}\left(-\omega^2 + (\Omega^2 - \Delta\Omega^2)\right) Q(\omega) - \sum_i f_i q_i(\omega) = F_{\text{ext}}(\omega), \\ (-\omega^2 + \omega_i^2) q_i(\omega) + f_i Q(\omega) = 0.\end{gathered} \tag{5.92}$$

Eliminating the q_i, we obtain

$$\left(-\omega^2 + (\Omega^2 - \Delta\Omega^2)\right) Q(\omega) - \sum_i \frac{f_i^2}{\omega_i^2 - \omega^2} Q(\omega) = F_{\text{ext}}(\omega). \tag{5.93}$$

As before, sums over the index i are replaced by integrals over the spectral function

$$\sum_i \frac{f_i^2}{\omega_i^2 - \omega^2} \to \frac{2}{\pi} \int_0^{\infty} \frac{\omega' J(\omega')}{\omega'^2 - \omega^2} d\omega', \tag{5.94}$$

and

$$-\Delta\Omega^2 \equiv \sum_i \left(\frac{f_i^2}{\omega_i^2}\right) \to \frac{2}{\pi} \int_0^{\infty} \frac{J(\omega')}{\omega'} d\omega'. \tag{5.95}$$

Then

$$Q(\omega) = \left(\frac{1}{\Omega^2 - \omega^2 + \Pi(\omega)}\right) F_{\text{ext}}(\omega), \tag{5.96}$$

where the *self-energy* $\Pi(\omega)$ is given by

$$\Pi(\omega) = \frac{2}{\pi} \int_0^{\infty} \left\{ \frac{J(\omega')}{\omega'} - \frac{\omega' J(\omega')}{\omega'^2 - \omega^2} \right\} d\omega' = -\omega^2 \frac{2}{\pi} \int_0^{\infty} \frac{J(\omega')}{\omega'(\omega'^2 - \omega^2)} d\omega'. \tag{5.97}$$

The expression

$$\mathcal{G}(\omega) \equiv \frac{1}{\Omega^2 - \omega^2 + \Pi(\omega)} \tag{5.98}$$

is a typical *response function*. Analogous objects occur in all branches of physics.

For viscous damping we know that $J(\omega) = \eta\omega$. Let us evaluate the integral occurring in $\Pi(\omega)$ for this case:

$$I(\omega) = \int_0^{\infty} \frac{d\omega'}{\omega'^2 - \omega^2}. \tag{5.99}$$

We will initially assume that ω is positive. Now,

$$\frac{1}{\omega'^2 - \omega^2} = \frac{1}{2\omega}\left(\frac{1}{\omega' - \omega} - \frac{1}{\omega' + \omega}\right), \tag{5.100}$$

so

$$I(\omega) = \left[\frac{1}{2\omega}\Big(\ln(\omega' - \omega) - \ln(\omega' + \omega)\Big)\right]_{\omega'=0}^{\infty}. \tag{5.101}$$

At the upper limit we have $\ln\Big((\infty - \omega)/(\infty + \omega)\Big) = \ln 1 = 0$. The lower limit contributes

$$-\frac{1}{2\omega}\Big(\ln(-\omega) - \ln(\omega)\Big). \tag{5.102}$$

To evaluate the logarithm of a negative quantity we must use

$$\ln \omega = \ln|\omega| + i \arg \omega, \tag{5.103}$$

where we will take $\arg \omega$ to lie in the range $-\pi < \arg \omega < \pi$.

To get an unambiguous answer, we need to give ω an infinitesimal imaginary part $\pm i\varepsilon$ (Figure 5.7). Depending on the sign of this imaginary part, we find that

$$I(\omega \pm i\varepsilon) = \pm\frac{i\pi}{2\omega}. \tag{5.104}$$

This formula remains true when the real part of ω is negative, and so

$$\Pi(\omega \pm i\varepsilon) = \mp i\eta\omega. \tag{5.105}$$

Now the frequency-space version of

$$\ddot{Q}(t) + \eta\dot{Q} + \Omega^2 Q = F_{\text{ext}}(t) \tag{5.106}$$

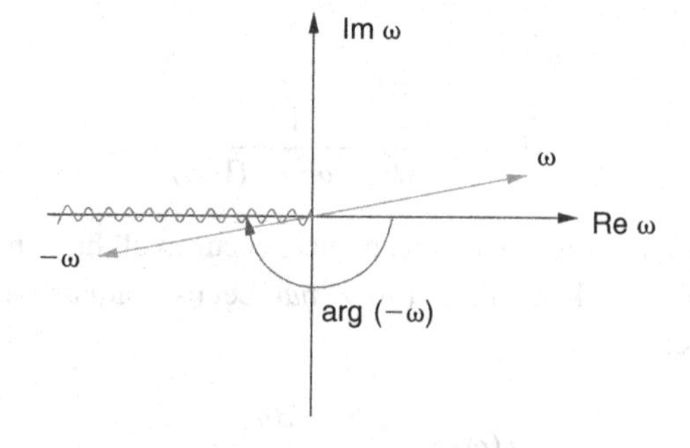

Figure 5.7 When ω has a small positive imaginary part, $\arg(-\omega) \approx -\pi$.

is

$$(-\omega^2 - i\eta\omega + \Omega^2)Q(\omega) = F_{\text{ext}}(\omega), \tag{5.107}$$

so we must opt for the small shift in ω that leads to $\Pi(\omega) = -i\eta\omega$. This means that we must regard ω as having a *positive* infinitesimal imaginary part, $\omega \to \omega + i\varepsilon$. This imaginary part is a good and needful thing: it effects the replacement of the ill-defined singular integrals

$$G(t) \stackrel{?}{=} \int_0^\infty \frac{1}{\omega_i^2 - \omega^2} e^{-i\omega t}\, d\omega, \tag{5.108}$$

which arise as we transform back to real time, with the unambiguous expressions

$$G_\varepsilon(t) = \int_0^\infty \frac{1}{\omega_i^2 - (\omega + i\varepsilon)^2} e^{-i\omega t}\, d\omega. \tag{5.109}$$

The latter, we know, give rise to properly causal real-time Green functions.

5.5.2 Plemelj formulæ

The functions we are meeting can all be cast in the form

$$f(\omega) = \frac{1}{\pi} \int_a^b \frac{\rho(\omega')}{\omega' - \omega}\, d\omega'. \tag{5.110}$$

If ω lies in the integration range $[a, b]$, then we divide by zero as we integrate over $\omega' = \omega$. We ought to avoid doing this, but this interval is often exactly where we desire to evaluate f. As before, we evade the division by zero by giving ω an infintesimally small imaginary part: $\omega \to \omega \pm i\varepsilon$. We can then apply the *Plemelj formulæ*, named for the Slovenian mathematician Josip Plemelj, which say that

$$\begin{aligned} \frac{1}{2}\Big(f(\omega + i\varepsilon) - f(\omega - i\varepsilon)\Big) &= i\rho(\omega), \quad \omega \in [a, b] \\ \frac{1}{2}\Big(f(\omega + i\varepsilon) + f(\omega - i\varepsilon)\Big) &= \frac{1}{\pi} P \int_a^b \frac{\rho(\omega')}{\omega' - \omega}\, d\omega'. \end{aligned} \tag{5.111}$$

As explained in Section 2.3.2, the "P" in front of the integral stands for *principal part*. Recall that it means that we are to delete an infinitesimal segment of the ω' integral lying symmetrically about the singular point $\omega' = \omega$.

The Plemelj formulæ mean that the otherwise smooth and analytic function $f(\omega)$ is discontinuous across the real axis between a and b (see Figure 5.8). If the discontinuity $\rho(\omega)$ is itself an analytic function then the line joining the points a and b is a *branch cut*, and the endpoints of the integral are *branch-point* singularities of $f(\omega)$.

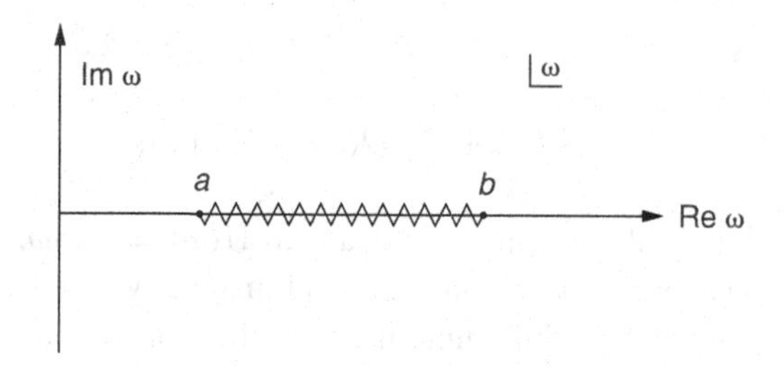

Figure 5.8 The analytic function $f(\omega)$ is discontinuous across the real axis between a and b.

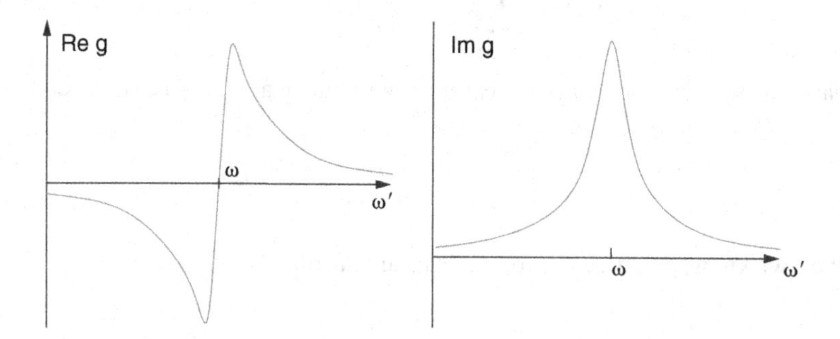

Figure 5.9 Sketch of the real and imaginary parts of $g(\omega') = 1/(\omega' - (\omega + i\varepsilon))$.

The reason for the discontinuity may be understood by considering Figure 5.9. The singular integrand is a product of $\rho(\omega')$ with

$$\frac{1}{\omega' - (\omega \pm i\varepsilon)} = \frac{\omega' - \omega}{(\omega' - \omega)^2 + \varepsilon^2} \pm \frac{i\varepsilon}{(\omega' - \omega)^2 + \varepsilon^2}. \tag{5.112}$$

The first term on the right is a symmetrically cut-off version $1/(\omega' - \omega)$ and provides the principal-part integral. The second term sharpens and tends to the delta function $\pm i\pi\delta(\omega' - \omega)$ as $\varepsilon \to 0$, and so gives $\pm i\pi\rho(\omega)$. Because of this explanation, the Plemelj equations are commonly encoded in physics papers via the "$i\varepsilon$" cabbala

$$\frac{1}{\omega' - (\omega \pm i\varepsilon)} = P\left(\frac{1}{\omega' - \omega}\right) \pm i\pi\delta(\omega' - \omega). \tag{5.113}$$

If ρ is real, as it often is, then $f(\omega + i\eta) = \Big(f(\omega - i\eta)\Big)^*$. The discontinuity across the real axis is then purely imaginary, and

$$\frac{1}{2}\Big(f(\omega + i\varepsilon) + f(\omega - i\varepsilon)\Big) \tag{5.114}$$

is the real part of f. In this case we can write (5.110) as

$$\mathrm{Re}f(\omega) = \frac{1}{\pi} P \int_a^b \frac{\mathrm{Im} f(\omega')}{\omega' - \omega} d\omega'. \tag{5.115}$$

This formula is typical of the relations linking the real and imaginary parts of causal response functions.

A practical example of such a relation is provided by the complex, frequency-dependent, *refractive index*, $n(\omega)$, of a medium. This is defined so that a travelling electromagnetic wave takes the form

$$\mathbf{E}(x, t) = \mathbf{E}_0 \, e^{in(\omega)kx - i\omega t}. \tag{5.116}$$

Here, $k = \omega/c$ is the *in vacuo* wavenumber. We can decompose n into its real and imaginary parts:

$$\begin{aligned} n(\omega) &= n_R + i n_I \\ &= n_R(\omega) + \frac{i}{2|k|}\gamma(\omega), \end{aligned} \tag{5.117}$$

where γ is the extinction coefficient, defined so that the intensity falls off as $I = I_0 \exp(-\gamma x)$. A non-zero γ can arise from either energy absorbtion or scattering out of the forward direction. For the refractive index, the function $f(\omega) = n(\omega) - 1$ can be written in the form of (5.110), and, using $n(-\omega) = n^*(\omega)$, this leads to the *Kramers–Kronig relation*

$$n_R(\omega) = 1 + \frac{c}{\pi} P \int_0^\infty \frac{\gamma(\omega')}{\omega'^2 - \omega^2} d\omega'. \tag{5.118}$$

Formulæ like this will be rigorously derived in Chapter 18 by the use of contour-integral methods.

5.5.3 Resolvent operator

Given a differential operator L, we define the *resolvent operator* to be $R_\lambda \equiv (L - \lambda I)^{-1}$. The resolvent is an analytic function of λ, except when λ lies in the spectrum of L.

We expand R_λ in terms of the eigenfunctions as

$$R_\lambda(x, \xi) = \sum_n \frac{\varphi_n(x)\varphi_n^*(\xi)}{\lambda_n - \lambda}. \tag{5.119}$$

When the spectrum is discrete, the resolvent has *poles* at the eigenvalues L. When the operator L has a continuous spectrum, the sum becomes an integral:

$$R_\lambda(x, \xi) = \int_{\mu \in \sigma(L)} \rho(\mu) \frac{\varphi_\mu(x)\varphi_\mu^*(\xi)}{\mu - \lambda} d\mu, \tag{5.120}$$

where $\rho(\mu)$ is the eigenvalue density of states. This is of the form that we saw in connection with the Plemelj formulæ. Consequently, when the spectrum comprises segments of the real axis, the resulting analytic function R_λ will be discontinuous across the real axis within them. The endpoints of the segments will be branch point singularities of R_λ, and the segments themselves, considered as subsets of the complex plane, are the branch cuts.

The trace of the resolvent $\mathrm{Tr}\, R_\lambda$ is defined by

$$\begin{aligned}\mathrm{Tr}\, R_\lambda &= \int dx\,\{R_\lambda(x,x)\}\\ &= \int dx \left\{\sum_n \frac{\varphi_n(x)\varphi_n^*(x)}{\lambda_n-\lambda}\right\}\\ &= \sum_n \frac{1}{\lambda_n-\lambda}\\ &\to \int \frac{\rho(\mu)}{\mu-\lambda}\,d\mu. \end{aligned} \tag{5.121}$$

Applying Plemelj to R_λ, we have

$$\mathrm{Im}\left[\lim_{\varepsilon\to 0}\left\{\mathrm{Tr}\, R_{\lambda+i\varepsilon}\right\}\right] = \pi\rho(\lambda). \tag{5.122}$$

Here, we have used the fact that ρ is real, so

$$\mathrm{Tr}\, R_{\lambda-i\varepsilon} = \left(\mathrm{Tr}\, R_{\lambda+i\varepsilon}\right)^*. \tag{5.123}$$

The non-zero imaginary part therefore shows that R_λ is discontinuous across the real axis at points lying in the continuous spectrum.

Example: Consider

$$L = -\partial_x^2 + m^2, \quad \mathcal{D}(L) = \{y, Ly \in L^2[-\infty,\infty]\}. \tag{5.124}$$

As we know, this operator has a continuous spectrum, with eigenfunctions

$$\varphi_k = \frac{1}{\sqrt{L}} e^{ikx}. \tag{5.125}$$

Here, L is the (very large) length of the interval. The eigenvalues are $E = k^2 + m^2$, so the spectrum is all positive numbers greater than m^2. The momentum density of states is

$$\rho(k) = \frac{L}{2\pi}. \tag{5.126}$$

The completeness relation is

$$\int_{-\infty}^{\infty} \frac{dk}{2\pi} e^{ik(x-\xi)} = \delta(x-\xi), \tag{5.127}$$

which is just the Fourier integral formula for the delta function.

The Green function for L is

$$G(x-y) = \int_{-\infty}^{\infty} dk \left(\frac{dn}{dk}\right) \frac{\varphi_k(x)\varphi_k^*(y)}{k^2+m^2} = \int_{-\infty}^{\infty} \frac{dk}{2\pi} \frac{e^{ik(x-y)}}{k^2+m^2} = \frac{1}{2m} e^{-m|x-y|}. \tag{5.128}$$

We can use the same calculation to look at the resolvent $R_\lambda = (-\partial_x^2 - \lambda)^{-1}$. Replacing m^2 by $-\lambda$, we have

$$R_\lambda(x,y) = \frac{1}{2\sqrt{-\lambda}} e^{-\sqrt{-\lambda}|x-y|}. \tag{5.129}$$

To appreciate this expression, we need to know how to evaluate $\sqrt{z}$ where z is complex. We write $z = |z|e^{i\phi}$ where we require $-\pi < \phi < \pi$. We now define

$$\sqrt{z} = \sqrt{|z|} e^{i\phi/2}. \tag{5.130}$$

When we evaluate $\sqrt{z}$ for z just below the negative real axis then this definition gives $-i\sqrt{|z|}$ (see Figure 5.10), and just above the axis we find $+i\sqrt{|z|}$. The discontinuity means that the negative real axis is a branch cut for the square-root function. The $\sqrt{-\lambda}$'s appearing in R_λ therefore mean that the *positive* real axis will be a branch cut for R_λ. This branch cut therefore coincides with the spectrum of L, as promised earlier.

If λ is positive and we shift $\lambda \to \lambda + i\varepsilon$ then

$$\frac{1}{2\sqrt{-\lambda}} e^{-\sqrt{-\lambda}|x-y|} \to \frac{i}{2\sqrt{\lambda}} e^{+i\sqrt{\lambda}|x-y| - \varepsilon|x-y|/2\sqrt{\lambda}}. \tag{5.131}$$

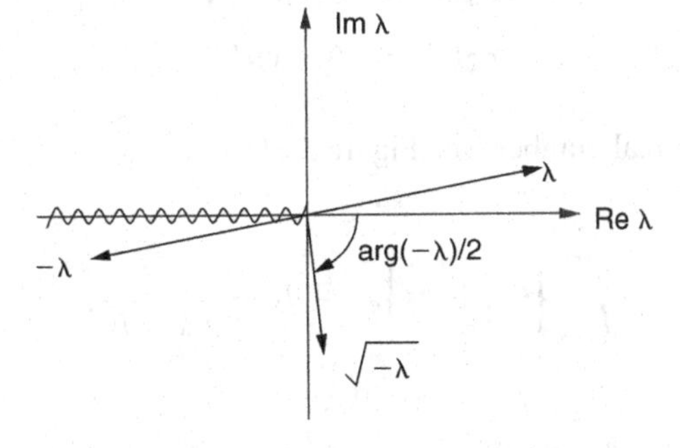

Figure 5.10 If $\text{Im}\,\lambda > 0$, and with the branch cut for $\sqrt{z}$ in its usual place along the negative real axis, then $\sqrt{-\lambda}$ has negative imaginary part and positive real part.

Notice that this decays away as $|x-y| \to \infty$. The square root retains a positive real part when λ is shifted to $\lambda - i\varepsilon$, and so the decay is still present:

$$\frac{1}{2\sqrt{-\lambda}} e^{-\sqrt{-\lambda}|x-y|} \to -\frac{i}{2\sqrt{\lambda}} e^{-i\sqrt{\lambda}|x-y| - \varepsilon|x-y|/2\sqrt{\lambda}}. \tag{5.132}$$

In each case, with λ either immediately above or immediately below the cut, the small imaginary part tempers the oscillatory behaviour of the Green function so that $\chi(x) = G(x,y)$ is square integrable and remains an element of $L^2[\mathbb{R}]$.

We now take the trace of R by setting $x = y$ and integrating:

$$\operatorname{Tr} R_{\lambda+i\varepsilon} = i\pi \frac{L}{2\pi\sqrt{|\lambda|}}. \tag{5.133}$$

Thus,

$$\rho(\lambda) = \theta(\lambda) \frac{L}{2\pi\sqrt{|\lambda|}}, \tag{5.134}$$

which coincides with our direct calculation.

Example: Let

$$L = -i\partial_x, \quad \mathcal{D}(L) = \{y, Ly \in L^2[\mathbb{R}]\}. \tag{5.135}$$

This has eigenfunctions e^{ikx} with eigenvalues k. The spectrum is therefore the entire real line. The local eigenvalue density of states is $1/2\pi$. The resolvent is therefore

$$(-i\partial_x - \lambda)^{-1}_{x,\xi} = \frac{1}{2\pi} \int_{-\infty}^{\infty} e^{ik(x-\xi)} \frac{1}{k-\lambda} dk. \tag{5.136}$$

To evaluate this, first consider the Fourier transforms of

$$\begin{aligned} F_1(x) &= \theta(x) e^{-\kappa x}, \\ F_2(x) &= -\theta(-x) e^{\kappa x}, \end{aligned} \tag{5.137}$$

where κ is a positive real number (see Figure 5.11).

We have

$$\int_{-\infty}^{\infty} \left\{ \theta(x) e^{-\kappa x} \right\} e^{-ikx}\, dx = \frac{1}{i} \frac{1}{k - i\kappa}, \tag{5.138}$$

$$\int_{-\infty}^{\infty} \left\{ -\theta(-x) e^{\kappa x} \right\} e^{-ikx}\, dx = \frac{1}{i} \frac{1}{k + i\kappa}. \tag{5.139}$$

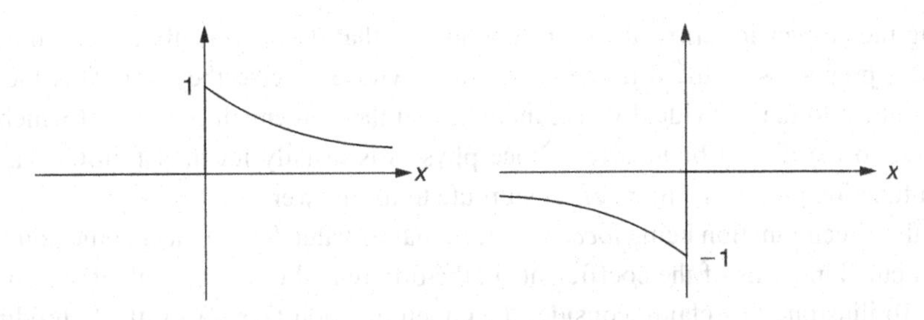

Figure 5.11 The functions $F_1(x) = \theta(x)e^{-\kappa x}$ and $F_2(x) = -\theta(-x)e^{\kappa x}$.

Inverting the transforms gives

$$\begin{aligned}\theta(x)e^{-\kappa x} &= \frac{1}{2\pi i}\int_{-\infty}^{\infty}\frac{1}{k-i\kappa}e^{ikx}\,dk,\\ -\theta(-x)e^{\kappa x} &= \frac{1}{2\pi i}\int_{-\infty}^{\infty}\frac{1}{k+i\kappa}e^{ikx}\,dk.\end{aligned} \tag{5.140}$$

These are important formulæ in their own right, and you should take care to understand them. Now we apply them to evaluating the integral defining R_λ.

If we write $\lambda = \mu + i\nu$, we find

$$\frac{1}{2\pi}\int_{-\infty}^{\infty} e^{ik(x-\xi)}\frac{1}{k-\lambda}\,dk = \begin{cases} i\theta(x-\xi)e^{i\mu(x-\xi)}e^{-\nu(x-\xi)}, & \nu > 0,\\ -i\theta(\xi-x)e^{i\mu(x-\xi)}e^{-\nu(x-\xi)}, & \nu < 0.\end{cases} \tag{5.141}$$

In each case, the resolvent is $\propto e^{i\lambda x}$ away from ξ, and has a jump of $+i$ at $x = \xi$ so as to produce the delta function. It decays either to the right or to the left, depending on the sign of ν. The Heaviside factor ensures that it is multiplied by zero on the exponentially growing side of $e^{-\nu x}$, so as to satisfy the requirement of square integrability.

Taking the trace of this resolvent is a little problematic. We are to set $x = \xi$ and integrate – but what value do we associate with $\theta(0)$? Remembering that Fourier transforms always give the mean of the two values at a jump discontinuity, it seems reasonable to set $\theta(0) = \frac{1}{2}$. With this definition, we have

$$\operatorname{Tr} R_\lambda = \begin{cases} \frac{i}{2}L, & \operatorname{Im}\lambda > 0,\\ -\frac{i}{2}L, & \operatorname{Im}\lambda < 0.\end{cases} \tag{5.142}$$

Our choice is therefore compatible with $\operatorname{Tr} R_{\lambda+i\varepsilon} = \pi\rho = L/2\pi$. We have been lucky. The ambiguous expression $\theta(0)$ is not always safely evaluated as $1/2$.

5.6 Locality and the Gelfand–Dikii equation

The answers to many quantum physics problems can be expressed either as sums over wavefunctions or as expressions involving Green functions. One of the advantages of writing the answer in terms of Green functions is that these typically depend only on the local properties of the differential operator whose inverse they are. This locality is in contrast to the individual wavefunctions and their eigenvalues, both of which are sensitive to the distant boundaries. Since physics is usually local, it follows that the Green function provides a more efficient route to the answer.

By the Green function being *local* we mean that its value for x, ξ near some point can be computed in terms of the coefficients in the differential operator evaluated near this point. To illustrate this claim, consider the Green function $G(x, \xi)$ for the Schrödinger operator $-\partial_x^2 + q(x) + \lambda$ on the entire real line. We will show that there is a not-exactly-obvious (but easy to obtain once you know the trick) local gradient expansion for the diagonal elements $D(x) \equiv G(x,x)$. These elements are often all that is needed in physics. We begin by recalling that we can write

$$G(x, \xi) \propto u(x)v(\xi)$$

where $u(x)$, $v(x)$ are solutions of $(-\partial_x^2 + q(x) + \lambda)y = 0$ satisfying suitable boundary conditions to the right and left respectively. We set $D(x) = G(x,x)$ and differentiate three times with respect to x. We find

$$\begin{aligned}\partial_x^3 D(x) &= u^{(3)}v + 3u''v' + 3u'v'' + uv^{(3)} \\ &= (\partial_x(q+\lambda)u)\, v + 3(q+\lambda)\partial_x(uv) + (\partial_x(q+\lambda)v)\, u.\end{aligned}$$

Here, in passing from the first to the second line, we have used the differential equation obeyed by u and v. We can re-express the second line as

$$(q\partial_x + \partial_x q - \frac{1}{2}\partial_x^3)D(x) = -2\lambda\partial_x D(x). \tag{5.143}$$

This relation is known as the *Gelfand–Dikii equation*. Using it we can find an expansion for the diagonal element $D(x)$ in terms of q and its derivatives. We begin by observing that for $q(x) \equiv 0$ we know that $D(x) = 1/(2\sqrt{\lambda})$. We therefore conjecture that we can expand

$$D(x) = \frac{1}{2\sqrt{\lambda}}\left(1 - \frac{b_1(x)}{2\lambda} + \frac{b_2(x)}{(2\lambda)^2} + \cdots + (-1)^n \frac{b_n(x)}{(2\lambda)^n} + \cdots\right).$$

If we insert this expansion into (5.143) we see that we get the recurrence relation

$$(q\partial_x + \partial_x q - \frac{1}{2}\partial_x^3)b_n = \partial_x b_{n+1}. \tag{5.144}$$

We can therefore find b_{n+1} from b_n by differentiation followed by a single integration. Remarkably, $\partial_x b_{n+1}$ is always the exact derivative of a polynomial in q and its derivatives.

Further, the integration constants must be zero so that we recover the $q \equiv 0$ result. If we carry out this process, we find

$$\begin{aligned}
b_1(x) &= q(x), \\
b_2(x) &= \frac{3\,q(x)^2}{2} - \frac{q''(x)}{2}, \\
b_3(x) &= \frac{5\,q(x)^3}{2} - \frac{5\,q'(x)^2}{4} - \frac{5\,q(x)\,q''(x)}{2} + \frac{q^{(4)}(x)}{4}, \\
b_4(x) &= \frac{35\,q(x)^4}{8} - \frac{35\,q(x)\,q'(x)^2}{4} - \frac{35\,q(x)^2\,q''(x)}{4} + \frac{21\,q''(x)^2}{8} \\
&\quad + \frac{7\,q'(x)\,q^{(3)}(x)}{2} + \frac{7\,q(x)\,q^{(4)}(x)}{4} - \frac{q^{(6)}(x)}{8},
\end{aligned} \tag{5.145}$$

and so on. (Note how the terms in the expansion are graded: each b_n is homogeneous in powers of q and its derivatives, provided we count two x derivatives as being worth one $q(x)$.) Keeping a few terms in this series expansion can provide an effective approximation for $G(x,x)$, but, in general, the series is not convergent, being only an *asymptotic expansion* for $D(x)$.

A similar strategy produces expansions for the diagonal element of the Green function of other one-dimensional differential operators. Such gradient expansions also exist in higher dimensions but the higher-dimensional *Seeley-coefficient* functions are not as easy to compute. Gradient expansions for the off-diagonal elements also exist, but, again, they are harder to obtain.

5.7 Further exercises and problems

Here are some further exercises that are intended to illustrate the material of this chapter:

Exercise 5.1: *Fredholm alternative*. A heavy elastic bar with uniform mass m per unit length lies almost horizontally. It is supported by a distribution of upward forces $F(x)$; see Figure 5.12.

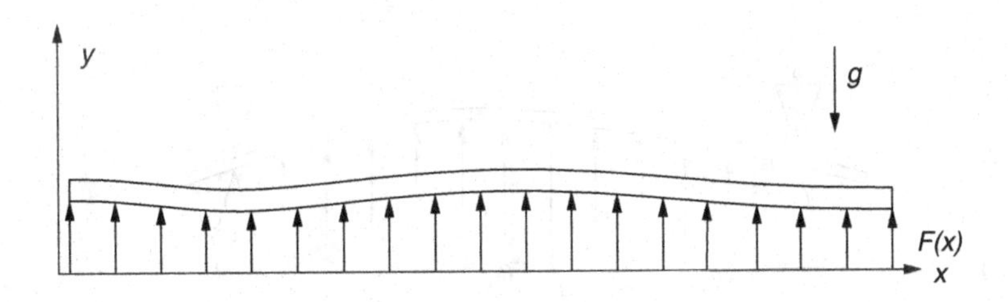

Figure 5.12 Elastic bar.

The shape of the bar, $y(x)$, can be found by minimizing the energy

$$U[y] = \int_0^L \left\{ \frac{1}{2}\kappa (y'')^2 - (F(x) - mg)y \right\} dx.$$

- Show that this minimization leads to the equation

$$\widehat{L}y \equiv \kappa \frac{d^4 y}{dx^4} = F(x) - mg, \quad y'' = y''' = 0 \quad \text{at} \quad x = 0, L.$$

- Show that the boundary conditions are such that the operator $\widehat{L}$ is self-adjoint with respect to an inner product with weight function 1.
- Find the zero modes which span the null space of $\widehat{L}$.
- If there are n linearly independent zero modes, then the codimension of the range of $\widehat{L}$ is also n. Using your explicit solutions from the previous part, find the conditions that must be obeyed by $F(x)$ for a solution of $\widehat{L}y = F - mg$ to exist. What is the physical meaning of these conditions?
- The solution to the equation and boundary conditions is not unique. Is this non-uniqueness physically reasonable? Explain.

Exercise 5.2: *Flexible rod again.* A flexible rod is supported near its ends by means of knife edges that constrain its position, but not its slope or curvature (Figure 5.13). It is acted on by a force $F(x)$.

The deflection of the rod is found by solving the boundary value problem

$$\frac{d^4 y}{dx^4} = F(x), \quad y(0) = y(1) = 0, \quad y''(0) = y''(1) = 0.$$

We wish to find the Green function $G(x, \xi)$ that facilitates the solution of this problem.

(a) If the differential operator and domain (boundary conditions) above is denoted by L, what is the operator and domain for $L^\dagger$? Is the problem self-adjoint?
(b) Are there any zero modes? Does F have to satisfy any conditions for the solution to exist?
(c) Write down the conditions, if any, obeyed by $G(x, \xi)$ and its derivatives $\partial_x G(x, \xi)$, $\partial^2_{xx} G(x, \xi)$, $\partial^3_{xxx} G(x, \xi)$ at $x = 0$, $x = \xi$ and $x = 1$.

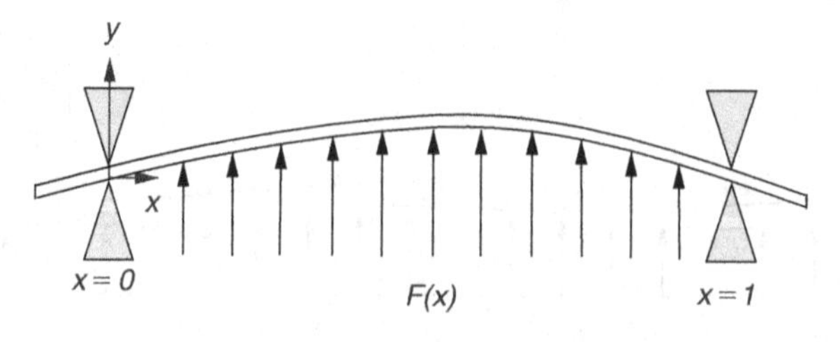

Figure 5.13 Simply supported rod.

(d) Using the conditions above, find $G(x,\xi)$. (This requires some boring algebra – but if you start from the "jump condition" and work down, it can be completed in under a page.)
(e) Is your Green function symmetric $\big(G(x,x) = G(\xi,x)\big)$? Is this in accord with the self-adjointness or not of the problem? (You can use this property as a check of your algebra.)
(f) Write down the integral giving the general solution of the boundary value problem. Assume, if necessary, that $F(x)$ is in the range of the differential operator. Differentiate your answer and see if it does indeed satisfy the differential equation and boundary conditions.

Exercise 5.3: *Hot ring*. The equation governing the steady state heat flow on a thin ring of unit circumference is

$$-y'' = f, \quad 0 < x < 1, \quad y(0) = y(1), \quad y'(0) = y'(1).$$

(a) This problem has a zero mode. Find the zero mode and the consequent condition on $f(x)$ for a solution to exist.
(b) Verify that a suitable modified Green function for the problem is

$$g(x,\xi) = \frac{1}{2}(x-\xi)^2 - \frac{1}{2}|x-\xi|.$$

You will need to verify that $g(x,\xi)$ satisfies both the differential equation *and* the boundary conditions.

Exercise 5.4: By using the observation that the left-hand side is 2π times the eigenfunction expansion of a modified Green function $G(x,0)$ for $L = -\partial_x^2$ on a circle of unit radius, show that

$$\sum_{n=-\infty}^{\infty} \frac{e^{inx}}{n^2} = \frac{1}{2}(x-\pi)^2 - \frac{\pi^2}{6}, \quad x \in [0, 2\pi).$$

The term with $n = 0$ is to be omitted from the sum.

Exercise 5.5: Seek a solution to the equation

$$-\frac{d^2y}{dx^2} = f(x), \quad x \in [0,1]$$

with inhomogeneous boundary conditions $y'(0) = F_0$, $y'(1) = F_1$. Observe that the corresponding homogeneous boundary condition problem has a zero mode. Therefore the solution, if one exists, cannot be unique.

(a) Show that there can be no solution to the differential equation and inhomogeneous boundary condition unless $f(x)$ satisfies the condition

$$\int_0^1 f(x)\,dx = F_0 - F_1. \qquad (\star)$$

(b) Let $G(x,\xi)$ denote the modified Green function (5.57)

$$G(x,\xi) = \begin{cases} \frac{1}{3} - \xi + \frac{x^2+\xi^2}{2}, & 0 < x < \xi \\ \frac{1}{3} - x + \frac{x^2+\xi^2}{2}, & \xi < x < 1. \end{cases}$$

Use the Lagrange-identity method for inhomogeneous boundary conditions to deduce that if a solution exists then it necessarily obeys

$$y(x) = \int_0^1 y(\xi)\,d\xi + \int_0^1 G(\xi,x)f(\xi)\,d\xi + G(1,x)F_1 - G(0,x)F_0.$$

(c) By differentiating with respect to x, show that

$$y_{\text{tentative}}(x) = \int_0^1 G(\xi,x)f(\xi)\,d\xi + G(1,x)F_1 - G(0,x)F_0 + C,$$

where C is an arbitrary constant, obeys the boundary conditions.

(d) By differentiating a second time with respect to x, show that $y_{\text{tentative}}(x)$ is a solution of the differential equation if, and only if, the condition $(\star)$ is satisfied.

Exercise 5.6: *Lattice Green functions*. The $k \times k$ matrices

$$\mathbf{T}_1 = \begin{pmatrix} 2 & -1 & 0 & 0 & 0 & \dots & 0 \\ -1 & 2 & -1 & 0 & 0 & \dots & 0 \\ 0 & -1 & 2 & -1 & 0 & \dots & 0 \\ \vdots & \vdots & \ddots & \ddots & \ddots & \vdots & \vdots \\ 0 & \dots & 0 & -1 & 2 & -1 & 0 \\ 0 & \dots & 0 & 0 & -1 & 2 & -1 \\ 0 & \dots & 0 & 0 & 0 & -1 & 2 \end{pmatrix},$$

$$\mathbf{T}_2 = \begin{pmatrix} 2 & -1 & 0 & 0 & 0 & \dots & 0 \\ -1 & 2 & -1 & 0 & 0 & \dots & 0 \\ 0 & -1 & 2 & -1 & 0 & \dots & 0 \\ \vdots & \vdots & \ddots & \ddots & \ddots & \vdots & \vdots \\ 0 & \dots & 0 & -1 & 2 & -1 & 0 \\ 0 & \dots & 0 & 0 & -1 & 2 & -1 \\ 0 & \dots & 0 & 0 & 0 & -1 & 1 \end{pmatrix}$$

represent two discrete lattice approximations to $-\partial_x^2$ on a finite interval.

(a) What are the boundary conditions defining the domains of the corresponding continuum differential operators? [They are either Dirichlet ($y = 0$) or Neumann ($y' = 0$) boundary conditions.] Make sure you explain your reasoning.

(b) Verify that

$$[\mathbf{T}_1^{-1}]_{ij} = \min(i,j) - \frac{ij}{k+1},$$
$$[\mathbf{T}_2^{-1}]_{ij} = \min(i,j).$$

(c) Find the continuum Green functions for the boundary value problems approximated by the matrix operators. Compare each of the matrix inverses with its corresponding continuum Green function. Are they similar?

Exercise 5.7: *Eigenfunction expansion.* The resolvent (Green function) $R_\lambda(x,\xi) = (L-\lambda)^{-1}_{x\xi}$ can be expanded as

$$(L-\lambda)^{-1}_{x\xi} = \sum_{\lambda_n} \frac{\varphi_n(x)\varphi_n(\xi)}{\lambda_n - \lambda},$$

where $\varphi_n(x)$ is the normalized eigenfunction corresponding to the eigenvalue λ_n. The resolvent therefore has a *pole* whenever λ approaches λ_n. Consider the case

$$R_{\omega^2}(x,\xi) = \left(-\frac{d^2}{dx^2} - \omega^2\right)^{-1}_{x\xi},$$

with boundary conditions $y(0) = y(L) = 0$.

(a) Show that

$$\begin{aligned} R_{\omega^2}(x,\xi) &= \frac{1}{\omega \sin \omega L} \sin \omega x \sin \omega (L-\xi), \quad x < \xi, \\ &= \frac{1}{\omega \sin \omega L} \sin \omega (L-x) \sin \omega \xi, \quad \xi < x. \end{aligned}$$

(b) Confirm that R_{ω^2} becomes singular at exactly those values of ω^2 corresponding to eigenvalues ω_n^2 of $-\frac{d^2}{dx^2}$.

(c) Find the associated eigenfunctions $\varphi_n(x)$ and, by taking the limit of R_{ω^2} as $\omega^2 \to \omega_n^2$, confirm that the *residue* of the pole (the coefficient of $1/(\omega_n^2 - \omega^2)$) is precisely the product of the *normalized* eigenfunctions $\varphi_n(x)\varphi_n(\xi)$.

Exercise 5.8: In this exercise we will investigate the self-adjointness of the operator $T = -i\partial/\partial x$ on the interval $[a,b]$ by using the resolvent operator $R_\lambda = (T - \lambda I)^{-1}$.

(a) The integral kernel $R_\lambda(x,\xi)$ is a Green function obeying

$$\left(-i\frac{\partial}{\partial x} - \lambda\right) R_\lambda(x,\xi) = \delta(x-\xi).$$

Use standard methods to show that

$$R_\lambda(x,\xi) = \frac{1}{2}\left(K_\lambda(\xi) + i\,\mathrm{sgn}\,(x-\xi)\right)e^{i\lambda(x-\xi)},$$

where $K_\lambda(\xi)$ is a number that depends on the boundary conditions imposed at the endpoints, a, b, of the interval.

(b) If T is to be self-adjoint then the Green function must be hermitian, i.e. $R_\lambda(x,\xi) = [R_\lambda(\xi,x)]^*$. Find the condition on K_λ for this to be true, and show that it implies that

$$\frac{R_\lambda(b,\xi)}{R_\lambda(a,\xi)} = e^{i\theta_\lambda},$$

where θ_λ is some real angle. Deduce that the range of R_λ is the set of functions

$$\mathcal{D}_\lambda = \{y(x) : y(b) = e^{i\theta_\lambda}y(a)\}.$$

Now the range of R_λ is the domain of $(T - \lambda I)$, which should be the same as the domain of T and therefore not depend on λ. We therefore require that θ_λ not depend on λ. Deduce that T will be self-adjoint only for boundary conditions $y(b) = e^{i\theta}y(a)$ – i.e. for twisted periodic boundary conditions.

(c) Show that with the twisted periodic boundary conditions of part (b), we have

$$K_\lambda = -\cot\left(\frac{\lambda(b-a)-\theta}{2}\right).$$

From this, show that $R_\lambda(x,\xi)$ has simple poles at $\lambda = \lambda_n$, where λ_n are the eigenvalues of T.

(d) Compute the residue of the pole of $R_\lambda(x,\xi)$ at the eigenvalue λ_n, and confirm that it is a product of the corresponding normalized eigenfunctions.

Problem 5.9: Consider the one-dimensional Dirac Hamiltonian

$$\begin{aligned}\widehat{H} &= \begin{pmatrix} -i\partial_x & m_1 - im_2 \\ m_1 + im_2 & +i\partial_x \end{pmatrix} \\ &= -i\widehat{\sigma}_3\partial_x + m_1(x)\widehat{\sigma}_1 + m_2(x)\widehat{\sigma}_2.\end{aligned}$$

Here $m_1(x)$, $m_2(x)$ are real functions, and the $\widehat{\sigma}_i$ are the Pauli matrices. $\widehat{H}$ acts on a two-component "spinor"

$$\Psi(x) = \begin{pmatrix}\psi_1(x) \\ \psi_2(x)\end{pmatrix}.$$

Impose self-adjoint boundary conditions

$$\frac{\psi_1(a)}{\psi_2(a)} = \exp\{i\theta_a\}, \quad \frac{\psi_1(b)}{\psi_2(b)} = \exp\{i\theta_b\}$$

at the ends of the interval $[a, b]$. Let $\Psi_L(x)$ be a solution of $\widehat{H}\Psi = \lambda\Psi$ obeying the boundary condition at $x = a$, and $\Psi_R(x)$ be a solution obeying the boundary condition at $x = b$. Define the "Wronskian" of these solutions to be

$$W(\Psi_L, \Psi_R) = \Psi_L^\dagger \widehat{\sigma}_3 \Psi_R.$$

(a) Show that, for real λ and the given boundary conditions, the Wronskian $W(\Psi_L, \Psi_R)$ is independent of position. Show also that $W(\Psi_L, \Psi_L) = W(\Psi_R, \Psi_R) = 0$.

(b) Show that the matrix-valued Green function $\widehat{G}(x, \xi)$ obeying

$$(\widehat{H} - \lambda I)\widehat{G}(x, \xi) = I\delta(x - \xi),$$

and the given boundary conditions, has entries

$$G_{\alpha\beta}(x, \xi) = \begin{cases} -\dfrac{i}{W^*}\psi_{L,\alpha}(x)\psi^*_{R,\beta}(\xi), & x < \xi, \\ +\dfrac{i}{W}\psi_{R,\alpha}(x)\psi^*_{L,\beta}(\xi), & x > \xi. \end{cases}$$

Observe that $G_{\alpha\beta}(x, \xi) = G^*_{\beta\alpha}(\xi, x)$, as befits the inverse of a self-adjoint operator.

(c) The Green function is discontinuous at $x = \xi$, but we can define a "position-diagonal" part by taking the average

$$G_{\alpha\beta}(x) \stackrel{\text{def}}{=} \frac{1}{2}\left(\frac{i}{W}\psi_{R,\alpha}(x)\psi^*_{L,\beta}(x) - \frac{i}{W^*}\psi_{L,\alpha}(x)\psi^*_{R,\beta}(x)\right).$$

Show that if we define the matrix $\widehat{g}(x)$ by setting $\widehat{g}(x) = \widehat{G}(x)\widehat{\sigma}_3$, then $\operatorname{tr}\widehat{g}(x) = 0$ and $\widehat{g}^2(x) = -\frac{1}{4}I$. Show further that

$$i\partial_x \widehat{g} = [\widehat{g}, \widehat{K}], \qquad (\star)$$

where $\widehat{K}(x) = \widehat{\sigma}_3\,(\lambda I - m_1(x)\widehat{\sigma}_1 - m_2(x)\widehat{\sigma}_2)$.

The equation $(\star)$ obtained in part (c) is the analogue of the Gelfand–Dikii equation for the Dirac Hamiltonian. It has applications in the theory of superconductivity, where $(\star)$ is known as the *Eilenberger equation*.

6
Partial differential equations

Most differential equations of physics involve quantities depending on both space and time. Inevitably they involve partial derivatives, and so are partial differential equations (PDEs). Although PDEs are inherently more complicated that ODEs, many of the ideas from the previous chapters – in particular the notion of self-adjointness and the resulting completeness of the eigenfunctions – carry over to the partial differential operators that occur in these equations.

6.1 Classification of PDEs

We focus on second-order equations in two variables, such as the wave equation

$$\frac{\partial^2 \varphi}{\partial x^2} - \frac{1}{c^2}\frac{\partial^2 \varphi}{\partial t^2} = f(x,t) \qquad \text{(Hyperbolic)}, \tag{6.1}$$

the Laplace or Poisson equation

$$\frac{\partial^2 \varphi}{\partial x^2} + \frac{\partial^2 \varphi}{\partial y^2} = f(x,y) \qquad \text{(Elliptic)}, \tag{6.2}$$

or Fourier's heat equation

$$\frac{\partial^2 \varphi}{\partial x^2} - \kappa\frac{\partial \varphi}{\partial t} = f(x,t) \qquad \text{(Parabolic)}. \tag{6.3}$$

What do the names hyperbolic, elliptic and parabolic mean? In high-school coordinate geometry we learned that a real quadratic curve

$$ax^2 + 2bxy + cy^2 + fx + gy + h = 0 \tag{6.4}$$

is a hyperbola, an ellipse or a parabola depending on whether the *discriminant*, $ac - b^2$, is less than zero, greater than zero or equal to zero, these being the conditions for the matrix

$$\begin{bmatrix} a & b \\ b & c \end{bmatrix} \tag{6.5}$$

to have signature $(+,-)$, $(+,+)$ or $(+,0)$.

By analogy, the equation

$$a(x,y)\frac{\partial^2\varphi}{\partial x^2} + 2b(x,y)\frac{\partial^2\varphi}{\partial x\partial y} + c(x,y)\frac{\partial^2\varphi}{\partial y^2} + \text{(lower orders)} = 0 \tag{6.6}$$

is said to be hyperbolic, elliptic or parabolic at a point (x,y) if

$$\begin{vmatrix} a(x,y) & b(x,y) \\ b(x,y) & c(x,y) \end{vmatrix} = (ac - b^2)|_{(x,y)} \tag{6.7}$$

is less than, greater than or equal to zero, respectively. This classification helps us understand what sort of initial or boundary data we need to specify the problem.

There are three broad classes of boundary conditions:

(a) **Dirichlet boundary conditions:** The value of the dependent variable is specified on the boundary.
(b) **Neumann boundary conditions:** The normal derivative of the dependent variable is specified on the boundary.
(c) **Cauchy boundary conditions:** Both the value and the normal derivative of the dependent variable are specified on the boundary.

Less commonly met are *Robin* boundary conditions, where the value of a linear combination of the dependent variable and the normal derivative of the dependent variable is specified on the boundary.

Cauchy boundary conditions are analogous to the initial conditions for a second-order ordinary differential equation. These are given at one end of the interval only. The other two classes of boundary condition are higher-dimensional analogues of the conditions we impose on an ODE at both ends of the interval.

Each class of PDEs requires a different class of boundary conditions in order to have a unique, stable solution.

(1) **Elliptic** equations require either Dirichlet or Neumann boundary conditions on a closed boundary surrounding the region of interest. Other boundary conditions are insufficient to determine a unique solution, are overly restrictive or lead to instabilities.
(2) **Hyperbolic** equations require Cauchy boundary conditions on an open surface. Other boundary conditions are either too restrictive for a solution to exist, or insufficient to determine a unique solution.
(3) **Parabolic** equations require Dirichlet or Neumann boundary conditions on an open surface. Other boundary conditions are too restrictive.

6.2 Cauchy data

Given a second-order ordinary differential equation

$$p_0 y'' + p_1 y' + p_2 y = f \tag{6.8}$$

with initial data $y(a), y'(a)$ we can construct the solution incrementally. We take a step $\delta x = \varepsilon$ and use the initial slope to find $y(a+\varepsilon) = y(a) + \varepsilon y'(a)$. Next we find $y''(a)$ from the differential equation

$$y''(a) = -\frac{1}{p_0}(p_1 y'(a) + p_2 y(a) - f(a)), \tag{6.9}$$

and use it to obtain $y'(a+\varepsilon) = y'(a) + \varepsilon y''(a)$. We now have initial data, $y(a+\varepsilon)$, $y'(a+\varepsilon)$, at the point $a+\varepsilon$, and can play the same game to proceed to $a+2\varepsilon$, and onwards.

Suppose now that we have the analogous situation of a second-order partial differential equation

$$a_{\mu\nu}(x)\frac{\partial^2 \varphi}{\partial x^\mu \partial x^\nu} + (\text{lower orders}) = 0 \tag{6.10}$$

in $\mathbb{R}^n$. We are also given initial data on a surface, Γ, of codimension one in $\mathbb{R}^n$ (see Figure 6.1).

At each point p on Γ we erect a basis $\mathbf{n}, \mathbf{t}_1, \mathbf{t}_2, \ldots, \mathbf{t}_{n-1}$, consisting of the normal to Γ and $n-1$ tangent vectors. The information we have been given consists of the value of φ at every point p together with

$$\frac{\partial \varphi}{\partial n} \stackrel{\text{def}}{=} n^\mu \frac{\partial \varphi}{\partial x^\mu}, \tag{6.11}$$

the normal derivative of φ at p. We want to know if these *Cauchy data* are sufficient to find the second derivative in the normal direction, and so construct similar Cauchy data

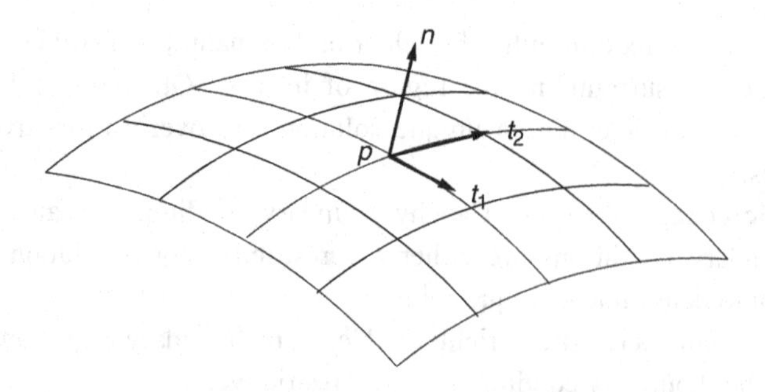

Figure 6.1 The surface Γ on which we are given Cauchy data.

on the adjacent surface $\Gamma + \varepsilon\mathbf{n}$. If so, we can repeat the process and systematically propagate the solution forward through $\mathbb{R}^n$.

From the given data, we can construct

$$\begin{aligned}\frac{\partial^2\varphi}{\partial n\partial t_i} &\stackrel{\text{def}}{=} n^\mu t_i^\nu \frac{\partial^2\varphi}{\partial x^\mu \partial x^\nu},\\ \frac{\partial^2\varphi}{\partial t_i\partial t_j} &\stackrel{\text{def}}{=} t_i^\nu t_j^\nu \frac{\partial^2\varphi}{\partial x^\mu \partial x^\nu},\end{aligned} \tag{6.12}$$

but we do not yet have enough information to determine

$$\frac{\partial^2\varphi}{\partial n\partial n} \stackrel{\text{def}}{=} n^\mu n^\nu \frac{\partial^2\varphi}{\partial x^\mu \partial x^\nu}. \tag{6.13}$$

Can we fill the data gap by using the differential equation (6.10)? Suppose that

$$\frac{\partial^2\varphi}{\partial x^\mu \partial x^\nu} = \phi_0^{\mu\nu} + n^\mu n^\nu \Phi \tag{6.14}$$

where $\phi_0^{\mu\nu}$ is a guess that is consistent with (6.12), and Φ is as yet unknown, and, because of the factor of $n^\mu n^\nu$, does not affect the derivatives (6.12). We plug into

$$a_{\mu\nu}(x_i)\frac{\partial^2\varphi}{\partial x^\mu \partial x^\nu} + (\text{known lower orders}) = 0 \tag{6.15}$$

and get

$$a_{\mu\nu}n^\mu n^\nu \Phi + (\text{known}) = 0. \tag{6.16}$$

We can therefore find Φ provided that

$$a_{\mu\nu}n^\mu n^\nu \neq 0. \tag{6.17}$$

If this expression *is* zero, we are stuck. It is like having $p_0(x) = 0$ in an ordinary differential equation. On the other hand, knowing Φ tells us the second normal derivative, and we can proceed to the adjacent surface where we play the same game.

Definition: A *characteristic surface* is a surface Σ such that $a_{\mu\nu}n^\mu n^\nu = 0$ at all points on Σ. We can therefore propagate our data forward, provided that the initial-data surface Γ is nowhere tangent to a characteristic surface. In two dimensions the characteristic surfaces become one-dimensional curves. An equation in two dimensions is hyperbolic, parabolic, or elliptic at a point (x, y) if it has two, one or zero characteristic curves through that point, respectively.

Characteristics are both a *curse* and *blessing*. They are a barrier to Cauchy data, but, as we see in the next two subsections, they are also the curves along which information is transmitted.

6.2.1 Characteristics and first-order equations

Suppose we have a linear first-order partial differential equation

$$a(x,y)\frac{\partial u}{\partial x} + b(x,y)\frac{\partial u}{\partial y} + c(x,y)u = f(x,y). \tag{6.18}$$

We can write this in vector notation as $(\mathbf{v}\cdot\nabla)u + cu = f$, where $\mathbf{v}$ is the vector field $\mathbf{v} = (a,b)$. If we define the *flow* of the vector field to be the family of parametrized curves $x(t), y(t)$ satisfying

$$\frac{dx}{dt} = a(x,y), \quad \frac{dy}{dt} = b(x,y), \tag{6.19}$$

then the partial differential equation (6.18) reduces to an ordinary linear differential equation

$$\frac{du}{dt} + c(t)u(t) = f(t) \tag{6.20}$$

along each flow line. Here,

$$\begin{aligned} u(t) &\equiv u(x(t),y(t)),\\ c(t) &\equiv c(x(t),y(t)),\\ f(t) &\equiv f(x(t),y(t)). \end{aligned} \tag{6.21}$$

Provided that $a(x,y)$ and $b(x,y)$ are never simultaneously zero, there will be one flow-line curve passing through each point in $\mathbb{R}^2$. If we have been given the initial value of u on a curve Γ that is nowhere tangent to any of these flow lines then we can propagate these data forward along the flow by solving (6.20). On the other hand, if the curve Γ does become tangent to one of the flow lines at some point then the data will generally be inconsistent with (6.18) at that point, and no solution can exist. The flow lines therefore play a role analogous to the characteristics of a second-order partial differential equation, and are therefore also called characteristics (see Figure 6.2). The trick of reducing the partial differential equation to a collection of ordinary differential equations along each of its flow lines is called the *method of characteristics*.

Exercise 6.1: Show that the general solution to the equation

$$\frac{\partial\varphi}{\partial x} - \frac{\partial\varphi}{\partial y} - (x-y)\varphi = 0$$

is

$$\varphi(x,y) = e^{-xy} f(x+y),$$

where f is an arbitrary function.

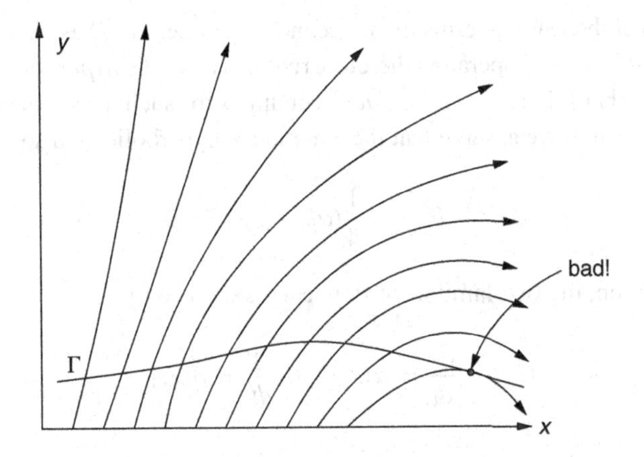

Figure 6.2 Initial data curve Γ, and flow-line characteristics.

6.2.2 Second-order hyperbolic equations

Consider a second-order equation containing the operator

$$D = a(x,y)\frac{\partial^2}{\partial x^2} + 2b(x,y)\frac{\partial^2}{\partial x \partial y} + c(x,y)\frac{\partial^2}{\partial y^2}. \tag{6.22}$$

We can always factorize

$$aX^2 + 2bXY + cY^2 = (\alpha X + \beta Y)(\gamma X + \delta Y), \tag{6.23}$$

and from this obtain

$$\begin{aligned} a\frac{\partial^2}{\partial x^2} + 2b\frac{\partial^2}{\partial x \partial y} + c\frac{\partial^2}{\partial y^2} &= \left(\alpha\frac{\partial}{\partial x} + \beta\frac{\partial}{\partial y}\right)\left(\gamma\frac{\partial}{\partial x} + \delta\frac{\partial}{\partial y}\right) + \text{lower}, \\ &= \left(\gamma\frac{\partial}{\partial x} + \delta\frac{\partial}{\partial y}\right)\left(\alpha\frac{\partial}{\partial x} + \beta\frac{\partial}{\partial y}\right) + \text{lower}. \end{aligned} \tag{6.24}$$

Here "lower" refers to terms containing only first-order derivatives such as

$$\alpha\left(\frac{\partial \gamma}{\partial x}\right)\frac{\partial}{\partial x}, \quad \beta\left(\frac{\partial \delta}{\partial y}\right)\frac{\partial}{\partial y}, \quad \text{etc.}$$

A necessary condition, however, for the coefficients $\alpha, \beta, \gamma, \delta$ to be *real* is that

$$\begin{aligned} ac - b^2 &= \alpha\beta\gamma\delta - \frac{1}{4}(\alpha\delta + \beta\gamma)^2 \\ &= -\frac{1}{4}(\alpha\delta - \beta\gamma)^2 \leq 0. \end{aligned} \tag{6.25}$$

A factorization of the leading terms in the second-order operator D as the product of two real first-order differential operators therefore requires that D be *hyperbolic* or *parabolic*. It is easy to see that this is also a *sufficient* condition for such a real factorization. For the rest of this section we assume that the equation is hyperbolic, and so

$$ac - b^2 = -\frac{1}{4}(\alpha\delta - \beta\gamma)^2 < 0. \tag{6.26}$$

With this condition, the two families of flow curves defined by

$$C_1 : \quad \frac{dx}{dt} = \alpha(x,y), \quad \frac{dy}{dt} = \beta(x,y), \tag{6.27}$$

and

$$C_2 : \quad \frac{dx}{dt} = \gamma(x,y), \quad \frac{dy}{dt} = \delta(x,y), \tag{6.28}$$

are distinct, and are the characteristics of D.

A hyperbolic second-order differential equation $Du = 0$ can therefore be written in either of two ways:

$$\left(\alpha\frac{\partial}{\partial x} + \beta\frac{\partial}{\partial y}\right) U_1 + F_1 = 0, \tag{6.29}$$

or

$$\left(\gamma\frac{\partial}{\partial x} + \delta\frac{\partial}{\partial y}\right) U_2 + F_2 = 0, \tag{6.30}$$

where

$$\begin{aligned} U_1 &= \gamma\frac{\partial u}{\partial x} + \delta\frac{\partial u}{\partial y}, \\ U_2 &= \alpha\frac{\partial u}{\partial x} + \beta\frac{\partial u}{\partial y}, \end{aligned} \tag{6.31}$$

and $F_{1,2}$ contain only $\partial u/\partial x$ and $\partial u/\partial y$. Given suitable Cauchy data, we can solve the two first-order partial differential equations by the method of characteristics described in the previous subsection, and so find $U_1(x,y)$ and $U_2(x,y)$. Because the hyperbolicity condition (6.26) guarantees that the determinant

$$\begin{vmatrix} \gamma & \delta \\ \alpha & \beta \end{vmatrix} = \gamma\beta - \alpha\delta$$

is not zero, we can solve (6.31) and so extract from $U_{1,2}$ the individual derivatives $\partial u/\partial x$ and $\partial u/\partial y$. From these derivatives and the initial values of u, we can determine $u(x,y)$.

6.3 Wave equation

The wave equation provides the paradigm for hyperbolic equations that can be solved by the method of characteristics.

6.3.1 d'Alembert's solution

Let $\varphi(x,t)$ obey the wave equation

$$\frac{\partial^2\varphi}{\partial x^2} - \frac{1}{c^2}\frac{\partial^2\varphi}{\partial t^2} = 0, \qquad -\infty < x < \infty. \tag{6.32}$$

We use the method of characteristics to propagate Cauchy data $\varphi(x,0) = \varphi_0(x)$ and $\dot{\varphi}(x,0) = v_0(x)$, given on the curve $\Gamma = \{x \in \mathbb{R}, t = 0\}$, forward in time.

We begin by factoring the wave equation as

$$0 = \left(\frac{\partial^2\varphi}{\partial x^2} - \frac{1}{c^2}\frac{\partial^2\varphi}{\partial t^2}\right) = \left(\frac{\partial}{\partial x} + \frac{1}{c}\frac{\partial}{\partial t}\right)\left(\frac{\partial\varphi}{\partial x} - \frac{1}{c}\frac{\partial\varphi}{\partial t}\right). \tag{6.33}$$

Thus,

$$\left(\frac{\partial}{\partial x} + \frac{1}{c}\frac{\partial}{\partial t}\right)(U - V) = 0, \tag{6.34}$$

where

$$U = \varphi' = \frac{\partial\varphi}{\partial x}, \qquad V = \frac{1}{c}\dot{\varphi} = \frac{1}{c}\frac{\partial\varphi}{\partial t}. \tag{6.35}$$

The quantity $U - V$ is therefore constant along the characteristic curves

$$x - ct = \text{const}. \tag{6.36}$$

Writing the linear factors in the reverse order yields the equation

$$\left(\frac{\partial}{\partial x} - \frac{1}{c}\frac{\partial}{\partial t}\right)(U + V) = 0. \tag{6.37}$$

This implies that $U + V$ is constant along the characteristics

$$x + ct = \text{const}. \tag{6.38}$$

Putting these two facts together tells us that

$$\begin{aligned} V(x,t') &= \frac{1}{2}[V(x,t') + U(x,t')] + \frac{1}{2}[V(x,t') - U(x,t')] \\ &= \frac{1}{2}[V(x+ct',0) + U(x+ct',0)] + \frac{1}{2}[V(x-ct',0) - U(x-ct',0)]. \end{aligned} \tag{6.39}$$

The value of the variable V at the point (x, t') has therefore been computed in terms of the values of U and V on the initial curve Γ. After changing variables from t' to $\xi = x \pm ct'$ as appropriate, we can integrate up to find that

$$\begin{aligned}
\varphi(x,t) &= \varphi(x,0) + c\int_0^t V(x,t')dt' \\
&= \varphi(x,0) + \frac{1}{2}\int_x^{x+ct} \varphi'(\xi,0)\,d\xi + \frac{1}{2}\int_x^{x-ct} \varphi'(\xi,0)\,d\xi \\
&\quad + \frac{1}{2c}\int_{x-ct}^{x+ct} \dot{\varphi}(\xi,0)\,d\xi \\
&= \frac{1}{2}\{\varphi(x+ct,0) + \varphi(x-ct,0)\} + \frac{1}{2c}\int_{x-ct}^{x+ct} \dot{\varphi}(\xi,0)\,d\xi. \qquad (6.40)
\end{aligned}$$

This result

$$\varphi(x,t) = \frac{1}{2}\{\varphi_0(x+ct) + \varphi_0(x-ct)\} + \frac{1}{2c}\int_{x-ct}^{x+ct} v_0(\xi)\,d\xi \qquad (6.41)$$

is usually known as *d'Alembert's solution* of the wave equation. It was actually obtained first by Euler in 1748.

The value of φ at x, t, is determined by only a finite interval of the initial Cauchy data. In more generality, $\varphi(x,t)$ depends only on what happens in the past *light-cone* of the point, which is bounded by a pair of characteristic curves. This is illustrated in Figure 6.3.

D'Alembert and Euler squabbled over whether φ_0 and v_0 had to be twice differentiable for the solution (6.41) to make sense. Euler wished to apply (6.41) to a plucked string, which has a discontinuous slope at the plucked point, but d'Alembert argued that the wave equation, with its second derivative, could not be applied in this case. This was a dispute that could not be resolved (in Euler's favour) until the advent of the theory of distributions.

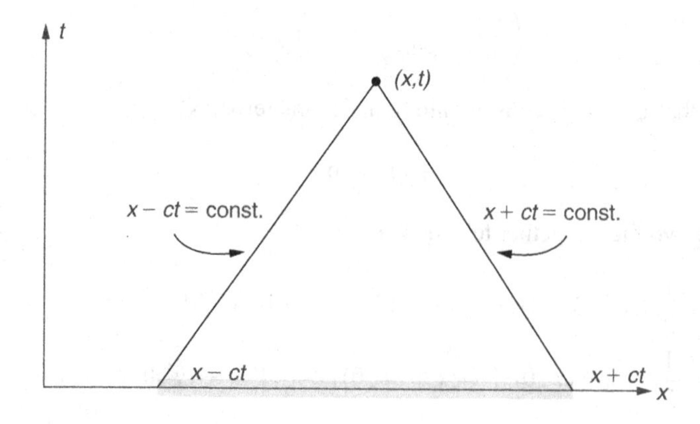

Figure 6.3 Range of Cauchy data influencing $\varphi(x,t)$.

It highlights an important difference between ordinary and partial differential equations: an ODE with smooth coefficients has smooth solutions; a PDE with smooth coefficients can admit discontinuous or even distributional solutions.

An alternative route to d'Alembert's solution uses a method that applies most effectively to PDEs with constant coefficients. We first seek a *general solution* to the PDE involving two arbitrary functions. Begin with a change of variables. Let

$$\begin{aligned} \xi &= x + ct, \\ \eta &= x - ct \end{aligned} \tag{6.42}$$

be *light-cone coordinates*. In terms of them, we have

$$\begin{aligned} x &= \frac{1}{2}(\xi + \eta), \\ t &= \frac{1}{2c}(\xi - \eta). \end{aligned} \tag{6.43}$$

Now,

$$\frac{\partial}{\partial \xi} = \frac{\partial x}{\partial \xi}\frac{\partial}{\partial x} + \frac{\partial t}{\partial \xi}\frac{\partial}{\partial t} = \frac{1}{2}\left(\frac{\partial}{\partial x} + \frac{1}{c}\frac{\partial}{\partial t}\right). \tag{6.44}$$

Similarly

$$\frac{\partial}{\partial \eta} = \frac{1}{2}\left(\frac{\partial}{\partial x} - \frac{1}{c}\frac{\partial}{\partial t}\right). \tag{6.45}$$

Thus

$$\left(\frac{\partial^2}{\partial x^2} - \frac{1}{c^2}\frac{\partial^2}{\partial t^2}\right) = \left(\frac{\partial}{\partial x} + \frac{1}{c}\frac{\partial}{\partial t}\right)\left(\frac{\partial}{\partial x} - \frac{1}{c}\frac{\partial}{\partial t}\right) = 4\frac{\partial^2}{\partial \xi \partial \eta}. \tag{6.46}$$

The characteristics of the equation

$$4\frac{\partial^2 \varphi}{\partial \xi \partial \eta} = 0 \tag{6.47}$$

are $\xi =$ const. or $\eta =$ const. There are two characteristics curves through each point, so the equation is still hyperbolic.

With light-cone coordinates it is easy to see that a solution to

$$\left(\frac{\partial^2}{\partial x^2} - \frac{1}{c^2}\frac{\partial^2}{\partial t^2}\right)\varphi = 4\frac{\partial^2 \varphi}{\partial \xi \partial \eta} = 0 \tag{6.48}$$

is

$$\varphi(x,t) = f(\xi) + g(\eta) = f(x+ct) + g(x-ct). \tag{6.49}$$

It is this expression that was obtained by d'Alembert (1746).

Following Euler, we use d'Alembert's general solution to propagate the Cauchy data $\varphi(x,0) \equiv \varphi_0(x)$ and $\dot{\varphi}(x,0) \equiv v_0(x)$ by using this information to determine the functions f and g. We have

$$\begin{aligned} f(x) + g(x) &= \varphi_0(x), \\ c(f'(x) - g'(x)) &= v_0(x). \end{aligned} \tag{6.50}$$

Integration of the second line with respect to x gives

$$f(x) - g(x) = \frac{1}{c}\int_0^x v_0(\xi)\, d\xi + A, \tag{6.51}$$

where A is an unknown (but irrelevant) constant. We can now solve for f and g, and find

$$\begin{aligned} f(x) &= \frac{1}{2}\varphi_0(x) + \frac{1}{2c}\int_0^x v_0(\xi)\, d\xi + \frac{1}{2}A, \\ g(x) &= \frac{1}{2}\varphi_0(x) - \frac{1}{2c}\int_0^x v_0(\xi)\, d\xi - \frac{1}{2}A, \end{aligned} \tag{6.52}$$

and so

$$\varphi(x,t) = \frac{1}{2}\left\{\varphi_0(x+ct) + \varphi_0(x-ct)\right\} + \frac{1}{2c}\int_{x-ct}^{x+ct} v_0(\xi)\, d\xi. \tag{6.53}$$

The unknown constant A has disappeared in the end result, and again we find "d'Alembert's" solution.

Exercise 6.2: Show that when the operator D in a constant-coefficient second-order PDE $D\varphi = 0$ is *reducible*, meaning that it can be factored into two distinct first-order factors $D = P_1 P_2$, where

$$P_i = \alpha_i \frac{\partial}{\partial x} + \beta_i \frac{\partial}{\partial y} + \gamma_i,$$

then the general solution to $D\varphi = 0$ can be written as $\varphi = \phi_1 + \phi_2$, where $P_1\phi_1 = 0$, $P_2\phi_2 = 0$. Hence, or otherwise, show that the general solution to the equation

$$\frac{\partial^2 \varphi}{\partial x \partial y} + 2\frac{\partial^2 \varphi}{\partial y^2} - \frac{\partial \varphi}{\partial x} - 2\frac{\partial \varphi}{\partial y} = 0$$

is

$$\varphi(x,y) = f(2x - y) + e^y g(x),$$

where f, g, are arbitrary functions.

Exercise 6.3: Show that when the constant-coefficient operator D is of the form

$$D = P^2 = \left(\alpha\frac{\partial}{\partial x} + \beta\frac{\partial}{\partial y} + \gamma\right)^2,$$

with $\alpha \neq 0$, then the general solution to $D\varphi = 0$ is given by $\varphi = \phi_1 + x\phi_2$, where $P\phi_{1,2} = 0$. (If $\alpha = 0$ and $\beta \neq 0$, then $\varphi = \phi_1 + y\phi_2$.)

6.3.2 Fourier's solution

In 1755 Daniel Bernoulli proposed solving for the motion of a finite length L of transversely vibrating string by setting

$$y(x,t) = \sum_{n=1}^{\infty} A_n \sin\left(\frac{n\pi x}{L}\right)\cos\left(\frac{n\pi ct}{L}\right), \tag{6.54}$$

but he did not know how to find the coefficients A_n (and perhaps did not care that his cosine time dependence restricted his solution to the initial condition $\dot{y}(x,0) = 0$). Bernoulli's idea was dismissed out of hand by Euler and d'Alembert as being too restrictive. They simply refused to believe that (almost) any chosen function could be represented by a trigonometric series expansion. It was only 50 years later, in a series of papers starting in 1807, that Joseph Fourier showed how to compute the A_n and insisted that indeed "any" function could be expanded in this way. Mathematicians have expended much effort in investigating the extent to which Fourier's claim is true.

We now try our hand at Bernoulli's game. Because we are solving the wave equation on the infinite line, we seek a solution as a Fourier *integral*. A sufficiently general form is

$$\varphi(x,t) = \int_{-\infty}^{\infty} \frac{dk}{2\pi}\left\{a(k)e^{ikx-i\omega_k t} + a^*(k)e^{-ikx+i\omega_k t}\right\}, \tag{6.55}$$

where $\omega_k \equiv c|k|$ is the *positive* root of $\omega^2 = c^2k^2$. The terms being summed by the integral are each individually of the form $f(x-ct)$ or $f(x+ct)$, and so $\varphi(x,t)$ is indeed a solution of the wave equation. The positive-root convention means that positive k corresponds to right-going waves, and negative k to left-going waves.

We find the amplitudes $a(k)$ by fitting to the Fourier transforms

$$\begin{aligned}
\Phi(k) &\stackrel{\text{def}}{=} \int_{-\infty}^{\infty} \varphi(x,t=0)e^{-ikx}dx,\\
\chi(k) &\stackrel{\text{def}}{=} \int_{-\infty}^{\infty} \dot{\varphi}(x,t=0)e^{-ikx}dx,
\end{aligned} \tag{6.56}$$

of the Cauchy data. Comparing

$$\begin{aligned}\varphi(x,t=0) &= \int_{-\infty}^{\infty}\frac{dk}{2\pi}\Phi(k)e^{ikx},\\ \dot\varphi(x,t=0) &= \int_{-\infty}^{\infty}\frac{dk}{2\pi}\chi(k)e^{ikx},\end{aligned}\tag{6.57}$$

with (6.55) shows that

$$\begin{aligned}\Phi(k) &= a(k)+a^*(-k),\\ \chi(k) &= i\omega_k\Big(a^*(-k)-a(k)\Big).\end{aligned}\tag{6.58}$$

Solving, we find

$$\begin{aligned}a(k) &= \frac{1}{2}\left(\Phi(k)+\frac{i}{\omega_k}\chi(k)\right),\\ a^*(k) &= \frac{1}{2}\left(\Phi(-k)-\frac{i}{\omega_k}\chi(-k)\right).\end{aligned}\tag{6.59}$$

The accumulated wisdom of 200 years of research on Fourier series and Fourier integrals shows that, when appropriately interpreted, this solution is equivalent to d'Alembert's.

6.3.3 Causal Green function

We now add a source term:

$$\frac{1}{c^2}\frac{\partial^2\varphi}{\partial t^2}-\frac{\partial^2\varphi}{\partial x^2}=q(x,t).\tag{6.60}$$

We solve this equation by finding a Green function such that

$$\left(\frac{1}{c^2}\frac{\partial^2}{\partial t^2}-\frac{\partial^2}{\partial x^2}\right)G(x,t;\xi,\tau)=\delta(x-\xi)\delta(t-\tau).\tag{6.61}$$

If the only waves in the system are those produced by the source, we should demand that the Green function be *causal*, in that $G(x,t;\xi,\tau)=0$ if $t<\tau$ (see Figure 6.4).

To construct the causal Green function, we integrate the equation over an infinitesimal time interval from $\tau-\varepsilon$ to $\tau+\varepsilon$ and so find Cauchy data

$$\begin{aligned}G(x,\tau+\varepsilon;\xi,\tau) &= 0,\\ \frac{d}{dt}G(x,\tau+\varepsilon;\xi,\tau) &= c^2\delta(x-\xi).\end{aligned}\tag{6.62}$$

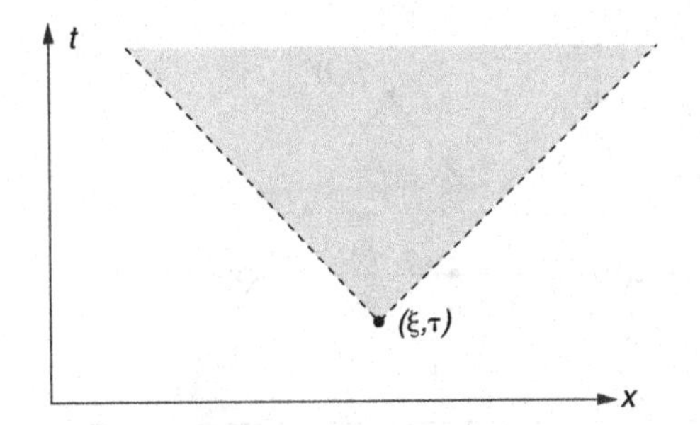

Figure 6.4 Support of $G(x,t;\xi,\tau)$ for fixed ξ,τ, or the "domain of influence".

We insert this data into d'Alembert's solution to get

$$\begin{aligned} G(x,t;\xi,\tau) &= \theta(t-\tau)\frac{c}{2}\int_{x-c(t-\tau)}^{x+c(t-\tau)} \delta(\zeta-\xi)d\zeta \\ &= \frac{c}{2}\theta(t-\tau)\left\{\theta\Big(x-\xi+c(t-\tau)\Big)-\theta\Big(x-\xi-c(t-\tau)\Big)\right\}. \end{aligned} \tag{6.63}$$

We can now use the Green function to write the solution to the inhomogeneous problem as

$$\varphi(x,t) = \iint G(x,t;\xi,\tau)q(\xi,\tau)\,d\tau d\xi. \tag{6.64}$$

The step-function form of $G(x,t;\xi,\tau)$ allows us to obtain

$$\begin{aligned} \varphi(x,t) &= \iint G(x,t;\xi,\tau)q(\xi,\tau)\,d\tau d\xi, \\ &= \frac{c}{2}\int_{-\infty}^{t} d\tau \int_{x-c(t-\tau)}^{x+c(t-\tau)} q(\xi,\tau)\,d\xi \\ &= \frac{c}{2}\iint_{\Omega} q(\xi,\tau)\,d\tau d\xi, \end{aligned} \tag{6.65}$$

where the domain of integration Ω is shown in Figure 6.5.

We can write the causal Green function in the form of Fourier's solution of the wave equation. We claim that

$$G(x,t;\xi,\tau) = c^2\int_{-\infty}^{\infty}\frac{d\omega}{2\pi}\int_{-\infty}^{\infty}\frac{dk}{2\pi}\left\{\frac{e^{ik(x-\xi)}e^{-i\omega(t-\tau)}}{c^2k^2-(\omega+i\varepsilon)^2}\right\}, \tag{6.66}$$

where the $i\varepsilon$ plays the same role in enforcing causality as it does for the harmonic oscillator in one dimension. This is only to be expected. If we decompose a vibrating string into normal modes, then each mode is an independent oscillator with $\omega_k^2 = c^2k^2$,

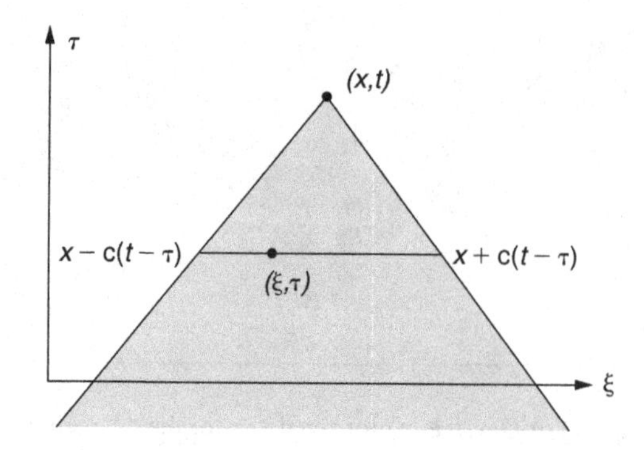

Figure 6.5 The region Ω, or the "domain of dependence".

and the Green function for the PDE is simply the sum of the ODE Green functions for each k mode. To confirm our claim, we exploit our previous results for the single-oscillator Green function to evaluate the integral over ω, and we find

$$G(x,t;0,0) = \theta(t)c^2 \int_{-\infty}^{\infty} \frac{dk}{2\pi} e^{ikx} \frac{1}{c|k|} \sin(|k|ct). \tag{6.67}$$

Despite the factor of $1/|k|$, there is no singularity at $k = 0$, so no $i\varepsilon$ is needed to make the integral over k well defined. We can do the k integral by recognizing that the integrand is nothing but the Fourier representation, $\frac{2}{k}\sin ak$, of a square-wave pulse. We end up with

$$G(x,t;0,0) = \theta(t)\frac{c}{2}\left\{\theta(x+ct) - \theta(x-ct)\right\}, \tag{6.68}$$

the same expression as from our direct construction. We can also write

$$G(x,t;0,0) = \frac{c}{2}\int_{-\infty}^{\infty} \frac{dk}{2\pi}\left(\frac{i}{|k|}\right)\left\{e^{ikx-ic|k|t} - e^{-ikx+ic|k|t}\right\}, \quad t > 0, \tag{6.69}$$

which is in explicit Fourier-solution form with $a(k) = ic/2|k|$.

Illustration: Radiation damping. Figure 6.6 shows a bead of mass M that slides without friction on the y-axis. The bead is attached to an infinite string which is initially undisturbed and lying along the x-axis. The string has tension T, and a density ρ, so the speed of waves on the string is $c = \sqrt{T/\rho}$. We show that either d'Alembert or Fourier can be used to compute the effect of the string on the motion of the bead.

We first use d'Alembert's general solution to show that wave energy emitted by the moving bead gives rise to an effective viscous damping force on it.

The string tension acting on the bead leads to the equation of motion $M\dot{v} = Ty'(0,t)$, and from the condition of no incoming waves we know that

$$y(x,t) = y(x-ct). \tag{6.70}$$

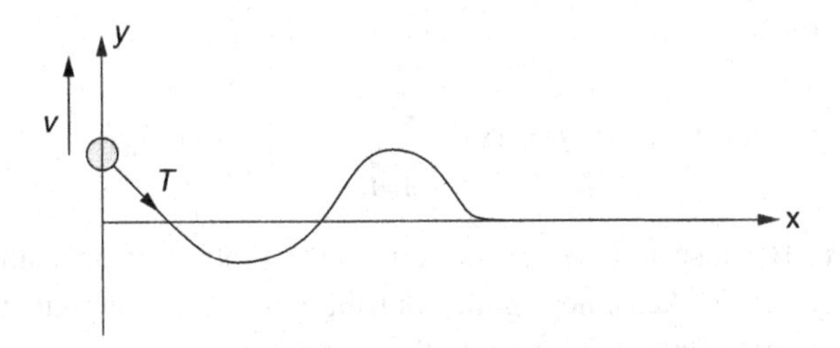

Figure 6.6 A bead connected to a string.

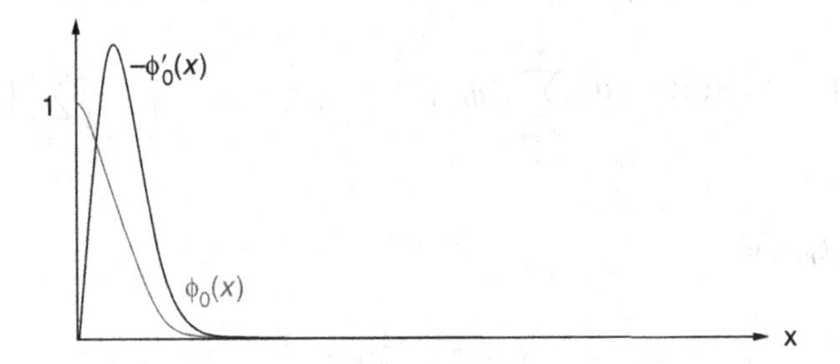

Figure 6.7 The function $\phi_0(x)$ and its derivative.

Thus $y'(0,t) = -\dot{y}(0,t)/c$. But the bead is attached to the string, so $v(t) = \dot{y}(0,t)$, and therefore

$$M\dot{v} = -\left(\frac{T}{c}\right)v. \tag{6.71}$$

The emitted radiation therefore generates a velocity-dependent drag force with friction coefficient $\eta = T/c$.

We need an infinitely long string for (6.71) to be true for all time. If the string had a finite length L, then, after a period of $2L/c$, energy will be reflected back to the bead and this will complicate matters.

We now show that Fourier's mode-decomposition of the string motion, combined with the Caldeira–Leggett analysis of Chapter 5, yields the same expression for the radiation damping as the d'Alembert solution. Our bead–string contraption has Lagrangian

$$L = \frac{M}{2}[\dot{y}(0,t)]^2 - V[y(0,t)] + \int_0^L \left\{\frac{\rho}{2}\dot{y}^2 - \frac{T}{2}y'^2\right\} dx. \tag{6.72}$$

Here, $V[y]$ is some potential energy for the bead.

To deal with the motion of the bead, we introduce a function $\phi_0(x)$ such that $\phi_0(0) = 1$ and $\phi_0(x)$ decreases rapidly to zero as x increases (see Figure 6.7). We therefore have $-\phi_0'(x) \approx \delta(x)$. We expand $y(x,t)$ in terms of $\phi_0(x)$ and the normal modes of a string with fixed ends as

$$y(x,t) = y(0,t)\phi_0(x) + \sum_{n=1}^{\infty} q_n(t)\sqrt{\frac{2}{L\rho}} \sin k_n x. \tag{6.73}$$

Here $k_n L = n\pi$. Because $y(0,t)\phi_0(x)$ describes the motion of only an infinitesimal length of string, $y(0,t)$ makes a negligible contribution to the string kinetic energy, but it provides a linear coupling of the bead to the string normal modes, $q_n(t)$, through the $Ty'^2/2$ term. Inserting the mode expansion into the Lagrangian, and after about half a page of arithmetic, we end up with

$$L = \frac{M}{2}[\dot{y}(0)]^2 - V[y(0)] + y(0)\sum_{n=1}^{\infty} f_n q_n + \sum_{n=1}^{\infty}\left(\frac{1}{2}\dot{q}_n^2 - \omega_n^2 q_n^2\right) - \frac{1}{2}\sum_{n=1}^{\infty}\left(\frac{f_n^2}{\omega_n^2}\right) y(0)^2, \tag{6.74}$$

where $\omega_n = ck_n$, and

$$f_n = T\sqrt{\frac{2}{L\rho}} k_n. \tag{6.75}$$

This is exactly the Caldeira–Leggett Lagrangian – including their frequency-shift counter-term that reflects that fact that a static displacement of an infinite string results in no additional force on the bead.[1] When L becomes large, the eigenvalue density of states

$$\rho(\omega) = \sum_n \delta(\omega - \omega_n) \tag{6.76}$$

becomes

$$\rho(\omega) = \frac{L}{\pi c}. \tag{6.77}$$

The Caldeira–Leggett spectral function

$$J(\omega) = \frac{\pi}{2}\sum_n \left(\frac{f_n^2}{\omega_n}\right)\delta(\omega - \omega_n), \tag{6.78}$$

is therefore

$$J(\omega) = \frac{\pi}{2}\cdot\frac{2T^2k^2}{L\rho}\cdot\frac{1}{kc}\cdot\frac{L}{\pi c} = \left(\frac{T}{c}\right)\omega, \tag{6.79}$$

[1] For a *finite* length of string that is fixed at the far end, the string tension *does* add $\frac{1}{2}Ty(0)^2/L$ to the static potential. In the mode expansion, this additional restoring force arises from the first term of $-\phi_0'(x) \approx 1/L + (2/L)\sum_{n=1}^{\infty} \cos k_n x$ in $\frac{1}{2}Ty(0)^2 \int (\phi_0')^2\, dx$. The subsequent terms provide the Caldeira–Leggett counter-term. The first-term contribution has been omitted in (6.74) as being unimportant for large L.

where we have used $c = \sqrt{T/\rho}$. Comparing with Caldeira and Leggett's $J(\omega) = \eta\omega$, we see that the effective viscosity is given by $\eta = T/c$, as before. The necessity of having an infinitely long string here translates into the requirement that we must have a *continuum* of oscillator modes. It is only after the sum over discrete modes ω_i is replaced by an integral over the continuum of ω's that no energy is ever returned to the system being damped.

For our bead and string, the mode-expansion approach is more complicated than d'Alembert's. In the important problem of the drag forces induced by the emission of radiation from an accelerated charged particle, however, the mode-expansion method leads to an informative resolution[2] of the pathologies of the Abraham–Lorentz equation,

$$M(\dot{\mathbf{v}} - \tau\ddot{\mathbf{v}}) = \mathbf{F}_{\text{ext}}, \quad \tau = \frac{2}{3}\frac{e^2}{Mc^3}\frac{1}{4\pi\varepsilon_0} \tag{6.80}$$

which is plagued by runaway, or apparently acausal, solutions.

6.3.4 Odd *vs.* even dimensions

Consider the wave equation for sound in three dimensions. We have a velocity potential ϕ which obeys the wave equation

$$\frac{\partial^2\phi}{\partial x^2} + \frac{\partial^2\phi}{\partial y^2} + \frac{\partial^2\phi}{\partial z^2} - \frac{1}{c^2}\frac{\partial^2\phi}{\partial t^2} = 0, \tag{6.81}$$

and from which the velocity, density and pressure fluctuations can be extracted as

$$\begin{aligned} v_1 &= \nabla\phi, \\ \rho_1 &= -\frac{\rho_0}{c^2}\dot{\phi}, \\ P_1 &= c^2\rho_1. \end{aligned} \tag{6.82}$$

In three dimensions, and considering only spherically symmetric waves, the wave equation becomes

$$\frac{\partial^2(r\phi)}{\partial r^2} - \frac{1}{c^2}\frac{\partial^2(r\phi)}{\partial t^2} = 0, \tag{6.83}$$

with solution

$$\phi(r,t) = \frac{1}{r}f\left(t - \frac{r}{c}\right) + \frac{1}{r}g\left(t + \frac{r}{c}\right). \tag{6.84}$$

[2] G. W. Ford, R. F. O'Connell, *Phys. Lett.* A, **157** (1991) 217.

Consider what happens if we put a point volume source at the origin (the sudden conversion of a negligible volume of solid explosive to a large volume of hot gas, for example). Let the rate at which volume is being intruded be $\dot{q}$. The gas velocity very close to the origin will be

$$v(r,t) = \frac{\dot{q}(t)}{4\pi r^2}. \tag{6.85}$$

Matching this to an outgoing wave gives

$$\frac{\dot{q}(t)}{4\pi r^2} = v_1(r,t) = \frac{\partial \phi}{\partial r} = -\frac{1}{r^2} f\left(t - \frac{r}{c}\right) - \frac{1}{rc} f'\left(t - \frac{r}{c}\right). \tag{6.86}$$

Close to the origin, in the *near field*, the term $\propto f/r^2$ will dominate, and so

$$-\frac{1}{4\pi}\dot{q}(t) = f(t). \tag{6.87}$$

Further away, in the *far field* or *radiation field*, only the second term will survive, and so

$$v_1 = \frac{\partial \phi}{\partial r} \approx -\frac{1}{rc} f'\left(t - \frac{r}{c}\right). \tag{6.88}$$

The far-field velocity-pulse profile v_1 is therefore the derivative of the near-field v_1 pulse profile (Figure 6.8).

The pressure pulse

$$P_1 = -\rho_0 \dot{\phi} = \frac{\rho_0}{4\pi r}\ddot{q}\left(t - \frac{r}{c}\right) \tag{6.89}$$

is also of this form. Thus, a sudden localized expansion of gas produces an outgoing pressure pulse which is first positive and then negative.

This phenomenon can be seen in (old, we hope) news footage of bomb blasts in tropical regions. A spherical vapour condensation wave can been seen spreading out from the explosion. The condensation cloud is caused by the air cooling below the dew-point in the low-pressure region which tails the over-pressure blast.

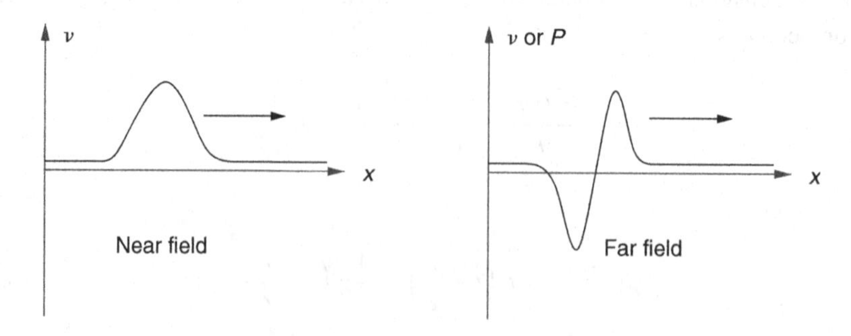

Figure 6.8 Three-dimensional blast wave.

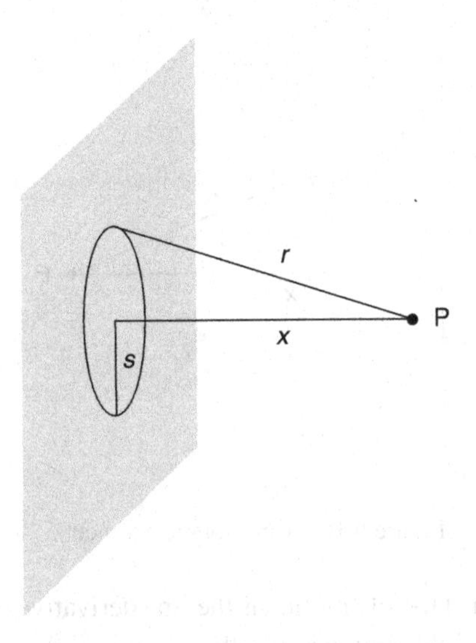

Figure 6.9 Sheet-source geometry.

Now consider what happens if we have a sheet of explosive, the simultaneous detonation of every part of which gives us a one-dimensional plane-wave pulse. We can obtain the plane wave by adding up the individual spherical waves from each point on the sheet.

Using the notation defined in Figure 6.9, we have

$$\phi(x,t) = 2\pi \int_0^\infty \frac{1}{\sqrt{x^2+s^2}} f\left(t - \frac{\sqrt{x^2+s^2}}{c}\right) s ds \tag{6.90}$$

with $f(t) = -\dot{q}(t)/4\pi$, where now $\dot{q}$ is the rate at which volume is being intruded per unit area of the sheet. We can write this as

$$\begin{aligned} 2\pi \int_0^\infty f\left(t - \frac{\sqrt{x^2+s^2}}{c}\right) d\sqrt{x^2+s^2} \\ = 2\pi c \int_{-\infty}^{t-x/c} f(\tau)\, d\tau, \\ = -\frac{c}{2} \int_{-\infty}^{t-x/c} \dot{q}(\tau)\, d\tau. \end{aligned} \tag{6.91}$$

In the second line we have defined $\tau = t - \sqrt{x^2+s^2}/c$, which, *inter alia*, interchanged the role of the upper and lower limits on the integral.

Thus, $v_1 = \phi'(x,t) = \frac{1}{2}\dot{q}(t - x/c)$. Since the near-field motion produced by the intruding gas is $v_1(r) = \frac{1}{2}\dot{q}(t)$, the far-field displacement exactly reproduces the initial motion, suitably delayed of course. (The factor $1/2$ is because half the intruded volume goes towards producing a pulse in the negative direction.)

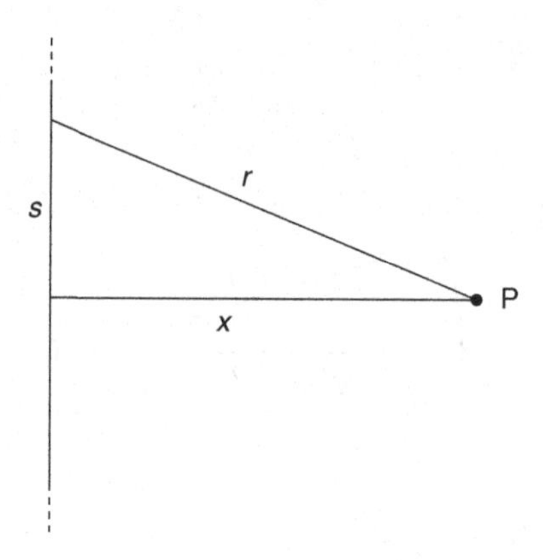

Figure 6.10 Line-source geometry.

In three dimensions, the far-field motion is the first derivative of the near-field motion. In one dimension, the far-field motion is exactly the same as the near-field motion. In two dimensions the far-field motion should therefore be the half-derivative of the near-field motion – but how do you half-differentiate a function? An answer is suggested by the theory of Laplace transformations as

$$\left(\frac{d}{dt}\right)^{\frac{1}{2}} F(t) \stackrel{\text{def}}{=} \frac{1}{\sqrt{\pi}} \int_{-\infty}^{t} \frac{\dot{F}(\tau)}{\sqrt{t-\tau}} d\tau. \tag{6.92}$$

Let us now repeat the explosive sheet calculation for an exploding wire.

Using the geometry shown in Figure 6.10, we have

$$ds = d\left(\sqrt{r^2 - x^2}\right) = \frac{r\,dr}{\sqrt{r^2 - x^2}}, \tag{6.93}$$

and combining the contributions of the two parts of the wire that are the same distance from p, we can write

$$\begin{aligned} \phi(x,t) &= \int_{x}^{\infty} \frac{1}{r} f\left(t - \frac{r}{c}\right) \frac{2r\,dr}{\sqrt{r^2 - x^2}} \\ &= 2\int_{x}^{\infty} f\left(t - \frac{r}{c}\right) \frac{dr}{\sqrt{r^2 - x^2}}, \end{aligned} \tag{6.94}$$

with $f(t) = -\dot{q}(t)/4\pi$, where now $\dot{q}$ is the volume intruded per unit length. We may approximate $r^2 - x^2 \approx 2x(r - x)$ for the near parts of the wire where $r \approx x$, since these make the dominant contribution to the integral. We also set $\tau = t - r/c$, and then have

$$\begin{aligned} \phi(x,t) &= \frac{2c}{\sqrt{2x}} \int_{-\infty}^{(t-x/c)} f(\tau) \frac{dr}{\sqrt{(ct - x) - c\tau}}, \\ &= -\frac{1}{2\pi}\sqrt{\frac{2c}{x}} \int_{-\infty}^{(t-x/c)} \dot{q}(\tau) \frac{d\tau}{\sqrt{(t - x/c) - \tau}}. \end{aligned} \tag{6.95}$$

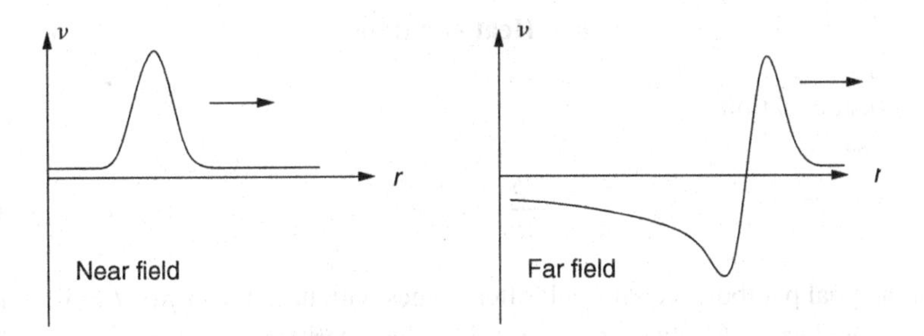

Figure 6.11 In two dimensions the far-field pulse has a long tail.

The far-field velocity is the x gradient of this,

$$v_1(r,t) = \frac{1}{2\pi c}\sqrt{\frac{2c}{x}}\int_{-\infty}^{(t-x/c)} \dddot{q}(\tau)\,\frac{d\tau}{\sqrt{(t-x/c)-\tau}}, \tag{6.96}$$

and is therefore proportional to the $1/2$-derivative of $\dot{q}(t-r/c)$.

A plot of near-field and far-field motions in Figure 6.11 shows how the far-field pulse never completely dies away to zero. This long tail means that one cannot use digital signalling in two dimensions.

Moral tale: One of our colleagues was performing numerical work on earthquake propagation. The source of his waves was a long, deep linear fault, so he used the two-dimensional wave equation. Not wanting to be troubled by the actual creation of the wave pulse, he took as initial data an outgoing finite-width pulse. After a short propagation time his numerical solution appeared to misbehave. New pulses were being emitted from the fault long after the initial one. He wasted several months in a vain attempt to improve the stability of his code before he realized that what he was seeing was real. The lack of a long tail on his pulse meant that it could not have been created by a briefly active line source. The new "unphysical" waves were a consequence of the source striving to suppress the long tail of the initial pulse. *Moral*: Always check that a solution of the form you seek actually exists before you waste your time trying to compute it.

Exercise 6.4: Use the calculus of improper integrals to show that, provided $F(-\infty)=0$, we have

$$\frac{d}{dt}\left(\frac{1}{\sqrt{\pi}}\int_{-\infty}^{t}\frac{\dot{F}(\tau)}{\sqrt{t-\tau}}d\tau\right) = \frac{1}{\sqrt{\pi}}\int_{-\infty}^{t}\frac{\ddot{F}(\tau)}{\sqrt{t-\tau}}d\tau. \tag{6.97}$$

This means that

$$\frac{d}{dt}\left(\frac{d}{dt}\right)^{\frac{1}{2}} F(t) = \left(\frac{d}{dt}\right)^{\frac{1}{2}}\frac{d}{dt}F(t). \tag{6.98}$$

6.4 Heat equation

Fourier's heat equation

$$\frac{\partial \phi}{\partial t} = \kappa \frac{\partial^2 \phi}{\partial x^2} \tag{6.99}$$

is the archetypal parabolic equation. It often comes with initial data $\phi(x, t = 0)$, but this is not Cauchy data, as the curve $t =$ const. is a characteristic.

The heat equation is also known as the *diffusion equation*.

6.4.1 Heat kernel

If we Fourier transform the initial data

$$\phi(x, t = 0) = \int_{-\infty}^{\infty} \frac{dk}{2\pi} \tilde{\phi}(k) e^{ikx}, \tag{6.100}$$

and write

$$\phi(x, t) = \int_{-\infty}^{\infty} \frac{dk}{2\pi} \tilde{\phi}(k, t) e^{ikx}, \tag{6.101}$$

we can plug this into the heat equation and find that

$$\frac{\partial \tilde{\phi}}{\partial t} = -\kappa k^2 \tilde{\phi}. \tag{6.102}$$

Hence,

$$\begin{aligned} \phi(x, t) &= \int_{-\infty}^{\infty} \frac{dk}{2\pi} \tilde{\phi}(k, t) e^{ikx} \\ &= \int_{-\infty}^{\infty} \frac{dk}{2\pi} \tilde{\phi}(k, 0) e^{ikx - \kappa k^2 t}. \end{aligned} \tag{6.103}$$

We may now express $\tilde{\phi}(k, 0)$ in terms of $\phi(x, 0)$ and rearrange the order of integration to get

$$\begin{aligned} \phi(x, t) &= \int_{-\infty}^{\infty} \frac{dk}{2\pi} \left(\int_{-\infty}^{\infty} \phi(\xi, 0) e^{ik\xi} \, d\xi \right) e^{ikx - \kappa k^2 t} \\ &= \int_{-\infty}^{\infty} \left(\int_{-\infty}^{\infty} \frac{dk}{2\pi} e^{ik(x - \xi) - \kappa k^2 t} \right) \phi(\xi, 0) d\xi \\ &= \int_{-\infty}^{\infty} G(x, \xi, t) \phi(\xi, 0) \, d\xi, \end{aligned} \tag{6.104}$$

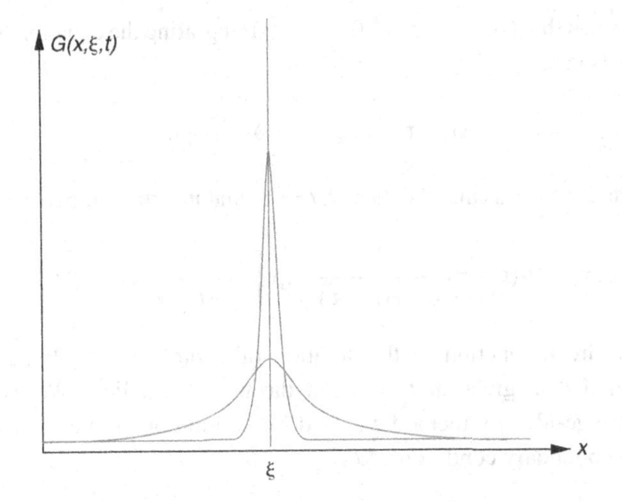

Figure 6.12 The heat kernel at three successive times.

where

$$G(x,\xi,t)=\int_{-\infty}^{\infty}\frac{dk}{2\pi}e^{ik(x-\xi)-\kappa k^2 t}=\frac{1}{\sqrt{4\pi\kappa t}}\exp\left\{-\frac{1}{4\kappa t}(x-\xi)^2\right\}. \tag{6.105}$$

Here, $G(x,\xi,t)$ is the *heat kernel*. It represents the spreading of a unit blob of heat.

As the heat spreads, the total amount of heat, represented by the area under the curve in Figure 6.12, remains constant:

$$\int_{-\infty}^{\infty}\frac{1}{\sqrt{4\pi\kappa t}}\exp\left\{-\frac{1}{4\kappa t}(x-\xi)^2\right\}dx=1. \tag{6.106}$$

The heat kernel possesses a *semigroup property*

$$G(x,\xi,t_1+t_2)=\int_{-\infty}^{\infty}G(x,\eta,t_2)G(\eta,\xi,t_1)d\eta. \tag{6.107}$$

Exercise: Prove this.

6.4.2 Causal Green function

Now we consider the inhomogeneous heat equation

$$\frac{\partial u}{\partial t}-\frac{\partial^2 u}{\partial x^2}=q(x,t), \tag{6.108}$$

with initial data $u(x,0)=u_0(x)$. We define a causal Green function by

$$\left(\frac{\partial}{\partial t}-\frac{\partial^2}{\partial x^2}\right)G(x,t;\xi,\tau)=\delta(x-\xi)\delta(t-\tau) \tag{6.109}$$

and the requirement that $G(x,t;\xi,\tau)=0$ if $t<\tau$. Integrating the equation from $t=\tau-\varepsilon$ to $t=\tau+\varepsilon$ tells us that

$$G(x,\tau+\varepsilon;\xi,\tau)=\delta(x-\xi). \tag{6.110}$$

Taking this delta function as initial data $\phi(x,t=\tau)$ and inserting into (6.104) we read off

$$G(x,t;\xi,\tau)=\theta(t-\tau)\frac{1}{\sqrt{4\pi(t-\tau)}}\exp\left\{-\frac{1}{4(t-\tau)}(x-\xi)^2\right\}. \tag{6.111}$$

We apply this Green function to the solution of a problem involving both a heat source and initial data given at $t=0$ on the entire real line. We exploit a variant of the Lagrange-identity method we used for solving one-dimensional ODEs with inhomogeneous boundary conditions. Let

$$D_{x,t}\equiv\frac{\partial}{\partial t}-\frac{\partial^2}{\partial x^2}, \tag{6.112}$$

and observe that its formal adjoint,

$$D^\dagger_{x,t}\equiv-\frac{\partial}{\partial t}-\frac{\partial^2}{\partial x^2} \tag{6.113}$$

is a "backward" heat-equation operator. The corresponding "backward" Green function

$$G^\dagger(x,t;\xi,\tau)=\theta(\tau-t)\frac{1}{\sqrt{4\pi(\tau-t)}}\exp\left\{-\frac{1}{4(\tau-t)}(x-\xi)^2\right\} \tag{6.114}$$

obeys

$$D^\dagger_{x,t}G^\dagger(x,t;\xi,\tau)=\delta(x-\xi)\delta(t-\tau), \tag{6.115}$$

with adjoint boundary conditions. These make $G^\dagger$ *anti-causal*, in that $G^\dagger(t-\tau)$ vanishes when $t>\tau$. Now we make use of the two-dimensional Lagrange identity

$$\begin{aligned}\int_{-\infty}^{\infty}dx\int_0^T dt\,&\left\{u(x,t)D^\dagger_{x,t}G^\dagger(x,t;\xi,\tau)-\left(D_{x,t}u(x,t)\right)G^\dagger(x,t;\xi,\tau)\right\}\\ &=\int_{-\infty}^{\infty}dx\left\{u(x,0)G^\dagger(x,0;\xi,\tau)\right\}-\int_{-\infty}^{\infty}dx\left\{u(x,T)G^\dagger(x,T;\xi,\tau)\right\}.\end{aligned} \tag{6.116}$$

Assume that (ξ,τ) lies within the region of integration. Then the left-hand side is equal to

$$u(\xi,\tau)-\int_{-\infty}^{\infty}dx\int_0^T dt\left\{q(x,t)G^\dagger(x,t;\xi,\tau)\right\}. \tag{6.117}$$

On the right-hand side, the second integral vanishes because $G^\dagger$ is zero on $t = T$. Thus,

$$u(\xi,\tau) = \int_{-\infty}^{\infty} dx \int_0^T dt \Big\{ q(x,t) G^\dagger(x,t;\xi,\tau) \Big\} + \int_{-\infty}^{\infty} \Big\{ u(x,0) G^\dagger(x,0;\xi,\tau) \Big\} \, dx. \tag{6.118}$$

Rewriting this by using

$$G^\dagger(x,t;\xi,\tau) = G(\xi,\tau;x,t), \tag{6.119}$$

and relabelling $x \leftrightarrow \xi$ and $t \leftrightarrow \tau$, we have

$$u(x,t) = \int_{-\infty}^{\infty} G(x,t;\xi,0) u_0(\xi) \, d\xi + \int_{-\infty}^{\infty} \int_0^t G(x,t;\xi,\tau) q(\xi,\tau) d\xi d\tau. \tag{6.120}$$

Note how the effects of any heat source $q(x,t)$ active prior to the initial-data epoch at $t = 0$ have been subsumed into the evolution of the initial data.

6.4.3 Duhamel's principle

Often, the temperature of the spatial boundary of a region is specified in addition to the initial data. Dealing with this type of problem leads us to a new strategy.

Suppose we are required to solve

$$\frac{\partial u}{\partial t} = \kappa \frac{\partial^2 u}{\partial x^2} \tag{6.121}$$

for the semi-infinite rod shown in Figure 6.13. We are given a specified temperature, $u(0,t) = h(t)$, at the end $x = 0$, and for all other points $x > 0$ we are given an initial condition $u(x,0) = 0$.

We begin by finding a solution $w(x,t)$ that satisfies the heat equation with $w(0,t) = 1$ and initial data $w(x,0) = 0$, $x > 0$. This solution is constructed in Problem 6.14, and is

$$w = \theta(t) \left\{ 1 - \operatorname{erf}\left(\frac{x}{2\sqrt{t}} \right) \right\}. \tag{6.122}$$

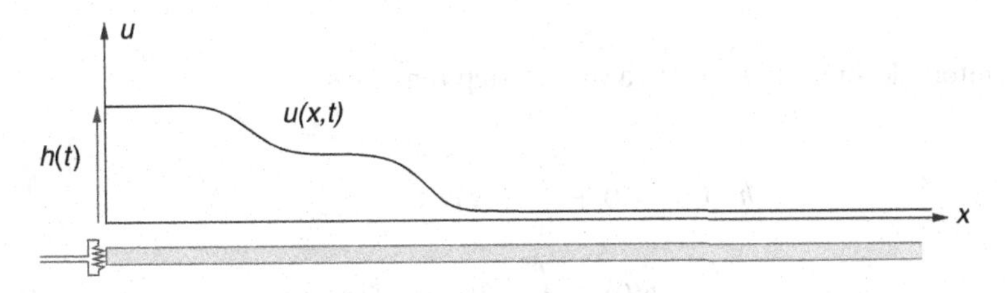

Figure 6.13 Semi-infinite rod heated at one end.

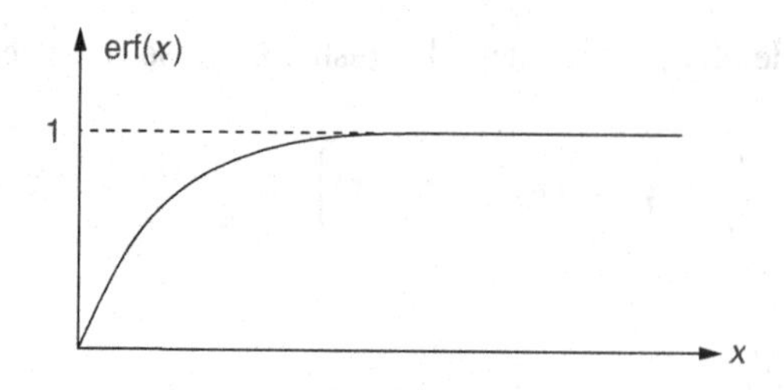

Figure 6.14 Error function.

Here erf (x) is the *error function*

$$\operatorname{erf}(x) = \frac{2}{\sqrt{\pi}} \int_0^x e^{-z^2} dz \tag{6.123}$$

which has the properties that erf $(0) = 0$ and erf $(x) \to 1$ as $x \to \infty$. See Figure 6.14.

If we were given

$$h(t) = h_0 \theta(t - t_0), \tag{6.124}$$

then the desired solution would be

$$u(x, t) = h_0 w(x, t - t_0). \tag{6.125}$$

For a sum

$$h(t) = \sum_n h_n \theta(t - t_n), \tag{6.126}$$

the principle of superposition (i.e. the linearity of the problem) tells us that the solution is the corresponding sum

$$u(x, t) = \sum_n h_n w(x, t - t_n). \tag{6.127}$$

We therefore decompose $h(t)$ into a sum of step functions

$$\begin{aligned} h(t) &= h(0) + \int_0^t \dot{h}(\tau)\, d\tau \\ &= h(0) + \int_0^\infty \theta(t - \tau) \dot{h}(\tau)\, d\tau. \end{aligned} \tag{6.128}$$

It should now be clear that

$$\begin{aligned} u(x,t) &= \int_0^t w(x,t-\tau)\dot{h}(\tau)\,d\tau + h(0)w(x,t) \\ &= -\int_0^t \left(\frac{\partial}{\partial\tau}w(x,t-\tau)\right) h(\tau)\,d\tau \\ &= \int_0^t \left(\frac{\partial}{\partial t}w(x,t-\tau)\right) h(\tau)\,d\tau. \end{aligned} \tag{6.129}$$

This is called *Duhamel's solution*, and the trick of expressing the data as a sum of Heaviside step functions is called Duhamel's principle.

We do not need to be as clever as Duhamel. We could have obtained this result by using the method of images to find a suitable causal Green function for the half-line, and then using the same Lagrange-identity method as before.

6.5 Potential theory

The study of boundary value problems involving the Laplacian is usually known as "potential theory". We seek solutions to these problems in some region Ω, whose boundary we denote by the symbol $\partial\Omega$.

Poisson's equation, $-\nabla^2\chi(\mathbf{r}) = f(\mathbf{r})$, $\mathbf{r} \in \Omega$, and the Laplace equation to which it reduces when $f(\mathbf{r}) \equiv 0$, come along with various boundary conditions, of which the commonest are

$$\begin{aligned} \chi &= g(\mathbf{r}) \quad \text{on} \quad \partial\Omega \qquad \text{(Dirichlet)}, \\ (\mathbf{n}\cdot\nabla)\chi &= g(\mathbf{r}) \quad \text{on} \quad \partial\Omega \qquad \text{(Neumann)}. \end{aligned} \tag{6.130}$$

A function for which $\nabla^2\chi = 0$ in some region Ω is said to be *harmonic* there.

6.5.1 Uniqueness and existence of solutions

We begin by observing that we need to be a little more precise about what it means for a solution to "take" a given value on a boundary. If we ask for a solution to the problem $\nabla^2\varphi = 0$ within $\Omega = \{(x,y) \in \mathbb{R}^2 : x^2 + y^2 < 1\}$ and $\varphi = 1$ on $\partial\Omega$, someone might claim that the function defined by setting $\varphi(x,y) = 0$ for $x^2 + y^2 < 1$ and $\varphi(x,y) = 1$ for $x^2 + y^2 = 1$ does the job – but such a discontinuous "solution" is hardly what we had in mind when we stated the problem. We must interpret the phrase "takes a given value on the boundary" as meaning that the boundary data is the limit, as we approach the boundary, of the solution within Ω.

With this understanding, we assert that a function harmonic in a bounded subset Ω of $\mathbb{R}^n$ is uniquely determined by the values it takes on the boundary of Ω. To see that this is so, suppose that φ_1 and φ_2 both satisfy $\nabla^2\varphi = 0$ in Ω, and coincide on the boundary.

Then $\chi = \varphi_1 - \varphi_2$ obeys $\nabla^2 \chi = 0$ in Ω, and is zero on the boundary. Integrating by parts we find that

$$\int_\Omega |\nabla \chi|^2 d^n r = \int_{\partial\Omega} \chi(\mathbf{n}\cdot\nabla)\chi \, dS = 0. \tag{6.131}$$

Here dS is the element of area on the boundary and $\mathbf{n}$ the outward-directed normal. Now, because the second derivatives exist, the partial derivatives entering into $\nabla\chi$ must be continuous, and so the vanishing of integral of $|\nabla\chi|^2$ tells us that $\nabla\chi$ is zero everywhere within Ω. This means that χ is constant – and because it is zero on the boundary it is zero everywhere.

An almost identical argument shows that if Ω is a bounded *connected* region, and if φ_1 and φ_2 both satisfy $\nabla^2\varphi = 0$ within Ω and take the same values of $(\mathbf{n}\cdot\nabla)\varphi$ on the boundary, then $\varphi_1 = \varphi_2 + \text{const}$. We have therefore shown that, if it exists, the solution of the Dirichlet boundary value problem is unique, and the solution of the Neumann problem is unique up to the addition of an arbitrary constant.

In the Neumann case, with boundary condition $(\mathbf{n}\cdot\nabla)\varphi = g(\mathbf{r})$, integration by parts gives

$$\int_\Omega \nabla^2\varphi \, d^n r = \int_{\partial\Omega} (\mathbf{n}\cdot\nabla)\varphi \, dS = \int_{\partial\Omega} g \, dS, \tag{6.132}$$

and so the boundary data $g(\mathbf{r})$ must satisfy $\int_{\partial\Omega} g\, dS = 0$ if a solution to $\nabla^2\varphi = 0$ is to exist. This is an example of the Fredholm alternative that relates the existence of a non-trivial null space to constraints on the source terms. For the inhomogeneous equation $-\nabla^2\varphi = f$, the Fredholm constraint becomes

$$\int_{\partial\Omega} g\, dS + \int_\Omega f \, d^n r = 0. \tag{6.133}$$

Given that we have satisfied any Fredholm constraint, do solutions to the Dirichlet and Neumann problem always exist? That solutions *should* exist is suggested by physics: the Dirichlet problem corresponds to an electrostatic problem with specified boundary potentials and the Neumann problem corresponds to finding the electric potential within a resistive material with prescribed current sources on the boundary. The Fredholm constraint says that if we drive current into the material, we must let it out somewhere. Surely solutions always exist to these physics problems? In the Dirichlet case we can even make a mathematically plausible argument for existence: we observe that the boundary value problem

$$\begin{aligned} \nabla^2\varphi &= 0, \quad \mathbf{r}\in\Omega \\ \varphi &= f, \quad \mathbf{r}\in\partial\Omega \end{aligned} \tag{6.134}$$

is solved by taking φ to be the χ that minimizes the functional

$$J[\chi] = \int_\Omega |\nabla\chi|^2 d^n r \tag{6.135}$$

over the set of continuously differentiable functions taking the given boundary values. Since $J[\chi]$ is positive, and hence bounded below, it seems intuitively obvious that there must be some function χ for which $J[\chi]$ is a minimum. The appeal of this *Dirichlet principle* argument led even Riemann astray. The fallacy was exposed by Weierstrass who provided counter-examples.

Consider, for example, the problem of finding a function $\varphi(x,y)$ obeying $\nabla^2\varphi = 0$ within the punctured disc $D' = \{(x,y) \in \mathbb{R}^2 : 0 < x^2 + y^2 < 1\}$ with boundary data $\varphi(x,y) = 1$ on the outer boundary at $x^2 + y^2 = 1$ and $\varphi(0,0) = 0$ on the inner boundary at the origin. We substitute the trial functions

$$\chi_\alpha(x,y) = (x^2 + y^2)^\alpha, \quad \alpha > 0, \tag{6.136}$$

all of which satisfy the boundary data, into the positive functional

$$J[\chi] = \int_{D'} |\nabla\chi|^2 \, dxdy \tag{6.137}$$

to find $J[\chi_\alpha] = 2\pi\alpha$. This number can be made as small as we like, and so the infimum of the functional $J[\chi]$ is zero. But if there is a minimizing φ, then $J[\varphi] = 0$ implies that φ is a constant, and a constant cannot satisfy the boundary conditions.

An analogous problem reveals itself in three dimensions when the boundary of Ω has a sharp re-entrant spike that is held at a different potential from the rest of the boundary. In this case we can again find a sequence of trial functions $\chi(\mathbf{r})$ for which $J[\chi]$ becomes arbitrarily small, but the sequence of χ's has no limit satisfying the boundary conditions. The physics argument also fails: if we tried to create a physical realization of this situation, the electric field would become infinite near the spike, and the charge would leak off and thwart our attempts to establish the potential difference. For reasonably smooth boundaries, however, a minimizing function *does* exist.

The Dirichlet–Poisson problem

$$\begin{aligned} -\nabla^2\varphi(\mathbf{r}) &= f(\mathbf{r}), \quad \mathbf{r} \in \Omega, \\ \varphi(\mathbf{r}) &= g(\mathbf{r}), \quad \mathbf{r} \in \partial\Omega, \end{aligned} \tag{6.138}$$

and the Neumann–Poisson problem

$$\begin{aligned} -\nabla^2\varphi(\mathbf{r}) &= f(\mathbf{r}), \quad x \in \Omega, \\ (\mathbf{n}\cdot\nabla)\varphi(\mathbf{r}) &= g(\mathbf{r}), \quad x \in \partial\Omega, \end{aligned}$$

supplemented with the Fredholm constraint

$$\int_\Omega f \, d^n r + \int_{\partial\Omega} g \, dS = 0 \tag{6.139}$$

also have solutions when $\partial\Omega$ is reasonably smooth. For the Neumann–Poisson problem, with the Fredholm constraint as stated, the region Ω must be connected, but its boundary need not be. For example, Ω can be the region between two nested spherical shells.

Exercise 6.5: Why did we insist that the region Ω be connected in our discussion of the Neumann problem? (Hint: how must we modify the Fredholm constraint when Ω consists of two or more disconnected regions?)

Exercise 6.6: *Neumann variational principles.* Let Ω be a bounded and connected three-dimensional region with a smooth boundary. Given a function f defined on Ω and such that $\int_\Omega f\, d^3r = 0$, define the functional

$$J[\chi] = \int_\Omega \left\{ \frac{1}{2}|\nabla\chi|^2 - \chi f \right\} d^3r.$$

Suppose that φ is a solution of the Neumann problem

$$-\nabla^2\varphi(\mathbf{r}) = f(\mathbf{r}), \quad \mathbf{r}\in\Omega,$$
$$(\mathbf{n}\cdot\nabla)\varphi(\mathbf{r}) = 0, \quad \mathbf{r}\in\partial\Omega.$$

Show that

$$J[\chi] = J[\varphi] + \int_\Omega \frac{1}{2}|\nabla(\chi-\varphi)|^2\, d^3r \geq J[\varphi]$$
$$= -\int_\Omega \frac{1}{2}|\nabla\varphi|^2\, d^3r = -\frac{1}{2}\int_\Omega \varphi f\, d^3r.$$

Deduce that φ is determined, up to the addition of a constant, as the function that minimizes $J[\chi]$ over the space of all continuously differentiable χ (and not just over functions satisfying the Neumann boundary condition).

Similarly, for g a function defined on the boundary $\partial\Omega$ and such that $\int_{\partial\Omega} g\, dS = 0$, set

$$K[\chi] = \int_\Omega \frac{1}{2}|\nabla\chi|^2\, d^3r - \int_{\partial\Omega} \chi g\, dS.$$

Now suppose that ϕ is a solution of the Neumann problem

$$-\nabla^2\phi(\mathbf{r}) = 0, \quad \mathbf{r}\in\Omega,$$
$$(\mathbf{n}\cdot\nabla)\phi(\mathbf{r}) = g(\mathbf{r}), \quad \mathbf{r}\in\partial\Omega.$$

Show that

$$K[\chi] = K[\phi] + \int_\Omega \frac{1}{2}|\nabla(\chi-\phi)|^2\, d^3r \geq K[\phi]$$
$$= -\int_\Omega \frac{1}{2}|\nabla\phi|^2\, d^3r = -\frac{1}{2}\int_{\partial\Omega} \phi g\, dS.$$

Deduce that ϕ is determined up to a constant as the function that minimizes $K[\chi]$ over the space of all continuously differentiable χ (and, again, not just over functions satisfying the Neumann boundary condition).

Show that when f and g fail to satisfy the integral conditions required for the existence of the Neumann solution, the corresponding functionals are not bounded below, and so no minimizing function can exist.

Exercise 6.7: Helmholtz decomposition. Let Ω be a bounded connected three-dimensional region with smooth boundary $\partial\Omega$.

(a) Cite the conditions for the existence of a solution to a suitable Neumann problem to show that if $\mathbf{u}$ is a smooth vector field defined in Ω, then there exist a unique solenoidal (i.e having zero divergence) vector field $\mathbf{v}$ with $\mathbf{v}\cdot\mathbf{n}=\mathbf{0}$ on the boundary $\partial\Omega$, and a unique (up to the addition of a constant) scalar field ϕ such that

$$\mathbf{u} = \mathbf{v} + \nabla\phi.$$

Here $\mathbf{n}$ is the outward normal to the (assumed smooth) bounding surface of Ω.

(b) In many cases (but not always) we can write a solenoidal vector field $\mathbf{v}$ as $\mathbf{v} = \operatorname{curl}\mathbf{w}$. Again by appealing to the conditions for existence and uniqueness of a Neumann problem solution, show that if we *can* write $\mathbf{v} = \operatorname{curl}\mathbf{w}$, then $\mathbf{w}$ is not unique, and we can always demand that it obey the conditions $\operatorname{div}\mathbf{w} = 0$ and $\mathbf{w}\cdot\mathbf{n} = 0$.

(c) Appeal to the Helmholtz decomposition of part (a) with $\mathbf{u} \to (\mathbf{v}\cdot\nabla)\mathbf{v}$ to show that in the Euler equation

$$\frac{\partial\mathbf{v}}{\partial t} + (\mathbf{v}\cdot\nabla)\mathbf{v} = -\nabla P, \quad \mathbf{v}\cdot\mathbf{n} = 0 \text{ on } \partial\Omega$$

governing the motion of an incompressible ($\operatorname{div}\mathbf{v} = 0$) fluid the instantaneous flow field $\mathbf{v}(x,y,z,t)$ uniquely determines $\partial\mathbf{v}/\partial t$, and hence the time evolution of the flow. (This observation provides the basis of practical algorithms for computing incompressible flows.)

We can always write the solenoidal field as $\mathbf{v} = \operatorname{curl}\mathbf{w} + \mathbf{h}$, where $\mathbf{h}$ obeys $\nabla^2\mathbf{h} = 0$ with suitable boundary conditions. See Exercise 6.16.

6.5.2 Separation of variables

Cartesian coordinates

When the region of interest is a square or a rectangle, we can solve Laplace boundary problems by separating the Laplace operator in cartesian coordinates. Let

$$\frac{\partial^2\varphi}{\partial x^2} + \frac{\partial^2\varphi}{\partial y^2} = 0, \tag{6.140}$$

and write

$$\varphi = X(x)Y(y), \tag{6.141}$$

so that

$$\frac{1}{X}\frac{\partial^2 X}{\partial x^2} + \frac{1}{Y}\frac{\partial^2 Y}{\partial y^2} = 0. \tag{6.142}$$

Since the first term is a function of x only, and the second of y only, both must be constants and the sum of these constants must be zero. Therefore

$$\begin{aligned} \frac{1}{X}\frac{\partial^2 X}{\partial x^2} &= -k^2, \\ \frac{1}{Y}\frac{\partial^2 Y}{\partial y^2} &= \quad k^2, \end{aligned} \tag{6.143}$$

or, equivalently,

$$\begin{aligned} \frac{\partial^2 X}{\partial x^2} + k^2 X = 0, \\ \frac{\partial^2 Y}{\partial y^2} - k^2 Y = 0. \end{aligned} \tag{6.144}$$

The number that we have, for later convenience, written as k^2 is called a *separation constant*. The solutions are $X = e^{\pm ikx}$ and $Y = e^{\pm ky}$. Thus

$$\varphi = e^{\pm ikx} e^{\pm ky}, \tag{6.145}$$

or a sum of such terms where the allowed k's are determined by the boundary conditions.

How do we know that the separated form $X(x)Y(y)$ captures all possible solutions? We can be confident that we have them all if we can use the separated solutions to solve boundary value problems with arbitrary boundary data.

We can use our separated solutions to construct the unique harmonic function taking given values on the sides of a square of side L shown in Figure 6.15. To see how to do this, consider the four families of functions

$$\begin{aligned} \varphi_{1,n} &= \sqrt{\frac{2}{L}} \frac{1}{\sinh n\pi} \sin \frac{n\pi x}{L} \sinh \frac{n\pi y}{L}, \\ \varphi_{2,n} &= \sqrt{\frac{2}{L}} \frac{1}{\sinh n\pi} \sinh \frac{n\pi x}{L} \sin \frac{n\pi y}{L}, \\ \varphi_{3,n} &= \sqrt{\frac{2}{L}} \frac{1}{\sinh n\pi} \sin \frac{n\pi x}{L} \sinh \frac{n\pi (L-y)}{L}, \\ \varphi_{4,n} &= \sqrt{\frac{2}{L}} \frac{1}{\sinh n\pi} \sinh \frac{n\pi (L-x)}{L} \sin \frac{n\pi y}{L}. \end{aligned} \tag{6.146}$$

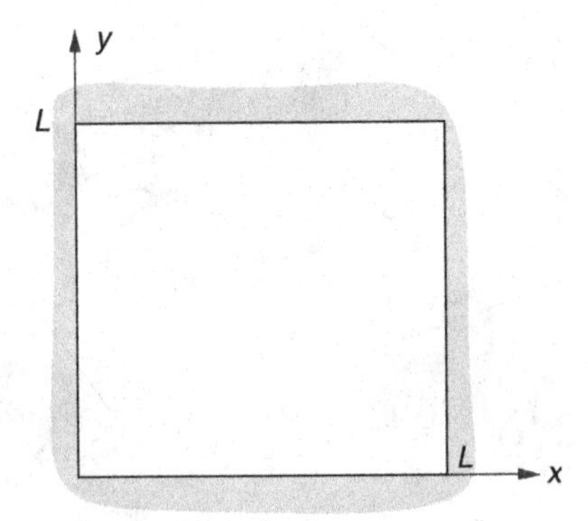

Figure 6.15 Square region.

Each of these comprises solutions to $\nabla^2\varphi = 0$. The family $\varphi_{1,n}(x,y)$ has been constructed so that every member is zero on three sides of the square, but on the side $y = L$ it becomes $\varphi_{1,n}(x,L) = \sqrt{2/L}\sin(n\pi x/L)$. The $\varphi_{1,n}(x,L)$ therefore constitute a complete orthonormal set in terms of which we can expand the boundary data on the side $y = L$. Similarly, the other families are non-zero on only one side, and are complete there. Thus, any boundary data can be expanded in terms of these four function sets, and the solution to the boundary value problem is given by a sum

$$\varphi(x,y) = \sum_{m=1}^{4}\sum_{n=1}^{\infty} a_{m,n}\varphi_{m,n}(x,y). \tag{6.147}$$

The solution to $\nabla^2\varphi = 0$ in the unit square with $\varphi = 1$ on the side $y = 1$ and zero on the other sides is, for example (see Figure 6.16)

$$\varphi(x,y) = \sum_{n=0}^{\infty} \frac{4}{(2n+1)\pi}\frac{1}{\sinh(2n+1)\pi}\sin\Big((2n+1)\pi x\Big)\sinh\Big((2n+1)\pi y\Big). \tag{6.148}$$

For cubes, and higher dimensional hypercubes, we can use similar boundary expansions. For the unit cube in three dimensions we would use

$$\varphi_{1,nm}(x,y,x) = \frac{1}{\sinh\left(\pi\sqrt{n^2+m^2}\right)}\sin(n\pi x)\sin(m\pi y)\sinh\left(\pi z\sqrt{n^2+m^2}\right),$$

to expand the data on the face $z = 1$, together with five other solution families, one for each of the other five faces of the cube.

If some of the boundaries are at infinity, we may need only some of these functions.

Example: Figure 6.17 shows three conducting sheets, each infinite in the z-direction. The central one has width a, and is held at voltage V_0. The outer two extend to infinity

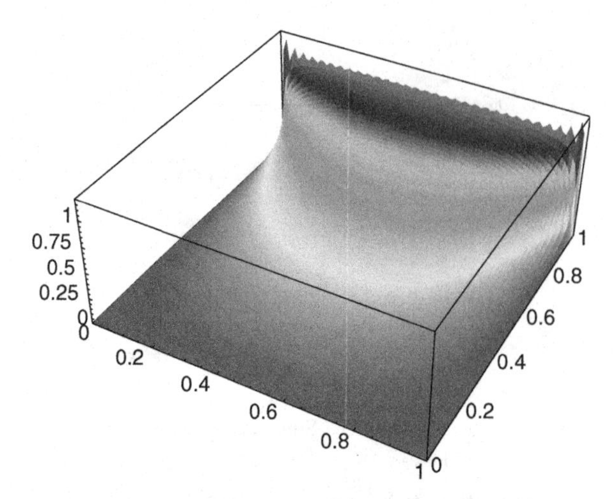

Figure 6.16 Plot of first 30 terms in Equation (6.148).

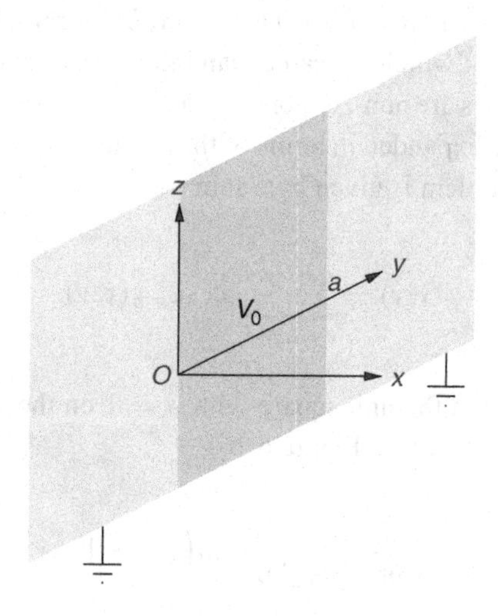

Figure 6.17 Conducting sheets.

also in the y-direction, and are grounded. The resulting potential should tend to zero as $|x|, |y| \to \infty$.

The voltage in the $x = 0$ plane is

$$\varphi(0, y, z) = \int_{-\infty}^{\infty} \frac{dk}{2\pi} a(k) e^{-iky}, \tag{6.149}$$

where

$$a(k) = V_0 \int_{-a/2}^{a/2} e^{iky}\, dy = \frac{2V_0}{k} \sin(ka/2). \tag{6.150}$$

Then, taking into account the boundary condition at large x, the solution to $\nabla^2\varphi = 0$ is

$$\varphi(x,y,z) = \int_{-\infty}^{\infty} \frac{dk}{2\pi} a(k) e^{-iky} e^{-|k||x|}. \tag{6.151}$$

The evaluation of this integral, and finding the charge distribution on the sheets, is left as an exercise.

The Cauchy problem is ill-posed

Although the Laplace equation has no characteristics, the Cauchy data problem is *ill-posed*, meaning that the solution is not a continuous function of the data. To see this, suppose we are given $\nabla^2\varphi = 0$ with Cauchy data on $y = 0$:

$$\begin{aligned} \varphi(x,0) &= 0, \\ \left.\frac{\partial\varphi}{\partial y}\right|_{y=0} &= \varepsilon \sin kx. \end{aligned} \tag{6.152}$$

Then

$$\varphi(x,y) = \frac{\varepsilon}{k} \sin(kx) \sinh(ky). \tag{6.153}$$

Provided k is large enough – even if ε is tiny – the exponential growth of the hyperbolic sine will make this arbitrarily large. Any infinitesimal uncertainty in the high-frequency part of the initial data will be vastly amplified, and the solution, although formally correct, is useless in practice.

Polar coordinates

We can use the separation of variables method in polar coordinates. Here,

$$\nabla^2\chi = \frac{\partial^2\chi}{\partial r^2} + \frac{1}{r}\frac{\partial\chi}{\partial r} + \frac{1}{r^2}\frac{\partial^2\chi}{\partial\theta^2}. \tag{6.154}$$

Set

$$\chi(r,\theta) = R(r)\Theta(\theta). \tag{6.155}$$

Then $\nabla^2\chi = 0$ implies

$$\begin{aligned} 0 &= \frac{r^2}{R}\left(\frac{\partial^2 R}{\partial r^2} + \frac{1}{r}\frac{\partial R}{\partial r}\right) + \frac{1}{\Theta}\frac{\partial^2\Theta}{\partial\theta^2} \\ &= m^2 - m^2, \end{aligned} \tag{6.156}$$

where in the second line we have written the separation constant as m^2. Therefore,

$$\frac{d^2\Theta}{d\theta^2} + m^2\Theta = 0, \tag{6.157}$$

implying that $\Theta = e^{im\theta}$, where m must be an integer if Θ is to be single-valued, and

$$r^2\frac{d^2R}{dr^2} + r\frac{dR}{dr} - m^2R = 0, \tag{6.158}$$

whose solutions are $R = r^{\pm m}$ when $m \neq 0$, and 1 or $\ln r$ when $m = 0$. The general solution is therefore a sum of these

$$\chi = A_0 + B_0 \ln r + \sum_{m\neq 0}(A_m r^{|m|} + B_m r^{-|m|})e^{im\theta}. \tag{6.159}$$

The singular terms, $\ln r$ and $r^{-|m|}$, are not solutions at the origin, and should be omitted when that point is part of the region where $\nabla^2\chi = 0$.

Example: *Dirichlet problem in the interior of the unit circle* (Figure 6.18). Solve $\nabla^2\chi = 0$ in $\Omega = \{\mathbf{r} \in \mathbb{R}^2 : |\mathbf{r}| < 1\}$ with $\chi = f(\theta)$ on $\partial\Omega \equiv \{|\mathbf{r}| = 1\}$.

We expand

$$\chi(r.\theta) = \sum_{m=-\infty}^{\infty} A_m r^{|m|} e^{im\theta}, \tag{6.160}$$

and read off the coefficients from the boundary data as

$$A_m = \frac{1}{2\pi}\int_0^{2\pi} e^{-im\theta'} f(\theta')\, d\theta'. \tag{6.161}$$

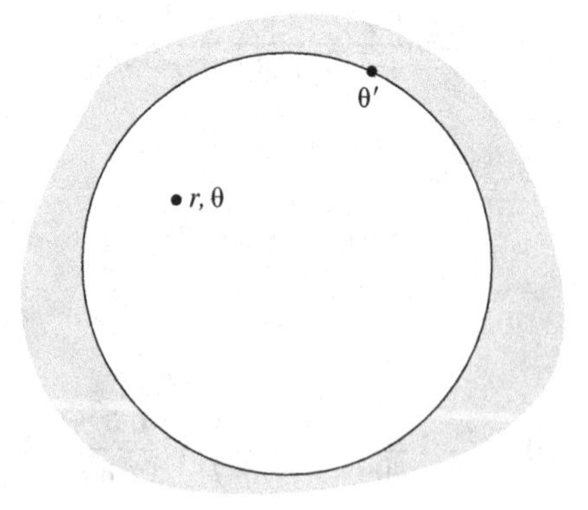

Figure 6.18 Dirichlet problem in the unit circle.

Thus,

$$\chi = \frac{1}{2\pi}\int_0^{2\pi}\left[\sum_{m=-\infty}^{\infty} r^{|m|}e^{im(\theta-\theta')}\right] f(\theta')\,d\theta'. \tag{6.162}$$

We can sum the geometric series

$$\begin{aligned}\sum_{m=-\infty}^{\infty} r^{|m|}e^{im(\theta-\theta')} &= \left(\frac{1}{1-re^{i(\theta-\theta')}} + \frac{re^{-i(\theta-\theta')}}{1-re^{-i(\theta-\theta')}}\right)\\ &= \frac{1-r^2}{1-2r\cos(\theta-\theta')+r^2}.\end{aligned} \tag{6.163}$$

Therefore,

$$\chi(r,\theta) = \frac{1}{2\pi}\int_0^{2\pi}\left(\frac{1-r^2}{1-2r\cos(\theta-\theta')+r^2}\right) f(\theta')\,d\theta'. \tag{6.164}$$

This expression is known as the *Poisson kernel formula*. Observe how the integrand sharpens towards a delta function as r approaches unity, and so ensures that the limiting value of $\chi(r,\theta)$ is consistent with the boundary data.

If we set $r = 0$ in the Poisson formula, we find

$$\chi(0,\theta) = \frac{1}{2\pi}\int_0^{2\pi} f(\theta')\,d\theta'. \tag{6.165}$$

We deduce that if $\nabla^2\chi = 0$ in some domain then the value of χ at a point in the domain is the average of its values on any circle centred on the chosen point and lying wholly in the domain.

This average-value property means that χ can have no local maxima or minima within Ω. The same result holds in $\mathbb{R}^n$, and a formal theorem to this effect can be proved:

Theorem: *(The mean-value theorem for harmonic functions): If χ is harmonic ($\nabla^2\chi = 0$) within the bounded (open, connected) domain $\Omega \in \mathbb{R}^n$, and is continuous on its closure $\overline{\Omega}$, and if $m \le \chi \le M$ on $\partial\Omega$, then $m < \chi < M$ within Ω – unless, that is, $m = M$, when $\chi = m$ is constant.*

Pie-shaped regions

Electrostatics problems involving regions with corners can often be understood by solving Laplace's equation within a pie-shaped region.

Figure 6.19 shows a pie-shaped region of opening angle α and radius R. If the boundary value of the potential is zero on the wedge and non-zero on the boundary arc, we can

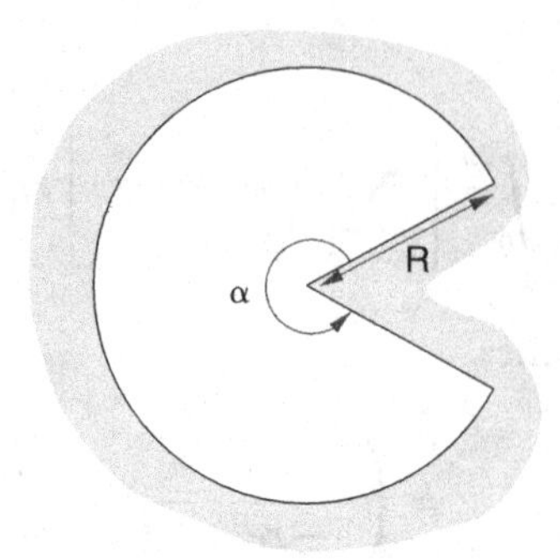

Figure 6.19 A pie-shaped region of opening angle α.

seek solutions as a sum of r, θ separated terms

$$\varphi(r,\theta)=\sum_{n=1}^{\infty} a_n r^{n\pi/\alpha}\sin\left(\frac{n\pi\theta}{\alpha}\right). \tag{6.166}$$

Here the trigonometric function is not 2π periodic, but instead has been constructed so as to make φ vanish at $\theta=0$ and $\theta=\alpha$. These solutions show that close to the edge of a conducting wedge of external opening angle α, the surface charge density σ usually varies as $\sigma(r)\propto r^{\alpha/\pi-1}$.

If we have non-zero boundary data on the edge of the wedge at $\theta=\alpha$, but have $\varphi=0$ on the edge at $\theta=0$ and on the curved arc $r=R$, then the solutions can be expressed as a continuous sum of r, θ separated terms

$$\begin{aligned}\varphi(r,\theta)&=\frac{1}{2i}\int_0^\infty a(\nu)\left(\left(\frac{r}{R}\right)^{i\nu}-\left(\frac{r}{R}\right)^{-i\nu}\right)\frac{\sinh(\nu\theta)}{\sinh(\nu\alpha)}\,d\nu,\\ &=\int_0^\infty a(\nu)\sin[\nu\ln(r/R)]\frac{\sinh(\nu\theta)}{\sinh(\nu\alpha)}\,d\nu.\end{aligned} \tag{6.167}$$

The Mellin sine transformation can be used to compute the coefficient function $a(\nu)$. This transformation lets us write

$$f(r)=\frac{2}{\pi}\int_0^\infty F(\nu)\sin(\nu\ln r)\,d\nu,\quad 0<r<1, \tag{6.168}$$

where

$$F(\nu)=\int_0^1 \sin(\nu\ln r)f(r)\,\frac{dr}{r}. \tag{6.169}$$

The Mellin sine transformation is a disguised version of the Fourier sine transform of functions on $[0,\infty)$. We simply map the positive x-axis onto the interval $(0,1]$ by the change of variables $x=-\ln r$.

Despite its complexity when expressed in terms of these formulae, the simple solution $\varphi(r,\theta) = a\theta$ is often the physically relevant one when the two sides of the wedge are held at different potentials and the potential is allowed to vary on the curved arc.

Example: Consider a pie-shaped region of opening angle π and radius $R = \infty$. This region can be considered to be the upper half-plane. Suppose that we are told that the positive x-axis is held at potential $+1/2$ and the negative x-axis is at potential $-1/2$, and are required to find the potential for positive y. If we separate Laplace's equation in cartesian coordinates and match to the boundary data on the x-axes, we end up with

$$\varphi_{xy}(x,y) = \frac{1}{\pi}\int_0^\infty \frac{1}{k} e^{-ky} \sin(kx)\, dk.$$

On the other hand, the function

$$\varphi_{r\theta}(r,\theta) = \frac{1}{\pi}(\pi/2 - \theta)$$

satisfies both Laplace's equation and the boundary data. At this point we ought to worry that we do not have enough data to determine the solution uniquely – nothing was said in the statement of the problem about the behaviour of φ on the boundary arc at infinity – but a little effort shows that

$$\begin{aligned} \frac{1}{\pi}\int_0^\infty \frac{1}{k} e^{-ky} \sin(kx)\, dk &= \frac{1}{\pi}\tan^{-1}\left(\frac{x}{y}\right), \quad y > 0 \\ &= \frac{1}{\pi}(\pi/2 - \theta), \end{aligned} \tag{6.170}$$

and so the two expressions for $\varphi(x,y)$ are equal.

6.5.3 Eigenfunction expansions

Elliptic operators are the natural analogues of the one-dimensional linear differential operators we studied in earlier chapters.

The operator $L = -\nabla^2$ is formally self-adjoint with respect to the inner product

$$\langle \phi, \chi \rangle = \iint \phi^* \chi \, dxdy. \tag{6.171}$$

This property follows from Green's identity

$$\iint_\Omega \left\{ \phi^*(-\nabla^2 \chi) - (-\nabla^2 \phi)^* \chi \right\} dxdy = \int_{\partial\Omega} \left\{ \phi^*(-\nabla \chi) - (-\nabla \phi)^* \chi \right\} \cdot \mathbf{n} ds \tag{6.172}$$

where $\partial\Omega$ is the boundary of the region Ω and $\mathbf{n}$ is the outward normal on the boundary.

The method of separation of variables also allows us to solve eigenvalue problems involving the Laplace operator. For example, the Dirichlet eigenvalue problem requires us to find the eigenfunctions and eigenvalues of the operator

$$L = -\nabla^2, \qquad \mathcal{D}(L) = \{\phi \in L^2[\Omega] : \phi = 0, \text{ on } \partial\Omega\}. \tag{6.173}$$

Suppose Ω is the rectangle $0 \le x \le L_x$, $0 \le y \le L_y$. The normalized eigenfunctions are

$$\phi_{n,m}(x,y) = \sqrt{\frac{4}{L_x L_y}} \sin\left(\frac{n\pi x}{L_x}\right) \sin\left(\frac{m\pi y}{L_y}\right), \tag{6.174}$$

with eigenvalues

$$\lambda_{n,m} = \left(\frac{n^2\pi^2}{L_x^2}\right) + \left(\frac{m^2\pi^2}{L_y^2}\right). \tag{6.175}$$

The eigenfunctions are orthonormal,

$$\int \phi_{n,m}\phi_{n',m'}\, dxdy = \delta_{nn'}\delta_{mm'}, \tag{6.176}$$

and complete. Thus, any function in $L^2[\Omega]$ can be expanded as

$$f(x,y) = \sum_{m,n=1}^{\infty} A_{nm}\phi_{n,m}(x,y), \tag{6.177}$$

where

$$A_{nm} = \iint \phi_{n,m}(x,y) f(x,y)\, dxdy. \tag{6.178}$$

We can find a complete set of eigenfunctions in product form whenever we can separate the Laplace operator in a system of coordinates ξ_i such that the boundary becomes $\xi_i = \text{const}$. Completeness in the multidimensional space is then guaranteed by the completeness of the eigenfunctions of each one-dimensional differential operator. For other than rectangular coordinates, however, the separated eigenfunctions are not elementary functions.

The Laplacian has a complete set of Dirichlet eigenfunctions in any region, but in general these eigenfunctions cannot be written as separated products of one-dimensional functions.

6.5.4 Green functions

Once we know the eigenfunctions φ_n and eigenvalues λ_n for $-\nabla^2$ in a region Ω, we can write down the Green function as

$$g(\mathbf{r},\mathbf{r}') = \sum_n \frac{1}{\lambda_n}\varphi_n(\mathbf{r})\varphi_n^*(\mathbf{r}').$$

For example, the Green function for the Laplacian in the entire $\mathbb{R}^n$ is given by the sum over eigenfunctions

$$g(\mathbf{r},\mathbf{r}') = \int \frac{d^n k}{(2\pi)^n}\frac{e^{i\mathbf{k}\cdot(\mathbf{r}-\mathbf{r}')}}{k^2}. \tag{6.179}$$

Thus

$$-\nabla_{\mathbf{r}}^2 g(\mathbf{r},\mathbf{r}') = \int \frac{d^n k}{(2\pi)^n} e^{i\mathbf{k}\cdot(\mathbf{r}-\mathbf{r}')} = \delta^n(\mathbf{r}-\mathbf{r}'). \tag{6.180}$$

We can evaluate the integral for any n by using *Schwinger's trick* to turn the integrand into a Gaussian:

$$\begin{aligned} g(\mathbf{r},\mathbf{r}') &= \int_0^\infty ds \int \frac{d^n k}{(2\pi)^n} e^{i\mathbf{k}\cdot(\mathbf{r}-\mathbf{r}')} e^{-sk^2} \\ &= \int_0^\infty ds \left(\sqrt{\frac{\pi}{s}}\right)^n \frac{1}{(2\pi)^n} e^{-\frac{1}{4s}|\mathbf{r}-\mathbf{r}'|^2} \\ &= \frac{1}{2^n \pi^{n/2}} \int_0^\infty dt\, t^{\frac{n}{2}-2} e^{-t|\mathbf{r}-\mathbf{r}'|^2/4} \\ &= \frac{1}{2^n \pi^{n/2}} \Gamma\left(\frac{n}{2}-1\right)\left(\frac{|\mathbf{r}-\mathbf{r}'|^2}{4}\right)^{1-n/2} \\ &= \frac{1}{(n-2)S_{n-1}}\left(\frac{1}{|\mathbf{r}-\mathbf{r}'|}\right)^{n-2}. \end{aligned} \tag{6.181}$$

Here, $\Gamma(x)$ is Euler's Gamma function:

$$\Gamma(x) = \int_0^\infty dt\, t^{x-1} e^{-t}, \tag{6.182}$$

and

$$S_{n-1} = \frac{2\pi^{n/2}}{\Gamma(n/2)} \tag{6.183}$$

is the surface area of the n-dimensional unit ball.

For three dimensions we find

$$g(\mathbf{r},\mathbf{r}') = \frac{1}{4\pi}\frac{1}{|\mathbf{r}-\mathbf{r}'|}, \qquad n=3. \tag{6.184}$$

In two dimensions the Fourier integral is divergent for small k. We may control this divergence by using *dimensional regularization*. We pretend that n is a continuous variable and use

$$\Gamma(x) = \frac{1}{x}\Gamma(x+1) \tag{6.185}$$

together with

$$a^x = e^{a\ln x} = 1 + a\ln x + \cdots \tag{6.186}$$

to examine the behaviour of $g(\mathbf{r},\mathbf{r}')$ near $n=2$:

$$\begin{aligned} g(\mathbf{r},\mathbf{r}') &= \frac{1}{4\pi}\frac{\Gamma(n/2)}{(n/2-1)}\left(1-(n/2-1)\ln(\pi|\mathbf{r}-\mathbf{r}'|^2)+O\left[(n-2)^2\right]\right)\\ &= \frac{1}{4\pi}\left(\frac{1}{n/2-1}-2\ln|\mathbf{r}-\mathbf{r}'|-\ln\pi-\gamma+\cdots\right). \end{aligned} \tag{6.187}$$

Here $\gamma = -\Gamma'(1) = 0.57721\ldots$ is the *Euler–Mascheroni constant.* Although the pole $1/(n-2)$ blows up at $n=2$, it is independent of position. We simply absorb it, and the $-\ln\pi-\gamma$, into an undetermined additive constant. Once we have done this, the limit $n\to 2$ can be taken and we find

$$g(\mathbf{r},\mathbf{r}') = -\frac{1}{2\pi}\ln|\mathbf{r}-\mathbf{r}'| + \text{const.}, \qquad n=2. \tag{6.188}$$

The constant does not affect the Green-function property, so we can choose any convenient value for it.

Although we have managed to sweep the small-k divergence of the Fourier integral under a rug, the hidden infinity still has the capacity to cause problems. The Green function in $\mathbb{R}^3$ allows us to solve for $\varphi(\mathbf{r})$ in the equation

$$-\nabla^2\varphi = q(\mathbf{r}),$$

with the boundary condition $\varphi(\mathbf{r})\to 0$ as $|\mathbf{r}|\to\infty$, as

$$\varphi(\mathbf{r}) = \int g(\mathbf{r},\mathbf{r}')q(\mathbf{r}')\,d^3r.$$

In two dimensions, however we try to adjust the arbitrary constant in (6.188), the divergence of the logarithm at infinity means that there can be no solution to the corresponding boundary-value problem unless $\int q(\mathbf{r})\, d^3r = 0$. This is not a Fredholm-alternative constraint because once the constraint is satisfied the solution is unique. The two-dimensional problem is therefore pathological from the viewpoint of Fredholm theory. This pathology is of the same character as the non-existence of solutions to the three-dimensional Dirichlet boundary value problem with boundary spikes. The Fredholm alternative applies, in general, only to operators possessing a discrete spectrum.

Exercise 6.8: Evaluate our formula for the $\mathbb{R}^n$ Laplace Green function,

$$g(\mathbf{r}, \mathbf{r}') = \frac{1}{(n-2)S_{n-1}|\mathbf{r} - \mathbf{r}'|^{n-2}}$$

with $S_{n-1} = 2\pi^{n/2}/\Gamma(n/2)$, for the case $n = 1$. Show that the resulting expression for $g(x, x')$ is not divergent, and obeys

$$-\frac{d^2}{dx^2} g(x, x') = \delta(x - x').$$

Our formula therefore makes sense as a Green function – even though the original integral (6.179) is linearly divergent at $k = 0$! We must defer an explanation of this miracle until we discuss *analytic continuation* in the context of complex analysis.
(Hint: recall that $\Gamma(1/2) = \sqrt{\pi}$).

6.5.5 Boundary value problems

We now look at how the Green function can be used to solve the interior Dirichlet boundary-value problem in regions where the method of separation of variables is not available. Figure 6.20 shows a bounded region Ω possessing a smooth boundary $\partial\Omega$.

We wish to solve $-\nabla^2\varphi = q(\mathbf{r})$ for $\mathbf{r} \in \Omega$ and with $\varphi(\mathbf{r}) = f(\mathbf{r})$ for $\mathbf{r} \in \partial\Omega$. Suppose we have found a Green function that obeys

$$-\nabla^2_{\mathbf{r}} g(\mathbf{r}, \mathbf{r}') = \delta^n(\mathbf{r} - \mathbf{r}'), \quad \mathbf{r}, \mathbf{r}' \in \Omega, \qquad g(\mathbf{r}, \mathbf{r}') = 0, \quad \mathbf{r} \in \partial\Omega. \tag{6.189}$$

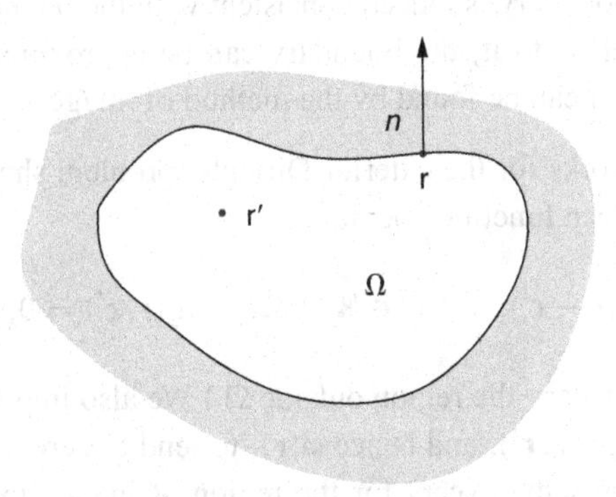

Figure 6.20 Interior Dirichlet problem.

We first show that $g(\mathbf{r},\mathbf{r}') = g(\mathbf{r}',\mathbf{r})$ by the same methods we used for one-dimensional self-adjoint operators. Next we follow the strategy that we used for one-dimensional inhomogeneous differential equations: we use Lagrange's identity (in this context called Green's theorem) to write

$$\int_\Omega d^n r \left\{ g(\mathbf{r},\mathbf{r}')\nabla_\mathbf{r}^2\varphi(\mathbf{r}) - \varphi(\mathbf{r})\nabla_\mathbf{r}^2 g(\mathbf{r},\mathbf{r}')\right\}$$
$$= \int_{\partial\Omega} d\mathbf{S}_\mathbf{r} \cdot \{g(\mathbf{r},\mathbf{r}')\nabla_\mathbf{r}\varphi(\mathbf{r}) - \varphi(\mathbf{r})\nabla_\mathbf{r} g(\mathbf{r},\mathbf{r}')\}, \tag{6.190}$$

where $d\mathbf{S}_\mathbf{r} = \mathbf{n}\, dS_\mathbf{r}$, with $\mathbf{n}$ the outward normal to $\partial\Omega$ at the point $\mathbf{r}$. The left-hand side is

$$\begin{aligned} \text{LHS} &= \int_\Omega d^n r\{-g(\mathbf{r},\mathbf{r}')q(\mathbf{r}) + \varphi(\mathbf{r})\delta^n(\mathbf{r}-\mathbf{r}')\}, \\ &= -\int_\Omega d^n r\, g(\mathbf{r},\mathbf{r}')\, q(\mathbf{r}) + \varphi(\mathbf{r}'), \\ &= -\int_\Omega d^n r\, g(\mathbf{r}',\mathbf{r})\, q(\mathbf{r}) + \varphi(\mathbf{r}'). \end{aligned} \tag{6.191}$$

On the right-hand side, the boundary condition on $g(\mathbf{r},\mathbf{r}')$ makes the first term zero, so

$$\text{RHS} = -\int_{\partial\Omega} dS_\mathbf{r} f(\mathbf{r})(\mathbf{n}\cdot\nabla_\mathbf{r})g(\mathbf{r},\mathbf{r}'). \tag{6.192}$$

Therefore,

$$\varphi(\mathbf{r}') = \int_\Omega g(\mathbf{r}',\mathbf{r})\, q(\mathbf{r})\, d^n r - \int_{\partial\Omega} f(\mathbf{r})(\mathbf{n}\cdot\nabla_\mathbf{r})g(\mathbf{r},\mathbf{r}')\, dS_\mathbf{r}. \tag{6.193}$$

In the language of Chapter 3, the first term is a particular integral and the second (the boundary integral term) is the complementary function.

Exercise 6.9: Assume that the boundary is a smooth surface. Show that the limit of $\varphi(\mathbf{r}')$ as $\mathbf{r}'$ approaches the boundary is indeed consistent with the boundary data $f(\mathbf{r}')$. (Hint: when $\mathbf{r}$, $\mathbf{r}'$ are very close to it, the boundary can be approximated by a straight-line segment, and so $g(\mathbf{r},\mathbf{r}')$ can be found by the method of images.)

A similar method works for the exterior Dirichlet problem shown in Figure 6.21. In this case we seek a Green function obeying

$$-\nabla_\mathbf{r}^2 g(\mathbf{r},\mathbf{r}') = \delta^n(\mathbf{r}-\mathbf{r}'), \quad \mathbf{r},\mathbf{r}' \in \mathbb{R}^n \setminus \Omega \qquad g(\mathbf{r},\mathbf{r}') = 0, \quad \mathbf{r} \in \partial\Omega. \tag{6.194}$$

(The notation $\mathbb{R}^n \setminus \Omega$ means the region outside Ω.) We also impose a further boundary condition by requiring $g(\mathbf{r},\mathbf{r}')$, and hence $\varphi(\mathbf{r})$, to tend to zero as $|\mathbf{r}| \to \infty$. The final formula for $\varphi(\mathbf{r})$ is the same except for the region of integration and the sign of the boundary term.

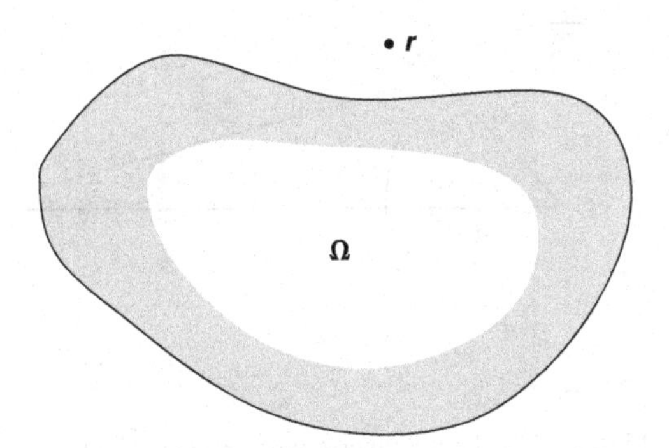

Figure 6.21 Exterior Dirichlet problem.

The hard part of both the interior and exterior problems is to find the Green function for the given domain.

Exercise 6.10: Suppose that $\varphi(x,y)$ is harmonic in the half-plane $y > 0$, tends to zero as $y \to \infty$ and takes the values $f(x)$ on the boundary $y = 0$. Show that

$$\varphi(x,y) = \frac{1}{\pi}\int_{-\infty}^{\infty} \frac{y}{(x-x')^2 + y^2} f(x')\, dx', \quad y > 0.$$

Deduce that the "energy" functional

$$S[f] \stackrel{\text{def}}{=} \frac{1}{2}\int_{y>0} |\nabla\varphi|^2\, dxdy \equiv -\frac{1}{2}\int_{-\infty}^{\infty} f(x) \left.\frac{\partial\varphi}{\partial y}\right|_{y=0} dx$$

can be expressed as

$$S[f] = \frac{1}{4\pi}\int_{-\infty}^{\infty}\int_{-\infty}^{\infty} \left\{\frac{f(x) - f(x')}{x - x'}\right\}^2 dx'dx.$$

The non-local functional $S[f]$ appears in the quantum version of the Caldeira–Leggett model. See also Exercise 2.24.

Method of images

When $\partial\Omega$ is a sphere or a circle we can find the Dirichlet Green functions for the region Ω by using the *method of images*.

Figure 6.22 shows a circle of radius R. Given a point B outside the circle, and a point X on the circle, we construct A inside and on the line OB, so that ∠OBX = ∠OXA. We now observe that △XOA is similar to △BOX, and so

$$\frac{\text{OA}}{\text{OX}} = \frac{\text{OX}}{\text{OB}}. \tag{6.195}$$

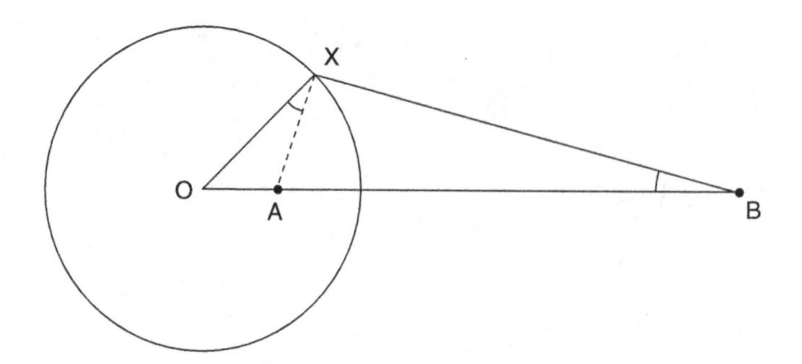

Figure 6.22 Points inverse with respect to a circle.

Thus, OA × OB = $(\mathrm{OX})^2 \equiv R^2$. The points A and B are therefore *mutually inverse* with respect to the circle. In particular, the point A does not depend on which point X was chosen.

Now let AX$= r_i$, BX$= r_0$ and OB$= B$. Then, using similar triangles again, we have

$$\frac{\mathrm{AX}}{\mathrm{OX}} = \frac{\mathrm{BX}}{\mathrm{OB}}, \tag{6.196}$$

or

$$\frac{R}{r_i} = \frac{B}{r_0}, \tag{6.197}$$

and so

$$\frac{1}{r_i}\left(\frac{R}{B}\right) - \frac{1}{r_0} = 0. \tag{6.198}$$

Interpreting the figure as a slice through the centre of a sphere of radius R, we see that if we put a unit charge at B, then the insertion of an *image charge* of magnitude $q = -R/B$ at A serves to keep the entire surface of the sphere at zero potential.

Thus, in three dimensions, and with Ω the region exterior to the sphere, the Dirichlet Green function is

$$g_\Omega(\mathbf{r}, \mathbf{r}_\mathrm{B}) = \frac{1}{4\pi}\left(\frac{1}{|\mathbf{r} - \mathbf{r}_\mathrm{B}|} - \left(\frac{R}{|\mathbf{r}_\mathrm{B}|}\right)\frac{1}{|\mathbf{r} - \mathbf{r}_\mathrm{A}|}\right). \tag{6.199}$$

In two dimensions, we find similarly that

$$g_\Omega(\mathbf{r}, \mathbf{r}_\mathrm{B}) = -\frac{1}{2\pi}\Big(\ln|\mathbf{r} - \mathbf{r}_\mathrm{B}| - \ln|\mathbf{r} - \mathbf{r}_\mathrm{A}| - \ln\left(|\mathbf{r}_\mathrm{B}|/R\right)\Big), \tag{6.200}$$

has $g_\Omega(\mathbf{r}, \mathbf{r}_\mathrm{B}) = 0$ for $\mathbf{r}$ on the circle. Thus, this is the Dirichlet Green function for Ω, the region exterior to the circle.

We can use the same method to construct the interior Green functions for the sphere and circle.

6.5.6 *Kirchhoff* vs. *Huygens*

Even if we do not have a Green function tailored for the specific region in which we are interested, we can still use the whole-space Green function to convert the differential equation into an *integral equation*, and so make progress. An example of this technique is provided by Kirchhoff's partial justification of Huygens' construction.

The Green function $G(\mathbf{r}, \mathbf{r}')$ for the elliptic Helmholtz equation

$$(-\nabla^2 + \kappa^2)G(\mathbf{r}, \mathbf{r}') = \delta^3(\mathbf{r} - \mathbf{r}') \tag{6.201}$$

in $\mathbb{R}^3$ is given by

$$\int \frac{d^3k}{(2\pi)^3} \frac{e^{i\mathbf{k}\cdot(\mathbf{r}-\mathbf{r}')}}{k^2 + \kappa^2} = \frac{1}{4\pi|\mathbf{r} - \mathbf{r}'|} e^{-\kappa|\mathbf{r}-\mathbf{r}'|}. \tag{6.202}$$

Exercise 6.11: Perform the k integration and confirm this.

For solutions of the wave equation with $e^{-i\omega t}$ time dependence, we want a Green function such that

$$\left[-\nabla^2 - \left(\frac{\omega^2}{c^2}\right)\right] G(\mathbf{r}, \mathbf{r}') = \delta^3(\mathbf{r} - \mathbf{r}'), \tag{6.203}$$

and so we have to take κ^2 negative. We therefore have two possible Green functions

$$G_{\pm}(\mathbf{r}, \mathbf{r}') = \frac{1}{4\pi|\mathbf{r} - \mathbf{r}'|} e^{\pm ik|\mathbf{r}-\mathbf{r}'|}, \tag{6.204}$$

where $k = |\omega|/c$. These correspond to taking the real part of κ^2 negative, but giving it an infinitesimal imaginary part, as we did when discussing resolvent operators in Chapter 5. If we want outgoing waves, we must take $G \equiv G_+$.

Now suppose we want to solve

$$(\nabla^2 + k^2)\psi = 0 \tag{6.205}$$

in an arbitrary region Ω. As before, we use Green's theorem to write

$$\int_\Omega \left\{G(\mathbf{r}, \mathbf{r}')(\nabla^2_\mathbf{r} + k^2)\psi(\mathbf{r}) - \psi(\mathbf{r})(\nabla^2_\mathbf{r} + k^2)G(\mathbf{r}, \mathbf{r}')\right\} d^nx$$
$$= \int_{\partial\Omega} \{G(\mathbf{r}, \mathbf{r}')\nabla_\mathbf{r}\psi(\mathbf{r}) - \psi(\mathbf{r})\nabla_\mathbf{r}G(\mathbf{r}, \mathbf{r}')\} \cdot d\mathbf{S}_\mathbf{r} \tag{6.206}$$

where $d\mathbf{S}_\mathbf{r} = \mathbf{n}\, dS_\mathbf{r}$, with $\mathbf{n}$ the outward normal to $\partial\Omega$ at the point $\mathbf{r}$. The left-hand side is

$$\int_\Omega \psi(\mathbf{r})\delta^n(\mathbf{r} - \mathbf{r}')\, d^nx = \begin{cases} \psi(\mathbf{r}'), & \mathbf{r}' \in \Omega \\ 0, & \mathbf{r}' \notin \Omega \end{cases} \tag{6.207}$$

and so

$$\psi(\mathbf{r}') = \int_{\partial\Omega} \{G(\mathbf{r},\mathbf{r}')(\mathbf{n}\cdot\nabla_x)\psi(\mathbf{r}) - \psi(\mathbf{r})(\mathbf{n}\cdot\nabla_{\mathbf{r}})G(\mathbf{r},\mathbf{r}')\}\, dS_{\mathbf{r}}, \quad \mathbf{r}' \in \Omega. \quad (6.208)$$

This must *not* be thought of as a solution to the wave equation in terms of an integral over the boundary, analogous to the solution (6.193) of the Dirichlet problem that we found in the last section. Here, unlike that earlier case, $G(\mathbf{r},\mathbf{r}')$ knows nothing of the boundary $\partial\Omega$, and so both terms in the surface integral contribute to ψ. We therefore have a formula for $\psi(\mathbf{r})$ in the interior in terms of both Dirichlet *and* Neumann data on the boundary $\partial\Omega$, and giving *both* over-prescribes the problem. If we take arbitrary values for ψ and $(\mathbf{n}\cdot\nabla)\psi$ on the boundary, and plug them into (6.208) so as to compute $\psi(\mathbf{r})$ within Ω then there is no reason for the resulting $\psi(\mathbf{r})$ to reproduce, as $\mathbf{r}$ approaches the boundary, the values ψ and $(\mathbf{n}\cdot\nabla)\psi$ appearing in the integral. If we demand that the output $\psi(\mathbf{r})$ *does* reproduce the input boundary data, then this is equivalent to demanding that the boundary data come from a solution of the differential equation in a region encompassing Ω.

The mathematical inconsistency of assuming arbitrary boundary data notwithstanding, this is exactly what we do when we follow Kirchhoff and use (6.208) to provide a justification of Huygens' construction as used in optics. Consider the problem of a plane wave, $\psi = e^{ikx}$, incident on a screen from the left and passing though the aperture labelled AB in Figure 6.23.

We take as the region Ω everything to the right of the obstacle. The Kirchhoff approximation consists of assuming that the values of ψ and $(\mathbf{n}\cdot\nabla)\psi$ on the surface AB are e^{ikx} and $-ike^{ikx}$, the same as they would be if the obstacle were not there, and that they are identically zero on all other parts of the boundary. In other words, we completely ignore any scattering by the material in which the aperture resides. We can then use our

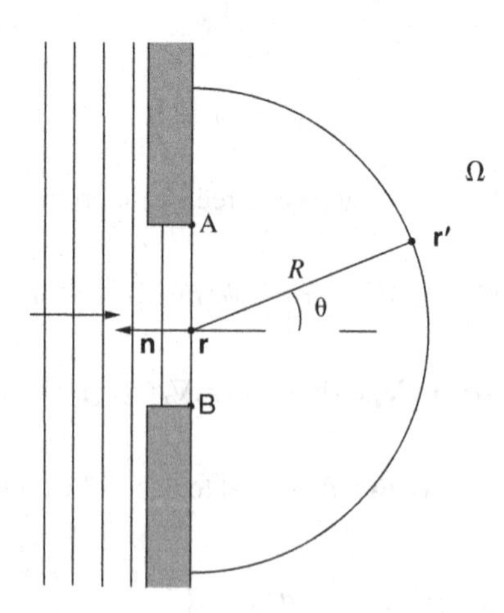

Figure 6.23 Huygens' construction.

formula to estimate ψ in the region to the right of the aperture. If we further set

$$\nabla_{\mathbf{r}} G(\mathbf{r},\mathbf{r}') \approx ik \frac{(\mathbf{r}-\mathbf{r}')}{|\mathbf{r}-\mathbf{r}'|^2} e^{ik|\mathbf{r}-\mathbf{r}'|}, \tag{6.209}$$

which is a good approximation provided we are more than a few wavelengths away from the aperture, we find

$$\psi(\mathbf{r}') \approx \frac{k}{4\pi i} \int_{\text{aperture}} \frac{e^{ik|\mathbf{r}-\mathbf{r}'|}}{|\mathbf{r}-\mathbf{r}'|} (1+\cos\theta) dS_{\mathbf{r}}. \tag{6.210}$$

Thus, each part of the wavefront on the surface AB acts as a source for the diffracted wave in Ω.

This result, although still an approximation, provides two substantial improvements to the naïve form of Huygens' construction as presented in elementary courses:

(i) There is factor of $(1+\cos\theta)$ which suppresses backward propagating waves. The traditional exposition of Huygens construction takes no notice of which way the wave is going, and so provides no explanation as to why a wavefront does not act as a source for a backward wave.
(ii) There is a factor of $i^{-1} = e^{-i\pi/2}$ which corrects a 90° error in the phase made by the naïve Huygens construction. For two-dimensional slit geometry we must use the more complicated two-dimensional Green function (it is a Bessel function), and this provides an $e^{-i\pi/4}$ factor which corrects for the 45° phase error that is manifest in the Cornu spiral of Fresnel diffraction.

For this reason the Kirchhoff approximation is widely used.

Problem 6.12: Use the method of images to construct (i) the Dirichlet, and (ii) the Neumann, Green function for the region Ω, consisting of everything to the right of the screen. Use your Green functions to write the solution to the diffraction problem in this region (a) in terms of the values of ψ on the aperture surface AB, and (b) in terms of the values of $(\mathbf{n}\cdot\nabla)\psi$ on the aperture surface. In each case, assume that the boundary data are identically zero on the dark side of the screen. Your expressions should coincide with the *Rayleigh–Sommerfeld diffraction integrals* of the first and second kind, respectively.[3] Explore the differences between the predictions of these two formulæ and that of Kirchhoff for the case of the diffraction of a plane wave incident on the aperture from the left.

[3] M. Born, E. Wolf, *Principles of Optics* Section 8.11.

6.6 Further exercises and problems

Problem 6.13: *Critical mass*. An infinite slab of fissile material has thickness L. The neutron density $n(x)$ in the material obeys the equation

$$\frac{\partial n}{\partial t} = D\frac{\partial^2 n}{\partial x^2} + \lambda n + \mu,$$

where $n(x,t)$ is zero at the surface of the slab at $x = 0, L$. Here, D is the neutron diffusion constant, the term λn describes the creation of new neutrons by induced fission and the constant μ is the rate of production per unit volume of neutrons by spontaneous fission.

(a) Expand $n(x,t)$ as a series,

$$n(x,t) = \sum_m a_m(t)\varphi_m(x),$$

where the $\varphi_m(x)$ are a complete set of functions you think suitable for solving the problem.

(b) Find an explicit expression for the coefficients $a_m(t)$ in terms of their intial values $a_m(0)$.

(c) Determine the critical thickness L_{crit} above which the slab will explode.

(d) Assuming that $L < L_{\text{crit}}$, find the equilibrium distribution $n_{\text{eq}}(x)$ of neutrons in the slab. (You may either sum your series expansion to get an explicit closed-form answer, or use another (Green function?) method.)

Problem 6.14: *Semi-infinite rod*. Consider the heat equation

$$\frac{\partial \theta}{\partial t} = D\nabla^2\theta, \quad 0 < x < \infty,$$

with the temperature $\theta(x,t)$ obeying the initial condition $\theta(x,0) = \theta_0$ for $0 < x < \infty$, and the boundary condition $\theta(0,t) = 0$.

(a) Show that the boundary condition at $x = 0$ may be satisfied at all times by introducing a suitable mirror image of the initial data in the region $-\infty < x < 0$, and then applying the heat kernel for the entire real line to this extended initial data. Show that the resulting solution of the semi-infinite rod problem can be expressed in terms of the *error function*

$$\text{erf}\,(x) \stackrel{\text{def}}{=} \frac{2}{\sqrt{\pi}}\int_0^x e^{-\xi^2}\,d\xi,$$

as

$$\theta(x,t) = \theta_0\,\text{erf}\left(\frac{x}{\sqrt{4t}}\right).$$

(b) Solve the same problem by using a Fourier integral expansion in terms of $\sin kx$ on the half-line $0 < x < \infty$ and obtaining the time evolution of the Fourier coefficients. Invert the transform and show that your answer reduces to that of part (a). (Hint: replace the initial condition by $\theta(x, 0) = \theta_0 e^{-\epsilon x}$, so that the Fourier transform converges, and then take the limit $\epsilon \to 0$ at the end of your calculation.)

Exercise 6.15: *Seasonal heat waves*. Suppose that the measured temperature of the air above the arctic permafrost at time t is expressed as a Fourier series

$$\theta(t) = \theta_0 + \sum_{n=1}^{\infty} \theta_n \cos n\omega t,$$

where the period $T = 2\pi/\omega$ is one year. Solve the heat equation for the soil temperature,

$$\frac{\partial \theta}{\partial t} = \kappa \frac{\partial^2 \theta}{\partial z^2}, \qquad 0 < z < \infty$$

with this boundary condition, and find the temperature $\theta(z, t)$ at a depth z below the surface as a function of time. Observe that the subsurface temperature fluctuates with the same period as that of the air, but with a phase lag that depends on the depth. Also observe that the longest-period temperature fluctuations penetrate the deepest into the ground. (Hint: for each Fourier component, write θ as $\mathrm{Re}[A_n(z) \exp in\omega t]$, where A_n is a complex function of z.)

The next problem is an illustration of a Dirichlet principle.

Exercise 6.16: *Helmholtz–Hodge decomposition.* Given a three-dimensional region Ω with smooth boundary $\partial\Omega$, introduce the real Hilbert space $L^2_{\mathrm{vec}}(\Omega)$ of finite-norm vector fields, with inner product

$$\langle \mathbf{u}, \mathbf{v} \rangle = \int_\Omega \mathbf{u} \cdot \mathbf{v} \, d^3x.$$

Consider the spaces $\mathcal{L} = \{\mathbf{v} : \mathbf{v} = \nabla\phi\}$ and $\mathcal{T} = \{\mathbf{v} : \mathbf{v} = \operatorname{curl} \mathbf{w}\}$ consisting of vector fields in $L^2_{\mathrm{vec}}(\Omega)$ that can be written as gradients and curls, respectively. (Strictly speaking, we should consider the completions of these spaces.)

(a) Show that if we demand that either (or both) of ϕ and the tangential component of $\mathbf{w}$ vanish on $\partial\Omega$, then the two spaces $\mathcal{L}$ and $\mathcal{T}$ are mutually orthogonal with respect to the $L^2_{\mathrm{vec}}(\Omega)$ inner product.

Let $\mathbf{u} \in L^2_{\mathrm{vec}}(\Omega)$. We will try to express $\mathbf{u}$ as the sum of a gradient and a curl by seeking to make the distance functional

$$\begin{aligned} F_{\mathbf{u}}[\phi, \mathbf{w}] &= \|\mathbf{u} - \nabla\phi - \operatorname{curl} \mathbf{w}\|^2 \\ &\stackrel{\text{def}}{=} \int_\Omega |\mathbf{u} - \nabla\phi - \operatorname{curl} \mathbf{w}|^2 \, d^3x \end{aligned}$$

equal to zero.

(b) Show that if we find a $\mathbf{w}$ and ϕ that minimize $F_{\mathbf{u}}[\phi, \mathbf{w}]$, then the residual vector field

$$\mathbf{h} \stackrel{\text{def}}{=} \mathbf{u} - \nabla\phi - \text{curl}\,\mathbf{w}$$

obeys $\text{curl}\,\mathbf{h} = 0$ and $\text{div}\,\mathbf{h} = 0$, together with boundary conditions determined by the constraints imposed on ϕ and $\mathbf{w}$:
 (i) If ϕ is unconstrained on $\partial\Omega$, but the tangential boundary component of $\mathbf{w}$ is required to vanish, then the component of $\mathbf{h}$ normal to the boundary must be zero.
 (ii) If $\phi = 0$ on $\partial\Omega$, but the tangential boundary component of $\mathbf{w}$ is unconstrained, then the tangential boundary component of $\mathbf{h}$ must be zero.
 (iii) If $\phi = 0$ on $\partial\Omega$ and also the tangential boundary component of $\mathbf{w}$ is required to vanish, then $\mathbf{h}$ need satisfy no boundary condition.

(c) Assuming that we can find suitable minimizing ϕ and $\mathbf{w}$, deduce that under each of the three boundary conditions of the previous part, we have a *Helmholtz–Hodge decomposition*

$$\mathbf{u} = \nabla\phi + \text{curl}\,\mathbf{w} + \mathbf{h}$$

into unique parts that are mutually $L^2_{\text{vec}}(\Omega)$ orthogonal. Observe that the residual vector field $\mathbf{h}$ is *harmonic* – i.e. it satisfies the equation $\nabla^2\mathbf{h} = 0$, where

$$\nabla^2\mathbf{h} \stackrel{\text{def}}{=} \nabla(\text{div}\,\mathbf{h}) - \text{curl}\,(\text{curl}\,\mathbf{h})$$

is the *vector Laplacian* acting on $\mathbf{h}$.

If $\mathbf{u}$ is sufficiently smooth, there will exist ϕ and $\mathbf{w}$ that minimize the distance $\|\mathbf{u} - \nabla\phi - \text{curl}\,\mathbf{w}\|$ and satisfy the boundary conditions. Whether or not $\mathbf{h}$ is needed in the decomposition is another matter. It depends both on how we constrain ϕ and $\mathbf{w}$, and on the topology of Ω. At issue is whether or not the boundary conditions imposed on $\mathbf{h}$ are sufficient to force it to be zero. If Ω is the interior of a torus, for example, then $\mathbf{h}$ can be non-zero whenever its tangential component is unconstrained.

The Helmholtz–Hodge decomposition is closely related to the vector-field eigenvalue problems commonly met with in electromagnetism or elasticity. The next few exercises lead up to this connection.

Exercise 6.17: *Self-adjointness and the vector Laplacian.* Consider the vector Laplacian (defined in the previous problem) as a linear operator on the Hilbert space $L^2_{\text{vec}}(\Omega)$.

(a) Show that

$$\int_\Omega d^3x \left\{\mathbf{u}\cdot(\nabla^2\mathbf{v}) - \mathbf{v}\cdot(\nabla^2\mathbf{u})\right\} = \int_{\partial\Omega} \{(\mathbf{n}\cdot\mathbf{u})\,\text{div}\,\mathbf{v} - (\mathbf{n}\cdot\mathbf{v})\,\text{div}\,\mathbf{u}$$
$$-\mathbf{u}\cdot(\mathbf{n}\times\text{curl}\,\mathbf{v}) + \mathbf{v}\cdot(\mathbf{n}\times\text{curl}\,\mathbf{u})\}\,dS$$

(b) Deduce from the identity in part (a) that the domain of ∇^2 coincides with the domain of $(\nabla^2)^\dagger$, and hence the vector Laplacian defines a truly self-adjoint operator with a complete set of mutually orthogonal eigenfunctions, when we take as boundary conditions one of the following:
   (i) Dirichlet–Dirichlet: $\mathbf{n}\cdot\mathbf{u} = 0$ and $\mathbf{n}\times\mathbf{u} = 0$ on $\partial\Omega$;
   (ii) Dirichlet–Neumann: $\mathbf{n}\cdot\mathbf{u} = 0$ and $\mathbf{n}\times \operatorname{curl}\mathbf{u} = 0$ on $\partial\Omega$;
   (iii) Neumann–Dirichlet: $\operatorname{div}\mathbf{u} = 0$ and $\mathbf{n}\times\mathbf{u} = 0$ on $\partial\Omega$; .
   (iv) Neumann–Neumann: $\operatorname{div}\mathbf{u} = 0$ and $\mathbf{n}\times\operatorname{curl}\mathbf{u} = 0$ on $\partial\Omega$:

(c) Show that the more general *Robin* boundary conditions

$$\alpha(n\cdot\mathbf{u}) + \beta \operatorname{div}\mathbf{u} = 0,$$
$$\lambda(\mathbf{n}\times\mathbf{u}) + \mu(\mathbf{n}\times\operatorname{curl}\mathbf{u}) = 0,$$

where $\alpha\ \beta$, $\mu\ \nu$ can be position dependent, also give rise to a truly self-adjoint operator.

Problem 6.18: *Cavity electrodynamics and the Hodge–Weyl decomposition.* Each of the self-adjoint boundary conditions in the previous problem gives rise to a complete set of mutually orthogonal vector eigenfunctions obeying

$$-\nabla^2\mathbf{u}_n = k_n^2\mathbf{u}_n.$$

For these eigenfunctions to describe the normal modes of the electric field $\mathbf{E}$ and the magnetic field $\mathbf{B}$ (which we identify with $\mathbf{H}$ as we will use units in which $\mu_0 = \epsilon_0 = 1$) within a cavity bounded by a perfect conductor, we need to additionally impose the Maxwell equations $\operatorname{div}\mathbf{B} = \operatorname{div}\mathbf{E} = 0$ everywhere within Ω, and to satisfy the perfect-conductor boundary conditions $\mathbf{n}\times\mathbf{E} = \mathbf{n}\cdot\mathbf{B} = 0$.

(a) For each eigenfunction $\mathbf{u}_n$ corresponding to a non-zero eigenvalue k_n^2, define

$$\mathbf{v}_n = \frac{1}{k_n^2}\operatorname{curl}(\operatorname{curl}\mathbf{u}_n), \quad \mathbf{w}_n = -\frac{1}{k_n^2}\nabla(\operatorname{div}\mathbf{u}_n),$$

so that $\mathbf{u}_n = \mathbf{v}_n + \mathbf{w}_n$. Show that $\mathbf{v}_n$ and $\mathbf{w}_m$ are, if non-zero, each eigenfunctions of $-\nabla^2$ with eigenvalue k_n^2. The vector eigenfunctions that are not in the null-space of ∇^2 can therefore be decomposed into their *transverse* (the $\mathbf{v}_n$, which obey $\operatorname{div}\mathbf{v}_n = 0$) and *longitudinal* (the $\mathbf{w}_n$, which obey $\operatorname{curl}\mathbf{w}_n = 0$) parts. However, it is not immediately clear what boundary conditions the $\mathbf{v}_n$ and $\mathbf{w}_n$ separately obey.

(b) The boundary-value problems of relevance to electromagnetism are:

$$\text{i)}\begin{cases} -\nabla^2\mathbf{h}_n = k_n^2\mathbf{h}_n, & \text{within } \Omega,\\ \mathbf{n}\cdot\mathbf{h}_n = 0, \quad \mathbf{n}\times\operatorname{curl}\mathbf{h}_n = 0, & \text{on } \partial\Omega;\end{cases}$$

$$\text{ii)} \begin{cases} -\nabla^2 \mathbf{e}_n = k_n^2 \mathbf{e}_n, & \text{within } \Omega, \\ \operatorname{div} \mathbf{e}_n = 0, \quad \mathbf{n} \times \mathbf{e}_n = 0, & \text{on } \partial\Omega; \end{cases}$$

$$\text{iii)} \begin{cases} -\nabla^2 \mathbf{b}_n = k_n^2 \mathbf{b}_n, & \text{within } \Omega, \\ \operatorname{div} \mathbf{b}_n = 0, \quad \mathbf{n} \times \operatorname{curl} \mathbf{b}_n = 0, & \text{on } \partial\Omega. \end{cases}$$

These problems involve, respectively, the Dirichlet–Neumann, Neumann–Dirichlet and Neumann–Neumann boundary conditions from the previous problem.
Show that the divergence-free transverse eigenfunctions

$$\mathbf{H}_n \stackrel{\text{def}}{=} \frac{1}{k_n^2} \operatorname{curl}\,(\operatorname{curl} \mathbf{h}_n), \quad \mathbf{E}_n \stackrel{\text{def}}{=} \frac{1}{k_n^2} \operatorname{curl}\,(\operatorname{curl} \mathbf{e}_n), \quad \mathbf{B}_n \stackrel{\text{def}}{=} \frac{1}{k_n^2} \operatorname{curl}\,(\operatorname{curl} \mathbf{b}_n)$$

obey $\mathbf{n} \cdot \mathbf{H}_n = \mathbf{n} \times \mathbf{E}_n = \mathbf{n} \times \operatorname{curl} \mathbf{B}_n = 0$ on the boundary, and that from these and the eigenvalue equations we can deduce that $\mathbf{n} \times \operatorname{curl} \mathbf{H}_n = \mathbf{n} \cdot \mathbf{B}_n = \mathbf{n} \cdot \operatorname{curl} \mathbf{E}_n = 0$ on the boundary. The perfect-conductor boundary conditions are therefore satisfied. Also show that the corresponding longitudinal eigenfunctions

$$\boldsymbol{\eta}_n \stackrel{\text{def}}{=} \frac{1}{k_n^2} \nabla(\operatorname{div} \mathbf{h}_n), \quad \boldsymbol{\epsilon}_n \stackrel{\text{def}}{=} \frac{1}{k_n^2} \nabla(\operatorname{div} \mathbf{e}_n), \quad \boldsymbol{\beta}_n \stackrel{\text{def}}{=} \frac{1}{k_n^2} \nabla(\operatorname{div} \mathbf{b}_n)$$

obey the boundary conditions $\mathbf{n} \cdot \boldsymbol{\eta}_n = \mathbf{n} \times \boldsymbol{\epsilon}_n = \mathbf{n} \times \boldsymbol{\beta}_n = 0$.

(c) By considering the counter-example provided by a rectangular box, show that the Dirichlet–Dirichlet boundary condition is not compatible with a longitudinal+transverse decomposition. (A purely transverse wave incident on such a boundary will, on reflection, acquire a longitudinal component.)

(d) Show that

$$0 = \int_\Omega \boldsymbol{\eta}_n \cdot \mathbf{H}_m \, d^3x = \int_\Omega \boldsymbol{\epsilon}_n \cdot \mathbf{E}_m \, d^3x = \int_\Omega \boldsymbol{\beta}_n \cdot \mathbf{B}_m \, d^3x,$$

but that the $\mathbf{v}_n$ and $\mathbf{w}_n$ obtained from the Dirichlet–Dirichlet boundary condition $\mathbf{u}_n$'s are not in general orthogonal to each other. Use the continuity of the $L^2_{\text{vec}}(\Omega)$ inner product

$$\mathbf{x}_n \to \mathbf{x} \quad \Rightarrow \quad \langle \mathbf{x}_n, \mathbf{y} \rangle \to \langle \mathbf{x}, \mathbf{y} \rangle$$

to show that this individual-eigenfunction orthogonality is retained by limits of sums of the eigenfunctions. Deduce that, for each of the boundary conditions (i)–(iii) (but *not* for the Dirichlet–Dirichlet case), we have the *Hodge–Weyl* decomposition of $L^2_{\text{vec}}(\Omega)$ as the orthogonal direct sum

$$L^2_{\text{vec}}(\Omega) = \mathcal{L} \oplus \mathcal{T} \oplus \mathcal{N},$$

where $\mathcal{L}$, $\mathcal{T}$ are respectively the spaces of functions representable as infinite sums of the longitudinal and transverse eigenfunctions, and $\mathcal{N}$ is the finite-dimensional space of harmonic (null-space) eigenfunctions.

Complete sets of vector eigenfunctions for the interior of a rectangular box, and for each of the four sets of boundary conditions we have considered, can be found in Morse and Feshbach §13.1.

Problem 6.19: Hodge–Weyl and Helmholtz–Hodge. In this exercise we consider the problem of what classes of vector-valued functions can be expanded in terms of the various families of eigenfunctions of the previous problem. It is tempting (but wrong) to think that we are restricted to expanding functions that obey the same boundary conditions as the eigenfunctions themselves. Thus, we might erroniously expect that the $\mathbf{E}_n$ are good only for expanding functions whose divergence vanishes and have vanishing tangential boundary components, or that the $\boldsymbol{\eta}_n$ can expand out only curl-free vector fields with vanishing normal boundary component. That this supposition can be false was exposed in section 2.2.3, where we showed that functions that are zero at the endpoints of an interval can be used to expand out functions that are not zero there. The key point is that each of our four families of $\mathbf{u}_n$ constitute a *complete orthonormal set* in $L^2_{\text{vec}}(\Omega)$, and can therefore be used expand *any* vector field. As a consequence, the infinite sum $\sum a_n \mathbf{E}_n \in \mathcal{T}$ can, for example, represent any vector-valued function $\mathbf{u} \in L^2_{\text{vec}}(\Omega)$ provided only that $\mathbf{u}$ possesses no component lying either in the subspace $\mathcal{L}$ of the longitudinal eigenfunctions $\boldsymbol{\epsilon}_n$, or in the nullspace $\mathcal{N}$.

(a) Let $\mathcal{T} = \langle \mathbf{E}_n \rangle$ be space of functions representable as infinite sums of the $\mathbf{E}_n$. Show that

$$\langle \mathbf{E}_n \rangle^{\perp} = \{\mathbf{u} : \operatorname{curl} \mathbf{u} = 0 \text{ within } \Omega,\ \mathbf{n} \times \mathbf{u} = 0 \text{ on } \partial\Omega\}.$$

Similarly show that

$$\begin{aligned}
\langle \boldsymbol{\epsilon}_n \rangle^{\perp} &= \{\mathbf{u} : \operatorname{div} \mathbf{u} = 0 \text{ within } \Omega, \quad \text{no condition on } \partial\Omega\},\\
\langle \mathbf{H}_n \rangle^{\perp} &= \{\mathbf{u} : \operatorname{curl} \mathbf{u} = 0 \text{ within } \Omega, \quad \text{no condition on } \partial\Omega\},\\
\langle \boldsymbol{\eta}_n \rangle^{\perp} &= \{\mathbf{u} : \operatorname{div} \mathbf{u} = 0 \text{ within } \Omega,\ \ \mathbf{n} \cdot \mathbf{u} = 0 \text{ on } \partial\Omega\},\\
\langle \mathbf{B}_n \rangle^{\perp} &= \{\mathbf{u} : \operatorname{curl} \mathbf{u} = 0 \text{ within } \Omega,\ \ \text{no condition on } \partial\Omega\}.\\
\langle \boldsymbol{\beta}_n \rangle^{\perp} &= \{\mathbf{u} : \operatorname{div} \mathbf{u} = 0 \text{ within } \Omega, \quad \text{no condition on } \partial\Omega\},
\end{aligned}$$

(b) Use the results of part (a) and the Helmholtz–Hodge decomposition to show that

$$\begin{aligned}
\langle \mathbf{E}_n \rangle &= \{\mathbf{u} \in L^2_{\text{vec}}(\Omega) : \mathbf{u} = \operatorname{curl} \mathbf{w},\ \text{no condition on } \mathbf{w} \text{ on } \partial\Omega\}.\\
\langle \boldsymbol{\epsilon}_n \rangle &= \{\mathbf{u} \in L^2_{\text{vec}}(\Omega) : \mathbf{u} = \nabla\phi,\ \text{where } \phi = 0 \text{ on } \partial\Omega\},\\
\langle \mathbf{H}_n \rangle &= \{\mathbf{u} \in L^2_{\text{vec}}(\Omega) : \mathbf{u} = \operatorname{curl} \mathbf{w},\ \text{where } \mathbf{n} \times \mathbf{w} = 0 \text{ on } \partial\Omega\},\\
\langle \boldsymbol{\eta}_n \rangle &= \{\mathbf{u} \in L^2_{\text{vec}}(\Omega) : \mathbf{u} = \nabla\phi,\ \text{no condition on } \phi \text{ on } \partial\Omega\},
\end{aligned}$$

$$\langle \mathbf{B}_n \rangle = \{\mathbf{u} \in L^2_{\text{vec}}(\Omega) : \mathbf{u} = \text{curl}\,\mathbf{w},\ \text{where } \mathbf{n} \times \mathbf{w} = 0 \text{ on } \partial\Omega\}.$$

$$\langle \boldsymbol{\beta}_n \rangle = \{\mathbf{u} \in L^2_{\text{vec}}(\Omega) : \mathbf{u} = \nabla\phi,\ \text{where } \phi = 0 \text{ on } \partial\Omega\}.$$

(c) Conclude from the previous part, that the Hodge-Weyl eigenspace decompositions

$$L^2_{\text{vec}}(\Omega) = \mathcal{L} \oplus \mathcal{T} \oplus \mathcal{N}$$

for each of the three vector Laplacian boundary condition families (i), (ii) and (iii) coincides the Helmholtz–Hodge decompositions under the conditions (i), (ii) and (iii) in problem 6. 6.16

(d) As an illustration of the practical distinctions between the decompositions in part (c), take Ω to be the unit cube in $\mathbb{R}^3$, and $\mathbf{u} = (1, 0, 0)$ a constant vector field. Show that with conditions (i) we have $\mathbf{u} \in \mathcal{L}$, but for (ii) we have $\mathbf{u} \in \mathcal{T}$, and for (iii) we have $\mathbf{u} \in \mathcal{N}$.

7
The mathematics of real waves

Waves are found everywhere in the physical world, but we often need more than the simple wave equation to understand them. The principal complications are nonlinearity and dispersion. In this chapter we will describe the mathematics lying behind some commonly observed, but still fascinating, phenomena.

7.1 Dispersive waves

In this section we will investigate the effects of *dispersion*, the dependence of the speed of propagation on the frequency of the wave. We will see that dispersion has a profound effect on the behaviour of a wavepacket.

7.1.1 Ocean waves

The most commonly seen dispersive waves are those on the surface of water. Although often used to illustrate wave motion in class demonstrations, these waves are not as simple as they seem.

In Chapter 1 we derived the equations governing the motion of water with a free surface. Now we will solve these equations. Recall that we described the flow by introducing a velocity potential ϕ such that $\mathbf{v} = \nabla\phi$, and a variable $h(x, t)$ which is the depth of the water at abscissa x (see Figure 7.1).

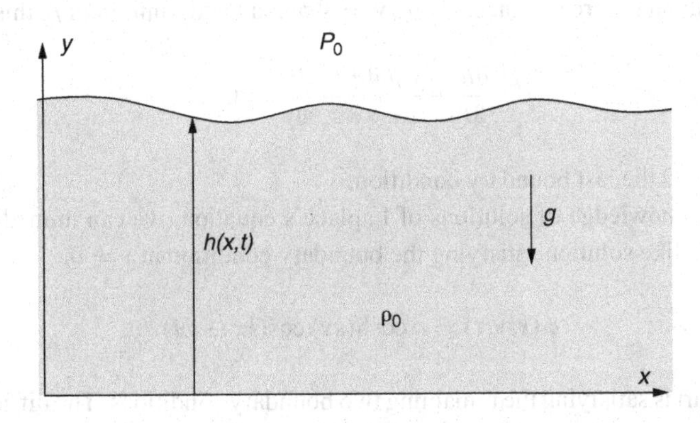

Figure 7.1 Water with a free surface.

Again looking back to Chapter 1, we see that the fluid motion is determined by imposing

$$\nabla^2 \phi = 0 \tag{7.1}$$

everywhere in the bulk of the fluid, together with boundary conditions

$$\frac{\partial \phi}{\partial y} = 0, \quad \text{on} \quad y = 0, \tag{7.2}$$

$$\frac{\partial \phi}{\partial t} + \frac{1}{2}(\nabla \phi)^2 + gy = 0, \quad \text{on the free surface } y = h, \tag{7.3}$$

$$\frac{\partial h}{\partial t} - \frac{\partial \phi}{\partial y} + \frac{\partial h}{\partial x}\frac{\partial \phi}{\partial x} = 0, \quad \text{on the free surface } y = h. \tag{7.4}$$

Recall the physical interpretation of these equations: the vanishing of the Laplacian of the velocity potential simply means that the bulk flow is incompressible

$$\operatorname{div} \mathbf{v} = \nabla^2 \phi = 0. \tag{7.5}$$

The first two of the boundary conditions are also easy to interpret: the first says that no water escapes through the lower boundary at $y = 0$. The second, a form of Bernoulli's equation, asserts that the free surface is everywhere at constant (atmospheric) pressure. The remaining boundary condition is more obscure. It states that a fluid particle initially on the surface stays on the surface. Remember that we set $f(x, y, t) = h(x, t) - y$, so the water surface is given by $f(x, y, t) = 0$. If the surface particles are carried with the flow then the convective derivative of f,

$$\frac{df}{dt} \stackrel{\text{def}}{=} \frac{\partial f}{\partial t} + (\mathbf{v} \cdot \nabla) f, \tag{7.6}$$

should vanish on the free surface. Using $\mathbf{v} = \nabla \phi$ and the definition of f, this reduces to

$$\frac{\partial h}{\partial t} + \frac{\partial \phi}{\partial x}\frac{\partial h}{\partial x} - \frac{\partial \phi}{\partial y} = 0, \tag{7.7}$$

which is indeed the last boundary condition.

Using our knowledge of solutions of Laplace's equation, we can immediately write down a wave-like solution satisfying the boundary condition at $y = 0$

$$\phi(x, y, t) = a \cosh(ky) \cos(kx - \omega t). \tag{7.8}$$

The tricky part is satisfying the remaining two boundary conditions. The difficulty is that they are nonlinear, and so couple modes with different wavenumbers. We will circumvent the difficulty by restricting ourselves to small-amplitude waves, for which the boundary conditions can be linearized. Suppressing all terms that contain a product of two or more small quantities, we are left with

$$\frac{\partial \phi}{\partial t} + gh = 0, \tag{7.9}$$

$$\frac{\partial h}{\partial t} - \frac{\partial \phi}{\partial y} = 0. \tag{7.10}$$

Because ϕ is already a small quantity, and the wave amplitude is a small quantity, linearization requires that these equations should be imposed at the equilibrium surface of the fluid $y = h_0$. It is convenient to eliminate h to get

$$\frac{\partial^2 \phi}{\partial t^2} + g\frac{\partial \phi}{\partial y} = 0, \qquad \text{on } y = h_0. \tag{7.11}$$

Inserting (7.8) into this boundary condition leads to the *dispersion equation*

$$\omega^2 = gk \tanh kh_0, \tag{7.12}$$

relating the frequency to the wavenumber.

Two limiting cases are of particular interest:

(i) *Long waves on shallow water:* Here $kh_0 \ll 1$, and, in this limit,

$$\omega = k\sqrt{gh_0}.$$

(ii) *Waves on deep water:* Here, $kh_0 \gg 1$, leading to $\omega = \sqrt{gk}$.

For deep water, the velocity potential becomes

$$\phi(x, y, t) = ae^{k(y-h_0)} \cos(kx - \omega t). \tag{7.13}$$

We see that the disturbance due to the surface wave dies away exponentially, and becomes very small only a few wavelengths below the surface.

Remember that the velocity of the fluid is $\mathbf{v} = \nabla\phi$. To follow the motion of individual particles of fluid we must solve the equations

$$\frac{dx}{dt} = v_x = -ake^{k(y-h_0)} \sin(kx - \omega t),$$

$$\frac{dy}{dt} = v_y = \quad ake^{k(y-h_0)} \cos(kx - \omega t). \tag{7.14}$$

This is a system of coupled nonlinear differential equations, but to find the small-amplitude motion of particles at the surface we may, to a first approximation, set

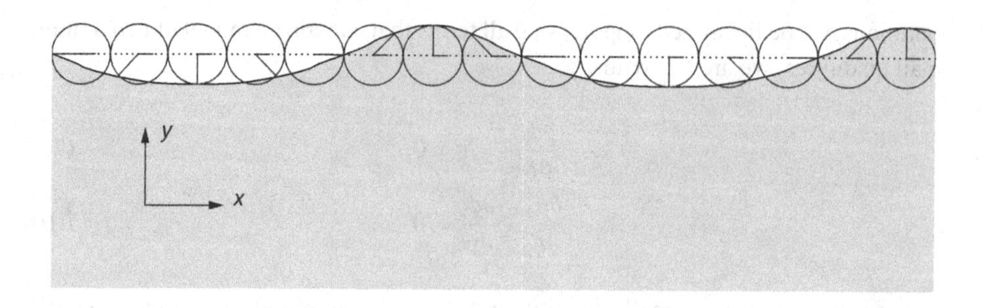

Figure 7.2 Circular orbits in deep water surface waves.

$x = x_0, y = h_0$ on the right-hand side. The orbits of the surface particles are therefore approximately

$$x(t) = x_0 - \frac{ak}{\omega}\cos(kx_0 - \omega t),$$

$$y(t) = y_0 - \frac{ak}{\omega}\sin(kx_0 - \omega t). \tag{7.15}$$

For right-moving waves, the particle orbits are clockwise circles. At the wave crest the particles move in the direction of the wave propagation; in the troughs they move in the opposite direction. Figure 7.2 shows that this motion results in an up-down-asymmetric cycloidal wave profile.

When the effect of the bottom becomes significant, the circular orbits deform into ellipses. For shallow water waves, the motion is principally back-and-forth with motion in the y-direction almost negligible.

7.1.2 Group velocity

The most important effect of dispersion is that the *group velocity* of the waves – the speed at which a wavepacket travels – differs from the *phase velocity* – the speed at which individual wave crests move. The group velocity is also the speed at which the *energy* associated with the waves travels.

Suppose that we have waves with dispersion equation $\omega = \omega(k)$. A right-going wavepacket of finite extent (Figure 7.3), and with initial profile $\varphi(x)$, can be Fourier analysed to give

$$\varphi(x) = \int_{-\infty}^{\infty} \frac{dk}{2\pi} A(k) e^{ikx}. \tag{7.16}$$

At later times this will evolve to

$$\varphi(x, t) = \int_{-\infty}^{\infty} \frac{dk}{2\pi} A(k) e^{ikx - i\omega(k)t}. \tag{7.17}$$

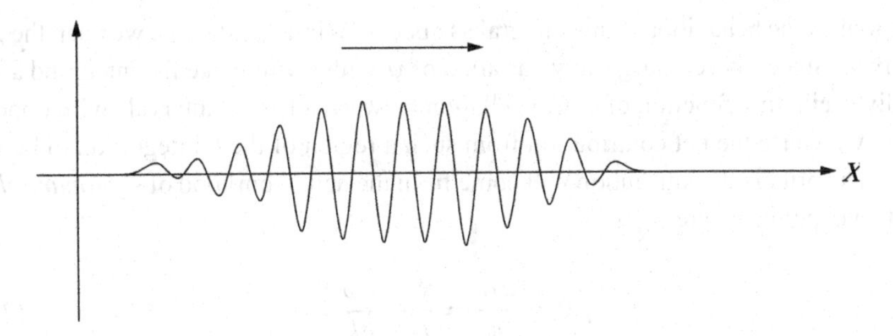

Figure 7.3 A right-going wavepacket.

Let us suppose for the moment that $A(k)$ is non-zero only for a narrow band of wavenumbers around k_0, and that, restricted to this narrow band, we can approximate the full $\omega(k)$ dispersion equation by

$$\omega(k) \approx \omega_0 + U(k - k_0). \tag{7.18}$$

Thus

$$\varphi(x,t) = \int_{-\infty}^{\infty} \frac{dk}{2\pi} A(k) e^{ik(x-Ut)-i(\omega_0-Uk_0)t}. \tag{7.19}$$

Comparing this with the Fourier expression for the initial profile, we find that

$$\varphi(x,t) = e^{-i(\omega_0-Uk_0)t}\varphi(x - Ut). \tag{7.20}$$

The pulse envelope therefore travels at speed U. This velocity

$$U \equiv \frac{\partial\omega}{\partial k} \tag{7.21}$$

is the *group velocity*. The individual wave crests, on the other hand, move at the *phase velocity* $\omega(k)/k$.

When the initial pulse contains a broad range of frequencies we can still explore its evolution. We make use of a powerful tool for estimating the behaviour of integrals that contain a large parameter. In this case the parameter is the time t. We begin by writing the Fourier representation of the wave as

$$\varphi(x,t) = \int_{-\infty}^{\infty} \frac{dk}{2\pi} A(k) e^{it\psi(k)} \tag{7.22}$$

where

$$\psi(k) = k\left(\frac{x}{t}\right) - \omega(k). \tag{7.23}$$

Now look at the behaviour of this integral as t becomes large, but while we keep the ratio x/t fixed. Since t is very large, any variation of ψ with k will make the integrand a very rapidly oscillating function of k. Cancellation between adjacent intervals with opposite phase will cause the net contribution from such a region of the k integration to be very small. The principal contribution will come from the neighbourhood of *stationary phase* points, i.e. points where

$$0 = \frac{d\psi}{dk} = \frac{x}{t} - \frac{\partial \omega}{\partial k}. \tag{7.24}$$

This means that, at points in space where $x/t = U$, we will only get contributions from the Fourier components with wavenumber satisfying

$$U = \frac{\partial \omega}{\partial k}. \tag{7.25}$$

The initial packet will therefore spread out, with those components of the wave having wavenumber k travelling at speed

$$v_{\text{group}} = \frac{\partial \omega}{\partial k}. \tag{7.26}$$

This is the same expression for the group velocity that we obtained in the narrow-band case. Again this speed of propagation should be contrasted with that of the wave crests, which travel at

$$v_{\text{phase}} = \frac{\omega}{k}. \tag{7.27}$$

The "stationary phase" argument may seem a little hand-waving, but it can be developed into a systematic approximation scheme. We will do this in Chapter 19.

Example: *Water waves*. The dispersion equation for waves on deep water is $\omega = \sqrt{gk}$. The phase velocity is therefore

$$v_{\text{phase}} = \sqrt{\frac{g}{k}}, \tag{7.28}$$

whilst the group velocity is

$$v_{\text{group}} = \frac{1}{2}\sqrt{\frac{g}{k}} = \frac{1}{2} v_{\text{phase}}. \tag{7.29}$$

This difference is easily demonstrated by tossing a stone into a pool and observing how individual wave crests overtake the circular wavepacket and die out at the leading edge, while new crests and troughs come into being at the rear and make their way to the front.

This result can be extended to three dimensions with

$$v^i_{\text{group}} = \frac{\partial \omega}{\partial k_i}. \tag{7.30}$$

Example: *de Broglie waves*. The plane-wave solutions of the time-dependent Schrödinger equation

$$i\frac{\partial \psi}{\partial t} = -\frac{1}{2m}\nabla^2\psi, \tag{7.31}$$

are

$$\psi = e^{i\mathbf{k}\cdot\mathbf{r} - i\omega t}, \tag{7.32}$$

with

$$\omega(k) = \frac{1}{2m}\mathbf{k}^2. \tag{7.33}$$

The group velocity is therefore

$$\mathbf{v}_{\text{group}} = \frac{1}{m}\mathbf{k}, \tag{7.34}$$

which is the classical velocity of the particle.

7.1.3 Wakes

There are many circumstances when waves are excited by an object moving at a constant velocity through a background medium, or by a stationary object immersed in a steady flow. The resulting *wakes* carry off energy, and therefore create *wave drag*. Wakes are involved, for example, in sonic booms, Čerenkov radiation, the Landau criterion for superfluidity and Landau damping of plasma oscillations. Here, we will consider some simple water-wave analogues of these effects. The common principle for all wakes is that the resulting wave pattern is time independent when observed from the object exciting it.

Example: *Obstacle in a stream*. Consider a log lying submerged in a rapidly flowing stream (Figure 7.4).

The obstacle disturbs the water and generates a train of waves. If the log lies athwart the stream, the problem is essentially one-dimensional and easy to analyse. The essential

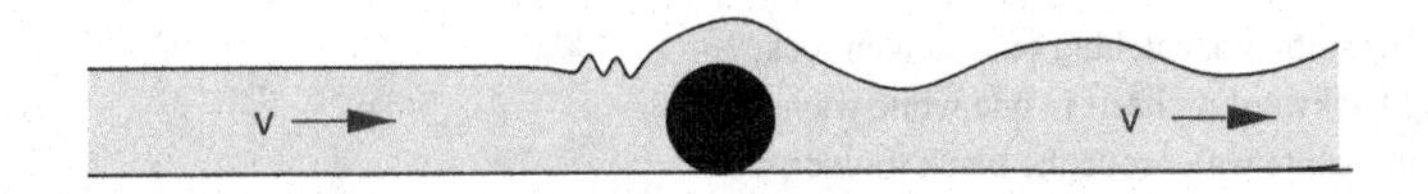

Figure 7.4 Log in a stream.

point is that the distance of the wave crests from the log does not change with time, and therefore the wavelength of the disturbance the log creates is selected by the condition that the *phase velocity* of the wave coincide with the velocity of the mean flow.[1] The group velocity does come into play, however. If the group velocity of the waves is *less* than the phase velocity, the energy being deposited in the wave train by the disturbance will be swept downstream, and the wake will lie *behind* the obstacle. If the group velocity is *higher* than the phase velocity, and this is the case with very short wavelength ripples on water where surface tension is more important than gravity, the energy will propagate against the flow, and so the ripples appear *upstream* of the obstacle.

Example: *Kelvin ship waves*. A more subtle problem is the pattern of waves left behind by a ship on deep water. The shape of the pattern is determined by the group velocity for deep-water waves being one-half that of the phase velocity.

How the wave pattern is formed can be understood from Figure 7.5. In order that the pattern of wave crests be time independent, the waves emitted in the direction AC must have phase velocity such that their crests travel from A to C while the ship goes from A to B. The crest of the wave emitted from the bow of the ship in the direction AC will therefore lie along the line BC – or at least there would be a wave crest on this line if the emitted wave energy travelled at the phase velocity. The angle at C must be a right angle because the direction of propagation is perpendicular to the wave crests. Euclid, by virtue of his angle-in-a-semicircle theorem, now tells us that the locus of all possible points C (for all directions of wave emission) is the larger circle. Because, however, the wave energy only travels at one-half the phase velocity, the waves going in the direction AC actually have significant amplitude only on the smaller circle, which has half the radius of the larger. The wake therefore lies on, and within, the Kelvin wedge, whose

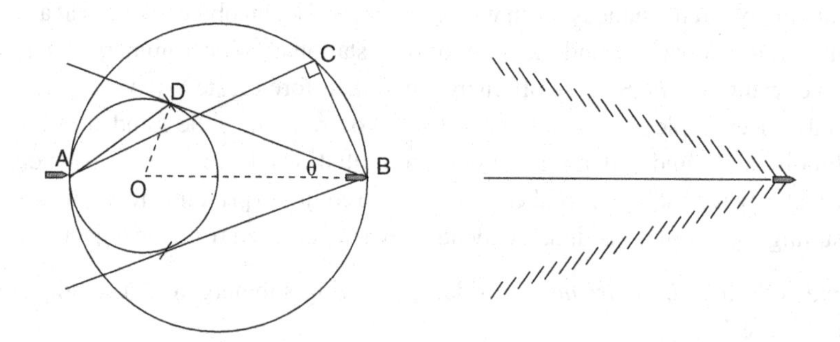

Figure 7.5 Kelvin's ship-wave construction.

[1] In his book *Waves in Fluids*, M. J. Lighthill quotes Robert Frost on this phenomenon:

> The black stream, catching on a sunken rock,
> Flung backward on itself in one white wave,
> And the white water rode the black forever,
> Not gaining but not losing.

boundary lies at an angle θ to the ship's path. This angle is determined by the ratio OD/OB=1/3 to be

$$\theta = \sin^{-1}(1/3) = 19.5^\circ. \tag{7.35}$$

Remarkably, this angle, and hence the width of the wake, is independent of the speed of the ship.

The waves actually on the edge of the wedge are usually the most prominent, and they will have crests perpendicular to the line AD. This orientation is indicated on the left-hand figure, and reproduced as the predicted pattern of wavecrests on the right. The prediction should be compared with the wave systems in Figures 7.6 and 7.7.

7.1.4 Hamilton's theory of rays

We have seen that wave packets travel at a frequency-dependent group velocity. We can extend this result to study the motion of waves in weakly inhomogeneous media, and so derive an analogy between the "geometric optics" limit of wave motion and classical dynamics.

Consider a packet composed of a roughly uniform train of waves spread out over a region that is substantially longer and wider than their mean wavelength. The essential feature of such a wave train is that at any particular point of space and time, $\mathbf{x}$ and t, it has a definite phase $\Theta(\mathbf{x}, t)$. Once we know this phase, we can define the local frequency

Figure 7.6 Large-scale Kelvin wakes. (Image source: US Navy).

Figure 7.7 Small-scale Kelvin wake. (Photograph by Fabrice Neyret).

ω and wavevector $\mathbf{k}$ by

$$\omega = -\left(\frac{\partial\Theta}{\partial t}\right)_x, \qquad k_i = \left(\frac{\partial\Theta}{\partial x_i}\right)_t. \tag{7.36}$$

These definitions are motivated by the idea that

$$\Theta(\mathbf{x}, t) \sim \mathbf{k}\cdot\mathbf{x} - \omega t, \tag{7.37}$$

at least locally.

We wish to understand how $\mathbf{k}$ changes as the wave propagates through a slowly varying medium. We introduce the inhomogeneity by assuming that the dispersion equation $\omega = \omega(\mathbf{k})$, which is initially derived for a uniform medium, can be extended to $\omega = \omega(\mathbf{k}, \mathbf{x})$, where the $\mathbf{x}$ dependence arises, for example, as a result of a position-dependent refractive index. This assumption is only an approximation, but it is a good approximation when the distance over which the medium changes is much larger than the distance between wave crests.

Applying the equality of mixed partials to the definitions of $\mathbf{k}$ and ω gives us

$$\left(\frac{\partial\omega}{\partial x_i}\right)_t = -\left(\frac{\partial k_i}{\partial t}\right)_{\mathbf{x}}, \quad \left(\frac{\partial k_i}{\partial x_j}\right)_{x_i} = \left(\frac{\partial k_j}{\partial x_i}\right)_{x_j}. \tag{7.38}$$

The subscripts indicate what is being left fixed when we differentiate. We must be careful about this, because we want to use the dispersion equation to express ω as a function of $\mathbf{k}$ and $\mathbf{x}$, and the wavevector $\mathbf{k}$ will itself be a function of $\mathbf{x}$ and t.

Taking this dependence into account, we write

$$\left(\frac{\partial \omega}{\partial x_i}\right)_t = \left(\frac{\partial \omega}{\partial x_i}\right)_{\mathbf{k}} + \left(\frac{\partial \omega}{\partial k_j}\right)_{\mathbf{x}} \left(\frac{\partial k_j}{\partial x_i}\right)_t. \tag{7.39}$$

We now use (7.38) to rewrite this as

$$\left(\frac{\partial k_i}{\partial t}\right)_{\mathbf{x}} + \left(\frac{\partial \omega}{\partial k_j}\right)_{\mathbf{x}} \left(\frac{\partial k_i}{\partial x_j}\right)_t = -\left(\frac{\partial \omega}{\partial x_i}\right)_{\mathbf{k}}. \tag{7.40}$$

Interpreting the left-hand side as a convective derivative

$$\frac{dk_i}{dt} = \left(\frac{\partial k_i}{\partial t}\right)_{\mathbf{x}} + (\mathbf{v}_g \cdot \nabla)k_i,$$

we read off that

$$\frac{dk_i}{dt} = -\left(\frac{\partial \omega}{\partial x_i}\right)_{\mathbf{k}} \tag{7.41}$$

provided we are moving at velocity

$$\frac{dx_i}{dt} = (\mathbf{v}_g)_i = \left(\frac{\partial \omega}{\partial k_i}\right)_{\mathbf{x}}. \tag{7.42}$$

Since this is the group velocity, the packet of waves is actually travelling at this speed. The last two equations therefore tell us how the orientation and wavelength of the wave train evolve if we ride along with the packet as it is refracted by the inhomogeneity.

The formulæ

$$\begin{aligned} \dot{\mathbf{k}} &= -\frac{\partial \omega}{\partial \mathbf{x}}, \\ \dot{\mathbf{x}} &= \frac{\partial \omega}{\partial \mathbf{k}}, \end{aligned} \tag{7.43}$$

are *Hamilton's ray equations*. These Hamilton equations are identical in form to Hamilton's equations for classical mechanics

$$\begin{aligned} \dot{\mathbf{p}} &= -\frac{\partial H}{\partial \mathbf{x}}, \\ \dot{\mathbf{x}} &= \frac{\partial H}{\partial \mathbf{p}}, \end{aligned} \tag{7.44}$$

except that $\mathbf{k}$ is playing the role of the canonical momentum, $\mathbf{p}$, and $\omega(\mathbf{k}, \mathbf{x})$ replaces the Hamiltonian, $H(\mathbf{p}, \mathbf{x})$. This formal equivalence of geometric optics and classical mechanics was a mystery in Hamilton's time. Today we understand that classical mechanics is nothing but the geometric optics limit of wave mechanics.

7.2 Making waves

Many waves occurring in nature are generated by the energy of some steady flow being stolen away to drive an oscillatory motion. Familiar examples include the music of a flute and the waves raised on the surface of water by the wind. The latter process is quite subtle and was not understood until the work of J. W. Miles in 1957. Miles showed that in order to excite waves the wind speed has to vary with the height above the water, and that waves of a given wavelength take energy only from the wind at that height where the windspeed matches the phase velocity of the wave. The resulting resonant energy transfer turns out to have analogues in many branches of science. In this section we will exhibit this phenomenon in the simpler situation where the varying flow is that of the water itself.

7.2.1 Rayleigh's equation

Consider water flowing in a shallow channel where friction forces prevent the water in contact with the stream-bed from moving. We will show that the resulting shear flow is unstable to the formation of waves on the water surface. The consequences of this instability are most often seen in a thin sheet of water running down the face of a dam. The sheet starts off flowing smoothly, but, as the water descends, waves form and break, and the water reaches the bottom in irregular pulses called *roll waves*.

It is easiest to describe what is happening from the vantage of a reference frame that rides along with the surface water. In this frame the velocity profile of the flow will be as shown in Figure 7.8.

Since the flow is incompressible but not irrotational, we will describe the motion by using a stream function Ψ, in terms of which the fluid velocity is given by

$$\begin{aligned} v_x &= -\partial_y \Psi, \\ v_y &= \partial_x \Psi. \end{aligned} \tag{7.45}$$

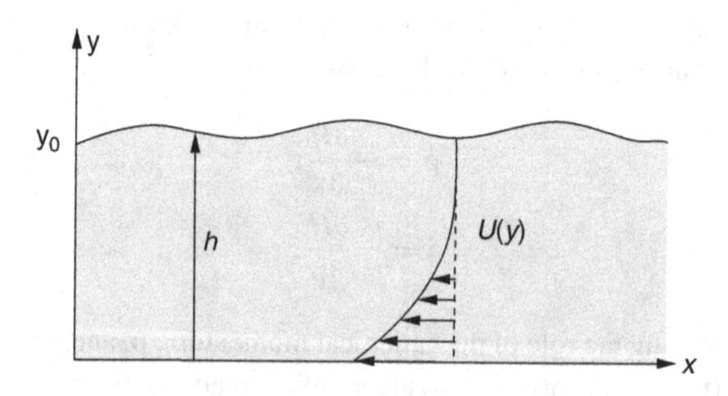

Figure 7.8 The velocity profile $U(y)$ in a frame at which the surface is at rest.

This parametrization automatically satisfies $\nabla \cdot \mathbf{v} = 0$, while (the z-component of) the vorticity becomes

$$\Omega \equiv \partial_x v_y - \partial_y v_x = \nabla^2 \Psi. \tag{7.46}$$

We will consider a stream function of the form[2]

$$\Psi(x,y,t) = \psi_0(y) + \psi(y) e^{ikx - i\omega t}, \tag{7.47}$$

where ψ_0 obeys $-\partial_y \psi_0 = v_x = U(y)$, and describes the horizontal mean flow. The term containing $\psi(y)$ represents a small-amplitude wave disturbance superposed on the mean flow. We will investigate whether this disturbance grows or decreases with time.

Euler's equation can be written as

$$\dot{\mathbf{v}} + \mathbf{v} \times \Omega = -\nabla \left(P + \frac{v^2}{2} + gy \right) = 0. \tag{7.48}$$

Taking the curl of this, and taking into account the two-dimensional character of the problem, we find that

$$\partial_t \Omega + (\mathbf{v} \cdot \nabla)\Omega = 0. \tag{7.49}$$

This, a general property of two-dimensional incompressible motion, says that vorticity is convected with the flow. We now express (7.49) in terms of Ψ, when it becomes

$$\nabla^2 \dot{\Psi} + (\mathbf{v} \cdot \nabla)\nabla^2 \Psi = 0. \tag{7.50}$$

Substituting the expression (7.47) into (7.50), and keeping only terms of first order in ψ, gives

$$-i\omega \left(\frac{d^2}{dy^2} - k^2 \right) \psi + iUk \left(\frac{d^2}{dy^2} - k^2 \right) \psi + ik\psi \partial_y(-\partial_y U) = 0,$$

or

$$\left(\frac{d^2}{dy^2} - k^2 \right) \psi - \left(\frac{\partial^2 U}{\partial y^2} \right) \frac{1}{(U - \omega/k)} \psi = 0. \tag{7.51}$$

This is *Rayleigh's equation*.[3] If only the first term were present, it would have solutions $\psi \propto e^{\pm ky}$, and we would have recovered the results of Section 7.1.1. The second term is significant, however. It will diverge if there is a point y_c such that $U(y_c) = \omega/k$. In other words, if there is a depth at which the flow speed coincides with the phase velocity

[2] The physical stream function is, of course, the real part of this expression.

[3] Lord Rayleigh, *Proc. Lond. Math. Soc.*, **11** (1879) 57.

of the wave disturbance, thus allowing a resonant interaction between the wave and flow. An actual infinity in (7.51) will be evaded, though, because ω will gain a small imaginary part $\omega \to \omega_R + i\gamma$. A positive imaginary part means that the wave amplitude is growing exponentially with time. A negative imaginary part means that the wave is being damped. With γ included, we then have

$$\frac{1}{(U - \omega/k)} \approx \frac{U - \omega_R/k}{(U - \omega_R/k)^2 + \gamma^2} + i\pi \operatorname{sgn}\left(\frac{\gamma}{k}\right) \delta\Big(U(y) - \omega_R/k\Big)$$
$$= \frac{U - \omega_R/k}{(U - \omega_R/k)^2 + \gamma^2} + i\pi \operatorname{sgn}\left(\frac{\gamma}{k}\right) \left|\frac{\partial U}{\partial y}\right|_{y_c}^{-1} \delta(y - y_c). \tag{7.52}$$

To specify the problem fully we need to impose boundary conditions on $\psi(y)$. On the lower surface we can set $\psi(0) = 0$, as this will keep the fluid at rest there. On the upper surface $y = h$ we apply Euler's equation

$$\dot{\mathbf{v}} + \mathbf{v} \times \Omega = -\nabla\left(P + \frac{v^2}{2} + gh\right) = 0. \tag{7.53}$$

We observe that P is constant, being atmospheric pressure, and the $v^2/2$ can be neglected as it is of second order in the disturbance. Then, considering the x-component, we have

$$-\nabla_x gh = -g\partial_x \int^t v_y dt = -g\left(\frac{k^2}{i\omega}\right)\psi \tag{7.54}$$

on the free surface. To lowest order we can apply the boundary condition on the equilibrium free surface $y = y_0$. The boundary condition is therefore

$$\frac{1}{\psi}\frac{d\psi}{dy} + \frac{k}{\omega}\frac{\partial U}{\partial y} = g\frac{k^2}{\omega^2}, \quad y = y_0. \tag{7.55}$$

We usually have $\partial U/\partial y = 0$ near the surface, so this simplifies to

$$\frac{1}{\psi}\frac{d\psi}{dy} = g\frac{k^2}{\omega^2}. \tag{7.56}$$

That this is sensible can be confirmed by considering the case of waves on still, deep water, where $\psi(y) = e^{|k|y}$. The boundary condition then reduces to $|k| = gk^2/\omega^2$, or $\omega^2 = g|k|$, which is the correct dispersion equation for such waves.

We find the corresponding dispersion equation for waves on shallow flowing water by computing

$$\left.\frac{1}{\psi}\frac{d\psi}{dy}\right|_{y_0}, \tag{7.57}$$

from Rayleigh's equation (7.51). Multiplying by ψ^* and integrating gives

$$0 = \int_0^{y_0} dy \left\{ \psi^* \left(\frac{d^2}{dy^2} - k^2 \right) \psi + k \left(\frac{\partial^2 U}{\partial y^2} \right) \frac{1}{(\omega - Uk)} |\psi|^2 \right\}. \tag{7.58}$$

An integration by parts then gives

$$\left[\psi^* \frac{d\psi}{dy} \right]_0^{y_0} = \int_0^{y_0} dy \left\{ \left| \frac{d\psi}{dy} \right| + k^2 |\psi|^2 + \left(\frac{\partial^2 U}{\partial y^2} \right) \frac{1}{(U - \omega/k)} |\psi|^2 \right\}. \tag{7.59}$$

The lower limit makes no contribution, since ψ^* is zero there. On using (7.52) and taking the imaginary part, we find

$$\operatorname{Im} \left(\psi^* \frac{d\psi}{dy} \right)_{y_0} = \operatorname{sgn} \left(\frac{\gamma}{k} \right) \pi \left(\frac{\partial^2 U}{\partial y^2} \right)_{y_c} \left| \frac{\partial U}{\partial y} \right|_{y_c}^{-1} |\psi(y_c)|, \tag{7.60}$$

or

$$\operatorname{Im} \left(\frac{1}{\psi} \frac{d\psi}{dy} \right)_{y_0} = \operatorname{sgn} \left(\frac{\gamma}{k} \right) \pi \left(\frac{\partial^2 U}{\partial y^2} \right)_{y_c} \left| \frac{\partial U}{\partial y} \right|_{y_c}^{-1} \frac{|\psi(y_c)|^2}{|\psi(y_0)|^2}. \tag{7.61}$$

This equation is most useful if the interaction with the flow does not substantially perturb $\psi(y)$ away from the still-water result $\psi(y) = \sinh(|k|y)$, and assuming this is so provides a reasonable first approximation.

If we insert (7.61) into (7.56), where we approximate,

$$g \left(\frac{k^2}{\omega^2} \right) \approx g \left(\frac{k^2}{\omega_R^2} \right) - 2ig \left(\frac{k^2}{\omega_R^3} \right) \gamma,$$

we find

$$\begin{aligned} \gamma &= \frac{\omega_R^3}{2gk^2} \operatorname{Im} \left(\frac{1}{\psi} \frac{d\psi}{dy} \right)_{y_0} \\ &= \operatorname{sgn} \left(\frac{\gamma}{k} \right) \pi \frac{\omega_R^3}{2gk^2} \left(\frac{\partial^2 U}{\partial y^2} \right)_{y_c} \left| \frac{\partial U}{\partial y} \right|_{y_c}^{-1} \frac{|\psi(y_c)|^2}{|\psi(y_0)|^2}. \end{aligned} \tag{7.62}$$

We see that either sign of γ is allowed by our analysis. Thus the resonant interaction between the shear flow and wave appears to lead to either exponential growth or damping of the wave. This is inevitable because our inviscid fluid contains no mechanism for dissipation, and its motion is necessarily time-reversal invariant. Nonetheless, as in our discussion of "friction without friction" in Section 5.2.2, only one sign of γ is actually observed. This sign is determined by the initial conditions, but a rigorous explanation of how this works mathematically is not easy, and is the subject of many papers. These show that the correct sign is given by

$$\gamma = -\pi \frac{\omega_R^3}{2gk^2} \left(\frac{\partial^2 U}{\partial y^2}\right)_{y_c} \left|\frac{\partial U}{\partial y}\right|_{y_c}^{-1} \frac{|\psi(y_c)|^2}{|\psi(y_0)|^2}. \quad (7.63)$$

Since our velocity profile has $\partial^2 U/\partial y^2 < 0$, this means that the waves grow in amplitude.

We can also establish the correct sign for γ by computing the change of momentum in the background flow due to the wave.[4] The crucial element is whether, in the neighbourhood of the critical depth, more fluid is overtaking the wave than lagging behind it. This is exactly what the quantity $\partial^2 U/\partial y^2$ measures.

7.3 Nonlinear waves

Nonlinear effects become important when some dimensionless measure of the amplitude of the disturbance, say $\Delta P/P$ for a sound wave, or $\Delta h/\lambda$ for a water wave, is no longer $\ll 1$.

7.3.1 Sound in air

The simplest nonlinear wave system is one-dimensional sound propagation in a gas. This problem was studied by Riemann.

The one-dimensional motion of a fluid is determined by the mass conservation equation

$$\partial_t \rho + \partial_x(\rho v) = 0, \quad (7.64)$$

and Euler's equation of motion

$$\rho(\partial_t v + v\partial_x v) = -\partial_x P. \quad (7.65)$$

In a fluid with equation of state $P = P(\rho)$, the speed of sound, c, is given by

$$c^2 = \frac{dP}{d\rho}. \quad (7.66)$$

It will in general depend on P, the speed of propagation being usually higher when the pressure is higher.

Riemann was able to simplify these equations by defining a new thermodynamic variable $\pi(P)$ as

$$\pi = \int_{P_0}^{P} \frac{1}{\rho c}\, dP, \quad (7.67)$$

[4] G. E. Vekstein, *Amer. J. Phys.*, **66** (1998) 886.

where P_0 is the equilibrium pressure of the undisturbed air. The quantity π obeys

$$\frac{d\pi}{dP} = \frac{1}{\rho c}. \tag{7.68}$$

In terms of π, Euler's equation divided by ρ becomes

$$\partial_t v + v\partial_x v + c\partial_x \pi = 0, \tag{7.69}$$

whilst the equation of mass conservation divided by ρ/c becomes

$$\partial_t \pi + v\partial_x \pi + c\partial_x v = 0. \tag{7.70}$$

Adding and subtracting, we get *Riemann's equations*

$$\begin{aligned} \partial_t(v+\pi) + (v+c)\partial_x(v+\pi) &= 0, \\ \partial_t(v-\pi) + (v-c)\partial_x(v-\pi) &= 0. \end{aligned} \tag{7.71}$$

These assert that the *Riemann invariants* $v \pm \pi$ are constant along the *characteristic curves*

$$\frac{dx}{dt} = v \pm c. \tag{7.72}$$

This tells us that signals travel at the speed $v \pm c$. In other words, they travel, with respect to the fluid, at the local speed of sound c. Using the Riemann equations, we can propagate initial data $v(x, t = 0)$, $\pi(x, t = 0)$ into the future by using the method of characteristics.

In Figure 7.9 the value of $v + \pi$ is constant along the characteristic curve C_+^A, which is the solution of

$$\frac{dx}{dt} = v + c \tag{7.73}$$

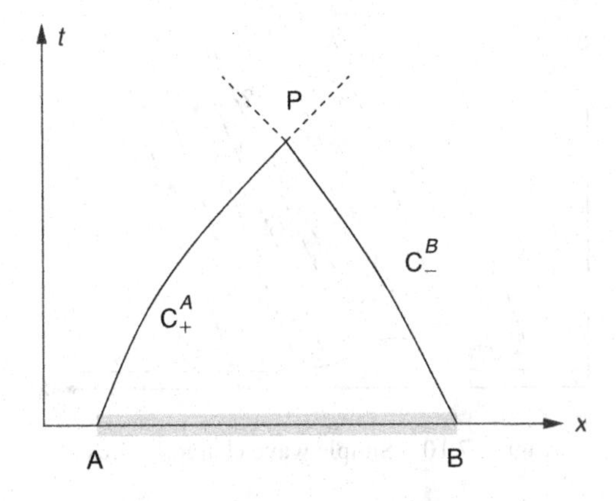

Figure 7.9 Nonlinear characteristic curves.

passing through A. The value of $v - \pi$ is constant along C_-^B, which is the solution of

$$\frac{dx}{dt} = v - c \tag{7.74}$$

passing through B. Thus the values of π and v at the point P can be found if we know the initial values of $v + \pi$ at the point A and $v - \pi$ at the point B. Having found v and π at P we can invert $\pi(P)$ to find the pressure P, and hence c, and so continue the characteristics into the future, as indicated by the dotted lines. We need, of course, to know v and c at every point along the characteristics C_+^A and C_-^B in order to construct them, and this requires us to treat every point as a "P". The values of the dynamical quantities at P therefore depend on the initial data at all points lying between A and B. This is the *domain of dependence* of P.

A sound wave caused by a localized excess of pressure will eventually break up into two distinct pulses, one going forwards and one going backwards. Once these pulses are sufficiently separated that they no longer interact with one another they are *simple waves*. Consider a forward-going pulse propagating into undisturbed air. The backward characteristics are coming from the undisturbed region where both π and v are zero. Clearly $\pi - v$ is zero everywhere on these characteristics, and so $\pi = v$. Now $\pi + v = 2v = 2\pi$ is constant on the forward characteristics, and so π and v are individually constant along them. Since π is constant, so is c. With v also being constant, this means that $c + v$ is constant. In other words, for a simple wave, *the characteristics are straight lines*.

This simple-wave simplification contains within it the seeds of its own destruction. Suppose we have a positive pressure pulse in a fluid whose speed of sound increases with the pressure. Figure 7.10 shows how, with this assumption, the straight-line characteristics travel faster in the high-pressure region, and eventually catch up with and intersect the slower-moving characteristics. When this happens the dynamical variables will become multivalued. How do we deal with this?

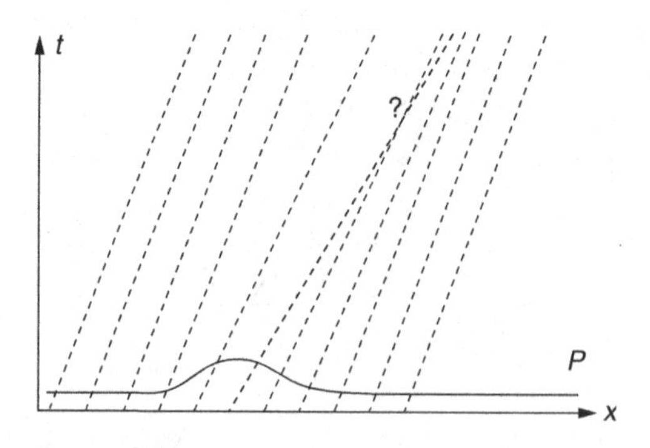

Figure 7.10 Simple wave characteristics.

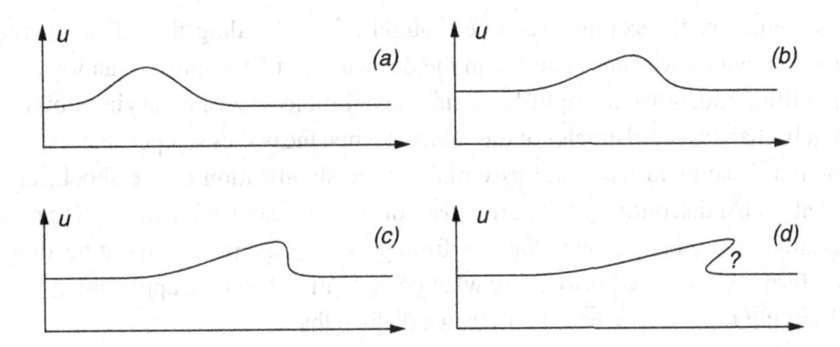

Figure 7.11 A breaking nonlinear wave.

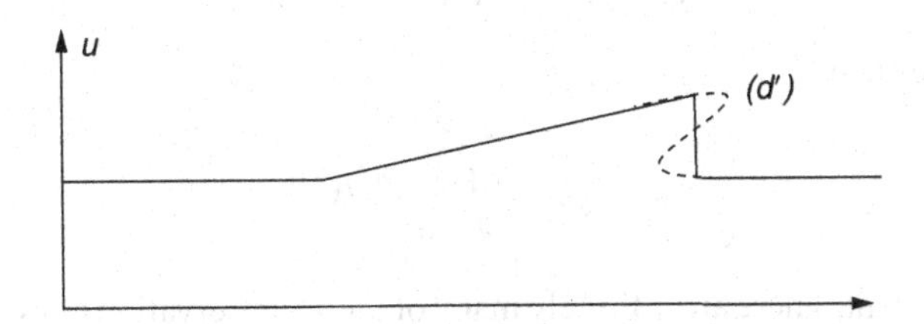

Figure 7.12 Formation of a shock.

7.3.2 Shocks

Let us untangle the multivaluedness by drawing another set of pictures. Suppose u obeys the nonlinear "half" wave equation

$$(\partial_t + u\partial_x)u = 0. \tag{7.75}$$

The velocity of propagation of the wave is therefore u itself, so the parts of the wave with large u will overtake those with smaller u, and the wave will "break", as shown in Figure 7.11.

Physics does not permit such multivalued solutions, and what usually happens is that the assumptions underlying the model which gave rise to the nonlinear equation will no longer be valid. New terms should be included in the equation which prevent the solution becoming multivalued, and instead a steep "shock" will form (Figure 7.12).

Examples of an equation with such additional terms are Burgers' equation

$$(\partial_t + u\partial_x)u = \nu\partial_{xx}^2 u, \tag{7.76}$$

and the Korteweg–de Vries (KdV) equation (4.11), which, by a suitable rescaling of x and t, we can write as

$$(\partial_t + u\partial_x)u = \delta\, \partial_{xxx}^3 u. \tag{7.77}$$

Burgers' equation, for example, can be thought of as including the effects of thermal conductivity, which was not included in the derivation of Riemann's equations. In both these modified equations, the right-hand side is negligible when u is varying slowly, but it completely changes the character of the solution when the waves steepen and try to break.

Although these extra terms are essential for the stabilization of the shock, once we know that such a discontinuous solution has formed, we can find many of its properties – for example the propagation velocity – from general principles, without needing their detailed form. All we need is to know what conservation laws are applicable.

Multiplying $(\partial_t + u\partial_x)u = 0$ by u^{n-1}, we deduce that

$$\partial_t \left\{ \frac{1}{n} u^n \right\} + \partial_x \left\{ \frac{1}{n+1} u^{n+1} \right\} = 0, \tag{7.78}$$

and this implies that

$$Q_n = \int_{-\infty}^{\infty} u^n \, dx \tag{7.79}$$

is time independent. There are infinitely many of these conservation laws, one for each n. Suppose that the n-th conservation law continues to hold even in the presence of the shock, and that the discontinuity is at $X(t)$. Then

$$\frac{d}{dt} \left\{ \int_{-\infty}^{X(t)} u^n \, dx + \int_{X(t)}^{\infty} u^n \, dx \right\} = 0. \tag{7.80}$$

This is equal to

$$u_-^n(X)\dot{X} - u_+^n(X)\dot{X} + \int_{-\infty}^{X(t)} \partial_t u^n \, dx + \int_{X(t)}^{\infty} \partial_t u^n \, dx = 0, \tag{7.81}$$

where $u_-^n(X) \equiv u^n(X - \epsilon)$ and $u_+^n(X) \equiv u^n(X + \epsilon)$. Now, using $(\partial_t + u\partial_x)u = 0$ in the regions away from the shock, where it is reliable, we can write this as

$$\begin{aligned} (u_+^n - u_-^n)\dot{X} &= -\frac{n}{n+1} \int_{-\infty}^{X(t)} \partial_x u^{n+1} \, dx - \frac{n}{n+1} \int_{X(t)}^{\infty} \partial_x u^{n+1} \, dx \\ &= \left(\frac{n}{n+1} \right) (u_+^{n+1} - u_-^{n+1}). \end{aligned} \tag{7.82}$$

The velocity at which the shock moves is therefore

$$\dot{X} = \left(\frac{n}{n+1} \right) \frac{(u_+^{n+1} - u_-^{n+1})}{(u_+^n - u_-^n)}. \tag{7.83}$$

Since the shock can only move at one velocity, only one of the infinitely many conservation laws can continue to hold in the modified theory!

Example: *Burgers' equation.* From

$$(\partial_t + u\partial_x)u = \nu\partial^2_{xx}u, \tag{7.84}$$

we deduce that

$$\partial_t u + \partial_x \left\{ \frac{1}{2}u^2 - \nu\partial_x u \right\} = 0, \tag{7.85}$$

so that $Q_1 = \int u\,dx$ is conserved, but further investigation shows that no other conservation law survives. The shock speed is therefore

$$\dot{X} = \frac{1}{2}\frac{(u_+^2 - u_-^2)}{(u_+ - u_-)} = \frac{1}{2}(u_+ + u_-). \tag{7.86}$$

Example: *KdV equation.* From

$$(\partial_t + u\partial_x)u = \delta\,\partial^3_{xxx}u, \tag{7.87}$$

we deduce that

$$\partial_t u + \partial_x \left\{ \frac{1}{2}u^2 - \delta\,\partial^2_{xx}u \right\} = 0,$$

$$\partial_t \left\{ \frac{1}{2}u^2 \right\} + \partial_x \left\{ \frac{1}{3}u^3 - \delta u\partial^2_{xx}u + \frac{1}{2}\delta(\partial_x u)^2 \right\} = 0$$

$$\vdots$$

where the dots refer to an infinite sequence of (not exactly obvious) conservation laws. Since more than one conservation law survives, the KdV equation cannot have shock-like solutions. Instead, the steepening wave breaks up into a sequence of *solitons*.

Example: *Hydraulic jump, or bore.*
A stationary hydraulic jump is a place in a stream where the fluid abruptly increases in depth from h_1 to h_2, and simultaneously slows down from supercritical (faster than wave speed) flow to subcritical (slower than wave speed) flow (Figure 7.13). Such jumps are

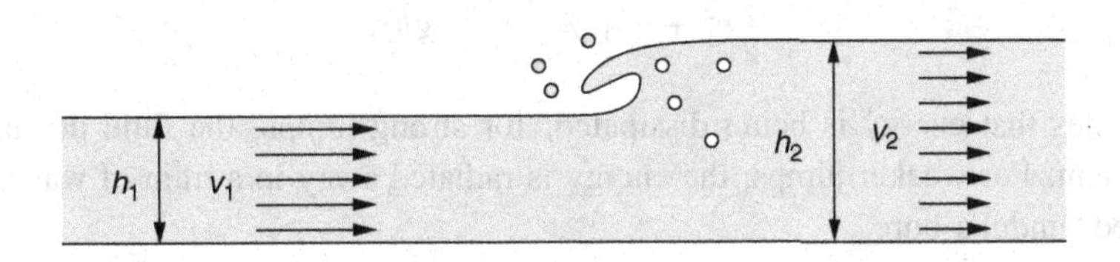

Figure 7.13 A hydraulic jump.

commonly seen near weirs, and white-water rapids.[5] A circular hydraulic jump is easily created in your kitchen sink. The moving equivalent is the *tidal bore*.

The equations governing uniform (meaning that v is independent of the depth) flow in channels are mass conservation

$$\partial_t h + \partial_x \{hv\} = 0, \tag{7.88}$$

and Euler's equation

$$\partial_t v + v \partial_x v = -\partial_x \{gh\}. \tag{7.89}$$

We could manipulate these into the Riemann form, and work from there, but it is more direct to combine them to derive the momentum conservation law

$$\partial_t \{hv\} + \partial_x \left\{ hv^2 + \frac{1}{2} gh^2 \right\} = 0. \tag{7.90}$$

From Euler's equation, assuming steady flow, $\dot{v} = 0$, we can also deduce Bernoulli's equation

$$\frac{1}{2} v^2 + gh = \text{const}, \tag{7.91}$$

which is an energy conservation law. At the jump, mass and momentum must be conserved:

$$\begin{aligned} h_1 v_1 &= h_2 v_2, \\ h_1 v_1^2 + \frac{1}{2} g h_1^2 &= h_2 v_2^2 + \frac{1}{2} g h_2^2, \end{aligned} \tag{7.92}$$

and v_2 may be eliminated to find

$$v_1^2 = \frac{1}{2} g \left(\frac{h_2}{h_1} \right) (h_1 + h_2). \tag{7.93}$$

A change of frame reveals that v_1 is the speed at which a wall of water of height $h = (h_2 - h_1)$ would propagate into stationary water of depth h_1.

Bernoulli's equation is inconsistent with the two equations we have used, and so

$$\frac{1}{2} v_1^2 + g h_1 \neq \frac{1}{2} v_2^2 + g h_2. \tag{7.94}$$

This means that energy is being dissipated: for strong jumps, the fluid downstream is turbulent. For weaker jumps, the energy is radiated away in a train of waves – the so-called "undular bore".

[5] The breaking crest of Frost's "white wave" is probably as much an example of a hydraulic jump as of a smooth downstream wake.

Example: *Shock wave in air.* At a shock wave in air we have conservation of mass

$$\rho_1 v_1 = \rho_2 v_2, \tag{7.95}$$

and momentum

$$\rho_1 v_1^2 + P_1 = \rho_2 v_2^2 + P_2. \tag{7.96}$$

In this case, however, Bernoulli's equation *does* hold,[6] so we also have

$$\frac{1}{2}v_1^2 + h_1 = \frac{1}{2}v_2^2 + h_2. \tag{7.97}$$

Here, h is the specific enthalpy ($U + PV$ per unit mass). Entropy, though, is not conserved, so we cannot use PV^γ = const. across the shock. From mass and momentum conservation alone we find

$$v_1^2 = \left(\frac{\rho_2}{\rho_1}\right)\frac{P_2 - P_1}{\rho_2 - \rho_1}. \tag{7.98}$$

For an ideal gas with $c_p/c_v = \gamma$, we can use energy conservation to eliminate the densities, and find

$$v_1 = c_0\sqrt{1 + \frac{\gamma+1}{2\gamma}\frac{P_2 - P_1}{P_1}}. \tag{7.99}$$

Here, c_0 is the speed of sound in the undisturbed gas.

7.3.3 Weak solutions

We want to make mathematically precise the sense in which a function u with a discontinuity can be a solution to the differential equation

$$\partial_t\left\{\frac{1}{n}u^n\right\} + \partial_x\left\{\frac{1}{n+1}u^{n+1}\right\} = 0, \tag{7.100}$$

even though the equation is surely meaningless if the functions to which the derivatives are being applied are not in fact differentiable.

[6] Recall that enthalpy is conserved in a *throttling process* even in the presence of dissipation. Bernoulli's equation for a gas is the generalization of this thermodynamic result to include the kinetic energy of the gas. The difference between the shock wave in air, where Bernoulli holds, and the hydraulic jump, where it does not, is that the enthalpy of the gas keeps track of the lost mechanical energy, which has been absorbed by the internal degrees of freedom. The Bernoulli equation for channel flow keeps track only of the mechanical energy of the mean flow.

We could play around with distributions like the Heaviside step function or the Dirac delta, but this is unsafe for nonlinear equations, because the product of two distributions is generally not meaningful. What we do is introduce a new concept. We say that u is a *weak solution* to (7.100) if

$$\int_{\mathbb{R}^2} dx\,dt \left\{ u^n \partial_t \varphi + \frac{n}{n+1} u^{n+1} \partial_x \varphi \right\} = 0, \tag{7.101}$$

for all test functions φ in some suitable space $\mathcal{T}$. This equation has formally been obtained from (7.100) by multiplying it by $\varphi(x, t)$, integrating over all space-time, and then integrating by parts to move the derivatives off u, and onto the smooth function φ. If u is assumed smooth then all these manipulations are legitimate and the new equation (7.101) contains no new information. A conventional solution to (7.100) is therefore also a weak solution. The new formulation (7.101), however, admits solutions in which u has shocks.

Let us see what is required of a weak solution if we assume that u is everywhere smooth except for a single jump from $u_-(t)$ to $u_+(t)$ at the point $X(t)$. Let $D_\pm$ be the regions to the left and right of the jump, as shown in Figure 7.14. Then the weak-solution condition (7.101) becomes

$$0 = \int_{D_-} dx\,dt \left\{ u^n \partial_t \varphi + \frac{n}{n+1} u^{n+1} \partial_x \varphi \right\} + \int_{D_+} dx\,dt \left\{ u^n \partial_t \varphi + \frac{n}{n+1} u^{n+1} \partial_x \varphi \right\}. \tag{7.102}$$

Let

$$\mathbf{n} = \left(\frac{1}{\sqrt{1 + |\dot{X}|^2}}, \frac{-\dot{X}}{\sqrt{1 + |\dot{X}|^2}} \right) \tag{7.103}$$

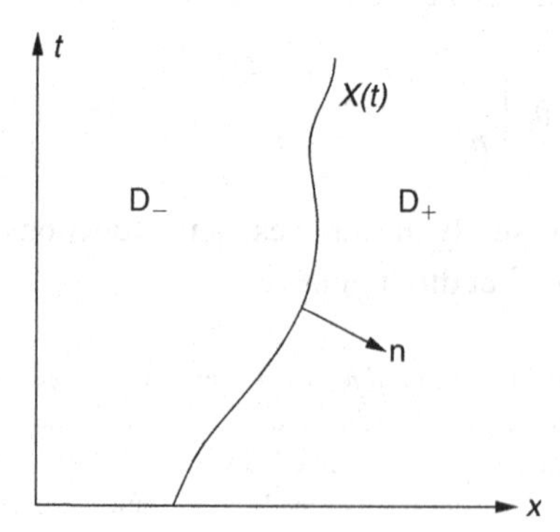

Figure 7.14 The geometry of the domains to the right and left of a jump.

be the unit outward normal to D_-; then, using the divergence theorem, we have

$$\begin{aligned}
\int_{D_-} dx\,dt \left\{u^n \partial_t \varphi + \frac{n}{n+1} u^{n+1} \partial_x \varphi\right\}
&= \int_{D_-} dx\,dt \left\{-\varphi \left(\partial_t u^n + \frac{n}{n+1} \partial_x u^{n+1}\right)\right\} \\
&+ \int_{\partial D_-} dt \left\{\varphi \left(-\dot{X}(t) u_-^n + \frac{n}{n+1} u_-^{n+1}\right)\right\} \qquad (7.104)
\end{aligned}$$

Here we have written the integration measure over the boundary as

$$ds = \sqrt{1 + |\dot{X}|^2}\, dt. \tag{7.105}$$

Performing the same manoeuvre for D_+, and observing that φ can be any smooth function, we deduce that

(i) $\partial_t u^n + \frac{n}{n+1} \partial_x u^{n+1} = 0$ within $D_\pm$.
(ii) $\dot{X}(u_+^n - u_-^n) = \frac{n}{n+1}(u_+^{n+1} - u_-^{n+1})$ on $X(t)$.

The reasoning here is identical to that in Chapter 1, where we considered variations at endpoints to obtain natural boundary conditions. We therefore end up with the same equations for the motion of the shock as before.

The notion of weak solutions is widely used in applied mathematics, and it is the principal ingredient of the *finite element* method of numerical analysis in continuum dynamics.

7.4 Solitons

A localized disturbance in a dispersive medium soon falls apart, since its various frequency components travel at differing speeds. At the same time, nonlinear effects will distort the wave profile. In some systems, however, these effects of dispersion and nonlinearity can compensate each other and give rise to *solitons* – stable solitary waves which propagate for long distances without changing their form. Not all equations possessing wave-like solutions also possess solitary wave solutions. The best known example of equations that do, are:

(1) The Korteweg–de Vries (KdV) equation, which in the form

$$\frac{\partial u}{\partial t} + u\frac{\partial u}{\partial x} = -\frac{\partial^3 u}{\partial x^3}, \tag{7.106}$$

has a solitary-wave solution

$$u(x,t) = 3\alpha^2 \text{sech}^2 \frac{1}{2}(\alpha x - \alpha^3 t) \tag{7.107}$$

which travels at speed α^2. The larger the amplitude, therefore, the faster the solitary wave travels. This equation applies to steep waves in shallow water.

(2) The nonlinear Shrödinger (NLS) equation with attractive interactions

$$i\frac{\partial \psi}{\partial t} = -\frac{1}{2m}\frac{\partial^2 \psi}{\partial x^2} - \lambda|\psi|^2\psi, \tag{7.108}$$

where $\lambda > 0$. It has solitary-wave solution

$$\psi = e^{ikx-i\omega t}\sqrt{\frac{\alpha}{m\lambda}}\operatorname{sech}\sqrt{\alpha}(x-Ut), \tag{7.109}$$

where

$$k = mU, \quad \omega = \frac{1}{2}mU^2 - \frac{\alpha}{2m}. \tag{7.110}$$

In this case, the speed is independent of the amplitude, and the moving solution can be obtained from a stationary one by means of a Galilean boost. The nonlinear equation for the stationary wavepacket may be solved by observing that

$$(-\partial_x^2 - 2\operatorname{sech}^2 x)\psi_0 = -\psi_0 \tag{7.111}$$

where $\psi_0(x) = \operatorname{sech} x$. This is the bound-state of the Pöschel–Teller equation that we have met several times before. The nonlinear Schrödinger equation describes many systems, including the dynamics of tornadoes, where the solitons manifest as the knot-like kinks sometimes seen winding their way up thin funnel clouds.[7]

(3) The sine-Gordon (SG) equation is

$$\frac{\partial^2 \varphi}{\partial t^2} - \frac{\partial^2 \varphi}{\partial x^2} + \frac{m^2}{\beta}\sin\beta\varphi = 0. \tag{7.112}$$

This has solitary-wave solutions

$$\varphi(x,t) = \frac{4}{\beta}\tan^{-1}\left\{e^{\pm m\gamma(x-Ut)}\right\}, \tag{7.113}$$

where $\gamma = (1-U^2)^{-\frac{1}{2}}$ and $|U| < 1$. The velocity is not related to the amplitude, and the moving soliton can again be obtained by boosting a stationary soliton. The boost is now a Lorentz transformation, and so we only get subluminal solitons, whose width is Lorentz contracted by the usual relativistic factor of γ. The sine-Gordon equation describes, for example, the evolution of light pulses whose frequency is in resonance with an atomic transition in the propagation medium.[8]

[7] H. Hasimoto, *J. Fluid Mech.*, **51** (1972) 477.

[8] See G. L. Lamb, *Rev. Mod. Phys.*, **43** (1971) 99, for a nice review.

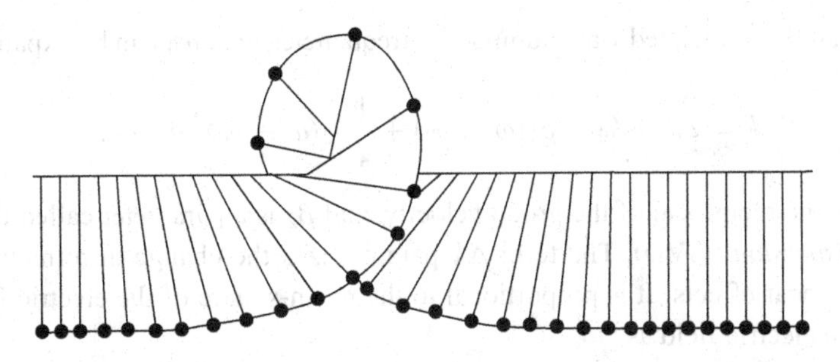

Figure 7.15 A sine-Gordon solitary wave as a twist in a ribbon of coupled pendulums.

In the case of the sine-Gordon soliton, the origin of the solitary wave is particularly easy to understand, as it can be realized as a "twist" in a chain of coupled pendulums (Figure 7.15). The handedness of the twist determines whether we take the $+$ or $-$ sign in the solution (7.113).

The existence of solitary-wave solutions is interesting in its own right. It was the fortuitous observation of such a wave by John Scott Russell on the Union Canal, near Hermiston in Scotland, that founded the subject.[9] Even more remarkable was Scott Russell's subsequent discovery (made in a specially constructed trough in his garden) of what is now called the *soliton property*: two colliding solitary waves interact in a complicated manner yet emerge from the encounter with their form unchanged, having suffered no more than a slight time delay. Each of the three equations given above has exact *multi-soliton* solutions which show this phenomenon.

After languishing for more than a century, soliton theory has grown to be a huge subject. It is, for example, studied by electrical engineers who use soliton pulses in fibre-optic communications. No other type of signal can propagate through thousands of kilometres of undersea cable without degradation. Solitons, or "quantum lumps" are also important in particle physics. The nucleon can be thought of as a knotted soliton (in this case called a "skyrmion") in the pion field, and gauge-field monopole solitons appear in many string and field theories. The soliton equations themselves are aristocrats among partial differential equations, with ties into almost every other branch of mathematics.

Practical illustration: Solitons in Optical Fibres. We wish to transmit picosecond pulses of light with a carrier frequency ω_0. Suppose that the dispersive properties of the fibre

[9] "I was observing the motion of a boat which was rapidly drawn along a narrow channel by a pair of horses, when the boat suddenly stopped – not so the mass of water in the channel which it had put in motion; it accumulated round the prow of the vessel in a state of violent agitation, then suddenly leaving it behind, rolled forward with great velocity, assuming the form of a large solitary elevation, a rounded, smooth and well-defined heap of water, which continued its course along the channel apparently without change of form or diminution of speed. I followed it on horseback, and overtook it still rolling on at a rate of some eight or nine miles an hour, preserving its original figure some thirty feet long and a foot to a foot and a half in height. Its height gradually diminished, and after a chase of one or two miles I lost it in the windings of the channel. Such, in the month of August 1834, was my first chance interview with that singular and beautiful phenomenon which I have called the Wave of Translation." – John Scott Russell, 1844.

are such that the associated wavenumber for frequencies near ω_0 can be expanded as

$$k = \Delta k + k_0 + \beta_1(\omega - \omega_0) + \frac{1}{2}\beta_2(\omega - \omega_0)^2 + \cdots . \tag{7.114}$$

Here, β_1 is the reciprocal of the group velocity, and β_2 is a parameter called the *group velocity dispersion* (GVD). The term Δk parametrizes the change in refractive index due to nonlinear effects. It is proportional to the mean-square of the electric field. Let us write the electric field as

$$E(x,t) = A(x,t)e^{ik_0 z - \omega_0 t}, \tag{7.115}$$

where $A(x,t)$ is a slowly varying envelope function. When we transform from Fourier variables to space and time we have

$$(\omega - \omega_0) \to i\frac{\partial}{\partial t}, \quad (k - k_0) \to -i\frac{\partial}{\partial z}, \tag{7.116}$$

and so the equation determining A becomes

$$-i\frac{\partial A}{\partial z} = i\beta_1\frac{\partial A}{\partial t} - \frac{\beta_2}{2}\frac{\partial^2 A}{\partial t^2} + \Delta k A. \tag{7.117}$$

If we set $\Delta k = \gamma|A^2|$, where γ is normally positive, we have

$$i\left(\frac{\partial A}{\partial z} + \beta_1\frac{\partial A}{\partial t}\right) = \frac{\beta_2}{2}\frac{\partial^2 A}{\partial t^2} - \gamma|A|^2 A. \tag{7.118}$$

We may get rid of the first-order time derivative by transforming to a frame moving at the group velocity. We do this by setting

$$\begin{aligned} \tau &= t - \beta_1 z, \\ \zeta &= z \end{aligned} \tag{7.119}$$

and using the chain rule. The equation for A ends up being

$$i\frac{\partial A}{\partial \zeta} = \frac{\beta_2}{2}\frac{\partial^2 A}{\partial \tau^2} - \gamma|A|^2 A. \tag{7.120}$$

This looks like our nonlinear Schrödinger equation, but with the role of space and time interchanged! Also, the coefficient of the second derivative has the wrong sign so, to make it coincide with the Schrödinger equation we studied earlier, we must have $\beta_2 < 0$. When this condition holds, we are said to be in the "anomalous dispersion" regime – although this is rather a misnomer since it is the *group refractive index*, $N_g = c/v_{\text{group}}$, that is decreasing with frequency, not the ordinary refractive index. For pure SiO_2 glass, β_2 is negative for wavelengths greater than 1.27 μm. We therefore have anomalous dispersion in the technologically important region near 1.55 μm, where the glass is most transparent. In the anomalous dispersion regime we have solitons with

$$A(\zeta,\tau) = e^{i\alpha|\beta_2|\zeta/2}\sqrt{\frac{\beta_2\alpha}{\gamma}}\operatorname{sech}\sqrt{\alpha}(\tau), \tag{7.121}$$

leading to

$$E(z,t) = \sqrt{\frac{\beta_2\alpha}{\gamma}}\operatorname{sech}\sqrt{\alpha}(t-\beta_1 z)e^{i\alpha|\beta_2|z/2}e^{ik_0 z - i\omega_0 t}. \tag{7.122}$$

This equation describes a pulse propagating at β_1^{-1}, which is the group velocity.

Exercise 7.1: Find the expression for the sine-Gordon soliton, by first showing that the static sine-Gordon equation

$$-\frac{\partial^2\varphi}{\partial x^2} + \frac{m^2}{\beta}\sin\beta\varphi = 0$$

implies that

$$\frac{1}{2}\varphi'^2 + \frac{m^2}{\beta^2}\cos\beta\varphi = \text{const.},$$

and solving this equation (for a suitable choice of the constant) by separation of variables. Next, show that if $f(x)$ is a solution of the static equation, then $f\big(\gamma(x-Ut)\big)$, $\gamma = (1-U^2)^{-1/2}$, $|U| < 1$ is a solution of the time-dependent equation.

Exercise 7.2: *Lax pair for the nonlinear Schrödinger equation.* Let L be the matrix differential operator

$$L = \begin{bmatrix} i\partial_x & \chi^* \\ \chi & i\partial_x \end{bmatrix},$$

and let P be the matrix

$$P = \begin{bmatrix} i|\chi|^2 & -\chi'^* \\ +\chi' & -i|\chi|^2 \end{bmatrix}.$$

Show that the equation

$$\dot{L} = [L, P]$$

is equivalent to the nonlinear Shrödinger equation

$$i\dot{\chi} = -\chi'' - 2|\chi|^2\chi.$$

7.5 Further exercises and problems

Here are some further problems on nonlinear and dispersive waves.

Problem 7.3: *The equation of telegraphy.* Oliver Heaviside's equations relating the voltage $v(x,t)$ and current $i(x,t)$ in a transmission line are

$$L\frac{\partial i}{\partial t} + Ri = -\frac{\partial v}{\partial x},$$

$$C\frac{\partial v}{\partial t} + Gv = -\frac{\partial i}{\partial x}.$$

Here R, C, L and G are respectively the resistance, capacitance, inductance and leakance of each unit length of the line.

(a) Show that Heaviside's equations lead to $v(x,t)$ obeying

$$LC\frac{\partial^2 v}{\partial t^2} + (LG+RC)\frac{\partial v}{\partial t} + RGv = \frac{\partial^2 v}{\partial x^2},$$

and also to a similar equation for $i(x,t)$.

(b) Seek a travelling-wave solution of the form

$$v(x,t) = v_0\, e^{i(kx+\omega t)},$$

$$i(x,t) = i_0\, e^{i(kx+\omega t)},$$

and find the dispersion equation relating ω and k. From this relation, show that signals propagate undistorted (i.e. with frequency-independent attenuation) at speed $1/\sqrt{LC}$ provided that the *Heaviside condition* $RC = LG$ is satisfied.

(c) Show that the *characteristic impedance* $Z \equiv v_0/i_0$ of the transmission line is given by

$$Z(\omega) = \sqrt{\frac{R+i\omega L}{G+i\omega C}}.$$

Deduce that the characteristic impedance is frequency independent if the Heaviside condition is satisfied.

In practical applications, the Heaviside condition can be satisfied by periodically inserting extra inductors – known as *loading coils* – into the line.

Problem 7.4: *Pantograph drag.* A high-speed train picks up its electrical power via a pantograph from an overhead line (Figure 7.16). The locomotive travels at speed U and the pantograph exerts a constant vertical force F on the power line.

We make the usual small-amplitude approximations and assume (not unrealistically) that the line is supported in such a way that its vertical displacement obeys an inhomogeneous Klein–Gordon equation

$$\rho\ddot{y} - Ty'' + \rho\Omega^2 y = F\delta(x-Ut),$$

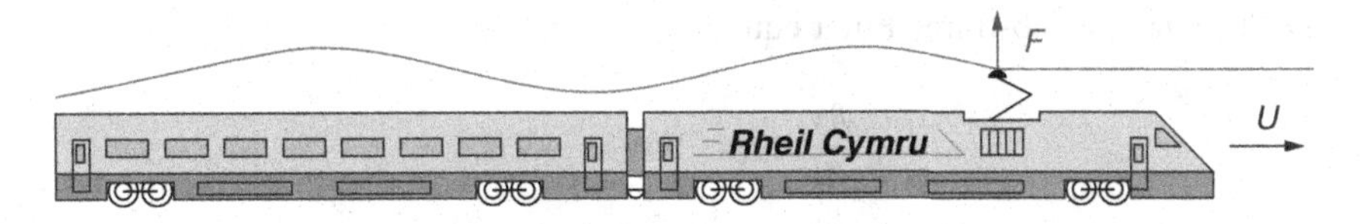

Figure 7.16 A high-speed train.

with $c = \sqrt{T/\rho}$, the velocity of propagation of short-wavelength transverse waves on the overhead cable.

(a) Assume that $U < c$ and solve for the steady state displacement of the cable about the pickup point. (Hint: the disturbance is time-independent when viewed from the train.)
(b) Now assume that $U > c$. Again find an expression for the displacement of the cable. (The same hint applies, but the physically appropriate boundary conditions are very different!)
(c) By equating the rate at which wave-energy

$$E = \int \left\{ \frac{1}{2}\rho\dot{y}^2 + \frac{1}{2}Ty'^2 + \frac{1}{2}\rho\Omega^2 y^2 \right\} dx$$

is being created to the rate at the which the locomotive is doing work, calculate the wave-drag on the train. In particular, show that there is no drag at all until U exceeds c. (Hint: while the front end of the wake is moving at speed U, the trailing end of the wake is moving forward at the *group velocity* of the wave-train.)
(d) By carefully considering the force the pantograph exerts on the overhead cable, again calculate the induced drag. You should get the same answer as in part (c) (Hint: to the order needed for the calculation, the tension in the cable is the same before and after the train has passed, but the direction in which the tension acts is different. The force F is therefore not *exactly* vertical, but has a small forward component. Don't forget that the resultant of the forces is accelerating the cable.)

This problem of wake formation and drag is related both to Čerenkov radiation and to the Landau criterion for superfluidity.

Exercise 7.5: *Inertial waves.* A rotating tank of incompressible ($\rho \equiv 1$) fluid can host waves whose restoring force is provided by angular momentum conservation. Suppose the fluid velocity at the point $\mathbf{r}$ is given by

$$\mathbf{v}(\mathbf{r}, t) = \mathbf{u}(\mathbf{r}, t) + \boldsymbol{\Omega} \times \mathbf{r},$$

where $\mathbf{u}$ is a perturbation imposed on the rigid rotation of the fluid at angular velocity $\boldsymbol{\Omega}$.

(a) Show that when viewed from a coordinate frame rotating with the fluid we have

$$\frac{\partial \mathbf{u}}{\partial t} = \left(\frac{\partial \mathbf{u}}{\partial t} - \boldsymbol{\Omega} \times \mathbf{u} + ((\boldsymbol{\Omega} \times \mathbf{r}) \cdot \nabla)\mathbf{u} \right)_{\text{lab}}.$$

Deduce that the lab-frame Euler equation

$$\frac{\partial \mathbf{v}}{\partial t} + (\mathbf{v} \cdot \nabla)\mathbf{v} = -\nabla P,$$

becomes, in the rotating frame,

$$\frac{\partial \mathbf{u}}{\partial t} + 2(\boldsymbol{\Omega} \times \mathbf{u}) + (\mathbf{u} \cdot \nabla)\mathbf{u} = -\nabla \left(P - \frac{1}{2}|\Omega \times \mathbf{r}|^2\right).$$

We see that in the non-inertial rotating frame the fluid experiences a $-2(\boldsymbol{\Omega} \times \mathbf{u})$ Coriolis and a $\nabla|\Omega \times \mathbf{r}|^2/2$ centrifugal force. By linearizing the rotating-frame Euler equation, show that for small $\mathbf{u}$ we have

$$\frac{\partial \boldsymbol{\omega}}{\partial t} - 2(\boldsymbol{\Omega} \cdot \nabla)\mathbf{u} = 0, \qquad (\star)$$

where $\boldsymbol{\omega} = \text{curl}\,\mathbf{u}$.

(b) Take $\boldsymbol{\Omega}$ to be directed along the z-axis. Seek plane-wave solutions to $(\star)$ in the form

$$\mathbf{u}(\mathbf{r}, t) = \mathbf{u}_0 e^{i(\mathbf{k}\cdot\mathbf{r} - \omega t)}$$

where $\mathbf{u}_0$ is a constant, and show that the dispersion equation for these small-amplitude *inertial waves* is

$$\omega = 2\Omega \sqrt{\frac{k_z^2}{k_x^2 + k_y^2 + k_z^2}}.$$

Deduce that the group velocity is directed *perpendicular* to $\mathbf{k}$ – i.e. at right-angles to the phase velocity. Conclude also that any slow flow that is steady (time independent) when viewed from the rotating frame is necessarily independent of the coordinate z. (This is the origin of the phenomenon of *Taylor columns*, which are columns of stagnant fluid lying above and below any obstacle immersed in such a flow.)

Exercise 7.6: *Nonlinear waves*. In this problem we will explore the Riemann invariants for a fluid with $P = \lambda^2 \rho^3/3$. This is the equation of state of a one-dimensional non-interacting Fermi gas.

(a) From the continuity equation

$$\partial_t \rho + \partial_x \rho v = 0,$$

and Euler's equation of motion

$$\rho(\partial_t v + v\partial_x v) = -\partial_x P,$$

deduce that

$$\left(\frac{\partial}{\partial t} + (\lambda\rho + v)\frac{\partial}{\partial x}\right)(\lambda\rho + v) = 0,$$

$$\left(\frac{\partial}{\partial t} + (-\lambda\rho + v)\frac{\partial}{\partial x}\right)(-\lambda\rho + v) = 0.$$

In what limit do these equations become equivalent to the wave equation for one-dimensional sound? What is the sound speed in this case?

(b) Show that the Riemann invariants $v \pm \lambda\rho$ are constant on suitably defined characteristic curves. What is the local speed of propagation of the waves moving to the right or left?

(c) The fluid starts from rest, $v = 0$, but with a region where the density is higher than elsewhere. Show that the Riemann equations will inevitably break down at some later time due to the formation of shock waves.

Exercise 7.7: *Burgers shocks.* As a simple mathematical model for the formation and decay of a shock wave consider *Burgers' equation*:

$$\partial_t u + u\partial_x u = \nu\, \partial_x^2 u.$$

Note its similarity to the Riemann equations of the previous exercise. The additional term on the right-hand side introduces dissipation and prevents the solution becoming multivalued.

(a) Show that if $\nu = 0$ any solution of Burgers' equation having a region where u decreases to the right will always eventually become multivalued.

(b) Show that the *Hopf–Cole* transformation, $u = -2\nu\, \partial_x \ln \psi$, leads to ψ obeying a heat diffusion equation

$$\partial_t \psi = \nu\, \partial_x^2 \psi.$$

(c) Show that

$$\psi(x,t) = Ae^{\nu a^2 t - ax} + Be^{\nu b^2 t - bx}$$

is a solution of this heat equation, and so deduce that Burgers' equation has a shock-wave-like solution which travels to the right at speed $C = \nu(a+b) = \frac{1}{2}(u_L + u_R)$, the mean of the wave speeds to the left and right of the shock. Show that the width of the shock is $\approx 4\nu/|u_L - u_R|$.

8
Special functions

In solving Laplace's equation by the method of separation of variables we come across the most important of the *special functions* of mathematical physics. These functions have been studied for many years, and books such as the Bateman manuscript project[1] summarize the results. Any serious student of theoretical physics needs to be familiar with this material, and should at least read the standard text: *A Course of Modern Analysis* by E. T. Whittaker and G. N. Watson (Cambridge University Press). Although it was originally published in 1902, nothing has superseded this book in its accessibility and usefulness.

8.1 Curvilinear coordinates

Laplace's equation can be separated in a number of coordinate systems. These are all *orthogonal* systems in that the local coordinate axes cross at right angles.

To any system of orthogonal curvilinear coordinates is associated a *metric* of the form

$$ds^2 = h_1^2(dx^1)^2 + h_2^2(dx^2)^2 + h_3^2(dx^3)^2. \tag{8.1}$$

This expression tells us the distance $\sqrt{ds^2}$ between the adjacent points $(x^1 + dx^1, x^2 + dx^2, x^3 + dx^3)$ and (x^1, x^2, x^3). In general, the h_i will depend on the coordinates x^i.

The most commonly used orthogonal curvilinear coordinate systems are plane polars, spherical polars and cylindrical polars. The Laplacian also separates in plane elliptic, or three-dimensional ellipsoidal coordinates and their degenerate limits, such as parabolic cylindrical coordinates – but these are not so often encountered, and for their properties we refer the reader to comprehensive treatises such as Morse and Feshbach's *Methods of Theoretical Physics.*

[1] The Bateman manuscript project contains the formulæ collected by Harry Bateman, who was professor of Mathematics, Theoretical Physics, and Aeronautics at the California Institute of Technology. After his death in 1946, several dozen shoe boxes full of file cards were found in his garage. These proved to be the index to a mountain of paper containing his detailed notes. A subset of the material was eventually published as the three-volume series *Higher Transcendental Functions*, and the two-volume *Tables of Integral Transformations*, A. Erdélyi *et al.* eds.

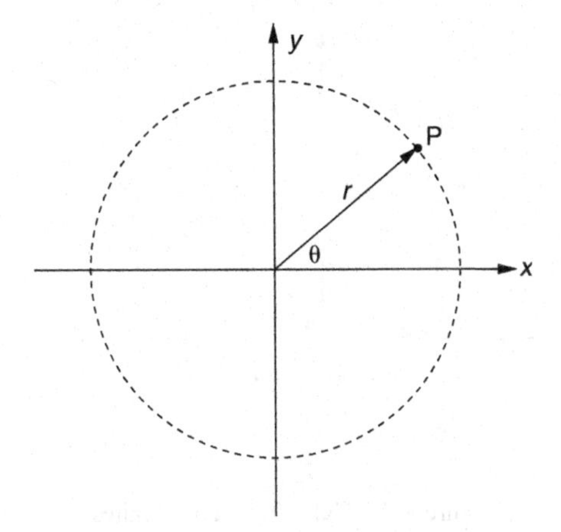

Figure 8.1 Plane polar coordinates.

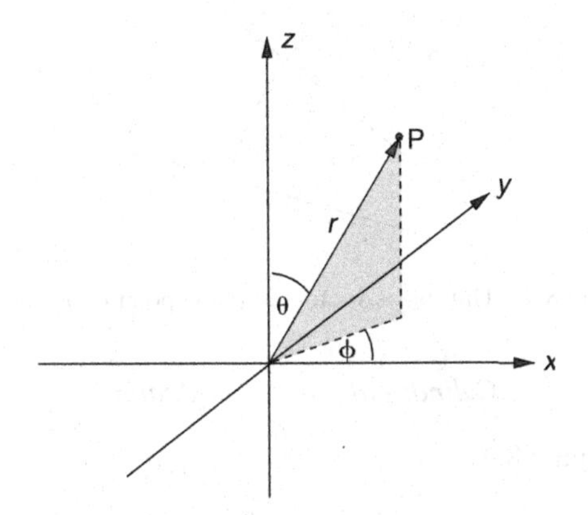

Figure 8.2 Spherical coordinates.

Plane polar coordinates

Plane polar coordinates (Figure 8.1) have metric

$$ds^2 = dr^2 + r^2 d\theta^2, \tag{8.2}$$

so $h_r = 1, h_\theta = r$.

Spherical polar coordinates

This system (Figure 8.2) has metric

$$ds^2 = dr^2 + r^2 d\theta^2 + r^2 \sin^2 \theta d\phi^2, \tag{8.3}$$

so $h_r = 1, h_\theta = r, h_\phi = r \sin\theta$.

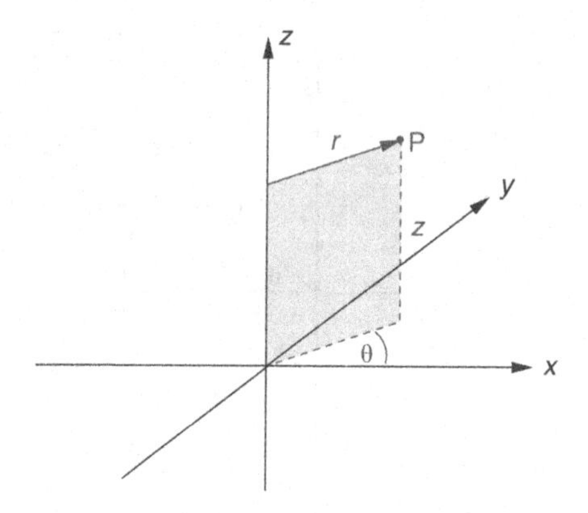

Figure 8.3 Cylindrical coordinates.

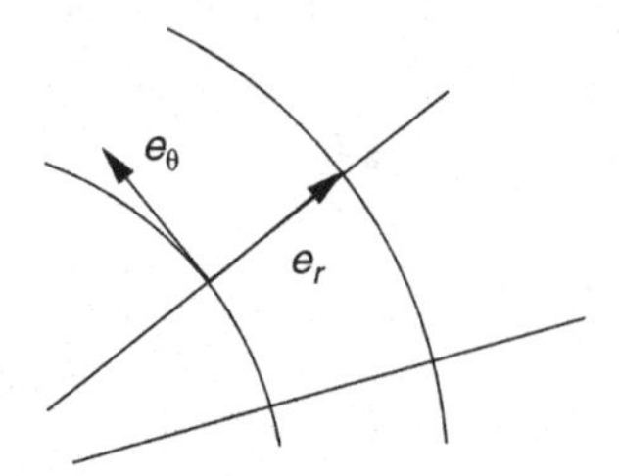

Figure 8.4 Unit basis vectors in plane polar coordinates.

Cylindrical polar coordinates

These have metric (Figure 8.3)

$$ds^2 = dr^2 + r^2 d\theta^2 + dz^2, \tag{8.4}$$

so $h_r = 1$, $h_\theta = r$, $h_z = 1$.

8.1.1 Div, grad and curl in curvilinear coordinates

It is very useful to know how to write the curvilinear coordinate expressions for the common operations of the vector calculus. Knowing these, we can then write down the expression for the Laplace operator.

The gradient operator

We begin with the gradient operator. This is a vector quantity, and to express it we need to understand how to associate a set of basis vectors with our coordinate system. The simplest thing to do is to take unit vectors $\mathbf{e}_i$ tangential to the local coordinate axes (Figure 8.4). Because the coordinate system is orthogonal, these unit vectors will then constitute an orthonormal system.

The vector corresponding to an infinitesimal coordinate displacement dx^i is then given by

$$d\mathbf{r} = h_1 dx^1 \mathbf{e}_1 + h_2 dx^2 \mathbf{e}_2 + h_3 dx^3 \mathbf{e}_3. \tag{8.5}$$

Using the orthonormality of the basis vectors, we find that

$$ds^2 \equiv |d\mathbf{r}|^2 = h_1^2 (dx^1)^2 + h_2^2 (dx^2)^2 + h_3^2 (dx^3)^2, \tag{8.6}$$

as before.

In the unit-vector basis, the gradient vector is

$$\text{grad}\,\phi \equiv \nabla\phi = \frac{1}{h_1}\left(\frac{\partial\phi}{\partial x_1}\right)\mathbf{e}_1 + \frac{1}{h_2}\left(\frac{\partial\phi}{\partial x_2}\right)\mathbf{e}_2 + \frac{1}{h_3}\left(\frac{\partial\phi}{\partial x_3}\right)\mathbf{e}_3, \tag{8.7}$$

so that

$$(\text{grad}\,\phi)\cdot d\mathbf{r} = \frac{\partial\phi}{\partial x^1}dx^1 + \frac{\partial\phi}{\partial x^2}dx^2 + \frac{\partial\phi}{\partial x^3}dx^3, \tag{8.8}$$

which is the change in the value ϕ due to the displacement.

The numbers $(h_1 dx^1, h_2 dx^2, h_3 dx^3)$ are often called the *physical components* of the displacement $d\mathbf{r}$, to distinguish them from the numbers (dx^1, dx^2, dx^3) which are the *coordinate components* of $d\mathbf{r}$. The physical components of a displacement vector all have the dimensions of length. The coordinate components may have different dimensions and units for each component. In plane polar coordinates, for example, the units will be meters and radians. This distinction extends to the gradient itself: the coordinate components of an electric field expressed in polar coordinates will have units of volts per metre and volts per radian for the radial and angular components, respectively. The factor $1/h_\theta = r^{-1}$ serves to convert the latter to volts per metre.

The divergence

The divergence of a vector field $\mathbf{A}$ is defined to be the flux of $\mathbf{A}$ out of an infinitesimal region, divided by volume of the region.

In Figure 8.5, the flux out of the two end faces is

$$dx^2 dx^3 \left[A_1 h_2 h_3|_{(x^1+dx^1,x^2,x^3)} - A_1 h_2 h_3|_{(x^1,x^2,x^3)}\right] \approx dx^1 dx^2 dx^3 \frac{\partial(A_1 h_2 h_3)}{\partial x^1}. \tag{8.9}$$

Adding the contributions from the other two pairs of faces, and dividing by the volume, $h_2 h_2 h_3 dx^1 dx^2 dx^3$, gives

$$\text{div}\,\mathbf{A} = \frac{1}{h_1 h_2 h_3}\left\{\frac{\partial}{\partial x_1}(h_2 h_3 A_1) + \frac{\partial}{\partial x_2}(h_1 h_3 A_2) + \frac{\partial}{\partial x_3}(h_1 h_2 A_3)\right\}. \tag{8.10}$$

Note that in curvilinear coordinates div $\mathbf{A}$ is no longer simply $\nabla\cdot\mathbf{A}$, although one often writes it as such.

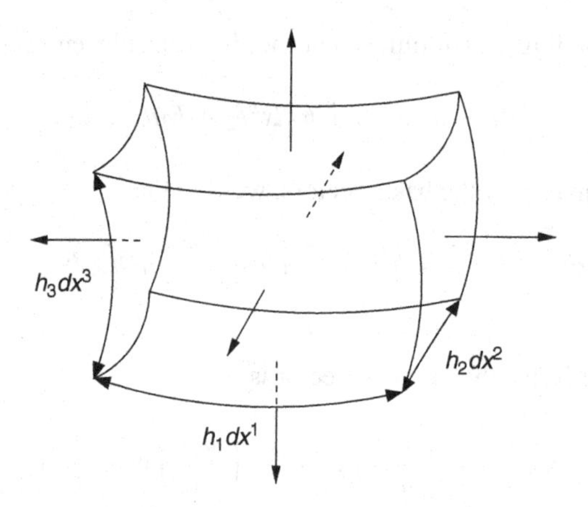

Figure 8.5 Flux out of an infinitesimal volume with sides of length h_1dx^1, h_2dx^2, h_3dx^3.

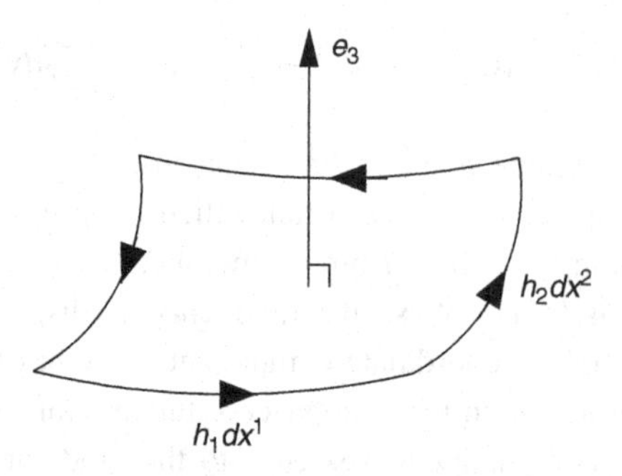

Figure 8.6 Line integral round an infinitesimal area with sides of length h_1dx^1, h_2dx^2 and normal $\mathbf{e}_3$.

The curl

The curl of a vector field **A** is a vector whose component in the direction of the normal to an infinitesimal area element is the line integral of **A** round the infinitesimal area, divided by the area (Figure 8.6).
The third component is, for example,

$$(\text{curl}\,\mathbf{A})_3 = \frac{1}{h_1h_2}\left(\frac{\partial h_2A_2}{\partial x^1} - \frac{\partial h_1A_1}{\partial x^2}\right). \tag{8.11}$$

The other two components are found by cyclically permuting $1 \to 2 \to 3 \to 1$ in this formula. The curl is thus is no longer equal to $\nabla \times \mathbf{A}$, although it is common to write it as if it were.

Note that the factors of h_i are disposed so that the vector identities

$$\text{curl}\,\text{grad}\,\varphi = 0, \tag{8.12}$$

and

$$\operatorname{div}\operatorname{curl}\mathbf{A} = 0, \tag{8.13}$$

continue to hold for any scalar field φ, and any vector field $\mathbf{A}$.

8.1.2 The Laplacian in curvilinear coordinates

The Laplacian acting on scalars is "div grad", and is therefore

$$\nabla^2\varphi = \frac{1}{h_1h_2h_3}\left\{\frac{\partial}{\partial x_1}\left(\frac{h_2h_3}{h_1}\frac{\partial\varphi}{\partial x_1}\right) + \frac{\partial}{\partial x_2}\left(\frac{h_1h_3}{h_2}\frac{\partial\varphi}{\partial x_2}\right) + \frac{\partial}{\partial x_3}\left(\frac{h_1h_2}{h_3}\frac{\partial\varphi}{\partial x_3}\right)\right\}. \tag{8.14}$$

This formula is worth committing to memory.

When the Laplacian is to act on a *vector field*, we must use the *vector Laplacian*

$$\nabla^2\mathbf{A} = \operatorname{grad}\operatorname{div}\mathbf{A} - \operatorname{curl}\operatorname{curl}\mathbf{A}. \tag{8.15}$$

In curvilinear coordinates this is no longer equivalent to the Laplacian acting on each component of $\mathbf{A}$, treating it as if it were a scalar. The expression (8.15) is the appropriate generalization of the vector Laplacian to curvilinear coordinates because it is defined in terms of the coordinate independent operators div, grad and curl, and reduces to the Laplacian on the individual components when the coordinate system is cartesian.

In spherical polars the Laplace operator acting on the scalar field φ is

$$\begin{aligned}
\nabla^2\varphi &= \frac{1}{r^2}\frac{\partial}{\partial r}\left(r^2\frac{\partial\varphi}{\partial r}\right) + \frac{1}{r^2\sin\theta}\frac{\partial}{\partial\theta}\left(\sin\theta\frac{\partial\varphi}{\partial\theta}\right) + \frac{1}{r^2\sin^2\theta}\frac{\partial^2\varphi}{\partial\phi^2}\\
&= \frac{1}{r}\frac{\partial^2(r\varphi)}{\partial r^2} + \frac{1}{r^2}\left\{\frac{1}{\sin\theta}\frac{\partial}{\partial\theta}\left(\sin\theta\frac{\partial\varphi}{\partial\theta}\right) + \frac{1}{\sin^2\theta}\frac{\partial^2\varphi}{\partial\phi^2}\right\}\\
&= \frac{1}{r}\frac{\partial^2(r\varphi)}{\partial r^2} - \frac{\hat{L}^2}{r^2}\varphi,
\end{aligned} \tag{8.16}$$

where

$$\hat{L}^2 = -\frac{1}{\sin\theta}\frac{\partial}{\partial\theta}\sin\theta\frac{\partial}{\partial\theta} - \frac{1}{\sin^2\theta}\frac{\partial^2}{\partial\phi^2}, \tag{8.17}$$

is (after multiplication by $\hbar^2$) the operator representing the square of the angular momentum in quantum mechanics.

In cylindrical polars the Laplacian is

$$\nabla^2 = \frac{1}{r}\frac{\partial}{\partial r}r\frac{\partial}{\partial r} + \frac{1}{r^2}\frac{\partial^2}{\partial\theta^2} + \frac{\partial^2}{\partial z^2}. \tag{8.18}$$

8.2 Spherical harmonics

We saw that Laplace's equation in spherical polars is

$$0 = \frac{1}{r}\frac{\partial^2 (r\varphi)}{\partial r^2} - \frac{\hat{L}^2}{r^2}\varphi. \tag{8.19}$$

To solve this by the method of separation of variables, we factorize

$$\varphi = R(r)Y(\theta, \phi), \tag{8.20}$$

so that

$$\frac{1}{Rr}\frac{d^2(rR)}{dr^2} - \frac{1}{r^2}\left(\frac{1}{Y}\hat{L}^2 Y\right) = 0. \tag{8.21}$$

Taking the separation constant to be $l(l+1)$, we have

$$r\frac{d^2(rR)}{dr^2} - l(l+1)(rR) = 0, \tag{8.22}$$

and

$$\hat{L}^2 Y = l(l+1)Y. \tag{8.23}$$

The solution for R is r^l or r^{-l-1}. The equation for Y can be further decomposed by setting $Y = \Theta(\theta)\Phi(\phi)$. Looking back at the definition of $\hat{L}^2$, we see that we can take

$$\Phi(\phi) = e^{im\phi} \tag{8.24}$$

with m an integer to ensure single-valuedness. The equation for Θ is then

$$\frac{1}{\sin\theta}\frac{d}{d\theta}\left(\sin\theta\frac{d\Theta}{d\theta}\right) - \frac{m^2}{\sin^2\theta}\Theta = -l(l+1)\Theta. \tag{8.25}$$

It is convenient to set $x = \cos\theta$; then

$$\left(\frac{d}{dx}(1-x^2)\frac{d}{dx} + l(l+1) - \frac{m^2}{1-x^2}\right)\Theta = 0. \tag{8.26}$$

8.2.1 Legendre polynomials

We first look at the axially symmetric case where $m = 0$. We are left with

$$\left(\frac{d}{dx}(1-x^2)\frac{d}{dx} + l(l+1)\right)\Theta = 0. \tag{8.27}$$

This is *Legendre's equation*. We can think of it as an eigenvalue problem

$$-\left(\frac{d}{dx}(1-x^2)\frac{d}{dx}\right)\Theta(x) = l(l+1)\Theta(x), \tag{8.28}$$

on the interval $-1 \le x \le 1$, this being the range of $\cos\theta$ for real θ. Legendre's equation is of Sturm–Liouville form, but with regular singular points at $x = \pm 1$. Because the endpoints of the interval are singular, we cannot impose as boundary conditions that Θ, Θ', or some linear combination of these, be zero there. We do need some boundary conditions, however, so as to have a self-adjoint operator and a complete set of eigenfunctions.

Given one or more singular endpoints, a possible route to a well-defined eigenvalue problem is to require solutions to be square-integrable, and so normalizable. This condition suffices for the harmonic-oscillator Schrödinger equation, for example, because at most one of the two solutions is square-integrable. For Legendre's equation with $l = 0$, the two independent solutions are $\Theta(x) = 1$ and $\Theta(x) = \ln(1+x) - \ln(1-x)$. Both of these solutions have finite $L^2[-1,1]$ norms, and this square integrability persists for all values of l. Thus, demanding normalizability is not enough to select a unique boundary condition. Instead, each endpoint possesses a one-parameter family of boundary conditions that lead to self-adjoint operators. We therefore make the more restrictive demand that the allowed eigenfunctions be *finite* at the endpoints. Because the north and south poles of the sphere are not special points, this is a physically reasonable condition. When l is an integer, then one of the solutions, $P_l(x)$, becomes a polynomial, and so is finite at $x = \pm 1$. The second solution $Q_l(x)$ is divergent at both ends, and so is not an allowed solution. When l is not an integer, neither solution is finite. The eigenvalues are therefore $l(l+1)$ with l zero or a positive integer. Despite its unfamiliar form, the "finite" boundary condition makes the Legendre operator self-adjoint, and the *Legendre polynomials* $P_l(x)$ form a complete orthogonal set for $L^2[-1,1]$.

Proving orthogonality is easy: we follow the usual strategy for Sturm–Liouville equations with non-singular boundary conditions to deduce that

$$[l(l+1) - m(m+1)]\int_{-1}^{1} P_l(x)P_m(x)\,dx = \left[(P_l P_m' - P_l' P_m)(1-x^2)\right]_{-1}^{1}. \tag{8.29}$$

Since the P_l's remain finite at ± 1, the right-hand side is zero because of the $(1-x^2)$ factor, and so $\int_{-1}^{1} P_l(x)P_m(x)\,dx$ is zero if $l \ne m$. (Observe that this last step differs from the usual argument where it is the vanishing of the eigenfunction or its derivative that makes the integrated-out term zero.)

Because they are orthogonal polynomials, the $P_l(x)$ can be obtained by applying the Gram–Schmidt procedure to the sequence $1, x, x^2, \ldots$ to obtain polynomials orthogonal with respect to the $w \equiv 1$ inner product, and then fixing the normalization constant. The result of this process can be expressed in closed form as

$$P_l(x) = \frac{1}{2^l l!}\frac{d^l}{dx^l}(x^2-1)^l. \tag{8.30}$$

This is called *Rodriguez 'formula*. It should be clear that this formula outputs a polynomial of degree l. The coefficient $1/2^l l!$ comes from the traditional normalization for the Legendre polynomials that makes $P_l(1) = 1$. This convention does not lead to an orthonormal set. Instead, we have

$$\int_{-1}^{1} P_l(x)P_m(x)\,dx = \frac{2}{2l+1}\delta_{lm}. \tag{8.31}$$

It is easy to show that this integral is zero if $l > m$ – simply integrate by parts l times so as to take the l derivatives off $(x^2-1)^l$ and onto $(x^2-1)^m$, which they kill. We will evaluate the $l = m$ integral in the next section.

We now show that the $P_l(x)$ given by Rodriguez' formula are indeed solutions of Legendre's equation: let $v = (x^2-1)^l$, then

$$(1-x^2)v' + 2lxv = 0. \tag{8.32}$$

We differentiate this $l+1$ times using Leibniz' theorem

$$\begin{aligned}[uv]^{(n)} &= \sum_{m=0}^{n} \binom{n}{m} u^{(m)} v^{(n-m)} \\ &= uv^{(n)} + nu'v^{(n-1)} + \frac{1}{2}n(n-1)u''v^{(n-2)} + \dots.\end{aligned} \tag{8.33}$$

We find that

$$\begin{aligned}[(1-x^2)v']^{(l+1)} &= (1-x^2)v^{(l+2)} - (l+1)2xv^{(l+1)} - l(l+1)v^{(l)}, \\ [2xnv]^{(l+1)} &= 2xlv^{(l+1)} + 2l(l+1)v^{(l)}.\end{aligned} \tag{8.34}$$

Putting these two terms together we obtain

$$\left((1-x^2)\frac{d^2}{dx^2} - 2x\frac{d}{dx} + l(l+1)\right)\frac{d^l}{dx^l}(x^2-1)^l = 0, \tag{8.35}$$

which is Legendre's equation.

The $P_l(x)$ have alternating parity

$$P_l(-x) = (-1)^l P_l(x), \tag{8.36}$$

and the first few are

$$\begin{aligned}P_0(x) &= 1, \\ P_1(x) &= x,\end{aligned}$$

$$P_2(x) = \frac{1}{2}(3x^2 - 1),$$

$$P_3(x) = \frac{1}{2}(5x^3 - 3x),$$

$$P_4(x) = \frac{1}{8}(35x^4 - 30x^2 + 3).$$

8.2.2 Axisymmetric potential problems

The essential property of the $P_l(x)$ is that the general axisymmetric solution of $\nabla^2\varphi = 0$ can be expanded in terms of them as

$$\varphi(r,\theta) = \sum_{l=0}^{\infty} \left(A_l r^l + B_l r^{-l-1}\right) P_l(\cos\theta). \tag{8.37}$$

You should memorize this formula. You should also know by heart the explicit expressions for the first four $P_l(x)$, and the factor of $2/(2l+1)$ in the orthogonality formula.

Example: *Point charge.* Put a unit charge at the point $\mathbf{R}$, and find an expansion for the potential as a Legendre polynomial series in a neighbourhood of the origin (Figure 8.7).

Let us start by assuming that $|\mathbf{r}| < |\mathbf{R}|$. We know that in this region the point charge potential $1/|\mathbf{r} - \mathbf{R}|$ is a solution of Laplace's equation, and so we can expand

$$\frac{1}{|\mathbf{r} - \mathbf{R}|} \equiv \frac{1}{\sqrt{r^2 + R^2 - 2rR\cos\theta}} = \sum_{l=0}^{\infty} A_l r^l P_l(\cos\theta). \tag{8.38}$$

We knew that the coefficients B_l were zero because φ is finite when $r = 0$. We can find the coefficients A_l by setting $\theta = 0$ and Taylor expanding

$$\frac{1}{|\mathbf{r} - \mathbf{R}|} = \frac{1}{R - r} = \frac{1}{R}\left(1 + \left(\frac{r}{R}\right) + \left(\frac{r}{R}\right)^2 + \cdots\right), \quad r < R. \tag{8.39}$$

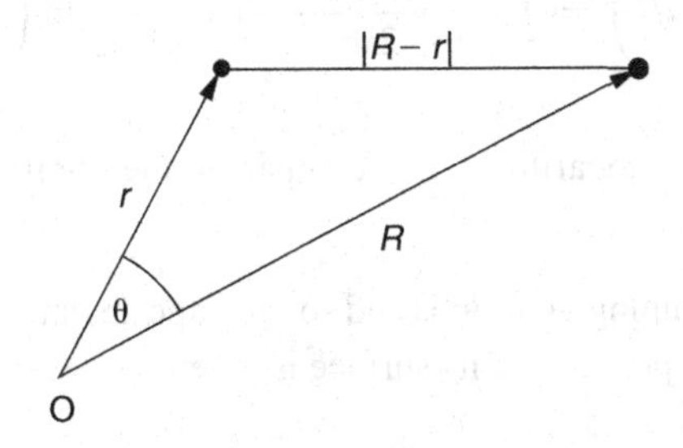

Figure 8.7 Geometry for generating function.

By comparing the two series and noting that $P_l(1) = 1$, we find that $A_l = R^{-l-1}$. Thus

$$\frac{1}{\sqrt{r^2+R^2-2rR\cos\theta}} = \frac{1}{R}\sum_{l=0}^{\infty}\left(\frac{r}{R}\right)^l P_l(\cos\theta), \quad r<R. \tag{8.40}$$

This last expression is the *generating function formula* for Legendre polynomials. It is also a useful formula to have in your long-term memory.

If $|\mathbf{r}| > |\mathbf{R}|$, then we must take

$$\frac{1}{|\mathbf{r}-\mathbf{R}|} \equiv \frac{1}{\sqrt{r^2+R^2-2rR\cos\theta}} = \sum_{l=0}^{\infty} B_l r^{-l-1} P_l(\cos\theta), \tag{8.41}$$

because we know that φ tends to zero when $r=\infty$. We now set $\theta = 0$ and compare with

$$\frac{1}{|\mathbf{r}-\mathbf{R}|} = \frac{1}{r-R} = \frac{1}{r}\left(1+\left(\frac{R}{r}\right)+\left(\frac{R}{r}\right)^2+\cdots\right), \quad R<r, \tag{8.42}$$

to get

$$\frac{1}{\sqrt{r^2+R^2-2rR\cos\theta}} = \frac{1}{r}\sum_{l=0}^{\infty}\left(\frac{R}{r}\right)^l P_l(\cos\theta), \quad R<r. \tag{8.43}$$

Observe that we made no use of the normalization integral

$$\int_{-1}^{1}\{P_l(x)\}^2\,dx = 2/(2l+1) \tag{8.44}$$

in deriving the generating function expansion for the Legendre polynomials. The following exercise shows that this expansion, taken together with their previously established orthogonality property, can be used to establish (8.44).

Exercise 8.1: Use the generating function for Legendre polynomials $P_l(x)$ to show that

$$\sum_{l=0}^{\infty} z^{2l}\left(\int_{-1}^{1}\{P_l(x)\}^2\,dx\right) = \int_{-1}^{1}\frac{1}{1-2xz+z^2}\,dx = -\frac{1}{z}\ln\left(\frac{1-z}{1+z}\right), \quad |z|<1.$$

By Taylor expanding the logarithm, and comparing the coefficients of z^{2l}, evaluate $\int_{-1}^{1}\{P_l(x)\}^2\,dx$.

Example: A planet is spinning on its axis and so its shape deviates slightly from a perfect sphere (Figure 8.8). The position of its surface is given by

$$R(\theta,\phi) = R_0 + \eta P_2(\cos\theta). \tag{8.45}$$

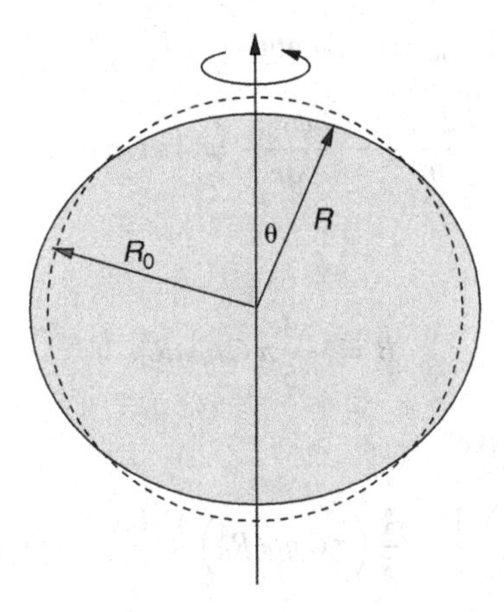

Figure 8.8 Deformed planet.

Observe that, to first order in η, this deformation does not alter the volume of the body. Assuming that the planet has a uniform density ρ_0, compute the external gravitational potential of the planet.

The gravitational potential obeys Poisson's equation

$$\nabla^2\phi = 4\pi G\rho(\mathbf{x}), \tag{8.46}$$

where G is Newton's gravitational constant. We expand ϕ as a power series in η

$$\phi(r,\theta) = \phi_0(r,\theta) + \eta\phi_1(r,\theta) + \dots \tag{8.47}$$

We also decompose the gravitating mass into a uniform undeformed sphere, which gives the external potential

$$\phi_{0,\text{ext}}(r,\theta) = -\left(\frac{4}{3}\pi R_0^3\rho_0\right)\frac{G}{r}, \quad r > R_0, \tag{8.48}$$

and a thin spherical shell of areal mass-density

$$\sigma(\theta) = \rho_0\eta P_2(\cos\theta). \tag{8.49}$$

The thin shell gives rise to the potential

$$\phi_{1,\text{int}}(r,\theta) = Ar^2 P_2(\cos\theta), \quad r < R_0, \tag{8.50}$$

and

$$\phi_{1,\text{ext}}(r,\theta) = B\frac{1}{r^3}P_2(\cos\theta), \quad r > R_0. \tag{8.51}$$

At the shell we must have $\phi_{1,\text{int}} = \phi_{1,\text{ext}}$ and

$$\frac{\partial \phi_{1,\text{ext}}}{\partial r} - \frac{\partial \phi_{1,\text{int}}}{\partial r} = 4\pi G\sigma(\theta). \tag{8.52}$$

Thus $A = BR_0^{-5}$, and

$$B = -\frac{4}{5}\pi G\eta\rho_0 R_0^4. \tag{8.53}$$

Putting this together, we have

$$\phi(r,\theta) = -\left(\frac{4}{3}\pi G\rho_0 R_0^3\right)\frac{1}{r} - \frac{4}{5}\left(\pi G\eta\rho_0 R_0^4\right)\frac{P_2(\cos\theta)}{r^3} + O(\eta^2), \quad r > R_0. \tag{8.54}$$

8.2.3 General spherical harmonics

When we do not have axisymmetry, we need the full set of spherical harmonics. These involve solutions of

$$\left(\frac{d}{dx}(1-x^2)\frac{d}{dx} + l(l+1) - \frac{m^2}{1-x^2}\right)\Phi = 0, \tag{8.55}$$

which is the *associated Legendre* equation. This looks like another complicated equation with singular endpoints, but its bounded solutions can be obtained by differentiating Legendre polynomials. On substituting $y = (1-x^2)^{m/2}z(x)$ into (8.55), and comparing the resulting equation for $z(x)$ with the m-th derivative of Legendre's equation, we find that

$$P_l^m(x) \stackrel{\text{def}}{=} (-1)^m(1-x^2)^{m/2}\frac{d^m}{dx^m}P_l(x) \tag{8.56}$$

is a solution of (8.55) that remains finite ($m = 0$) or goes to zero ($m > 0$) at the endpoints $x = \pm 1$. Since $P_l(x)$ is a polynomial of degree l, we must have $P_l^m(x) = 0$ if $m > l$. For each l, the allowed values of m in this formula are therefore $0, 1, \ldots, l$. Our definition (8.56) of the $P_l^m(x)$ can be extended to negative integer m by interpreting $d^{-|m|}/dx^{-|m|}$ as an instruction to integrate the Legendre polynomial m times, instead of differentiating it, but the resulting $P_l^{-|m|}(x)$ are proportional to $P_l^m(x)$, so nothing new is gained by this conceit.

The *spherical harmonics* are the normalized product of these *associated Legendre functions* with the corresponding $e^{im\phi}$:

$$Y_l^m(\theta,\phi) \propto P_l^{|m|}(\cos\theta)e^{im\phi}, \quad -l \le m \le l. \tag{8.57}$$

The first few are

$$l = 0 \quad Y_0^0 = \frac{1}{\sqrt{4\pi}} \tag{8.58}$$

$$l=1 \quad \begin{cases} Y_1^1 &= -\sqrt{\frac{3}{8\pi}}\sin\theta\, e^{i\phi}, \\ Y_1^0 &= \sqrt{\frac{3}{4\pi}}\cos\theta, \\ Y_1^{-1} &= \sqrt{\frac{3}{8\pi}}\sin\theta\, e^{-i\phi}, \end{cases} \tag{8.59}$$

$$l=2 \quad \begin{cases} Y_2^2 &= \frac{1}{4}\sqrt{\frac{15}{2\pi}}\sin^2\theta\, e^{2i\phi}, \\ Y_2^1 &= -\sqrt{\frac{15}{8\pi}}\sin\theta\cos\theta\, e^{i\phi}, \\ Y_2^0 &= \sqrt{\frac{5}{4\pi}}\left(\frac{3}{2}\cos^2\theta - \frac{1}{2}\right), \\ Y_2^{-1} &= \sqrt{\frac{15}{8\pi}}\sin\theta\cos\theta\, e^{-i\phi}, \\ Y_2^{-2} &= \frac{1}{4}\sqrt{\frac{15}{2\pi}}\sin^2\theta\, e^{-2i\phi}. \end{cases} \tag{8.60}$$

The spherical harmonics compose an orthonormal

$$\int_0^{2\pi} d\phi \int_0^{\pi} \sin\theta d\theta \left[Y_l^m(\theta,\phi)\right]^* Y_{l'}^{m'}(\theta,\phi) = \delta_{ll'}\delta_{mm'}, \tag{8.61}$$

and complete

$$\sum_{l=0}^{\infty}\sum_{m=-l}^{l} [Y_l^m(\theta',\phi')]^* Y_l^m(\theta,\phi) = \delta(\phi-\phi')\delta(\cos\theta' - \cos\theta) \tag{8.62}$$

set of functions on the unit sphere. In terms of them, the general solution to $\nabla^2\varphi = 0$ is

$$\varphi(r,\theta,\phi) = \sum_{l=0}^{\infty}\sum_{m=-l}^{l} \left(A_{lm}r^l + B_{lm}r^{-l-1}\right) Y_l^m(\theta,\phi). \tag{8.63}$$

This is definitely a formula to remember.

For $m=0$, the spherical harmonics are independent of the azimuthal angle ϕ, and so must be proportional to the Legendre polynomials. The exact relation is

$$Y_l^0(\theta,\phi) = \sqrt{\frac{2l+1}{4\pi}} P_l(\cos\theta). \tag{8.64}$$

If we use a unit vector $\mathbf{n}$ to denote a point on the unit sphere, we have the symmetry properties

$$[Y_l^m(\mathbf{n})]^* = (-1)^m Y_l^{-m}(\mathbf{n}), \qquad Y_l^m(-\mathbf{n}) = (-1)^l Y_l^m(\mathbf{n}). \tag{8.65}$$

These identities are useful when we wish to know how quantum mechanical wavefunctions transform under time reversal or parity.

There is an addition theorem

$$P_l(\cos\gamma) = \frac{4\pi}{2l+1} \sum_{m=-l}^{l} [Y_l^m(\theta',\phi')]^* Y_l^m(\theta,\phi), \tag{8.66}$$

where γ is the angle between the directions (θ,ϕ) and (θ',ϕ'), and is found from

$$\cos\gamma = \cos\theta\cos\theta' + \sin\theta\sin\theta'\cos(\phi-\phi'). \tag{8.67}$$

The addition theorem is established by first showing that the right-hand side is rotationally invariant, and then setting the direction (θ',ϕ') to point along the z-axis. Addition theorems of this sort are useful because they allow one to replace a simple function of an entangled variable by a sum of functions of unentangled variables. For example, the point-charge potential can be disentangled as

$$\frac{1}{|\mathbf{r}-\mathbf{r}'|} = \sum_{l=0}^{\infty}\sum_{m=-l}^{l} \frac{4\pi}{2l+1}\left(\frac{r_<^l}{r_>^{l+1}}\right)[Y_l^m(\theta',\phi')]^* Y_l^m(\theta,\phi), \tag{8.68}$$

where $r_<$ is the smaller of $|\mathbf{r}|$ or $|\mathbf{r}'|$, and $r_>$ is the greater and (θ,ϕ), (θ',ϕ') specify the direction of $\mathbf{r}$, $\mathbf{r}'$, respectively. This expansion is derived by combining the generating function for the Legendre polynomials with the addition formula. It is useful for defining and evaluating multipole expansions.

Exercise 8.2: Show that

$$\left.\begin{matrix} Y_1^1 \\ Y_1^0 \\ Y_1^{-1} \end{matrix}\right\} \propto \left\{\begin{matrix} x+iy, \\ z, \\ x-iy \end{matrix}\right.$$

$$\left.\begin{matrix} Y_2^2 \\ Y_2^1 \\ Y_2^0 \\ Y_2^{-1} \\ Y_2^{-2} \end{matrix}\right\} \propto \left\{\begin{matrix} (x+iy)^2, \\ (x+iy)z, \\ x^2+y^2-2z^2, \\ (x-iy)z, \\ (x-iy)^2, \end{matrix}\right.$$

where $x^2+y^2+z^2=1$ are the usual cartesian coordinates, restricted to the unit sphere.

8.3 Bessel functions

In cylindrical polar coordinates, Laplace's equation is

$$0 = \nabla^2\varphi = \frac{1}{r}\frac{\partial}{\partial r}r\frac{\partial\varphi}{\partial r} + \frac{1}{r^2}\frac{\partial^2\varphi}{\partial\theta^2} + \frac{\partial^2\varphi}{\partial z^2}. \tag{8.69}$$

If we set $\varphi = R(r)e^{im\phi}e^{\pm kx}$ we find that $R(r)$ obeys

$$\frac{d^2R}{dr^2} + \frac{1}{r}\frac{dR}{dr} + \left(k^2 - \frac{m^2}{r^2}\right)R = 0. \tag{8.70}$$

Now

$$\frac{d^2y}{dx^2} + \frac{1}{x}\frac{dy}{dx} + \left(1 - \frac{\nu^2}{x^2}\right)y = 0 \tag{8.71}$$

is *Bessel's equation* and its solutions are *Bessel functions* of order ν. The solutions for R will therefore be Bessel functions of order m, but with x replaced by kr.

8.3.1 Cylindrical Bessel functions

We now set about solving Bessel's equation,

$$\frac{d^2y}{dx^2} + \frac{1}{x}\frac{dy}{dx} + \left(1 - \frac{\nu^2}{x^2}\right)y(x) = 0. \tag{8.72}$$

This has a regular singular point at the origin, and an irregular singular point at infinity. We seek a series solution of the form

$$y = x^\lambda(1 + a_1x + a_2x^2 + \cdots), \tag{8.73}$$

and find from the indicial equation that $\lambda = \pm\nu$. Setting $\lambda = \nu$ and inserting the series into the equation, we find, with a conventional choice for normalization, that

$$y = J_\nu(x) \stackrel{\text{def}}{=} \left(\frac{x}{2}\right)^\nu \sum_{n=0}^{\infty} \frac{(-1)^n}{n!(n+\nu)!}\left(\frac{x}{2}\right)^{2n}. \tag{8.74}$$

Here $(n+\nu)! \equiv \Gamma(n+\nu+1)$. The functions $J_\nu(x)$ are called *cylindrical Bessel functions.*

If ν is an integer we find that $J_{-n}(x) = (-1)^n J_n(x)$, so we have only found one of the two independent solutions. It is therefore traditional to define the *Neumann function*

$$N_\nu(x) = \frac{J_\nu(x)\cos\nu\pi - J_{-\nu}(x)}{\sin\nu\pi}, \tag{8.75}$$

as this remains an independent second solution even as ν becomes integral. At short distance, and for ν not an integer

$$\begin{aligned} J_\nu(x) &= \left(\frac{x}{2}\right)^\nu \frac{1}{\Gamma(\nu+1)} + \cdots, \\ N_\nu(x) &= \frac{1}{\pi}\left(\frac{x}{2}\right)^{-\nu}\Gamma(\nu) + \cdots \end{aligned} \tag{8.76}$$

When ν tends to zero, we have

$$J_0(x) = 1 - \frac{1}{4}x^2 + \cdots$$
$$N_0(x) = \left(\frac{2}{\pi}\right)(\ln x/2 + \gamma) + \cdots, \tag{8.77}$$

where $\gamma = 0.57721\ldots$ denotes the Euler–Mascheroni constant. For fixed ν, and $x \gg \nu$, we have the asymptotic expansions

$$J_\nu(x) \sim \sqrt{\frac{2}{\pi x}}\cos\left(x - \frac{1}{2}\nu\pi - \frac{1}{4}\pi\right)\left(1 + O\left(\frac{1}{x}\right)\right), \tag{8.78}$$

$$N_\nu(x) \sim \sqrt{\frac{2}{\pi x}}\sin\left(x - \frac{1}{2}\nu\pi - \frac{1}{4}\pi\right)\left(1 + O\left(\frac{1}{x}\right)\right). \tag{8.79}$$

It is therefore natural to define the *Hankel functions*

$$H_\nu^{(1)}(x) = J_\nu(x) + iN_\nu(x) \sim \sqrt{\frac{2}{\pi x}}e^{i(x-\nu\pi/2-\pi/4)}, \tag{8.80}$$

$$H_\nu^{(2)}(x) = J_\nu(x) - iN_\nu(x) \sim \sqrt{\frac{2}{\pi x}}e^{-i(x-\nu\pi/2-\pi/4)}. \tag{8.81}$$

We will derive these asymptotic forms in Chapter 19.

Generating function

The two-dimensional wave equation

$$\left(\nabla^2 - \frac{1}{c^2}\frac{\partial^2}{\partial t^2}\right)\Phi(r,\theta,t) = 0 \tag{8.82}$$

has solutions

$$\Phi = e^{i\omega t}e^{in\theta}J_n(kr), \tag{8.83}$$

where $k = |\omega|/c$. Equivalently, the two-dimensional Helmholtz equation

$$(\nabla^2 + k^2)\Phi = 0, \tag{8.84}$$

has solutions $e^{in\theta}J_n(kr)$. It also has solutions with $J_n(kr)$ replaced by $N_n(kr)$, but these are not finite at the origin. Since the $e^{in\theta}J_n(kr)$ are the only solutions that are finite at the origin, any other finite solution should be expandable in terms of them. In particular, we should be able to expand a plane-wave solution:

$$e^{iky} = e^{ikr\sin\theta} = \sum_n a_n e^{in\theta}J_n(kr). \tag{8.85}$$

As we will see in a moment, the a_n's are all unity, so in fact

$$e^{ikr\sin\theta} = \sum_{n=-\infty}^{\infty} e^{in\theta} J_n(kr). \tag{8.86}$$

This *generating function* is the historical origin of the Bessel functions. They were introduced by the astronomer Wilhelm Bessel as a method of expressing the eccentric anomaly of a planetary position as a Fourier sine series in the mean anomaly – a modern version of Hipparchus' epicycles.

From the generating function we see that

$$J_n(x) = \frac{1}{2\pi} \int_0^{2\pi} e^{-in\theta + ix\sin\theta}\, d\theta. \tag{8.87}$$

Whenever you come across a formula like this, involving the Fourier integral of the exponential of a trigonometric function, you are probably dealing with a Bessel function.

The generating function can also be written as

$$e^{\frac{x}{2}\left(t-\frac{1}{t}\right)} = \sum_{n=-\infty}^{\infty} t^n J_n(x). \tag{8.88}$$

Expanding the left-hand side and using the binomial theorem, we find

$$\begin{aligned} LHS &= \sum_{m=0}^{\infty} \left(\frac{x}{2}\right)^m \frac{1}{m!} \left[\sum_{r+s=m} \frac{(r+s)!}{r!s!} (-1)^s t^r t^{-s} \right], \\ &= \sum_{r=0}^{\infty}\sum_{s=0}^{\infty} (-1)^s \left(\frac{x}{2}\right)^{r+s} \frac{t^{r-s}}{r!s!}, \\ &= \sum_{n=-\infty}^{\infty} t^n \left\{ \sum_{s=0}^{\infty} \frac{(-1)^s}{s!(s+n)!} \left(\frac{x}{2}\right)^{2s+n} \right\}. \end{aligned} \tag{8.89}$$

We recognize that the sum in the braces is the series expansion defining $J_n(x)$. This therefore proves the generating function formula.

Bessel identities

There are many identities and integrals involving Bessel functions. The standard reference is the monumental *Treatise on the Theory of Bessel Functions* by G. N. Watson. Here are just a few formulæ for your delectation:

(i) Starting from the generating function

$$\exp\left\{\tfrac{1}{2}x\left(t - \frac{1}{t}\right)\right\} = \sum_{n=-\infty}^{\infty} J_n(x) t^n, \tag{8.90}$$

we can, with a few lines of work, establish the recurrence relations

$$2J_n'(x) = J_{n-1}(x) - J_{n+1}(x), \tag{8.91}$$

$$\frac{2n}{x}J_n(x) = J_{n-1}(x) + J_{n+1}(x), \tag{8.92}$$

together with

$$J_0'(x) = -J_1(x), \tag{8.93}$$

$$J_n(x+y) = \sum_{r=-\infty}^{\infty} J_r(x)J_{n-r}(y). \tag{8.94}$$

(ii) From the series expansion for $J_n(x)$ we find

$$\frac{d}{dx}\left\{x^n J_n(x)\right\} = x^n J_{n-1}(x). \tag{8.95}$$

(iii) By similar methods, we find

$$\left(\frac{1}{x}\frac{d}{dx}\right)^m \left\{x^{-n}J_n(x)\right\} = (-1)^m x^{-n-m} J_{n+m}(x). \tag{8.96}$$

(iv) Again from the series expansion, we find

$$\int_0^\infty J_0(ax)e^{-px}dx = \frac{1}{\sqrt{a^2+p^2}}. \tag{8.97}$$

Semiclassical picture

The Schrödinger equation

$$-\frac{\hbar^2}{2m}\nabla^2\psi = E\psi \tag{8.98}$$

can be separated in cylindrical polar coordinates, and has eigenfunctions

$$\psi_{k,l}(r,\theta) = J_l(kr)e^{il\theta}. \tag{8.99}$$

The eigenvalues are $E = \hbar^2k^2/2m$. The quantity $L = \hbar l$ is the angular momentum of the Schrödinger particle about the origin. If we impose rigid-wall boundary conditions that $\psi_{k,l}(r,\theta)$ vanish on the circle $r = R$, then the allowed k form a discrete set $k_{l,n}$, where $J_l(k_{l,n}R) = 0$. To find the energy eigenvalues we therefore need to know the location of the zeros of $J_l(x)$. There is no closed form equation for these numbers, but they are tabulated. The zeros for $kR \gg l$ are also approximated by the zeros of the asymptotic expression

$$J_l(kR) \sim \sqrt{\frac{2}{\pi kR}} \cos\left(kR - \frac{1}{2}l\pi - \frac{1}{4}\pi\right), \tag{8.100}$$

which are located at

$$k_{l,n}R = \frac{1}{2}l\pi + \frac{1}{4}\pi + (2n+1)\frac{\pi}{2}. \tag{8.101}$$

If we let $R \to \infty$, then the spectrum becomes continuous and we are describing unconfined scattering states. Since the particles are free, their classical motion is in a straight line at constant velocity. A classical particle making a closest approach at a distance $r_{\rm min}$ has angular momentum $L = pr_{\rm min}$. Since $p = \hbar k$ is the particle's linear momentum, we have $l = kr_{\rm min}$. Because the classical particle is never closer than $r_{\rm min}$, the quantum mechanical wavefunction representing such a particle will become evanescent (i.e. tend rapidly to zero) as soon as r is smaller than $r_{\rm min}$. We therefore expect that $J_l(kr) \approx 0$ if $kr < l$. This effect is dramatically illustrated by the *Mathematica*™ plot in Figure 8.9.

Improved asymptotic expressions, which give a better estimate of the $J_l(kr)$ zeros, are the approximations

$$\begin{aligned} J_l(kr) &\approx \sqrt{\frac{2}{\pi kx}} \cos(kx - l\theta - \pi/4), \quad r \gg r_{\rm min}, \\ N_l(kr) &\approx \sqrt{\frac{2}{\pi kx}} \sin(kx - l\theta - \pi/4), \quad r \gg r_{\rm min}. \end{aligned} \tag{8.102}$$

Here $\theta = \cos^{-1}(r_{\rm min}/r)$ and $x = r\sin\theta$ are functions of r. They have a geometric origin in the right-angled triangle in Figure 8.10. The parameter x has the physical interpretation

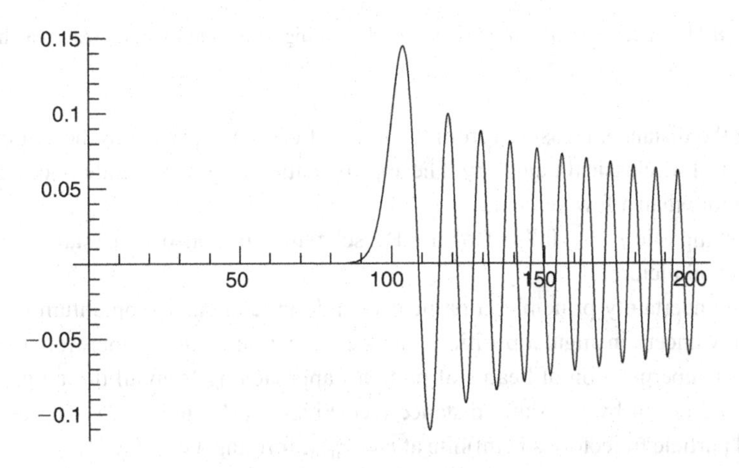

Figure 8.9 $J_{100}(x)$.

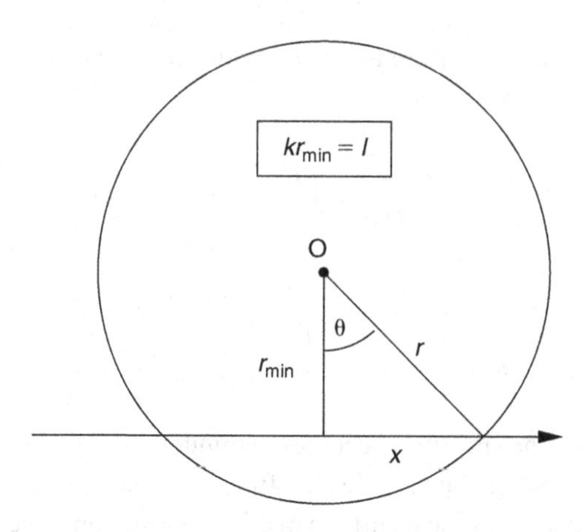

Figure 8.10 The geometric origin of $x(r)$ and $\theta(r)$ in (8.102).

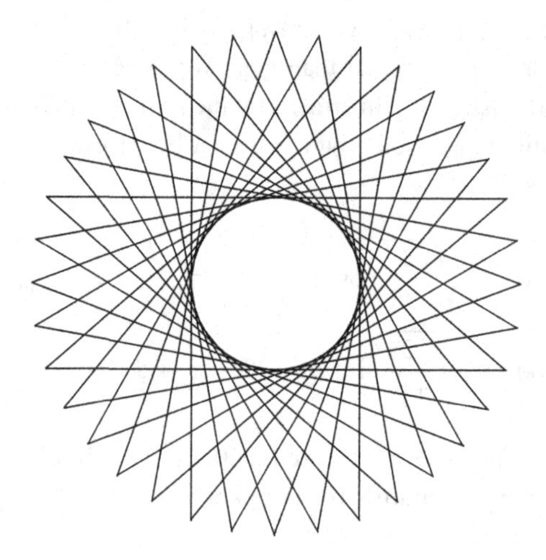

Figure 8.11 A collection of trajectories, each missing the origin by r_{min}, leaves a "hole".

of being the distance, measured from the point of closest approach to the origin, along the straight-line classical trajectory. The approximation is quite accurate once r exceeds r_{min} by more than a few percent.

The asymptotic $r^{-1/2}$ fall-off of the Bessel function is also understandable in the semiclassical picture.

By the uncertainty principle, a particle with definite angular momentum must have completely uncertain angular position. The wavefunction $J_l(kr)e^{il\theta}$ therefore represents a coherent superposition of beams of particles approaching from all directions, but all missing the origin by the same distance (see Figures 8.11 and 8.12). The density of classical particle trajectories is infinite at $r = r_{min}$, forming a caustic. By "conservation of lines", the particle density falls off as $1/r$ as we move outwards. The particle density

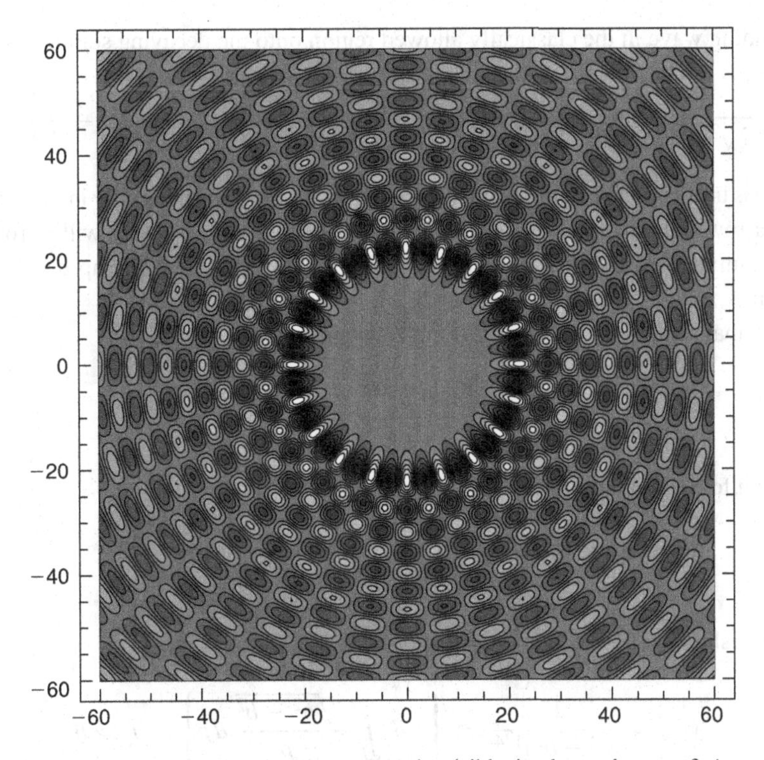

Figure 8.12 The hole appearing in Figure 8.11 is visible in the real part of $\psi_{k,20}(r\theta) = e^{i20\theta} J_{20}(kr)$.

is proportional to $|\psi|^2$, so ψ itself decreases as $r^{-1/2}$. In contrast to the classical particle density, the quantum mechanical wavefunction amplitude remains finite at the caustic – the "geometric optics" infinity being tempered by diffraction effects.

Exercise 8.3: The WKB (Wentzel–Kramers–Brillouin) approximation to a solution of the Schrödinger equation

$$-\frac{d^2\psi}{dx^2} + V(x)\psi(x) = E\psi(x)$$

sets

$$\psi(x) \approx \frac{1}{\sqrt{\kappa(x)}} \exp\left\{\pm i \int_a^x \kappa(\xi)\, d\xi\right\},$$

where $\kappa(x) = \sqrt{E - V(x)}$, and a is some conveniently chosen constant. This form of the approximation is valid in classically allowed regions, where κ is real, and away from "turning points" where κ goes to zero. In a classically forbidden region, where κ is imaginary, the solutions should decay exponentially. The *connection rule* that matches the standing wave in the classically allowed region onto the decaying solution is

$$\frac{1}{2\sqrt{|\kappa(x)|}}\exp\left\{-\left|\int_a^x \kappa(\xi)\,d\xi\right|\right\} \to \frac{1}{\sqrt{\kappa(x)}}\cos\left\{\left|\int_a^x \kappa(\xi)\,d\xi\right| - \frac{\pi}{4}\right\},$$

where a is the classical turning point. (The connection is safely made only in the direction of the arrow. This is because a small error in the phase of the cosine will introduce a small admixture of the growing solution, which will eventually swamp the decaying solution.)

Show that setting $y(r) = r^{-1/2}\psi(r)$ in Bessel's equation

$$-\frac{d^2y}{dr^2} - \frac{1}{r}\frac{dy}{dr} + \frac{l^2y}{r^2} = k^2y$$

reduces it to Schrödinger form

$$-\frac{d^2\psi}{dr^2} + \frac{(l^2 - 1/4)}{r^2}\psi = k^2\psi.$$

From this show that a WKB approximation to $y(r)$ is

$$\begin{aligned} y(r) &\approx \frac{1}{(r^2 - b^2)^{1/4}}\exp\left\{\pm ik\int_b^r \frac{\sqrt{\rho^2 - b^2}}{\rho}\,d\rho\right\}, \quad r \gg b \\ &= \frac{1}{\sqrt{x(r)}}\exp\{\pm i[kx(r) - l\theta(r)]\}, \end{aligned}$$

where $kb = \sqrt{l^2 - 1/4} \approx l$, and $x(r)$ and $\theta(r)$ were defined in connection with (8.102). Deduce that the expressions (8.102) are WKB approximations and are therefore accurate once we are away from the classical turning point at $r = b \equiv r_{\min}$.

8.3.2 Orthogonality and completeness

We can write the equation obeyed by $J_n(kr)$ in Sturm–Liouville form. We have

$$\frac{1}{r}\frac{d}{dr}\left(r\frac{dy}{dr}\right) + \left(k^2 - \frac{m^2}{r^2}\right)y = 0. \tag{8.103}$$

Comparison with the standard Sturm–Liouville equation shows that the weight function, $w(r)$, is r, and the eigenvalues are k^2.

From Lagrange's identity we obtain

$$\begin{aligned} (k_1^2 - k_2^2)\int_0^R J_m(k_1r)J_m(k_2r)r\,dr \\ = R\left[k_2J_m(k_1R)J_m'(k_2R) - k_1J_m(k_2R)J_m'(k_1R)\right]. \end{aligned} \tag{8.104}$$

We have no contribution from the origin on the right-hand side because all J_m Bessel functions except J_0 vanish there, whilst $J_0'(0) = 0$. For each m we get a set of orthogonal functions, $J_m(k_n x)$, provided the $k_n R$ are chosen to be roots of $J_m(k_n R) = 0$ or $J_m'(k_n R) = 0$.

We can find the normalization constants by differentiating (8.104) with respect to k_1 and then setting $k_1 = k_2$ in the result. We find

$$\begin{aligned}\int_0^R \left[J_m(kr)\right]^2 r\,dr &= \frac{1}{2}R^2\left[\left[J_m'(kR)\right]^2 + \left(1 - \frac{m^2}{k^2R^2}\right)\left[J_m(kR)\right]^2\right],\\ &= \frac{1}{2}R^2\left[[J_n(kR)]^2 - J_{n-1}(kR)J_{n+1}(kR)\right].\end{aligned} \tag{8.105}$$

(The second equality follows on applying the recurrence relations for the $J_n(kr)$, and provides an expression that is perhaps easier to remember.) For Dirichlet boundary conditions we will require $k_n R$ to be a zero of J_m, and so we have

$$\int_0^R \left[J_m(kr)\right]^2 r\,dr = \frac{1}{2}R^2\left[J_m'(kR)\right]^2. \tag{8.106}$$

For Neumann boundary conditions we require $k_n R$ to be a zero of J_m'. In this case

$$\int_0^R \left[J_m(kr)\right]^2 r\,dr = \frac{1}{2}R^2\left(1 - \frac{m^2}{k^2R^2}\right)\left[J_m(kR)\right]^2. \tag{8.107}$$

Example: Harmonic function in a cylinder. We wish to solve $\nabla^2 V = 0$ within a cylinder of height L and radius a (Figure 8.13). The voltage is prescribed on the upper surface of the cylinder: $V(r,\theta,L) = U(r,\theta)$. We are told that $V = 0$ on all other parts of the boundary.

The general solution of Laplace's equation will be a sum of terms such as

$$\left\{\begin{matrix}\sinh(kz)\\ \cosh(kz)\end{matrix}\right\} \times \left\{\begin{matrix}J_m(kr)\\ N_m(kr)\end{matrix}\right\} \times \left\{\begin{matrix}\sin(m\theta)\\ \cos(m\theta)\end{matrix}\right\}, \tag{8.108}$$

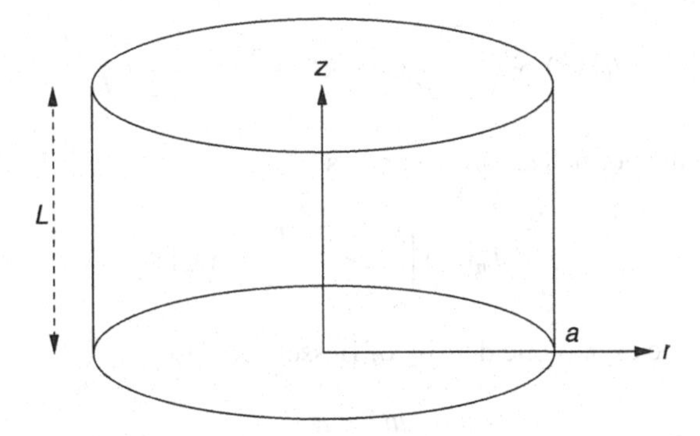

Figure 8.13 Cylinder geometry.

where the braces indicate a choice of upper or lower functions. We must take only the $\sinh(kz)$ terms because we know that $V = 0$ at $z = 0$, and only the $J_m(kr)$ terms because V is finite at $r = 0$. The k's are also restricted by the boundary condition on the sides of the cylinder to be such that $J_m(ka) = 0$. We therefore expand the prescribed voltage as

$$U(r,\theta) = \sum_{m,n} \sinh(k_{nm}L) J_m(k_{mn}r) \left[A_{nm}\sin(m\theta) + B_{nm}\cos(m\theta)\right], \tag{8.109}$$

and use the orthonormality of the trigonometric and Bessel function to find the coefficients to be

$$A_{nm} = \frac{2\mathrm{cosech}(k_{nm}L)}{\pi a^2 [J_m'(k_{nm}a)]^2} \int_0^{2\pi} d\theta \int_0^a U(r,\theta) J_m(k_{nm}r)\sin(m\theta)\, r dr, \tag{8.110}$$

$$B_{nm} = \frac{2\mathrm{cosech}(k_{nm}L)}{\pi a^2 [J_m'(k_{nm}a)]^2} \int_0^{2\pi} d\theta \int_0^a U(r,\theta) J_m(k_{nm}r)\cos(m\theta)\, r dr, \quad m \neq 0, \tag{8.111}$$

and

$$B_{n0} = \frac{1}{2}\frac{2\mathrm{cosech}(k_{n0}L)}{\pi a^2 [J_0'(k_{n0}a)]^2} \int_0^{2\pi} d\theta \int_0^a U(r,\theta) J_0(k_{n0}r)\, r dr. \tag{8.112}$$

Then we fit the boundary data expansion to the general solution, and so find

$$V(r,\theta,z) = \sum_{m,n} \sinh(k_{nm}z) J_m(k_{mn}r) \left[A_{nm}\sin(m\theta) + B_{nm}\cos(m\theta)\right]. \tag{8.113}$$

Hankel transforms

When the radius, R, of the region in which we are performing our eigenfunction expansion becomes infinite, the eigenvalue spectrum will become continuous, and the sum over the discrete k_n Bessel-function zeros must be replaced by an integral over k. By using the asymptotic approximation

$$J_n(kR) \sim \sqrt{\frac{2}{\pi kR}} \cos\left(kR - \frac{1}{2}n\pi - \frac{1}{4}\pi\right), \tag{8.114}$$

we may estimate the normalization integral as

$$\int_0^R \left[J_m(kr)\right]^2 r\, dr \sim \frac{R}{\pi k} + O(1). \tag{8.115}$$

We also find that the asymptotic density of Bessel zeros is

$$\frac{dn}{dk} = \frac{R}{\pi}. \tag{8.116}$$

Putting these two results together shows that the continuous-spectrum orthogonality and completeness relations are

$$\int_0^\infty J_n(kr)J_n(k'r)\,r dr = \frac{1}{k}\delta(k-k'), \tag{8.117}$$

$$\int_0^\infty J_n(kr)J_n(kr')\,k dk = \frac{1}{r}\delta(r-r'), \tag{8.118}$$

respectively. These two equations establish that the *Hankel transform* (also called the *Fourier–Bessel transform*) of a function $f(r)$, which is defined by

$$F(k) = \int_0^\infty J_n(kr)f(r)r\,dr, \tag{8.119}$$

has as its inverse

$$f(r) = \int_0^\infty J_n(kr)F(k)k\,dk. \tag{8.120}$$

(See exercise 8.14 for an alternative derivation of the Hankel-transform pair.)
Some Hankel transform pairs:

$$\begin{aligned}\int_0^\infty e^{-ar}J_0(kr)\,dr &= \frac{1}{\sqrt{k^2+a^2}},\\ \int_0^\infty \frac{J_0(kr)}{\sqrt{k^2+a^2}}\,kdk &= \frac{e^{-ar}}{r}.\end{aligned} \tag{8.121}$$

$$\begin{aligned}\int_0^\infty \cos(ar)J_0(kr)\,dr &= \begin{cases}0, & k<a,\\ 1/\sqrt{k^2-a^2}, & k>a.\end{cases}\\ \int_a^\infty \frac{J_0(kr)}{\sqrt{k^2-a^2}}\,kdk &= \frac{1}{r}\cos(ar).\end{aligned} \tag{8.122}$$

$$\begin{aligned}\int_0^\infty \sin(ar)J_0(kr)\,dr &= \begin{cases}1/\sqrt{a^2-k^2}, & k<a,\\ 0, & k>a.\end{cases}\\ \int_0^a \frac{J_0(kr)}{\sqrt{a^2-k^2}}\,kdk &= \frac{1}{r}\sin(ar).\end{aligned} \tag{8.123}$$

Example: Weber's disc problem. Consider a thin isolated conducting disc of radius a lying on the xy-plane in $\mathbb{R}^3$. The disc is held at potential V_0. We seek the potential V in the entirety of $\mathbb{R}^3$, such that $V \to 0$ at infinity.

It is easiest to first find V in the half-space $z \geq 0$, and then extend the solution to $z < 0$ by symmetry. Because the problem is axisymmetric, we will make use of cylindrical polar coordinates with their origin at the centre of the disc. In the region $z \geq 0$ the potential $V(r,z)$ obeys

$$
\begin{aligned}
\nabla^2 V(r,z) &= 0, \quad z > 0, \\
V(r,z) &\to 0 \quad |z| \to \infty, \\
V(r,0) &= V_0, \quad r < a, \\
\left.\frac{\partial V}{\partial z}\right|_{z=0} &= 0, \quad r > a.
\end{aligned}
\tag{8.124}
$$

This is a *mixed boundary value problem*. We have imposed Dirichlet boundary conditions on $r < a$ and Neumann boundary conditions for $r > a$.

We expand the axisymmetric solution of Laplace's equation in terms of Bessel functions as

$$
V(r,z) = \int_0^\infty A(k) e^{-k|z|} J_0(kr)\, dk, \tag{8.125}
$$

and so require the unknown coeffcient function $A(k)$ to obey

$$
\begin{aligned}
\int_0^\infty A(k) J_0(kr)\, dk &= V_0, \quad r < a \\
\int_0^\infty k A(k) J_0(kr)\, dk &= 0, \quad r > a.
\end{aligned}
\tag{8.126}
$$

No elementary algorithm for solving such a pair of *dual integral equations* exists. In this case, however, some inspired guesswork helps. By integrating the first equation of the transform pair (8.122) with respect to a, we discover that

$$
\int_0^\infty \frac{\sin(ar)}{r} J_0(kr)\, dr = \begin{cases} \pi/2, & k < a, \\ \sin^{-1}(a/k), & k > a. \end{cases} \tag{8.127}
$$

With this result in hand, we then observe that (8.123) tells us that the function

$$
A(k) = \frac{2V_0 \sin(ka)}{\pi k} \tag{8.128}
$$

satisfies both equations. Thus

$$
V(r,z) = \frac{2V_0}{\pi} \int_0^\infty e^{-k|z|} \sin(ka) J_0(kr) \frac{dk}{k}. \tag{8.129}
$$

The potential on the plane $z = 0$ can be evaluated explicitly to be

$$
V(r,0) = \begin{cases} V_0, & r < a, \\ (2V_0/\pi)\sin^{-1}(a/r), & r > a. \end{cases} \tag{8.130}
$$

The charge distribution on the disc can also be found as

$$\begin{aligned}\sigma(r) &= \left.\frac{\partial V}{\partial z}\right|_{z=0_-} - \left.\frac{\partial V}{\partial z}\right|_{z=0_+} \\ &= \frac{4V_0}{\pi}\int_0^\infty \sin(ak)J_0\,dk \\ &= \frac{4V_0}{\pi\sqrt{a^2-r^2}}, \quad r<a. \end{aligned} \tag{8.131}$$

8.3.3 Modified Bessel functions

When k is real the Bessel function $J_n(kr)$ and the Neumann $N_n(kr)$ function oscillate at large distance. When k is purely imaginary, it is convenient to combine them so as to have functions that grow or decay exponentially. These combinations are the *modified Bessel functions* $I_n(kr)$ and $K_n(kr)$.

These functions are initially defined for non-integer ν by

$$I_\nu(x) = i^{-\nu}J_\nu(ix), \tag{8.132}$$

$$K_\nu(x) = \frac{\pi}{2\sin\nu\pi}[I_{-\nu}(x) - I_\nu(x)]. \tag{8.133}$$

The factor of $i^{-\nu}$ in the definition of $I_\nu(x)$ is inserted to make I_ν real. Our definition of $K_\nu(x)$ is that in Abramowitz and Stegun's *Handbook of Mathematical Functions*. It differs from that of Whittaker and Watson, who divide by $\tan\nu\pi$ instead of $\sin\nu\pi$.

At short distance, and for $\nu > 0$,

$$I_\nu(x) = \left(\frac{x}{2}\right)^\nu \frac{1}{\Gamma(\nu+1)} + \cdots, \tag{8.134}$$

$$K_\nu(x) = \frac{1}{2}\Gamma(\nu)\left(\frac{x}{2}\right)^{-\nu} + \cdots \tag{8.135}$$

When ν becomes an integer we must take limits, and in particular

$$I_0(x) = 1 + \frac{1}{4}x^2 + \cdots, \tag{8.136}$$

$$K_0(x) = -(\ln x/2 + \gamma) + \cdots \tag{8.137}$$

The large x asymptotic behaviour is

$$I_\nu(x) \sim \frac{1}{\sqrt{2\pi x}}e^x, \quad x\to\infty, \tag{8.138}$$

$$K_\nu(x) \sim \frac{\pi}{\sqrt{2x}}e^{-x}, \quad x\to\infty. \tag{8.139}$$

From the expression for $J_n(x)$ as an integral, we have

$$I_n(x) = \frac{1}{2\pi}\int_0^{2\pi} e^{in\theta} e^{x\cos\theta}\, d\theta = \frac{1}{\pi}\int_0^{\pi} \cos(n\theta)e^{x\cos\theta}\, d\theta \tag{8.140}$$

for integer n. When n is not an integer we still have an expression for $I_\nu(x)$ as an integral, but now it is

$$I_\nu(x) = \frac{1}{\pi}\int_0^{\pi} \cos(\nu\theta)e^{x\cos\theta}\, d\theta - \frac{\sin\nu\pi}{\pi}\int_0^{\infty} e^{-x\cosh t - \nu t} dt. \tag{8.141}$$

Here we need $|\arg x| < \pi/2$ for the second integral to converge. The origin of the "extra" infinite integral must remain a mystery until we learn how to use complex integral methods for solving differential equations. From the definition of $K_\nu(x)$ in terms of I_ν we find

$$K_\nu(x) = \int_0^{\infty} e^{-x\cosh t}\cosh(\nu t)\, dt, \quad |\arg x| < \pi/2. \tag{8.142}$$

Physics illustration: Light propagation in optical fibres. Consider the propagation of light of frequency ω_0 down a straight section of optical fibre. Typical fibres are made of two materials: an outer layer, or *cladding*, with refractive index n_2, and an inner *core* with refractive index $n_1 > n_2$. The core of a fibre used for communication is usually less than 10 μm in diameter.

We will treat the light field E as a scalar. (This is not a particularly good approximation for real fibres, but the complications due to the vector character of the electromagnetic field are considerable.) We suppose that E obeys

$$\frac{\partial^2 E}{\partial x^2} + \frac{\partial^2 E}{\partial y^2} + \frac{\partial^2 E}{\partial z^2} - \frac{n^2(x,y)}{c^2}\frac{\partial^2 E}{\partial t^2} = 0. \tag{8.143}$$

Here $n(x, y)$ is the refractive index of the fibre, which is assumed to lie along the z-axis. We set

$$E(x, y, z, t) = \psi(x, y, z)e^{ik_0 z - i\omega_0 t} \tag{8.144}$$

where $k_0 = \omega_0/c$. The amplitude ψ is a (relatively) slowly varying envelope function. Plugging into the wave equation we find that

$$\frac{\partial^2 \psi}{\partial x^2} + \frac{\partial^2 \psi}{\partial y^2} + \frac{\partial^2 \psi}{\partial z^2} + 2ik_0\frac{\partial \psi}{\partial z} + \left(\frac{n^2(x,y)}{c^2}\omega_0^2 - k_0^2\right)\psi = 0. \tag{8.145}$$

Because ψ is slowly varying, we neglect the second derivative of ψ with respect to z, and this becomes

$$2ik_0\frac{\partial \psi}{\partial z} = -\left(\frac{\partial^2}{\partial x^2} + \frac{\partial^2}{\partial y^2}\right)\psi + k_0^2\left(1 - n^2(x,y)\right)\psi, \tag{8.146}$$

which is the two-dimensional time-dependent Schrödinger equation, but with t replaced by $z/2k_0$, where z is the distance down the fibre. The wave modes that will be trapped and guided by the fibre will be those corresponding to bound states of the axisymmetric potential

$$V(x,y) = k_0^2(1 - n^2(r)). \tag{8.147}$$

If these bound states have (negative) "energy" E_n, then $\psi \propto e^{-iE_n z/2k_0}$, and so the actual wavenumber for frequency ω_0 is

$$k = k_0 - E_n/2k_0. \tag{8.148}$$

In order to have a unique propagation velocity for signals on the fibre, it is therefore necessary that the potential support one, and only one, bound state.

If

$$\begin{aligned} n(r) &= n_1, \quad r < a, \\ &= n_2, \quad r > a, \end{aligned} \tag{8.149}$$

then the bound state solutions will be of the form

$$\psi(r,\theta) = \begin{cases} e^{in\theta} e^{i\beta z} J_n(\kappa r), & r < a, \\ A e^{in\theta} e^{i\beta z} K_n(\gamma r), & r > a, \end{cases} \tag{8.150}$$

where

$$\kappa^2 = (n_1^2 k_0^2 - \beta^2), \tag{8.151}$$

$$\gamma^2 = (\beta^2 - n_2^2 k_0^2). \tag{8.152}$$

To ensure that we have a solution decaying away from the core, we need β to be such that both κ and γ are real. We therefore require

$$n_1^2 > \frac{\beta^2}{k_0^2} > n_2^2. \tag{8.153}$$

At the interface both ψ and its radial derivative must be continuous, and so we will have a solution only if β is such that

$$\kappa \frac{J_n'(\kappa a)}{J_n(\kappa a)} = \gamma \frac{K_n'(\gamma a)}{K_n(\gamma a)}.$$

This Schrödinger approximation to the wave equation has other applications. It is called the *paraxial approximation*.

8.3.4 Spherical Bessel functions

Consider the wave equation

$$\left(\nabla^2 - \frac{1}{c^2}\frac{\partial^2}{\partial t^2}\right)\varphi(r,\theta,\phi,t) = 0 \tag{8.154}$$

in spherical polar coordinates. To apply separation of variables, we set

$$\varphi = e^{i\omega t} Y_l^m(\theta,\phi)\chi(r), \tag{8.155}$$

and find that

$$\frac{d^2\chi}{dr^2} + \frac{2}{r}\frac{d\chi}{dr} - \frac{l(l+1)}{r^2}\chi + \frac{\omega^2}{c^2}\chi = 0. \tag{8.156}$$

Substitute $\chi = r^{-1/2}R(r)$ and we have

$$\frac{d^2R}{dr^2} + \frac{1}{r}\frac{dR}{dr} + \left(\frac{\omega^2}{c^2} - \frac{(l+\frac{1}{2})^2}{r^2}\right)R = 0. \tag{8.157}$$

This is Bessel's equation with $\nu^2 \to (l+\frac{1}{2})^2$. Therefore the general solution is

$$R = AJ_{l+\frac{1}{2}}(kr) + BJ_{-l-\frac{1}{2}}(kr), \tag{8.158}$$

where $k = |\omega|/c$. Now inspection of the series definition of the J_ν reveals that

$$J_{\frac{1}{2}}(x) = \sqrt{\frac{2}{\pi x}}\sin x, \tag{8.159}$$

$$J_{-\frac{1}{2}}(x) = \sqrt{\frac{2}{\pi x}}\cos x, \tag{8.160}$$

so these Bessel functions are actually elementary functions. This is true of all Bessel functions of half-integer order, $\nu = \pm 1/2, \pm 3/2, \ldots$ We define the *spherical Bessel functions* by[2]

$$j_l(x) = \sqrt{\frac{\pi}{2x}}J_{l+\frac{1}{2}}(x), \tag{8.161}$$

$$n_l(x) = (-1)^{l+1}\sqrt{\frac{\pi}{2x}}J_{-(l+\frac{1}{2})}(x). \tag{8.162}$$

[2] We are using the definitions from Schiff's *Quantum Mechanics*.

The first few are

$$j_0(x) = \frac{1}{x}\sin x,$$

$$j_1(x) = \frac{1}{x^2}\sin x - \frac{1}{x}\cos x,$$

$$j_2(x) = \left(\frac{3}{x^3} - \frac{1}{x}\right)\sin x - \frac{3}{x^2}\cos x,$$

$$n_0(x) = -\frac{1}{x}\cos x,$$

$$n_1(x) = -\frac{1}{x^2}\cos x - \frac{1}{x}\sin x,$$

$$n_2(x) = -\left(\frac{3}{x^3} - \frac{1}{x}\right)\cos x - \frac{3}{x^2}\sin x.$$

Despite the appearance of negative powers of x, the $j_l(x)$ are all finite at $x = 0$. The $n_l(x)$ all diverge to $-\infty$ as $x \to 0$. In general

$$j_l(x) = f_l(x)\sin x + g_l(x)\cos(x), \tag{8.163}$$

$$n_l(x) = -f_l(x)\cos(x) + g_l(x)\sin x, \tag{8.164}$$

where $f_l(x)$ and $g_l(x)$ are polynomials in $1/x$.

We also define the spherical Hankel functions by

$$h_l^{(1)}(x) = j_l(x) + in_l(x), \tag{8.165}$$

$$h_l^{(2)}(x) = j_l(x) - in_l(x). \tag{8.166}$$

These behave like

$$h_l^{(1)}(x) \sim \frac{1}{x}e^{i(x-[l+1]\pi/2)}, \tag{8.167}$$

$$h_l^{(2)}(x) \sim \frac{1}{x}e^{-i(x-[l+1]\pi/2)}, \tag{8.168}$$

at large x.

The solution to the wave equation regular at the origin is therefore a sum of terms such as

$$\varphi_{k,l,m}(r,\theta,\phi,t) = j_l(kr)Y_l^m(\theta,\phi)e^{-i\omega t}, \tag{8.169}$$

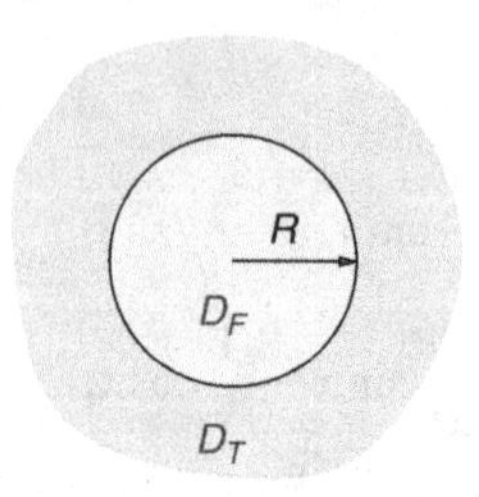

Figure 8.14 Fission core.

where $\omega = \pm ck$, with $k > 0$. For example, the plane wave e^{ikz} has expansion

$$e^{ikz} = e^{ikr\cos\theta} = \sum_{l=0}^{\infty}(2l+1)i^l j_l(kr)P_l(\cos\theta), \tag{8.170}$$

or equivalently, using (8.66),

$$e^{i\mathbf{k}\cdot\mathbf{r}} = 4\pi \sum_{l=0}^{\infty}\sum_{m=-l}^{l} i^l j_l(kr)\left[Y_l^m(\hat{\mathbf{k}})\right]^* Y_l^m(\hat{\mathbf{r}}) \tag{8.171}$$

where $\hat{\mathbf{k}}$, $\hat{\mathbf{r}}$ are unit vectors in the direction of $\mathbf{k}$ and $\mathbf{r}$, respectively, and are used as a shorthand notation to indicate the angles that should be inserted into the spherical harmonics. This angular-momentum-adapted expansion of a plane wave provides a useful tool in scattering theory.

Exercise 8.4: *Peierls' problem. Critical mass.* The core of a nuclear device consists of a sphere of fissile ^{235}U of radius R. It is surrounded by a thick shell of non-fissile material which acts as a neutron reflector, or *tamper* (Figure 8.14).

In the core, the fast neutron density $n(\mathbf{r}, t)$ obeys

$$\frac{\partial n}{\partial t} = \nu n + D_F \nabla^2 n. \tag{8.172}$$

Here the term with ν accounts for the production of additional neutrons due to induced fission. The term with D_F describes the diffusion of the fast neutrons. In the tamper the neutron flux obeys

$$\frac{\partial n}{\partial t} = D_T \nabla^2 n. \tag{8.173}$$

Both the neutron density n and flux $\mathbf{j} \equiv D_{F,T}\nabla n$, are continuous across the interface between the two materials. Find an equation determining the critical radius R_c above which the neutron density grows exponentially. Show that the critical radius for an assembly with a tamper consisting of ^{238}U ($D_T = D_F$) is one-half of that for a core surrounded only by air ($D_T = \infty$), and so the use of a thick ^{238}U tamper reduces the critical mass by a factor of eight.

Factorization and recurrence

The equation obeyed by the spherical Bessel function is

$$-\frac{d^2\chi_l}{dx^2} - \frac{2}{x}\frac{d\chi_l}{dx} + \frac{l(l+1)}{x^2}\chi_l = k^2\chi_l, \tag{8.174}$$

or, in Sturm–Liouville form,

$$-\frac{1}{x^2}\frac{d}{dx}\left(x^2\frac{d\chi_l}{dx}\right) + \frac{l(l+1)}{x^2}\chi_l = k^2\chi_l. \tag{8.175}$$

The corresponding differential operator is formally self-adjoint with respect to the inner product

$$\langle f, g\rangle = \int_0^\infty (f^*g)x^2dx. \tag{8.176}$$

Now, the operator

$$D_l = -\frac{d^2}{dx^2} - \frac{2}{x}\frac{d}{dx} + \frac{l(l+1)}{x^2} \tag{8.177}$$

factorizes as

$$D_l = \left(-\frac{d}{dx} + \frac{l-1}{x}\right)\left(\frac{d}{dx} + \frac{l+1}{x}\right), \tag{8.178}$$

or as

$$D_l = \left(\frac{d}{dx} + \frac{l+2}{x}\right)\left(-\frac{d}{dx} + \frac{l}{x}\right). \tag{8.179}$$

Since, with respect to the $w = x^2$ inner product, we have

$$\left(\frac{d}{dx}\right)^\dagger = -\frac{1}{x^2}\frac{d}{dx}x^2 = -\frac{d}{dx} - \frac{2}{x}, \tag{8.180}$$

we can write

$$D_l = A_l^\dagger A_l = A_{l+1}A_{l+1}^\dagger, \tag{8.181}$$

where

$$A_l = \left(\frac{d}{dx} + \frac{l+1}{x}\right). \tag{8.182}$$

From this we can deduce

$$A_l j_l \propto j_{l-1}, \tag{8.183}$$

$$A_{l+1}^{\dagger} j_l \propto j_{l+1}. \tag{8.184}$$

The constants of proportionality are in each case unity. The same recurrence formulæ hold for the spherical Neumann functions n_l.

8.4 Singular endpoints

In this section we will exploit our knowledge of the Laplace eigenfunctions in spherical and plane polar coordinates to illustrate Weyl's theory of self-adjoint boundary conditions at singular endpoints. We also connect Weyl's theory with concepts from scattering theory.

8.4.1 Weyl's theorem

Consider the Sturm–Liouville eigenvalue problem

$$Ly \equiv -\frac{1}{w}[p(r)y']' + q(r)y = \lambda y \tag{8.185}$$

on the interval $[0, R]$. Here $p(r)$, $q(r)$ and $w(r)$ are all supposed real, so the equation is formally self-adjoint with respect to the inner product

$$\langle u, v\rangle_w = \int_0^R wu^* v\, dr. \tag{8.186}$$

If $r = 0$ is a singular point of (8.185), then we will be unable to impose boundary conditions of our accustomed form

$$ay(0) + by'(0) = 0 \tag{8.187}$$

because one or both of the linearly independent solutions $y_1(r)$ and $y_2(r)$ will diverge as $r \to 0$. The range of possibilities was ennumerated by Weyl:

Theorem: *(Hermann Weyl, 1910): Suppose that $r = 0$ is a singular point and $r = R$ a regular point of the differential equation (8.185). Then*

I. Either:

- *(a)* ***Limit-circle case****: There exists a λ_0 such that both of the linearly independent solutions to (8.185) have a w norm that is convergent in the vicinity of $r = 0$. In this case both solutions have convergent w norm for* ***all*** *values of λ.*

Or

- *(b)* ***Limit-point case****: No more than one solution has convergent w norm for any λ.*

II. *In either case, whenever* Im $\lambda \neq 0$, *there is at least* one *finite-norm solution. When λ lies on the real axis there may or may not exist a finite norm solution.*

We will not attempt to prove Weyl's theorem. The proof is not difficult and may be found in many standard texts.[3] It is just a little more technical than the level of this book. We will instead illustrate it with enough examples to make the result plausible, and its practical consequences clear.

When we come to construct the resolvent $R_\lambda(r,r')$ obeying

$$(L - \lambda I)R_\lambda(r,r') = \delta(r - r') \tag{8.188}$$

by writing it as a product of $y_<$ and $y_>$ we are obliged to choose a normalizable function for $y_<$, the solution obeying the boundary condition at $r = 0$. We must do this so that the range of R_λ will be in $L^2[0,R]$. In the limit-point case, and when Im $\lambda \neq 0$, there is only one choice for $y_<$. There is therefore a unique resolvent, a unique self-adjoint operator $L - \lambda I$ of which R_λ is the inverse, and hence L is a uniquely specified differential operator.[4]

In the limit-circle case there is more than one choice for $y_<$ and hence more than one way of making L into a self-adjoint operator. To what boundary conditions do these choices correspond?

Suppose that the two normalizable solutions for $\lambda = \lambda_0$ are $y_1(r)$ and $y_2(r)$. The essence of Weyl's theorem is that once we are sufficiently close to $r = 0$ the exact value of λ is unimportant and all solutions behave as a linear combination of these two. We can therefore impose as a boundary condition that the allowed solutions be proportional to a specified real linear combination

$$y(r) \propto ay_1(r) + by_2(r), \quad r \to 0. \tag{8.189}$$

This is a natural generalization of the regular case where we have a solution $y_1(r)$ with boundary conditions $y_1(0) = 1$, $y_1'(0) = 0$, so $y_1(r) \sim 1$, and a solution $y_2(r)$ with $y_2(0) = 0$, $y_2'(0) = 1$, so $y_2(r) \sim r$. The regular self-adjoint boundary condition

$$ay(0) + by'(0) = 0 \tag{8.190}$$

with real a, b then forces $y(r)$ to be

$$y(r) \propto by_1(r) - ay_2(r) \sim b\,1 - a\,r, \quad r \to 0. \tag{8.191}$$

Example: Consider the radial part of the Laplace eigenvalue problem in two dimensions.

$$L\psi \equiv -\frac{1}{r}\frac{dr}{dr}\left(r\frac{d\psi}{dr}\right) + \frac{m^2}{r^2}\psi = k^2\psi. \tag{8.192}$$

[3] For example: Ivar Stackgold *Boundary Value Problems of Mathematical Physics*, Volume I (SIAM 2000).

[4] When λ is on the real axis then there may be no normalizable solution, and R_λ cannot exist. This will occur only when λ is in the continuous spectrum of the operator L, and is not a problem as the same operator L is obtained for any λ.

The differential operator L is formally self-adjoint with respect to the inner product

$$\langle \psi, \chi \rangle = \int_0^R \psi^* \chi \, r dr. \tag{8.193}$$

When $k^2 = 0$, the $m^2 \neq 0$ equation has solutions $\psi = r^{\pm m}$, and, of the normalization integrals

$$\int_0^R |r^m|^2 \, r dr, \quad \int_0^R |r^{-m}|^2 \, r dr, \tag{8.194}$$

only the first, containing the positive power of r, is convergent. For $m \neq 0$ we are therefore in Weyl's *limit-point* case. For $m^2 = 0$, however, the $k^2 = 0$ solutions are $\psi_1(r) = 1$ and $\psi_2(r) = \ln r$. Both normalization integrals

$$\int_0^R 1^2 \, r dr, \quad \int_0^R |\ln r|^2 \, r dr \tag{8.195}$$

converge and we are in the *limit-circle* case at $r = 0$. When $k^2 > 0$ these solutions become

$$\begin{aligned} J_0(kr) &= 1 - \frac{1}{4}(kr)^2 + \cdots \\ N_0(kr) &= \left(\frac{2}{\pi}\right)[\ln(kr/2) + \gamma] + \cdots \end{aligned} \tag{8.196}$$

Both remain normalizable, in conformity with Weyl's theorem. The self-adjoint boundary conditions at $r \to 0$ are therefore that near $r = 0$ the allowed functions become proportional to

$$1 + \alpha \ln r \tag{8.197}$$

with α some specified real constant.

Example: Consider the radial equation that arises when we separate the Laplace eigenvalue problem in spherical polar coordinates.

$$-\frac{1}{r^2}\left(\frac{d}{dr} r^2 \frac{d\psi}{dr}\right) + \frac{l(l+1)}{r^2}\psi = k^2 \psi. \tag{8.198}$$

When $k = 0$ this has solutions $\psi = r^l, r^{-l-1}$. For non-zero l only the first of the normalization integrals

$$\int_0^R r^{2l} \, r^2 dr, \quad \int_0^R r^{-2l-2} \, r^2 dr, \tag{8.199}$$

is finite. Thus, for $l \neq 0$, we are again in the limit-point case, and the boundary condition at the origin is uniquely determined by the requirement that the solution be normalizable.

When $l = 0$, however, the two $k^2 = 0$ solutions are $\psi_1(r) = 1$ and $\psi_2(r) = 1/r$. Both integrals

$$\int_0^R r^2\, dr, \quad \int_0^R r^{-2} r^2\, dr \tag{8.200}$$

converge, so we are again in the limit-circle case. For positive k^2, these solutions evolve into

$$\psi_{1,k}(r) = j_0(kr) = \frac{\sin kr}{kr}, \quad \psi_{2,k}(r) = -kn_0(kr) = \frac{\cos kr}{r}. \tag{8.201}$$

Near $r = 0$, we have $\psi_{1,k} \sim 1$ and $\psi_{2,k} \sim 1/r$, exactly the same behaviour as the $k^2 = 0$ solutions.

We obtain a self-adjoint operator if we choose a constant a_s and demand that all functions in the domain be proportional to

$$\psi(r) \sim 1 - \frac{a_s}{r} \tag{8.202}$$

as we approach $r = 0$. If we write the solution with this boundary condition as

$$\begin{aligned} \psi_k(r) = \frac{\sin(kr + \eta)}{r} &= \cos\eta \left(\frac{\sin(kr)}{r} + \tan\eta\, \frac{\cos(kr)}{r} \right) \\ &\sim k\cos\eta \left(1 + \frac{\tan\eta}{kr} \right), \end{aligned} \tag{8.203}$$

we can read off the phase shift η as

$$\tan\eta(k) = -ka_s. \tag{8.204}$$

These boundary conditions arise in quantum mechanics when we study the scattering of particles whose de Broglie wavelength is much larger than the range of the scattering potential. The incident wave is unable to resolve any of the internal structure of the potential and perceives its effect only as a singular boundary condition at the origin. In this context the constant a_s is called the *scattering length*. This physical model explains why only the $l = 0$ partial waves have a choice of boundary condition: classical particles with angular momentum $l \neq 0$ would miss the origin by a distance $r_{\min} = l/k$ and never see the potential.

The quantum picture also helps explain the physical origin of the distinction between the limit-point and limit-circle cases. A point potential can have a bound state that extends far beyond the short range of the potential. If the corresponding eigenfunction is normalizable, the bound particle has a significant amplitude to be found at non-zero r, and this amplitude must be included in the completeness relation and in the eigenfunction expansion of the Green function. When the state is not normalizable, however, the particle spends all its time very close to the potential, and its eigenfunction makes zero contribution to the Green function and completeness sum at any non-zero r. Any admixture of this non-normalizable state allowed by the boundary conditions can therefore be ignored, and, as far as the external world is concerned, all boundary conditions look alike. The next few exercises will illustrate this.

Exercise 8.5: The two-dimensional "delta function" potential. Consider the quantum mechanical problem in $\mathbb{R}^2$

$$\left(-\nabla^2 + V(|\mathbf{r}|)\right)\psi = E\psi$$

with V an attractive circular square well:

$$V(r) = \begin{cases} -\lambda/\pi a^2, & r < a \\ 0, & r > a. \end{cases}$$

The factor of πa^2 has been inserted to make this a regulated version of $V(\mathbf{r}) = -\lambda\delta^2(\mathbf{r})$. Let $\mu = \sqrt{\lambda/\pi a^2}$.

(i) By matching the functions

$$\psi(r) \propto \begin{cases} J_0(\mu r), & r < a \\ K_0(\kappa r), & r > a, \end{cases}$$

at $r = a$, show that as a becomes small, we can scale λ towards zero in such a way that the well becomes infinitely deep yet there remains a single bound state with finite binding energy

$$E_0 \equiv \kappa^2 = \frac{4}{a^2} e^{-2\gamma} e^{-4\pi/\lambda}.$$

It is only after scaling λ in this way that we have a well-defined quantum mechanical problem with a "point" potential.

(ii) Show that in the scaling limit, the associated wavefunction obeys the singular-endpoint boundary condition

$$\psi(r) \to 1 + \alpha \ln r, \quad r \to 0$$

where

$$\alpha = \frac{1}{\gamma + \ln \kappa/2}.$$

Observe that by varying κ^2 between 0 and ∞ we can make α be any real number. So the entire range of possible self-adjoint boundary conditions may be obtained by specifying the binding energy of an attractive potential.

(iii) Assume that we have fixed the boundary conditions by specifying κ, and consider the scattering of unbound particles off the short-range potential. It is natural to define the phase shift $\eta(k)$ so that

$$\psi_k(r) = \cos\eta J_0(kr) - \sin\eta N_0(kr)$$
$$\sim \sqrt{\frac{2}{\pi kr}}\cos(kr - \pi/4 + \eta), \quad r \to \infty.$$

Show that

$$\cot\eta = \left(\frac{2}{\pi}\right)\ln k/\kappa.$$

Exercise 8.6: *The three-dimensional "delta function" potential.* Repeat the calculation of the previous exercise for the case of a three-dimensional delta function potential

$$V(r) = \begin{cases} -\lambda/(4\pi a^3/3), & r < a \\ 0, & r > a. \end{cases}$$

(i) Show that as we take $a \to 0$, the delta function strength λ can be adjusted so that the scattering length becomes

$$a_s = \left(\frac{\lambda}{4\pi a^2} - \frac{1}{a}\right)^{-1}$$

and remains finite.

(ii) Show that when this a_s is positive, the attractive potential supports a single bound state with external wavefunction

$$\psi(r) \propto \frac{1}{r}e^{-\kappa r}$$

where $\kappa = a_s^{-1}$.

Exercise 8.7: *The pseudo-potential*. Consider a particle of mass μ confined in a large sphere of radius R. At the centre of the sphere is a singular potential whose effects can be parametrized by its scattering length a_s and the resultant phase shift

$$\eta(k) \approx \tan\eta(k) = -a_s k.$$

In the absence of the potential, the normalized $l = 0$ wavefunctions would be

$$\psi_n(r) = \sqrt{\frac{1}{2\pi R}}\frac{\sin k_n r}{r}$$

where $k_n = n\pi/R$.

(i) Show that the presence of the singular potential perturbs the ψ_n eigenstate so that its energy E_n changes by an amount

$$\Delta E_n = \frac{\hbar^2}{2\mu}\frac{2a_s k_n^2}{R}.$$

(ii) Show this energy shift can be written as if it were the result of applying first-order perturbation theory

$$\Delta E_n \approx \langle n|V_{ps}|n\rangle \equiv \int d^3r|\psi_n|^2 V_{ps}(r)$$

to an artificial *pseudo-potential*

$$V_{ps}(r) = \frac{2\pi a_s \hbar^2}{\mu}\delta^3(r).$$

Although the energy shift is small when R is large, it is not a first-order perturbation effect and the pseudo-potential is a convenient fiction which serves to parametrize the effect of the true potential. Even the sign of the pseudo-potential may differ from that of the actual short-distance potential. For our attractive "delta function", for example, the pseudo-potential changes from being attractive to being repulsive as the bound state is peeled off the bottom of the unbound continuum. The change of sign occurs not by a_s passing through zero, but by it passing through infinity. It is difficult to manipulate a single potential so as to see this dramatic effect, but when the particles have spin, and a spin-dependent interaction potential, it is possible to use a magnetic field to arrange for a bound state of one spin configuration to pass through the zero of energy of the other. The resulting *Feshbach resonance* has the same effect on the scattering length as the conceptually simpler *shape resonance* obtained by tuning the single potential.

The pseudo-potential formula is commonly used to describe the pairwise interaction of a dilute gas of particles of mass m, where it reads

$$V_{ps}(r) = \frac{4\pi a_s \hbar^2}{m}\delta^3(r). \tag{8.205}$$

The internal energy-density of the gas due to the two-body interaction then becomes

$$u(\rho) = \frac{1}{2}\frac{4\pi a_s \hbar^2}{m}\rho^2,$$

where ρ is the particle-number density.

The factor-of-two difference between the formula in the exercise and (8.205) arises because the μ in the exercise must be understood as the *reduced mass* $\mu = m^2/(m+m) = m/2$ of the pair of interacting particles.

Example: In n dimensions, the "$l = 0$" part of the Laplace operator is

$$\frac{d^2}{dr^2} + \frac{(n-1)}{r}\frac{d}{dr}.$$

This is formally self adjoint with respect to the natural inner product

$$\langle \psi, \chi \rangle_n = \int_0^\infty r^{n-1} \psi^* \chi \, dr. \tag{8.206}$$

The zero eigenvalue solutions are $\psi_1(r) = 1$ and $\psi_2(r) = r^{2-n}$. The second of these ceases to be normalizable once $n \geq 4$. In four space dimensions and above, therefore, we are always in the limit-point case. No point interaction – no matter how strong – can affect the physics. This non-interaction result extends, with slight modification, to the quantum field theory of relativistic particles. Here we find that contact interactions become *irrelevent* or *non-renormalizable* in more than four space-time dimensions.

8.5 Further exercises and problems

Here are some further problems involving Legendre polynomials, associated Legendre functions and Bessel functions.

Exercise 8.8: A sphere of radius a is made by joining two conducting hemispheres along their equators. The hemispheres are electrically insulated from one another and maintained at two different potentials V_1 and V_2.

(a) Starting from the general expression

$$V(r,\theta) = \sum_{l=0}^{\infty} \left(a_l r^l + \frac{b_l}{r^{l+1}} \right) P_l(\cos\theta)$$

find an integral expression for the coefficients a_l, b_l that are relevant to the electric field *outside* the sphere. Evaluate the integrals giving b_1, b_2 and b_3.

(b) Use your results from part (a) to compute the electric dipole moment of the sphere as a function of the potential difference $V_1 - V_2$.

(c) Now the two hemispheres are electrically connected and the entire surface is at one potential. The sphere is immersed in a uniform electric field $\mathbf{E}$. What is its dipole moment now?

Problem 8.9: Tides and gravity. The Earth is not exactly spherical. Two major causes of the deviation from sphericity are the Earth's rotation and the tidal forces it feels from the Sun and the Moon. In this problem we will study the effects of rotation and tides on a self-gravitating sphere of fluid of uniform density ρ_0.

(a) Consider the equilibrium of a nearly spherical body of fluid rotating homogeneously with angular velocity ω_0. Show that the effect of rotation can be accounted for by introducing an "effective gravitational potential"

$$\varphi_{\text{eff}} = \varphi_{\text{grav}} + \frac{1}{3}\omega_0^2 R^2 (P_2(\cos\theta) - 1),$$

where R, θ are spherical coordinates defined with their origin in the centre of the body and $\hat{z}$ along the axis of rotation.

(b) A small planet is in a circular orbit about a distant massive star. It rotates about an axis perpendicular to the plane of the orbit so that it always keeps the same face directed towards thc star. Show that the planet experiences an effective external potential

$$\varphi_{\text{tidal}} = -\Omega^2 R^2 P_2(\cos\theta),$$

together with a potential, of the same sort as in part (a), that arises from the once-per-orbit rotation. Here Ω is the orbital angular velocity, and R, θ are spherical coordinates defined with their origin at the centre of the planet and $\hat{z}$ pointing at the star.

(c) Each of the external potentials slightly deforms the initially spherical planet so that the surface is given by

$$R(\theta, \phi) = R_0 + \eta P_2(\cos\theta).$$

(with θ being measured with respect to different axes for the rotation and tidal effects). Show that, to first order in η, this deformation does not alter the volume of the body. Observe that positive η corresponds to a prolate spheroid and negative η to an oblate one.

(d) The gravitational field of the deformed spheroid can be found by approximating it as an undeformed homogeneous sphere of radius R_0, together with a thin spherical shell of radius R_0 and surface mass density $\sigma = \rho_0 \eta P_2(\cos\theta)$. Use the general axisymmetric solution

$$\varphi(R, \theta, \phi) = \sum_{l=0}^{\infty} \left(A_l R^l + \frac{B_l}{R^{l+1}} \right) P_l(\cos\theta)$$

of Laplace's equation, together with Poisson's equation

$$\nabla^2 \varphi = 4\pi G \rho(\mathbf{r})$$

for the gravitational potential, to obtain expressions for φ_{shell} in the regions $R > R_0$ and $R \leq R_0$.

(e) The surface of the fluid will be an equipotential of the combined potentials of the homogeneous sphere, the thin shell and the effective external potential of the tidal or centrifugal forces. Use this fact to find η (to lowest order in the angular velocities) for the two cases. Do not include the centrifugal potential from part (b) when computing the tidal distortion. We never include the variation of the centrifugal potential across a planet when calculating tidal effects. This is because this variation is due to the once-per-year rotation, and contributes to the oblate equatorial bulge and not to the prolate tidal bulge.[5] $\left(\text{Answer: } \eta_{\text{rot}} = -\frac{5}{2}\frac{\omega_0^2 R_0}{4\pi G\rho_0}, \text{ and } \eta_{\text{tide}} = \frac{15}{2}\frac{\Omega^2 R_0}{4\pi G\rho_0}.\right)$

Exercise 8.10: *Dielectric sphere.* Consider a solid dielectric sphere of radius a and permittivity ϵ. The sphere is placed in an electric field which takes the constant value $\mathbf{E} = E_0\hat{\mathbf{z}}$ a long distance from the sphere. Recall that Maxwell's equations require that $\mathbf{D}_\perp$ and $\mathbf{E}_\parallel$ be continuous across the surface of the sphere.

(a) Use the expansions

$$\Phi_{in} = \sum_l A_l r^l P_l(\cos\theta)$$

$$\Phi_{out} = \sum_l (B_l r^l + C_l r^{-l-1}) P_l(\cos\theta)$$

and find all non-zero coefficients A_l, B_l, C_l.

(b) Show that the $\mathbf{E}$ field inside the sphere is uniform and of magnitude $\frac{3\epsilon_0}{\epsilon+2\epsilon_0}E_0$.

(c) Show that the electric field is unchanged if the dielectric is replaced by the polarization-induced surface charge density

$$\sigma_{\text{induced}} = 3\epsilon_0 \left(\frac{\epsilon - \epsilon_0}{\epsilon + 2\epsilon_0}\right) E_0 \cos\theta.$$

(Some systems of units may require extra 4π's in this last expression. In SI units $\mathbf{D} \equiv \epsilon\mathbf{E} = \epsilon_0\mathbf{E} + \mathbf{P}$, and the polarization-induced charge density is $\rho_{\text{induced}} = -\nabla\cdot\mathbf{P}$.)

Exercise 8.11: *Hollow sphere.* The potential on a spherical surface of radius a is $\Phi(\theta, \phi)$. We want to express the potential inside the sphere as an integral over the surface in a manner analagous to the Poisson kernel in two dimensions.

(a) By using the generating function for Legendre polynomials, show that

$$\frac{1 - r^2}{(1 + r^2 - 2r\cos\theta)^{3/2}} = \sum_{l=0}^{\infty} (2l+1) r^l P_l(\cos\theta), \quad r < 1.$$

[5] Our Earth rotates about its axis $365\frac{1}{4} + 1$ times in a year, not $365\frac{1}{4}$ times. The "+1" is this effect.

(b) Starting from the expansion

$$\Phi_{in}(r,\theta,\phi) = \sum_{l=0}^{\infty}\sum_{m=-l}^{l} A_{lm} r^l Y_l^m(\theta,\phi)$$

$$A_{lm} = \frac{1}{a^l}\int_{S^2}\left[Y_l^m(\theta,\phi)\right]^* \Phi(\theta,\phi)\, d\cos\theta\, d\phi$$

and using the addition formula for spherical harmonics, show that

$$\Phi_{in}(r,\theta,\phi) = \frac{a(a^2-r^2)}{4\pi}\int_{S^2}\frac{\Phi(\theta',\phi')}{(r^2+a^2-2ar\cos\gamma)^{3/2}} d\cos\theta' d\phi'$$

where $\cos\gamma = \cos\theta\cos\theta' + \sin\theta\sin\theta'\cos(\phi-\phi')$.

(c) By setting $r = 0$, deduce that a three-dimensional harmonic function cannot have a local maximum or minimum.

Problem 8.12: We have several times met with the Pöschel–Teller eigenvalue problem

$$\left(-\frac{d^2}{dx^2} - n(n+1)\operatorname{sech}^2 x\right)\psi = E\psi, \qquad (\star)$$

in the particular case that $n = 1$. We now consider this problem for any positive integer n.

(a) Set $\xi = \tanh x$ in $(\star)$ and show that it becomes

$$\left(\frac{d}{d\xi}(1-\xi^2)\frac{d}{d\xi} + n(n+1) + \frac{E}{1-\xi^2}\right)\psi = 0.$$

(b) Compare the equation in part (a) with the associated Legendre equation and deduce that the bound-state eigenfunctions and eigenvalues of the original Pöschel–Teller equation are

$$\psi_m(x) = P_n^m(\tanh x), \quad E_m = -m^2, \quad m = 1,\ldots,n,$$

where $P_n^m(\xi)$ is the associated Legendre function. Observe that the list of bound states does not include $\psi_0 = P_n^0(\tanh x) \equiv P_n(\tanh x)$. This is because ψ_0 is not normalizable, being the lowest of the unbound $E \geq 0$ continuous-spectrum states.

(c) Now seek continuous spectrum solutions to $(\star)$ in the form

$$\psi_k(x) = e^{ikx} f(\tanh x),$$

and show if we take $E = k^2$, where k is any real number, then $f(\xi)$ obeys

$$(1-\xi^2)\frac{d^2 f}{d\xi^2} + 2(ik-\xi)\frac{df}{d\xi} + n(n+1)f = 0. \qquad (\star\star)$$

(d) Let us denote by $P_n^{(k)}(\xi)$ the solutions of ($\star\star$) that reduce to the Legendre polynomial $P_n(\xi)$ when $k = 0$. Show that the first few $P_n^{(k)}(\xi)$ are

$$P_0^{(k)}(\xi) = 1,$$
$$P_1^{(k)}(\xi) = \xi - ik,$$
$$P_2^{(k)}(\xi) = \frac{1}{2}(3\xi^2 - 1 - 3ik\xi - k^2).$$

Explore the properties of the $P_n^{(k)}(\xi)$, and show that they include

(i) $P_n^{(k)}(-\xi) = (-1)^n P_n^{(-k)}(\xi)$.
(ii) $(n+1)P_{n+1}^{(k)}(\xi) = (2n+1)xP_n^{(k)}(\xi) - (n + k^2/n)P_{n-1}^{(k)}(\xi)$.
(iii) $P_n^{(k)}(1) = (1-ik)(2-ik)\dots(n-ik)/n!$.

(The $P_n^{(k)}(\xi)$ are the $\nu = -\mu = ik$ special case of the Jacobi polynomials $P_n^{(\nu,\mu)}(\xi)$.)

Problem 8.13: Bessel functions and impact parameters. In two dimensions we can expand a plane wave as

$$e^{iky} = \sum_{n=-\infty}^{\infty} J_n(kr)e^{in\theta}.$$

(a) What do you think the resultant wave will look like if we take only a finite segment of this sum? For example

$$\phi(x) = \sum_{l=10}^{17} J_n(kr)e^{in\theta}.$$

Think about:

(i) The quantum interpretation of $\hbar l$ as angular momentum $= \hbar kd$, where d is the *impact parameter*, the amount by which the incoming particle misses the origin.
(ii) Diffraction: one cannot have a plane wave of finite width.

(b) After writing down your best guess for the previous part, confirm your understanding by using *Mathematica* or another package to plot the real part of ϕ as defined above. The following *Mathematica* code may work.

```
Clear[bit,tot]
bit[l_,x_,y_]:=Cos[l ArcTan[x,y]]BesselJ[l,Sqrt[x^2+y^2]]
tot[x_,y_] :=Sum[bit[l,x,y],{l,10,17}]
ContourPlot[tot[x,y],{x,-40,40},{y,-40,40},PlotPoints ->200].
Display["wave",\%,"EPS"]
```

Run it, or some similar code, as a batchfile. Try different ranges for the sum.

Exercise 8.14: Consider the two-dimensional Fourier transform

$$\widetilde{f}(\mathbf{k}) = \int e^{i\mathbf{k}\cdot\mathbf{x}} f(\mathbf{x})\, d^2x$$

of a function that in polar coordinates is of the form $f(r,\theta) = \exp\{-il\theta\} f(r)$.

(a) Show that

$$\widetilde{f}(\mathbf{k}) = 2\pi i^l e^{-il\theta_k} \int_0^\infty J_l(kr) f(r)\, r dr,$$

where k, θ_k are the polar coordinates of $\mathbf{k}$.

(b) Use the inversion formula for the two-dimensional Fourier transform to establish the inversion formula (8.120) for the Hankel transform

$$F(k) = \int_0^\infty J_l(kr) f(r)\, r dr.$$

9
Integral equations

A problem involving a differential equation can often be recast as one involving an integral equation. Sometimes this new formulation suggests a method of attack or approximation scheme that would not have been apparent in the original language. It is also usually easier to extract general properties of the solution when the problem is expressed as an integral equation.

9.1 Illustrations

Here are some examples.

A boundary-value problem: Consider the differential equation for the unknown $u(x)$

$$-u'' + \lambda V(x)u = 0 \tag{9.1}$$

with the boundary conditions $u(0) = u(L) = 0$. To turn this into an integral equation we introduce the Green function

$$G(x,y) = \begin{cases} \frac{1}{L}x(y-L), & 0 \le x \le y \le L, \\ \frac{1}{L}y(x-L), & 0 \le y \le x \le L, \end{cases} \tag{9.2}$$

so that

$$-\frac{d^2}{dx^2}G(x,y) = \delta(x-y). \tag{9.3}$$

Then we can pretend that $\lambda V(x)u(x)$ in the differential equation is a known source term, and substitute it for "$f(x)$" in the usual Green function solution. We end up with

$$u(x) + \lambda \int_0^L G(x,y)V(y)u(y)\,dx = 0. \tag{9.4}$$

This *integral equation* for u has not solved the problem, but is equivalent to the original problem. Note, in particular, that the boundary conditions are implicit in this formulation: if we set $x = 0$ or L in the second term, it becomes zero because the Green function is zero at those points. The integral equation then says that $u(0)$ and $u(L)$ are both zero.

An initial value problem: Consider essentially the same differential equation as before, but now with initial data:

$$-u'' + V(x)u = 0, \qquad u(0) = 0, \quad u'(0) = 1. \tag{9.5}$$

In this case, we claim that the inhomogeneous integral equation

$$u(x) - \int_0^x (x-t)V(t)u(t)\,dt = x \tag{9.6}$$

is equivalent to the given problem. Let us check the claim. First, the initial conditions. Rewrite the integral equation as

$$u(x) = x + \int_0^x (x-t)V(t)u(t)\,dt, \tag{9.7}$$

so it is manifest that $u(0) = 0$. Now differentiate to get

$$u'(x) = 1 + \int_0^x V(t)u(t)\,dt. \tag{9.8}$$

This shows that $u'(0) = 1$, as required. Differentiating once more confirms that $u'' = V(x)u$.

These examples reveal that one advantage of the integral equation formulation is that the boundary or initial value conditions are automatically encoded in the integral equation itself, and do not have to be added as riders.

9.2 Classification of integral equations

The classification of linear integral equations is best described by a list:

(A) (i) Limits on integrals fixed $\Rightarrow$ Fredholm equation.
  (ii) One integration limit is $x \Rightarrow$ Volterra equation.
(B) (i) Unknown under integral only $\Rightarrow$ Type I.
  (ii) Unknown also outside integral $\Rightarrow$ Type II.
(C) (i) Homogeneous.
  (ii) Inhomogeneous.

For example,

$$u(x) = \int_0^L G(x,y)u(y)\,dy \tag{9.9}$$

is a Type II homogeneous Fredholm equation, whilst

$$u(x) = x + \int_0^x (x-t)V(t)u(t)\,dt \tag{9.10}$$

is a Type II inhomogeneous Volterra equation.

The equation

$$f(x) = \int_a^b K(x,y)u(y)\,dy, \tag{9.11}$$

an inhomogeneous Type I Fredholm equation, is analogous to the matrix equation

$$\mathbf{Kx} = \mathbf{b}. \tag{9.12}$$

On the other hand, the equation

$$u(x) = \frac{1}{\lambda}\int_a^b K(x,y)u(y)\,dy, \tag{9.13}$$

a homogeneous Type II Fredholm equation, is analogous to the matrix eigenvalue problem

$$\mathbf{Kx} = \lambda\mathbf{x}. \tag{9.14}$$

Finally,

$$f(x) = \int_a^x K(x,y)u(y)\,dy, \tag{9.15}$$

an inhomogeneous Type I Volterra equation, is the analogue of a system of linear equations involving an upper triangular matrix.

The function $K(x,y)$ appearing in these expressions is called the *kernel*. The phrase "kernel of the integral operator" can therefore refer either to the function K or the null-space of the operator. The context should make clear which meaning is intended.

9.3 Integral transforms

When the kernel of the Fredholm equation is of the form $K(x-y)$, with x and y taking values on the entire real line, then it is translation invariant and we can solve the integral equation by using the Fourier transformation

$$\widetilde{u}(k) = \mathcal{F}(u) = \int_{-\infty}^{\infty} u(x)e^{ikx}\,dx \tag{9.16}$$

$$u(x) = \mathcal{F}^{-1}(\widetilde{u}) = \int_{-\infty}^{\infty} \widetilde{u}(k)e^{-ikx}\,\frac{dk}{2\pi}. \tag{9.17}$$

Integral equations involving translation-invariant Volterra kernels usually succumb to a Laplace transform

$$\widetilde{u}(p) = \mathcal{L}(u) = \int_0^\infty u(x)e^{-px}\,dx \tag{9.18}$$

$$u(x) = \mathcal{L}^{-1}(\widetilde{u}) = \frac{1}{2\pi i}\int_{\gamma-i\infty}^{\gamma+i\infty} \widetilde{u}(p)e^{px}\,dp. \tag{9.19}$$

The Laplace inversion formula is the *Bromwich contour integral*, where γ is chosen so that all the singularities of $\widetilde{u}(p)$ lie to the left of the contour. In practice one finds the inverse Laplace transform by using a table of Laplace transforms, such as the Bateman tables of integral transforms mentioned in the introduction to Chapter 8.

For kernels of the form $K(x/y)$ the Mellin transform,

$$\widetilde{u}(\sigma) = \mathcal{M}(u) = \int_0^\infty u(x)x^{\sigma-1}\,dx \tag{9.20}$$

$$u(x) = \mathcal{M}^{-1}(\widetilde{u}) = \frac{1}{2\pi i}\int_{\gamma-i\infty}^{\gamma+i\infty} \widetilde{u}(\sigma)x^{-\sigma}\,d\sigma, \tag{9.21}$$

is the tool of choice. Again the inversion formula requires a Bromwich contour integral, and so usually requires tables of Mellin transforms.

9.3.1 Fourier methods

The class of problems that succumb to a Fourier transform can be thought of as a continuous version of a matrix problem where the entries in the matrix depend only on their distance from the main diagonal (Figure 9.1).

Example: Consider the Type II Fredholm equation

$$u(x) - \lambda\int_{-\infty}^{\infty} e^{-|x-y|}u(y)\,dy = f(x), \tag{9.22}$$

where we will assume that $\lambda < 1/2$. Here the x-space kernel operator

$$K(x-y) = \delta(x-y) - \lambda e^{-|x-y|} \tag{9.23}$$

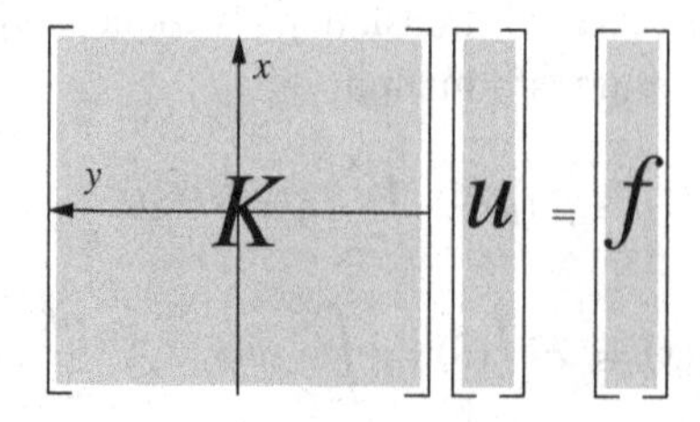

Figure 9.1 The matrix form of the equation $\int_{-\infty}^{\infty} K(x-y)u(y)\,dy = f(x)$.

has Fourier transform

$$\widetilde{K}(k) = 1 - \frac{2\lambda}{k^2+1} = \frac{k^2 + (1-2\lambda)}{k^2+1} = \frac{k^2+a^2}{k^2+1}, \tag{9.24}$$

where $a^2 = 1 - 2\lambda$. From

$$\left(\frac{k^2+a^2}{k^2+1}\right)\widetilde{u}(k) = \widetilde{f}(k) \tag{9.25}$$

we find

$$\begin{aligned}\widetilde{u}(k) &= \left(\frac{k^2+1}{k^2+a^2}\right)\widetilde{f}(k)\\ &= \left(1 + \frac{1-a^2}{k^2+a^2}\right)\widetilde{f}(k).\end{aligned} \tag{9.26}$$

Inverting the Fourier transform gives

$$\begin{aligned}u(x) &= f(x) + \frac{1-a^2}{2a}\int_{-\infty}^{\infty} e^{-a|x-y|} f(y)\,dy\\ &= f(x) + \frac{\lambda}{\sqrt{1-2\lambda}}\int_{-\infty}^{\infty} e^{-\sqrt{1-2\lambda}|x-y|} f(y)\,dy.\end{aligned} \tag{9.27}$$

This solution is no longer valid when the parameter λ exceeds $1/2$. This is because zero then lies in the spectrum of the operator we are attempting to invert. The spectrum is continuous and the Fredholm alternative does not apply.

9.3.2 Laplace transform methods

The Volterra problem

$$\int_0^x K(x-y)u(y)\,dy = f(x), \quad 0 < x < \infty \tag{9.28}$$

can also be solved by the application of an integral transform. In this case we observe that the value of $K(x)$ is only needed for positive x (see Figure 9.2), and this suggests that we take a *Laplace transform* over the positive real axis.

Abel's equation

As an example of Laplace methods, consider Abel's equation

$$f(x) = \int_0^x \frac{1}{\sqrt{x-y}} u(y)\,dy, \tag{9.29}$$

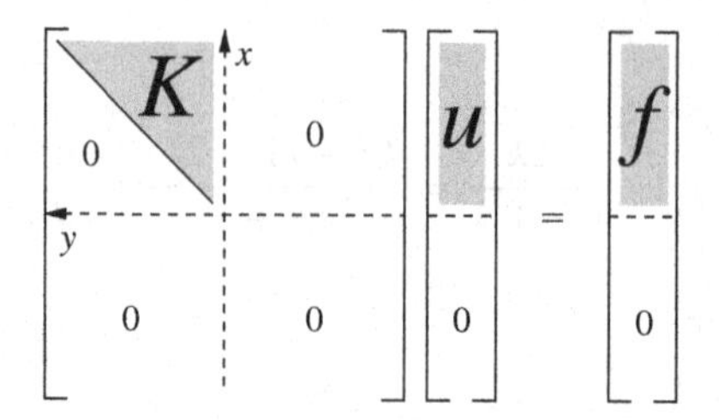

Figure 9.2 We only require the value of $K(x)$ for x positive, and u and f can be set to zero for $x < 0$.

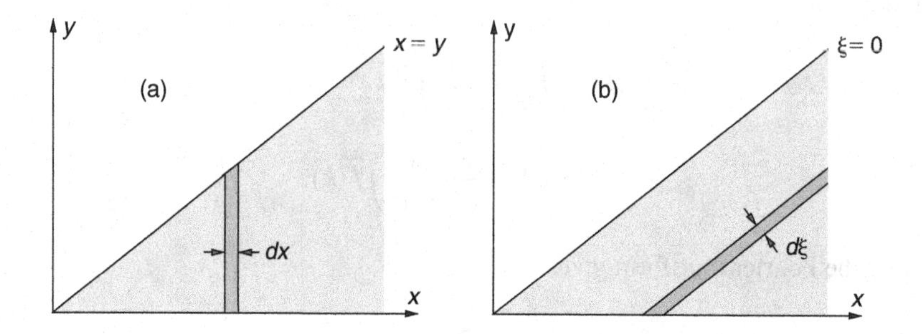

Figure 9.3 Regions of integration for the convolution theorem: (a) Integrating over y at fixed x, then over x; (b) integrating over η at fixed ξ, then over ξ.

where we are given $f(x)$ and wish to find $u(x)$. Here it is clear that we need $f(0) = 0$ for the equation to make sense. We have met this integral transformation before in the definition of the "half-derivative". It is an example of the more general equation of the form

$$f(x) = \int_0^x K(x-y)u(y)\,dy. \tag{9.30}$$

Let us take the Laplace transform of both sides of (9.30):

$$\begin{aligned}\mathcal{L}f(p) &= \int_0^\infty e^{-px}\left(\int_0^x K(x-y)u(y)\,dy\right)dx\\ &= \int_0^\infty dx\int_0^x dy\, e^{-px}K(x-y)u(y).\end{aligned} \tag{9.31}$$

Now we make the change of variables (see Figure 9.3)

$$\begin{aligned}x &= \xi + \eta,\\ y &= \eta.\end{aligned} \tag{9.32}$$

This has Jacobian

$$\frac{\partial(x,y)}{\partial(\xi,\eta)} = 1, \tag{9.33}$$

and the integral becomes

$$\begin{aligned}\mathcal{L}f(p) &= \int_0^\infty \int_0^\infty e^{-p(\xi+\eta)} K(\xi)u(\eta)\, d\xi\, d\eta \\ &= \int_0^\infty e^{-p\xi} K(\xi)\, d\xi \int_0^\infty e^{-p\eta} u(\eta)\, d\eta \\ &= \mathcal{L}K(p)\ \mathcal{L}u(p). \end{aligned} \tag{9.34}$$

Thus the Laplace transform of a Volterra convolution is the product of the Laplace transforms. We can now invert:

$$u = \mathcal{L}^{-1}\big(\mathcal{L}f/\mathcal{L}K\big). \tag{9.35}$$

For Abel's equation, we have

$$K(x) = \frac{1}{\sqrt{x}}, \tag{9.36}$$

the Laplace transform of which is

$$\mathcal{L}K(p) = \int_0^\infty x^{\frac{1}{2}-1} e^{-px}\, dx = p^{-1/2}\Gamma\left(\frac{1}{2}\right) = p^{-1/2}\sqrt{\pi}. \tag{9.37}$$

Therefore, the Laplace transform of the solution $u(x)$ is

$$\mathcal{L}u(p) = \frac{1}{\sqrt{\pi}} p^{1/2}(\mathcal{L}f) = \frac{1}{\pi}(\sqrt{\pi}p^{-1/2}p\mathcal{L}f). \tag{9.38}$$

Now $f(0) = 0$, and so

$$p\mathcal{L}f = \mathcal{L}\left(\frac{d}{dx}f\right), \tag{9.39}$$

as may be seen by an integration by parts in the definition. Using this observation, and depending on whether we put the p next to f or outside the parentheses, we conclude that the solution of Abel's equation can be written in two equivalent ways:

$$u(x) = \frac{1}{\pi}\frac{d}{dx}\int_0^x \frac{1}{\sqrt{x-y}} f(y)\, dy = \frac{1}{\pi}\int_0^x \frac{1}{\sqrt{x-y}} f'(y)\, dy. \tag{9.40}$$

Proving the equality of these two expressions was a problem we set ourselves in Chapter 6.

Here is another way of establishing the equality: assume for the moment that $K(0)$ is finite, and that, as we have already noted, $f(0) = 0$. Then,

$$\frac{d}{dx}\int_0^x K(x-y)f(y)\, dy \tag{9.41}$$

is equal to

$$\begin{aligned}
&K(0)f(x) + \int_0^x \partial_x K(x-y)f(y)\,dy,\\
&= K(0)f(x) - \int_0^x \partial_y K(x-y)f(y)\,dy\\
&= K(0)f(x) - \int_0^x \partial_y\Big(K(x-y)f(y)\Big)\,dy + \int_0^x K(x-y)f'(y)\,dy\\
&= K(0)f(x) - K(0)f(x) - K(x)f(0) + \int_0^x K(x-y)f'(y)\,dy\\
&= \int_0^x K(x-y)f'(y)\,dy.
\end{aligned} \tag{9.42}$$

Since $K(0)$ cancelled out, we need not worry that it is divergent! More rigorously, we should regularize the improper integral by raising the lower limit on the integral to a small positive quantity, and then taking the limit to zero at the end of the calculation.

Radon transforms

An Abel integral equation lies at the heart of the method for reconstructing the image in a computer aided tomography (CAT) scan. By rotating an X-ray source about a patient and recording the direction-dependent shadow, we measure the integral of his or her tissue density $f(x,y)$ along all lines in a slice (which we will take to be the xy-plane) through his or her body. The resulting information is the *Radon transform* F of the function f.

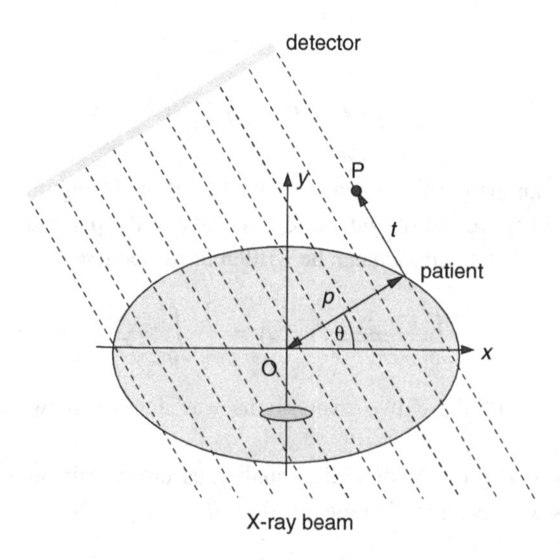

Figure 9.4 The geometry of the CAT scan Radon transformation, showing the location of the point P with coordinates $x = p\cos\theta - t\sin\theta, y = p\sin\theta + t\cos\theta$.

If we parametrize the family of lines by p and θ, as shown in Figure 9.4, we have

$$F(p,\theta) = \int_{-\infty}^{\infty} f(p\cos\theta - t\sin\theta, p\sin\theta + t\cos\theta)\, dt,$$
$$= \int_{\mathbb{R}^2} \delta(x\cos\theta + y\sin\theta - p) f(x,y)\, dxdy. \tag{9.43}$$

We will assume that f is zero outside some finite region (the patient), and so these integrals converge.

We wish to invert the transformation and recover f from the data $F(p,\theta)$. This problem was solved by Johann Radon in 1917. Radon made clever use of the Euclidean group to simplify the problem. He observed that we may take the point O at which we wish to find f to be the origin, and defined[1]

$$F_O(p) = \frac{1}{2\pi}\int_0^{2\pi} \left\{ \int_{\mathbb{R}^2} \delta(x\cos\theta + y\sin\theta - p) f(x,y)\, dxdy \right\} d\theta. \tag{9.44}$$

Thus $F_O(p)$ is the angular average over all lines tangent to a circle of radius p about the desired inversion point. Radon then observed that if he additionally defines

$$\bar{f}(r) = \frac{1}{2\pi}\int_0^{2\pi} f(r\cos\phi, r\sin\phi)\, d\phi \tag{9.45}$$

then he can substitute $\bar{f}(r)$ for $f(x,y)$ in (9.44) without changing the value of the integral. Furthermore $\bar{f}(0) = f(0,0)$. Hence, taking polar coordinates in the xy-plane, he has

$$F_O(p) = \frac{1}{2\pi}\int_0^{2\pi} \left\{ \int_{\mathbb{R}^2} \delta(r\cos\phi\cos\theta + r\sin\phi\sin\theta - p)\bar{f}(r)\, r d\phi dr \right\} d\theta. \tag{9.46}$$

We can now use

$$\delta(g(\phi)) = \sum_n \frac{1}{|g'(\phi_n)|}\delta(\phi - \phi_n), \tag{9.47}$$

where the sum is over the zeros ϕ_n of $g(\phi) = r\cos(\theta - \phi) - p$, to perform the ϕ integral. Any given point $x = r\cos\phi$, $y = r\sin\phi$ lies on two distinct lines if and only if $p < r$. Thus $g(\phi)$ has two zeros if $p < r$, but none if $r < p$. Consequently

$$F_O(p) = \frac{1}{2\pi}\int_0^{2\pi} \left\{ \int_p^{\infty} \frac{2}{\sqrt{r^2 - p^2}}\bar{f}(r)\, rdr \right\} d\theta. \tag{9.48}$$

[1] We trust that the reader will forgive the anachronism of our expressing Radon's formulæ in terms of Dirac's delta function.

Nothing in the inner integral depends on θ. The outer integral is therefore trivial, and so

$$F_O(p) = \int_p^\infty \frac{2}{\sqrt{r^2 - p^2}} \bar{f}(r)\, r dr. \tag{9.49}$$

We can extract $F_O(p)$ from the data. We could therefore solve the Abel equation (9.49) and recover the complete function $\bar{f}(r)$. We are only interested in $\bar{f}(0)$, however, and it is easier to verify a claimed solution. Radon asserts that

$$f(0,0) = \bar{f}(0) = -\frac{1}{\pi} \int_0^\infty \frac{1}{p} \left(\frac{\partial}{\partial p} F_O(p) \right) dp. \tag{9.50}$$

To prove that his claim is true we must first take the derivative of $F_O(p)$ and show that

$$\left(\frac{\partial}{\partial p} F_O(p) \right) = \int_p^\infty \frac{2p}{\sqrt{r^2 - p^2}} \left(\frac{\partial}{\partial r} \bar{f}(r) \right) dr. \tag{9.51}$$

The details of this computation are left as an exercise. It is little different from the differentiation of the integral transform at the end of the last section. We then substitute (9.51) into (9.50) and evaluate the resulting integral

$$I = -\frac{1}{\pi} \int_0^\infty \frac{1}{p} \left\{ \int_p^\infty \frac{2p}{\sqrt{r^2 - p^2}} \left(\frac{\partial}{\partial r} \bar{f}(r) \right) dr \right\} dp \tag{9.52}$$

by exchanging the order of the integrations, as shown in Figure 9.5.

After the interchange we have

$$I = -\frac{2}{\pi} \int_0^\infty \left\{ \int_0^r \frac{1}{\sqrt{r^2 - p^2}} dp \right\} \left(\frac{\partial}{\partial r} \bar{f}(r) \right) dr. \tag{9.53}$$

Since

$$\int_0^r \frac{1}{\sqrt{r^2 - p^2}} dp = \frac{\pi}{2}, \tag{9.54}$$

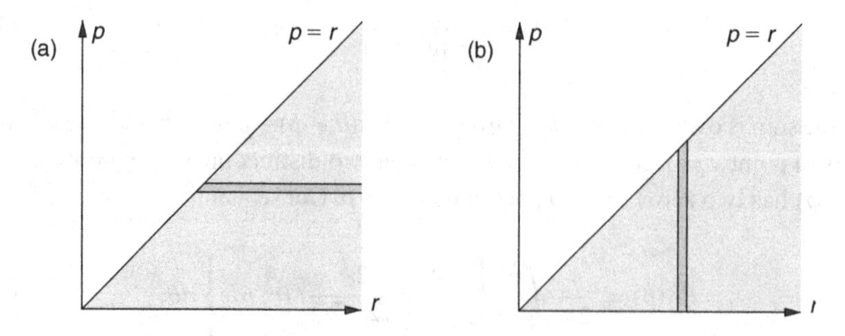

Figure 9.5 (a) In (9.52) we integrate first over r and then over p. The inner r integral is therefore from $r = p$ to $r = \infty$. (b) In (9.53) we integrate first over p and then over r. The inner p integral therefore runs from $p = 0$ to $p = r$.

the inner integral is independent of r. We thus obtain

$$I = -\int_0^\infty \left(\frac{\partial}{\partial r}\bar{f}(r)\right) dr = \bar{f}(0) = f(0,0). \tag{9.55}$$

Radon's inversion formula is therefore correct.

Although Radon found a closed-form inversion formula, the numerical problem of reconstructing the image from the partial and noisy data obtained from a practical CAT scanner is quite delicate, and remains an active area of research.

9.4 Separable kernels

Let

$$K(x,y) = \sum_{i=1}^{N} p_i(x)q_i(y), \tag{9.56}$$

where $\{p_i\}$ and $\{q_i\}$ are two linearly independent sets of functions. The range of K is therefore the span $\langle p_i\rangle$ of the set $\{p_i\}$. Such kernels are said to be *separable*. The theory of integral equations containing such kernels is especially transparent.

9.4.1 Eigenvalue problem

Consider the eigenvalue problem

$$\lambda u(x) = \int_D K(x,y)u(y)\, dy \tag{9.57}$$

for a separable kernel. Here, D is some range of integration, and $x \in D$. If $\lambda \neq 0$, we know that u has to be in the range of K, so we can write

$$u(x) = \sum_i \xi_i p_i(x). \tag{9.58}$$

Inserting this into the integral, we find that our problem reduces to the finite matrix eigenvalue equation

$$\lambda \xi_i = A_{ij}\xi_j, \tag{9.59}$$

where

$$A_{ij} = \int_D q_i(y)p_j(y)\, dy. \tag{9.60}$$

Matters are especially simple when $q_i = p_i^*$. In this case $A_{ij} = A_{ji}^*$, so the matrix A is hermitian and has N linearly independent eigenvectors. Further, none of the N

associated eigenvalues can be zero. To see that this is so suppose that $v(x) = \sum_i \zeta_i p_i(x)$ is an eigenvector with zero eigenvalue. In other words, suppose that

$$0 = \sum_i p_i(x) \int_D p_i^*(y) p_j(y) \zeta_j \, dy. \tag{9.61}$$

Since the $p_i(x)$ are linearly independent, we must have

$$0 = \int_D p_i^*(y) p_j(y) \zeta_j \, dy = 0, \tag{9.62}$$

for each i separately. Multiplying by ζ_i^* and summing we find

$$0 = \int_D \left| \sum_j p_j(y) \zeta_j \right|^2 dy = \int_D |v(y)|^2 \, dy, \tag{9.63}$$

and so $v(x)$ itself must have been zero. The remaining (infinite in number) eigenfunctions span $\langle q_i \rangle^\perp$ and have $\lambda = 0$.

9.4.2 Inhomogeneous problem

It is easiest to discuss inhomogeneous separable-kernel problems by example. Consider the equation

$$u(x) = f(x) + \mu \int_0^1 K(x,y) u(y) \, dy, \tag{9.64}$$

where $K(x,y) = xy$. Here, $f(x)$ and μ are given, and $u(x)$ is to be found. We know that $u(x)$ must be of the form

$$u(x) = f(x) + ax, \tag{9.65}$$

and the only task is to find the constant a. We plug u into the integral equation and, after cancelling a common factor of x, we find

$$a = \mu \int_0^1 y u(y) \, dy = \mu \int_0^1 y f(y) \, dy + a\mu \int_0^1 y^2 \, dy. \tag{9.66}$$

The last integral is equal to $\mu a/3$, so

$$a \left(1 - \frac{1}{3}\mu \right) = \mu \int_0^1 y f(y) \, dy, \tag{9.67}$$

and finally

$$u(x) = f(x) + x \frac{\mu}{(1 - \mu/3)} \int_0^1 y f(y) \, dy. \tag{9.68}$$

Notice that this solution is meaningless if $\mu = 3$. We can relate this to the eigenvalues of the kernel $K(x,y) = xy$. The eigenvalue problem for this kernel is

$$\lambda u(x) = \int_0^1 xyu(y)\,dy. \tag{9.69}$$

On substituting $u(x) = ax$, this reduces to $\lambda ax = ax/3$, and so $\lambda = 1/3$. All other eigenvalues are zero. Our inhomogeneous equation was of the form

$$(1 - \mu K)u = f \tag{9.70}$$

and the operator $(1 - \mu K)$ has an infinite set of eigenfunctions with eigenvalue 1, and a single eigenfunction, $u_0(x) = x$, with eigenvalue $(1 - \mu/3)$. The eigenvalue becomes zero, and hence the inverse ceases to exist, when $\mu = 3$.

A solution to the problem $(1 - \mu K)u = f$ *may* still exist even when $\mu = 3$. But now, applying the Fredholm alternative, we see that f must satisfy the condition that it be orthogonal to all solutions of $(1 - \mu K)^\dagger v = 0$. Since our kernel is hermitian, this means that f must be orthogonal to the zero mode $u_0(x) = x$. For the case of $\mu = 3$, the equation is

$$u(x) = f(x) + 3\int_0^1 xyu(y)\,dy, \tag{9.71}$$

and to have a solution f must obey $\int_0^1 yf(y)\,dy = 0$. We again set $u = f(x) + ax$, and find

$$a = 3\int_0^1 yf(y)\,dy + a3\int_0^1 y^2\,dy, \tag{9.72}$$

but now this reduces to $a = a$. The general solution is therefore

$$u = f(x) + ax \tag{9.73}$$

with a arbitrary.

9.5 Singular integral equations

Equations involving principal-part integrals, such as the *airfoil equation*

$$\frac{P}{\pi}\int_{-1}^1 \varphi(x)\frac{1}{x-y}\,dx = f(y), \tag{9.74}$$

in which f is given and we are to find φ, are called *singular integral equations*. Their solution depends on what conditions are imposed on the unknown function $\varphi(x)$ at the endpoints of the integration region. We will consider only this simplest example here.[2]

[2] The classic text is N. I. Muskhelishvili, *Singular Integral Equations*.

9.5.1 Solution via Tchebychef polynomials

Recall the definition of the Tchebychef polynomials from Chapter 2. We set

$$T_n(x) = \cos(n\cos^{-1} x), \tag{9.75}$$

$$U_{n-1}(x) = \frac{\sin(n\cos^{-1} x)}{\sin(\cos^{-1} x)} = \frac{1}{n} T_n'(x). \tag{9.76}$$

These are the Tchebychef polynomials of the first and second kind, respectively. The orthogonality of the functions $\cos n\theta$ and $\sin n\theta$ over the interval $[0, \pi]$ translates into

$$\int_{-1}^{1} \frac{1}{\sqrt{1-x^2}} T_n(x)\, T_m(x)\, dx = h_n\, \delta_{nm}, \quad n, m \geq 0, \tag{9.77}$$

where $h_0 = \pi$, $h_n = \pi/2$, $n > 0$ and

$$\int_{-1}^{1} \sqrt{1-x^2}\, U_{n-1}(x)\, U_{m-1}(x)\, dx = \frac{\pi}{2}\, \delta_{nm}, \quad n, m > 0. \tag{9.78}$$

The sets $\{T_n(x)\}$ and $\{U_n(x)\}$ are complete in $L^2_w[0, 1]$ with the weight functions $w = (1-x^2)^{-1/2}$ and $w = (1-x^2)^{1/2}$, respectively.

Rather less obvious are the principal-part integral identities (valid for $-1 < y < 1$)

$$P\int_{-1}^{1} \frac{1}{\sqrt{1-x^2}} \frac{1}{x-y}\, dx = 0, \tag{9.79}$$

$$P\int_{-1}^{1} \frac{1}{\sqrt{1-x^2}} T_n(x) \frac{1}{x-y}\, dx = \pi\, U_{n-1}(y), \quad n > 0, \tag{9.80}$$

and

$$P\int_{-1}^{1} \sqrt{1-x^2}\, U_{n-1}(x) \frac{1}{x-y}\, dx = -\pi\, T_n(y), \quad n > 0. \tag{9.81}$$

These correspond, after we set $x = \cos\theta$ and $y = \cos\phi$, to the trigonometric integrals

$$P\int_{0}^{\pi} \frac{\cos n\theta}{\cos\theta - \cos\phi}\, d\theta = \pi \frac{\sin n\phi}{\sin\phi}, \tag{9.82}$$

and

$$P\int_{0}^{\pi} \frac{\sin\theta \sin n\theta}{\cos\theta - \cos\phi}\, d\theta = -\pi \cos n\phi, \tag{9.83}$$

respectively. We will motivate and derive these formulæ at the end of this section.

Granted the validity of these principal-part integrals we can solve the integral equation

$$\frac{P}{\pi}\int_{-1}^{1}\varphi(x)\frac{1}{x-y}\,dx = f(y), \quad y \in [-1,1], \tag{9.84}$$

for φ in terms of f, subject to the condition that φ be bounded at $x = \pm 1$. We show that no solution exists unless f satisfies the condition

$$\int_{-1}^{1}\frac{1}{\sqrt{1-x^2}}f(x)\,dx = 0, \tag{9.85}$$

but if f does satisfy this condition then there is a unique solution

$$\varphi(y) = -\frac{\sqrt{1-y^2}}{\pi}P\int_{-1}^{1}\frac{1}{\sqrt{1-x^2}}f(x)\frac{1}{x-y}\,dx. \tag{9.86}$$

To understand why this is the solution, and why there is a condition on f, expand

$$f(x) = \sum_{n=1}^{\infty} b_n T_n(x). \tag{9.87}$$

Here, the condition on f translates into the absence of a term involving $T_0 \equiv 1$ in the expansion. Then,

$$\varphi(x) = -\sqrt{1-x^2}\sum_{n=1}^{\infty} b_n U_{n-1}(x), \tag{9.88}$$

with b_n the coefficients that appear in the expansion of f, solves the problem. That this is so may be seen on substituting this expansion for φ into the integral equation and using the second of the principal-part identities. This identity provides no way to generate a term with T_0; hence the constraint. Next we observe that the expansion for φ is generated term-by-term from the expansion for f by substituting this into the integral form of the solution and using the first principal-part identity.

Similarly, we solve for $\varphi(y)$ in

$$\frac{P}{\pi}\int_{-1}^{1}\varphi(x)\frac{1}{x-y}\,dx = f(y), \quad y \in [-1,1], \tag{9.89}$$

where now φ is permitted to be singular at $x = \pm 1$. In this case there is always a solution, but it is not unique. The solutions are

$$\varphi(y) = \frac{1}{\pi\sqrt{1-y^2}}P\int_{-1}^{1}\sqrt{1-x^2}f(x)\frac{1}{x-y}\,dx + \frac{C}{\sqrt{1-y^2}}, \tag{9.90}$$

where C is an arbitrary constant. To see this, expand

$$f(x) = \sum_{n=1}^{\infty} a_n U_{n-1}(x), \tag{9.91}$$

and then

$$\varphi(x) = \frac{1}{\sqrt{1-x^2}} \left(\sum_{n=1}^{\infty} a_n T_n(x) + C T_0 \right) \tag{9.92}$$

satisfies the equation for any value of the constant C. Again the expansion for φ is generated from that of f by use of the second principal-part identity.

Explanation of the principal-part identities

The principal-part identities can be extracted from the analytic properties of the resolvent operator $R_\lambda(n - n') \equiv (\hat{H} - \lambda I)^{-1}_{n,n'}$ for a tight-binding model of the conduction band in a one-dimensional crystal with nearest neighbour hopping. The eigenfunctions $u_E(n)$ for this problem obey

$$u_E(n+1) + u_E(n-1) = E\, u_E(n) \tag{9.93}$$

and are

$$u_E(n) = e^{in\theta}, \quad -\pi < \theta < \pi, \tag{9.94}$$

with energy eigenvalues $E = 2\cos\theta$.

The resolvent $R_\lambda(n)$ obeys

$$R_\lambda(n+1) + R_\lambda(n-1) - \lambda R_\lambda(n) = \delta_{n0}, \quad n \in \mathbb{Z}, \tag{9.95}$$

and can be expanded in terms of the energy eigenfunctions as

$$R_\lambda(n - n') = \sum_E \frac{u_E(n) u_E^*(n')}{E - \lambda} = \int_{-\pi}^{\pi} \frac{e^{i(n-n')\theta}}{2\cos\theta - \lambda} \frac{d\theta}{2\pi}. \tag{9.96}$$

If we set $\lambda = 2\cos\phi$, we observe that

$$\int_{-\pi}^{\pi} \frac{e^{in\theta}}{2\cos\theta - 2\cos\phi} \frac{d\theta}{2\pi} = \frac{1}{2i\sin\phi} e^{i|n|\phi}, \quad \operatorname{Im}\phi > 0. \tag{9.97}$$

That this integral is correct can be confirmed by observing that it is evaluating the Fourier coefficient of the double geometric series

$$\sum_{n=-\infty}^{\infty} e^{-in\theta} e^{i|n|\phi} = \frac{2i\sin\phi}{2\cos\theta - 2\cos\phi}, \quad \operatorname{Im}\phi > 0. \tag{9.98}$$

By writing $e^{in\theta} = \cos n\theta + i \sin n\theta$ and observing that the sine term integrates to zero, we find that

$$\int_0^\pi \frac{\cos n\theta}{\cos\theta - \cos\phi} d\theta = \frac{\pi}{i \sin\phi}(\cos n\phi + i \sin n\phi), \tag{9.99}$$

where $n > 0$, and again we have taken $\mathrm{Im}\,\phi > 0$. Now let ϕ approach the real axis from above, and apply the Plemelj formula. We find

$$P\int_0^\pi \frac{\cos n\theta}{\cos\theta - \cos\phi} d\theta = \pi \frac{\sin n\phi}{\sin\phi}. \tag{9.100}$$

This is the first principal-part integral identity. The second identity,

$$P\int_0^\pi \frac{\sin\theta \sin n\theta}{\cos\theta - \cos\phi} d\theta = -\pi \cos n\phi, \tag{9.101}$$

is obtained from the first by using the addition theorems for the sine and cosine.

9.6 Wiener–Hopf equations I

We have seen that Volterra equations of the form

$$\int_0^x K(x-y)\, u(y)\, dy = f(x), \quad 0 < x < \infty, \tag{9.102}$$

having translation invariant kernels, may be solved for u by using a Laplace transform. The apparently innocent modification (see Figure 9.6)

$$\int_0^\infty K(x-y)\, u(y)\, dy = f(x), \quad 0 < x < \infty \tag{9.103}$$

leads to an equation that is much harder to deal with. In these *Wiener–Hopf* equations, we are still only interested in the upper left quadrant of the continuous matrix $K(x-y)$ and $K(x-y)$ still has entries depending only on their distance from the main diagonal.

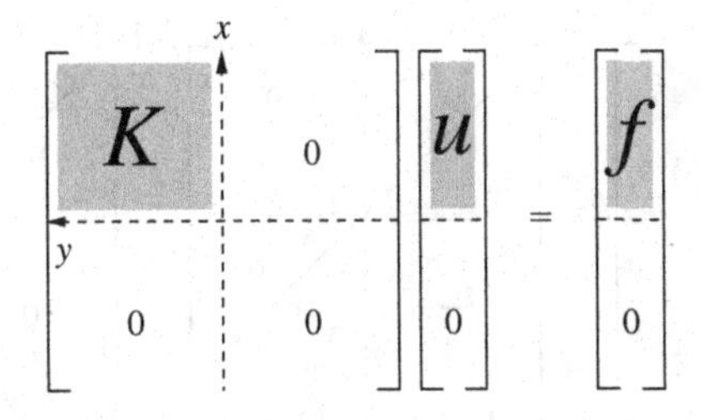

Figure 9.6 The matrix form of (9.103).

Now, however, we make use of the values of $K(x)$ for all of $-\infty < x < \infty$. This suggests the use of a Fourier transform. The problem is that, in order to Fourier transform, we must integrate over the entire real line on *both sides* of the equation and this requires us to know the values of $f(x)$ for negative values of x – but we have not been given this information (and do not really need it). We therefore make the replacement

$$f(x) \to f(x) + g(x), \tag{9.104}$$

where $f(x)$ is non-zero only for positive x, and $g(x)$ non-zero only for negative x. We then solve

$$\int_0^\infty K(x-y)u(y)\,dy = \begin{cases} f(x), & 0 < x < \infty, \\ g(x), & -\infty < x < 0, \end{cases} \tag{9.105}$$

so as to find u and g simultaneously. In other words, we extend the problem to one on the whole real line, but with the negative-x source term $g(x)$ chosen so that the solution $u(x)$ is non-zero only for positive x. We represent this pictorially in Figure 9.7.

To find u and g we try to make an "LU" decomposition of the matrix K into the product $K = L^{-1}U$ of an upper triangular matrix $U(x-y)$ and a lower triangular matrix $L^{-1}(x-y)$; see Figure 9.8. Written out in full, the product $L^{-1}U$ is

$$K(x-y) = \int_{-\infty}^{\infty} L^{-1}(x-t)U(t-y)\,dt. \tag{9.106}$$

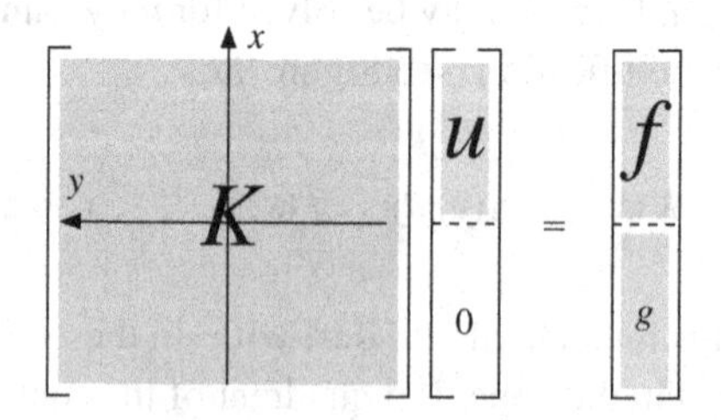

Figure 9.7 The matrix form of (9.105) with both f and g.

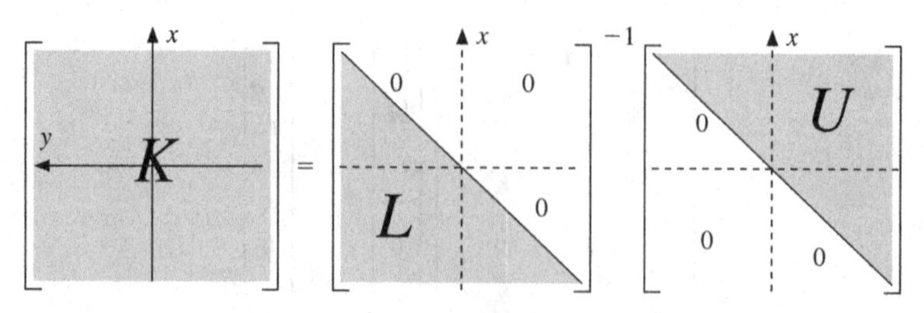

Figure 9.8 The matrix decomposition $K = L^{-1}U$.

Now the inverse of a lower triangular matrix is also lower triangular, and so $L(x-y)$ itself is lower triangular. This means that the function $U(x)$ is zero for negative x, whilst $L(x)$ is zero when x is positive.

If we can find such a decomposition, then on multiplying both sides by L, Equation (9.103) becomes

$$\int_0^x U(x-y)u(y)\,dy = h(x), \quad 0 < x < \infty, \tag{9.107}$$

where

$$h(x) \stackrel{\text{def}}{=} \int_x^\infty L(x-y)f(y)\,dy, \quad 0 < x < \infty. \tag{9.108}$$

These two equations come from the upper half of the full matrix equation represented in Figure 9.9.

The lower parts of the matrix equation have no influence on (9.107) and (9.108): the function $h(x)$ depends only on f, and while $g(x)$ should be chosen to give the column of zeros below h, we do not, in principle, need to know it. This is because we could solve the Volterra equation $Uu = h$ (9.107) *via* a Laplace transform. In practice (as we will see) it is easier to find $g(x)$, and then, knowing the (f, g) column vector, obtain $u(x)$ by solving (9.105). This we can do by Fourier transform.

The difficulty lies in finding the LU decomposition. For finite matrices this decomposition is a standard technique in numerical linear algebra. It is equivalent to the method of *Gaussian elimination*, which, although we were probably never told its name, is the strategy taught in high school for solving simultaneous equations. For continuously infinite matrices, however, making such a decomposition demands techniques far beyond those learned in school. It is a particular case of the scalar *Riemann–Hilbert problem*, and its solution requires the use of complex variable methods.

On taking the Fourier transform of (9.106) we see that we are being asked to factorize

$$\widetilde{K}(k) = [\widetilde{L}(k)]^{-1}\widetilde{U}(k) \tag{9.109}$$

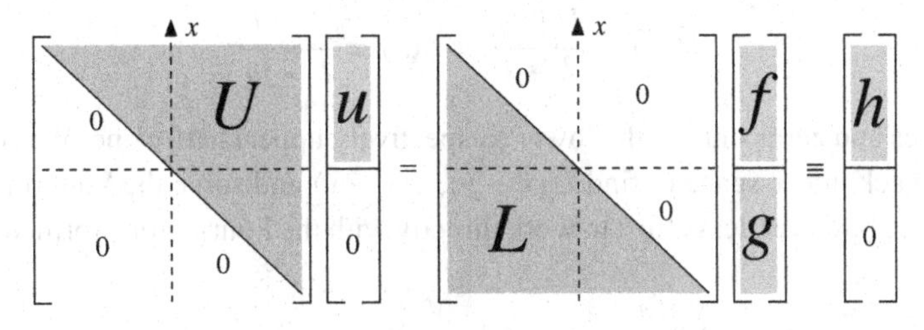

Figure 9.9 Equation (9.107) and the definition (9.108) correspond to the upper half of these two matrix equations.

where

$$\widetilde{U}(k) = \int_0^\infty e^{ikx} U(x)\, dx \tag{9.110}$$

is analytic (i.e. has no poles or other singularities) in the region $\mathrm{Im}\, k \geq 0$, and similarly

$$\widetilde{L}(k) = \int_{-\infty}^0 e^{ikx} L(x)\, dx \tag{9.111}$$

has no poles for $\mathrm{Im}\, k \leq 0$, these analyticity conditions being consequences of the vanishing conditions $U(x-y) = 0, x < y$ and $L(x-y) = 0, x > y$. There will be more than one way of factoring $\widetilde{K}$ into functions with these no-pole properties, but, because the inverse of an upper or lower triangular matrix is also upper or lower triangular, the matrices $U^{-1}(x-y)$ and $L^{-1}(x-y)$ have the same vanishing properties, and, because these inverse matrices correspond to the reciprocals of the Fourier transform, we must also demand that $\widetilde{U}(k)$ and $\widetilde{L}(k)$ have no *zeros* in the upper and lower half-plane, respectively. The combined no-poles, no-zeros conditions will usually determine the factors up to constants. If we are able to factorize $\widetilde{K}(k)$ in this manner, we have effected the LU decomposition. When $\widetilde{K}(k)$ is a rational function of k we can factorize by inspection. In the general case, more sophistication is required.

Example: Let us solve the equation

$$u(x) - \lambda \int_0^\infty e^{-|x-y|} u(y)\, dy = f(x), \tag{9.112}$$

where we will assume that $\lambda < 1/2$. Here the kernel function is

$$K(x,y) = \delta(x-y) - \lambda e^{-|x-y|}. \tag{9.113}$$

This has Fourier transform

$$\widetilde{K}(k) = 1 - \frac{2\lambda}{k^2+1} = \frac{k^2 + (1-2\lambda)}{k^2+1} = \left(\frac{k+ia}{k+i}\right)\left(\frac{k-i}{k-ia}\right)^{-1}, \tag{9.114}$$

where $a^2 = 1 - 2\lambda$. We were able to factorize this by inspection with

$$\widetilde{U}(k) = \frac{k+ia}{k+i}, \quad \widetilde{L}(k) = \frac{k-i}{k-ia} \tag{9.115}$$

having poles and zeros only in the lower (respectively upper) half-plane. We could now transform back into x-space to find $U(x-y)$, $L(x-y)$ and solve the Volterra equation $Uu = h$. It is, however, less effort to work directly with the Fourier transformed equation in the form

$$\left(\frac{k+ia}{k+i}\right)\widetilde{u}_+(k) = \left(\frac{k-i}{k-ia}\right)(\widetilde{f}_+(k) + \widetilde{g}_-(k)). \tag{9.116}$$

Here we have placed subscripts on $\widetilde{f}(k)$, $\widetilde{g}(k)$ and $\widetilde{u}(k)$ to remind us that these Fourier transforms are analytic in the upper (+) or lower (−) half-plane. Since the left-hand side of this equation is analytic in the upper half-plane, so must be the right-hand side. We therefore choose $\widetilde{g}_-(k)$ to eliminate the potential pole at $k = ia$ that might arise from the first term on the right. This we can do by setting

$$\left(\frac{k-i}{k-ia}\right) g_-(k) = \frac{\alpha}{k-ia} \tag{9.117}$$

for some as yet undetermined constant α. (Observe that the resultant $g_-(k)$ is indeed analytic in the lower half-plane. This analyticity ensures that $g(x)$ is zero for positive x.) We can now solve for $\widetilde{u}(k)$ as

$$\begin{aligned}\widetilde{u}(k) &= \left(\frac{k+i}{k+ia}\right)\left(\frac{k-i}{k-ia}\right)\widetilde{f}_+(k) + \left(\frac{k+i}{k+ia}\right)\frac{\alpha}{k-ia} \\ &= \frac{k^2+1}{k^2+a^2}\widetilde{f}_+(k) + \alpha\frac{k+i}{k^2+a^2} \\ &= \widetilde{f}_+(k) + \frac{1-a^2}{k^2+a^2}\widetilde{f}_+(k) + \alpha\frac{k+i}{k^2+a^2}.\end{aligned} \tag{9.118}$$

The inverse Fourier transform of

$$\frac{k+i}{k^2+a^2} \tag{9.119}$$

is

$$\frac{i}{2|a|}(1-|a|\,\mathrm{sgn(x)})e^{-|a||x|}, \tag{9.120}$$

and that of

$$\frac{1-a^2}{k^2+a^2} = \frac{2\lambda}{k^2+(1-2\lambda)} \tag{9.121}$$

is

$$\frac{\lambda}{\sqrt{1-2\lambda}}e^{-\sqrt{1-2\lambda}|x|}. \tag{9.122}$$

Consequently

$$\begin{aligned}u(x) &= f(x) + \frac{\lambda}{\sqrt{1-2\lambda}}\int_0^\infty e^{-\sqrt{1-2\lambda}|x-y|}f(y)\,dy \\ &\quad + \beta(1-\sqrt{1-2\lambda}\,\mathrm{sgn}\,x)e^{-\sqrt{1-2\lambda}|x|}.\end{aligned} \tag{9.123}$$

Here β is some multiple of α, and we have used the fact that $f(y)$ is zero for negative y to make the lower limit on the integral 0 instead of $-\infty$. We determine the as yet unknown β from the requirement that $u(x) = 0$ for $x < 0$. We find that this will be the case if we take

$$\beta = -\frac{\lambda}{a(a+1)} \int_0^\infty e^{-ay} f(y)\, dy. \tag{9.124}$$

The solution is therefore, for $x > 0$,

$$\begin{aligned} u(x) = f(x) &+ \frac{\lambda}{\sqrt{1-2\lambda}} \int_0^\infty e^{-\sqrt{1-2\lambda}|x-y|} f(y)\, dy \\ &+ \frac{\lambda(\sqrt{1-2\lambda}-1)}{1-2\lambda+\sqrt{1-2\lambda}} e^{-\sqrt{1-2\lambda}x} \int_0^\infty e^{-\sqrt{1-2\lambda}y} f(y)\, dy. \end{aligned} \tag{9.125}$$

Not every invertible n-by-n matrix has a plain LU decomposition. For a related reason not every Wiener–Hopf equation can be solved so simply. Instead there is a topological *index theorem* that determines whether solutions can exist, and, if solutions do exist, whether they are unique. We shall therefore return to this problem once we have aquired a deeper understanding of the interaction between topology and complex analysis.

9.7 Some functional analysis

We have hitherto avoided, as far as it is possible, the full rigours of mathematics. For most of us, and for most of the time, we can solve our physics problems by using calculus rather than analysis. It is worth, nonetheless, being familiar with the proper mathematical language so that when something tricky comes up we know where to look for help. The modern setting for the mathematical study of integral and differential equations is the discipline of *functional analysis*, and the classic text for the mathematically inclined physicist is the four-volume set *Methods of Modern Mathematical Physics* by Michael Reed and Barry Simon. We cannot summarize these volumes in a few paragraphs, but we can try to provide enough background for us to be able to explain a few issues that may have puzzled the alert reader.

This section requires the reader to have sufficient background in real analysis to know what it means for a set to be compact.

9.7.1 Bounded and compact operators

(i) A linear operator $K : L^2 \to L^2$ is *bounded* if there is a positive number M such that

$$\|Kx\| \le M\|x\|, \quad \forall x \in L^2. \tag{9.126}$$

If K is bounded then the smallest such M is the *norm* of K, which we denote by $\|K\|$. Thus

$$\|Kx\| \leq \|K\| \, \|x\|. \tag{9.127}$$

For a finite-dimensional matrix, $\|K\|$ is the largest eigenvalue of K. The function Kx is a continuous function of x if, and only if, it is bounded. "Bounded" and "continuous" are therefore synonyms. Linear differential operators are *never* bounded, and this is the source of most of the complications in their theory.

(ii) If the operators A and B are bounded, then so is AB and

$$\|AB\| \leq \|A\|\|B\|. \tag{9.128}$$

(iii) A linear operator $K : L^2 \to L^2$ is *compact* (or *completely continuous*) if it maps bounded sets in L^2 to relatively compact sets (sets whose closure is compact). Equivalently, K is compact if the image sequence Kx_n of every bounded sequence of functions x_n contains a convergent subsequence. Compact $\Rightarrow$ continuous, but not *vice versa*. One can show that, given any positive number M, a compact self-adjoint operator has only a finite number of eigenvalues with λ outside the interval $[-M, M]$. The eigenvectors u_n with non-zero eigenvalues span the range of the operator. Any vector can therefore be written

$$u = u_0 + \sum_i a_i u_i, \tag{9.129}$$

where u_0 lies in the null-space of K. The Green function of a linear differential operator defined on a finite interval is usually the integral kernel of a compact operator.

(iv) If K is compact then

$$H = I + K \tag{9.130}$$

is *Fredholm*. This means that H has a finite-dimensional kernel and co-kernel, and that the Fredholm alternative applies.

(v) An integral kernel is *Hilbert–Schmidt* if

$$\int |K(\xi, \eta)|^2 \, d\xi d\eta < \infty. \tag{9.131}$$

This means that K can be expanded in terms of a complete orthonormal set $\{\phi_m\}$ as

$$K(x, y) = \sum_{n,m=1}^{\infty} A_{nm} \phi_n(x) \phi_m^*(y) \tag{9.132}$$

in the sense that

$$\lim_{N,M\to\infty} \left\| \sum_{n,m=1}^{N,M} A_{nm}\phi_n\phi_m^* - K \right\| = 0. \tag{9.133}$$

Now the finite sum

$$\sum_{n,m=1}^{N,M} A_{nm}\phi_n(x)\phi_m^*(y) \tag{9.134}$$

is automatically compact since it is bounded and has finite-dimensional range. (The unit ball in a Hilbert space is relatively compact $\Leftrightarrow$ the space is finite dimensional.) Thus, Hilbert–Schmidt implies that K is approximated in norm by compact operators. But it is not hard to show that a norm-convergent limit of compact operators is compact, so K itself is compact. Thus

Hilbert–Schmidt $\Rightarrow$ compact.

It is easy to test a given kernel to see if it is Hilbert–Schmidt (simply use the definition) and therein lies the utility of the concept.

If we have a Hilbert–Schmidt Green function g, we can recast our differential equation as an integral equation with g as kernel, and this is why the Fredholm alternative works for a large class of linear differential equations.

Example: Consider the Legendre-equation operator

$$L = -\frac{d}{dx}(1-x^2)\frac{d}{dx} \tag{9.135}$$

acting on functions $u \in L^2[-1,1]$ with boundary conditions that u be finite at the endpoints. This operator has a normalized zero mode $u_0 = 1/\sqrt{2}$, so it cannot have an inverse. There exists, however, a modified Green function $g(x,x')$ that satisfies

$$Lu = \delta(x-x') - \frac{1}{2}. \tag{9.136}$$

It is

$$g(x,x') = \ln 2 - \frac{1}{2} - \frac{1}{2}\ln(1+x_>)(1-x_<), \tag{9.137}$$

where $x_>$ is the greater of x and x', and $x_<$ the lesser. We may verify that

$$\int_{-1}^{1}\int_{-1}^{1} |g(x,x')|^2\, dxdx' < \infty, \tag{9.138}$$

so g is Hilbert–Schmidt and therefore the kernel of a compact operator. The eigenvalue problem

$$Lu_n = \lambda_n u_n \tag{9.139}$$

can be recast as the integral equation

$$\mu_n u_n = \int_{-1}^{1} g(x,x')u_n(x')\,dx' \tag{9.140}$$

with $\mu_n = \lambda_n^{-1}$. The compactness of g guarantees that there is a complete set of eigenfunctions (these being the Legendre polynomials $P_n(x)$ for $n > 0$) having eigenvalues $\mu_n = 1/n(n+1)$. The operator g also has the eigenfunction P_0 with eigenvalue $\mu_0 = 0$. This example provides the justification for the claim that the "finite" boundary conditions we adopted for the Legendre equation in Chapter 8 give us a self-adjoint operator.

Note that $K(x,y)$ does not have to be bounded for K to be Hilbert–Schmidt.

Example: The kernel

$$K(x,y) = \frac{1}{(x-y)^{\alpha}}, \quad |x|,|y| < 1 \tag{9.141}$$

is Hilbert–Schmidt provided $\alpha < \frac{1}{2}$.

Example: The kernel

$$K(x,y) = \frac{1}{2m}e^{-m|x-y|}, \quad x,y \in \mathbb{R} \tag{9.142}$$

is not Hilbert–Schmidt because $|K(x-y)|$ is constant along the lines $x - y =$ constant, which lie parallel to the diagonal. K has a continuous spectrum consisting of all positive real numbers less than $1/m^2$. It cannot be compact, therefore, but it is bounded with $\|K\| = 1/m^2$. The integral equation (9.22) contains this kernel, and the Fredholm alternative does not apply to it.

9.7.2 Closed operators

One motivation for our including a brief account of functional analysis is that an attentive reader will have realized that some of the statements we have made in earlier chapters appear to be inconsistent. We have asserted in Chapter 2 that no significance can be attached to the value of an L^2 function at any particular point – only integrated averages matter. In later chapters, though, we have happily imposed boundary conditions that require these very functions to take specified values at the endpoints of our interval. In this section we will resolve this paradox. The apparent contradiction is intimately connected with our imposing boundary conditions only on derivatives of lower order than that of the differential equation, but understanding why this is so requires some function-analytic language.

Differential operators L are never continuous; we cannot deduce from $u_n \to u$ that $Lu_n \to Lu$. Differential operators can be *closed*, however. A closed operator is one for which whenever a sequence u_n converges to a limit u and at the same time the image sequence Lu_n also converges to a limit f, then u is in the domain of L and $Lu = f$. The name is not meant to imply that the domain of definition is closed, but indicates instead that the *graph* of L – this being the set $\{u, Lu\}$ considered as a subset of $L^2[a,b] \times L^2[a,b]$ – contains its limit points and so is a closed set.

Any self-adjoint operator is automatically closed. To see why this is so, recall that in defining the adjoint of an operator A, we say that y is in the domain of $A^\dagger$ if there is a z such that $\langle y, Ax\rangle = \langle z, x\rangle$ for all x in the domain of A. We then set $A^\dagger y = z$. Now suppose that $y_n \to y$ and $A^\dagger y_n = z_n \to z$. The Cauchy–Schwartz–Bunyakovski inequality shows that the inner product is a continuous function of its arguments. Consequently, if x is in the domain of A, we can take the limit of $\langle y_n, Ax\rangle = \langle A^\dagger y_n, x\rangle = \langle z_n, x\rangle$ to deduce that $\langle y, Ax\rangle = \langle z, x\rangle$. But this means that y is in the domain of $A^\dagger$, and $z = A^\dagger y$. The adjoint of any operator is therefore a closed operator. A self-adjoint operator, being its own adjoint, is therefore necessarily closed.

A deep result states that a closed operator defined on a closed domain is bounded. Since they are always unbounded, the domain of a closed differential operator can never be a closed set.

An operator may not be closed but may be *closable*, in that we can make it closed by including additional functions in its domain. The essential requirement for closability is that we never have two sequences u_n and v_n which converge to the **same** limit, w, while Lu_n and Lv_n both converge, but to **different** limits. Closability is equivalent to requiring that if $u_n \to 0$ and Lu_n converges, then Lu_n converges to zero.

Example: Let $L = d/dx$. Suppose that $u_n \to 0$ and $Lu_n \to f$. If φ is a smooth L^2 function that vanishes at 0, 1, then

$$\int_0^1 \varphi f\, dx = \lim_{n\to\infty} \int_0^1 \varphi \frac{du_n}{dx}\, dx = -\lim_{n\to\infty} \int_0^1 \phi' u_n\, dx = 0. \tag{9.143}$$

Here we have used the continuity of the inner product to justify the interchange of the order of limit and integral. By the same arguments we used when dealing with the calculus of variations, we deduce that $f = 0$. Thus d/dx is closable.

If an operator is closable, we may as well add the extra functions to its domain and make it closed. Let us consider what closure means for the operator

$$L = \frac{d}{dx}, \quad \mathcal{D}(L) = \{y \in C^1[0,1] : y'(0) = 0\}. \tag{9.144}$$

Here, in fixing the derivative at the endpoint, we are imposing a boundary condition of higher order than we ought.

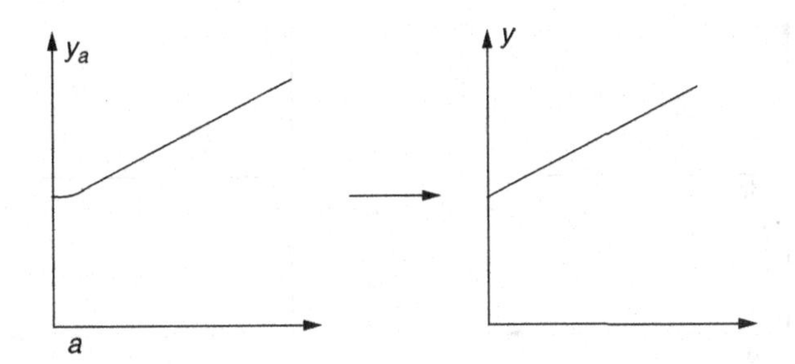

Figure 9.10 $\lim_{a\to 0} y_a = y$ in $L^2[0, 1]$.

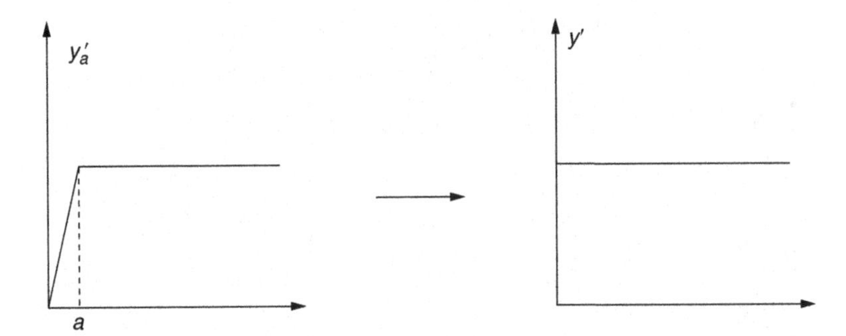

Figure 9.11 $y'_a \to y'$ in $L^2[0, 1]$.

Consider the sequence of differentiable functions y_a shown in Figure 9.10. These functions have vanishing derivative at $x = 0$, but tend in L^2 to a function y whose derivative is non-zero at $x = 0$.

Figure 9.11 shows that the derivative of these functions also converges in L^2.

If we want L to be closed, we should therefore extend the domain of definition of L to include functions with non-vanishing endpoint derivative. We can also use this method to add to the domain of L functions that are only *piecewise differentiable* – i.e. functions with a discontinuous derivative.

Now consider what happens if we try to extend the domain of

$$L = \frac{d}{dx}, \quad \mathcal{D}(L) = \{y, y' \in L^2 : y(0) = 0\}, \tag{9.145}$$

to include functions that do not vanish at the endpoint. Take the sequence of functions y_a shown in Figure 9.12. These functions vanish at the origin, and converge in L^2 to a function that does not vanish at the origin.

Now, as Figure 9.13 shows, the derivatives converge towards the derivative of the limit function – *together with a delta function near the origin*. The area under the functions $|y'_a(x)|^2$ grows without bound and the sequence Ly_a becomes infinitely far from the derivative of the limit function when distance is measured in the L^2 norm.

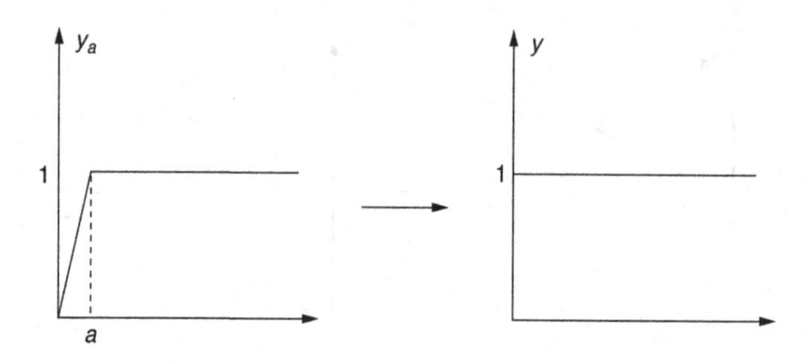

Figure 9.12 $\lim_{a\to 0} y_a = y$ in $L^2[0, 1]$.

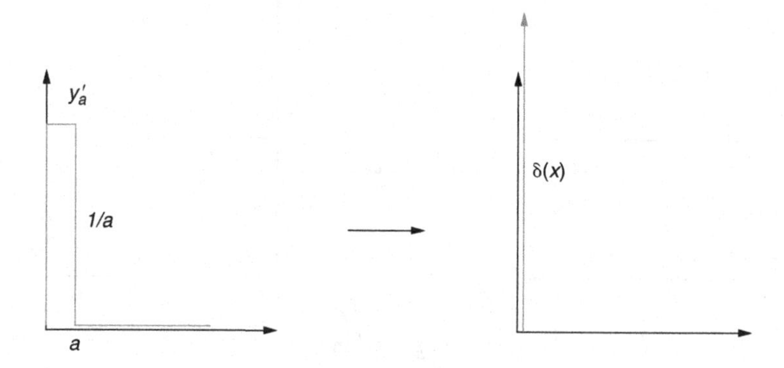

Figure 9.13 $y'_a \to \delta(x)$, but the delta function is not an element of $L^2[0, 1]$.

We therefore cannot use closure to extend the domain to include these functions. Another way of saying this is that in order for the weak derivative of y to be in L^2, and therefore for y to be in the domain of d/dx, the function y need not be classically differentiable, but its L^2 equivalence class must contain a *continuous* function – and continuous functions do have well-defined values. It is the values of this continuous representative that are constrained by the boundary conditions.

This story repeats for differential operators of any order: if we try to impose boundary conditions of too high an order, they are washed out in the process of closing the operator. Boundary conditions of lower order cannot be eliminated, however, and so make sense as statements involving functions in L^2.

9.8 Series solutions

One of the advantages of recasting a problem as an integral equation is that the equation often suggests a systematic approximation scheme. Usually we start from the solution of an exactly solvable problem and expand the desired solution about it as an infinite series in some small parameter. The terms in such a perturbation series may become progressively harder to evaluate, but, if we are lucky, the sum of the first few will prove adaquate for our purposes.

9.8.1 Liouville–Neumann–Born series

The geometric series

$$S = 1 - x + x^2 - x^3 + \cdots \tag{9.146}$$

converges to $1/(1+x)$ provided $|x| < 1$. Suppose we wish to solve

$$(I + \lambda K)\varphi = f \tag{9.147}$$

where K is an integral operator. It is then natural to write

$$\varphi = (I + \lambda K)^{-1} f = (1 - \lambda K + \lambda^2 K^2 - \lambda^3 K^3 + \cdots) f, \tag{9.148}$$

where

$$\begin{aligned} K^2(x,y) &= \int K(x,z)K(z,y)\,dz, \\ K^3(x,y) &= \int K(x,z_1)K(z_1,z_2)K(z_2,y)\,dz_1 dz_2, \end{aligned} \tag{9.149}$$

and so on. This *Liouville–Neumann series* will converge, and yield a solution to the problem, provided that $\lambda\|K\| < 1$. In quantum mechanics this series is known as the *Born series*.

9.8.2 Fredholm series

A familiar result from high-school algebra is *Cramer's rule*, which gives the solution of a set of linear equations in terms of ratios of determinants. For example, the system of equations

$$\begin{aligned} a_{11}x_1 + a_{12}x_2 + a_{13}x_3 &= b_1, \\ a_{21}x_1 + a_{22}x_2 + a_{23}x_3 &= b_2, \\ a_{31}x_1 + a_{32}x_2 + a_{33}x_3 &= b_3, \end{aligned} \tag{9.150}$$

has solution

$$x_1 = \frac{1}{D}\begin{vmatrix} b_1 & a_{12} & a_{13} \\ b_2 & a_{22} & a_{23} \\ b_3 & a_{32} & a_{33} \end{vmatrix}, \quad x_2 = \frac{1}{D}\begin{vmatrix} a_{11} & b_1 & a_{13} \\ a_{21} & b_2 & a_{23} \\ a_{31} & b_3 & a_{33} \end{vmatrix}, \quad x_3 = \frac{1}{D}\begin{vmatrix} a_{11} & a_{12} & b_1 \\ a_{21} & a_{22} & b_2 \\ a_{31} & a_{32} & b_3 \end{vmatrix}, \tag{9.151}$$

where

$$D = \begin{vmatrix} a_{11} & a_{12} & a_{13} \\ a_{21} & a_{22} & a_{23} \\ a_{31} & a_{32} & a_{33} \end{vmatrix}. \tag{9.152}$$

Although not as computationally efficient as standard Gaussian elimination, Cramer's rule is useful in that it is a closed-form solution. It is equivalent to the statement that the inverse of a matrix is given by the transposed matrix of the cofactors, divided by the determinant.

A similar formula for integral equations was given by Fredholm. The equations he considered were, in operator form,

$$(I + \lambda K)\varphi = f, \tag{9.153}$$

where I is the identity operator, K is an integral operator with kernel $K(x,y)$ and λ a parameter. We motivate Fredholm's formula by giving an expansion for the determinant of a finite matrix. Let $\mathbf{K}$ be an n-by-n matrix

$$D(\lambda) \stackrel{\text{def}}{=} \det(\mathbf{I} + \lambda\mathbf{K}) \equiv \begin{vmatrix} 1+\lambda K_{11} & \lambda K_{12} & \cdots & \lambda K_{1n} \\ \lambda K_{21} & 1+\lambda K_{22} & \cdots & \lambda K_{2n} \\ \vdots & \vdots & \ddots & \vdots \\ \lambda K_{n1} & \lambda K_{n2} & \cdots & 1+\lambda K_{nn} \end{vmatrix}. \tag{9.154}$$

Then

$$D(\lambda) = \sum_{m=0}^{n} \frac{\lambda^m}{m!} A_m, \tag{9.155}$$

where $A_0 = 1, A_1 = \operatorname{tr}\mathbf{K} \equiv \sum_i K_{ii}$,

$$A_2 = \sum_{i_1,i_2=1}^{n} \begin{vmatrix} K_{i_1i_1} & K_{i_1i_2} \\ K_{i_2i_1} & K_{i_2i_2} \end{vmatrix}, \quad A_3 = \sum_{i_1,i_2,i_3=1}^{n} \begin{vmatrix} K_{i_1i_1} & K_{i_1i_2} & K_{i_1i_3} \\ K_{i_2i_1} & K_{i_2i_2} & K_{i_2i_3} \\ K_{i_3i_1} & K_{i_3i_2} & K_{i_3i_3} \end{vmatrix}. \tag{9.156}$$

The pattern for the rest of the terms should be obvious, as should the proof.

As observed above, the inverse of a matrix is the reciprocal of the determinant of the matrix multiplied by the transposed matrix of the cofactors. So, if $D_{\mu\nu}$ is the cofactor of the term in $D(\lambda)$ associated with $K_{\nu\mu}$, then the solution of the matrix equation

$$(\mathbf{I} + \lambda\mathbf{K})\mathbf{x} = \mathbf{b} \tag{9.157}$$

is

$$x_\mu = \frac{D_{\mu 1}b_1 + D_{\mu 2}b_2 + \cdots + D_{\mu n}b_n}{D(\lambda)}. \tag{9.158}$$

If $\mu \neq \nu$ we have

$$D_{\mu\nu} = \lambda K_{\mu\nu} + \lambda^2 \sum_i \begin{vmatrix} K_{\mu\nu} & K_{\mu i} \\ K_{i\nu} & K_{ii} \end{vmatrix} + \lambda^3 \frac{1}{2!} \sum_{i_1 i_2} \begin{vmatrix} K_{\mu\nu} & K_{\mu i_1} & K_{\mu i_2} \\ K_{i_1\nu} & K_{i_1 i_1} & K_{i_1 i_2} \\ K_{i_2\nu} & K_{i_2 i_1} & K_{i_2 i_2} \end{vmatrix} + \cdots \tag{9.159}$$

When $\mu = \nu$ we have

$$D_{\mu\nu} = \delta_{\mu\nu} \widetilde{D}(\lambda), \tag{9.160}$$

where $\widetilde{D}(\lambda)$ is the expression analogous to $D(\lambda)$, but with the μ-th row and column deleted.

These elementary results suggest the definition of the *Fredholm determinant* of the integral kernel $K(x,y)$, $a < x,y < b$, as

$$D(\lambda) = \mathrm{Det}\,|I + \lambda K| \equiv \sum_{m=0}^{\infty} \frac{\lambda^m}{m!} A_m, \tag{9.161}$$

where $A_0 = 1$, $A_1 = \mathrm{Tr}\, K \equiv \int_a^b K(x,x)\, dx$,

$$A_2 = \int_a^b \int_a^b \begin{vmatrix} K(x_1,x_1) & K(x_1,x_2) \\ K(x_2,x_1) & K(x_2,x_2) \end{vmatrix} dx_1 dx_2,$$

$$A_3 = \int_a^b \int_a^b \int_a^b \begin{vmatrix} K(x_1,x_1) & K(x_1,x_2) & K(x_1,x_3) \\ K(x_2,x_1) & K(x_2,x_2) & K(x_2,x_3) \\ K(x_3,x_1) & K(x_3,x_2) & K(x_3,x_3) \end{vmatrix} dx_1 dx_2 dx_3, \tag{9.162}$$

etc. We also define

$$\begin{aligned} D(x,y,\lambda) = {} & \lambda K(x,y) + \lambda^2 \int_a^b \begin{vmatrix} K(x,y) & K(x,\xi) \\ K(\xi,y) & K(\xi,\xi) \end{vmatrix} d\xi \\ & + \lambda^3 \frac{1}{2!} \int_a^b \int_a^b \begin{vmatrix} K(x,y) & K(x,\xi_1) & K(x,\xi_2) \\ K(\xi_1,y) & K(\xi_1,\xi_1) & K(\xi_1,\xi_2) \\ K(\xi_2,y) & K(\xi_2,\xi_1) & K(\xi_2,\xi_2) \end{vmatrix} d\xi_1 d\xi_2 + \cdots, \end{aligned} \tag{9.163}$$

and then

$$\varphi(x) = f(x) + \frac{1}{D(\lambda)} \int_a^b D(x,y,\lambda) f(y)\, dy \tag{9.164}$$

is the solution of the equation

$$\varphi(x) + \lambda \int_a^b K(x,y) \varphi(y)\, dy = f(x). \tag{9.165}$$

If $|K(x,y)| < M$ in $[a,b] \times [a,b]$, the *Fredholm series* for $D(\lambda)$ and $D(x,y,\lambda)$ converge for all λ, and define entire functions. In this feature it is unlike the Neumann series, which has a finite radius of convergence.

The proof of these claims follows from the identity

$$D(x,y,\lambda) + \lambda D(\lambda)K(x,y) + \lambda \int_a^b D(x,\xi,\lambda)K(\xi,y)\,d\xi = 0, \tag{9.166}$$

or, more compactly with $G(x,y) = D(x,y,\lambda)/D(\lambda)$,

$$(I+G)(I+\lambda K) = I. \tag{9.167}$$

For details see Whitaker and Watson §11.2.

Example: The equation

$$\varphi(x) = x + \lambda \int_0^1 xy\varphi(y)\,dy \tag{9.168}$$

gives us

$$D(\lambda) = 1 - \frac{1}{3}\lambda, \quad D(x,y,\lambda) = \lambda xy \tag{9.169}$$

and so

$$\varphi(x) = \frac{3x}{3-\lambda}. \tag{9.170}$$

(We have considered this equation and solution before, in Section 9.4.)

9.9 Further exercises and problems

Exercise 9.1: The following problems should be relatively easy.

(a) Solve the inhomogeneous Type II Fredholm integral equation

$$u(x) = e^x + \lambda \int_0^1 xy\,u(y)\,dy\,.$$

(b) Solve the homogeneous Type II Fredholm integral equation

$$u(x) = \lambda \int_0^\pi \sin(x-y)\,u(y)\,dy\,.$$

(c) Solve the integral equation

$$u(x) = x + \lambda \int_0^1 (yx + y^2)\,u(y)\,dy$$

to second order in λ using

(i) the Neumann series; and
(ii) the Fredholm series.

(d) By differentiating, solve the integral equation: $u(x) = x + \int_0^x u(y)\,dy$.

(e) Solve the integral equation: $u(x) = x^2 + \int_0^1 xy\,u(y)\,dy$.

(f) Find the eigenfunction(s) and eigenvalue(s) of the integral equation

$$u(x) = \lambda \int_0^1 e^{x-y}\,u(y)\,dy\,.$$

(g) Solve the integral equation: $u(x) = e^x + \lambda \int_0^1 e^{x-y}\,u(y)\,dy$.

(h) Solve the integral equation

$$u(x) = x + \int_0^1 dy\,(1 + xy)\,u(y)$$

for the unknown function $u(x)$.

Exercise 9.2: Solve the integral equation

$$u(x) = f(x) + \lambda \int_0^1 x^3 y^3 u(y)dy, \quad 0 < x < 1$$

for the unknown $u(x)$ in terms of the given function $f(x)$. For what values of λ does a unique solution $u(x)$ exist without restrictions on $f(x)$? For what value $\lambda = \lambda_0$ does a solution exist only if $f(x)$ satisfies some condition? Using the language of the Fredholm alternative, and the range and null-space of the relevant operators, explain what is happening when $\lambda = \lambda_0$. For the case $\lambda = \lambda_0$ find explicitly the condition on $f(x)$ and, assuming this condition is satisfied, write down the corresponding general solution for $u(x)$. Check that this solution does indeed satisfy the integral equation.

Exercise 9.3: Use a Laplace transform to find the solution to the generalized Abel equation

$$f(x) = \int_0^x (x-t)^{-\mu}\,u(t)dt, \quad 0 < \mu < 1,$$

where $f(x)$ is given and $f(0) = 0$. Your solution will be of the form

$$u(x) = \int_0^x K(x-t)f'(t)dt,$$

and you should give an explicit expression for the kernel $K(x-t)$.

You will find the formula

$$\int_0^\infty t^{\mu-1}e^{-pt}dt = p^{-\mu}\,\Gamma(\mu), \quad \mu > 0$$

to be useful.

Exercise 9.4: *Translationally invariant kernels.*

(a) Consider the integral equation: $u(x) = g(x) + \lambda \int_{-\infty}^{\infty} K(x,y)\, u(y)\, dy$, with the translationally invariant kernel $K(x,y) = Q(x-y)$, in which g, λ and Q are known. Show that the Fourier transforms $\hat{u}$, $\hat{g}$ and $\hat{Q}$ satisfy $\hat{u}(q) = \hat{g}(q)/\{1 - \sqrt{2\pi}\lambda\hat{Q}(q)\}$. Expand this result to second order in λ to recover the second-order Liouville–Neumann–Born series.

(b) Use Fourier transforms to find a solution of the integral equation

$$u(x) = e^{-|x|} + \lambda \int_{-\infty}^{\infty} e^{-|x-y|}\, u(y)\, dy$$

that remains finite as $|x| \to \infty$.

(c) Use Laplace transforms to find a solution of the integral equation

$$u(x) = e^{-x} + \lambda \int_{0}^{x} e^{-|x-y|}\, u(y)\, dy \qquad x > 0.$$

Exercise 9.5: The integral equation

$$\frac{1}{\pi} \int_{0}^{\infty} dy \, \frac{\phi(y)}{x+y} = f(x), \quad x > 0,$$

relates the unknown function ϕ to the known function f.

(i) Show that the changes of variables

$$x = \exp 2\xi, \quad y = \exp 2\eta,$$

$$\phi(\exp 2\eta)\, \exp \eta = \psi(\eta), \quad f(\exp 2\xi)\, \exp \xi = g(\xi),$$

convert the integral equation into one that can be solved by an integral transform.

(ii) Hence, or otherwise, construct an explicit formula for $\phi(x)$ in terms of a double integral involving $f(y)$.

You may use without proof the integral

$$\int_{-\infty}^{\infty} d\xi \, \frac{e^{-is\xi}}{\cosh \xi} = \frac{\pi}{\cosh \pi s/2}.$$

Exercise 9.6: *Using Mellin transforms.* Recall that the Mellin transform $\tilde{f}(s)$ of the function $f(t)$ is defined by

$$\tilde{f}(s) = \int_{0}^{\infty} dt\, t^{s-1} f(t).$$

(a) Given two functions, $f(t)$ and $g(t)$, a Mellin convolution $f * g$ can be defined through

$$(f * g)(t) = \int_{0}^{\infty} f(tu^{-1})\, g(u)\, \frac{du}{u}.$$

Show that the Mellin transform of the Mellin convolution $f * g$ is

$$\widetilde{f * g}(s) = \int_0^\infty t^{s-1}(f * g)(t)\, dt = \widetilde{f}(s)\widetilde{g}(s).$$

Similarly find the Mellin transform of

$$(f\#g)(t) \stackrel{\text{def}}{=} \int_0^\infty f(tu)g(u)\, du.$$

(b) The unknown function $F(t)$ satisfies Fox's integral equation,

$$F(t) = G(t) + \int_0^\infty dv\, Q(tv)\, F(v),$$

in which G and Q are known. Solve for the Mellin transform $\widetilde{F}$ in terms of the Mellin transforms $\widetilde{G}$ and $\widetilde{Q}$.

Exercise 9.7: Some more easy problems:

(a) Solve the Lalesco–Picard integral equation

$$u(x) = \cos \mu x + \frac{1}{4} \int_{-\infty}^{\infty} dy\, \mathrm{e}^{-|x-y|}\, u(y)\, .$$

(b) For $\lambda \neq 3$, solve the integral equation

$$\phi(x) = 1 + \lambda \int_0^1 dy\, xy\, \phi(y)\, .$$

(c) By taking derivatives, show that the solution of the Volterra equation

$$x = \int_0^x dy\, \left(\mathrm{e}^x + \mathrm{e}^y\right) \psi(y)$$

satisfies a first-order differential equation. Hence, solve the integral equation.

Exercise 9.8: *Principal-part integrals.*

(a) If w is real, show that

$$P \int_{-\infty}^{\infty} e^{-u^2} \frac{1}{u - w} du = -2\sqrt{\pi} e^{-w^2} \int_0^w e^{u^2} du.$$

(This is easier than it looks.)

(b) If y is real, but *not* in the interval $(-1, 1)$, show that

$$\int_{-1}^{1} \frac{1}{(y - x)\sqrt{1 - x^2}} dx = \frac{\pi}{\sqrt{y^2 - 1}}.$$

Now let $y \in (-1, 1)$. Show that

$$P \int_{-1}^{1} \frac{1}{(y-x)\sqrt{1-x^2}} \, dx = 0.$$

(This is harder than it looks.)

Exercise 9.9: Consider the integral equation

$$u(x) = g(x) + \lambda \int_0^1 K(x,y)\, u(y)\, dy\,,$$

in which only u is unknown.

(a) Write down the solution $u(x)$ to second order in the Liouville–Neumann–Born series.
(b) Suppose $g(x) = x$ and $K(x,y) = \sin 2\pi xy$. Compute $u(x)$ to second order in the Liouville–Neumann–Born series.

Exercise 9.10: Show that the application of the Fredholm series method to the equation

$$\varphi(x) = x + \lambda \int_0^1 (xy + y^2)\varphi(y)\, dy$$

gives

$$D(\lambda) = 1 - \frac{2}{3}\lambda - \frac{1}{72}\lambda^2$$

and

$$D(x,y,\lambda) = \lambda(xy + y^2) + \lambda^2 \left(\frac{1}{2}xy^2 - \frac{1}{3}xy - \frac{1}{3}y^2 + \frac{1}{4}y \right).$$

10
Vectors and tensors

In this chapter we explain how a vector space V gives rise to a family of associated tensor spaces, and how mathematical objects such as linear maps or quadratic forms should be understood as being elements of these spaces. We then apply these ideas to physics. We make extensive use of notions and notations from the appendix on linear algebra, so it may help to review that material before we begin.

10.1 Covariant and contravariant vectors

When we have a vector space V over $\mathbb{R}$, and $\{\mathbf{e}_1, \mathbf{e}_2, \ldots, \mathbf{e}_n\}$ and $\{\mathbf{e}'_1, \mathbf{e}'_2, \ldots, \mathbf{e}'_n\}$ are both bases for V, then we may expand each of the basis vectors $\mathbf{e}_\mu$ in terms of the $\mathbf{e}'_\mu$ as

$$\mathbf{e}_\nu = a^\mu_\nu \mathbf{e}'_\mu. \tag{10.1}$$

We are here, as usual, using the Einstein summation convention that repeated indices are to be summed over. Written out in full for a three-dimensional space, the expansion would be

$$\begin{aligned}
\mathbf{e}_1 &= a^1_1 \mathbf{e}'_1 + a^2_1 \mathbf{e}'_2 + a^3_1 \mathbf{e}'_3, \\
\mathbf{e}_2 &= a^1_2 \mathbf{e}'_1 + a^2_2 \mathbf{e}'_2 + a^3_2 \mathbf{e}'_3, \\
\mathbf{e}_3 &= a^1_3 \mathbf{e}'_1 + a^2_3 \mathbf{e}'_2 + a^3_3 \mathbf{e}'_3.
\end{aligned}$$

We could also have expanded the $\mathbf{e}'_\mu$ in terms of the $\mathbf{e}_\mu$ as

$$\mathbf{e}'_\nu = (a^{-1})^\mu_\nu \mathbf{e}'_\mu. \tag{10.2}$$

As the notation implies, the matrices of coefficients a^μ_ν and $(a^{-1})^\mu_\nu$ are inverses of each other:

$$a^\mu_\nu (a^{-1})^\nu_\sigma = (a^{-1})^\mu_\nu a^\nu_\sigma = \delta^\mu_\sigma. \tag{10.3}$$

If we know the components x^μ of a vector $\mathbf{x}$ in the $\mathbf{e}_\mu$ basis then the components x'^μ of $\mathbf{x}$ in the $\mathbf{e}'_\mu$ basis are obtained from

$$\mathbf{x} = x'^\mu \mathbf{e}'_\mu = x^\nu \mathbf{e}_\nu = \left(x^\nu a^\mu_\nu\right) \mathbf{e}'_\mu \tag{10.4}$$

by comparing the coefficients of $\mathbf{e}'_\mu$. We find that $x'^\mu = a^\mu_\nu x^\nu$. Observe how the $\mathbf{e}_\mu$ and the x^μ transform in "opposite" directions. The components x^μ are therefore said to transform *contravariantly*.

Associated with the vector space V is its *dual space* V^*, whose elements are *covectors*, i.e. linear maps $\mathbf{f} : V \to \mathbb{R}$. If $\mathbf{f} \in V^*$ and $\mathbf{x} = x^\mu \mathbf{e}_\mu$, we use the linearity property to evaluate $\mathbf{f}(\mathbf{x})$ as

$$\mathbf{f}(\mathbf{x}) = \mathbf{f}(x^\mu \mathbf{e}_\mu) = x^\mu \mathbf{f}(\mathbf{e}_\mu) = x^\mu f_\mu. \tag{10.5}$$

Here, the set of numbers $f_\mu = \mathbf{f}(\mathbf{e}_\mu)$ are the components of the covector $\mathbf{f}$. If we change basis so that $\mathbf{e}_\nu = a^\mu_\nu \mathbf{e}'_\mu$ then

$$f_\nu = \mathbf{f}(\mathbf{e}_\nu) = \mathbf{f}(a^\mu_\nu \mathbf{e}'_\mu) = a^\mu_\nu \mathbf{f}(\mathbf{e}'_\mu) = a^\mu_\nu f'_\mu. \tag{10.6}$$

We conclude that $f_\nu = a^\mu_\nu f'_\mu$. The f_μ components transform in the same manner as the basis. They are therefore said to transform *covariantly*. In physics it is traditional to call the the set of numbers x^μ with upstairs indices (the components of) a *contravariant vector*. Similarly, the set of numbers f_μ with downstairs indices is called (the components of) a *covariant vector*. Thus, contravariant vectors are elements of V and covariant vectors are elements of V^*.

The relationship between V and V^* is one of mutual duality, and to mathematicians it is only a matter of convenience which space is V and which space is V^*. The evaluation of $\mathbf{f} \in V^*$ on $\mathbf{x} \in V$ is therefore often written as a "pairing" $(\mathbf{f}, \mathbf{x})$, which gives equal status to the objects being put together to get a number. A physics example of such a mutually dual pair is provided by the space of displacements $\mathbf{x}$ and the space of wavenumbers $\mathbf{k}$. The units of $\mathbf{x}$ and $\mathbf{k}$ are different (metres *versus* metres^{-1}). There is therefore no meaning to "$\mathbf{x} + \mathbf{k}$", and $\mathbf{x}$ and $\mathbf{k}$ are not elements of the same vector space. The "dot" in expressions such as

$$\psi(\mathbf{x}) = e^{i\mathbf{k}\cdot\mathbf{x}} \tag{10.7}$$

cannot be a true inner product (which requires the objects it links to be in the same vector space) but is instead a pairing

$$(\mathbf{k}, \mathbf{x}) \equiv \mathbf{k}(\mathbf{x}) = k_\mu x^\mu. \tag{10.8}$$

In describing the physical world we usually give priority to the space in which we live, breathe and move, and so treat it as being "V". The displacement vector $\mathbf{x}$ then becomes the contravariant vector, and the Fourier-space wave number $\mathbf{k}$, being the more abstract quantity, becomes the covariant covector.

Our vector space may come equipped with a *metric* that is derived from a non-degenerate inner product. We regard the inner product as being a bilinear form $\mathbf{g} : V \times V \to \mathbb{R}$, so the length $\|\mathbf{x}\|$ of a vector $\mathbf{x}$ is $\sqrt{\mathbf{g}(\mathbf{x}, \mathbf{x})}$. The set of numbers

$$g_{\mu\nu} = \mathbf{g}(\mathbf{e}_\mu, \mathbf{e}_\nu) \tag{10.9}$$

comprises (the components of) the *metric tensor*. In terms of them, the inner product $\langle \mathbf{x}, \mathbf{y} \rangle$ of the pair of vectors $\mathbf{x} = x^\mu \mathbf{e}_\mu$ and $\mathbf{y} = y^\mu \mathbf{e}_\mu$ becomes

$$\langle \mathbf{x}, \mathbf{y} \rangle \equiv \mathbf{g}(\mathbf{x}, \mathbf{y}) = g_{\mu\nu} x^\mu y^\nu. \tag{10.10}$$

Real-valued inner products are always symmetric, so $\mathbf{g}(\mathbf{x}, \mathbf{y}) = \mathbf{g}(\mathbf{y}, \mathbf{x})$ and $g_{\mu\nu} = g_{\nu\mu}$. As the product is non-degenerate, the matrix $g_{\mu\nu}$ has an inverse, which is traditionally written as $g^{\mu\nu}$. Thus

$$g_{\mu\nu} g^{\nu\lambda} = g^{\lambda\nu} g_{\nu\mu} = \delta^\lambda_\mu. \tag{10.11}$$

The additional structure provided by the metric permits us to identify V with V^*. The identification is possible, because, given any $\mathbf{f} \in V^*$, we can find a vector $\widetilde{\mathbf{f}} \in V$ such that

$$\mathbf{f}(\mathbf{x}) = \langle \widetilde{\mathbf{f}}, \mathbf{x} \rangle. \tag{10.12}$$

We obtain $\widetilde{\mathbf{f}}$ by solving the equation

$$f_\mu = g_{\mu\nu} \widetilde{f}^\nu \tag{10.13}$$

to get $\widetilde{f}^\nu = g^{\nu\mu} f_\mu$. We may now drop the tilde and identify $\mathbf{f}$ with $\widetilde{\mathbf{f}}$, and hence V with V^*. When we do this, we say that the covariant components f_μ are related to the contravariant components f^μ by *raising*

$$f^\mu = g^{\mu\nu} f_\nu, \tag{10.14}$$

or *lowering*

$$f_\mu = g_{\mu\nu} f^\nu, \tag{10.15}$$

the index μ using the metric tensor. Bear in mind that this $V \cong V^*$ identification depends crucially on the metric. A different metric will, in general, identify an $\mathbf{f} \in V^*$ with a completely different $\widetilde{\mathbf{f}} \in V$.

We may play this game in the Euclidean space $\mathbb{E}^n$ with its "dot" inner product. Given a vector $\mathbf{x}$ and a basis $\mathbf{e}_\mu$ for which $g_{\mu\nu} = \mathbf{e}_\mu \cdot \mathbf{e}_\nu$, we can define two sets of components for the same vector. Firstly the coefficients x^μ appearing in the basis expansion

$$\mathbf{x} = x^\mu \mathbf{e}_\mu, \tag{10.16}$$

and secondly the "components"

$$x_\mu = \mathbf{e}_\mu \cdot \mathbf{x} = \mathbf{g}(\mathbf{e}_\mu, \mathbf{x}) = \mathbf{g}(\mathbf{e}_\mu, x^\nu \mathbf{e}_\nu) = \mathbf{g}(\mathbf{e}_\mu, \mathbf{e}_\nu) x^\nu = g_{\mu\nu} x^\nu \tag{10.17}$$

of $\mathbf{x}$ along the basis vectors. These two sets of numbers are then respectively called the contravariant and covariant components of the vector $\mathbf{x}$. If the $\mathbf{e}_\mu$ constitute an orthonormal basis, where $g_{\mu\nu} = \delta_{\mu\nu}$, then the two sets of components (covariant and contravariant) are numerically coincident. In a non-orthogonal basis they will be different, and we must take care never to add contravariant components to covariant ones.

10.2 Tensors

We now introduce tensors in two ways: firstly as sets of numbers labelled by indices and equipped with transformation laws that tell us how these numbers change as we change basis; and secondly as basis-independent objects that are elements of a vector space constructed by taking multiple tensor products of the spaces V and V^*.

10.2.1 Transformation rules

After we change basis $\mathbf{e}_\mu \to \mathbf{e}'_\mu$, where $\mathbf{e}_\nu = a^\mu_\nu \mathbf{e}'_\mu$, the metric tensor will be represented by a new set of components

$$g'_{\mu\nu} = \mathbf{g}(\mathbf{e}'_\mu, \mathbf{e}'_\nu). \tag{10.18}$$

These are related to the old components by

$$g_{\mu\nu} = \mathbf{g}(\mathbf{e}_\mu, \mathbf{e}_\nu) = \mathbf{g}(a^\rho_\mu \mathbf{e}'_\rho, a^\sigma_\nu \mathbf{e}'_\sigma) = a^\rho_\mu a^\sigma_\nu \mathbf{g}(\mathbf{e}'_\rho, \mathbf{e}'_\sigma) = a^\rho_\mu a^\sigma_\nu \, g'_{\rho\sigma}. \tag{10.19}$$

This transformation rule for $g_{\mu\nu}$ has both of its subscripts behaving like the downstairs indices of a covector. We therefore say that $g_{\mu\nu}$ transforms as a *doubly covariant tensor*. Written out in full, for a two-dimensional space, the transformation law is

$$\begin{aligned}
g_{11} &= a^1_1 a^1_1 g'_{11} + a^1_1 a^2_1 g'_{12} + a^2_1 a^1_1 g'_{21} + a^2_1 a^2_1 g'_{22}, \\
g_{12} &= a^1_1 a^1_2 g'_{11} + a^1_1 a^2_2 g'_{12} + a^2_1 a^1_2 g'_{21} + a^2_1 a^2_2 g'_{22}, \\
g_{21} &= a^1_2 a^1_1 g'_{11} + a^1_2 a^2_1 g'_{12} + a^2_2 a^1_1 g'_{21} + a^2_2 a^2_1 g'_{22}, \\
g_{22} &= a^1_2 a^1_2 g'_{11} + a^1_2 a^2_2 g'_{12} + a^2_2 a^1_2 g'_{21} + a^2_2 a^2_2 g'_{22}.
\end{aligned}$$

In three dimensions each row would have nine terms, and sixteen in four dimensions.

A set of numbers $Q^{\alpha\beta}{}_{\gamma\delta\epsilon}$, whose indices range from 1 to the dimension of the space and that transforms as

$$Q^{\alpha\beta}{}_{\gamma\delta\epsilon} = (a^{-1})^\alpha_{\alpha'} (a^{-1})^\beta_{\beta'} \, a^{\gamma'}_\gamma a^{\delta'}_\delta a^{\epsilon'}_\epsilon \, Q'^{\alpha'\beta'}{}_{\gamma'\delta'\epsilon'}, \tag{10.20}$$

or conversely as

$$Q'^{\alpha\beta}{}_{\gamma\delta\epsilon} = a^\alpha_{\alpha'} a^\beta_{\beta'} (a^{-1})^{\gamma'}_\gamma (a^{-1})^{\delta'}_\delta (a^{-1})^{\epsilon'}_\epsilon \, Q^{\alpha'\beta'}{}_{\gamma'\delta'\epsilon'}, \tag{10.21}$$

comprises the components of a *doubly contravariant, triply covariant* tensor. More compactly, the $Q^{\alpha\beta}{}_{\gamma\delta\epsilon}$ are the components of a tensor of type $(2, 3)$. Tensors of type (p, q) are defined analogously. The total number of indices $p + q$ is called the *rank* of the tensor.

Note how the indices are wired up in the transformation rules (10.20) and (10.21): free (not summed over) upstairs indices on the left-hand side of the equations match to free upstairs indices on the right-hand side, similarly for the downstairs indices. Also upstairs indices are summed only with downstairs ones.

Similar conditions apply to equations relating tensors in any particular basis. If they are violated you do not have a valid tensor equation – meaning that an equation valid in one basis will not be valid in another basis. Thus an equation

$$A^{\mu}{}_{\nu\lambda} = B^{\mu\tau}{}_{\nu\lambda\tau} + C^{\mu}{}_{\nu\lambda} \tag{10.22}$$

is fine, but

$$A^{\mu}{}_{\nu\lambda} \stackrel{?}{=} B^{\nu}{}_{\mu\lambda} + C^{\mu}{}_{\nu\lambda\sigma\sigma} + D^{\mu}{}_{\nu\lambda\tau} \tag{10.23}$$

has something wrong in each term.

Incidentally, although not illegal, it is a good idea not to write tensor indices directly underneath one another – i.e. do not write Q^{ij}_{kjl} – because if you raise or lower indices using the metric tensor, and some pages later in a calculation try to put them back where they were, they might end up in the wrong order.

Tensor algebra

The sum of two tensors of a given type is also a tensor of that type. The sum of two tensors of different types is not a tensor. Thus each particular type of tensor constitutes a distinct vector space, but one derived from the common underlying vector space whose change-of-basis formula is being utilized.

Tensors can be combined by multiplication: if $A^{\mu}{}_{\nu\lambda}$ and $B^{\mu}{}_{\nu\lambda\tau}$ are tensors of type $(1,2)$ and $(1,3)$, respectively, then

$$C^{\alpha\beta}{}_{\nu\lambda\rho\sigma\tau} = A^{\alpha}{}_{\nu\lambda} B^{\beta}{}_{\rho\sigma\tau} \tag{10.24}$$

is a tensor of type $(2,5)$.

An important operation is *contraction*, which consists of setting one or more contravariant index equal to a covariant index and summing over the repeated indices. This reduces the rank of the tensor. So, for example,

$$D_{\rho\sigma\tau} = C^{\alpha\beta}{}_{\alpha\beta\rho\sigma\tau} \tag{10.25}$$

is a tensor of type $(0,3)$. Similarly $\mathbf{f}(\mathbf{x}) = f_{\mu}x^{\mu}$ is a type $(0,0)$ tensor, i.e. an *invariant* – a number that takes the same value in all bases. Upper indices can only be contracted with lower indices, and *vice versa*. For example, the array of numbers $A_{\alpha} = B_{\alpha\beta\beta}$ obtained from the type $(0,3)$ tensor $B_{\alpha\beta\gamma}$ is *not* a tensor of type $(0,1)$.

The contraction procedure outputs a tensor because setting an upper index and a lower index to a common value μ and summing over μ leads to the factor $\ldots(a^{-1})^{\mu}_{\alpha} a^{\beta}_{\mu}\ldots$ appearing in the transformation rule. Now

$$(a^{-1})^{\mu}_{\alpha} a^{\beta}_{\mu} = \delta^{\beta}_{\alpha}, \tag{10.26}$$

and the Kronecker delta effects a summation over the corresponding pair of indices in the transformed tensor.

Although often associated with general relativity, tensors occur in many places in physics. They are used, for example, in elasticity theory, where the word "tensor" in its modern meaning was introduced by Woldemar Voigt in 1898. Voigt, following Cauchy and Green, described the infinitesimal deformation of an elastic body by the *strain tensor* $e_{\alpha\beta}$, which is a tensor of type (0,2). The forces to which the strain gives rise are described by the *stress tensor* $\sigma^{\lambda\mu}$. A generalization of Hooke's law relates stress to strain via a tensor of elastic constants $c^{\alpha\beta\gamma\delta}$ as

$$\sigma^{\alpha\beta} = c^{\alpha\beta\gamma\delta} e_{\gamma\delta}. \tag{10.27}$$

We study stress and strain in more detail later in this chapter.

Exercise 10.1: Show that $g^{\mu\nu}$, the matrix inverse of the metric tensor $g_{\mu\nu}$, is indeed a doubly contravariant tensor, as the position of its indices suggests.

10.2.2 Tensor character of linear maps and quadratic forms

As an illustration of the tensor concept and of the need to distinguish between upstairs and downstairs indices, we contrast the properties of matrices representing linear maps and those representing quadratic forms.

A linear map $M : V \to V$ is an object that exists independently of any basis. Given a basis, however, it is represented by a matrix $M^\mu{}_\nu$ obtained by examining the action of the map on the basis elements:

$$M(\mathbf{e}_\mu) = \mathbf{e}_\nu M^\nu{}_\mu. \tag{10.28}$$

Acting on $\mathbf{x}$ we get a new vector $\mathbf{y} = M(\mathbf{x})$, where

$$y^\nu \mathbf{e}_\nu = \mathbf{y} = M(\mathbf{x}) = M(x^\mu \mathbf{e}_\mu) = x^\mu M(\mathbf{e}_\mu) = x^\mu M^\nu{}_\mu \mathbf{e}_\nu = M^\nu{}_\mu x^\mu \, \mathbf{e}_\nu. \tag{10.29}$$

We therefore have

$$y^\nu = M^\nu{}_\mu x^\mu, \tag{10.30}$$

which is the usual matrix multiplication $\mathbf{y} = \mathbf{M}\mathbf{x}$. When we change basis, $\mathbf{e}_\nu = a^\mu_\nu \mathbf{e}'_\mu$, then

$$\mathbf{e}_\nu M^\nu{}_\mu = M(\mathbf{e}_\mu) = M(a^\rho_\mu \mathbf{e}'_\rho) = a^\rho_\mu M(\mathbf{e}'_\rho) = a^\rho_\mu \mathbf{e}'_\sigma M'^\sigma{}_\rho = a^\rho_\mu (a^{-1})^\nu_\sigma \mathbf{e}_\nu M'^\sigma{}_\rho. \tag{10.31}$$

Comparing coefficients of $\mathbf{e}_\nu$, we find

$$M^\nu{}_\mu = a^\rho_\mu (a^{-1})^\nu_\sigma M'^\sigma{}_\rho, \tag{10.32}$$

or, conversely,

$$M'^{\nu}{}_{\mu} = (a^{-1})^{\rho}_{\mu} a^{\nu}_{\sigma} M^{\sigma}{}_{\rho}. \tag{10.33}$$

Thus a matrix representing a linear map has the tensor character suggested by the position of its indices, i.e. it transforms as a type $(1, 1)$ tensor. We can derive the same formula in matrix notation. In the new basis the vectors $\mathbf{x}$ and $\mathbf{y}$ have new components $\mathbf{x}' = \mathbf{A}\mathbf{x}$, and $\mathbf{y}' = \mathbf{A}\mathbf{y}$. Consequently $\mathbf{y} = \mathbf{M}\mathbf{x}$ becomes

$$\mathbf{y}' = \mathbf{A}\mathbf{y} = \mathbf{A}\mathbf{M}\mathbf{x} = \mathbf{A}\mathbf{M}\mathbf{A}^{-1}\mathbf{x}', \tag{10.34}$$

and the matrix representing the map M has new components

$$\mathbf{M}' = \mathbf{A}\mathbf{M}\mathbf{A}^{-1}. \tag{10.35}$$

Now consider the quadratic form $Q : V \to \mathbb{R}$ that is obtained from a symmetric bilinear form $Q : V \times V \to \mathbb{R}$ by setting $Q(\mathbf{x}) = Q(\mathbf{x}, \mathbf{x})$. We can write

$$Q(\mathbf{x}) = Q_{\mu\nu} x^{\mu} x^{\nu} = x^{\mu} Q_{\mu\nu} x^{\nu} = \mathbf{x}^T \mathbf{Q}\mathbf{x}, \tag{10.36}$$

where $Q_{\mu\nu} \equiv Q(\mathbf{e}_{\mu}, \mathbf{e}_{\nu})$ are the entries in the symmetric matrix $\mathbf{Q}$, the suffix T denotes transposition, and $\mathbf{x}^T \mathbf{Q}\mathbf{x}$ is standard matrix-multiplication notation. Just as does the metric tensor, the coefficients $Q_{\mu\nu}$ transform as a type $(0, 2)$ tensor:

$$Q_{\mu\nu} = a^{\alpha}_{\mu} a^{\beta}_{\nu} Q'_{\alpha\beta}. \tag{10.37}$$

In matrix notation the vector $\mathbf{x}$ again transforms to have new components $\mathbf{x}' = \mathbf{A}\mathbf{x}$, but $\mathbf{x}'^T = \mathbf{x}^T \mathbf{A}^T$. Consequently

$$\mathbf{x}'^T \mathbf{Q}'\mathbf{x}' = \mathbf{x}^T \mathbf{A}^T \mathbf{Q}'\mathbf{A}\mathbf{x}. \tag{10.38}$$

Thus

$$\mathbf{Q} = \mathbf{A}^T \mathbf{Q}'\mathbf{A}. \tag{10.39}$$

The message is that linear maps and quadratic forms can both be represented by matrices, but these matrices correspond to distinct types of tensor and transform differently under a change of basis.

A matrix representing a linear map has a basis-independent determinant. Similarly the *trace* of a matrix representing a linear map

$$\operatorname{tr} \mathbf{M} \stackrel{\text{def}}{=} M^{\mu}{}_{\mu} \tag{10.40}$$

is a tensor of type $(0, 0)$, i.e. a scalar, and therefore basis independent. On the other hand, while you can certainly compute the determinant or the trace of the matrix representing a quadratic form in some particular basis, when you change basis and calculate the determinant or trace of the transformed matrix, you will get a different number.

It *is* possible to make a quadratic form out of a linear map, but this requires using the metric to lower the contravariant index on the matrix representing the map:

$$Q(\mathbf{x}) = x^\mu g_{\mu\nu} Q^\nu{}_\lambda x^\lambda = \mathbf{x} \cdot \mathbf{Q}\mathbf{x}. \tag{10.41}$$

Be careful, therefore: the matrices "**Q**" in $\mathbf{x}^T\mathbf{Q}\mathbf{x}$ and in $\mathbf{x} \cdot \mathbf{Q}\mathbf{x}$ are representing different mathematical objects.

Exercise 10.2: In this problem we will use the distinction between the transformation law of a quadratic form and that of a linear map to resolve the following "paradox":

- In quantum mechanics we are taught that the matrices representing two operators can be simultaneously diagonalized only if they commute.
- In classical mechanics we are taught how, given the Lagrangian

$$L = \sum_{ij} \left(\frac{1}{2}\dot{q}_i M_{ij} \dot{q}_j - \frac{1}{2} q_i V_{ij} q_j \right),$$

to construct normal coordinates Q_i such that L becomes

$$L = \sum_i \left(\frac{1}{2}\dot{Q}_i^2 - \frac{1}{2}\omega_i^2 Q_i^2 \right).$$

We have apparantly managed to simultaneously diagonalize the matrices $M_{ij} \to \text{diag}\,(1, \dots, 1)$ and $V_{ij} \to \text{diag}\,(\omega_1^2, \dots, \omega_n^2)$, even though there is no reason for them to commute with each other!

Show that when **M** and **V** are a pair of symmetric matrices, with **M** being positive definite, then there exists an invertible matrix **A** such that $\mathbf{A}^T\mathbf{M}\mathbf{A}$ and $\mathbf{A}^T\mathbf{V}\mathbf{A}$ are simultaneously diagonal. (Hint: consider **M** as defining an inner product, and use the Gramm–Schmidt procedure to first find an orthonormal frame in which $M'_{ij} = \delta_{ij}$. Then show that the matrix corresponding to **V** in this frame can be diagonalized by a further transformation that does not perturb the already diagonal M'_{ij}.)

10.2.3 Tensor product spaces

We may regard the set of numbers $Q^{\alpha\beta}{}_{\gamma\delta\epsilon}$ as being the components of an object **Q** that is an element of the vector space of type $(2, 3)$ tensors. We denote this vector space by the symbol $V \otimes V \otimes V^* \otimes V^* \otimes V^*$, the notation indicating that it is derived from the original V and its dual V^* by taking *tensor products* of these spaces. The tensor **Q** is to be thought of as existing as an element of $V \otimes V \otimes V^* \otimes V^* \otimes V^*$ independently of any basis, but given a basis $\{\mathbf{e}_\mu\}$ for V, and the dual basis $\{\mathbf{e}^{*\nu}\}$ for V^*, we expand it as

$$\mathbf{Q} = Q^{\alpha\beta}{}_{\gamma\delta\epsilon}\, \mathbf{e}_\alpha \otimes \mathbf{e}_\beta \otimes \mathbf{e}^{*\gamma} \otimes \mathbf{e}^{*\delta} \otimes \mathbf{e}^{*\epsilon}. \tag{10.42}$$

Here the tensor product symbol "⊗" is distributive

$$\mathbf{a} \otimes (\mathbf{b} + \mathbf{c}) = \mathbf{a} \otimes \mathbf{b} + \mathbf{a} \otimes \mathbf{c},$$
$$(\mathbf{a} + \mathbf{b}) \otimes \mathbf{c} = \mathbf{a} \otimes \mathbf{c} + \mathbf{b} \otimes \mathbf{c}, \tag{10.43}$$

and associative

$$(\mathbf{a} \otimes \mathbf{b}) \otimes \mathbf{c} = \mathbf{a} \otimes (\mathbf{b} \otimes \mathbf{c}), \tag{10.44}$$

but is not commutative

$$\mathbf{a} \otimes \mathbf{b} \neq \mathbf{b} \otimes \mathbf{a}. \tag{10.45}$$

Everything commutes with the field, however,

$$\lambda(\mathbf{a} \otimes \mathbf{b}) = (\lambda \mathbf{a}) \otimes \mathbf{b} = \mathbf{a} \otimes (\lambda \mathbf{b}). \tag{10.46}$$

If we change basis $\mathbf{e}_\alpha = a^\beta_\alpha \mathbf{e}'_\beta$ then these rules lead, for example, to

$$\mathbf{e}_\alpha \otimes \mathbf{e}_\beta = a^\lambda_\alpha a^\mu_\beta \, \mathbf{e}'_\lambda \otimes \mathbf{e}'_\mu. \tag{10.47}$$

From this change-of-basis formula, we deduce that

$$T^{\alpha\beta} \mathbf{e}_\alpha \otimes \mathbf{e}_\beta = T^{\alpha\beta} a^\lambda_\alpha a^\mu_\beta \, \mathbf{e}'_\lambda \otimes \mathbf{e}'_\mu = T'^{\lambda\mu} \, \mathbf{e}'_\lambda \otimes \mathbf{e}'_\mu, \tag{10.48}$$

where

$$T'^{\lambda\mu} = T^{\alpha\beta} a^\lambda_\alpha a^\mu_\beta. \tag{10.49}$$

The analogous formula for $\mathbf{e}_\alpha \otimes \mathbf{e}_\beta \otimes \mathbf{e}^{*\gamma} \otimes \mathbf{e}^{*\delta} \otimes \mathbf{e}^{*\epsilon}$ reproduces the transformation rule for the components of $\mathbf{Q}$.

The meaning of the tensor product of a collection of vector spaces should now be clear: if $\mathbf{e}_\mu$ consititute a basis for V, the space $V \otimes V$ is, for example, the space of all linear combinations[1] of the abstract symbols $\mathbf{e}_\mu \otimes \mathbf{e}_\nu$, which we declare by *fiat* to constitute a basis for this space. There is no geometric significance (as there is with a vector product $\mathbf{a} \times \mathbf{b}$) to the tensor product $\mathbf{a} \otimes \mathbf{b}$, so the $\mathbf{e}_\mu \otimes \mathbf{e}_\nu$ are simply useful place-keepers. Remember that these are *ordered* pairs, $\mathbf{e}_\mu \otimes \mathbf{e}_\nu \neq \mathbf{e}_\nu \otimes \mathbf{e}_\mu$.

[1] Do not confuse the tensor-product space $V \otimes W$ with the cartesian product $V \times W$. The latter is the set of all ordered pairs $(\mathbf{x}, \mathbf{y})$, $\mathbf{x} \in V$, $\mathbf{y} \in W$. The tensor product includes also *formal sums* of such pairs. The cartesian product of two vector spaces can be given the structure of a vector space by defining an addition operation $\lambda(\mathbf{x}_1, \mathbf{y}_1) + \mu(\mathbf{x}_2, \mathbf{y}_2) = (\lambda \mathbf{x}_1 + \mu \mathbf{x}_2, \lambda \mathbf{y}_1 + \mu \mathbf{y}_2)$, but this construction does not lead to the tensor product. Instead it defines the *direct sum* $V \oplus W$.

Although there is no *geometric* meaning, it is possible, however, to give an *algebraic* meaning to a product like $\mathbf{e}^{*\lambda}\otimes\mathbf{e}^{*\mu}\otimes\mathbf{e}^{*\nu}$ by viewing it as a multilinear form $V\times V\times V :\to \mathbb{R}$. We define

$$\mathbf{e}^{*\lambda}\otimes\mathbf{e}^{*\mu}\otimes\mathbf{e}^{*\nu}\,(\mathbf{e}_\alpha,\mathbf{e}_\beta,\mathbf{e}_\gamma)=\delta^\lambda_\alpha\,\delta^\mu_\beta\,\delta^\nu_\gamma. \tag{10.50}$$

We may also regard it as a linear map $V\otimes V\otimes V :\to \mathbb{R}$ by defining

$$\mathbf{e}^{*\lambda}\otimes\mathbf{e}^{*\mu}\otimes\mathbf{e}^{*\nu}\,(\mathbf{e}_\alpha\otimes\mathbf{e}_\beta\otimes\mathbf{e}_\gamma)=\delta^\lambda_\alpha\,\delta^\mu_\beta\,\delta^\nu_\gamma \tag{10.51}$$

and extending the definition to general elements of $V\otimes V\otimes V$ by linearity. In this way we establish an isomorphism

$$V^*\otimes V^*\otimes V^*\cong (V\otimes V\otimes V)^*. \tag{10.52}$$

This multiple personality is typical of tensor spaces. We have already seen that the metric tensor is simultaneously an element of $V^*\otimes V^*$ and a map $\mathbf{g}: V\to V^*$.

Tensor products and quantum mechanics

When we have two quantum-mechanical systems having Hilbert spaces $\mathcal{H}^{(1)}$ and $\mathcal{H}^{(2)}$, the Hilbert space for the combined system is $\mathcal{H}^{(1)}\otimes\mathcal{H}^{(2)}$. Quantum mechanics books usually denote the vectors in these spaces by the Dirac "bra-ket" notation in which the basis vectors of the separate spaces are denoted by[2] $|n_1\rangle$ and $|n_2\rangle$, and that of the combined space by $|n_1,n_2\rangle$. In this notation, a state in the combined system is a linear combination

$$|\Psi\rangle=\sum_{n_1,n_2}|n_1,n_2\rangle\langle n_1,n_2|\Psi\rangle. \tag{10.53}$$

This is the tensor product in disguise. To unmask it, we simply make the notational translation

$$\begin{aligned}|\Psi\rangle &\to \mathbf{\Psi}\\ \langle n_1,n_2|\Psi\rangle &\to \psi^{n_1,n_2}\\ |n_1\rangle &\to \mathbf{e}^{(1)}_{n_1}\\ |n_2\rangle &\to \mathbf{e}^{(2)}_{n_2}\\ |n_1,n_2\rangle &\to \mathbf{e}^{(1)}_{n_1}\otimes\mathbf{e}^{(2)}_{n_2}.\end{aligned} \tag{10.54}$$

Then (10.53) becomes

$$\mathbf{\Psi}=\psi^{n_1,n_2}\,\mathbf{e}^{(1)}_{n_1}\otimes\mathbf{e}^{(2)}_{n_2}. \tag{10.55}$$

[2] We assume for notational convenience that the Hilbert spaces are finite dimensional.

Entanglement: Suppose that $\mathcal{H}^{(1)}$ has basis $\mathbf{e}_1^{(1)}, \dots, \mathbf{e}_m^{(1)}$ and $\mathcal{H}^{(2)}$ has basis $\mathbf{e}_1^{(2)}, \dots, \mathbf{e}_n^{(2)}$. The Hilbert space $\mathcal{H}^{(1)} \otimes \mathcal{H}^{(2)}$ is then nm dimensional. Consider a state

$$\boldsymbol{\Psi} = \psi^{ij} \mathbf{e}_i^{(1)} \otimes \mathbf{e}_j^{(2)} \in \mathcal{H}^{(1)} \otimes \mathcal{H}^{(2)}. \tag{10.56}$$

If we can find vectors

$$\begin{aligned} \boldsymbol{\Phi} &\equiv \phi^i \mathbf{e}_i^{(1)} \in \mathcal{H}^{(1)}, \\ \mathbf{X} &\equiv \chi^j \mathbf{e}_j^{(2)} \in \mathcal{H}^{(2)}, \end{aligned} \tag{10.57}$$

such that

$$\boldsymbol{\Psi} = \boldsymbol{\Phi} \otimes \mathbf{X} \equiv \phi^i \chi^j \mathbf{e}_i^{(1)} \otimes \mathbf{e}_j^{(2)} \tag{10.58}$$

then the tensor $\boldsymbol{\Psi}$ is said to be *decomposable* and the two quantum systems are said to be *unentangled*. If there are no such vectors then the two systems are *entangled* in the sense of the Einstein–Podolski–Rosen (EPR) paradox.

Quantum states are really in one-to-one correspondence with *rays* in the Hilbert space, rather than vectors. If we denote the n-dimensional vector space over the field of the complex numbers as $\mathbb{C}^n$, the space of rays, in which we do not distinguish between the vectors $\mathbf{x}$ and $\lambda\mathbf{x}$ when $\lambda \neq 0$, is denoted by $\mathbb{C}P^{n-1}$ and is called *complex projective space*. Complex projective space is where *algebraic geometry* is studied. The set of decomposable states may be thought of as a subset of the complex projective space $\mathbb{C}P^{nm-1}$, and, since, as the following exercise shows, this subset is defined by a finite number of homogeneous polynomial equations, it forms what algebraic geometers call a *variety*. This particular subset is known as the *Segre variety*.

Exercise 10.3: The Segre conditions for a state to be decomposable.

(i) By counting the number of independent components that are at our disposal in $\boldsymbol{\Psi}$, and comparing that number with the number of free parameters in $\boldsymbol{\Phi} \otimes \mathbf{X}$, show that the coefficients ψ^{ij} must satisfy $(n-1)(m-1)$ relations if the state is to be decomposable.

(ii) If the state is decomposable, show that

$$0 = \begin{vmatrix} \psi^{ij} & \psi^{il} \\ \psi^{kj} & \psi^{kl} \end{vmatrix}$$

for all sets of indices i, j, k, l.

(iii) Assume that ψ^{11} is not zero. Using your count from part (i) as a guide, find a subset of the relations from part (ii) that constitute a necessary and sufficient set of conditions for the state Ψ to be decomposable. Include a proof that your set is indeed sufficient.

10.2.4 Symmetric and skew-symmetric tensors

By examining the transformation rule you may see that if a pair of upstairs or downstairs indices is *symmetric* (say $Q^{\mu\nu}{}_{\rho\sigma\tau} = Q^{\nu\mu}{}_{\rho\sigma\tau}$) or *skew-symmetric* ($Q^{\mu\nu}{}_{\rho\sigma\tau} = -Q^{\nu\mu}{}_{\rho\sigma\tau}$) in one basis, it remains so after the basis has been changed. (This is **not** true of a pair composed of one upstairs and one downstairs index.) It makes sense, therefore, to define symmetric and skew-symmetric tensor product spaces. Thus skew-symmetric doubly-contravariant tensors can be regarded as belonging to the space denoted by $\bigwedge^2 V$ and expanded as

$$\mathbf{A} = \frac{1}{2}A^{\mu\nu}\,\mathbf{e}_\mu \wedge \mathbf{e}_\nu, \tag{10.59}$$

where the coefficients are skew-symmetric, $A^{\mu\nu} = -A^{\nu\mu}$, and the *wedge product* of the basis elements is associative and distributive, as is the tensor product, but in addition obeys $\mathbf{e}_\mu \wedge \mathbf{e}_\nu = -\mathbf{e}_\nu \wedge \mathbf{e}_\mu$. The "1/2" (replaced by $1/p!$ when there are p indices) is convenient in that each independent component only appears once in the sum. For example, in three dimensions,

$$\frac{1}{2}A^{\mu\nu}\,\mathbf{e}_\mu \wedge \mathbf{e}_\nu = A^{12}\,\mathbf{e}_1 \wedge \mathbf{e}_2 + A^{23}\,\mathbf{e}_2 \wedge \mathbf{e}_3 + A^{31}\,\mathbf{e}_3 \wedge \mathbf{e}_1. \tag{10.60}$$

Symmetric doubly contravariant tensors can be regarded as belonging to the space $\mathrm{sym}^2 V$ and expanded as

$$\mathbf{S} = S^{\alpha\beta}\,\mathbf{e}_\alpha \odot \mathbf{e}_\beta \tag{10.61}$$

where $\mathbf{e}_\alpha \odot \mathbf{e}_\beta = \mathbf{e}_\beta \odot \mathbf{e}_\alpha$ and $S^{\alpha\beta} = S^{\beta\alpha}$. (We do not insert a "1/2" here because including it leads to no particular simplification in any consequent equations.)

We can treat these symmetric and skew-symmetric products as symmetric or skew multilinear forms. Define, for example,

$$\mathbf{e}^{*\alpha} \wedge \mathbf{e}^{*\beta}\,(\mathbf{e}_\mu, \mathbf{e}_\nu) = \delta^\alpha_\mu \delta^\beta_\nu - \delta^\alpha_\nu \delta^\beta_\mu, \tag{10.62}$$

and

$$\mathbf{e}^{*\alpha} \wedge \mathbf{e}^{*\beta}\,(\mathbf{e}_\mu \wedge \mathbf{e}_\nu) = \delta^\alpha_\mu \delta^\beta_\nu - \delta^\alpha_\nu \delta^\beta_\mu. \tag{10.63}$$

We need two terms on the right-hand side of these examples because the skew-symmetry of $\mathbf{e}^{*\alpha} \wedge \mathbf{e}^{*\beta}(\ ,\ )$ in its slots does not allow us the luxury of demanding that the $\mathbf{e}_\mu$ be inserted in the exact order of the $\mathbf{e}^{*\alpha}$ to get a non-zero answer. Because the p-th order analogue of (10.62) form has $p!$ terms on its right-hand side, some authors like to divide the right-hand side by $p!$ in this definition. We prefer the one above, though. With our definition, and with $\mathbf{A} = \frac{1}{2}A_{\mu\nu}\mathbf{e}^{*\mu} \wedge \mathbf{e}^{*\nu}$ and $\mathbf{B} = \frac{1}{2}B^{\alpha\beta}\mathbf{e}_\alpha \wedge \mathbf{e}_\beta$, we have

$$\mathbf{A}(\mathbf{B}) = \frac{1}{2}A_{\mu\nu}B^{\mu\nu} = \sum_{\mu<\nu} A_{\mu\nu}B^{\mu\nu}, \tag{10.64}$$

so the sum is only over independent terms.

The wedge ($\wedge$) product notation is standard in mathematics wherever skew-symmetry is implied.[3] The "sym" and $\odot$ are not. Different authors use different notations for spaces of symmetric tensors. This reflects the fact that skew-symmetric tensors are extremely useful and appear in many different parts of mathematics, while symmetric ones have fewer special properties (although they are common in physics). Compare the relative usefulness of determinants and permanents.

Exercise 10.4: Show that in d dimensions:

(i) the dimension of the space of skew-symmetric covariant tensors with p indices is $d!/p!(d-p)!$;
(ii) the dimension of the space of symmetric covariant tensors with p indices is $(d+p-1)!/p!(d-1)!$.

Bosons and fermions

Spaces of symmetric and skew-symmetric tensors appear whenever we deal with the quantum mechanics of many indistinguishable particles possessing Bose or Fermi statistics. If we have a Hilbert space $\mathcal{H}$ of single-particle states with basis $\mathbf{e}_i$ then the N-boson space is $\mathrm{Sym}^N\mathcal{H}$ which consists of states

$$\mathbf{\Phi} = \Phi^{i_1 i_2 \ldots i_N} \mathbf{e}_{i_1} \odot \mathbf{e}_{i_2} \odot \cdots \odot \mathbf{e}_{i_N}, \tag{10.65}$$

and the N-fermion space is $\bigwedge^N \mathcal{H}$, which contains states

$$\mathbf{\Psi} = \frac{1}{N!} \Psi^{i_1 i_2 \ldots i_N} \mathbf{e}_{i_1} \wedge \mathbf{e}_{i_2} \wedge \cdots \wedge \mathbf{e}_{i_N}. \tag{10.66}$$

The symmetry of the Bose wavefunction

$$\Phi^{i_1 \ldots i_\alpha \ldots i_\beta \ldots i_N} = \Phi^{i_1 \ldots i_\beta \ldots i_\alpha \ldots i_N}, \tag{10.67}$$

and the skew-symmetry of the Fermion wavefunction

$$\Psi^{i_1 \ldots i_\alpha \ldots i_\beta \ldots i_N} = -\Psi^{i_1 \ldots i_\beta \ldots i_\alpha \ldots i_N}, \tag{10.68}$$

under the interchange of the particle labels α, β is then natural.

Slater determinants and the Plücker relations: Some N-fermion states can be decomposed into a product of single-particle states

$$\begin{aligned} \mathbf{\Psi} &= \boldsymbol{\psi}_1 \wedge \boldsymbol{\psi}_2 \wedge \cdots \wedge \boldsymbol{\psi}_N \\ &= \psi_1^{i_1} \psi_2^{i_2} \cdots \psi_N^{i_N} \mathbf{e}_{i_1} \wedge \mathbf{e}_{i_2} \wedge \cdots \wedge \mathbf{e}_{i_N}. \end{aligned} \tag{10.69}$$

[3] Skew products and abstract vector spaces were introduced simultaneously in Hermann Grassmann's *Ausdehnungslehre* (1844). Grassmann's mathematics was not appreciated in his lifetime. In his disappointment he turned to other fields, making significant contributions to the theory of colour mixtures (Grassmann's law), and to the philology of Indo-European languages (another Grassmann's law).

Comparing the coefficients of $\mathbf{e}_{i_1} \wedge \mathbf{e}_{i_2} \wedge \cdots \wedge \mathbf{e}_{i_N}$ in (10.66) and (10.69) shows that the many-body wavefunction can then be written as

$$\Psi^{i_1 i_2 \ldots i_N} = \begin{vmatrix} \psi_1^{i_1} & \psi_1^{i_2} & \cdots & \psi_1^{i_N} \\ \psi_2^{i_1} & \psi_2^{i_2} & \cdots & \psi_2^{i_N} \\ \vdots & \vdots & \ddots & \vdots \\ \psi_N^{i_1} & \psi_N^{i_2} & \cdots & \psi_N^{i_N} \end{vmatrix}. \tag{10.70}$$

The wavefunction is therefore given by a single *Slater determinant*. Such wavefunctions correspond to a very special class of states. The general many-fermion state is not decomposable, and its wavefunction can only be expressed as a sum of many Slater determinants. The Hartree–Fock method of quantum chemistry is a variational approximation that takes such a single Slater determinant as its trial wavefunction and varies only the one-particle wavefunctions $\langle i|\psi_a\rangle \equiv \psi_a^i$. It is a remarkably successful approximation, given the very restricted class of wavefunctions it explores.

As with the Segre condition for two distinguishable quantum systems to be unentangled, there is a set of necessary and sufficient conditions on the $\Psi^{i_1 i_2 \ldots i_N}$ for the state $\boldsymbol{\Psi}$ to be decomposable into single-particle states. The conditions are that

$$\Psi^{i_1 i_2 \ldots i_{N-1}[j_1}\Psi^{j_2 j_3 \ldots j_{N+1}]} = 0 \tag{10.71}$$

for any choice of indices $i_1, \ldots i_{N-1}$ and $j_1, \ldots, j_{N+1}$. The square brackets [...] indicate that the expression is to be antisymmetrized over the indices enclosed in the brackets. For example, a three-particle state is decomposable if and only if

$$\Psi^{i_1 i_2 j_1}\Psi^{j_2 j_3 j_4} - \Psi^{i_1 i_2 j_2}\Psi^{j_1 j_3 j_4} + \Psi^{i_1 i_2 j_3}\Psi^{j_1 j_2 j_4} - \Psi^{i_1 i_2 j_4}\Psi^{j_1 j_2 j_3} = 0. \tag{10.72}$$

These conditions are called the *Plücker relations* after Julius Plücker who discovered them long before the advent of quantum mechanics.[4] It is easy to show that Plücker's relations are necessary conditions for decomposability. It takes more sophistication to show that they are sufficient. We will therefore defer this task to the exercises at the end of the chapter. As far as we are aware, the Plücker relations are not exploited by quantum chemists, but, in disguise as the *Hirota bilinear equations*, they constitute the geometric condition underpinning the many-soliton solutions of the Korteweg–de Vries and other soliton equations.

[4] As well as his extensive work in algebraic geometry, Plücker (1801–68) made important discoveries in experimental physics. He was, for example, the first person to observe the deflection of cathode rays – beams of electrons – by a magnetic field, and the first to point out that each element had its characteristic emission spectrum.

10.2.5 Kronecker and Levi-Civita tensors

Suppose the tensor δ^μ_ν is defined, with respect to some basis, to be unity if $\mu = \nu$ and zero otherwise. In a new basis it will transform to

$$\delta'^\mu_\nu = a^\mu_\rho (a^{-1})^\sigma_\nu \delta^\rho_\sigma = a^\mu_\rho (a^{-1})^\rho_\nu = \delta^\mu_\nu. \tag{10.73}$$

In other words the Kronecker delta symbol of type $(1, 1)$ has the same numerical components in all coordinate systems. This is not true of the Kronecker delta symbol of type $(0, 2)$, i.e. of $\delta_{\mu\nu}$.

Now consider an n-dimensional space with a tensor $\eta_{\mu_1\mu_2\ldots\mu_n}$ whose components, in some basis, coincide with the Levi-Civita symbol $\epsilon_{\mu_1\mu_2\ldots\mu_n}$. We find that in a new frame the components are

$$\begin{aligned}\eta'_{\mu_1\mu_2\ldots\mu_n} &= (a^{-1})^{\nu_1}_{\mu_1} (a^{-1})^{\nu_2}_{\mu_2} \cdots (a^{-1})^{\nu_n}_{\mu_n} \epsilon_{\nu_1\nu_2\ldots\nu_n} \\ &= \epsilon_{\mu_1\mu_2\ldots\mu_n} (a^{-1})^{\nu_1}_1 (a^{-1})^{\nu_2}_2 \cdots (a^{-1})^{\nu_n}_n \epsilon_{\nu_1\nu_2\ldots\nu_n} \\ &= \epsilon_{\mu_1\mu_2\ldots\mu_n} \det \mathbf{A}^{-1} \\ &= \eta_{\mu_1\mu_2\ldots\mu_n} \det \mathbf{A}^{-1}. \end{aligned} \tag{10.74}$$

Thus, unlike the δ^μ_ν, the Levi-Civita symbol is not quite a tensor.

Consider also the quantity

$$\sqrt{g} \stackrel{\text{def}}{=} \sqrt{\det [g_{\mu\nu}]}. \tag{10.75}$$

Here we assume that the metric is positive-definite, so that the square root is real, and that we have taken the positive square root. Since

$$\det [g'_{\mu\nu}] = \det [(a^{-1})^\rho_\mu (a^{-1})^\sigma_\nu g_{\rho\sigma}] = (\det \mathbf{A})^{-2} \det [g_{\mu\nu}], \tag{10.76}$$

we see that

$$\sqrt{g'} = |\det \mathbf{A}|^{-1} \sqrt{g}. \tag{10.77}$$

Thus $\sqrt{g}$ is also not quite an invariant. This is only to be expected, because $\mathbf{g}(\ ,\ )$ is a quadratic form and we know that there is no basis-independent meaning to the determinant of such an object.

Now define

$$\varepsilon_{\mu_1\mu_2\ldots\mu_n} = \sqrt{g}\, \epsilon_{\mu_1\mu_2\ldots\mu_n}, \tag{10.78}$$

and assume that $\varepsilon_{\mu_1\mu_2\ldots\mu_n}$ has the type $(0, n)$ tensor character implied by its indices. When we look at how this transforms, and restrict ourselves to *orientation preserving* changes of basis, i.e. ones for which $\det \mathbf{A}$ is positive, we see that factors of $\det \mathbf{A}$ conspire to give

$$\varepsilon'_{\mu_1\mu_2\dots\mu_n} = \sqrt{g'}\,\epsilon_{\mu_1\mu_2\dots\mu_n}. \tag{10.79}$$

A similar exercise indicates that if we define $\epsilon^{\mu_1\mu_2\dots i_n}$ to be numerically equal to $\epsilon_{i_1 i_2\dots\mu_n}$ then

$$\varepsilon^{\mu_1\mu_2\dots\mu_n} = \frac{1}{\sqrt{g}}\,\epsilon^{\mu_1\mu_2\dots\mu_n} \tag{10.80}$$

also transforms as a tensor – in this case a type $(n, 0)$ contravariant one – provided that the factor of $1/\sqrt{g}$ is always calculated with respect to the current basis.

If the dimension n is even and we are given a skew-symmetric tensor $F_{\mu\nu}$, we can therefore construct an invariant

$$\varepsilon^{\mu_1\mu_2\dots\mu_n}F_{\mu_1\mu_2}\cdots F_{\mu_{n-1}\mu_n} = \frac{1}{\sqrt{g}}\epsilon^{\mu_1\mu_2\dots\mu_n}F_{\mu_1\mu_2}\cdots F_{\mu_{n-1}\mu_n}. \tag{10.81}$$

Similarly, given a skew-symmetric covariant tensor $F_{\mu_1\dots\mu_m}$ with m ($\le n$) indices we can form its *dual*, denoted by F^*, an $(n-m)$-contravariant tensor with components

$$(F^*)^{\mu_{m+1}\dots\mu_n} = \frac{1}{m!}\varepsilon^{\mu_1\mu_2\dots\mu_n}F_{\mu_1\dots\mu_m} = \frac{1}{\sqrt{g}}\frac{1}{m!}\epsilon^{\mu_1\mu_2\dots\mu_n}F_{\mu_1\dots\mu_m}. \tag{10.82}$$

We meet this "dual" tensor again, when we study differential forms.

10.3 Cartesian tensors

If we restrict ourselves to cartesian coordinate systems having orthonormal basis vectors, so that $g_{ij} = \delta_{ij}$, then there are considerable simplifications. In particular, we do not have to make a distinction between co- and contravariant indices. We shall usually write their indices as roman-alphabet suffixes.

A change of basis from one orthogonal n-dimensional basis $\mathbf{e}_i$ to another $\mathbf{e}'_i$ will set

$$\mathbf{e}'_i = O_{ij}\mathbf{e}_j, \tag{10.83}$$

where the numbers O_{ij} are the entries in an *orthogonal* matrix $\mathbf{O}$, i.e. a real matrix obeying $\mathbf{O}^T\mathbf{O} = \mathbf{O}\mathbf{O}^T = \mathbf{I}$, where T denotes the transpose. The set of n-by-n orthogonal matrices constitutes the *orthogonal group* $\mathrm{O}(n)$.

10.3.1 Isotropic tensors

The Kronecker δ_{ij} with both indices downstairs is unchanged by $\mathrm{O}(n)$ transformations,

$$\delta'_{ij} = O_{ik}O_{jl}\delta_{kl} = O_{ik}O_{jk} = O_{ik}O^T_{kj} = \delta_{ij}, \tag{10.84}$$

and has the same components in any cartesian frame. We say that its components are *numerically invariant*. A similar property holds for tensors made up of products of δ_{ij}, such as

$$T_{ijklmn} = \delta_{ij}\delta_{kl}\delta_{mn}. \tag{10.85}$$

It is possible to show[5] that any tensor whose components are numerically invariant under all orthogonal transformations is a sum of products of this form. The most general O(n) invariant tensor of rank four is, for example.

$$\alpha\delta_{ij}\delta_{kl} + \beta\delta_{ik}\delta_{lj} + \gamma\delta_{il}\delta_{jk}. \tag{10.86}$$

The determinant of an orthogonal transformation must be ± 1. If we only allow orientation-preserving changes of basis then we restrict ourselves to orthogonal transformations O_{ij} with $\det \mathbf{O} = 1$. These are the *proper* orthogonal transformations. In n dimensions they constitute the group SO(n). Under SO(n) transformations, both δ_{ij} and $\epsilon_{i_1 i_2 \dots i_n}$ are numerically invariant and the most general SO(n) invariant tensors consist of sums of products of δ_{ij}'s and $\epsilon_{i_1 i_2 \dots i_n}$'s. The most general SO(4)-invariant rank-four tensor is, for example,

$$\alpha\delta_{ij}\delta_{kl} + \beta\delta_{ik}\delta_{lj} + \gamma\delta_{il}\delta_{jk} + \lambda\epsilon_{ijkl}. \tag{10.87}$$

Tensors that are numerically invariant under SO(n) are known as *isotropic tensors*.

As there is no longer any distinction between co- and contravariant indices, we can now contract any pair of indices. In three dimensions, for example,

$$B_{ijkl} = \epsilon_{nij}\epsilon_{nkl} \tag{10.88}$$

is a rank-four isotropic tensor. Now $\epsilon_{i_1 \dots i_n}$ is *not* invariant when we transform via an orthogonal transformation with $\det \mathbf{O} = -1$, but the product of two ϵ's *is* invariant under such transformations. The tensor B_{ijkl} is therefore numerically invariant under the larger group O(3) and must be expressible as

$$B_{ijkl} = \alpha\delta_{ij}\delta_{kl} + \beta\delta_{ik}\delta_{lj} + \gamma\delta_{il}\delta_{jk} \tag{10.89}$$

for some coefficients α, β and γ. The following exercise explores some consequences of this and related facts.

Exercise 10.5: We defined the n-dimensional Levi-Civita symbol by requiring that $\epsilon_{i_1 i_2 \dots i_n}$ be antisymmetric in all pairs of indices, and $\epsilon_{12\dots n} = 1$.

[5] The proof is surprisingly complicated. See, for example, M. Spivak, *A Comprehensive Introduction to Differential Geometry* (second edition) Vol. V, pp. 466–481.

(a) Show that $\epsilon_{123} = \epsilon_{231} = \epsilon_{312}$, but that $\epsilon_{1234} = -\epsilon_{2341} = \epsilon_{3412} = -\epsilon_{4123}$.
(b) Show that

$$\epsilon_{ijk}\epsilon_{i'j'k'} = \delta_{ii'}\delta_{jj'}\delta_{kk'} + \text{five other terms},$$

where you should write out all six terms explicitly.
(c) Show that $\epsilon_{ijk}\epsilon_{ij'k'} = \delta_{jj'}\delta_{kk'} - \delta_{jk'}\delta_{kj'}$.
(d) For dimension $n = 4$, write out $\epsilon_{ijkl}\epsilon_{ij'k'l'}$ as a sum of products of δ's similar to the one in part (c).

Exercise 10.6: *Vector products*. The vector product of two three-vectors may be written in cartesian components as $(\mathbf{a} \times \mathbf{b})_i = \epsilon_{ijk}a_j b_k$. Use this and your results about ϵ_{ijk} from the previous exercise to show that

(i) $\mathbf{a} \cdot (\mathbf{b} \times \mathbf{c}) = \mathbf{b} \cdot (\mathbf{c} \times \mathbf{a}) = \mathbf{c} \cdot (\mathbf{a} \times \mathbf{b})$,
(ii) $\mathbf{a} \times (\mathbf{b} \times \mathbf{c}) = (\mathbf{a} \cdot \mathbf{c})\mathbf{b} - (\mathbf{a} \cdot \mathbf{b})\mathbf{c}$,
(iii) $(\mathbf{a} \times \mathbf{b}) \cdot (\mathbf{c} \times \mathbf{d}) = (\mathbf{a} \cdot \mathbf{c})(\mathbf{b} \cdot \mathbf{d}) - (\mathbf{a} \cdot \mathbf{d})(\mathbf{b} \cdot \mathbf{c})$.
(iv) If we take **a**, **b**, **c** and **d**, with $\mathbf{d} \equiv \mathbf{b}$, to be unit vectors, show that the identities (i) and (iii) become the sine and cosine rule, respectively, of spherical trigonometry. (Hint: for the spherical sine rule, begin by showing that $\mathbf{a} \cdot [(\mathbf{a} \times \mathbf{b}) \times (\mathbf{a} \times \mathbf{c})] = \mathbf{a} \cdot (\mathbf{b} \times \mathbf{c})$.)

10.3.2 Stress and strain

As an illustration of the utility of cartesian tensors, we consider their application to elasticity.

Suppose that an elastic body is slightly deformed so that the particle that was originally at the point with cartesian coordinates x_i is moved to $x_i + \eta_i$. We define the (infinitesimal) *strain tensor* e_{ij} by

$$e_{ij} = \frac{1}{2}\left(\frac{\partial \eta_j}{\partial x_i} + \frac{\partial \eta_i}{\partial x_j}\right). \tag{10.90}$$

It is automatically symmetric: $e_{ij} = e_{ji}$. We will leave for later (Exercise 11.3) a discussion of why this is the natural definition of strain, and also the modifications necessary were we to employ a non-cartesian coordinate system.

To define the *stress tensor* σ_{ij} we consider the portion Ω of the body in Figure 10.1, and an element of area $d\mathbf{S} = \mathbf{n}\, d|S|$ on its boundary. Here, $\mathbf{n}$ is the unit normal vector pointing out of Ω. The force $\mathbf{F}$ exerted on this surface element by the parts of the body exterior to Ω has components

$$F_i = \sigma_{ij} n_j \, d|S|. \tag{10.91}$$

That $\mathbf{F}$ is a linear function of $\mathbf{n}\, d|S|$ can be seen by considering the forces on a small tetrahedron, three of whose sides coincide with the coordinate planes, the fourth side having $\mathbf{n}$ as its normal. In the limit that the lengths of the sides go to zero as ϵ, the mass

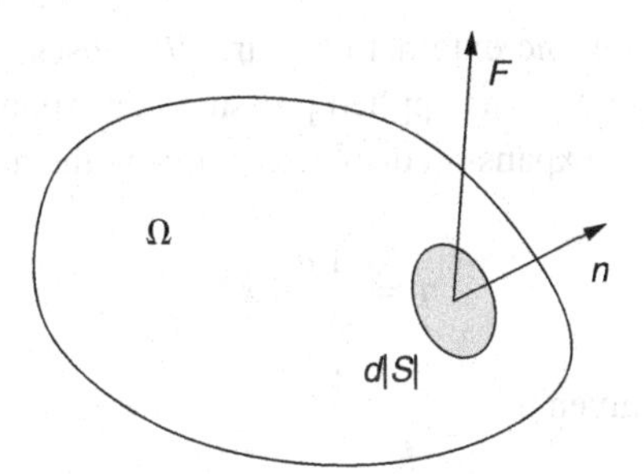

Figure 10.1 Stress forces.

of the body scales to zero as ϵ^3, but the forces are proprtional to the areas of the sides and go to zero only as ϵ^2. Only if the linear relation holds true can the acceleration of the tetrahedron remain finite. A similar argument applied to torques and the moment of inertia of a small cube shows that $\sigma_{ij} = \sigma_{ji}$.

A generalization of Hooke's law,

$$\sigma_{ij} = c_{ijkl} e_{kl}, \tag{10.92}$$

relates the stress to the strain via the tensor of *elastic constants* c_{ijkl}. This rank-four tensor has the symmetry properties

$$c_{ijkl} = c_{klij} = c_{jikl} = c_{ijlk}. \tag{10.93}$$

In other words, the tensor is symmetric under the interchange of the first and second pairs of indices, and also under the interchange of the individual indices in either pair.

For an isotropic material – a material whose properties are invariant under the rotation group SO(3) – the tensor of elastic constants must be an isotropic tensor. The most general such tensor with the required symmetries is

$$c_{ijkl} = \lambda \delta_{ij} \delta_{kl} + \mu(\delta_{ik}\delta_{jl} + \delta_{il}\delta_{jk}). \tag{10.94}$$

An isotropic material is therefore characterized by only two independent parameters, λ and μ. These are called the *Lamé* constants after the mathematical engineer Gabriel Lamé. In terms of them the generalized Hooke's law becomes

$$\sigma_{ij} = \lambda \delta_{ij} e_{kk} + 2\mu e_{ij}. \tag{10.95}$$

By considering particular deformations, we can express the more directly measurable *bulk modulus*, *shear modulus*, *Young's modulus* and *Poisson's ratio* in terms of λ and μ.

The bulk modulus κ is defined by

$$dP = -\kappa \frac{dV}{V}, \tag{10.96}$$

where an infinitesimal isotropic external pressure dP causes a change $V \to V + dV$ in the volume of the material. This applied pressure corresponds to a surface stress of $\sigma_{ij} = -\delta_{ij}\, dP$. An isotropic expansion displaces points in the material so that

$$\eta_i = \frac{1}{3}\frac{dV}{V}x_i. \tag{10.97}$$

The strains are therefore given by

$$e_{ij} = \frac{1}{3}\delta_{ij}\frac{dV}{V}. \tag{10.98}$$

Inserting this strain into the stress–strain relation gives

$$\sigma_{ij} = \delta_{ij}\left(\lambda + \frac{2}{3}\mu\right)\frac{dV}{V} = -\delta_{ij}dP. \tag{10.99}$$

Thus

$$\kappa = \lambda + \frac{2}{3}\mu. \tag{10.100}$$

To define the shear modulus, we assume a deformation $\eta_1 = \theta x_2$, so $e_{12} = e_{21} = \theta/2$, with all other e_{ij} vanishing (see Figure 10.2).

The applied shear stress is $\sigma_{12} = \sigma_{21}$. The shear modulus is defined to be σ_{12}/θ. Inserting the strain components into the stress–strain relation gives

$$\sigma_{12} = \mu\theta, \tag{10.101}$$

and so the shear modulus is equal to the Lamé constant μ. We can therefore write the generalized Hooke's law as

$$\sigma_{ij} = 2\mu\left(e_{ij} - \tfrac{1}{3}\delta_{ij}e_{kk}\right) + \kappa e_{kk}\delta_{ij}, \tag{10.102}$$

which reveals that the shear modulus is associated with the traceless part of the strain tensor, and the bulk modulus with the trace.

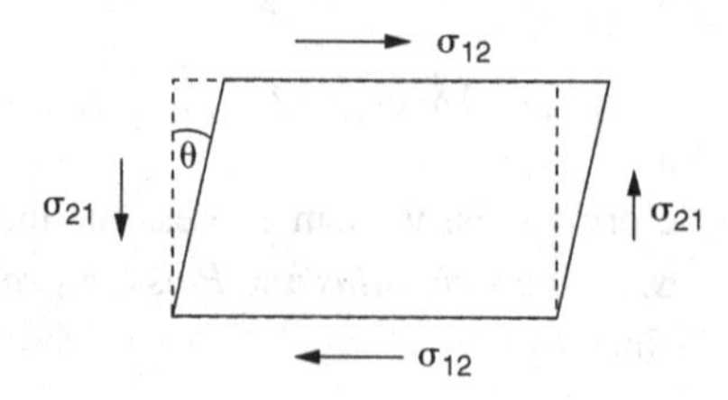

Figure 10.2 Shear strain. The arrows show the direction of the applied stresses. The σ_{21} on the vertical faces are necessary to stop the body rotating.

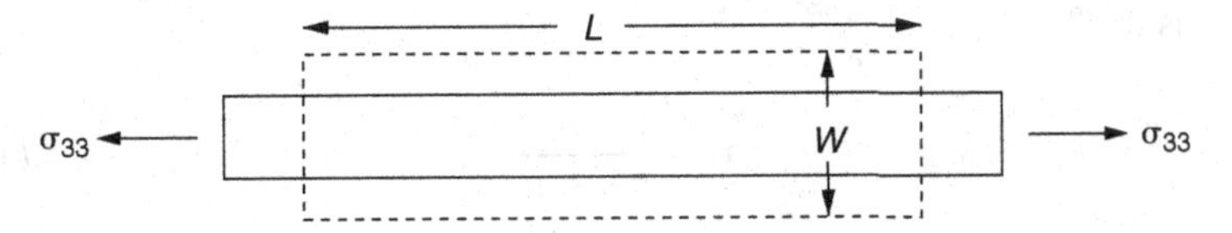

Figure 10.3 Forces on a stretched wire.

Young's modulus Y is measured by stretching a wire of initial length L and square cross-section of side W under a tension $T = \sigma_{33} W^2$.

We define Y so that

$$\sigma_{33} = Y \frac{dL}{L}. \tag{10.103}$$

At the same time as the wire stretches, its width changes $W \to W + dW$. Poisson's ratio σ is defined by

$$\frac{dW}{W} = -\sigma \frac{dL}{L}, \tag{10.104}$$

so that σ is positive if the wire gets thinner as it gets longer. The displacements are

$$\begin{aligned} \eta_3 &= z \left(\frac{dL}{L} \right), \\ \eta_1 &= x \left(\frac{dW}{W} \right) = -\sigma x \left(\frac{dL}{L} \right), \\ \eta_2 &= y \left(\frac{dW}{W} \right) = -\sigma y \left(\frac{dL}{L} \right), \end{aligned} \tag{10.105}$$

so the strain components are

$$e_{33} = \frac{dL}{L}, \quad e_{11} = e_{22} = \frac{dW}{W} = -\sigma e_{33}. \tag{10.106}$$

We therefore have

$$\sigma_{33} = (\lambda(1 - 2\sigma) + 2\mu) \left(\frac{dL}{L} \right), \tag{10.107}$$

leading to

$$Y = \lambda(1 - 2\sigma) + 2\mu. \tag{10.108}$$

Now, the side of the wire is a free surface with no forces acting on it, so

$$0 = \sigma_{22} = \sigma_{11} = (\lambda(1 - 2\sigma) - 2\sigma\mu) \left(\frac{dL}{L} \right). \tag{10.109}$$

This tells us that[6]

$$\sigma = \frac{1}{2}\frac{\lambda}{\lambda + \mu}, \tag{10.110}$$

and

$$Y = \mu\left(\frac{3\lambda + 2\mu}{\lambda + \mu}\right). \tag{10.111}$$

Other relations, following from those above, are

$$\begin{aligned} Y &= 3\kappa(1 - 2\sigma), \\ &= 2\mu(1 + \sigma). \end{aligned} \tag{10.112}$$

Exercise 10.7: Show that the symmetries

$$c_{ijkl} = c_{klij} = c_{jikl} = c_{ijlk}$$

imply that a general homogeneous material has 21 independent elastic constants. (This result was originally obtained by George Green, of Green function fame.)

Exercise 10.8: A steel beam is forged so that its cross-section has the shape of a region $\Gamma \in \mathbb{R}^2$. When undeformed, it lies along the z-axis. The centroid O of each cross-section is defined so that

$$\int_\Gamma x\,dxdy = \int_\Gamma y\,dxdy = 0,$$

when the coordinates x, y are taken with the centroid O as the origin. The beam is slightly bent away from the z-axis so that the line of centroids remains in the yz-plane (see Figure 10.4). At a particular cross-section with centroid O, the line of centroids has radius of curvature R.

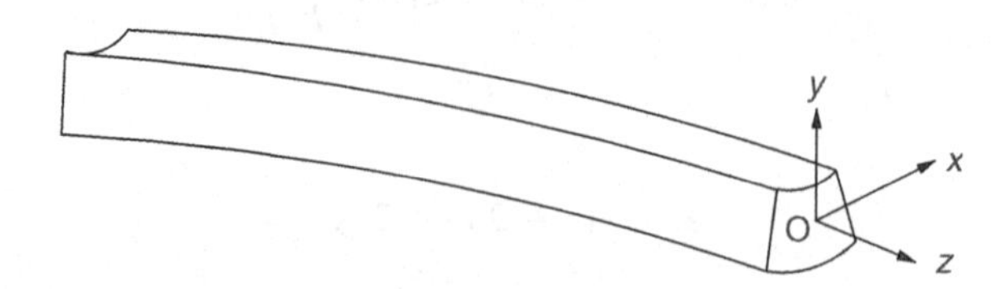

Figure 10.4 Bent beam.

[6] Poisson and Cauchy erroneously believed that $\lambda = \mu$, and hence that $\sigma = 1/4$.

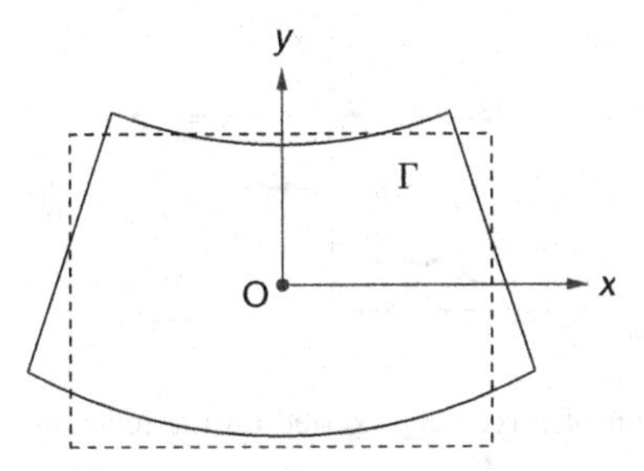

Figure 10.5 The original (dashed) and anticlastically deformed (full) cross-section.

Assume that the deformation in the vicinity of O is such that

$$\eta_x = -\frac{\sigma}{R}xy,$$

$$\eta_y = \frac{1}{2R}\left\{\sigma(x^2 - y^2) - z^2\right\},$$

$$\eta_z = \frac{1}{R}yz.$$

Observe that for this assumed deformation, and for a positive Poisson ratio, the cross-section deforms *anticlastically* – the sides bend *up* as the beam bends *down*. This is shown in Figure 10.5.

Compute the strain tensor resulting from the assumed deformation, and show that its only non-zero components are

$$e_{xx} = -\frac{\sigma}{R}y, \qquad e_{yy} = -\frac{\sigma}{R}y, \qquad e_{zz} = \frac{1}{R}y.$$

Next, show that

$$\sigma_{zz} = \left(\frac{Y}{R}\right)y,$$

and that all other components of the stress tensor vanish (Figure 10.6). Deduce from this vanishing that the assumed deformation satisfies the free-surface boundary condition, and so is indeed the way the beam responds when it is bent by forces applied at its ends.

The work done in bending the beam

$$\int_{\text{beam}} \frac{1}{2} e_{ij} c_{ijkl} e_{kl}\, d^3x$$

is stored as elastic energy. Show that for our bent rod this energy is equal to

$$\int \frac{YI}{2}\left(\frac{1}{R^2}\right) ds \approx \int \frac{YI}{2}(y'')^2 dz,$$

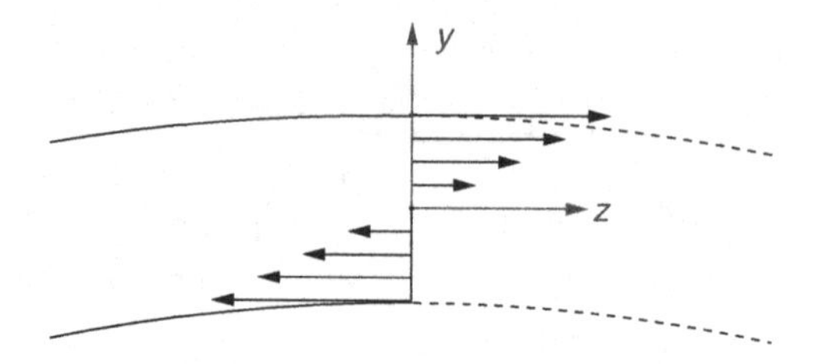

Figure 10.6 The distribution of forces σ_{zz} exerted on the left-hand part of the bent rod by the material to its right.

where s is the arc-length taken along the line of centroids of the beam,

$$I = \int_{\Gamma} y^2 \, dxdy$$

is the moment of inertia of the region Γ about the x axis and y'' denotes the second derivative of the deflection of the beam with respect to z (which approximates the arc-length). This last formula for the strain energy has been used in a number of our calculus of variations problems.

10.3.3 Maxwell stress tensor

Consider a small cubical element of an elastic body. If the stress tensor were position independent, the external forces on each pair of opposing faces of the cube would be equal in magnitude but pointing in opposite directions. There would therefore be no net external force on the cube. When σ_{ij} is *not* constant then we claim that the total force acting on an infinitesimal element of volume dV is

$$F_i = \partial_j \sigma_{ij} \, dV. \tag{10.113}$$

To see that this assertion is correct, consider a finite region Ω with boundary $\partial\Omega$, and use the divergence theorem to write the total force on Ω as

$$F_i^{\text{tot}} = \int_{\partial\Omega} \sigma_{ij} n_j d|S| = \int_{\Omega} \partial_j \sigma_{ij} dV. \tag{10.114}$$

Whenever the force-per-unit-volume f_i acting on a body can be written in the form $f_i = \partial_j \sigma_{ij}$, we refer to σ_{ij} as a "stress tensor", by analogy with stress in an elastic solid. As an example, let $\mathbf{E}$ and $\mathbf{B}$ be electric and magnetic fields. For simplicity, initially assume them to be static. The force per unit volume exerted by these fields on a distribution of charge ρ and current $\mathbf{j}$ is

$$\mathbf{f} = \rho\mathbf{E} + \mathbf{j} \times \mathbf{B}. \tag{10.115}$$

From Gauss' law $\rho = \operatorname{div}\mathbf{D}$, and with $\mathbf{D} = \epsilon_0\mathbf{E}$, we find that the force per unit volume due to the electric field has components

$$\begin{aligned}\rho E_i = (\partial_j D_j)E_i &= \epsilon_0\Big(\partial_j(E_iE_j) - E_j\,\partial_j E_i\Big)\\ &= \epsilon_0\Big(\partial_j(E_iE_j) - E_j\,\partial_i E_j\Big)\\ &= \epsilon_0\partial_j\left(E_iE_j - \frac{1}{2}\delta_{ij}|E|^2\right).\end{aligned} \tag{10.116}$$

Here, in passing from the first line to the second, we have used the fact that $\operatorname{curl}\mathbf{E}$ is zero for static fields, and so $\partial_j E_i = \partial_i E_j$. Similarly, using $\mathbf{j} = \operatorname{curl}\mathbf{H}$, together with $\mathbf{B} = \mu_0\mathbf{H}$ and $\operatorname{div}\mathbf{B} = 0$, we find that the force per unit volume due to the magnetic field has components

$$(\mathbf{j}\times\mathbf{B})_i = \mu_0\partial_j\left(H_iH_j - \frac{1}{2}\delta_{ij}|H|^2\right). \tag{10.117}$$

The quantity

$$\sigma_{ij} = \epsilon_0\left(E_iE_j - \frac{1}{2}\delta_{ij}|E|^2\right) + \mu_0\left(H_iH_j - \frac{1}{2}\delta_{ij}|H|^2\right) \tag{10.118}$$

is called the *Maxwell stress tensor*. Its utility lies in the fact that the total electromagnetic force on an isolated body is the integral of the Maxwell stress over its surface. We do not need to know the fields within the body.

Michael Faraday was the first to intuit a picture of electromagnetic stresses and attributed both a longitudinal tension and a mutual lateral repulsion to the field lines. Maxwell's tensor expresses this idea mathematically.

Exercise 10.9: Allow the fields in the preceding calculation to be time dependent. Show that Maxwell's equations

$$\begin{aligned}\operatorname{curl}\mathbf{E} &= -\frac{\partial\mathbf{B}}{\partial t}, & \operatorname{div}\mathbf{B} &= 0,\\ \operatorname{curl}\mathbf{H} &= \mathbf{j} + \frac{\partial\mathbf{D}}{\partial t}, & \operatorname{div}\mathbf{D} &= \rho,\end{aligned}$$

with $\mathbf{B} = \mu_0\mathbf{H}$, $\mathbf{D} = \epsilon_0\mathbf{E}$ and $c = 1/\sqrt{\mu_0\epsilon_0}$, lead to

$$(\rho\mathbf{E} + \mathbf{j}\times\mathbf{B})_i + \frac{\partial}{\partial t}\left\{\frac{1}{c^2}(\mathbf{E}\times\mathbf{H})_i\right\} = \partial_j\sigma_{ij}.$$

The left-hand side is the time rate of change of the mechanical (first term) and electromagnetic (second term) momentum density. Observe that we can equivalently write

$$\frac{\partial}{\partial t}\left\{\frac{1}{c^2}(\mathbf{E}\times\mathbf{H})_i\right\} + \partial_j(-\sigma_{ij}) = -(\rho\mathbf{E} + \mathbf{j}\times\mathbf{B})_i,$$

and think of this as a local field-momentum conservation law. In this interpretation $-\sigma_{ij}$ is thought of as the *momentum flux* tensor, its entries being the flux in direction j of the component of field momentum in direction i. The term on the right-hand side is the rate at which momentum is being supplied to the electromagnetic field by the charges and currents.

10.4 Further exercises and problems

Exercise 10.10: *Quotient theorem.* Suppose that you have come up with some recipe for generating an array of numbers T^{ijk} in any coordinate frame, and want to know whether these numbers are the components of a triply contravariant tensor. Suppose further that you know that, given the components a_{ij} of an arbitrary doubly covariant tensor, the numbers

$$T^{ijk}a_{jk} = v^i$$

transform as the components of a contravariant vector. Show that T^{ijk} does indeed transform as a triply contravariant tensor. (The natural generalization of this result to arbitrary tensor types is known as the *quotient theorem*.)

Exercise 10.11: Let $T^i{}_j$ be the 3-by-3 array of components of a tensor. Show that the quantities

$$a = T^i{}_i, \quad b = T^i{}_jT^j{}_i, \quad c = T^i{}_jT^j{}_kT^k{}_i$$

are invariant. Further show that the eigenvalues of the linear map represented by the matrix $T^i{}_j$ can be found by solving the cubic equation

$$\lambda^3 - a\lambda^2 + \frac{1}{2}(a^2 - b)\lambda - \frac{1}{6}(a^3 - 3ab + 2c) = 0.$$

Exercise 10.12: Let the covariant tensor R_{ijkl} possess the following symmetries:

(i) $R_{ijkl} = -R_{jikl}$,
(ii) $R_{ijkl} = -R_{ijlk}$,
(iii) $R_{ijkl} + R_{iklj} + R_{iljk} = 0$.

Use the properties (i), (ii), (iii) to show that:

(a) $R_{ijkl} = R_{klij}$.
(b) If $R_{ijkl}x^iy^jx^ky^l = 0$ for all vectors x^i, y^i, then $R_{ijkl} = 0$.
(c) If B_{ij} is a symmetric covariant tensor and we set $A_{ijkl} = B_{ik}B_{jl} - B_{il}B_{jk}$, then A_{ijkl} has the same symmetries as R_{ijkl}.

Exercise 10.13: Write out Euler's equation for fluid motion

$$\dot{\mathbf{v}} + (\mathbf{v}\cdot\nabla)\mathbf{v} = -\nabla h$$

in cartesian tensor notation. Transform it into

$$\dot{\mathbf{v}} - \mathbf{v} \times \boldsymbol{\omega} = -\nabla \left(\frac{1}{2} \mathbf{v}^2 + h \right),$$

where $\boldsymbol{\omega} = \nabla \times \mathbf{v}$ is the vorticity. Deduce Bernoulli's theorem, that for steady ($\dot{\mathbf{v}} = 0$) flow the quantity $\frac{1}{2}\mathbf{v}^2 + h$ is constant along streamlines.

Exercise 10.14: The elastic properties of an infinite homogeneous and isotropic solid of density ρ are described by Lamé constants λ and μ. Show that the equation of motion for small-amplitude vibrations is

$$\rho \frac{\partial^2 \eta_i}{\partial t^2} = (\lambda + \mu) \frac{\partial^2 \eta_j}{\partial x_i \partial x_j} + \mu \frac{\partial^2 \eta_i}{\partial x_j^2}.$$

Here η_i are the cartesian components of the displacement vector $\boldsymbol{\eta}(\mathbf{x}, t)$ of the particle initially at the point $\mathbf{x}$. Seek plane-wave solutions of the form

$$\boldsymbol{\eta} = \mathbf{a} \exp\{i\mathbf{k} \cdot \mathbf{x} - i\omega t\},$$

and deduce that there are two possible types of wave: longitudinal "P-waves", which have phase velocity

$$v_{\mathrm{P}} = \sqrt{\frac{\lambda + 2\mu}{\rho}},$$

and transverse "S-waves", which have phase velocity

$$v_{\mathrm{S}} = \sqrt{\frac{\mu}{\rho}}.$$

Exercise 10.15: *Symmetric integration*. Show that the n-dimensional integral

$$I_{\alpha\beta\gamma\delta} = \int \frac{d^n k}{(2\pi)^n} (k_\alpha k_\beta k_\gamma k_\delta) f(k^2),$$

is equal to

$$A(\delta_{\alpha\beta}\delta_{\gamma\delta} + \delta_{\alpha\gamma}\delta_{\beta\delta} + \delta_{\alpha\delta}\delta_{\beta\gamma}),$$

where

$$A = \frac{1}{n(n+2)} \int \frac{d^n k}{(2\pi)^n} (k^2)^2 f(k^2).$$

Similarly evaluate

$$I_{\alpha\beta\gamma\delta\epsilon} = \int \frac{d^n k}{(2\pi)^n} (k_\alpha k_\beta k_\gamma k_\delta k_\epsilon) f(k^2).$$

Exercise 10.16: Write down the most general three-dimensional isotropic tensors of rank two and three.

In piezoelectric materials, the application of an electric field E_i induces a mechanical strain that is described by a rank-two symmetric tensor

$$e_{ij} = d_{ijk} E_k,$$

where d_{ijk} is a third-rank tensor that depends only on the material. Show that e_{ij} can only be non-zero in an anisotropic material.

Exercise 10.17: In three dimensions, a rank-five isotropic tensor T_{ijklm} is a linear combination of expressions of the form $\epsilon_{i_1 i_2 i_3} \delta_{i_4 i_5}$ for some assignment of the indices i, j, k, l, m to the $i_1, \ldots, i_5$. Show that, on taking into account the symmetries of the Kronecker and Levi-Civita symbols, we can construct *ten* distinct products $\epsilon_{i_1 i_2 i_3} \delta_{i_4 i_5}$. Only *six* of these are linearly independent, however. Show, for example, that

$$\epsilon_{ijk} \delta_{lm} - \epsilon_{jkl} \delta_{im} + \epsilon_{kli} \delta_{jm} - \epsilon_{lij} \delta_{km} = 0,$$

and find the three other independent relations of this sort.[7]

(Hint: begin by showing that, in three dimensions,

$$\delta^{i_1 i_2 i_3 i_4}_{i_5 i_6 i_7 i_8} \stackrel{\text{def}}{=} \begin{vmatrix} \delta_{i_1 i_5} & \delta_{i_1 i_6} & \delta_{i_1 i_7} & \delta_{i_1 i_8} \\ \delta_{i_2 i_5} & \delta_{i_2 i_6} & \delta_{i_2 i_7} & \delta_{i_2 i_8} \\ \delta_{i_3 i_5} & \delta_{i_3 i_6} & \delta_{i_3 i_7} & \delta_{i_3 i_8} \\ \delta_{i_4 i_5} & \delta_{i_4 i_6} & \delta_{i_4 i_7} & \delta_{i_4 i_8} \end{vmatrix} = 0,$$

and contract with $\epsilon_{i_6 i_7 i_8}$.)

Problem 10.18: *The Plücker relations.* This problem provides a challenging test of your understanding of linear algebra. It leads you through the task of deriving the necessary and sufficient conditions for

$$\mathbf{A} = A^{i_1 \ldots i_k}\, \mathbf{e}_{i_1} \wedge \ldots \wedge \mathbf{e}_{i_k} \in \bigwedge\nolimits^k V$$

to be decomposable as

$$\mathbf{A} = \mathbf{f}_1 \wedge \mathbf{f}_2 \wedge \ldots \wedge \mathbf{f}_k.$$

The trick is to introduce two subspaces of V,

[7] Such relations are called *syzygies*. A recipe for constructing linearly independent basis sets of isotropic tensors can be found in: G. F. Smith, *Tensor*, **19** (1968) 79.

(i) W, the smallest subspace of V such that $\mathbf{A} \in \bigwedge^k W$,
(ii) $W' = \{\mathbf{v} \in V : \mathbf{v} \wedge \mathbf{A} = 0\}$,

and explore their relationship.

(a) Show that if $\{\mathbf{w}_1, \mathbf{w}_2, \dots, \mathbf{w}_n\}$ constitute a basis for W', then

$$\mathbf{A} = \mathbf{w}_1 \wedge \mathbf{w}_2 \wedge \cdots \wedge \mathbf{w}_n \wedge \varphi$$

for some $\varphi \in \bigwedge^{k-n} V$. Conclude that $W' \subseteq W$, and that equality holds if and only if $\mathbf{A}$ is decomposable, in which case $W = W' = \text{span}\{\mathbf{f}_1 \dots \mathbf{f}_k\}$.
(b) Now show that W is the image space of $\bigwedge^{k-1} V^*$ under the map that takes

$$\Xi = \Xi_{i_1 \dots i_{k-1}} \mathbf{e}^{*i_1} \wedge \dots \wedge \mathbf{e}^{*i_{k-1}} \in \bigwedge{}^{k-1} V^*$$

to

$$i(\Xi)\mathbf{A} \stackrel{\text{def}}{=} \Xi_{i_1 \dots i_{k-1}} A^{i_1 \dots i_{k-1} j} \mathbf{e}_j \in V.$$

Deduce that the condition $W \subseteq W'$ is that

$$\left(i(\Xi)\mathbf{A}\right) \wedge \mathbf{A} = 0, \quad \forall\, \Xi \in \bigwedge{}^{k-1} V^*.$$

(c) By taking

$$\Xi = \mathbf{e}^{*i_1} \wedge \dots \wedge \mathbf{e}^{*i_{k-1}},$$

show that the condition in part (b) can be written as

$$A^{i_1 \dots i_{k-1} j_1} A^{j_2 j_3 \dots j_{k+1}} \mathbf{e}_{j_1} \wedge \dots \wedge \mathbf{e}_{j_{k+1}} = 0.$$

Deduce that the necessary and sufficient conditions for decomposability are that

$$A^{i_1 \dots i_{k-1} [j_1} A^{j_2 j_3 \dots j_{k+1}]} = 0,$$

for all possible index sets $i_1, \dots, i_{k-1}, j_1, \dots j_{k+1}$. Here $[\dots]$ denotes anti-symmetrization of the enclosed indices.

11
Differential calculus on manifolds

In this section we will apply what we have learned about vectors and tensors in linear algebra to vector and tensor *fields* in a general curvilinear coordinate system. Our aim is to introduce the reader to the modern language of advanced calculus, and in particular to the calculus of differential forms on surfaces and manifolds.

11.1 Vector and covector fields

Vector fields – electric, magnetic, velocity fields, and so on – appear everywhere in physics. After perhaps struggling with it in introductory courses, we rather take the field concept for granted. There remain subtleties, however. Consider an electric field. It makes sense to add two field vectors at a single point, but there is no physical meaning to the sum of field vectors $\mathbf{E}(x_1)$ and $\mathbf{E}(x_2)$ at two distinct points. We should therefore regard all possible electric fields at a single point as living in a vector space, but each different point in space comes with its own field-vector space.

This view seems even more reasonable when we consider velocity vectors describing motion on a curved surface. A velocity vector lives in the *tangent space* to the surface at each point, and each of these spaces is a differently oriented subspace of the higher-dimensional ambient space (see Figure 11.1).

Mathematicians call such a collection of vector spaces – one for each of the points in a surface – a *vector bundle* over the surface. Thus, the *tangent bundle* over a surface is the totality of all vector spaces tangent to the surface. Why a *bundle*? This word is used because the individual tangent spaces are not completely independent, but are tied together in a rather non-obvious way. Try to construct a smooth field of unit vectors tangent to the surface of a sphere. However hard you work you will end up in trouble

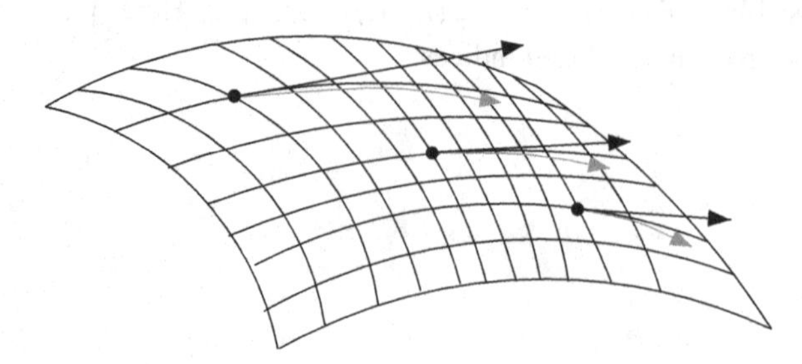

Figure 11.1 Each point on a surface has its own vector space of tangents.

somewhere. You cannot comb a hairy ball. On the surface of a torus you will have no such problems. You can comb a hairy doughnut. The tangent spaces collectively know something about the surface they are tangent to.

Although we spoke in the previous paragraph of vectors tangent to a curved surface, it is useful to generalize this idea to vectors lying in the tangent space of an n-dimensional *manifold*, or n-manifold. An n-manifold M is essentially a space that locally looks like a part of $\mathbb{R}^n$. This means that some open neighbourhood of each point $x \in M$ can be parametrized by an n-dimensional coordinate system. A coordinate parametrization is called a *chart*. Unless M is $\mathbb{R}^n$ itself (or part of it), a chart will cover only part of M, and more than one will be required for complete coverage. Where a pair of charts overlap, we demand that the transformation formula giving one set of coordinates as a function of the other be a smooth (C^∞) function, and possess a smooth inverse.[1] A collection of such smoothly related coordinate charts covering all of M is called an *atlas*. The advantage of thinking in terms of manifolds is that we do not have to understand their properties as arising from some embedding in a higher dimensional space. Whatever structure they have, they possess in, and of, themselves.

Classical mechanics provides a familiar illustration of these ideas. Except in pathological cases, the configuration space M of a mechanical system is a manifold. When the system has n degrees of freedom we use generalized coordinates q^i, $i = 1, \ldots, n$ to parametrize M. The tangent bundle of M then provides the setting for Lagrangian mechanics. This bundle, denoted by TM, is the $2n$-dimensional space each of whose points consists of a point $q = (q^1, \ldots, q^n)$ in M paired with a tangent vector lying in the tangent space TM_q at that point. If we think of the tangent vector as a velocity, the natural coordinates on TM become $(q^1, q^2, \ldots, q^n; \dot{q}^1, \dot{q}^2, \ldots, \dot{q}^n)$, and these are the variables that appear in the Lagrangian of the system.

If we consider a vector tangent to some curved surface, it will stick out of it. If we have a vector tangent to a manifold, it is a straight arrow lying atop bent coordinates. Should we restrict the length of the vector so that it does not stick out too far? Are we restricted to only infinitesimal vectors? It is best to avoid all this by adopting a clever notion of what a vector in a tangent space is. The idea is to focus on a well-defined object such as a derivative. Suppose that our space has coordinates x^μ. (These are *not* the contravariant components of some vector.) A *directional derivative* is an object such as $X^\mu \partial_\mu$, where ∂_μ is shorthand for $\partial/\partial x^\mu$. When the components X^μ are functions of the coordinates x^σ, this object is called a tangent-vector field, and we write[2]

$$X = X^\mu \partial_\mu. \tag{11.1}$$

[1] A formal definition of a manifold contains some further technical restrictions (that the space be *Hausdorff* and *paracompact*) that are designed to eliminate pathologies. We are more interested in doing calculus than in proving theorems, and so we will ignore these niceties.

[2] We are going to stop using bold symbols to distinguish between intrinsic objects and their components, because from now on almost everything will be something other than a number, and too much black ink would just be confusing.

We regard the ∂_μ at a point x as a basis for TM_x, the tangent-vector space at x, and the $X^\mu(x)$ as the (contravariant) components of the vector X at that point. Although they are not little arrows, what the ∂_μ are is mathematically clear, and so we know perfectly well how to deal with them.

When we change coordinate system from x^μ to z^ν by regarding the x^μ's as invertable functions of the z^ν's, i.e.

$$\begin{aligned} x^1 &= x^1(z^1, z^2, \ldots, z^n), \\ x^2 &= x^2(z^1, z^2, \ldots, z^n), \\ &\vdots \\ x^n &= x^n(z^1, z^2, \ldots, z^n), \end{aligned} \tag{11.2}$$

then the chain rule for partial differentiation gives

$$\partial_\mu \equiv \frac{\partial}{\partial x^\mu} = \frac{\partial z^\nu}{\partial x^\mu}\frac{\partial}{\partial z^\nu} = \left(\frac{\partial z^\nu}{\partial x^\mu}\right)\partial'_\nu, \tag{11.3}$$

where ∂'_ν is shorthand for $\partial/\partial z^\nu$. By demanding that

$$X = X^\mu \partial_\mu = X'^\nu \partial'_\nu \tag{11.4}$$

we find the components in the z^ν coordinate frame to be

$$X'^\nu = \left(\frac{\partial z^\nu}{\partial x^\mu}\right) X^\mu. \tag{11.5}$$

Conversely, using

$$\frac{\partial x^\sigma}{\partial z^\nu}\frac{\partial z^\nu}{\partial x^\mu} = \frac{\partial x^\sigma}{\partial x^\mu} = \delta^\sigma_\mu, \tag{11.6}$$

we have

$$X^\nu = \left(\frac{\partial x^\nu}{\partial z^\mu}\right) X'^\mu. \tag{11.7}$$

This, then, is the transformation law for a contravariant vector.

It is worth pointing out that the basis vectors ∂_μ are *not* unit vectors. As we have no metric, and therefore no notion of length anyway, we cannot try to normalize them. If you insist on drawing (small?) arrows, think of ∂_1 as starting at a point $(x^1, x^2, \ldots, x^n)$ and with its head at $(x^1 + 1, x^2, \ldots, x^n)$; see Figure 11.2. Of course this is only a good picture if the coordinates are not too "curvy".

Example: The surface of the unit sphere is a manifold. It is usually denoted by S^2. We may label its points with spherical polar coordinates, θ measuring the co-latitude and ϕ

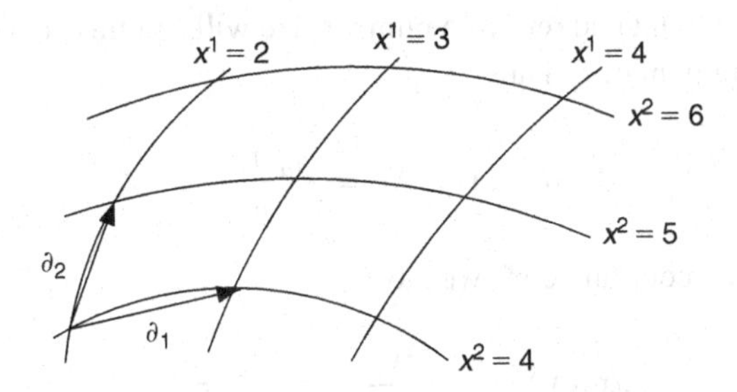

Figure 11.2 Approximate picture of the vectors ∂_1 and ∂_2 at the point $(x^1, x^2) = (2, 4)$.

measuring the longitude. These will be useful everywhere except at the north and south poles, where they become singular because at $\theta = 0$ or π all values of the longitude ϕ correspond to the same point. In this coordinate basis, the tangent vector representing the velocity field due to a rigid rotation of one radian per second about the z-axis is

$$V_z = \partial_\phi. \tag{11.8}$$

Similarly

$$\begin{aligned} V_x &= -\sin\phi\, \partial_\theta - \cot\theta \cos\phi\, \partial_\phi, \\ V_y &= \quad \cos\phi\, \partial_\theta - \cot\theta \sin\phi \partial_\phi, \end{aligned} \tag{11.9}$$

respectively represent rigid rotations about the x- and y-axes.

We now know how to think about vectors. What about their dual-space partners, the covectors? These live in the *cotangent bundle* T^*M, and for them a cute notational game, due to Élie Cartan, is played. We write the basis vectors dual to the ∂_μ as $dx^\mu(\ )$. Thus

$$dx^\mu(\partial_\nu) = \delta^\mu_\nu. \tag{11.10}$$

When evaluated on a vector field $X = X^\mu \partial_\mu$, the basis covectors dx^μ return its components:

$$dx^\mu(X) = dx^\mu(X^\nu \partial_\nu) = X^\nu dx^\mu(\partial_\nu) = X^\nu \delta^\mu_\nu = X^\mu. \tag{11.11}$$

Now, any smooth function $f \in C^\infty(M)$ will give rise to a field of covectors in T^*M. This is because a vector field X acts on the scalar function f as

$$Xf = X^\mu \partial_\mu f \tag{11.12}$$

and Xf is another scalar function. This new function gives a number – and thus an element of the field $\mathbb{R}$ – at each point $x \in M$. But this is exactly what a covector does: it takes in a vector at a point and returns a number. We will call this covector field "df". It is essentially the gradient of f. Thus

$$df(X) \stackrel{\text{def}}{=} Xf = X^\mu \frac{\partial f}{\partial x^\mu}. \tag{11.13}$$

If we take f to be the coordinate x^ν, we have

$$dx^\nu(X) = X^\mu \frac{\partial x^\nu}{\partial x^\mu} = X^\mu \delta^\nu_\mu = X^\nu, \tag{11.14}$$

so this viewpoint is consistent with our previous definition of dx^ν. Thus

$$df(X) = \frac{\partial f}{\partial x^\mu} X^\mu = \frac{\partial f}{\partial x^\mu} dx^\mu(X) \tag{11.15}$$

for any vector field X. In other words, we can expand df as

$$df = \frac{\partial f}{\partial x^\mu} dx^\mu. \tag{11.16}$$

This is *not* some approximation to a change in f, but is an exact expansion of the covector field df in terms of the basis covectors dx^μ.

We may retain something of the notion that dx^μ represents the (contravariant) components of a small displacement in x provided that we think of dx^μ as a machine into which we insert the small displacement (a vector) and have it spit out the numerical components δx^μ. This is the same distinction that we make between sin( ) as a function into which one can plug x, and $\sin x$, the number that results from inserting in this particular value of x. Although seemingly innocent, we know that it is a distinction of great power.

The change of coordinates transformation law for a covector field f_μ is found from

$$f_\mu \, dx^\mu = f'_\nu \, dz^\nu, \tag{11.17}$$

by using

$$dx^\mu = \left(\frac{\partial x^\mu}{\partial z^\nu}\right) dz^\nu. \tag{11.18}$$

We find

$$f'_\nu = \left(\frac{\partial x^\mu}{\partial z^\nu}\right) f_\mu. \tag{11.19}$$

A general tensor such as $Q^{\lambda\mu}{}_{\rho\sigma\tau}$ transforms as

$$Q'^{\lambda\mu}{}_{\rho\sigma\tau}(z) = \frac{\partial z^\lambda}{\partial x^\alpha} \frac{\partial z^\mu}{\partial x^\beta} \frac{\partial x^\gamma}{\partial z^\rho} \frac{\partial x^\delta}{\partial z^\sigma} \frac{\partial x^\epsilon}{\partial z^\tau} Q^{\alpha\beta}{}_{\gamma\delta\epsilon}(x). \tag{11.20}$$

Observe how the indices are wired up: Those for the new tensor coefficients in the new coordinates, z, are attached to the new z's, and those for the old coefficients are attached to the old x's. Upstairs indices go in the numerator of each partial derivative, and downstairs ones are in the denominator.

The language of bundles and sections

At the beginning of this section, we introduced the notion of a vector bundle. This is a particular example of the more general concept of a *fibre bundle*, where the vector space at each point in the manifold is replaced by a "fibre" *over* that point. The fibre can be any mathematical object, such as a set, tensor space or another manifold. Mathematicians visualize the bundle as a collection of fibres growing out of the manifold, much as stalks of wheat grow out of the soil. When one slices through a patch of wheat with a scythe, the blade exposes a cross-section of the stalks. By analogy, a choice of an element of the the fibre over each point in the manifold is called a *cross-section*, or, more commonly, a *section* of the bundle. In this language, a tangent-vector field becomes a section of the tangent bundle, and a field of covectors becomes a section of the cotangent bundle.

We provide a more detailed account of bundles in Chapter 16.

11.2 Differentiating tensors

If f is a function then $\partial_\mu f$ are components of the covariant vector df. Suppose that a^μ is a contravariant vector. Are $\partial_\nu a^\mu$ the components of a type $(1, 1)$ tensor? The answer is *no*! In general, differentiating the components of a tensor does not give rise to another tensor. One can see why at two levels:

(a) Consider the transformation laws. They contain expressions of the form $\partial x^\mu/\partial z^\nu$. If we differentiate both sides of the transformation law of a tensor, these factors are also differentiated, but tensor transformation laws never contain second derivatives, such as $\partial^2 x^\mu/\partial z^\nu \partial z^\sigma$.

(b) Differentiation requires subtracting vectors or tensors at different points – but vectors at different points are in different vector spaces, so their difference is not defined.

These two reasons are really one and the same. We need to be cleverer to get new tensors by differentiating old ones.

11.2.1 Lie bracket

One way to proceed is to note that the vector field X is an *operator*. It makes sense, therefore, to try to compose two of them to make another. Look at XY, for example:

$$XY = X^\mu \partial_\mu (Y^\nu \partial_\nu) = X^\mu Y^\nu \partial^2_{\mu\nu} + X^\mu \left(\frac{\partial Y^\nu}{\partial x^\mu} \right) \partial_\nu. \tag{11.21}$$

What are we to make of this? Not much! There is no particular interpretation for the second derivative, and as we saw above, it does not transform nicely. But suppose we take a *commutator*:

$$[X, Y] = XY - YX = \big(X^\mu(\partial_\mu Y^\nu) - Y^\mu(\partial_\mu X^\nu)\big)\,\partial_\nu. \tag{11.22}$$

The second derivatives have cancelled, and what remains is a directional derivative and so a *bona fide* vector field. The components

$$[X, Y]^\nu \equiv X^\mu(\partial_\mu Y^\nu) - Y^\mu(\partial_\mu X^\nu) \tag{11.23}$$

are the components of a new contravariant vector field made from the two old vector fields. This new vector field is called the *Lie bracket* of the two fields, and has a geometric interpretation.

To understand the geometry of the Lie bracket, we first define the *flow* associated with a tangent-vector field X. This is the map that takes a point x_0 and maps it to $x(t)$ by solving the family of equations

$$\frac{dx^\mu}{dt} = X^\mu(x^1, x^2, \dots), \tag{11.24}$$

with initial condition $x^\mu(0) = x_0^\mu$. In words, we regard X as the velocity field of a flowing fluid, and let x ride along with the fluid.

Now envisage X and Y as two velocity fields. Suppose we flow along X for a brief time t, then along Y for another brief interval s. Next we switch back to X, but with a minus sign, for time t, and then to $-Y$ for a final interval of s. We have tried to retrace our path, but a short exercise with Taylor's theorem shows that we will fail to return to our exact starting point. We will miss by $\delta x^\mu = st[X, Y]^\mu$, plus corrections of cubic order in s and t (see Figure 11.3).

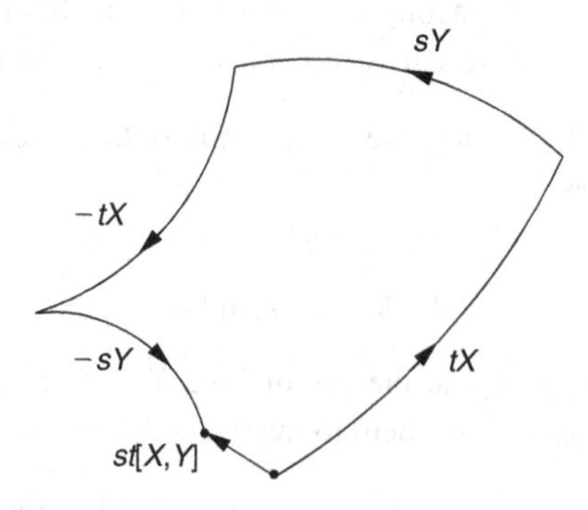

Figure 11.3 We try to retrace our steps but fail to return by a distance proportional to the Lie bracket.

Example: Let

$$V_x = -\sin\phi\,\partial_\theta - \cot\theta\cos\phi\,\partial_\phi,$$
$$V_y = \cos\phi\,\partial_\theta - \cot\theta\sin\phi\,\partial_\phi,$$

be two vector fields in $T(S^2)$. We find that

$$[V_x, V_y] = -V_z,$$

where $V_z = \partial_\phi$.

Frobenius' theorem

Suppose that in some region of a d-dimensional manifold M we are given $n < d$ linearly independent tangent-vector fields X_i. Such a set is called a *distribution* by differential geometers. (The concept has nothing to do with probability, or with objects like "$\delta(x)$" which are also called "distributions".) At each point x, the span $\langle X_i(x)\rangle$ of the field vectors forms a subspace of the tangent space TM_x, and we can picture this subspace as a fragment of an n-dimensional surface passing through x. It is possible that these surface fragments fit together to make a stack of smooth surfaces – called a *foliation* (see Figure 11.4) – that fill out the d-dimensional space, and have the given X_i as their tangent vectors.

If this is the case then starting from x and taking steps only along the X_i we find ourselves restricted to the n-surface, or *n-submanifold*, N passing though the original point x.

Alternatively, the surface fragments may form such an incoherent jumble that starting from x and moving only along the X_i we can find our way to any point in the neighbourhood of x. It is also possible that some intermediate case applies, so that moving along the X_i restricts us to an m-surface, where $d > m > n$. The Lie bracket provides us with the appropriate tool with which to investigate these possibilities.

First a definition: if there are functions $c_{ij}{}^{k}(x)$ such that

$$[X_i, X_j] = c_{ij}{}^{k}(x) X_k, \tag{11.25}$$

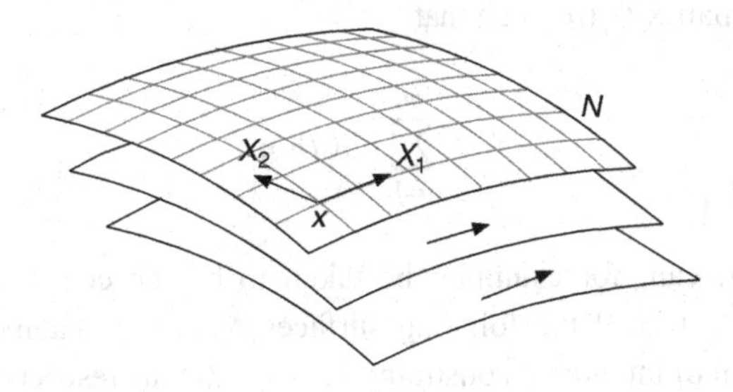

Figure 11.4 A local foliation.

i.e. the Lie brackets close within the set $\{X_i\}$ at each point x, then the distribution is said to be *involutive*, and the vector fields are said to be "in involution" with each other. When our given distribution is involutive, then the first case holds, and, at least locally, there is a foliation by n-submanifolds N. A formal statement of this is:

Theorem: *(Frobenius): A smooth (C^∞) involutive distribution is completely integrable: locally, there are coordinates $x^\mu, \mu = 1,\ldots,d$ such that $X_i = \sum_{\mu=1}^{n} X_i^\mu \partial_\mu$, and the surfaces N through each point are in the form $x^\mu = const.$ for $\mu = n+1,\ldots,d$. Conversely, if such coordinates exist then the distribution is involutive.*

A half-proof: If such coordinates exist then it is obvious that the Lie bracket of any pair of vectors in the form $X_i = \sum_{\mu=1}^{n} X_i^\mu \partial_\mu$ can also be expanded in terms of the first n basis vectors. A logically equivalent statement exploits the geometric interpretation of the Lie bracket: if the Lie brackets of the fields X_i do *not* close within the n-dimensional span of the X_i, then a sequence of back-and-forth manœvres along the X_i allows us to escape into a new direction, and so the X_i *cannot* be tangent to an n-surface. Establishing the converse – that closure implies the existence of the foliation – is rather more technical, and we will not attempt it.

Involutive and non-involutive distributions appear in classical mechanics under the guise of *holonomic* and *anholonomic* constraints. In mechanics, constraints are not usually given as a list of the directions (vector fields) in which we are free to move, but instead as a list of restrictions imposed on the permitted motion. In a d-dimensional mechanical system we might have a set of m independent constraints of the form $\omega^i_\mu(q)\dot{q}^\mu = 0$, $i = 1,\ldots,m$. Such restrictions are most naturally expressed in terms of the covector fields

$$\omega^i = \sum_{\mu=1}^{d} \omega^i_\mu(q) dq^\mu, \quad i = 1 \leq i \leq m. \tag{11.26}$$

We can write the constraints as the m conditions $\omega^i(\dot{q}) = 0$ that must be satisfied if $\dot{q} \equiv \dot{q}^\mu \partial_\mu$ is to be an allowed motion. The list of constraints is known a *Pfaffian* system of equations. These equations indirectly determine an $n = d - m$ dimensional distribution of permitted motions. The Pfaffian system is said to be *integrable* if this distribution is involutive, and hence integrable. In this case there is a set of m functions $g^i(q)$ and an invertible m-by-m matrix $f^i{}_j(q)$ such that

$$\omega^i = \sum_{j=1}^{m} f^i{}_j(q) dg^j. \tag{11.27}$$

The functions $g^i(q)$ can, for example, be taken to be the coordinate functions x^μ, $\mu = n+1,\ldots,d$, that label the foliating surfaces N in the statement of Frobenius' theorem. The system of integrable constraints $\omega^i(\dot{q}) = 0$ thus restricts us to the surfaces $g^i(q) = \text{constant}$.

For example, consider a particle moving in three dimensions. If we are told that the velocity vector is constrained by $\omega(\dot{q}) = 0$, where

$$\omega = x\,dx + y\,dy + z\,dz \tag{11.28}$$

we realize that the particle is being forced to move on a sphere passing through the initial point. In spherical coordinates the associated distribution is the set $\{\partial_\theta, \partial_\phi\}$, which is clearly involutive because $[\partial_\theta, \partial_\phi] = 0$. The functions $f(x,y,z)$ and $g(x,y,z)$ from the previous paragraph can be taken to be $r = \sqrt{x^2+y^2+z^2}$, and the constraint covector written as $\omega = f\,dg = r\,dr$.

The foliation is the family of nested spheres whose centre is the origin. (The foliation is not global because it becomes singular at $r = 0$.) Constraints like this, which restrict the motion to a surface, are said to be *holonomic*.

Suppose, on the other hand, we have a ball rolling on a table. Here, we have a five-dimensional configuration manifold $M = \mathbb{R}^2 \times S^3$, parametrized by the centre of mass $(x,y) \in \mathbb{R}^2$ of the ball and the three Euler angles $(\theta, \phi, \psi) \in S^3$ defining its orientation. Three no-slip rolling conditions

$$\begin{aligned} \dot{x} &= \quad \dot{\psi}\sin\theta\sin\phi + \dot{\theta}\cos\phi, \\ \dot{y} &= -\dot{\psi}\sin\theta\cos\phi + \dot{\theta}\sin\phi, \\ 0 &= \quad \dot{\psi}\cos\theta + \dot{\phi}, \end{aligned} \tag{11.29}$$

(see Exercise 11.17) link the rate of change of the Euler angles to the velocity of the centre of mass. At each point in this five-dimensional manifold we are free to roll the ball in two directions, and so we might expect that the reachable configurations constitute a two-dimensional surface embedded in the full five-dimensional space. The two vector fields

$$\begin{aligned} \mathbf{roll_x} &= \partial_x - \sin\phi\cot\theta\,\partial_\phi + \cos\phi\,\partial_\theta + \operatorname{cosec}\theta\sin\phi\,\partial_\psi, \\ \mathbf{roll_y} &= \partial_y + \cos\phi\cot\theta\,\partial_\phi + \sin\phi\,\partial_\theta - \operatorname{cosec}\theta\cos\phi\,\partial_\psi, \end{aligned} \tag{11.30}$$

describing the permitted x- and y-direction rolling motion are not in involution, however. By calculating enough Lie brackets we eventually obtain five linearly independent velocity vector fields, and starting from one configuration we can reach any other. The no-slip rolling condition is said to be *non-integrable*, or *anholonomic*. Such systems are tricky to deal with in Lagrangian dynamics.

The following exercise provides a familiar example of the utility of non-holonomic constraints:

Exercise 11.1: *Parallel parking using Lie brackets*. The configuration space of a car is four dimensional, and parametrized by coordinates (x, y, θ, ϕ), as shown in Figure 11.5. Define the following vector fields:

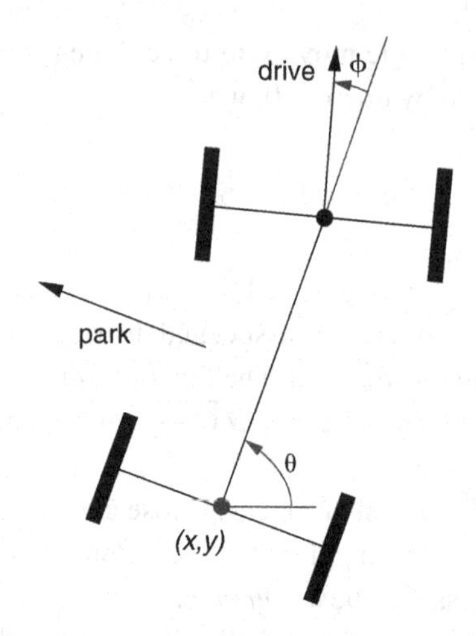

Figure 11.5 Coordinates for car parking.

(a) (front wheel) **drive** $= \cos\phi(\cos\theta\,\partial_x + \sin\theta\,\partial_y) + \sin\phi\,\partial_\theta$.
(b) **steer** $= \partial_\phi$.
(c) (front wheel) **skid** $= -\sin\phi(\cos\theta\,\partial_x + \sin\theta\,\partial_y) + \cos\phi\,\partial_\theta$.
(d) **park** $= -\sin\theta\,\partial_x + \cos\theta\,\partial_y$.

Explain why these are apt names for the vector fields, and compute the six Lie brackets:

$$[\mathbf{steer}, \mathbf{drive}],\ [\mathbf{steer}, \mathbf{skid}],\ [\mathbf{skid}, \mathbf{drive}],$$

$$[\mathbf{park}, \mathbf{drive}],\ [\mathbf{park}, \mathbf{park}],\ [\mathbf{park}, \mathbf{skid}].$$

The driver can use only the operations (±) **drive** and (±) **steer** to manœvre the car. Use the geometric interpretation of the Lie bracket to explain how a suitable sequence of motions (forward, reverse and turning the steering wheel) can be used to manoeuvre a car sideways into a parking space.

11.2.2 Lie derivative

Another derivative that we can define is the *Lie derivative* along a vector field X. It is defined by its action on a scalar function f as

$$\mathcal{L}_X f \stackrel{\text{def}}{=} Xf, \tag{11.31}$$

on a vector field by

$$\mathcal{L}_X Y \stackrel{\text{def}}{=} [X, Y], \tag{11.32}$$

and on anything else by requiring it to be a *derivation*, meaning that it obeys Leibniz' rule. For example, let us compute the Lie derivative of a covector F. We first introduce an arbitrary vector field Y and plug it into F to get the scalar function $F(Y)$. Leibniz' rule is then the statement that

$$\mathcal{L}_X F(Y) = (\mathcal{L}_X F)(Y) + F(\mathcal{L}_X Y). \tag{11.33}$$

Since $F(Y)$ is a function and Y is a vector, both of whose derivatives we know how to compute, we know the first and third of the three terms in this equation. From $\mathcal{L}_X F(Y) = XF(Y)$ and $F(\mathcal{L}_X Y) = F([X,Y])$, we have

$$XF(Y) = (\mathcal{L}_X F)(Y) + F([X,Y]), \tag{11.34}$$

and so

$$(\mathcal{L}_X F)(Y) = XF(Y) - F([X,Y]). \tag{11.35}$$

In components, this becomes

$$\begin{aligned}(\mathcal{L}_X F)(Y) &= X^\nu \partial_\nu (F_\mu Y^\mu) - F_\nu (X^\mu \partial_\mu Y^\nu - Y^\mu \partial_\mu X^\nu)\\ &= (X^\nu \partial_\nu F_\mu + F_\nu \partial_\mu X^\nu) Y^\mu.\end{aligned} \tag{11.36}$$

Note how all the derivatives of Y^μ have cancelled, so $\mathcal{L}_X F(\ )$ depends only on the local value of Y. The Lie derivative of F is therefore still a covector field. This is true in general: the Lie derivative does not change the tensor character of the objects on which it acts. Dropping the passive spectator field Y^ν, we have a formula for $\mathcal{L}_X F$ in components:

$$(\mathcal{L}_X F)_\mu = X^\nu \partial_\nu F_\mu + F_\nu \partial_\mu X^\nu. \tag{11.37}$$

Another example is provided by the Lie derivative of a type $(0,2)$ tensor, such as a metric tensor. This is

$$(\mathcal{L}_X g)_{\mu\nu} = X^\alpha \partial_\alpha g_{\mu\nu} + g_{\mu\alpha} \partial_\nu X^\alpha + g_{\alpha\nu} \partial_\mu X^\alpha. \tag{11.38}$$

The Lie derivative of a metric measures the extent to which the displacement $x^\alpha \to x^\alpha + \epsilon X^\alpha(x)$ deforms the geometry. If we write the metric as

$$g(\ ,\ ) = g_{\mu\nu}(x)\, dx^\mu \otimes dx^\nu, \tag{11.39}$$

we can understand both this geometric interpretation and the origin of the three terms appearing in the Lie derivative. We simply make the displacement $x^\alpha \to x^\alpha + \epsilon X^\alpha$ in the coefficients $g_{\mu\nu}(x)$ and in the two dx^α. In the latter we write

$$d(x^\alpha + \epsilon X^\alpha) = dx^\alpha + \epsilon \frac{\partial X^\alpha}{\partial x^\beta} dx^\beta. \tag{11.40}$$

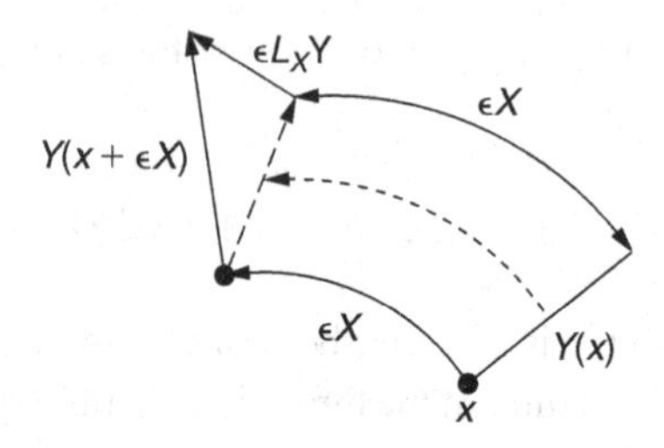

Figure 11.6 Computing the Lie derivative of a vector.

Then we see that

$$g_{\mu\nu}(x)\, dx^\mu \otimes dx^\nu \to \left[g_{\mu\nu}(x) + \epsilon(X^\alpha \partial_\alpha g_{\mu\nu} + g_{\mu\alpha}\partial_\nu X^\alpha + g_{\alpha\nu}\partial_\mu X^\alpha)\right] dx^\mu \otimes dx^\nu$$
$$= [g_{\mu\nu} + \epsilon(\mathcal{L}_X g)_{\mu\nu}]\, dx^\mu \otimes dx^\nu. \tag{11.41}$$

A displacement field X that does not change distances between points, i.e. one that gives rise to an *isometry*, must therefore satisfy $\mathcal{L}_X g = 0$. Such an X is said to be a *Killing field* after Wilhelm Killing who introduced them in his study of non-Euclidean geometries.

The geometric interpretation of the Lie derivative of a vector field is as follows: in order to compute the X directional derivative of a vector field Y, we need to be able to subtract the vector $Y(x)$ from the vector $Y(x+\epsilon X)$, divide by ϵ and take the limit $\epsilon \to 0$. To do this we have somehow to get the vector $Y(x)$ from the point x, where it normally resides, to the new point $x + \epsilon X$, so both vectors are elements of the same vector space. The Lie derivative achieves this by carrying the old vector to the new point along the field X (see Figure 11.6).

Imagine the vector Y as drawn in ink in a flowing fluid whose velocity field is X. Initially the tail of Y is at x and its head is at $x + Y$. After flowing for a time ϵ, its tail is at $x + \epsilon X$ – i.e. exactly where the tail of $Y(x + \epsilon X)$ lies. Where the head of the transported vector ends up depends on how the flow has stretched and rotated the ink, but it is this distorted vector that is subtracted from $Y(x + \epsilon X)$ to get $\epsilon \mathcal{L}_X Y = \epsilon[X, Y]$.

Exercise 11.2: The metric on the unit sphere equipped with polar coordinates is

$$g(\ ,\ ) = d\theta \otimes d\theta + \sin^2\theta\, d\phi \otimes d\phi.$$

Consider

$$V_x = -\sin\phi\, \partial_\theta - \cot\theta \cos\phi\, \partial_\phi,$$

which is the vector field of a rigid rotation about the x-axis. Compute the Lie derivative $\mathcal{L}_{V_x} g$, and show that it is zero.

Exercise 11.3: Suppose we have an unstrained block of material in real space. A coordinate system ξ^1, ξ^2, ξ^3, is attached to the material of the body. The point with coordinate ξ is located at $(x^1(\xi), x^2(\xi), x^3(\xi))$ where x^1, x^2, x^3 are the usual $\mathbf{R}^3$ cartesian coordinates.

(a) Show that the induced metric in the ξ coordinate system is

$$g_{\mu\nu}(\xi) = \sum_{a=1}^{3} \frac{\partial x^a}{\partial \xi^\mu} \frac{\partial x^a}{\partial \xi^\nu}.$$

(b) The body is now deformed by an infinitesimal strain vector field $\eta(\xi)$. The atom with coordinate ξ^μ is moved to what was $\xi^\mu + \eta^\mu(\xi)$, or, equivalently, the atom initially at cartesian coordinate $x^a(\xi)$ is moved to $x^a + \eta^\mu \partial x^a / \partial \xi^\mu$. Show that the new induced metric is

$$g_{\mu\nu} + \delta g_{\mu\nu} = g_{\mu\nu} + \mathcal{L}_\eta g_{\mu\nu}.$$

(c) Define the *strain tensor* to be 1/2 of the Lie derivative of the metric with respect to the deformation. If the original ξ coordinate system coincided with the cartesian one, show that this definition reduces to the familiar form

$$e_{ab} = \frac{1}{2}\left(\frac{\partial \eta_a}{\partial x^b} + \frac{\partial \eta_b}{\partial x^a}\right),$$

all tensors being cartesian.

(d) Part (c) gave us the geometric definitition of *infinitesimal strain*. If the body is deformed substantially, the *Cauchy–Green finite strain tensor* is defined as

$$E_{\mu\nu}(\xi) = \frac{1}{2}\left(g_{\mu\nu} - g^{(0)}_{\mu\nu}\right),$$

where $g^{(0)}_{\mu\nu}$ is the metric in the undeformed body and $g_{\mu\nu}$ the metric in the deformed body. Explain why this is a reasonable definition.

11.3 Exterior calculus

11.3.1 Differential forms

The objects we introduced in Section 11.1, the dx^μ, are called 1-forms, or differential 1-forms. They are fields living in the cotangent bundle T^*M of M. More precisely, they are *sections* of the cotangent bundle. Sections of the bundle whose fibre above $x \in M$ is the p-th skew-symmetric tensor power $\bigwedge^p(T^*M_x)$ of the cotangent space are known as p-forms.

For example,

$$A = A_\mu dx^\mu = A_1 dx^1 + A_2 dx^2 + A_3 dx^3 \tag{11.42}$$

is a 1-form,

$$F = \frac{1}{2} F_{\mu\nu} dx^\mu \wedge dx^\nu = F_{12} dx^1 \wedge dx^2 + F_{23} dx^2 \wedge dx^3 + F_{31} dx^3 \wedge dx^1 \tag{11.43}$$

is a 2-form, and

$$\Omega = \frac{1}{3!}\Omega_{\mu\nu\sigma}\, dx^\mu \wedge dx^\nu \wedge dx^\sigma = \Omega_{123}\, dx^1 \wedge dx^2 \wedge dx^3 \tag{11.44}$$

is a 3-form. All the coefficients are skew-symmetric tensors, so, for example,

$$\Omega_{\mu\nu\sigma} = \Omega_{\nu\sigma\mu} = \Omega_{\sigma\mu\nu} = -\Omega_{\nu\mu\sigma} = -\Omega_{\mu\sigma\nu} = -\Omega_{\sigma\nu\mu}. \tag{11.45}$$

In each example we have explicitly written out all the independent terms for the case of three dimensions. Note how the $p!$ disappears when we do this and keep only distinct components. In d dimensions the space of p-forms is $d!/p!(d-p)!$ dimensional, and all p-forms with $p > d$ vanish identically.

As with the wedge products in Chapter 1, we regard a p-form as a p-linear skew-symetric function with p slots into which we can drop vectors to get a number. For example the basis two-forms give

$$dx^\mu \wedge dx^\nu(\partial_\alpha, \partial_\beta) = \delta^\mu_\alpha \delta^\nu_\beta - \delta^\mu_\beta \delta^\nu_\alpha. \tag{11.46}$$

The analogous expression for a p-form would have $p!$ terms. We can define an algebra of differential forms by "wedging" them together in the obvious way, so that the product of a p-form with a q-form is a $(p+q)$-form. The wedge product is associative and distributive but not, of course, commutative. Instead, if a is a p-form and b a q-form, then

$$a \wedge b = (-1)^{pq}\, b \wedge a. \tag{11.47}$$

Actually it is customary in this game to suppress the "$\wedge$" and simply write $F = \frac{1}{2}F_{\mu\nu}\, dx^\mu dx^\nu$, it being assumed that you know that $dx^\mu dx^\nu = -dx^\nu dx^\mu$ – what else could it be?

11.3.2 The exterior derivative

These p-forms may seem rather complicated, so it is perhaps surprising that all the vector calculus (div, grad, curl, the divergence theorem and Stokes' theorem, etc.) that you have learned in the past reduce, in terms of them, to two simple formulæ! Indeed Élie Cartan's calculus of p-forms is slowly supplanting traditional vector calculus, much as Willard Gibbs' and Oliver Heaviside's vector calculus supplanted the tedious component-by-component formulæ you find in Maxwell's *Treatise on Electricity and Magnetism*.

The basic tool is the *exterior derivative* "d", which we now define axiomatically:

(i) If f is a function (0-form), then df coincides with the previous definition, i.e. $df(X) = Xf$ for any vector field X.
(ii) d is an *anti-derivation*: if a is a p-form and b a q-form then

$$d(a \wedge b) = da \wedge b + (-1)^p a \wedge db. \tag{11.48}$$

(iii) *Poincaré's lemma*: $d^2 = 0$, meaning that $d(da) = 0$ for any p-form a.
(iv) d is linear. That $d(\alpha a) = \alpha da$, for constant α follows already from (i) and (ii), so the new fact is that $d(a+b) = da + db$.

It is not immediately obvious that axioms (i), (ii) and (iii) are compatible with one another. If we use axiom (i), (ii) and $d(dx^i) = 0$ to compute the d of $\Omega = \frac{1}{p!}\Omega_{i_1,\ldots,i_p} dx^{i_1}\cdots dx^{i_p}$, we find

$$\begin{aligned} d\Omega &= \frac{1}{p!}(d\Omega_{i_1,\ldots,i_p})\, dx^{i_1}\cdots dx^{i_p} \\ &= \frac{1}{p!}\partial_k\Omega_{i_1,\ldots,i_p}\, dx^k dx^{i_1}\cdots dx^{i_p}. \end{aligned} \tag{11.49}$$

Now compute

$$d(d\Omega) = \frac{1}{p!}\left(\partial_l\partial_k\Omega_{i_1,\ldots,i_p}\right) dx^l dx^k dx^{i_1}\cdots dx^{i_p}. \tag{11.50}$$

Fortunately this is zero because $\partial_l\partial_k\Omega = \partial_k\partial_l\Omega$, while $dx^l dx^k = -dx^k dx^l$.

As another example let $A = A_1 dx^1 + A_2 dx^2 + A_3 dx^3$. Then

$$\begin{aligned} dA &= \left(\frac{\partial A_2}{\partial x^1} - \frac{\partial A_1}{\partial x^2}\right) dx^1 dx^2 + \left(\frac{\partial A_1}{\partial x^3} - \frac{\partial A_3}{\partial x^1}\right) dx^3 dx^1 \\ &\quad + \left(\frac{\partial A_3}{\partial x^2} - \frac{\partial A_2}{\partial x^3}\right) dx^2 dx^3 \\ &= \frac{1}{2}F_{\mu\nu} dx^\mu dx^\nu, \end{aligned} \tag{11.51}$$

where

$$F_{\mu\nu} = \partial_\mu A_\nu - \partial_\nu A_\mu. \tag{11.52}$$

You will recognize the components of curl **A** hiding in here.

Again, if $F = F_{12}dx^1dx^2 + F_{23}dx^2dx^3 + F_{31}dx^3dx^1$ then

$$dF = \left(\frac{\partial F_{23}}{\partial x^1} + \frac{\partial F_{31}}{\partial x^2} + \frac{\partial F_{12}}{\partial x^3}\right) dx^1 dx^2 dx^3. \tag{11.53}$$

This looks like a divergence.

The axiom $d^2 = 0$ encompasses both "curl grad $= 0$" and "div curl $= 0$", together with an infinite number of higher-dimensional analogues. The familiar "curl $=\nabla\times$", meanwhile, is only defined in three-dimensional space.

The exterior derivative takes p-forms to $(p+1)$-forms, i.e. skew-symmetric type $(0,p)$ tensors to skew-symmetric $(0,p+1)$ tensors. How does "d" get around the fact that the derivative of a tensor is not a tensor? Well, if you apply the transformation law for A_μ, and the chain rule to $\frac{\partial}{\partial x^\mu}$ to find the transformation law for $F_{\mu\nu} = \partial_\mu A_\nu - \partial_\nu A_\mu$, you will see why: all the derivatives of the $\frac{\partial z^\nu}{\partial x^\mu}$ cancel, and $F_{\mu\nu}$ is a *bona fide* tensor of type $(0, 2)$. This sort of cancellation is why skew-symmetric objects are useful, and symmetric ones less so.

Exercise 11.4: Use axiom (ii) to compute $d(d(a \wedge b))$ and confirm that it is zero.

Closed and exact forms

The Poincaré lemma, $d^2 = 0$, leads to some important terminology:

(i) A p-form ω is said to be *closed* if $d\omega = 0$.
(ii) A p-form ω is said to *exact* if $\omega = d\eta$ for some $(p-1)$-form η.

An exact form is necessarily closed, but a closed form is not necessarily exact. The question of when closed $\Rightarrow$ exact is one involving the global topology of the space in which the forms are defined, and will be the subject of Chapter 13.

Cartan's formulæ

It is sometimes useful to have expressions for the action of d coupled with the evaluation of the subsequent $(p + 1)$ forms.

If f, η, ω are $0, 1, 2$-forms, respectively, then $df, d\eta, d\omega$ are $1, 2, 3$-forms. When we plug in the appropriate number of vector fields X, Y, Z, then, after some labour, we will find

$$df(X) = Xf. \tag{11.54}$$

$$d\eta(X, Y) = X\eta(Y) - Y\eta(X) - \eta([X, Y]). \tag{11.55}$$

$$\begin{aligned} d\omega(X, Y, Z) = {} & X\omega(Y, Z) + Y\omega(Z, X) + Z\omega(X, Y) \\ & - \omega([X, Y], Z) - \omega([Y, Z], X) - \omega([Z, X], Y). \end{aligned} \tag{11.56}$$

These formulæ, and their higher-p analogues, express d in terms of geometric objects, and so make it clear that the exterior derivative is itself a geometric object, independent of any particular coordinate choice.

Let us demonstrate the correctness of the second formula. With $\eta = \eta_\mu dx^\mu$, the left-hand side, $d\eta(X, Y)$, is equal to

$$\partial_\mu \eta_\nu \, dx^\mu dx^\nu (X, Y) = \partial_\mu \eta_\nu (X^\mu Y^\nu - X^\nu Y^\mu). \tag{11.57}$$

The right-hand side is equal to

$$X^\mu \partial_\mu (\eta_\nu Y^\nu) - Y^\mu \partial_\mu (\eta_\nu X^\nu) - \eta_\nu (X^\mu \partial_\mu Y^\nu - Y^\mu \partial_\mu X^\nu). \tag{11.58}$$

On using the product rule for the derivatives in the first two terms, we find that all derivatives of the components of X and Y cancel, and we are left with exactly those terms appearing on the left.

Exercise 11.5: Let ω^i, $i = 1, \dots, r$, be a linearly independent set of 1-forms defining a Pfaffian system (see Section 11.2.1) in d dimensions.

(i) Use Cartan's formulæ to show that the corresponding $(d - r)$-dimensional distribution is involutive if and only if there is an r-by-r matrix of 1-forms $\theta^i{}_j$ such that

$$d\omega^i = \sum_{j=1}^{r} \theta^i{}_j \wedge \omega^j.$$

(ii) Show that the conditions in part (i) are satisfied if there are r functions g^i and an invertible r-by-r matrix of functions $f^i{}_j$ such that

$$\omega^i = \sum_{j=1}^{r} f^i{}_j dg^i.$$

In this case foliation surfaces are given by the conditions $g^i(x) = \text{const.}$, $i = 1, \dots, r$.

It is also possible, but considerably harder, to show that (i) $\Rightarrow$ (ii). Doing so would constitute a proof of Frobenius' theorem.

Exercise 11.6: Let ω be a closed 2-form, and let $\text{Null}(\omega)$ be the space of vector fields X such that $\omega(X, \;) = 0$. Use the Cartan formulæ to show that if $X, Y \in \text{Null}(\omega)$, then $[X, Y] \in \text{Null}(\omega)$.

Lie derivative of forms

Given a p-form ω and a vector field X, we can form a $(p-1)$-form called $i_X\omega$ by writing

$$i_X\omega(\underbrace{\ldots\ldots}_{p-1 \text{ slots}}) = \omega(\overbrace{X, \underbrace{\ldots\ldots}_{p-1 \text{ slots}}}^{p \text{ slots}}). \tag{11.59}$$

Acting on a 0-form, i_X is defined to be 0. This procedure is called the *interior multiplication* by X. It is simply a contraction

$$\omega_{j i_2 \ldots j_p} \to \omega_{k j_2 \ldots j_p} X^k, \tag{11.60}$$

but it is convenient to have a special symbol for this operation. It is perhaps surprising that i_X turns out to be an anti-derivation, just as is d. If η and ω are p and q forms respectively, then

$$i_X(\eta \wedge \omega) = (i_X\eta) \wedge \omega + (-1)^p \eta \wedge (i_X\omega), \tag{11.61}$$

even though i_X involves no differentiation. For example, if $X = X^\mu \partial_\mu$, then

$$\begin{aligned} i_X(dx^\mu \wedge dx^\nu) &= dx^\mu \wedge dx^\nu (X^\alpha \partial_\alpha, \ ), \\ &= X^\mu dx^\nu - dx^\mu X^\nu, \\ &= (i_X dx^\mu) \wedge (dx^\nu) - dx^\mu \wedge (i_X dx^\nu). \end{aligned} \tag{11.62}$$

One reason for introducing i_X is that there is a nice (and profound) formula for the Lie derivative of a p-form in terms of i_X. The formula is called the *infinitesimal homotopy relation*. It reads

$$\mathcal{L}_X \omega = (d\, i_X + i_X d)\omega. \tag{11.63}$$

This formula is proved by verifying that it is true for functions and 1-forms, and then showing that it is a derivation – in other words that it satisfies Leibniz' rule. From the derivation property of the Lie derivative, we immediately deduce that the formula works for any p-form.

That the formula is true for functions should be obvious: since $i_X f = 0$ by definition, we have

$$(d\, i_X + i_X d)f = i_X df = df(X) = Xf = \mathcal{L}_X f. \tag{11.64}$$

To show that the formula works for one forms, we evaluate

$$\begin{aligned} (d\, i_X + i_X d)(f_\nu\, dx^\nu) &= d(f_\nu X^\nu) + i_X(\partial_\mu f_\nu\, dx^\mu dx^\nu) \\ &= \partial_\mu (f_\nu X^\nu) dx^\mu + \partial_\mu f_\nu (X^\mu dx^\nu - X^\nu dx^\mu) \\ &= (X^\nu \partial_\nu f_\mu + f_\nu \partial_\mu X^\nu) dx^\mu. \end{aligned} \tag{11.65}$$

In going from the second to the third line, we have interchanged the dummy labels $\mu \leftrightarrow \nu$ in the term containing dx^ν. We recognize that the 1-form in the last line is indeed $\mathcal{L}_X f$.

To show that $di_X + i_X d$ is a derivation we must apply $d\, i_X + i_X d$ to $a \wedge b$ and use the anti-derivation property of i_x and d. This is straightforward once we recall that d takes a p-form to a $(p+1)$-form while i_X takes a p-form to a $(p-1)$-form.

Exercise 11.7: Let

$$\omega = \frac{1}{p!}\omega_{i_1 \ldots i_p}\, dx^{i_1} \cdots dx^{i_p}.$$

Use the anti-derivation property of i_X to show that

$$i_X \omega = \frac{1}{(p-1)!}\, \omega_{\alpha i_2 \ldots i_p} X^\alpha dx^{i_2} \cdots dx^{i_p},$$

and so verify the equivalence of (11.59) and (11.60).

Exercise 11.8: Use the infinitesimal homotopy relation to show that $\mathcal{L}$ and d commute, i.e. for ω a p-form, we have

$$d\left(\mathcal{L}_X\omega\right) = \mathcal{L}_X(d\omega).$$

11.4 Physical applications

11.4.1 Maxwell's equations

In relativistic[3] four-dimensional tensor notation the two source-free Maxwell's equations

$$\begin{aligned} \operatorname{curl}\mathbf{E} &= -\frac{\partial \mathbf{B}}{\partial t}, \\ \operatorname{div}\mathbf{B} &= 0, \end{aligned} \tag{11.66}$$

reduce to the single equation

$$\frac{\partial F_{\mu\nu}}{\partial x^\lambda} + \frac{\partial F_{\nu\lambda}}{\partial x^\mu} + \frac{\partial F_{\lambda\mu}}{\partial x^\nu} = 0, \tag{11.67}$$

where

$$F_{\mu\nu} = \begin{pmatrix} 0 & -E_x & -E_y & -E_z \\ E_x & 0 & B_z & -B_y \\ E_y & -B_z & 0 & B_x \\ E_z & B_y & -B_x & 0 \end{pmatrix}. \tag{11.68}$$

The "F" is traditional, for Michael Faraday. In form language, the relativistic equation becomes the even more compact expression $dF = 0$, where

$$\begin{aligned} F &\equiv \frac{1}{2} F_{\mu\nu} dx^\mu dx^\nu \\ &= B_x dydz + B_y dzdx + B_z dxdy + E_x dxdt + E_y dydt + E_z dzdt, \end{aligned} \tag{11.69}$$

is a Minkowski-space 2-form.

Exercise 11.9: Verify that the source-free Maxwell equations are indeed equivalent to $dF = 0$.

The equation $dF = 0$ is automatically satisfied if we introduce a 4-vector 1-form potential $A = -\phi dt + A_x dx + A_y dy + A_z dz$ and set $F = dA$.

[3] In this section we will use units in which $c = \epsilon_0 = \mu_0 = 1$. We take the Minkowski metric to be $g_{\mu\nu} = \operatorname{diag}(-1, 1, 1, 1)$ where $x^0 = t$, $x^1 = x$, etc.

The two Maxwell equations with sources

$$\begin{aligned}\operatorname{div}\mathbf{D} &= \rho,\\ \operatorname{curl}\mathbf{H} &= \mathbf{j} + \frac{\partial \mathbf{D}}{\partial t},\end{aligned} \tag{11.70}$$

reduce in 4-tensor notation to the single equation

$$\partial_\mu F^{\mu\nu} = J^\nu. \tag{11.71}$$

Here $J^\mu = (\rho, \mathbf{j})$ is the current 4-vector.

This source equation takes a little more work to express in form language, but it can be done. We need a new concept: the *Hodge "star" dual* of a form. In d dimensions the "$\star$" map takes a p-form to a $(d-p)$-form. It depends on both the metric and the *orientation*. The latter means a canonical choice of the order in which to write our basis forms, with orderings that differ by an even permutation being counted as the same. The full d-dimensional definition involves the Levi-Civita duality operation of Chapter 10, combined with the use of the metric tensor to raise indices. Recall that $\sqrt{g} = \sqrt{\det g_{\mu\nu}}$. (In Minkowski-signature metrics we should replace $\sqrt{g}$ by $\sqrt{-g}$.) We define "$\star$" to be a linear map

$$\star : \bigwedge^p (T^*M) \to \bigwedge^{(d-p)} (T^*M) \tag{11.72}$$

such that

$$\star\, dx^{i_1} \dots dx^{i_p} \stackrel{\text{def}}{=} \frac{1}{(d-p)!} \sqrt{g} g^{i_1 j_1} \dots g^{i_p j_p} \epsilon_{j_1 \cdots j_p j_{p+1} \cdots j_d} dx^{j_{p+1}} \dots dx^{j_d}. \tag{11.73}$$

Although this definition looks a trifle involved, computations involving it are not so intimidating. The trick is to work, whenever possible, with oriented orthonormal frames. If we are in Euclidean space and $\{\mathbf{e}^{*i_1}, \mathbf{e}^{*i_2}, \dots, \mathbf{e}^{*i_d}\}$ is an ordering of the orthonormal basis for $(T^*M)_x$ whose orientation is equivalent to $\{\mathbf{e}^{*1}, \mathbf{e}^{*2}, \dots, \mathbf{e}^{*d}\}$ then

$$\star\,(\mathbf{e}^{*i_1} \wedge \mathbf{e}^{*i_2} \wedge \dots \wedge \mathbf{e}^{*i_p}) = \mathbf{e}^{*i_{p+1}} \wedge \mathbf{e}^{*i_{p+2}} \wedge \dots \wedge \mathbf{e}^{*i_d}. \tag{11.74}$$

For example, in three dimensions, and with x, y, z our usual cartesian coordinates, we have

$$\begin{aligned}\star\, dx &= dydz,\\ \star\, dy &= dzdx,\\ \star\, dz &= dxdy.\end{aligned} \tag{11.75}$$

An analogous method works for Minkowski-signature $(-,+,+,+)$ metrics, except that now we must include a minus sign for each negatively normed dt factor in the form being "starred". Taking $\{dt, dx, dy, dz\}$ as our oriented basis, we therefore find[4]

$$\begin{aligned} \star\, dxdy &= -dzdt, \\ \star\, dydz &= -dxdt, \\ \star\, dzdx &= -dydt, \\ \star\, dxdt &= dydz, \\ \star\, dydt &= dzdx, \\ \star\, dzdt &= dxdy. \end{aligned} \tag{11.76}$$

For example, the first of these equations is derived by observing that $(dxdy)(-dzdt) = dtdxdydz$, and that there is no "dt" in the product $dxdy$. The fourth follows from observing that $(dxdt)(-dydx) = dtdxdydz$, but there is a negative-normed "dt" in the product $dxdt$.

The $\star$ map is constructed so that if

$$\alpha = \frac{1}{p!}\alpha_{i_1 i_2 \ldots i_p} dx^{i_1} dx^{i_2} \cdots dx^{i_p}, \tag{11.77}$$

and

$$\beta = \frac{1}{p!}\beta_{i_1 i_2 \ldots i_p} dx^{i_1} dx^{i_2} \cdots dx^{i_p}, \tag{11.78}$$

then

$$\alpha \wedge (\star\beta) = \beta \wedge (\star\alpha) = \langle \alpha, \beta \rangle\, \sigma, \tag{11.79}$$

where the inner product $\langle \alpha, \beta \rangle$ is defined to be the invariant

$$\langle \alpha, \beta \rangle = \frac{1}{p!} g^{i_1 j_1} g^{i_2 j_2} \cdots g^{i_p j_p} \alpha_{i_1 i_2 \ldots i_p} \beta_{j_1 j_2 \ldots j_p}, \tag{11.80}$$

and σ is the *volume form*

$$\sigma = \sqrt{g}\, dx^1 dx^2 \cdots dx^d. \tag{11.81}$$

In future we will write $\alpha \star \beta$ for $\alpha \wedge (\star\beta)$. Bear in mind that the "$\star$" in this expression is acting on β and is not some new kind of binary operation.

We now apply these ideas to Maxwell. From the field-strength 2-form

$$F = B_x dydz + B_y dzdx + B_z dxdy + E_x dxdt + E_y dydt + E_z dzdt, \tag{11.82}$$

[4] See for example: C. W. Misner, K. S. Thorn and J. A. Wheeler, *Gravitation* (MTW) p. 108.

we get a dual 2-form

$$\star F = -B_x dxdt - B_y dydt - B_z dzdt + E_x dydz + E_y dzdx + E_z dxdy. \tag{11.83}$$

We can check that we have correctly computed the Hodge star of F by taking the wedge product, for which we find

$$F \star F = \frac{1}{2}(F_{\mu\nu}F^{\mu\nu})\sigma = (B_x^2 + B_y^2 + B_z^2 - E_x^2 - E_y^2 - E_z^2)dtdxdydz. \tag{11.84}$$

Observe that the expression $B^2 - E^2$ is a Lorentz scalar. Similarly, from the current 1-form

$$J \equiv J_\mu dx^\mu = -\rho\, dt + j_x dx + j_y dy + j_z dz, \tag{11.85}$$

we derive the dual current 3-form

$$\star J = \rho\, dxdydz - j_x dtdydz - j_y dtdzdx - j_z dtdxdy, \tag{11.86}$$

and check that

$$J \star J = (J_\mu J^\mu)\sigma = (-\rho^2 + j_x^2 + j_y^2 + j_z^2)dtdxdydz. \tag{11.87}$$

Observe that

$$d \star J = \left(\frac{\partial \rho}{\partial t} + \operatorname{div} \mathbf{j}\right) dtdxdydz = 0 \tag{11.88}$$

expresses the charge conservation law.

Writing out the terms explicitly shows that the source-containing Maxwell equations reduce to $d \star F = \star J$. All four Maxwell equations are therefore very compactly expressed as

$$\boxed{dF = 0, \quad d \star F = \star J.}$$

Observe that current conservation $d \star J = 0$ follows from the second Maxwell equation as a consequence of $d^2 = 0$.

Exercise 11.10: Show that for a p-form ω in d Euclidean dimensions we have

$$\star\star\, \omega = (-1)^{p(d-p)}\omega.$$

Show, further, that for a Minkowski metric an additional minus sign has to be inserted. (For example, $\star\star F = -F$, even though $(-1)^{2(4-2)} = +1$.)

11.4.2 Hamilton's equations

Hamiltonian dynamics take place in *phase space*, a manifold with coordinates $(q^1, \ldots, q^n, p^1, \ldots, p^n)$. Since momentum is a naturally covariant vector,[5] phase space is usually the *cotangent bundle* T^*M of the configuration manifold M. We are writing the indices on the p's upstairs though, because we are considering them as coordinates in T^*M.

We expect that you are familiar with Hamilton's equations in their q, p setting. Here, we shall describe them as they appear in a modern book on Mechanics, such as Abrahams and Marsden's *Foundations of Mechanics*, or V. I. Arnold's *Mathematical Methods of Classical Mechanics*.

Phase space is an example of a *symplectic manifold*, a manifold equipped with a *symplectic form* – a closed, non-degenerate, 2-form field

$$\omega = \frac{1}{2}\omega_{ij} dx^i dx^j. \tag{11.89}$$

Recall that the word *closed* means that $d\omega = 0$. *Non-degenerate* means that for any point x the statement that $\omega(X, Y) = 0$ for all vectors $Y \in TM_x$ implies that $X = 0$ at that point (or equivalently that for all x the matrix $\omega_{ij}(x)$ has an inverse $\omega^{ij}(x)$).

Given a *Hamiltonian* function H on our symplectic manifold, we define a velocity vector-field v_H by solving

$$dH = -i_{v_H}\omega = -\omega(v_H,) \tag{11.90}$$

for v_H. If the symplectic form is $\omega = dp^1 dq^1 + dp^2 dq^2 + \cdots + dp^n dq^n$, this is nothing but a fancy form of Hamilton's equations. To see this, we write

$$dH = \frac{\partial H}{\partial q^i} dq^i + \frac{\partial H}{\partial p^i} dp^i \tag{11.91}$$

and use the customary notation $(\dot{q}^i, \dot{p}^i)$ for the velocity-in-phase-space components, so that

$$v_H = \dot{q}^i \frac{\partial}{\partial q^i} + \dot{p}^i \frac{\partial}{\partial p^i}. \tag{11.92}$$

Now we work out

$$\begin{aligned} i_{v_H}\omega &= dp^i dq^i (\dot{q}^j \partial_{q^j} + \dot{p}^j \partial_{p^j},) \\ &= \dot{p}^i dq^i - \dot{q}^i dp^i, \end{aligned} \tag{11.93}$$

[5] To convince yourself of this, remember that in quantum mechanics $\hat{p}_\mu = -i\hbar \frac{\partial}{\partial x^\mu}$, and the gradient of a function is a covector.

so, comparing coefficients of dp^i and dq^i on the two sides of $dH = -i_{v_H}\omega$, we read off

$$\dot{q}^i = \frac{\partial H}{\partial p^i}, \quad \dot{p}^i = -\frac{\partial H}{\partial q^i}. \tag{11.94}$$

Darboux' theorem, which we will not try to prove, says that for any point x we can always find coordinates p, q, valid in some neighbourhood of x, such that $\omega = dp^1 dq^1 + dp^2 dq^2 + \cdots dp^n dq^n$. Given this fact, it is not unreasonable to think that there is little to be gained by using the abstract differential-form language. In simple cases this is so, and the traditional methods are fine. It may be, however, that the neighbourhood of x where the Darboux coordinates work is not the entire phase space, and we need to cover the space with overlapping p, q coordinate charts. Then, what is a p in one chart will usually be a combination of p's and q's in another. In this case, the traditional form of Hamilton's equations loses its appeal in comparison to the coordinate-free $dH = -i_{v_H}\omega$.

Given two functions H_1, H_2 we can define their *Poisson bracket* $\{H_1, H_2\}$. Its importance lies in Dirac's observation that the passage from classical mechanics to quantum mechanics is accomplished by replacing the Poisson bracket of two quantities, A and B, with the commutator of the corresponding operators $\hat{A}$, and $\hat{B}$:

$$i[\hat{A}, \hat{B}] \quad \longleftrightarrow \quad \hbar\{A, B\} + O\left(\hbar^2\right). \tag{11.95}$$

We define the Poisson bracket by[6]

$$\{H_1, H_2\} \stackrel{\text{def}}{=} \left.\frac{dH_2}{dt}\right|_{H_1} = v_{H_1} H_2. \tag{11.96}$$

Now, $v_{H_1} H_2 = dH_2(v_{H_1})$, and Hamilton's equations say that $dH_2(v_{H_1}) = \omega(v_{H_1}, v_{H_2})$. Thus,

$$\{H_1, H_2\} = \omega(v_{H_1}, v_{H_2}). \tag{11.97}$$

The skew symmetry of $\omega(v_{H_1}, v_{H_2})$ shows that despite the asymmetrical appearance of the definition we have skew symmetry: $\{H_1, H_2\} = -\{H_2, H_1\}$.

Moreover, since

$$v_{H_1}(H_2 H_3) = (v_{H_1} H_2) H_3 + H_2 (v_{H_1} H_3), \tag{11.98}$$

the Poisson bracket is a derivation:

$$\{H_1, H_2 H_3\} = \{H_1, H_2\} H_3 + H_2 \{H_1, H_3\}. \tag{11.99}$$

[6] Our definition differs in sign from the traditional one, but has the advantage of minimizing the number of minus signs in subsequent equations.

Neither the skew symmetry nor the derivation property require the condition that $d\omega = 0$. What does need ω to be closed is the *Jacobi identity*:

$$\{\{H_1, H_2\}, H_3\} + \{\{H_2, H_3\}, H_1\} + \{\{H_3, H_1\}, H_2\} = 0. \tag{11.100}$$

We establish Jacobi by using Cartan's formula in the form

$$\begin{aligned} d\omega(v_{H_1}, v_{H_2}, v_{H_3}) &= v_{H_1}\omega(v_{H_2}, v_{H_3}) + v_{H_2}\omega(v_{H_3}, v_{H_1}) + v_{H_3}\omega(v_{H_1}, v_{H_2}) \\ &\quad - \omega([v_{H_1}, v_{H_2}], v_{H_3}) - \omega([v_{H_2}, v_{H_3}], v_{H_1}) \\ &\quad - \omega([v_{H_3}, v_{H_1}], v_{H_2}). \end{aligned} \tag{11.101}$$

It is relatively straightforward to interpret each term in the first of these lines as Poisson brackets. For example,

$$v_{H_1}\omega(v_{H_2}, v_{H_3}) = v_{H_1}\{H_2, H_3\} = \{H_1, \{H_2, H_3\}\}. \tag{11.102}$$

Relating the terms in the second line to Poisson brackets requires a little more effort. We proceed as follows:

$$\begin{aligned} \omega([v_{H_1}, v_{H_2}], v_{H_3}) &= -\omega(v_{H_3}, [v_{H_1}, v_{H_2}]) \\ &= dH_3([v_{H_1}, v_{H_2}]) \\ &= [v_{H_1}, v_{H_2}]H_3 \\ &= v_{H_1}(v_{H_2}H_3) - v_{H_2}(v_{H_1}H_3) \\ &= \{H_1, \{H_2, H_3\}\} - \{H_2, \{H_1, H_3\}\} \\ &= \{H_1, \{H_2, H_3\}\} + \{H_2, \{H_3, H_1\}\}. \end{aligned} \tag{11.103}$$

Adding everything together now shows that

$$\begin{aligned} 0 &= d\omega(v_{H_1}, v_{H_2}, v_{H_3}) \\ &= -\{\{H_1, H_2\}, H_3\} - \{\{H_2, H_3\}, H_1\} - \{\{H_3, H_1\}, H_2\}. \end{aligned} \tag{11.104}$$

If we rearrange the Jacobi identity as

$$\{H_1, \{H_2, H_3\}\} - \{H_2, \{H_1, H_3\}\} = \{\{H_1, H_2\}, H_3\}, \tag{11.105}$$

we see that it is equivalent to

$$[v_{H_1}, v_{H_2}] = v_{\{H_1, H_2\}}.$$

The algebra of Poisson brackets is therefore *homomorphic* to the algebra of the Lie brackets. The correspondence is not an *isomorphism*, however: the assignment $H \mapsto v_H$ fails to be one-to-one because constant functions map to the zero vector field.

Exercise 11.11: Use the infinitesimal homotopy relation, to show that $\mathcal{L}_{v_H}\omega = 0$, where v_H is the vector field corresponding to H. Suppose now that the phase space is $2n$ dimensional. Show that in local Darboux coordinates the $2n$-form $\omega^n/n!$ is, up to a sign, the phase-space volume element $d^n p\, d^n q$. Show that $\mathcal{L}_{v_H}\omega^n/n! = 0$ and that this result is *Liouville's theorem* on the conservation of phase-space volume.

The classical mechanics of spin

It is sometimes said in books on quantum mechanics that the spin of an electron, or other elementary particle, is a purely quantum concept and cannot be described by classical mechanics. This statement is false, but spin *is* the simplest system in which traditional physicists' methods become ugly and it helps to use the modern symplectic language. A "spin" $\mathbf{S}$ can be regarded as a fixed length vector that can point in any direction in $\mathbb{R}^3$. We will take it to be of unit length so that its components are

$$
\begin{aligned}
S_x &= \sin\theta\cos\phi, \\
S_y &= \sin\theta\sin\phi, \\
S_z &= \cos\theta,
\end{aligned} \tag{11.106}
$$

where θ and ϕ are polar coordinates on the 2-sphere S^2.

The surface of the sphere turns out to be both the configuration space and the phase space. In particular the phase space for a spin is *not* the cotangent bundle of the configuration space. This has to be so: we learned from Niels Bohr that a $2n$-dimensional phase space contains roughly one quantum state for every $\hbar^n$ of phase-space volume. A cotangent bundle always has infinite volume, so its corresponding Hilbert space is necessarily infinite dimensional. A quantum spin, however, has a *finite-dimensional* Hilbert space so its classical phase space must have a finite total volume. This finite-volume phase space seems unnatural in the traditional view of mechanics, but it fits comfortably into the modern symplectic picture.

We want to treat all points on the sphere alike, and so it is natural to take the symplectic 2-form to be proportional to the element of area. Suppose that $\omega = \sin\theta\, d\theta d\phi$. We could write $\omega = -d\cos\theta\, d\phi$ and regard ϕ as "q" and $-\cos\theta$ as "p" (Darboux' theorem in action!), but this identification is singular at the north and south poles of the sphere, and, besides, it obscures the spherical symmetry of the problem, which is manifest when we think of ω as d(area).

Let us take our Hamiltonian to be $H = BS_x$, corresponding to an applied magnetic field in the x-direction, and see what Hamilton's equations give for the motion. First we take the exterior derivative

$$
d(BS_x) = B(\cos\theta\cos\phi d\theta - \sin\theta\sin\phi d\phi). \tag{11.107}
$$

This is to be set equal to

$$
-\omega(v_{BS_x},\ ) = v^\theta(-\sin\theta)d\phi + v^\phi\sin\theta d\theta. \tag{11.108}
$$

Comparing coefficients of $d\theta$ and $d\phi$, we get

$$v_{(BS_x)} = v^\theta \partial_\theta + v^\phi \partial_\phi = B(\sin\phi\partial_\theta + \cos\phi\cot\theta\partial_\phi), \tag{11.109}$$

i.e. B times the velocity vector for a rotation about the x-axis. This velocity field therefore describes a steady Larmor precession of the spin about the applied field. This is exactly the motion predicted by quantum mechanics. Similarly, setting $B = 1$, we find

$$\begin{aligned} v_{S_y} &= -\cos\phi\partial_\theta + \sin\phi\cot\theta\partial_\phi, \\ v_{S_z} &= -\partial_\phi. \end{aligned} \tag{11.110}$$

From these velocity fields we can compute the Poisson brackets:

$$\begin{aligned} \{S_x, S_y\} &= \omega(v_{S_x}, v_{S_y}) \\ &= \sin\theta d\theta d\phi(\sin\phi\partial_\theta + \cos\phi\cot\theta\partial_\phi, -\cos\phi\partial_\theta + \sin\phi\cot\theta\partial_\phi) \\ &= \sin\theta(\sin^2\phi\cot\theta + \cos^2\phi\cot\theta) \\ &= \cos\theta = S_z. \end{aligned}$$

Repeating the exercise leads to

$$\begin{aligned} \{S_x, S_y\} &= S_z, \\ \{S_y, S_z\} &= S_x, \\ \{S_z, S_x\} &= S_y. \end{aligned} \tag{11.111}$$

These Poisson brackets for our classical "spin" are to be compared with the commutation relations $[\hat{S}_x, \hat{S}_y] = i\hbar\hat{S}_z$ etc. for the quantum spin operators $\hat{S}_i$.

11.5 Covariant derivatives

Covariant derivatives are a general class of derivatives that act on sections of a vector or tensor bundle over a manifold. We will begin by considering derivatives on the tangent bundle, and in the exercises indicate how the idea generalizes to other bundles.

11.5.1 Connections

The Lie and exterior derivatives require no structure beyond that which comes for free with our manifold. Another type of derivative that can act on tangent-space vectors and tensors is the *covariant derivative* $\nabla_X \equiv X^\mu \nabla_\mu$. This requires an additional mathematical object called an *affine connection*.

The covariant derivative is defined by:

(i) Its action on scalar functions as

$$\nabla_X f = Xf. \tag{11.112}$$

(ii) Its action on a basis set of tangent-vector fields $\mathbf{e}_a(x) = e_a^{\mu}(x)\partial_\mu$ (a local frame, or *vielbein*[7]) by introducing a set of functions $\omega^i{}_{jk}(x)$ and setting

$$\nabla_{\mathbf{e}_k}\mathbf{e}_j = \mathbf{e}_i\omega^i{}_{jk}. \tag{11.113}$$

(iii) Extending this definition to any other type of tensor by requiring ∇_X to be a derivation.

(iv) Requiring that the result of applying ∇_X to a tensor is a tensor of the same type.

The set of functions $\omega^i{}_{jk}(x)$ is the *connection*. In any local coordinate chart we can choose them at will, and different choices define different covariant derivatives. (There may be global compatibility constraints, however, which appear when we assemble the charts into an atlas.)

Warning: Despite having the appearance of one, $\omega^i{}_{jk}$ is *not* a tensor. It transforms inhomogeneously under a change of frame or coordinates – see Equation (11.132).

We can, of course, take as our basis vectors the coordinate vectors $\mathbf{e}_\mu \equiv \partial_\mu$. When we do this it is traditional to use the symbol Γ for the coordinate frame connection instead of ω. Thus,

$$\nabla_\mu\mathbf{e}_\nu \equiv \nabla_{\mathbf{e}_\mu}\mathbf{e}_\nu = \mathbf{e}_\lambda\Gamma^\lambda{}_{\nu\mu}. \tag{11.114}$$

The numbers $\Gamma^\lambda{}_{\nu\mu}$ are often called *Christoffel symbols*.

As an example consider the covariant derivative of a vector $f^\nu\mathbf{e}_\nu$. Using the derivation property we have

$$\begin{aligned}\nabla_\mu(f^\nu\mathbf{e}_\nu) &= (\partial_\mu f^\nu)\mathbf{e}_\nu + f^\nu\nabla_\mu\mathbf{e}_\nu\\ &= (\partial_\mu f^\nu)\mathbf{e}_\nu + f^\nu\mathbf{e}_\lambda\Gamma^\lambda{}_{\nu\mu}\\ &= \mathbf{e}_\nu\left\{\partial_\mu f^\nu + f^\lambda\Gamma^\nu{}_{\lambda\mu}\right\}.\end{aligned} \tag{11.115}$$

In the first line we have used the defining property that $\nabla_{\mathbf{e}_\mu}$ acts on the functions f^ν as ∂_μ, and in the last line we interchanged the dummy indices ν and λ. We often abuse the notation by writing only the components, and set

$$\nabla_\mu f^\nu = \partial_\mu f^\nu + f^\lambda\Gamma^\nu{}_{\lambda\mu}. \tag{11.116}$$

Similarly, acting on the components of a mixed tensor, we would write

$$\nabla_\mu A^\alpha{}_{\beta\gamma} = \partial_\mu A^\alpha{}_{\beta\gamma} + \Gamma^\alpha{}_{\lambda\mu}A^\lambda{}_{\beta\gamma} - \Gamma^\lambda{}_{\beta\mu}A^\alpha{}_{\lambda\gamma} - \Gamma^\lambda{}_{\gamma\mu}A^\alpha{}_{\beta\lambda}. \tag{11.117}$$

When we use this notation, we are no longer regarding the tensor components as "functions".

[7] In practice *viel*, "many", is replaced by the appropriate German numeral: *ein-*, *zwei-*, *drei-*, *vier-*, *fünf-*, ..., indicating the dimension. The word *bein* means "leg".

Observe that the plus and minus signs in (11.117) are required so that, for example, the covariant derivative of the scalar function $f_\alpha g^\alpha$ is

$$\begin{aligned}\nabla_\mu \left(f_\alpha g^\alpha\right) &= \partial_\mu \left(f_\alpha g^\alpha\right)\\ &= \left(\partial_\mu f_\alpha\right) g^\alpha + f_\alpha \left(\partial_\mu g^\alpha\right)\\ &= \left(\partial_\mu f_\alpha - f_\lambda \Gamma^\lambda{}_{\alpha\mu}\right) g^\alpha + f_\alpha \left(\partial_\mu g^\alpha + g^\lambda \Gamma^\alpha{}_{\lambda\mu}\right)\\ &= \left(\nabla_\mu f_\alpha\right) g^\alpha + f_\alpha \left(\nabla_\mu g^\alpha\right),\end{aligned} \tag{11.118}$$

and so satisfies the derivation property.

Parallel transport

We have defined the covariant derivative *via* its formal calculus properties. It has, however, a geometrical interpretation. As with the Lie derivative, in order to compute the derivative along X of the vector field Y, we have to somehow carry the vector $Y(x)$ from the tangent space TM_x to the tangent space $TM_{x+\epsilon X}$, where we can subtract it from $Y(x+\epsilon X)$. The Lie derivative carries Y along with the X flow. The covariant derivative implicitly carries Y by "parallel transport". If $\gamma : s \mapsto x^\mu(s)$ is a parametrized curve with tangent vector $X^\mu \partial_\mu$, where

$$X^\mu = \frac{dx^\mu}{ds}, \tag{11.119}$$

then we say that the vector field $Y(x^\mu(s))$ is *parallel transported* along the curve γ if

$$\nabla_X Y = 0, \tag{11.120}$$

at each point $x^\mu(s)$. Thus, a vector that in the vielbein frame $\mathbf{e}_i$ at x has components Y^i will, after being parallel transported to $x + \epsilon X$, end up with components

$$Y^i - \epsilon \omega^i{}_{jk} Y^j X^k. \tag{11.121}$$

In a coordinate frame, after parallel transport through an infinitesimal displacement δx^μ, the vector $Y^\nu \partial_\nu$ will have components

$$Y^\nu \to Y^\nu - \Gamma^\nu{}_{\lambda\mu} Y^\lambda \delta x^\mu, \tag{11.122}$$

and so

$$\begin{aligned}\delta x^\mu \nabla_\mu Y^\nu &= Y^\nu(x^\mu + \delta x^\mu) - \{Y^\nu(x) - \Gamma^\nu{}_{\lambda\mu} Y^\lambda \delta x^\mu\}\\ &= \delta x^\mu \{\partial_\mu Y^\nu + \Gamma^\nu{}_{\lambda\mu} Y^\lambda\}.\end{aligned} \tag{11.123}$$

Curvature and torsion

As we said earlier, the connection $\omega^i{}_{jk}(x)$ is not itself a tensor. Two important quantities which *are* tensors, are associated with ∇_X:

(i) The *torsion*

$$T(X,Y) = \nabla_X Y - \nabla_Y X - [X,Y]. \tag{11.124}$$

The quantity $T(X,Y)$ is a vector depending linearly on X, Y, so T at the point x is a map $TM_x \times TM_x \to TM_x$, and so a tensor of type (1,2). In a coordinate frame it has components

$$T^\lambda{}_{\mu\nu} = \Gamma^\lambda{}_{\mu\nu} - \Gamma^\lambda{}_{\nu\mu}. \tag{11.125}$$

(ii) The *Riemann curvature tensor*

$$R(X,Y)Z = \nabla_X \nabla_Y Z - \nabla_Y \nabla_Z Z - \nabla_{[X,Y]} Z. \tag{11.126}$$

The quantity $R(X,Y)Z$ is also a vector, so $R(X,Y)$ is a linear map $TM_x \to TM_x$, and thus R itself is a tensor of type (1,3). Written out in a coordinate frame, we have

$$R^\alpha{}_{\beta\mu\nu} = \partial_\mu \Gamma^\alpha{}_{\beta\nu} - \partial_\nu \Gamma^\alpha{}_{\beta\mu} + \Gamma^\alpha{}_{\lambda\mu}\Gamma^\lambda{}_{\beta\nu} - \Gamma^\alpha{}_{\lambda\nu}\Gamma^\lambda{}_{\beta\mu}. \tag{11.127}$$

If our manifold comes equipped with a metric tensor $g_{\mu\nu}$ (and is thus a *Riemann manifold*), and if we require both that $T = 0$ and $\nabla_\mu g_{\alpha\beta} = 0$, then the connection is uniquely determined, and is called the *Riemann*, or *Levi-Civita*, connection. In a coordinate frame it is given by

$$\Gamma^\alpha{}_{\mu\nu} = \frac{1}{2} g^{\alpha\lambda}\left(\partial_\mu g_{\lambda\nu} + \partial_\nu g_{\mu\lambda} - \partial_\lambda g_{\mu\nu}\right). \tag{11.128}$$

This is the connection that appears in general relativity.

The curvature tensor measures the degree of path dependence in parallel transport: if $Y^\nu(x)$ is parallel transported along a path $\gamma : s \mapsto x^\mu(s)$ from a to b, and if we deform γ so that $x^\mu(s) \to x^\mu(s) + \delta x^\mu(s)$ while keeping the endpoints a, b fixed, then

$$\delta Y^\alpha(b) = -\int_a^b R^\alpha{}_{\beta\mu\nu}(x) Y^\beta(x) \delta x^\mu \, dx^\nu. \tag{11.129}$$

If $R^\alpha{}_{\beta\mu\nu} \equiv 0$ then the effect of parallel transport from a to b will be independent of the route taken.

The geometric interpretation of $T_{\mu\nu}$ is less transparent. On a two-dimensional surface a connection is torsion free when the tangent space "rolls without slipping" along the curve γ.

Exercise 11.12: *Metric compatibility*. Show that the Riemann connection

$$\Gamma^{\alpha}{}_{\mu\nu} = \frac{1}{2} g^{\alpha\lambda} \left(\partial_\mu g_{\lambda\nu} + \partial_\nu g_{\mu\lambda} - \partial_\lambda g_{\mu\nu} \right)$$

follows from the torsion-free condition $\Gamma^{\alpha}{}_{\mu\nu} = \Gamma^{\alpha}{}_{\nu\mu}$ together with the *metric compatibility condition*

$$\nabla_\mu g_{\alpha\beta} \equiv \partial_\mu g_{\alpha\beta} - \Gamma^{\nu}{}_{\alpha\mu} g_{\nu\beta} - \Gamma^{\nu}{}_{\alpha\mu} g_{\alpha\nu} = 0.$$

Show that "metric compatibility" means that the operation of raising or lowering indices commutes with covariant derivation.

Exercise 11.13: *Geodesic equation*. Let $\gamma : s \mapsto x^\mu(s)$ be a parametrized path from a to b. Show that the Euler–Lagrange equation that follows from minimizing the distance functional

$$S(\gamma) = \int_a^b \sqrt{g_{\mu\nu} \dot{x}^\mu \dot{x}^\nu}\, ds,$$

where the dots denote differentiation with respect to the parameter s, is

$$\frac{d^2 x^\mu}{ds^2} + \Gamma^{\mu}{}_{\alpha\beta} \frac{dx^\alpha}{ds} \frac{dx^\beta}{ds} = 0.$$

Here $\Gamma^{\mu}{}_{\alpha\beta}$ is the Riemann connection (11.128).

Exercise 11.14: Show that if A^μ is a vector field then, for the Riemann connection,

$$\nabla_\mu A^\mu = \frac{1}{\sqrt{g}} \frac{\partial \sqrt{g} A^\mu}{\partial x^\mu}.$$

In other words, show that

$$\Gamma^{\alpha}{}_{\alpha\mu} = \frac{1}{\sqrt{g}} \frac{\partial \sqrt{g}}{\partial x^\mu}.$$

Deduce that the Laplacian acting on a scalar field ϕ can be defined by setting either

$$\nabla^2 \phi = g_{\mu\nu} \nabla_\mu \nabla_\nu \phi,$$

or

$$\nabla^2 \phi = \frac{1}{\sqrt{g}} \frac{\partial}{\partial x^\mu} \left(\sqrt{g} g^{\mu\nu} \frac{\partial \phi}{\partial x^\nu} \right),$$

the two definitions being equivalent.

11.5.2 Cartan's form viewpoint

Let $\mathbf{e}^{*j}(x) = e^{*j}{}_{\mu}(x)dx^{\mu}$ be the basis of 1-forms dual to the vielbein frame $\mathbf{e}_i(x) = e_i^{\mu}(x)\partial_{\mu}$. Since

$$\delta^i_j = \mathbf{e}^{*i}(\mathbf{e}_j) = e^{*j}{}_{\mu}e_i^{\mu}, \tag{11.130}$$

the matrices $e^{*j}{}_{\mu}$ and e_i^{μ} are inverses of one another. We can use them to change from roman vielbein indices to Greek coordinate frame indices. For example:

$$g_{ij} = g(\mathbf{e}_i, \mathbf{e}_j) = e_i^{\mu} g_{\mu\nu} e_j^{\nu}, \tag{11.131}$$

and

$$\omega^i{}_{jk} = e^{*i}{}_{\nu}(\partial_{\mu} e_j^{\nu})e_k^{\mu} + e^{*i}{}_{\lambda} e_j^{\nu} e_k^{\mu}\, \Gamma^{\lambda}{}_{\nu\mu}. \tag{11.132}$$

Cartan regards the connection as being a matrix Ω of 1-forms with matrix entries $\omega^i{}_j = \omega^i{}_{j\mu}dx^{\mu}$. In this language Equation (11.113) becomes

$$\nabla_X \mathbf{e}_j = \mathbf{e}_i \omega^i{}_j(X). \tag{11.133}$$

Cartan's viewpoint separates off the index μ, which refers to the direction $\delta x^{\mu} \propto X^{\mu}$ in which we are differentiating, from the matrix indices i and j that act on the components of the vector or tensor being differentiated. This separation becomes very natural when the vector space spanned by the $\mathbf{e}_i(x)$ is no longer the tangent space, but some other "internal" vector space attached to the point x. Such internal spaces are common in physics, an important example being the "colour space" of gauge field theories. Physicists, following Hermann Weyl, call a connection on an internal space a "gauge potential". To mathematicians it is simply a connection on the vector bundle that has the internal spaces as its fibres.

Cartan also regards the torsion $\mathbf{T}$ and curvature $\mathbf{R}$ as forms; in this case vector- and matrix-valued 2-forms, respectively, with entries

$$T^i = \frac{1}{2} T^i{}_{\mu\nu} dx^{\mu} dx^{\nu}, \tag{11.134}$$

$$R^i{}_k = \frac{1}{2} R^i{}_{k\mu\nu} dx^{\mu} dx^{\nu}. \tag{11.135}$$

In his form language the equations defining the torsion and curvature become *Cartan's structure equations*:

$$d\mathbf{e}^{*i} + \omega^i{}_j \wedge \mathbf{e}^{*j} = T^i, \tag{11.136}$$

and

$$d\omega^i{}_k + \omega^i{}_j \wedge \omega^j{}_k = R^i{}_k. \tag{11.137}$$

The last equation can be written more compactly as

$$d\Omega + \Omega \wedge \Omega = \mathbf{R}. \tag{11.138}$$

From this, by taking the exterior derivative, we obtain the *Bianchi identity*

$$d\mathbf{R} - \mathbf{R} \wedge \Omega + \Omega \wedge \mathbf{R} = 0. \tag{11.139}$$

On a Riemann manifold, we can take the vielbein frame $\mathbf{e}_i$ to be orthonormal. In this case the roman-index metric $g_{ij} = g(\mathbf{e}_i, \mathbf{e}_j)$ becomes δ_{ij}. There is then no distinction between covariant and contravariant roman indices, and the connection and curvature forms, Ω, $\mathbf{R}$, being infinitesimal rotations, become skew symmetric matrices:

$$\omega_{ij} = -\omega_{ji}, \quad R_{ij} = -R_{ji}. \tag{11.140}$$

11.6 Further exercises and problems

Exercise 11.15: Consider the vector fields $X = y\partial_x$, $Y = \partial_y$ in $\mathbb{R}^2$. Find the flows associated with these fields, and use them to verify the statements made in Section 11.2.1 about the geometric interpretation of the Lie bracket.

Exercise 11.16: Show that the pair of vector fields $L_z = x\partial_y - y\partial_x$ and $L_y = z\partial_x - x\partial_z$ in $\mathbb{R}^3$ is in involution wherever they are both non-zero. Show further that the general solution of the system of partial differential equations

$$\begin{aligned} (x\partial_y - y\partial_x)f &= 0, \\ (x\partial_z - z\partial_x)f &= 0, \end{aligned}$$

in $\mathbb{R}^3$ is $f(x,y,z) = F(x^2 + y^2 + z^2)$, where F is an arbitrary function.

Exercise 11.17: In the rolling conditions (11.29) we are using the "Y" convention for Euler angles. In this convention θ and ϕ are the usual spherical polar coordinate angles with respect to the space-fixed xyz-axes. They specify the direction of the body-fixed Z-axis about which we make the final ψ rotation – see Figure 11.7.

(a) Show that (11.29) are indeed the no-slip rolling conditions

$$\begin{aligned} \dot{x} &= \omega_y, \\ \dot{y} &= -\omega_x, \\ 0 &= \omega_z, \end{aligned}$$

where $(\omega_x, \omega_y, \omega_z)$ are the components of the ball's angular velocity in the xyz space-fixed frame.

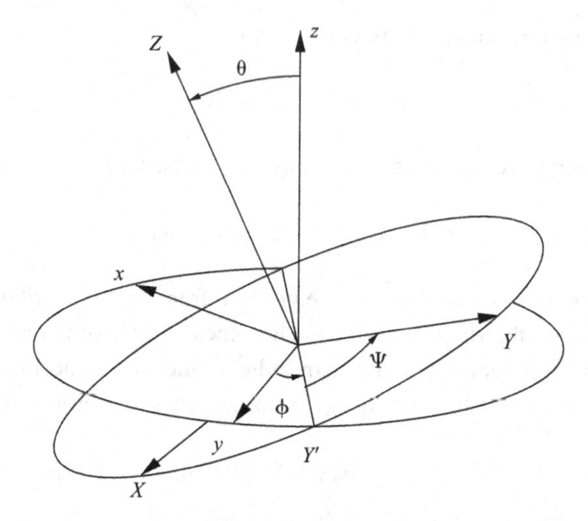

Figure 11.7 The "Y" convention for Euler angles. The XYZ axes are fixed to the ball, and the xyz-axes are fixed in space. We first rotate the ball through an angle ϕ about the z-axis, thus taking $y \to Y'$, then through θ about Y', and finally through ψ about Z, so taking $Y' \to Y$.

(b) Solve the three constraints in (11.29) so as to obtain the vector fields $\mathbf{roll_x}$, $\mathbf{roll_y}$ of (11.30).

(c) Show that

$$[\mathbf{roll_x}, \mathbf{roll_y}] = -\mathbf{spin_z},$$

where $\mathbf{spin_z} \equiv \partial_\phi$, corresponds to a rotation about a vertical axis through the point of contact. This is a new motion, being forbidden by the $\omega_z = 0$ condition.

(d) Show that

$$[\mathbf{spin_z}, \mathbf{roll_x}] = \mathbf{spin_x},$$
$$[\mathbf{spin_z}, \mathbf{roll_y}] = \mathbf{spin_y},$$

where the new vector fields

$$\mathbf{spin_x} \equiv -(\mathbf{roll_y} - \partial_y),$$
$$\mathbf{spin_y} \equiv \ \ (\mathbf{roll_x} - \partial_x),$$

correspond to rotations of the ball about the space-fixed x- and y-axes through its centre, and with the centre of mass held fixed.

We have generated five independent vector fields from the original two. Therefore, by sufficient rolling to-and-fro, we can position the ball anywhere on the table, and in any orientation.

Exercise 11.18: The semiclassical dynamics of charge $-e$ electrons in a magnetic solid are governed by the equations[8]

$$
\begin{aligned}
\dot{\mathbf{r}} &= \frac{\partial \epsilon(\mathbf{k})}{\partial \mathbf{k}} - \dot{\mathbf{k}} \times \Omega, \\
\dot{\mathbf{k}} &= -\frac{\partial V}{\partial \mathbf{r}} - e\dot{\mathbf{r}} \times \mathbf{B}.
\end{aligned}
$$

Here $\mathbf{k}$ is the Bloch momentum of the electron, $\mathbf{r}$ is its position, $\epsilon(\mathbf{k})$ its band energy (in the extended-zone scheme) and $\mathbf{B}(\mathbf{r})$ is the external magnetic field. The components Ω_i of the *Berry curvature* $\Omega(\mathbf{k})$ are given in terms of the periodic part $|u(\mathbf{k})\rangle$ of the Bloch wavefunctions of the band by

$$
\Omega_i = i\epsilon_{ijk}\frac{1}{2}\left(\left\langle \frac{\partial u}{\partial k_j}\middle| \frac{\partial u}{\partial k_k}\right\rangle - \left\langle \frac{\partial u}{\partial k_k}\middle| \frac{\partial u}{\partial k_j}\right\rangle\right).
$$

The only property of $\Omega(\mathbf{k})$ needed for the present problem, however, is that $\mathrm{div}_{\mathbf{k}}\Omega = 0$.

(a) Show that these equations are Hamiltonian, with

$$
H(\mathbf{r}, \mathbf{k}) = \epsilon(\mathbf{k}) + V(\mathbf{r})
$$

and with

$$
\omega = dk_i dx_i - \frac{e}{2}\epsilon_{ijk}B_i(\mathbf{r})dx_j dx_k + \frac{1}{2}\epsilon_{ijk}\Omega_i(\mathbf{k})dk_j dk_k
$$

as the symplectic form.[9]

(b) Confirm that the ω defined in part (b) is closed, and that the Poisson brackets are given by

$$
\begin{aligned}
\{x_i, x_j\} &= -\frac{\epsilon_{ijk}\Omega_k}{(1 + e\mathbf{B}\cdot\Omega)}, \\
\{x_i, k_j\} &= -\frac{\delta_{ij} + eB_i\Omega_j}{(1 + e\mathbf{B}\cdot\Omega)}, \\
\{k_i, k_j\} &= \frac{\epsilon_{ijk}eB_k}{(1 + e\mathbf{B}\cdot\Omega)}.
\end{aligned}
$$

(c) Show that the conserved phase-space volume $\omega^3/3!$ is equal to

$$
(1 + e\mathbf{B}\cdot\Omega)d^3k d^3x,
$$

instead of the naïvely expected $d^3k d^3x$.

[8] M. C. Chang, Q. Niu, *Phys. Rev. Lett.*, **75** (1995) 1348.

[9] C. Duval, Z. Horváth, P. A. Horváthy, L. Martina, P. C. Stichel, *Mod. Phys. Lett., B* **20** (2006) 373.

The following two exercises show that Cartan's expression for the curvature tensor remains valid for covariant differentiation in "internal" spaces. There is, however, no natural concept analogous to the torsion tensor for internal spaces.

Exercise 11.19: *Non-abelian gauge fields as matrix-valued forms*. In a non-abelian Yang–Mills gauge theory, such as QCD, the vector potential

$$A = A_\mu dx^\mu$$

is matrix-valued, meaning that the components A_μ are matrices which do not necessarily commute with each other. (These matrices are elements of the Lie algebra of the gauge group, but we won't need this fact here.) The matrix-valued curvature, or field-strength, 2-form F is defined by

$$F = dA + A^2 = \frac{1}{2} F_{\mu\nu} dx^\mu dx^\nu.$$

Here a combined matrix and wedge product is to be understood:

$$(A^2)^a{}_b \equiv A^a{}_c \wedge A^c{}_b = A^a{}_{c\mu} A^c{}_{b\nu}\, dx^\mu dx^\nu.$$

(i) Show that $A^2 = \frac{1}{2}[A_\mu, A_\nu] dx^\mu dx^\nu$, and hence show that

$$F_{\mu\nu} = \partial_\mu A_\nu - \partial_\nu A_\mu + [A_\mu, A_\nu].$$

(ii) Define the *gauge-covariant derivatives*

$$\nabla_\mu = \partial_\mu + A_\mu,$$

and show that the commutator $[\nabla_\mu, \nabla_\nu]$ of two of these is equal to $F_{\mu\nu}$. Show further that if X, Y are two vector fields with Lie bracket $[X, Y]$ and $\nabla_X \equiv X^\mu \nabla_\mu$, then

$$F(X, Y) = [\nabla_X, \nabla_Y] - \nabla_{[X,Y]}.$$

(iii) Show that F obeys the Bianchi identity

$$dF - FA + AF = 0.$$

Again wedge and matrix products are to be understood. This equation is the non-abelian version of the source-free Maxwell equation $dF = 0$.

(iv) Show that, in any number of dimensions, the Bianchi identity implies that the 4-form $\operatorname{tr}(F^2)$ is closed, i.e. that $d\operatorname{tr}(F^2) = 0$. Similarly show that the $2n$-form $\operatorname{tr}(F^n)$ is closed. (Here the "tr" means a trace over the roman matrix indices, and not over the Greek space-time indices.)

(v) Show that,

$$\operatorname{tr}(F^2) = d\left\{\operatorname{tr}\left(AdA + \frac{2}{3}A^3\right)\right\}.$$

The 3-form $\operatorname{tr}(AdA + \frac{2}{3}A^3)$ is called a *Chern–Simons* form.

Exercise 11.20: *Gauge transformations*. Here we consider how the matrix-valued vector potential transforms when we make a change of gauge. In other words, we seek the non-abelian version of $A_\mu \to A_\mu + \partial_\mu \phi$.

(i) Let g be an invertible matrix, and δg a matrix describing a small change in g. Show that the corresponding change in the inverse matrix is given by $\delta(g^{-1}) = -g^{-1}(\delta g)g^{-1}$.

(ii) Show that under the *gauge transformation*

$$A \to A^g \equiv g^{-1}Ag + g^{-1}dg,$$

we have $F \to g^{-1}Fg$. (Hint: the labour is minimized by exploiting the covariant derivative identity in part (ii) of the previous exercise.)

(iii) Deduce that $\operatorname{tr}(F^n)$ is *gauge invariant*.

(iv) Show that a necessary condition for the matrix-valued gauge field A to be "pure gauge", i.e. for there to be a position-dependent matrix $g(x)$ such that $A = g^{-1}dg$, is that $F = 0$, where F is the curvature 2-form of the previous exercise. (If we are working in a simply connected region, then $F = 0$ is also a *sufficient* condition for there to be a g such that $A = g^{-1}dg$, but this is a little harder to prove.)

In a gauge theory based on a Lie group G, the matrices g will be elements of the group, or, more generally, they will form a matrix representation of the group.

12

Integration on manifolds

One usually thinks of integration as requiring *measure* – a notion of volume, and hence of size and length, and so a *metric*. A metric, however, is not required for integrating differential forms. They come pre-equipped with whatever notion of length, area or volume is required.

12.1 Basic notions

12.1.1 Line integrals

Consider, for example, the form df. We want to try to give a meaning to the symbol

$$I_1 = \int_\Gamma df. \tag{12.1}$$

Here, Γ is a path in our space starting at some point P_0 and ending at the point P_1. Any reasonable definition of I_1 should end up with the answer that we would immediately write down if we saw an expression like I_1 in an elementary calculus class. This answer is

$$I_1 = \int_\Gamma df = f(P_1) - f(P_0). \tag{12.2}$$

No notion of a metric is needed here. There is, however, a geometric picture of what we have done. We draw in our space the surfaces $\ldots, f(x) = -1, f(x) = 0, f(x) = 1, \ldots$, and perhaps fill in intermediate values if necessary. We then start at P_0 and travel from there to P_1, keeping track of how many of these surfaces we pass through (with sign -1, if we pass back through them). The integral of df is this number. Figure 12.1 illustrates a case in which $\int_\Gamma df = 5.5 - 1.5 = 4$.

What we have defined is a *signed integral*. If we parametrize the path as $x(s)$, $0 \le s \le 1$, and with $x(0) = P_0$, $x(1) = P_1$ we have

$$I_1 = \int_0^1 \left(\frac{df}{ds}\right) ds \tag{12.3}$$

where the right-hand side is an ordinary one-variable integral. It is important that we did not write $\left|\frac{df}{ds}\right|$ in this integral. The absence of the modulus sign ensures that if we partially

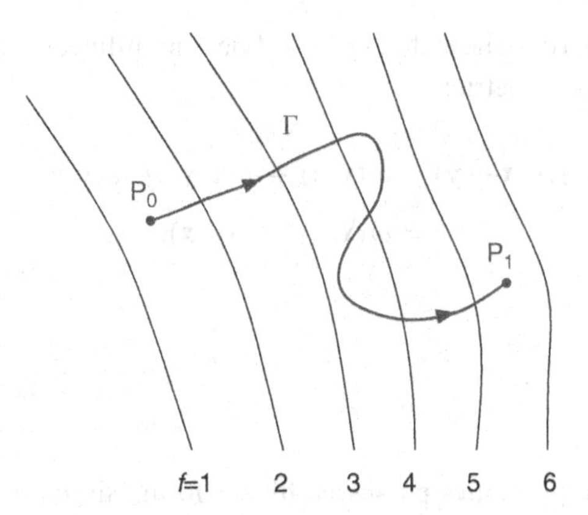

Figure 12.1 The integral of a 1-form.

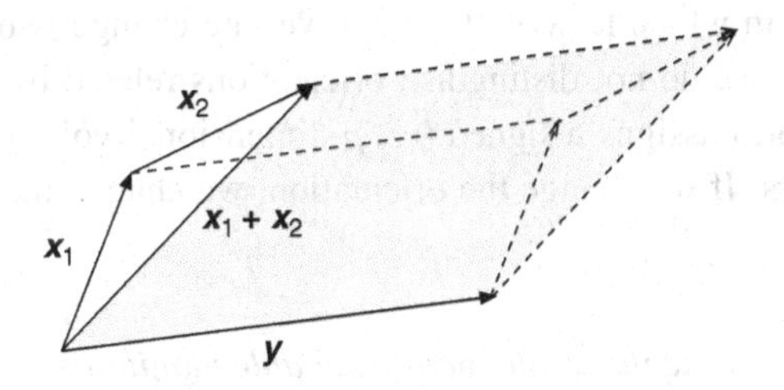

Figure 12.2 Additivity of $\omega(\mathbf{x}, \mathbf{y})$.

retrace our route, so that we pass over some part of Γ three times – twice forward and once back – we obtain the same answer as if we went only forward.

12.1.2 Skew-symmetry and orientations

What about integrating 2- and 3-forms? Why the skew-symmetry? To answer these questions, think about assigning some sort of "area" in $\mathbb{R}^2$ to the parallelogram defined by the two vectors $\mathbf{x}, \mathbf{y}$. This is going to be some function of the two vectors. Let us call it $\omega(\mathbf{x}, \mathbf{y})$. What properties do we demand of this function? There are at least three:

(i) *Scaling*: If we double the length of one of the vectors, we expect the area to double. Generalizing this, we demand that $\omega(\lambda\mathbf{x}, \mu\mathbf{y}) = (\lambda\mu)\omega(\mathbf{x}, \mathbf{y})$. (Note that we are not putting modulus signs on the lengths, so we are allowing negative "areas", and allowing for the sign to change when we reverse the direction of a vector.)

(ii) *Additivity*: The drawing in Figure 12.2 shows that we ought to have

$$\omega(\mathbf{x}_1 + \mathbf{x}_2, \mathbf{y}) = \omega(\mathbf{x}_1, \mathbf{y}) + \omega(\mathbf{x}_2, \mathbf{y}), \tag{12.4}$$

similarly for the second slots.

(iii) *Degeneration*: If the two sides coincide, the area should be zero. Thus $\omega(\mathbf{x}, \mathbf{x}) = 0$.

The first two properties show that ω should be a multilinear form. The third shows that it must be skew-symmetric:

$$
\begin{aligned}
0 = \omega(\mathbf{x}+\mathbf{y},\mathbf{x}+\mathbf{y}) &= \omega(\mathbf{x},\mathbf{x}) + \omega(\mathbf{x},\mathbf{y}) + \omega(\mathbf{y},\mathbf{x}) + \omega(\mathbf{y},\mathbf{y}) \\
&= \omega(\mathbf{x},\mathbf{y}) + \omega(\mathbf{y},\mathbf{x}).
\end{aligned} \tag{12.5}
$$

So we have

$$
\omega(\mathbf{x},\mathbf{y}) = -\omega(\mathbf{y},\mathbf{x}). \tag{12.6}
$$

These are exactly the properties possessed by a 2-form. Similarly, a 3-form outputs a volume element.

These volume elements are *oriented*. Remember that an orientation of a set of vectors is a choice of order in which to write them. If we interchange two vectors, the orientation changes sign. We do not distinguish orientations related by an even number of interchanges. A p-form assigns a signed ($\pm$) p-dimensional volume element to an orientated set of vectors. If we change the orientation, we change the sign of the volume element.

Orientable and non-orientable manifolds

In the classic video game *Asteroids*, you could select periodic boundary conditions so that your spaceship would leave the right-hand side of the screen and re-appear on the left (Figure 12.3). The game universe was topologically a torus T^2. Suppose that we modify the game code so that each bit of the spaceship re-appears at the point *diametrically opposite* the point it left. This does not seem like a drastic change until you play a game with a left-hand drive (US) spaceship. If you send the ship off the screen and watch as it

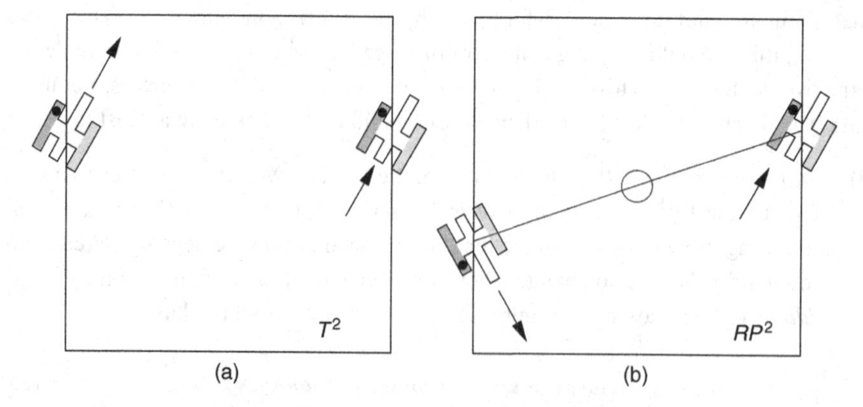

Figure 12.3 A spaceship leaves one side of the screen and returns on the other with (a) torus boundary conditions, (b) projective-plane boundary conditions. Observe how, in case (b), the spaceship has changed from being left-handed to being right-handed.

re-appears on the opposite side, you will observe the ship transmogrify into a right-hand-drive (British) craft. If we ourselves made such an excursion, we would end up starving to death because all our left-handed digestive enzymes would have been converted to right-handed ones. The game-space we have constructed is topologically equivalent to the *real projective plane* $\mathbb{R}P^2$. The lack of a global notion of being left- or right-handed makes it an example of a *non-orientable* manifold.

A manifold or surface is *orientable* if we can choose a global orientation for the tangent bundle. The simplest way to do this would be to find a smoothly varying set of basis-vector fields, $\mathbf{e}_\mu(x)$, on the surface and define the orientation by choosing an order, $\mathbf{e}_1(x), \mathbf{e}_2(x), \ldots, \mathbf{e}_d(x)$, in which to write them. In general, however, a globally defined smooth basis will not exist (try to construct one for the two-sphere, S^2!). We will, however, be able to find a continuously varying orientated basis $\mathbf{e}_1^{(i)}(x), \mathbf{e}_2^{(i)}(x), \ldots, \mathbf{e}_d^{(i)}(x)$ for each member, labelled by (i), of an atlas of coordinate charts. We should choose the charts so that the intersection of any pair forms a connected set. Assuming that this has been done, the orientation of a pair of overlapping charts is said to coincide if the determinant, $\det A$, of the map $\mathbf{e}_\mu^{(i)} = A^\nu_{\ \mu}\mathbf{e}_\nu^{(j)}$ relating the bases in the region of overlap, is positive.[1] If bases can be chosen so that all overlap determinants are positive, the manifold is *orientable* and the selected bases define the orientation. If bases cannot be so chosen, the manifold or surface is *non-orientable*.

Exercise 12.1: Consider a *three-dimensional* ball B^3 with diametrically opposite points of its surface identified. What would happen to an aircraft flying through the surface of the ball? Would it change handedness, turn inside out or simply turn upside down? Is this ball an orientable 3-manifold?

12.2 Integrating p-forms

A p-form is naturally integrated over an oriented p-dimensional surface or manifold. Rather than start with an abstract definition, we will first explain this pictorially, and then translate the pictures into mathematics.

12.2.1 Counting boxes

To visualize integrating 2-forms let us try to make sense of

$$\int_\Omega df dg, \tag{12.7}$$

where Ω is an oriented two-dimensional surface embedded in three dimensions. The surfaces $f = \text{const.}$ and $g = \text{const.}$ break the space up into a series of tubes. The oriented

[1] The determinant will have the same sign in the entire overlap region. If it did not, continuity and connectedness would force it to be zero somewhere, implying that one of the putative bases was not linearly independent there.

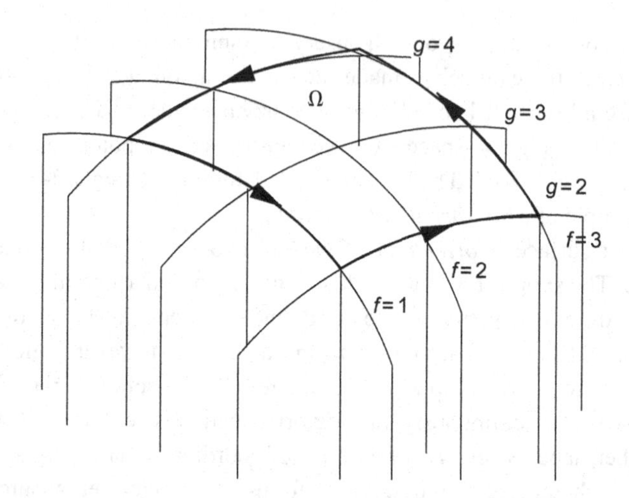

Figure 12.4 The integration region cuts the tubes into parallelograms.

surface Ω cuts these tubes in a two-dimensional mesh of (oriented) parallelograms as shown in Figure 12.4.

We define an integral by counting how many parallelograms (including fractions of a parallelogram) there are, taking the number to be positive if the parallelogram given by the mesh is oriented in the same way as the surface, and negative otherwise. To compute

$$\int_{\Omega} h\, df dg \tag{12.8}$$

we do the same, but weight each parallelogram, by the value of h at that point. The integral $\int_{\Omega} f dx dy$, over a region in $\mathbb{R}^2$, thus ends up being the number we would compute in a multivariate calculus class, but the integral $\int_{\Omega} f dy dx$ would be minus this. Similarly we compute

$$\int_{\Xi} df\, dg\, dh \tag{12.9}$$

of the 3-form $df\, dg\, dh$ over the oriented volume Ξ, by counting how many boxes defined by the level surfaces of f, g, h are included in Ξ.

An equivalent way of thinking of the integral of a p-form uses its definition as a skew-symmetric p-linear function. Accordingly we evaluate

$$I_2 = \int_{\Omega} \omega, \tag{12.10}$$

where ω is a 2-form, and Ω is an oriented 2-surface, by plugging vectors into ω. In Figure 12.5 we show a tiling of the surface Ω by a collection of (infinitesimal) parallelograms, each defined by an oriented pair of vector $\mathbf{v}_1$ and $\mathbf{v}_2$ that lie in the tangent space at one corner point x of the parallelogram. At each point x, we insert these vectors into

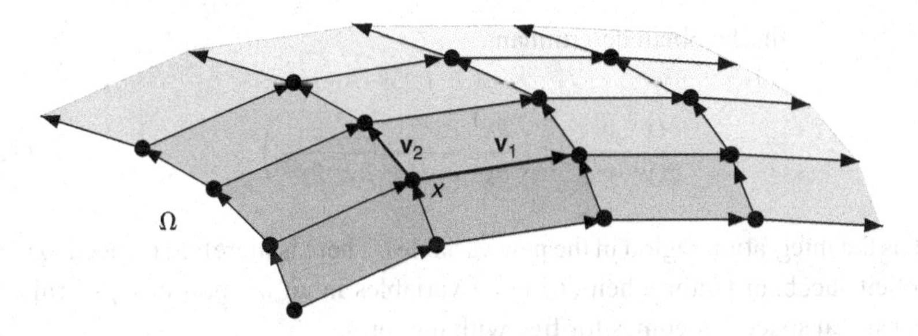

Figure 12.5 A tiling of Ω with small oriented parallelograms.

the 2-form (in the order specified by their orientation) to get $\omega(\mathbf{v}_1, \mathbf{v}_2)$, and then sum the resulting numbers over all the parallelograms to get I_2. Similarly, we integrate a p-form over an oriented p-dimensional region by decomposing the region into infinitesimal p-dimensional oriented parallelepipeds, inserting their defining vectors into the form, and summing their contributions.

12.2.2 Relation to conventional integrals

In the previous section we explained how to think pictorially about the integral. Here, we interpret the pictures as multivariable calculus.

We begin by motivating our recipe by considering a change of variables in an integral in $\mathbb{R}^2$. Suppose we set $x_1 = x(y_1, y_2)$, $x_2 = x_2(y_1, y_2)$ in

$$I_4 = \int_\Omega f(x) dx^1 dx^2, \tag{12.11}$$

and use

$$\begin{aligned} dx^1 &= \frac{\partial x^1}{\partial y^1} dy^1 + \frac{\partial x^1}{\partial y^2} dy^2, \\ dx^2 &= \frac{\partial x^2}{\partial y^1} dy^1 + \frac{\partial x^2}{\partial y^2} dy^2. \end{aligned} \tag{12.12}$$

Since $dy^1 dy^2 = -dy^2 dy^1$, we have

$$dx^1 dx^2 = \left(\frac{\partial x^1}{\partial y^1} \frac{\partial x^2}{\partial y^2} - \frac{\partial x^2}{\partial y^1} \frac{\partial x^1}{\partial y^2} \right) dy^1 dy^2. \tag{12.13}$$

Thus

$$\int_\Omega f(x) dx^1 dx^2 = \int_{\Omega'} f(x(y)) \frac{\partial(x^1, x^2)}{\partial(y^1, y^2)} dy^1 dy^2 \tag{12.14}$$

where $\frac{\partial(x^1,y^1)}{\partial(y^1,y^2)}$ is the Jacobian determinant

$$\frac{\partial(x^1,y^1)}{\partial(y^1,y^2)} = \left(\frac{\partial x^1}{\partial y^1}\frac{\partial x^2}{\partial y^2} - \frac{\partial x^2}{\partial y^1}\frac{\partial x^1}{\partial y^2}\right), \tag{12.15}$$

and Ω' is the integration region in the new variables. There is therefore no need to include an explicit Jacobian factor when changing variables in an integral of a p-form over a p-dimensional space – it comes for free with the form.

This observation leads us to the general prescription: to evaluate $\int_\Omega \omega$, the integral of a p-form

$$\omega = \frac{1}{p!}\omega_{\mu_1\mu_2\dots\mu_p} dx^{\mu_1}\cdots dx^{\mu_p} \tag{12.16}$$

over the region Ω of a p-dimensional surface in a $d \geq p$ dimensional space, substitute a parametrization

$$\begin{aligned} x^1 &= x^1(\xi^1,\xi^2,\dots,\xi^p),\\ &\vdots\\ x^d &= x^d(\xi^1,\xi^2,\dots,\xi^p), \end{aligned} \tag{12.17}$$

of the surface into ω. Next, use

$$dx^\mu = \frac{\partial x^\mu}{\partial \xi^i} d\xi^i, \tag{12.18}$$

so that

$$\omega \to \omega(x(\xi))_{i_1 i_2 \dots i_p} \frac{\partial x^{i_1}}{\partial \xi^1}\cdots\frac{\partial x^{i_p}}{\partial \xi^p} d\xi^1\cdots d\xi^p, \tag{12.19}$$

which we regard as a p-form on Ω. (Our customary $1/p!$ is absent here because we have chosen a particular order for the $d\xi$'s.) Then

$$\int_\Omega \omega \overset{\text{def}}{=} \int_\Omega \omega(x(\xi))_{i_1 i_2 \dots i_p} \frac{\partial x^{i_1}}{\partial \xi^1}\cdots\frac{\partial x^{i_p}}{\partial \xi^p} d\xi^1\cdots d\xi^p, \tag{12.20}$$

where the right-hand side is an ordinary multiple integral. This recipe is a generalization of the formula (12.3), which reduced the integral of a 1-form to an ordinary single-variable integral. Because the appropriate Jacobian factor appears automatically, the numerical value of the integral does not depend on the choice of parametrization of the surface.

Example: To integrate the 2-form $x\,dydz$ over the surface of a two-dimensional sphere of radius R, we parametrize the surface with polar angles as

$$\begin{aligned} x &= R\sin\phi\sin\theta, \\ y &= R\cos\phi\sin\theta, \\ z &= R\cos\theta. \end{aligned} \tag{12.21}$$

Then

$$\begin{aligned} dy &= -R\sin\phi\sin\theta d\phi + R\cos\phi\cos\theta d\theta, \\ dz &= -R\sin\theta d\theta, \end{aligned} \tag{12.22}$$

and so

$$x\,dydz = R^3\sin^2\phi\,\sin^3\theta\; d\phi d\theta. \tag{12.23}$$

We therefore evaluate

$$\begin{aligned} \int_{\text{sphere}} x\,dydz &= R^3\int_0^{2\pi}\int_0^{\pi} \sin^2\phi\,\sin^3\theta\; d\phi d\theta \\ &= R^3\int_0^{2\pi}\sin^2\phi\, d\phi\int_0^{\pi}\sin^3\theta\, d\theta \\ &= R^3\pi\int_{-1}^{1}(1-\cos^2\theta)\, d\cos\theta \\ &= \frac{4}{3}\pi R^3. \end{aligned} \tag{12.24}$$

The volume form

Although we do not need any notion of length to integrate a differential form, a p-dimensional surface embedded or immersed in $\mathbb{R}^d$ does inherit a distance scale from the $\mathbb{R}^d$ Euclidean metric, and this can be used to define the area or volume of the surface. When the cartesian coordinates $x^1,\ldots,x^d$ of a point in the surface are given as $x^a(\xi^1,\ldots,\xi^p)$, where the $\xi^1,\ldots,\xi^p$, are coordinates on the surface, then the inherited, or *induced*, metric is

$$\text{“}ds^2\text{''} \equiv g(\ ,\ ) \equiv g_{\mu\nu}\, d\xi^\mu \otimes d\xi^\nu, \tag{12.25}$$

where

$$g_{\mu\nu} = \sum_{a=1}^{d}\frac{\partial x^a}{\partial \xi^\mu}\frac{\partial x^a}{\partial \xi^\nu}. \tag{12.26}$$

The *volume form* associated with the induced metric is

$$d(\text{Volume}) = \sqrt{g}\, d\xi^1 \cdots d\xi^p, \tag{12.27}$$

where $g = \det(g_{\mu\nu})$. The integral of this p-form over a p-dimensional region gives the area, or p-dimensional volume, of the region.

If we change the parametrization of the surface from ξ^μ to ζ^μ, neither the $d\xi^1 \cdots d\xi^p$ nor the $\sqrt{g}$ are separately invariant, but the Jacobian arising from the change of the p-form, $d\xi^1 \cdots d\xi^p \to d\zeta^1 \cdots d\zeta^p$ cancels against the factor coming from the transformation law of the metric tensor $g_{\mu\nu} \to g'_{\mu\nu}$, leading to

$$\sqrt{g}\, d\xi^1 \cdots d\xi^p = \sqrt{g'} d\zeta^1 \cdots d\zeta^p. \tag{12.28}$$

The volume of the surface is therefore independent of the coordinate system used to evaluate it.

Example: The induced metric on the surface of a unit-radius two-sphere embedded in $\mathbb{R}^3$ is, expressed in polar angles,

$$\text{“}ds^2\text{''} = \mathbf{g}(\ ,\ ) = d\theta \otimes d\theta + \sin^2\theta\, d\phi \otimes d\phi.$$

Thus

$$g = \begin{vmatrix} 1 & 0 \\ 0 & \sin^2\theta \end{vmatrix} = \sin^2\theta,$$

and

$$d(\text{Area}) = \sin\theta\, d\theta d\phi.$$

12.3 Stokes' theorem

All of the integral theorems of classical vector calculus are special cases of

Stokes' theorem: If $\partial\Omega$ denotes the (oriented) boundary of the (oriented) region Ω, then

$$\boxed{\int_\Omega d\omega = \int_{\partial\Omega} \omega.}$$

We will not provide a detailed proof. Apart from notation, it would parallel the proof of Stokes' or Green's theorems in ordinary vector calculus: the exterior derivative d has been *defined* so that the theorem holds for an infinitesimal square, cube or hypercube. We therefore divide Ω into many such small regions. We then observe that the contributions of the interior boundary faces cancel because all interior faces are shared between two adjacent regions, and so occur twice with opposite orientations. Only the contribution of the outer boundary remains.

Example: If Ω is a region of $\mathbb{R}^2$, then from

$$d\left[\frac{1}{2}(x\,dy - y\,dx)\right] = dxdy,$$

we have

$$\text{Area}\,(\Omega) = \int_\Omega dxdy = \frac{1}{2}\int_{\partial\Omega}(x\,dy - y\,dx).$$

Example: Again, if Ω is a region of $\mathbb{R}^2$, then from $d[r^2 d\theta/2] = r\,drd\theta$ we have

$$\text{Area}\,(\Omega) = \int_\Omega r\,drd\theta = \frac{1}{2}\int_{\partial\Omega} r^2 d\theta.$$

Example: If Ω is the interior of a sphere of radius R, then

$$\int_\Omega dxdydz = \int_{\partial\Omega} x\,dydz = \frac{4}{3}\pi R^3.$$

Here we have referred back to (12.24) to evaluate the surface integral.

Example: *Archimedes' tombstone.* Archimedes of Syracuse gave instructions that his tombstone should have displayed on it a diagram consisting of a sphere and circumscribed cylinder. Cicero, while serving as quæstor in Sicily, had the stone restored.[2] Cicero himself suggested that this act was the only significant contribution by a Roman to the history of pure mathematics. The carving on the stone was to commemorate Archimedes' results about the areas and volumes of spheres, including the one illustrated in Figure 12.6, that the area of the spherical cap cut off by slicing through the cylinder is equal to the area cut off on the cylinder.

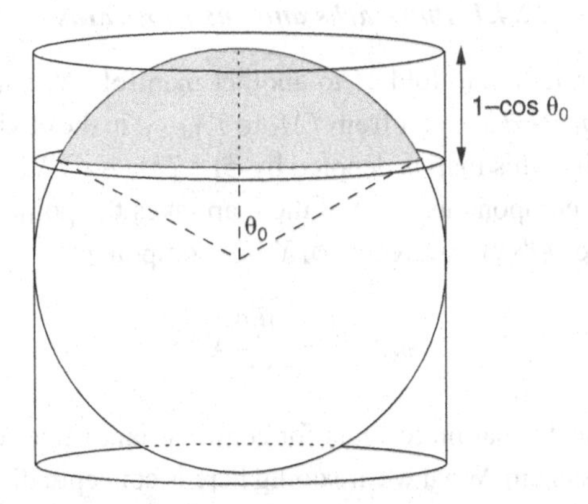

Figure 12.6 Sphere and circumscribed cylinder.

[2] Marcus Tullius Cicero, *Tusculan Disputations*, Book V, Sections 64–66.

We can understand this result via Stokes' theorem: if the 2-sphere S^2 is parametrized by spherical polar coordinates θ, ϕ, and Ω is a region on the sphere, then

$$\text{Area}\,(\Omega) = \int_\Omega \sin\theta d\theta d\phi = \int_{\partial\Omega} (1 - \cos\theta) d\phi,$$

and applying this to the figure, where the cap is defined by $\theta < \theta_0$, gives

$$\text{Area}\,(\text{cap}) = 2\pi(1 - \cos\theta_0),$$

which is indeed the area cut off on the cylinder.

Exercise 12.2: The sphere S^n can be thought of as the locus of points in $\mathbb{R}^{n+1}$ obeying $\sum_{i=1}^{n+1}(x^i)^2 = 1$. Use its invariance under orthogonal transformations to show that the element of surface "volume" of the n-sphere can be written as

$$d(\text{Volume on } S^n) = \frac{1}{n!}\epsilon_{\alpha_1\alpha_2\ldots\alpha_{n+1}} x^{\alpha_1}\, dx^{\alpha_2} \ldots dx^{\alpha_{n+1}}.$$

Use Stokes' theorem to relate the integral of this form over the *surface* of the sphere to the volume of the *solid* unit sphere. Confirm that we get the correct proportionality between the volume of the solid unit sphere and the volume or area of its surface.

12.4 Applications

We now know how to integrate forms. What sort of forms should we seek to integrate? For a physicist working with a classical or quantum field, a plentiful supply of interesting forms is obtained by using the field to *pull back* geometric objects.

12.4.1 Pull-backs and push-forwards

If we have a map ϕ from a manifold M to another manifold N, and we choose a point $x \in M$, we can *push forward* a vector from TM_x to $TN_{\phi(x)}$, in the obvious way (map head-to-head and tail-to-tail). This map is denoted by $\phi_* : TM_x \to TN_{\phi(x)}$; see Figure 12.7.

If the vector X has components X^μ and the map takes the point with coordinates x^μ to one with coordinates $\xi^\mu(x)$, the vector $\phi_* X$ has components

$$(\phi_* X)^\mu = \frac{\partial \xi^\mu}{\partial x^\nu} X^\nu. \tag{12.29}$$

This looks like the transformation formula for contravariant vector components under a change of coordinate system. What we are doing here is conceptually different, however. A change of coordinates produces a *passive* transformation – i.e. a new description for an unchanging vector. A push-forward is an *active* transformation – we are changing a vector into a different one. Furthermore, the map from $M \to N$ is not being assumed to

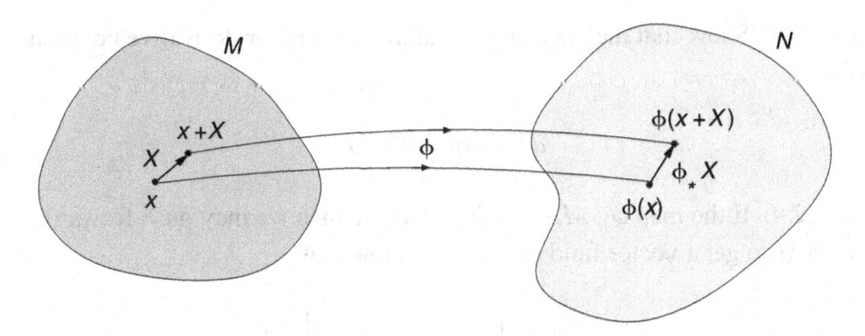

Figure 12.7 Pushing forward a vector X from TM_x to $TN_{\phi(x)}$.

be one-to-one, so, contrary to the requirement imposed on a coordinate transformation, it may not be possible to invert the functions $\xi^\mu(x)$ and write the x^ν's as functions of the ξ^μ's.

While we can push forward individual vectors, we cannot always push forward a vector *field* X from TM to TN. If two distinct points, x_1 and x_2, should happen to map to the same point $\xi \in N$, and $X(x_1) \neq X(x_2)$, we would not know whether to choose $\phi_*[X(x_1)]$ or $\phi_*[X(x_2)]$ as $[\phi_* X](\xi)$. This problem does not occur for differential forms. A map $\phi : M \to N$ induces a natural, and always well-defined, *pull-back* map $\phi^* : \bigwedge^p(T^*N) \to \bigwedge^p(T^*M)$ which works as follows: given a form $\omega \in \bigwedge^p(T^*N)$, we define $\phi^*\omega$ as a form on M by specifying what we get when we plug the vectors $X_1, X_2, \dots, X_p \in TM$ into it. We evaluate the form at $x \in M$ by pushing the vectors $X_i(x)$ forward from TM_x to $TN_{\phi(x)}$, plugging them into ω at $\phi(x)$ and declaring the result to be the evaluation of $\phi^*\omega$ on the X_i at x. Symbolically

$$[\phi^*\omega](X_1, X_2, \dots, X_p) = \omega(\phi_* X_1, \phi_* X_2, \dots, \phi_* X_p). \tag{12.30}$$

This may seem rather abstract, but the idea is in practice quite simple: if the map takes $x \in M \to \xi(x) \in N$, and if

$$\omega = \frac{1}{p!}\omega_{i_1 \dots i_p}(\xi) d\xi^{i_1} \dots d\xi^{i_p}, \tag{12.31}$$

then

$$\begin{aligned}\phi^*\omega &= \frac{1}{p!}\omega_{i_1 i_2 \dots i_p}[\xi(x)] d\xi^{i_1}(x) d\xi^{i_2}(x) \cdots d\xi^{i_p}(x) \\ &= \frac{1}{p!}\omega_{i_1 i_2 \dots i_p}[\xi(x)] \frac{\partial \xi^{i_1}}{\partial x^{\mu_1}} \frac{\partial \xi^{i_2}}{\partial x^{\mu_2}} \cdots \frac{\partial \xi^{i_p}}{\partial x^{\mu_1}} dx^{\mu_p} \cdots dx^{\mu_p}.\end{aligned} \tag{12.32}$$

Computationally, the process of pulling back a form is so transparent that it is easy to confuse it with a simple change of variable. That it is not the same operation will become clear in the next few sections where we consider maps that are many-to-one.

Exercise 12.3: Show that the operation of taking an exterior derivative commutes with a pull-back:

$$d\left[\phi^*\omega\right] = \phi^*(d\omega).$$

Exercise 12.4: If the map $\phi : M \to N$ is invertible then we may push forward a vector field X on M to get a vector field $\phi_* X$ on N. Show that

$$\mathcal{L}_X[\phi^*\omega] = \phi^*\left[\mathcal{L}_{\phi_* X}\omega\right].$$

Exercise 12.5: Again assume that $\phi : M \to N$ is invertible. By using the coordinate expressions for the Lie bracket along with the effect of a push-forward, show that if X, Y are vector fields on TM then

$$\phi_*([X,Y]) = [\phi_* X, \phi_* Y],$$

as vector fields on TN.

12.4.2 Spin textures

As an application of pull-backs we consider some of the topological aspects of *spin textures* which are fields of unit vectors $\mathbf{n}$, or "spins", in two or three dimensions.

Consider a smooth map $\varphi : \mathbb{R}^2 \to S^2$ that assigns $x \mapsto \mathbf{n}(x)$, where $\mathbf{n}$ is a three-dimensional unit vector whose tip defines a point on the 2-sphere S^2. A physical example of such an $\mathbf{n}(x)$ would be the local direction of the spin polarization in a ferromagnetically coupled two-dimensional electron gas.

In terms of $\mathbf{n}$, the area 2-form on the 2-sphere becomes

$$\Omega = \frac{1}{2}\mathbf{n}\cdot(d\mathbf{n}\times d\mathbf{n}) \equiv \frac{1}{2}\epsilon_{ijk}n^i dn^j dn^k. \tag{12.33}$$

The φ map pulls this area-form back to

$$F \equiv \varphi^*\Omega = \frac{1}{2}(\epsilon_{ijk}n^i\partial_\mu n^j\partial_\nu n^k)dx^\mu dx^\nu = (\epsilon_{ijk}n^i\partial_1 n^j\partial_2 n^k)\,dx^1 dx^2 \tag{12.34}$$

which is a differential form in $\mathbb{R}^2$. We will call it the *topological charge density*. It measures the area on the 2-sphere swept out by the $\mathbf{n}$ vectors as we explore a square in $\mathbb{R}^2$ of side dx^1 by dx^2.

Suppose now that the $\mathbf{n}$ tends to the same unit vector $\mathbf{n}(\infty)$ at large distance in all directions. This allows us to think of "infinity" as a single point, and the assignment $\varphi : x \mapsto \mathbf{n}(x)$ as a map from S^2 to S^2. Such maps are characterized topologically by their "*topological charge*", or *winding number* N which counts the number of times the image of the originating x-sphere wraps round the target $\mathbf{n}$-sphere. A mathematician would call this number the *Brouwer degree* of the map φ. It is intuitively plausible that a continuous map from a sphere to itself will wrap a whole number of times, and so we expect

$$N = \frac{1}{4\pi}\int_{\mathbb{R}^2}\left\{\epsilon_{ijk}n^i\partial_1 n^j\partial_2 n^k\right\}dx^1dx^2, \tag{12.35}$$

to be an integer. We will soon show that this is indeed so, but first we will demonstrate that N is a *topological invariant*.

In two dimensions the form $F = \varphi^*\Omega$ is automatically closed because the exterior derivative of any 2-form is zero – there being no 3-forms in two dimensions. Even if we consider a field $\mathbf{n}(x^1,\ldots,x^m)$ in $m > 2$ dimensions, however, we still have $dF = 0$. This is because

$$dF = \frac{1}{2}\epsilon^{ijk}\partial_\sigma n^i\partial_\mu n^j\partial_\nu n^k dx^\sigma dx^\mu dx^\nu. \tag{12.36}$$

If we plug infinitesimal vectors into the dx^μ to get their components δx^μ, we have to evaluate the triple-product of three vectors $\delta n^i = \partial_\mu n^i\delta x^\mu$, each of which is tangent to the 2-sphere. But the tangent space of S^2 is two-dimensional, so any three tangent vectors $\mathbf{t}_1, \mathbf{t}_2, \mathbf{t}_3$, are linearly dependent and their triple-product $\mathbf{t}_1\cdot(\mathbf{t}_2\times\mathbf{t}_3)$ is therefore zero.

Although it is closed, $F = \varphi^*\Omega$ will not generally be the d of a globally defined 1-form. Suppose, however, that we vary the map so that $\mathbf{n}(x) \to \mathbf{n}(x) + \delta\mathbf{n}(x)$. The corresponding change in the topological charge density is

$$\delta F = \varphi^*[\mathbf{n}\cdot(d(\delta\mathbf{n})\times d\mathbf{n})], \tag{12.37}$$

and this variation *can* be written as a total derivative:

$$\delta F = d\{\varphi^*[\mathbf{n}\cdot(\delta\mathbf{n}\times d\mathbf{n})]\} \equiv d\{\epsilon_{ijk}n^i\delta n^j\partial_\mu n^k dx^\mu\}. \tag{12.38}$$

In these manipulations we have used $\delta\mathbf{n}\cdot(d\mathbf{n}\times d\mathbf{n}) = d\mathbf{n}\cdot(\delta\mathbf{n}\times d\mathbf{n}) = 0$, the triple-products being zero for the linear-dependence reason adduced earlier. From Stokes' theorem, we have

$$\delta N = \int_{S^2}\delta F = \int_{\partial S^2}\epsilon_{ijk}n^i\delta n^j\partial_\mu n^k dx^\mu. \tag{12.39}$$

Because the sphere has no boundary, i.e. $\partial S^2 = \emptyset$, the last integral vanishes, so we conclude that $\delta N = 0$ under any smooth deformation of the map $\mathbf{n}(x)$. This is what we mean when we say that N is a topological invariant. Equivalently, on $\mathbb{R}^2$, with $\mathbf{n}$ constant at infinity, we have

$$\delta N = \int_{\mathbb{R}^2}\delta F = \int_\Gamma\epsilon_{ijk}n^i\delta n^j\partial_\mu n^k dx^\mu, \tag{12.40}$$

where Γ is a curve surrounding the origin at large distance. Again $\delta N = 0$, this time because $\partial_\mu n^k = 0$ everywhere on Γ.

In some physical applications, the field $\mathbf{n}$ winds in localized regions called *skyrmions*. These knots in the spin field behave very much as elementary particles, retaining their identity as they move through the system. The winding number counts how many skyrmions (minus the number of anti-skyrmions, which wind with opposite orientation) there are. To construct a smooth multi-skyrmion map $\varphi : \mathbb{R}^2 \to S^2$ with positive winding number N, take a set of $N+1$ complex numbers $\lambda, a_1, \ldots, a_N$ and another set of N complex numbers $b_1, \ldots, b_N$ such that no b coincides with any a. Then put

$$e^{i\phi} \tan \frac{\theta}{2} = \lambda \frac{(z-a_1)\ldots(z-a_N)}{(z-b_1)\ldots(z-b_N)} \tag{12.41}$$

where $z = x^1 + ix^2$, and θ and ϕ are spherical polar coordinates specifying the direction $\mathbf{n}$ at the point (x^1, x^2). At the points $z = a_i$ the vector $\mathbf{n}$ points straight up, and at the points $z = b_i$ it points straight down. You will show in Exercise 12.12 that this particular $\mathbf{n}$-field configuration minimizes the energy functional

$$\begin{aligned} E[\mathbf{n}] &= \frac{1}{2} \int (\partial_1 \mathbf{n} \cdot \partial_1 \mathbf{n} + \partial_2 \mathbf{n} \cdot \partial_2 \mathbf{n}) \, dx^1 dx^2 \\ &= \frac{1}{2} \int \left(|\nabla n^1|^2 + |\nabla n^2|^2 + |\nabla n^3|^2 \right) dx^1 dx^2 \end{aligned} \tag{12.42}$$

for the given winding number N. In the next section we will explain the geometric origin of the mysterious combination $e^{i\phi} \tan \theta/2$.

12.4.3 The Hopf map

You may recall that in Section 10.2.3 we defined *complex projective space* $\mathbb{C}P^n$ to be the set of *rays* in a complex $(n+1)$-dimensional vector space. A ray is an equivalence class of vectors $[\zeta_1, \zeta_2, \ldots, \zeta_{n+1}]$, where the ζ_i are not all zero, and where we do not distinguish between $[\zeta_1, \zeta_2, \ldots, \zeta_{n+1}]$ and $[\lambda\zeta_1, \lambda\zeta_2, \ldots, \lambda\zeta_{n+1}]$ for non-zero complex λ. The space of rays is a $2n$-dimensional real manifold: in a region where ζ_{n+1} does not vanish, we can take as coordinates the real numbers $\xi_1, \ldots, \xi_n, \eta_1, \ldots, \eta_n$ where

$$\xi_1 + i\eta_1 = \frac{\zeta_1}{\zeta_{n+1}}, \quad \xi_2 + i\eta_2 = \frac{\zeta_2}{\zeta_{n+1}}, \ldots, \xi_n + i\eta_n = \frac{\zeta_n}{\zeta_{n+1}}. \tag{12.43}$$

Similar coordinate charts can be constructed in the regions where other ζ_i are non-zero. Every point in $\mathbb{C}P^n$ lies in at least one of these coordinate charts, and the coordinate transformation rules for going from one chart to another are smooth.

The simplest complex projective space, $\mathbb{C}P^1$, is the real 2-sphere S^2 in disguise. This rather non-obvious fact is revealed by the use of a *stereographic map* to make the equivalence class $[\zeta_1, \zeta_2] \in \mathbb{C}P^1$ correspond to a point $\mathbf{n}$ on the sphere. When ζ_1 is non-zero, the class $[\zeta_1, \zeta_2]$ is uniquely determined by the ratio $\zeta_2/\zeta_1 = |\zeta_2/\zeta_1| e^{i\phi}$, which we plot on the complex plane. We think of this copy of $\mathbb{C}$ as being the xy-plane in $\mathbb{R}^3$. We then draw a straight line connecting the plotted point to the south pole of a unit sphere

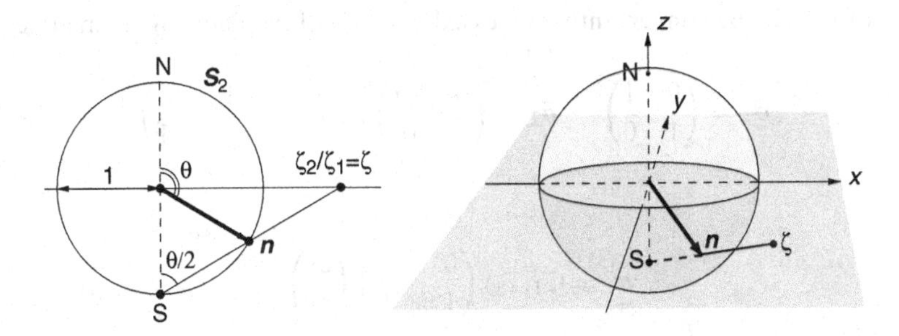

Figure 12.8 Two views of the stereographic map between the 2-sphere and the complex plane. The point $\zeta = \zeta_2/\zeta_1 \in \mathbb{C}$ corresponds to the unit vector $\mathbf{n} \in S^2$.

circumscribed about the origin in $\mathbb{R}^3$. The point where this line (continued, if necessary) intersects the sphere is the tip of the unit vector $\mathbf{n}$.

If ζ_2 were zero, $\mathbf{n}$ would end up at the north pole, where the $\mathbb{R}^3$ coordinate z takes the value $z = 1$. If ζ_1 goes to zero with ζ_2 fixed, $\mathbf{n}$ moves smoothly to the south pole $z = -1$. We therefore extend the definition of our map to the case $\zeta_1 = 0$ by making the equivalence class $[0, \zeta_2]$ correspond to the south pole. We can find an explicit formula for this map. Figure 12.8 shows that $\zeta_2/\zeta_1 = e^{i\phi} \tan\theta/2$, and this relation suggests the use of the "t"-substitution formulæ:

$$\sin\theta = \frac{2t}{1+t^2}, \quad \cos\theta = \frac{1-t^2}{1+t^2}, \tag{12.44}$$

where $t = \tan\theta/2$. Since the x, y, z components of $\mathbf{n}$ are given by

$$\begin{aligned} n^1 &= \sin\theta\cos\phi, \\ n^2 &= \sin\theta\sin\phi, \\ n^3 &= \cos\theta, \end{aligned} \tag{12.45}$$

we find that

$$n^1 + in^2 = \frac{2(\zeta_2/\zeta_1)}{1+|\zeta_2/\zeta_1|^2}, \quad n^3 = \frac{1-|\zeta_2/\zeta_1|^2}{1+|\zeta_2/\zeta_1|^2}. \tag{12.46}$$

We can multiply through by $|\zeta_1|^2 = \overline{\zeta}_1\zeta_1$, and so write this correspondence in a more symmetrical manner:

$$\begin{aligned} n^1 &= \frac{\overline{\zeta}_1\zeta_2 + \overline{\zeta}_2\zeta_1}{|\zeta_1|^2+|\zeta_2|^2}, \\ n^2 &= \frac{1}{i}\left(\frac{\overline{\zeta}_1\zeta_2 - \overline{\zeta}_2\zeta_1}{|\zeta_1|^2+|\zeta_2|^2}\right), \\ n^3 &= \frac{|\zeta_1|^2-|\zeta_2|^2}{|\zeta_1|^2+|\zeta_2|^2}. \end{aligned} \tag{12.47}$$

This last form can be conveniently expressed in terms of the Pauli sigma matrices

$$\widehat{\sigma}_1 = \begin{pmatrix} 0 & 1 \\ 1 & 0 \end{pmatrix}, \quad \widehat{\sigma}_2 = \begin{pmatrix} 0 & -i \\ i & 0 \end{pmatrix}, \quad \widehat{\sigma}_3 = \begin{pmatrix} 1 & 0 \\ 0 & -1 \end{pmatrix} \tag{12.48}$$

as

$$\begin{aligned} n^1 &= (\bar{z}_1, \bar{z}_2) \begin{pmatrix} 0 & 1 \\ 1 & 0 \end{pmatrix} \begin{pmatrix} z_1 \\ z_2 \end{pmatrix}, \\ n^2 &= (\bar{z}_1, \bar{z}_2) \begin{pmatrix} 0 & -i \\ i & 0 \end{pmatrix} \begin{pmatrix} z_1 \\ z_2 \end{pmatrix}, \\ n^3 &= (\bar{z}_1, \bar{z}_2) \begin{pmatrix} 1 & 0 \\ 0 & -1 \end{pmatrix} \begin{pmatrix} z_1 \\ z_2 \end{pmatrix}, \end{aligned} \tag{12.49}$$

where

$$\begin{pmatrix} z_1 \\ z_2 \end{pmatrix} = \frac{1}{\sqrt{|\zeta_1|^2 + |\zeta_2|^2}} \begin{pmatrix} \zeta_1 \\ \zeta_2 \end{pmatrix} \tag{12.50}$$

is a normalized 2-vector, which we think of as a *spinor*.

The correspondence $\mathbb{C}P^1 \simeq S^2$ now has a quantum-mechanical interpretation: any unit 3-vector $\mathbf{n}$ can be obtained as the expectation value of the $\hat{\sigma}$ matrices in a normalized spinor state. Conversely, any normalized spinor $\psi = (z_1, z_2)^T$ gives rise to a unit vector *via*

$$n^i = \psi^\dagger \hat{\sigma}^i \psi. \tag{12.51}$$

Now, since

$$1 = |z_1|^2 + |z_2|^2, \tag{12.52}$$

the normalized spinor can be thought of as defining a point in S^3. This means that the one-to-one correspondence $[z_1, z_2] \leftrightarrow \mathbf{n}$ also gives rise to a map from $S^3 \to S^2$. This is called the *Hopf map*:

$$\text{Hopf} : S^3 \to S^2. \tag{12.53}$$

The dimension reduces from three to two, so the Hopf map cannot be one-to-one. Even after we have normalized $[\zeta_1, \zeta_2]$, we are still left with a choice of overall phase. Both (z_1, z_2) and $(z_1 e^{i\theta}, z_2 e^{i\theta})$, although distinct points in S^3, correspond to the same point in $\mathbb{C}P^1$, and hence in S^2. The inverse image of a point in S^2 is a geodesic circle in S^3. Later, we will show that any two such geodesic circles are linked, and this makes the Hopf map topologically non-trivial, in that it cannot be continuously deformed to a constant map – i.e. to a map that takes all of S^3 to a single point in S^2.

Exercise 12.6: We have seen that the stereographic map relates the point with spherical polar coordinates θ, ϕ to the complex number

$$\zeta = e^{i\phi} \tan\theta/2.$$

We can therefore set $\zeta = \xi + i\eta$ and take ξ, η as *stereographic coordinates* on the sphere. Show that in these coordinates the sphere metric is given by

$$\begin{aligned} g(\ ,\ ) &\equiv d\theta \otimes d\theta + \sin^2\theta\, d\phi \otimes d\phi \\ &= \frac{2}{(1+|\zeta|^2)^2}(d\bar{\zeta} \otimes d\zeta + d\zeta \otimes d\bar{\zeta}) \\ &= \frac{4}{(1+\xi^2+|\eta|^2)^2}(d\xi \otimes d\xi + d\eta \otimes d\eta), \end{aligned}$$

and that the area 2-form becomes

$$\begin{aligned} \Omega &\equiv \sin\theta\, d\theta \wedge d\phi \\ &= \frac{2i}{(1+|\zeta|^2)^2}\, d\zeta \wedge d\bar{\zeta} \\ &= \frac{4}{(1+\xi^2+\eta^2)^2}\, d\xi \wedge d\eta. \end{aligned} \tag{12.54}$$

12.4.4 Homotopy and the Hopf map

We can use the Hopf map to factor the map $\varphi : x \mapsto \mathbf{n}(x)$ *via* the 3-sphere by specifying the spinor ψ at each point, instead of the vector $\mathbf{n}$, and so mapping indirectly

$$\varphi : \mathbb{R}^2 \xrightarrow{\psi} S^3 \xrightarrow{\text{Hopf}} S^2.$$

It might seem that for a given spin-field $\mathbf{n}(x)$ we can choose the overall phase of $\psi(x) \equiv (z_1(x), z_2(x))^T$ as we like; however, if we demand that the z_i's be *continuous* functions of x then there is a rather non-obvious topological restriction which has important physical consequences. To see how this comes about, we first express the winding number in terms of the z_i. We find (after a page or two of algebra) that

$$F = (\epsilon_{ijk} n^i \partial_1 n^j \partial_2 n^k)\, dx^1 dx^2 = \frac{2}{i}\sum_{i=1}^{2}(\partial_1 \bar{z}_i \partial_2 z_i - \partial_2 \bar{z}_i \partial_1 z_i)\, dx^1 dx^2, \tag{12.55}$$

and so the topological charge N is given by

$$N = \frac{1}{2\pi i}\int \sum_{i=1}^{2}(\partial_1 \bar{z}_i \partial_2 z_i - \partial_2 \bar{z}_i \partial_1 z_i)\, dx^1 dx^2. \tag{12.56}$$

Now, when written in terms of the z_i variables, the form F becomes a total derivative:

$$F = \frac{2}{i}\sum_{i=1}^{2}(\partial_1\bar{z}_i\partial_2 z_i - \partial_2\bar{z}_i\partial_1 z_i)\,dx^1dx^2$$
$$= d\left\{\frac{1}{i}\sum_{i=1}^{2}\left(\bar{z}_i\partial_\mu z_i - (\partial_\mu\bar{z}_i)z_i\right)dx^\mu\right\}. \tag{12.57}$$

Furthermore, because $\mathbf{n}$ is fixed at large distance, we have $(z_1, z_2) = e^{i\theta}(c_1, c_2)$ near infinity, where c_1, c_2 are constants with $|c_1|^2 + |c_2|^2 = 1$. Thus, near infinity,

$$\frac{1}{2i}\sum_{i=1}^{2}\left(\bar{z}_i\partial_\mu z_i - (\partial_\mu\bar{z}_i)z_i\right) \to (|c_1|^2 + |c_2|^2)d\theta = d\theta. \tag{12.58}$$

We combine this observation with Stokes' theorem to obtain

$$N = \frac{1}{2\pi i}\int_\Gamma \frac{1}{2}\sum_{i=1}^{2}\left(\bar{z}_i\partial_\mu z_i - (\partial_\mu\bar{z}_i)z_i\right)dx^\mu = \frac{1}{2\pi}\int_\Gamma d\theta. \tag{12.59}$$

Here, as in the previous section, Γ is a curve surrounding the origin at large distance. Now $\int d\theta$ is the total change in θ as we circle the boundary. While the phase $e^{i\theta}$ has to return to its original value after a round trip, the angle θ can increase by an integer multiple of 2π. The *winding number* $\oint d\theta/2\pi$ can therefore be non-zero, but must be an integer.

We have uncovered the rather surprising fact that the topological charge of the map $\varphi : S^2 \to S^2$ is equal to the winding number of the phase angle θ at infinity. This is the topological restriction referred to in the preceding paragraph. As a byproduct, we have confirmed our conjecture that the topological charge N is an integer. The existence of this integer invariant shows that the smooth maps $\varphi : S^2 \to S^2$ fall into distinct *homotopy classes* labelled by N. Maps with different values of N cannot be continuously deformed into one another, and, while we have not shown that it is so, two maps with the same value of N can be deformed into each other.

Maps that can be continuously deformed one into the other are said to be *homotopic*. The set of homotopy classes of the maps of the n-sphere into a manifold M is denoted by $\pi_n(M)$. In the present case $M = S^2$. We are therefore claiming that

$$\pi_2(S^2) = \mathbb{Z}, \tag{12.60}$$

where we are identifying the homotopy class with its winding number $N \in \mathbb{Z}$.

12.4.5 The Hopf index

We have so far discussed maps from S^2 to S^2. It is perhaps not too surprising that such maps are classified by a winding number. What is rather more surprising is that maps

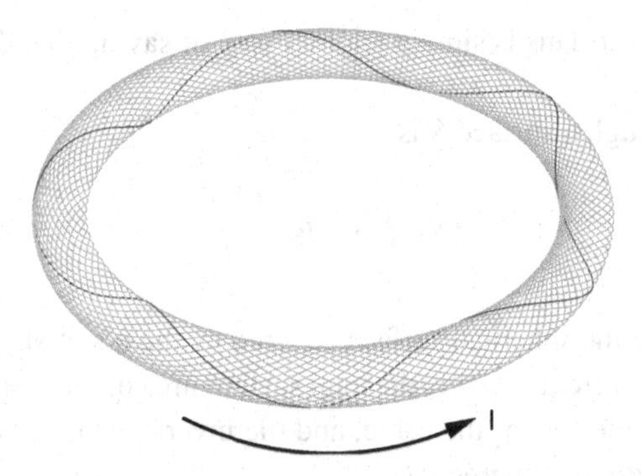

Figure 12.9 A twisted cable with $N = 5$.

$\varphi : S^3 \to S^2$ also have an associated topological number. If we continue to assume that $\mathbf{n}$ tends to a constant vector at infinity so that we can think of $\mathbb{R}^3 \cup \{\infty\}$ as being S^3, this number will label the homotopy classes $\pi_3(S^2)$ of fields of unit vectors $\mathbf{n}$ in *three* dimensions. We will think of the third dimension as being time. In this situation an interesting set of $\mathbf{n}$ fields to consider are the $\mathbf{n}(x, t)$ corresponding moving skyrmions. The world-lines of these skyrmions will be tubes outside of which $\mathbf{n}$ is constant, and such that on any slice through the tube, $\mathbf{n}$ will cover the target $\mathbf{n}$-sphere once.

To motivate the formula we will find for the topological number, we begin with a problem from magnetostatics. Suppose we are given a cable originally made up of a bundle of many parallel wires. The cable is then twisted N times about its axis and bent into a closed loop, the end of each individual wire being attached to its beginning to make a continuous circuit (Figure 12.9). A current I flows in the cable in such a manner that each individual wire carries only an infinitesimal part δI_i of the total. The sense of the current is such that as we flow with it around the cable each wire wraps N times anticlockwise about all the others. The current produces a magnetic field $\mathbf{B}$. Can we determine the integer twisting number N knowing only this $\mathbf{B}$ field?

The answer is *yes*. We use Ampère's law in integral form,

$$\oint_\Gamma \mathbf{B} \cdot d\mathbf{r} = (\text{current encircled by } \Gamma). \tag{12.61}$$

We also observe that the current density $\nabla \times \mathbf{B} = \mathbf{J}$ at a point is directed along the tangent to the wire passing through that point. We therefore integrate along each individual wire as it encircles the others, and sum over the wires to find

$$NI^2 = \sum_{\text{wires } i} \delta I_i \oint \mathbf{B} \cdot d\mathbf{r}_i = \int \mathbf{B} \cdot \mathbf{J}\, d^3x = \int \mathbf{B} \cdot (\nabla \times \mathbf{B})\, d^3x. \tag{12.62}$$

We now apply this insight to our three-dimensional field of unit vectors $\mathbf{n}(x)$. The quantity playing the role of the current density $\mathbf{J}$ is the *topological current*

$$J^\sigma = \frac{1}{2}\epsilon^{\sigma\mu\nu}\epsilon_{ijk} n^i \partial_\mu n^j \partial_\nu n^k. \tag{12.63}$$

Observe that div $\mathbf{J} = 0$. This is simply another way of saying that the 2-form $F = \varphi^*\Omega$ is closed.

The flux of $\mathbf{J}$ through a surface S is

$$I = \int_S \mathbf{J} \cdot d\mathbf{S} = \int_S F \tag{12.64}$$

and this is the area of the spherical surface covered by the $\mathbf{n}$'s. A skyrmion, for example, has total topological current $I = 4\pi$, the total surface area of the 2-sphere. The skyrmion world-line will play the role of the cable, and the inverse images in $\mathbb{R}^3$ of points on S^2 correspond to the individual wires.

In form language, the field corresponding to $\mathbf{B}$ can be any 1-form A such that $dA = F$. Thus

$$N_{\text{Hopf}} = \frac{1}{I^2} \int_{\mathbb{R}^3} \mathbf{B} \cdot \mathbf{J}\, d^3x = \frac{1}{16\pi^2} \int_{\mathbb{R}^3} AF \tag{12.65}$$

will be an integer. This integer is the *Hopf linking number*, or *Hopf index*, and counts the number of times the skyrmion twists before it bites its tail to form a closed-loop world-line.

There is another way of obtaining this formula, and of understanding the number $16\pi^2$. We observe that the two-form F and the one-form A are the pull-back from S^3 to $\mathbb{R}^3$ along ψ of the forms

$$\mathcal{F} = \frac{1}{i} \sum_{i=1}^{2} (d\bar{z}_i dz_i - dz_i d\bar{z}_i),$$

$$\mathcal{A} = \frac{1}{i} \sum_{i=1}^{2} (\bar{z}_i dz_i - z_i d\bar{z}_i), \tag{12.66}$$

respectively. If we substitute $z_{1,2} = \xi_{1,2} + i\eta_{1,2}$, we find that

$$\mathcal{A}\mathcal{F} = 8(\xi_1 d\eta_1 d\xi_2 d\eta_2 - \eta_1 d\xi_1 d\xi_2 d\eta_2 + \xi_2 d\eta_2 d\xi_1 d\eta_1 - \eta_2 d\xi_2 d\xi_1 d\eta_1). \tag{12.67}$$

We know from Exercise 12.2 that this expression is eight times the volume 3-form on the 3-sphere. Now the total volume of the unit 3-sphere is $2\pi^2$, and so, from our factored map $x \mapsto \psi \mapsto \mathbf{n}$ we have that

$$N_{\text{Hopf}} = \frac{1}{16\pi^2} \int_{\mathbb{R}^3} \psi^*(\mathcal{A}\mathcal{F}) = \frac{1}{2\pi^2} \int_{\mathbb{R}^3} \psi^* d(\text{Volume on } S^3) \tag{12.68}$$

is the number of times the normalized spinor $\psi(x)$ covers S^3 as x covers $\mathbb{R}^3$. For the Hopf map itself, this number is unity, and so the loop in S^3 that is the inverse image of a point in S^2 will twist once around any other such inverse image loop.

We have now established that

$$\pi_3(S^2) = \mathbb{Z}. \tag{12.69}$$

This result, implying that there are many maps from the 3-sphere to the 2-sphere that are not smoothly deformable to a constant map, was a great surprise when Hopf discovered it.

One of the principal physics consequences of the existence of the Hopf index is that "quantum lump" quasi-particles such as the skyrmion can be fermions, even though they are described by commuting (and therefore bosonic) fields.

To understand how this can be, we first explain that the collection of homotopy classes $\pi_n(M)$ is not just a *set*. It has the additional structure of being a *group*: we can compose two homotopy classes to get a third, the composition is associative, and each homotopy class has an inverse. To define the group composition law, we think of S^n as the interior of an n-dimensional cube with the map $f : S^n \to M$ taking a fixed value $m_0 \in M$ at all points on the boundary of the cube. The boundary can then be considered to be a single point on S^n. We then take one of the n dimensions as being "time" and place two cubes and their maps f_1, f_2 into contact, with f_1 being "earlier" and f_2 being "later". We thus get a continuous map from a bigger box into M. The homotopy class of this map, after we relax the condition that the map takes the value m_0 on the common boundary, defines the composition $[f_2] \circ [f_1]$ of the two homotopy classes corresponding to f_1 and f_2. The composition may be shown to be independent of the choice of representative functions in the two classes. The inverse of a homotopy class $[f]$ is obtained by reversing the direction of "time" for each of the maps in the class. This group structure appears to depend on the fixed point m_0. As long as M is arcwise connected, however, the groups obtained from different m_0's are *isomorphic*, or equivalent. In the case of $\pi_2(S^2) = \mathbb{Z}$ and $\pi_3(S^2) = \mathbb{Z}$, the composition law is simply the addition of the integers $N \in \mathbb{Z}$ that label the classes. A useful exposition of homotopy theory for physicists is to be found in a review article by David Mermin.[3]

When we quantize using Feynman's "sum over histories" path integral, we have the option of multiplying the contributions of histories f that are not deformable into one another by distinct phase factors $\exp\{i\phi([f])\}$. The choice of phases must, however, be compatible with the composition of histories by concatenating one after the other – the same operation as composing homotopy classes. This means that the product $\exp\{i\phi([f_1])\}\exp\{i\phi([f_2])\}$ of the phase factors for two possible histories must be the phase factor $\exp\{i\phi([f_2] \circ [f_1])\}$ assigned to the composition of their homotopy classes. If our quantum system consists of spins $\mathbf{n}$ in two space and one time dimension we can consistently assign a phase factor $\exp(i\pi N_{\text{Hopf}})$ to a history. The rotation of a single skyrmion twists the world-line cable through 2π and so makes $N_{\text{Hopf}} = 1$. The rotation therefore causes the wavefunction to change sign. We will show, in the next section, that a history where two particles change places can be continuously deformed into a history where they do not interchange, but instead one of them is twisted through 2π. The

[3] N. D. Mermin, *Rev. Mod. Phys.*, **51** (1979) 591.

wavefunction of a pair of skyrmions therefore changes sign when they are interchanged. This means that the quantized skyrmion is a fermion.

12.4.6 Twist and writhe

Consider two oriented non-intersecting closed curves γ_1 and γ_2 in $\mathbb{R}^3$. Imagine that γ_2 carries a unit current in the direction of its orientation and so gives rise to a magnetic field. Ampère's law then tells us that the number of times γ_1 encircles γ_2 is

$$\begin{aligned}\mathrm{Lk}(\gamma_1,\gamma_2) &= \oint_{\gamma_1} \mathbf{B}(\mathbf{r}_1)\cdot d\mathbf{r}_1 \\ &= \frac{1}{4\pi}\oint_{\gamma_1}\oint_{\gamma_2} \frac{(\mathbf{r}_1-\mathbf{r}_2)\cdot(d\mathbf{r}_1\times d\mathbf{r}_2)}{|\mathbf{r}_1-\mathbf{r}_2|^3}. \qquad (12.70)\end{aligned}$$

Here the second expression follows from the first by an application of the Biot–Savart law to compute the $\mathbf{B}$ field due to the current. This expression also shows that $\mathrm{Lk}(\gamma_1,\gamma_2)$, which is called the *Gauss linking number*, is symmetric under the interchange $\gamma_1 \leftrightarrow \gamma_2$ of the two curves. It changes sign, however, if one of the curves changes orientation, or if the pair of curves is reflected in a mirror.

We can relate the Gauss linking number to the Brouwer degree of a map. Introduce parameters t_1, t_2 with $0 < t_1, t_2 \leq 1$ to label points on the two curves. The curves are closed, so $\mathbf{r}_1(0) = \mathbf{r}_1(1)$, and similarly for $\mathbf{r}_2$. Let us also define a unit vector

$$\mathbf{n}(t_1,t_2) = \frac{\mathbf{r}_1(t_1)-\mathbf{r}_2(t_2)}{|\mathbf{r}_1(t_1)-\mathbf{r}_2(t_2)|}. \qquad (12.71)$$

Then

$$\begin{aligned}\mathrm{Lk}(\gamma_1,\gamma_2) &= \frac{1}{4\pi}\oint_{\gamma_1}\oint_{\gamma_2} \frac{\mathbf{r}_1(t_1)-\mathbf{r}_2(t_2)}{|\mathbf{r}_1(t_1)-\mathbf{r}_2(t_2)|^3}\cdot\left(\frac{\partial \mathbf{r}_1}{\partial t_1}\times\frac{\partial \mathbf{r}_2}{\partial t_2}\right) dt_1 dt_2 \\ &= -\frac{1}{4\pi}\int_{T^2} \mathbf{n}\cdot\left(\frac{\partial \mathbf{n}}{\partial t_1}\times\frac{\partial \mathbf{n}}{\partial t_2}\right) dt_1 dt_2 \qquad (12.72)\end{aligned}$$

is seen to be (minus) the winding number of the map

$$\mathbf{n} : [0,1]\times[0,1] \to S^2 \qquad (12.73)$$

of the 2-torus into the sphere. Our previous results on maps into the 2-sphere therefore confirm our Ampère-law intuition that $\mathrm{Lk}(\gamma_1,\gamma_2)$ is an integer. The linking number is also topological invariant, being unchanged under any deformation of the curves that does not cause one to pass through the other.

An important application of these ideas occurs in biology, where the curves are the two complementary strands of a closed loop of DNA. We can think of two such parallel curves as forming the edges of a *ribbon* $\{\gamma_1,\gamma_2\}$ of width ϵ. Let us denote by γ the curve

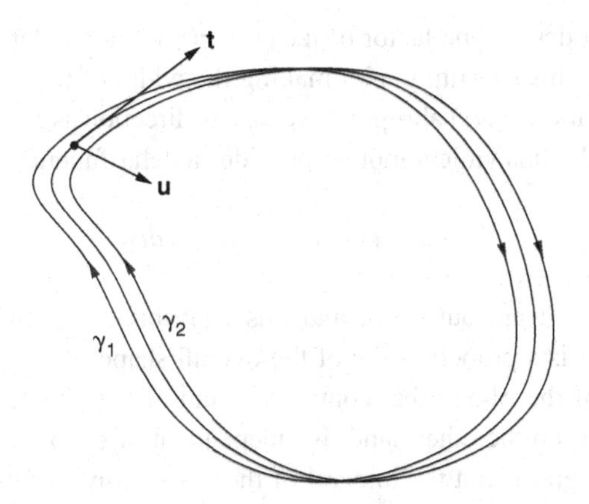

Figure 12.10 An oriented ribbon $\{\gamma_1, \gamma_2\}$ showing the vectors **t** and **u**.

$\mathbf{r}(t)$ running along the axis of the ribbon midway between γ_1 and γ_2. The unit tangent to γ at the point $\mathbf{r}(t)$ is

$$\mathbf{t}(t) = \frac{\dot{\mathbf{r}}(t)}{|\dot{\mathbf{r}}(t)|}, \tag{12.74}$$

where, as usual, the dots denote differentiation with respect to t. We also introduce a unit vector $\mathbf{u}(t)$ that is perpendicular to $\mathbf{t}(t)$ and lies in the ribbon, pointing from $\mathbf{r}_1(t)$ to $\mathbf{r}_2(t)$; see Figure 12.10.

We will assign a common value of the parameter t to a point on γ and the points nearest to $\mathbf{r}(t)$ on γ_1 and γ_2. Consequently

$$\begin{aligned}\mathbf{r}_1(t) &= \mathbf{r}(t) - \frac{1}{2}\epsilon\, \mathbf{u}(t)\\ \mathbf{r}_2(t) &= \mathbf{r}(t) + \frac{1}{2}\epsilon\, \mathbf{u}(t).\end{aligned} \tag{12.75}$$

We can express $\dot{\mathbf{u}}$ as

$$\dot{\mathbf{u}} = \boldsymbol{\omega} \times \mathbf{u} \tag{12.76}$$

for some angular-velocity vector $\boldsymbol{\omega}(t)$. The quantity

$$\mathrm{Tw} = \frac{1}{2\pi} \oint_\gamma (\boldsymbol{\omega} \cdot \mathbf{t})\, dt \tag{12.77}$$

is called the *twist* of the ribbon. It is not usually an integer, and is a property of the ribbon $\{\gamma_1, \gamma_2\}$ itself, being independent of the choice of parametrization t.

If we set $\mathbf{r}_1(t)$ and $\mathbf{r}_2(t)$ equal to the single axis curve $\mathbf{r}(t)$ in the integrand of (12.70), the resulting "self-linking" integral, or *writhe*,

$$\mathrm{Wr} \stackrel{\text{def}}{=} \frac{1}{4\pi} \oint_\gamma \oint_\gamma \frac{(\mathbf{r}(t_1) - \mathbf{r}(t_2)) \cdot (\dot{\mathbf{r}}(t_1) \times \dot{\mathbf{r}}(t_2))}{|\mathbf{r}(t_1) - \mathbf{r}(t_2)|^3}\, dt_1 dt_2 \tag{12.78}$$

remains convergent despite the factor of $|\mathbf{r}(t_1) - \mathbf{r}(t_2)|^3$ in the denominator. However, if we try to achieve this substitution by making the width of the ribbon ϵ tend to zero, we find that the vector $\mathbf{n}(t_1, t_2)$ abruptly reverses its direction as t_1 passes t_2. In the limit of infinitesimal width this violent motion provides a delta-function contribution

$$-(\boldsymbol{\omega} \cdot \mathbf{t})\delta(t_1 - t_2)\, dt_1 \wedge dt_2 \tag{12.79}$$

to the 2-sphere area swept out by $\mathbf{n}$, and this contribution is invisible to the writhe integral. The writhe is a property only of the overall shape of the axis curve γ, and is independent both of the ribbon that contains it, and of the choice of parametrization. The linking number, on the other hand, is independent of ϵ, so the $\epsilon \to 0$ limit of the linking-number integral is not the integral of the $\epsilon \to 0$ limit of its integrand. Instead we have

$$\mathrm{Lk}(\gamma_1, \gamma_2) = \frac{1}{2\pi}\oint_\gamma (\boldsymbol{\omega} \cdot \mathbf{t})\, dt + \frac{1}{4\pi}\oint_\gamma \oint_\gamma \frac{(\mathbf{r}(t_1) - \mathbf{r}(t_2)) \cdot (\dot{\mathbf{r}}(t_1) \times \dot{\mathbf{r}}(t_2))}{|\mathbf{r}(t_1) - \mathbf{r}(t_2)|^3}\, dt_1 dt_2. \tag{12.80}$$

This formula

$$\mathrm{Lk} = \mathrm{Tw} + \mathrm{Wr} \tag{12.81}$$

is known as the *Calugareanu–White–Fuller* relation, and is the basis for the claim, made in the previous section, that the world-line of an extended particle with an exchange ($\mathrm{Wr} = \pm 1$) can be deformed into a world-line with a 2π rotation ($\mathrm{Tw} = \pm 1$) without changing the topologically invariant linking number.

By setting

$$\mathbf{n}(t_1, t_2) = \frac{\mathbf{r}(t_1) - \mathbf{r}(t_2)}{|\mathbf{r}(t_1) - \mathbf{r}(t_2)|} \tag{12.82}$$

we can express the writhe as

$$\mathrm{Wr} = -\frac{1}{4\pi}\int_{T^2} \mathbf{n} \cdot \left(\frac{\partial \mathbf{n}}{\partial t_1} \times \frac{\partial \mathbf{n}}{\partial t_2}\right) dt_1 dt_2, \tag{12.83}$$

but we must take care to recognize that this new $\mathbf{n}(t_1, t_2)$ is discontinuous across the line $t = t_1 = t_2$. It is equal to $\mathbf{t}(t)$ for t_1 infinitesimally larger than t_2, and equal to $-\mathbf{t}(t)$ when t_1 is infinitesimally smaller than t_2. By cutting the square domain of integration and reassembling it into a rhomboid, as shown in Figure 12.11, we obtain a continuous integrand and see that the writhe is (minus) the 2-sphere area (counted with multiplicities and divided by 4π) of a region whose boundary is composed of two curves Γ, the *tangent indicatrix*, or *tantrix*, on which $\mathbf{n} = \mathbf{t}(t)$, and its oppositely oriented antipodal counterpart Γ' on which $\mathbf{n} = -\mathbf{t}(t)$.

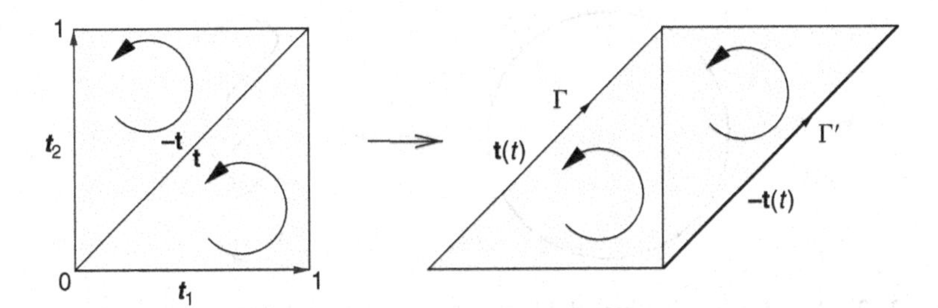

Figure 12.11 Cutting and reassembling the domain of integration in (12.83).

The 2-sphere area $\Omega(\Gamma)$ bounded by Γ is only determined by Γ up to the addition of integer multiples of 4π. Taking note that the "wrong" orientation of the boundary Γ (see Figure 12.11 again) compensates for the minus sign before the integral in (12.83), we have

$$4\pi \,\mathrm{Wr} = 2\Omega(\Gamma) + 4\pi n. \tag{12.84}$$

Thus,

$$\mathrm{Wr} = \frac{1}{2\pi}\Omega(\Gamma), \mod 1. \tag{12.85}$$

We can do better than (12.85) once we realize that by allowing crossings we can continuously deform any closed curve into a perfect circle. Each self-crossing causes Lk and Wr (but not Tw which, being a local functional, does not care about crossings) to jump by ± 2. For a perfect circle $\mathrm{Wr} = 0$ whilst $\Omega = 2\pi$. We therefore have an improved estimate of the additive integer that is left undetermined by Γ, and from it we obtain

$$\mathrm{Wr} = 1 + \frac{1}{2\pi}\Omega(\Gamma), \mod 2. \tag{12.86}$$

This result is due to Brock Fuller.[4]

We can use our ribbon language to describe conformational transitions in long molecules. The elastic energy of a closed rod (or DNA molecule) can be approximated by

$$E = \int_\gamma \left\{ \frac{1}{2}\alpha(\boldsymbol{\omega}\cdot\mathbf{t})^2 + \frac{1}{2}\beta\kappa^2 \right\} ds. \tag{12.87}$$

Here we are parametrizing the curve by its arc-length s. The constant α is the torsional stiffness coefficient, β is the flexural stiffness and

$$\kappa(s) = \left| \frac{d^2\mathbf{r}(s)}{ds^2} \right| = \left| \frac{d\mathbf{t}(s)}{ds} \right| \tag{12.88}$$

[4] F. Brock Fuller, *Proc. Natl. Acad. Sci. USA*, **75** (1978) 3557.

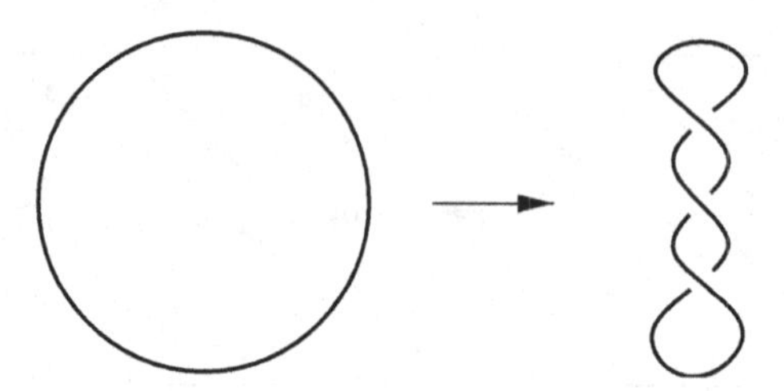

Figure 12.12 A molecule initially with Lk = 3, Tw = 3, Wr = 0 writhes to a new configuration with Lk = 3, Tw = 0, Wr = 3.

is the local curvature. Suppose that our molecule has linking number n, i.e. it was twisted n times before the ends were joined together to make a loop.

When $\beta \gg \alpha$ the molecule will minimize its bending energy by forming a planar circle with Wr ≈ 0 and Tw $\approx n$. If we increase α, or decrease β, there will come a point at which the molecule will seek to save torsional energy at the expense of bending, and will suddenly writhe into a new configuration with Wr $\approx n$ and Tw ≈ 0 (Figure 12.12). Such twist-to-writhe transformations will be familiar to anyone who has struggled to coil a garden hose or electric cable.

12.5 Further exercises and problems

Exercise 12.7: A 2-form is expressed in cartesian coordinates as

$$\omega = \frac{1}{r^3}(zdxdy + xdydz + ydzdx)$$

where $r = \sqrt{x^2 + y^2 + z^2}$.

(a) Evaluate $d\omega$ for $r \neq 0$.
(b) Evaluate the integral

$$\Phi = \int_P \omega,$$

over the infinite plane $P = \{-\infty < x < \infty, -\infty < y < \infty, z = 1\}$.
(c) A sphere is embedded into $\mathbb{R}^3$ by the map φ, which takes the point $(\theta, \phi) \in S^2$ to the point $(x, y, z) \in \mathbb{R}^3$, where

$$x = R\cos\phi\sin\theta,$$

$$y = R\sin\phi\sin\theta,$$

$$z = R\cos\theta.$$

Pull back ω and find the 2-form $\varphi^*\omega$ on the sphere. (Hint: the form $\varphi^*\omega$ is both familiar and simple. If you end up with an intractable mess of trigonometric functions, you have made an algebraic error.)

(d) By exploiting the result of part (c), or otherwise, evaluate the integral

$$\Phi = \int_{S^2(R)} \omega$$

where $S^2(R)$ is the surface of a 2-sphere of radius R centred at the origin.

The following four exercises all explore the same geometric facts relating to Stokes' theorem and the area 2-form of a sphere, but in different physical settings.

Exercise 12.8: A flywheel of moment of inertia I can rotate without friction about an axle whose direction is specified by a unit vector $\mathbf{n}$ (Figure 12.13). The flywheel and axle are initially stationary. The direction $\mathbf{n}$ of the axle is made to describe a simple closed curve $\gamma = \partial\Omega$ on the unit sphere, and is then left stationary.

Show that once the axle has returned to rest in its initial direction, the flywheel has also returned to rest, but has rotated through an angle $\theta = \text{Area}(\Omega)$ when compared with its initial orientation. The area of Ω is to be counted as positive if the path γ surrounds it in a clockwise sense, and negative otherwise. Observe that the path γ bounds two regions with opposite orientations. Taking into account the fact that we cannot define the rotation angle at intermediate steps, show that the area of either region can be used to compute θ, the results being physically indistinguishable. (Hint: show that the component $L_Z = I(\dot{\psi} + \dot{\phi}\cos\theta)$ of the flywheel's angular momentum along the axle is a constant of the motion.)

Exercise 12.9: A ball of unit radius rolls without slipping on a table. The ball moves in such a way that the point in contact with table describes a closed path $\gamma = \partial\Omega$ *on the ball*. (The corresponding path on the *table* will not necessarily be closed.) Show that the final orientation of the ball will be such that it has rotated, when compared with its initial

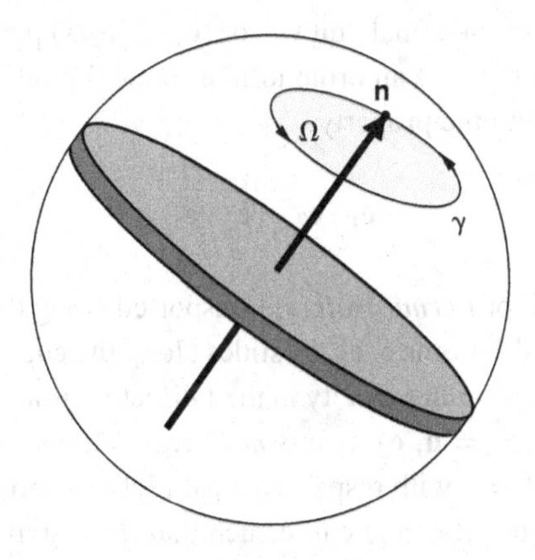

Figure 12.13 Flywheel.

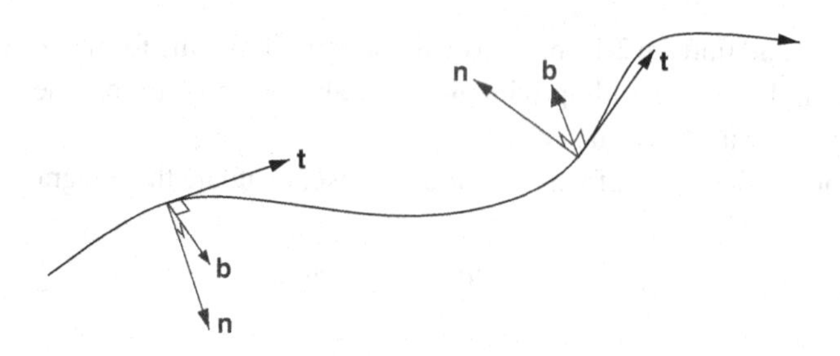

Figure 12.14 Serret–Frenet frames.

orientation, through an angle $\phi = \text{Area}(\Omega)$ about a vertical axis through its centre. As in the previous problem, the area is counted positive if γ encircles Ω in an anticlockwise sense. (Hint: recall the no-slip rolling condition $\dot{\phi} + \dot{\psi}\cos\theta = 0$ from (11.29).)

Exercise 12.10: Let a curve in $\mathbb{R}^3$ be parametrized by its arc-length s as $\mathbf{r}(s)$. Then the unit tangent to the curve is given by

$$\mathbf{t}(s) = \dot{\mathbf{r}} \stackrel{\text{def}}{=} \frac{d\mathbf{r}}{ds}.$$

The *principal normal* $\mathbf{n}(s)$ and the *binormal* $\mathbf{b}(s)$ to the curve are defined by the requirement that $\dot{\mathbf{t}} = \kappa\mathbf{n}$ with the *curvature* $\kappa(s)$ positive, and that $\mathbf{t}$, $\mathbf{n}$ and $\mathbf{b} = \mathbf{t}\times\mathbf{n}$ form a right-handed orthonormal frame (Figure 12.14).

(a) Show that there exists a scalar $\tau(s)$, the *torsion* of the curve, such that $\mathbf{t}$, $\mathbf{n}$ and $\mathbf{b}$ obey the *Serret–Frenet* relations

$$\begin{pmatrix}\dot{\mathbf{t}}\\ \dot{\mathbf{n}}\\ \dot{\mathbf{b}}\end{pmatrix} = \begin{pmatrix}0 & \kappa & 0\\ -\kappa & 0 & \tau\\ 0 & -\tau & 0\end{pmatrix}\begin{pmatrix}\mathbf{t}\\ \mathbf{n}\\ \mathbf{b}\end{pmatrix}.$$

(b) Any pair of mutually orthogonal unit vectors $\mathbf{e}_1(s)$, $\mathbf{e}_2(s)$ perpendicular to $\mathbf{t}$ and such that $\mathbf{e}_1\times\mathbf{e}_2 = \mathbf{t}$ can serve as an orthonormal frame for vectors in the normal plane. A basis pair $\mathbf{e}_1$, $\mathbf{e}_2$ with the property

$$\dot{\mathbf{e}}_1\cdot\mathbf{e}_2 - \dot{\mathbf{e}}_2\cdot\mathbf{e}_1 = 0$$

is said to be *parallel*, or *Fermi–Walker*, transported along the curve. In other words, a parallel-transported 3-frame $\mathbf{t}$, $\mathbf{e}_1$, $\mathbf{e}_2$ slides along the curve $\mathbf{r}(s)$ in such a way that the component of its angular velocity in the $\mathbf{t}$ direction is always zero. Show that the Serret–Frenet frame $\mathbf{e}_1 = \mathbf{n}$, $\mathbf{e}_2 = \mathbf{b}$ is *not* parallel transported, but instead rotates at angular velocity $\dot{\theta} = \tau$ with respect to a parallel-transported frame.

(c) Consider a finite segment of the curve such that the initial and final Serret–Frenet frames are parallel, and so $\mathbf{t}(s)$ defines a closed path $\gamma = \partial\Omega$ on the unit sphere. Fill in the line-by-line justifications for the following sequence of manipulations:

$$\begin{aligned}
\int_\gamma \tau\, ds &= \frac{1}{2}\int_\gamma (\mathbf{b}\cdot\dot{\mathbf{n}} - \mathbf{n}\cdot\dot{\mathbf{b}})\, ds \\
&= \frac{1}{2}\int_\gamma (\mathbf{b}\cdot d\mathbf{n} - \mathbf{n}\cdot d\mathbf{b}) \\
&= \frac{1}{2}\int_\Omega (d\mathbf{b}\cdot d\mathbf{n} - d\mathbf{n}\cdot d\mathbf{b}) \qquad (*) \\
&= \frac{1}{2}\int_\Omega \{(d\mathbf{b}\cdot\mathbf{t})(\mathbf{t}\cdot d\mathbf{n}) - (d\mathbf{n}\cdot\mathbf{t})(\mathbf{t}\cdot d\mathbf{b})\} \\
&= \frac{1}{2}\int_\Omega \{(\mathbf{b}\cdot d\mathbf{t})(d\mathbf{t}\cdot\mathbf{n}) - (\mathbf{n}\cdot d\mathbf{t})(d\mathbf{t}\cdot\mathbf{b})\} \\
&= -\frac{1}{2}\int_\Omega \mathbf{t}\cdot(d\mathbf{t}\times d\mathbf{t}) \\
&= -\mathrm{Area}(\Omega).
\end{aligned}$$

(The line marked '$(*)$' is the one that requires most thought. How can we define "**b**" and "**n**" in the interior of Ω?)

(d) Conclude that a Fermi–Walker transported frame will have rotated through an angle $\theta = \mathrm{Area}(\Omega)$, compared to its initial orientation, by the time it reaches the end of the curve.

The plane of transversely polarized light propagating in a monomode optical fibre is Fermi–Walker transported, and this rotation can be studied experimentally.[5]

Exercise 12.11: *Foucault's pendulum (in disguise).* A particle of mass m is constrained by a pair of frictionless plates to move in a plane Π that passes through the origin O. The particle is attracted to O by a force $-\kappa\mathbf{r}$, and it therefore executes harmonic motion within Π. The orientation of the plane, specified by a normal vector $\mathbf{n}$, can be altered in such a way that Π continues to pass through the centre of attraction O.

(a) Show that the constrained motion is described by the equation

$$m\ddot{\mathbf{r}} + \kappa\mathbf{r} = \lambda(t)\mathbf{n},$$

and determine $\lambda(t)$ in terms of m, $\mathbf{n}$ and $\dot{\mathbf{r}}$.

(b) Initially the particle motion is given by

$$\mathbf{r}(t) = \mathbf{A}\cos(\omega t + \phi).$$

[5] A. Tomita, R. Y. Chao, *Phys. Rev. Lett.*, **57** (1986) 937.

Now assume that $\mathbf{n}$ changes direction slowly compared to the frequency $\omega = \sqrt{\kappa/m}$. Seek a solution in the form

$$\mathbf{r}(t) = \mathbf{A}(t)\cos(\omega t + \phi),$$

and show that $\dot{\mathbf{A}} = -\mathbf{n}(\dot{\mathbf{n}} \cdot \mathbf{A})$. Deduce that $|\mathbf{A}|$ remains constant, and so $\dot{\mathbf{A}} = \boldsymbol{\omega} \times \mathbf{A}$ for some angular velocity vector $\boldsymbol{\omega}$. Show that $\boldsymbol{\omega}$ is perpendicular to $\mathbf{n}$.

(c) Show that the results of part (b) imply that the direction of oscillation $\mathbf{A}$ is "parallel transported", in the sense of the previous problem. Conclude that if $\mathbf{n}$ slowly describes a closed loop $\gamma = \partial\Omega$ on the unit sphere, then the direction of oscillation $\mathbf{A}$ ends up rotated through an angle $\theta = \text{Area}(\Omega)$.

The next exercise introduces a clever trick for solving some of the nonlinear partial differential equations of field theory. The class of equations to which it and its generalizations are applicable is rather restricted, but when they work they provide a complete multi-soliton solution.

Problem 12.12: In this problem you will find the spin field $\mathbf{n}(x)$ that minimizes the energy functional

$$E[\mathbf{n}] = \frac{1}{2}\int_{\mathbb{R}^2} \left(|\nabla n^1|^2 + |\nabla n^2|^2 + |\nabla n^3|^2\right) dx^1 dx^2$$

for a given positive winding number N.

(a) Use the results of Exercise 12.6 to write the winding number N, defined in (12.35), and the energy functional $E[\mathbf{n}]$ as

$$4\pi N = \int \frac{4}{(1+\xi^2+\eta^2)^2}\left(\partial_1\xi\,\partial_2\eta - \partial_1\eta\,\partial_2\xi\right) dx^1 dx^2,$$

$$E[\mathbf{n}] = \frac{1}{2}\int \frac{4}{(1+\xi^2+\eta^2)^2}\left((\partial_1\xi)^2 + (\partial_2\xi)^2 + (\partial_1\eta)^2 + (\partial_2\eta)^2\right) dx^1 dx^2,$$

where ξ and η are stereographic coordinates on S^2 specifying the direction of the unit vector $\mathbf{n}$.

(b) Deduce the inequality

$$E - 4\pi N \stackrel{\text{def}}{=} \frac{1}{2}\int \frac{4}{(1+\xi^2+\eta^2)^2}|(\partial_1 + i\partial_2)(\xi + i\eta)|^2\, dx^1 dx^2 \geq 0.$$

(c) Deduce that, for winding number $N > 0$, the minimum-energy solutions have energy $E = 4\pi N$ and are obtained by solving the *first-order* linear partial differential equation

$$\left(\frac{\partial}{\partial x^1} + i\frac{\partial}{\partial x^2}\right)(\xi + i\eta) = 0.$$

(d) Solve the partial differential equation in part (c), and hence show that the minimal-energy solutions with winding number $N > 0$ are given by

$$\xi + i\eta = \lambda \frac{(z - a_1) \dots (z - a_N)}{(z - b_1) \dots (z - b_N)},$$

where $z = x^1 + ix^2$, and $\lambda, a_1, \dots, a_N$ and $b_1, \dots, b_N$ are arbitrary complex numbers – except that no a may coincide with any b. This is the solution that we displayed at the end of Section 12.4.2.

(e) Repeat the analysis for $N < 0$. Show that the solutions are given in terms of rational functions of $\bar{z} = x^1 - ix^2$.

The idea of combining the energy functional and the topological charge into a single, manifestly positive, functional is due to Evgueny Bogomol'nyi. The resulting first-order linear equation is therefore called a *Bogomolnyi equation*. If we had tried to find a solution directly in terms of $\mathbf{n}$, we would have ended up with a horribly nonlinear second-order partial differential equation.

Exercise 12.13: *Lobachevski space*. The hyperbolic plane of Lobachevski geometry can be realized by embedding the $Z \geq R$ branch of the two-sheeted hyperboloid $Z^2 - X^2 - Y^2 = R^2$ into a Minkowski space with metric $ds^2 = -dZ^2 + dX^2 + dY^2$.

We can parametrize the embedded surface by making an "imaginary radius" version of the stereographic map, in which the point P on the hyperboloid is labelled by the coordinates of the point Q on the XY-plane (see Figure 12.15).

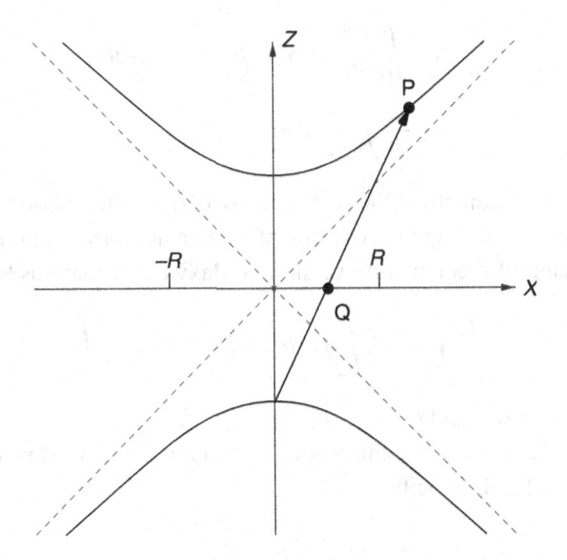

Figure 12.15 A slice through the embedding of two-dimensional Lobachevski space into three-dimensional Minkowski space, showing the stereographic parametrization of the embedded space by the Poincaré disc $X^2 + Y^2 < R^2$.

(i) Show that this embedding induces the metric

$$g(\ ,\ ) = \frac{4R^4}{(R^2 - X^2 - Y^2)^2}(dX \otimes dX + dY \otimes dY), \quad X^2 + Y^2 < R^2,$$

of the Poincaré disc model (see Problem 1.7) on the hyperboloid.

(ii) Use the induced metric to show that the area of a disc of hyperbolic radius ρ is given by

$$Area = 4\pi R^2 \sinh^2\left(\frac{\rho}{2R}\right) = 2\pi R^2(\cosh(\rho/R) - 1),$$

and so is only given by $\pi\rho^2$ when ρ is small compared to the scale R of the hyperbolic space. It suffices to consider circles with their centres at the origin. You will first need to show that the hyperbolic distance ρ from the centre of the disc to a point at Euclidean distance r is

$$\rho = R \ln\left(\frac{R+r}{R-r}\right).$$

Exercise 12.14: Faraday's "flux rule" for computing the electromotive force $\mathcal{E}$ in a circuit containing a thin moving wire is usually derived by the following manipulations:

$$\begin{aligned}
\mathcal{E} &\equiv \oint_{\partial\Omega} (\mathbf{E} + \mathbf{v} \times \mathbf{B}) \cdot d\mathbf{r} \\
&= \int_{\Omega} \operatorname{curl} \mathbf{E} \cdot d\mathbf{S} - \oint_{\partial\Omega} \mathbf{B} \cdot (\mathbf{v} \times d\mathbf{r}) \\
&= -\int_{\Omega} \frac{\partial \mathbf{B}}{\partial t} \cdot d\mathbf{S} - \oint_{\partial\Omega} \mathbf{B} \cdot (\mathbf{v} \times d\mathbf{r}) \\
&= -\frac{d}{dt}\int_{\Omega} \mathbf{B} \cdot d\mathbf{S}.
\end{aligned}$$

(a) Show that if we parametrize the surface Ω as $x^\mu(u, v, \tau)$, with u, v labelling points on Ω, and τ parametrizing the evolution of Ω, then the corresponding manipulations in the covariant differential-form version of Maxwell's equations lead to

$$\frac{d}{d\tau}\int_{\Omega} F = \int_{\Omega} \mathcal{L}_V F = \int_{\partial\Omega} i_V F = -\int_{\partial\Omega} f,$$

where $V^\mu = \partial x^\mu / \partial \tau$ and $f = -i_V F$.

(b) Show that if we take τ to be the proper time along the world-line of each element of Ω, then V is the 4-velocity

$$V^\mu = \frac{1}{\sqrt{1 - \mathbf{v}^2}}(1, \mathbf{v}),$$

and $f = -i_V F$ becomes the 1-form corresponding to the Lorentz-force 4-vector.

It is not clear that the terms in this covariant form of Faraday's law can be given any physical interpretation outside the low-velocity limit. When parts of $\partial\Omega$ have different velocities, the relation of the integrals to measurements made at fixed coordinate time requires thought.[6]

The next pair of exercises explores some physics appearances of the continuum Hopf linking number (12.65).

Exercise 12.15: The equations governing the motion of an incompressible inviscid fluid are $\nabla \cdot \mathbf{v} = 0$ and Euler's equation

$$\frac{D\mathbf{v}}{Dt} \stackrel{\text{def}}{=} \frac{\partial \mathbf{v}}{\partial t} + (\mathbf{v} \cdot \nabla)\mathbf{v} = -\nabla P.$$

Recall that the operator $\partial/\partial t + \mathbf{v} \cdot \nabla$, here written as D/Dt, is called the *convective derivative*.

(a) Take the curl of Euler's equation to show that if $\boldsymbol{\omega} = \nabla \times \mathbf{v}$ is the *vorticity* then

$$\frac{D\boldsymbol{\omega}}{Dt} \equiv \frac{\partial \boldsymbol{\omega}}{\partial t} + (\mathbf{v} \cdot \nabla)\boldsymbol{\omega} = (\boldsymbol{\omega} \cdot \nabla)\mathbf{v}.$$

(b) Combine Euler's equation with part (a) to show that

$$\frac{D}{Dt}(\mathbf{v} \cdot \boldsymbol{\omega}) = \nabla \cdot \left\{ \boldsymbol{\omega} \left(\frac{1}{2}\mathbf{v}^2 - P \right) \right\}.$$

(c) Show that if Ω is a volume moving with the fluid, and f is a scalar function, then

$$\frac{d}{dt}\int_\Omega f(\mathbf{r}, t)\, dV = \int_\Omega \frac{Df}{Dt}\, dV.$$

(d) Conclude that when $\boldsymbol{\omega}$ is zero at infinity the *helicity*

$$I = \int \mathbf{v} \cdot (\nabla \times \mathbf{v})\, dV = \int \mathbf{v} \cdot \boldsymbol{\omega}\, dV$$

is a constant of the motion.

The helicity measures the Hopf linking number of the vortex lines. The discovery[7] of its conservation launched the field of *topological fluid dynamics*.

Exercise 12.16: Let $\mathbf{B} = \nabla \times \mathbf{A}$ and $\mathbf{E} = -\partial \mathbf{A}/\partial t - \nabla \phi$ be the electric and magnetic fields in an incompressible and perfectly conducting fluid. In such a fluid, the co-moving electromotive force $\mathbf{E} + \mathbf{v} \times \mathbf{B}$ must vanish everywhere.

[6] See E. Marx, *J. Franklin Inst.*, **300** (1975) 353.
[7] H. K. Moffatt, *J. Fluid Mech.*, **35** (1969) 117.

(a) Use Maxwell's equations to show that

$$\frac{\partial \mathbf{A}}{\partial t} = \mathbf{v} \times (\nabla \times \mathbf{A}) - \nabla \phi,$$
$$\frac{\partial \mathbf{B}}{\partial t} = \nabla \times (\mathbf{v} \times \mathbf{B}).$$

(b) From part (a) show that the convective derivative of $\mathbf{A} \cdot \mathbf{B}$ is given by

$$\frac{D}{Dt} (\mathbf{A} \cdot \mathbf{B}) = \nabla \cdot \{\mathbf{B} (\mathbf{A} \cdot \mathbf{v} - \phi)\}.$$

(c) By using the same reasoning as in the previous problem, and assuming that $\mathbf{B}$ is zero at infinity, conclude that *Woltjer's invariant*,

$$\int (\mathbf{A} \cdot \mathbf{B}) \, dV = \int \epsilon_{ijk} A_i \partial_j A_k d^3x = \int AF,$$

is a constant of the motion.

This result shows that the Hopf linking number of the magnetic field lines is independent of time. It is an essential ingredient in the geodynamo theory of the Earth's magnetic field.

13

An introduction to differential topology

Topology is the study of the consequences of continuity. We all know that a continuous real function defined on a connected interval and positive at one point and negative at another must take the value zero at some point between. This fact seems obvious – although a course of real analysis will convince you of the need for a proof. A less obvious fact, but one that follows from the previous one, is that a continuous function defined on the unit circle must posses two diametrically opposite points at which it takes the same value. To see that this is so, consider $f(\theta+\pi)-f(\theta)$. This difference (if not initially zero, in which case there is nothing further to prove) changes sign as θ is advanced through π, because the two terms exchange roles. It was therefore zero somewhere. This observation has practical application in daily life: our local coffee shop contains four-legged tables that wobble because the floor is not level. They are round tables, however, and because they possess no misguided levelling screws all four legs have the same length. We are therefore guaranteed that by rotating the table about its centre through an angle of less than $\pi/2$ we will find a stable location. A ninety-degree rotation interchanges the pair of legs that are both on the ground with the pair that are rocking, and at the change-over point all four legs must be simultaneously on the ground.

Similar effects with a practical significance for physics appear when we try to extend our vector and tensor calculus from a local region to an entire manifold. A smooth field of vectors tangent to the sphere S^2 will always possess a zero – i.e. a point at which the vector field vanishes. On the torus T^2, however, we can construct a nowhere-zero vector field. This shows that the global topology of the manifold influences the way in which the tangent spaces are glued together to form the tangent bundle. To study this influence in a systematic manner we need first to understand how to characterize the global structure of a manifold, and then to see how this structure affects the mathematical and physical objects that live on it.

13.1 Homeomorphism and diffeomorphism

In the previous chapter we met with a number of *topological invariants* associated with mappings. These *homotopy* invariants were unaffected by continuous deformations of a map, and served to distinguish between topologically distinct mappings. Similarly, *homology* invariants help classify topologically distinct manifolds. The analogue of the winding number is the set of *Betti numbers* of a manifold. If two manifolds have different Betti numbers they are certainly distinct. Unfortunately, if two manifolds have the same

Betti numbers, we cannot be sure that they are topologically identical. It is a Holy Grail of topology to find a complete set of invariants such that having them all coincide would be enough to say that two manifolds were topologically the same.

In the previous paragraph we were deliberately vague in our use of the terms "distinct" and the "same". Two topological spaces (spaces equipped with a definition of what is to be considered an open set) are regarded as being the "same", or *homeomorphic*, if there is a one-to-one, onto, continuous map between them whose inverse is also continuous. Manifolds come with the additional structure of differentiability: we may therefore talk of "smooth" maps, meaning that their expression in coordinates is infinitely (C^∞) differentiable. We regard two manifolds as being the "same", or *diffeomorphic*, if there is a one-to-one onto C^∞ map between them whose inverse is also C^∞. The distinction between homeomorphism and diffeomorphism sounds like a mere technical nicety, but it has consequences for physics. Edward Witten discovered[1] that there are 992 distinct 11-spheres. These are manifolds that are all homeomorphic to the 11-sphere, but diffeomorphically inequivalent. This fact is crucial for the cancellation of global gravitational anomalies in the $E_8 \times E_8$ or SO(32) symmetric superstring theories.

Since we are interested in the consequences of topology for calculus, we shall restrict ourselves to the interpretation "same" = diffeomorphic.

13.2 Cohomology

Betti numbers arise in answer to what seems like a simple calculus problem: when can a vector field whose divergence vanishes be written as the curl of something? We shall see that the answer depends on the global structure of the space the field inhabits.

13.2.1 Retractable spaces: Converse of Poincaré's lemma

Poincaré's lemma asserts that $d^2 = 0$. In traditional vector-calculus language this reduces to the statements curl (grad ϕ) $= 0$ and div (curl $\mathbf{w}$) $= 0$. We often assume that the converse is true: if curl $\mathbf{v} = 0$, we expect that we can find a ϕ such that $\mathbf{v} = \operatorname{grad}\phi$, and if div $\mathbf{v} = 0$ that we can find a $\mathbf{w}$ such that $\mathbf{v} = \operatorname{curl}\mathbf{w}$. You know a formula for the first case:

$$\phi(x) = \int_{x_0}^{x} \mathbf{v} \cdot d\mathbf{x}, \tag{13.1}$$

but you probably do not know the corresponding formula for $\mathbf{w}$. Using differential forms, and provided the space in which these forms live has suitable *topological* properties, it is straightforward to find a solution for the general problem: If ω is closed, meaning that $d\omega = 0$, find χ such that $\omega = d\chi$.

[1] E. Witten, *Comm. Math. Phys.*, **117** (1986) 197.

The "suitable topological properties" referred to in the previous paragraph is that the space be *retractable*. Suppose that the closed form ω is defined in a domain Ω. We say that Ω is retractable to the point O if there exists a smooth map $\varphi_t : \Omega \to \Omega$ which depends continuously on a parameter $t \in [0, 1]$ and for which $\varphi_1(x) = x$ and $\varphi_0(x) = \mathrm{O}$. Applying this retraction map to the form, we will then have $\varphi_1^*\omega = \omega$ and $\varphi_0^*\omega = 0$. Let us set $\varphi_t(x^\mu) = x^\mu(t)$. Define $\eta(x,t)$ to be the velocity-vector field that corresponds to the coordinate flow:

$$\frac{dx^\mu}{dt} = \eta^\mu(x,t). \tag{13.2}$$

An easy exercise, using the interpretation of the Lie derivative in (11.41), shows that

$$\frac{d}{dt}(\varphi_t^*\omega) = \mathcal{L}_\eta(\varphi_t^*\omega). \tag{13.3}$$

We now use the infinitesimal homotopy relation and our assumption that $d\omega = 0$, and hence (from Exercise 12.3) that $d(\varphi_t^*\omega) = 0$, to write

$$\mathcal{L}_\eta(\varphi_t^*\omega) = (i_\eta d + d i_\eta)(\varphi_t^*\omega) = d[i_\eta(\varphi_t^*\omega)]. \tag{13.4}$$

Using this, we can integrate up with respect to t to find

$$\omega = \varphi_1^*\omega - \varphi_0^*\omega = d\left(\int_0^1 i_\eta(\varphi_t^*\omega)dt\right). \tag{13.5}$$

Thus

$$\chi = \int_0^1 i_\eta(\varphi_t^*\omega)dt \tag{13.6}$$

solves our problem.

This magic formula for χ makes use of nearly all the "calculus on manifolds" concepts that we have introduced so far. The notation is so powerful that it has also suppressed nearly everything that a traditionally educated physicist would find familiar. We will therefore unpack the symbols by means of a concrete example. Let us take Ω to be the whole of $\mathbb{R}^3$. This can be retracted to the origin via the map $\varphi_t(x^\mu) = x^\mu(t) = tx^\mu$. The velocity field whose flow gives

$$x^\mu(t) = t\,x^\mu(1)$$

is $\eta^\mu(x,t) = x^\mu/t$. To verify this, compute

$$\frac{dx^\mu(t)}{dt} = x^\mu(1) = \frac{1}{t}x^\mu(t),$$

so $x^\mu(t)$ is indeed the solution to

$$\frac{dx^\mu}{dt} = \eta^\mu(x(t), t).$$

Now let us apply this retraction to $\omega = Adydz + Bdzdx + Cdxdy$ with

$$d\omega = \left(\frac{\partial A}{\partial x} + \frac{\partial B}{\partial y} + \frac{\partial C}{\partial z}\right) dxdydz = 0. \tag{13.7}$$

The pull-back φ_t^* gives

$$\varphi_t^* \omega = A(tx, ty, tz)d(ty)d(tz) + \text{(two similar terms)}. \tag{13.8}$$

The interior product with

$$\eta = \frac{1}{t}\left(x\frac{\partial}{\partial x} + y\frac{\partial}{\partial y} + z\frac{\partial}{\partial z}\right) \tag{13.9}$$

then gives

$$i_\eta \varphi_t^* \omega = tA(tx, ty, tz)(y\,dz - z\,dy) + \text{(two similar terms)}. \tag{13.10}$$

Finally we form the ordinary integral over t to get

$$\begin{aligned}
\chi = & \int_0^1 i_\eta(\varphi_t^*\omega)dt \\
= & \left[\int_0^1 A(tx, ty, tz)t\,dt\right](ydz - zdy) \\
& + \left[\int_0^1 B(tx, ty, tz)t\,dt\right](zdx - xdz) \\
& + \left[\int_0^1 C(tx, ty, tz)t\,dt\right](xdy - ydx).
\end{aligned} \tag{13.11}$$

In this expression the integrals in the square brackets are just numerical coefficients, i.e. the "dt" is not part of the 1-form. It is instructive, because not entirely trivial, to let "d" act on χ and verify that the construction works. If we focus first on the term involving A, we find that $d[\int_0^1 A(tx, ty, tz)t\,dt](ydz - zdy)$ can be grouped as

$$\begin{aligned}
& \left[\int_0^1 \left\{2tA + t^2\left(x\frac{\partial A}{\partial x} + y\frac{\partial A}{\partial y} + z\frac{\partial A}{\partial z}\right)\right\} dt\right] dydz \\
& - \int_0^1 t^2 \frac{\partial A}{\partial x} dt\,(xdydz + ydzdx + zdxdy).
\end{aligned} \tag{13.12}$$

The first of these terms is equal to

$$\left[\int_0^1 \frac{d}{dt}\left\{t^2 A(tx, ty, tz)\right\} dt\right] dydz = A(x,y,x)\, dydz, \tag{13.13}$$

which is part of ω. The second term will combine with the terms involving B, C, to become

$$-\int_0^1 t^2\left(\frac{\partial A}{\partial x}+\frac{\partial B}{\partial y}+\frac{\partial C}{\partial z}\right) dt\,(xdydz+ydzdx+zdxdy), \tag{13.14}$$

which is zero by our hypothesis. Putting together the A, B, C terms does, therefore, reconstitute ω.

13.2.2 Obstructions to exactness

The condition that Ω be retractable plays an essential role in the converse to Poincaré's lemma. In its absence, $d\omega = 0$ does not guarantee that there is an χ such that $\omega = d\chi$. Consider, for example, a vector field $\mathbf{v}$ with $\operatorname{curl}\mathbf{v} \equiv 0$ in a two-dimensional annulus $\Omega = \{R_0 < |\mathbf{r}| < R_1\}$. In the annulus (a non-retractable space) the condition that $\operatorname{curl}\mathbf{v} \equiv 0$ does not prohibit $\oint_\Gamma \mathbf{v}\cdot d\mathbf{r}$ being non-zero for some closed path Γ encircling the central hole. When this line integral is non-zero then there can be no single-valued χ such that $\mathbf{v} = \operatorname{grad}\chi$. If there were such a χ, then

$$\oint_\Gamma \mathbf{v}\cdot d\mathbf{r} = \chi(0) - \chi(0) = 0. \tag{13.15}$$

A non-zero value for $\oint_\Gamma \mathbf{v}\cdot d\mathbf{r}$ therefore constitutes an *obstruction* to the existence of a χ such that $\mathbf{v} = \operatorname{grad}\ \chi$.

Example: The sphere S^2 is not retractable: any attempt to pull its points back to the north pole will necessarily tear a hole in the surface somewhere. Related to this fact is that whilst the area 2-form $\sin\theta d\theta d\phi$ is closed, it cannot be written as the d of something. We can try to write

$$\sin\theta d\theta d\phi = d[(1-\cos\theta)d\phi], \tag{13.16}$$

but the 1-form $(1-\cos\theta)d\phi$ is singular at the south pole, $\theta = \pi$. We could try

$$\sin\theta d\theta d\phi = d[(-1-\cos\theta)d\phi], \tag{13.17}$$

but this is singular at the north pole, $\theta = 0$. There is no escape. We know that

$$\int_{S^2} \sin\theta d\theta d\phi = 4\pi, \tag{13.18}$$

but if $\sin\theta d\theta d\phi = d\chi$ then Stokes' theorem says that

$$\int_{S^2} \sin\theta d\theta d\phi \stackrel{?}{=} \int_{\partial S^2} \chi = 0 \tag{13.19}$$

because $\partial S^2 = \emptyset$. Again, a non-zero value for $\int \omega$ over some boundary-less region has provided an obstruction to finding an χ such that $\omega = d\chi$.

13.2.3 De Rham cohomology

We have seen that, sometimes, the condition $d\omega = 0$ allows us to find a χ such that $\omega = d\chi$, and sometimes it does not. If the region in which we seek χ is retractable, we can always construct it. If the region is not retractable there may be an obstruction to the existence of χ. In order to describe the various possibilities we introduce the language of *cohomology*, or more precisely *de Rham cohomology*, named for the Swiss mathematician Georges de Rham who did the most to create it.

For simplicity, suppose that we are working in a compact manifold M without boundary. Let $\Omega^p(M) = \bigwedge^p(T^*M)$ be the space of all smooth p-form fields. It is a vector space over $\mathbb{R}$: we can add p-form fields and multiply them by real constants, but, as is the vector space $C^\infty(M)$ of smooth functions on M, it is infinite dimensional. The subspace $Z^p(M)$ of *closed* forms – those with $d\omega = 0$ – is also an infinite-dimensional vector space, and the same is true of the space $B^p(M)$ of *exact* forms – those that can be written as $\omega = d\chi$ for some globally defined $(p-1)$-form χ. Now consider the space $H^p = Z^p/B^p$, which is the space of closed forms *modulo* exact forms. In this space we do not distinguish between two forms, ω_1 and ω_2 when there is a χ, such that $\omega_1 = \omega_2 + d\chi$. We say that ω_1 and ω_2 are *cohomologous* in $H^p(M)$, and write $\omega_1 \sim \omega_2$. We will use the symbol $[\omega]$ to denote the equivalence class of forms cohomologous to ω. Now a miracle happens! For a compact manifold M, the space $H^p(M)$ is *finite* dimensional! It is called the p-th (de Rham) cohomology space of the manifold, and depends only on the global topology of M. In particular, it does not depend on any metric we may have chosen for M.

Sometimes we write $H^p_{\text{dR}}(M,\mathbb{R})$ to make clear that we are dealing with de Rham cohomology, and that we are working with vector spaces over the real numbers. This is because there is also a valuable space $H^p_{\text{dR}}(M,\mathbb{Z})$, where we only allow multiplication by integers.

The cohomology space $H^p_{\text{dR}}(M,\mathbb{R})$ codifies all potential obstructions to solving the problem of finding a $(p-1)$-form χ such that $d\chi = \omega$: we can find such a χ if and only if ω is cohomologous to zero in $H^p_{\text{dR}}(M,\mathbb{R})$. If $H^p_{\text{dR}}(M,\mathbb{R}) = \{0\}$, which is the case if M is retractable, then all closed p-forms are cohomologous to zero. If $H^p_{\text{dR}}(M,\mathbb{R}) \neq \{0\}$, then some closed p-forms ω will not be cohomologous to zero. We can test whether $\omega \sim 0 \in H^p_{\text{dR}}(M,\mathbb{R})$ by forming suitable integrals.

13.3 Homology

To understand what the suitable integrals, of the last section are, we need to think about the spaces that are the cohomology spaces' vector-space duals. These *homology* spaces are simple to understand pictorially.

The basic idea is that, given a region Ω, we can find its boundary $\partial\Omega$. Inspection of a few simple cases will soon lead to the conclusion that the "boundary of a boundary" consists of nothing. In symbols, $\partial^2 = 0$. The statement "$\partial^2 = 0$" is clearly analogous to "$d^2 = 0$", and, pursuing the analogy, we can construct a vector space of "regions" and define two "regions" as being *homologous* if they differ by the boundary of another "region".

13.3.1 Chains, cycles and boundaries

We begin by making precise the vague notions of region and boundary.

Simplicial complexes

The set of all curves and surfaces in a manifold M is infinite dimensional, but the homology spaces are finite dimensional. Life would be much easier if we could use finite-dimensional spaces throughout. Mathematicians therefore do what any computationally minded physicist would do: they approximate the smooth manifold by a discrete polygonal grid.[2] Were they interested in distances, they would necessarily use many small polygons so as to obtain a good approximation to the detailed shape of the manifold. The global topology, though, can often be captured by a rather coarse discretization. The result of this process is to reduce a complicated problem in differential geometry to one of simple algebra. The resulting theory is therefore known as *algebraic topology*.

It turns out to be convenient to approximate the manifold by generalized triangles. We therefore dissect M into line segments (if one-dimensional), triangles (if two-dimensional), tetrahedra (if three-dimensional) or higher-dimensional *p-simplices* (singular: *simplex*). The rules for the dissection are (see Figure 13.1):

(a) Every point must belong to at least one simplex.
(b) A point can belong to only a finite number of simplices.
(c) Two different simplices either have no points in common, or

[2] This discrete approximation leads to what is known as *simplicial homology*. Simplicial homology is rather primitive and old fashioned, having been supplanted by *singular homology* and the theory of *CW complexes*. The modern definitions are superior for proving theorems, but are less intuitive, and for smooth manifolds lead to the same conclusions as the simpler-to-describe simplicial theory.

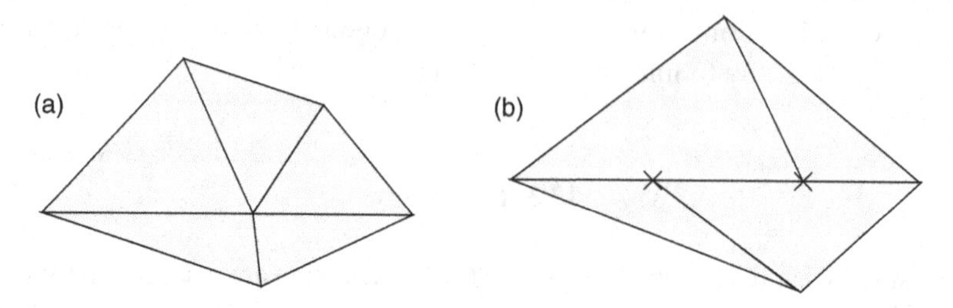

Figure 13.1 Triangles, or 2-simplices, that are (a) allowed, (b) not allowed in a dissection. In (b) the problem is that only parts of edges are in common.

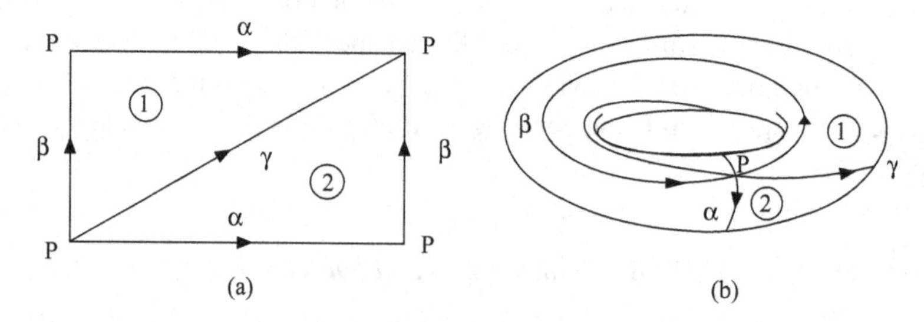

Figure 13.2 A triangulation of the 2-torus. (a) The torus as a rectangle with periodic boundary conditions: the two edges labelled α will be glued together point-by-point along the arrows when we reassemble the torus, and so are to be regarded as a single edge. The two sides labelled β will be glued similarly. (b) The assembled torus: all four P's are now in the same place, and correspond to a single point.

(i) one is a face (or edge, or vertex) of the other;
(ii) the set of points in common is the whole of a shared face (or edge, or vertex).

The collection of simplices composing the dissected space is called a *simplicial complex*. We will denote it by S.

We may not need many triangles to capture the global topology. For example, Figure 13.2 shows how a two-dimensional torus can be decomposed into two 2-simplices (triangles) bounded by three 1-simplices (edges) α, β, γ, and with only a single 0-simplex (vertex) P. Computations are easier to describe, however, if each simplex in the decomposition is uniquely specified by its vertices. For this, we usually need a slightly finer dissection. Figure 13.3 shows a decomposition of the torus into 18 triangles, each of which is uniquely labelled by three points drawn from a set of nine vertices. In this figure vertices with identical labels are to be regarded as the same vertex, as are the corresponding sides of triangles. Thus, each of the edges P_1P_2, P_2P_3, P_3P_1 at the top of the figure are to be glued point-by-point to the corresponding edges on the bottom of the figure; similarly along the sides. The resulting simplicial complex then has 27 edges.

We may triangulate the sphere S^2 as a tetrahedron with vertices P_1, P_2, P_3, P_4. This dissection has six edges: P_1P_2, P_1P_3, P_1P_4, P_2P_3, P_2P_4, P_3P_4, and four faces: $P_2P_3P_4$, $P_1P_3P_4$, $P_1P_2P_4$ and $P_1P_2P_3$ (see Figure 13.4).

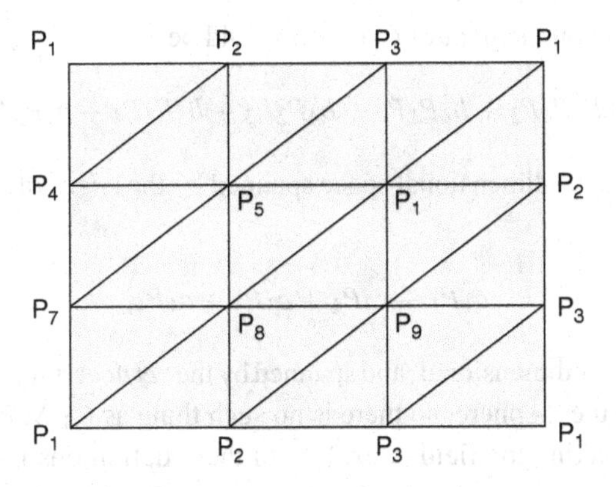

Figure 13.3 A second triangulation of the 2-torus.

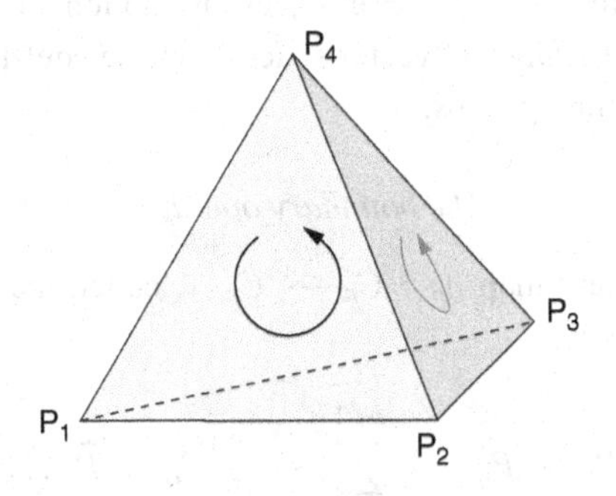

Figure 13.4 A tetrahedral triangulation of the 2-sphere. The circulating arrows on the faces indicate the choice of orientation $P_1P_2P_4$ and $P_2P_3P_4$.

p-chains

We assign to simplices an orientation defined by the order in which we write their defining vertices. The interchange of any pair of vertices reverses the orientation, and we consider there to be a relative minus sign between oppositely oriented but otherwise identical simplices: $P_2P_1P_3P_4 = -P_1P_2P_3P_4$.

We now construct abstract vector spaces $C_p(S,\mathbb{R})$ of *p-chains* which have oriented p-simplices as their basis vectors. The most general elements of $C_2(S,\mathbb{R})$, with S being the tetrahedral triangulation of the sphere S^2, would be

$$a_1P_2P_3P_4 + a_2P_1P_3P_4 + a_3P_1P_2P_4 + a_4P_1P_2P_3, \tag{13.20}$$

where the coefficients $a_1, \ldots, a_4$ are real numbers. We regard the distinct faces as being linearly independent basis elements for $C_2(S,\mathbb{R})$. The space is therefore four dimensional. If we had triangulated the sphere so that it had 16 triangular faces, the space C_2 would be 16 dimensional.

Similarly, the general element of $C_1(S, \mathbb{R})$ would be

$$b_1P_1P_2 + b_2P_1P_3 + b_3P_1P_4 + b_4P_2P_3 + b_5P_2P_4 + b_6P_3P_4, \tag{13.21}$$

and so $C_1(S, \mathbb{R})$ is a six-dimensional space spanned by the *edges* of the tetrahedron. For $C_0(S, \mathbb{R})$ we have

$$c_1P_1 + c_2P_2 + c_3P_3 + c_4P_4, \tag{13.22}$$

and so $C_0(S, \mathbb{R})$ is four dimensional, and spanned by the *vertices*. Our manifold comprises only the *surface* of the 2-sphere, so there is no such thing as $C_3(S, \mathbb{R})$.

The reason for making the field $\mathbb{R}$ explicit in these definitions is that we sometimes gain more information about the topology if we allow only integer coefficients. The space of such p-chains is then denoted by $C_p(S, \mathbb{Z})$. Because a vector space requires that coefficients be drawn from a field, these objects are no longer vector spaces. They can be thought of as either *modules* – "vector spaces" whose coefficients are drawn from a ring – or as additive abelian groups.

The boundary operator

We now introduce a linear map $\partial_p : C_p \to C_{p-1}$, called the *boundary operator*. Its action on a p-simplex is

$$\partial_p P_{i_1} P_{i_2} \cdots P_{i_{p+1}} = \sum_{j=1}^{p+1} (-1)^{j-1} P_{i_1} \ldots \widehat{P}_{i_j} \ldots P_{i_{p+1}}, \tag{13.23}$$

where the "hat" indicates that P_{i_j} is to be omitted. The resulting $(p-1)$-chain is called the *boundary* of the simplex. For example (see Figure 13.5)

$$\begin{aligned} \partial_2(P_2P_3P_4) &= P_3P_4 - P_2P_4 + P_2P_3, \\ &= P_3P_4 + P_4P_2 + P_2P_3. \end{aligned} \tag{13.24}$$

The boundary of a line segment is the difference of its endpoints

$$\partial_1(P_1P_2) = P_2 - P_1. \tag{13.25}$$

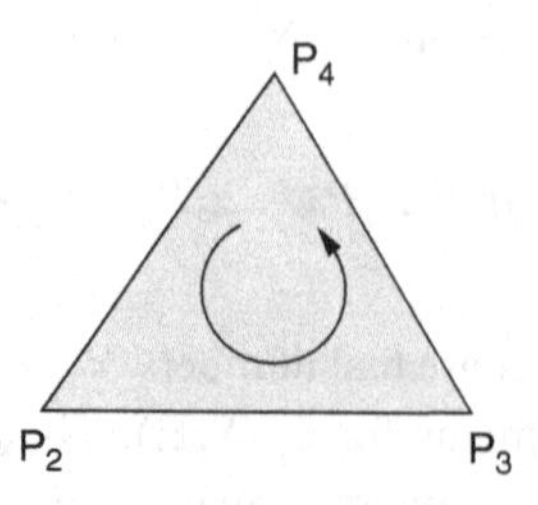

Figure 13.5 The oriented triangle $P_2P_3P_4$ has boundary $P_3P_4 + P_4P_2 + P_2P_3$.

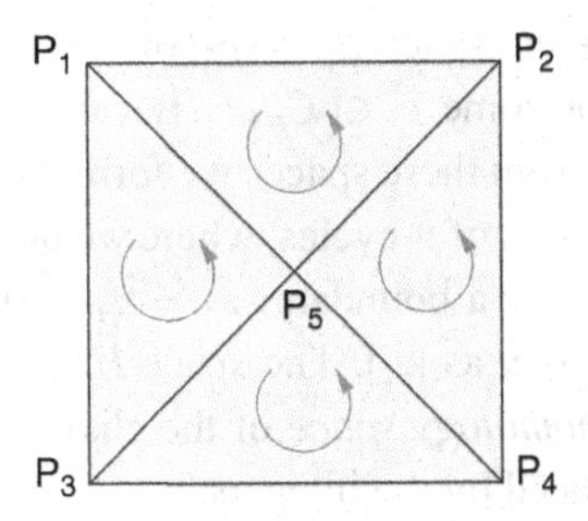

Figure 13.6 Compatibly oriented simplices.

Finally, for any point,

$$\partial_0 P_i = 0. \tag{13.26}$$

Because ∂_p is defined to be a linear map, when it is applied to a p-chain $c = a_1 s_1 + a_2 s_2 + \cdots + a_n s_n$, where the s_i are p-simplices, we have $\partial_p c = a_1 \partial_p s_1 + a_2 \partial_p s_2 + \cdots + a_n \partial_p s_n$.

When we take the "∂" of a chain of compatibly oriented simplices that together make up some region, the internal boundaries cancel in pairs, and the "boundary" of the chain really is the oriented geometric boundary of the region. For example, in Figure 13.6 we find that

$$\partial(P_1P_5P_2 + P_2P_5P_4 + P_3P_4P_5 + P_1P_3P_5) = P_1P_3 + P_3P_4 + P_4P_2 + P_2P_1, \tag{13.27}$$

which is the anticlockwise directed boundary of the square.

For each of the examples above, we find that $\partial_{p-1}\partial_p s = 0$. From the definition (13.23) we can easily establish that this identity holds for any p-simplex s. As chains are sums of simplices and ∂_p is linear, it remains true for any $c \in C_p$. Thus $\partial_{p-1}\partial_p = 0$. We will usually abbreviate this statement as $\partial^2 = 0$.

Cycles, boundaries and homology

A *chain complex* is a doubly infinite sequence of spaces (these can be vector spaces, modules, abelian groups, or many other mathematical objects) such as $\ldots, C_{-2}, C_{-1}, C_0, C_1, C_2 \ldots$, together with structure-preserving maps

$$\ldots \xrightarrow{\partial_{p+1}} C_p \xrightarrow{\partial_p} C_{p-1} \xrightarrow{\partial_{p-1}} C_{p-2} \xrightarrow{\partial_{p-2}} \ldots, \tag{13.28}$$

possessing the property that $\partial_{p-1}\partial_p = 0$. The finite sequence of C_p's that we constructed from our simplicial complex is an example of a chain complex where C_p is zero-dimensional for $p < 0$ or $p > d$. Chain complexes are a useful tool in mathematics, and the ideas that we explain in this section have many applications.

Given any chain complex we can define two important linear subspaces of each of the C_p's. The first is the space Z_p of *p-cycles*. This consists of those $z \in C_p$ such that $\partial_p z = 0$. The second is the space B_p of *p-boundaries*, and consists of those $b \in C_p$ such that $b = \partial_{p+1} c$ for some $c \in C_{p+1}$. Because $\partial^2 = 0$, the boundaries B_p constitute a subspace of Z_p. From these spaces we form the quotient space $H_p = Z_p/B_p$, consisting of *equivalence classes* of p-cycles, where we deem z_1 and z_2 to be equivalent, or *homologous*, if they differ by a boundary: $z_2 = z_1 + \partial c$. We write the equivalence class of cycles homologous to z_i as $[z_i]$. The space H_p, or, more accurately, $H_p(\mathbb{R})$, is called the p-th (simplicial) *homology space* of the chain complex. It becomes the p-th homology *group* if $\mathbb{R}$ is replaced by the integers.

We can construct these homology spaces for any chain complex. When the chain complex is derived from a simplicial complex decomposition of a manifold M a remarkable thing happens. The spaces C_p, Z_p and B_p all depend on the details of how the manifold M has been dissected to form the simplicial complex S. The homology space H_p, however, is *independent* of the dissection. This is neither obvious nor easy to prove. We will rely on examples to make it plausible. Granted this independence, we will write $H_p(M)$, or $H_p(M, \mathbb{R})$, so as to make it clear that H_p is a property of M. The dimension b_p of $H_p(M)$ is called the p-th *Betti number* of the manifold:

$$b_p \stackrel{\text{def}}{=} \dim H_p(M). \tag{13.29}$$

Example: *The 2-sphere.* For the tetrahedral dissection of the 2-sphere, any vertex is P_i homologous to any other, as $P_i - P_j = \partial(P_j P_i)$ and all $P_j P_i$ belong to C_2. Furthermore, $\partial P_i = 0$, so $H_0(S^2)$ is one-dimensional. In general, the dimension of $H_0(M)$ is the number of disconnected pieces making up M. We will write $H_0(S^2) = \mathbb{R}$, regarding $\mathbb{R}$ as the archetype of a one-dimensional vector space.

Now let us consider $H_1(S^2)$. We first find the space of 1-cycles Z_1. An element of C_1 will be in Z_1 only if each vertex that is the beginning of an edge is also the end of an edge, and that these edges have the same coefficient. Thus,

$$z_1 = P_2P_3 + P_3P_4 + P_4P_2$$

is a cycle, as is

$$z_2 = P_1P_4 + P_4P_2 + P_2P_1.$$

These are both boundaries of faces of the tetrahedron. It should be fairly easy to convince yourself that Z_1 is the space of linear combinations of these together with boundaries of the other faces

$$z_3 = P_1P_4 + P_4P_3 + P_3P_1,$$
$$z_4 = P_1P_3 + P_3P_2 + P_2P_1.$$

Any three of these are linearly independent, and so Z_1 is three-dimensional. Because all of the cycles are boundaries, every element of Z_1 is homologous to $\mathbf{0}$, and so $H_1(S^2) = \{\mathbf{0}\}$.

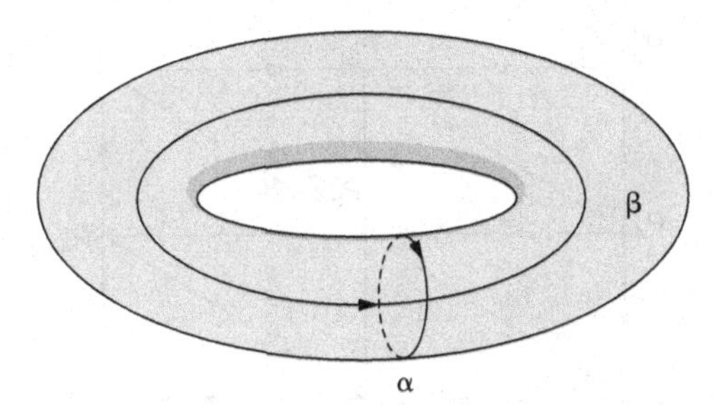

Figure 13.7 A basis of 1-cycles on the 2-torus.

We also see that $H_2(S^2) = \mathbb{R}$. Here the basis element is

$$P_2P_3P_4 - P_1P_3P_4 + P_1P_2P_4 - P_1P_2P_3, \tag{13.30}$$

which is the 2-chain corresponding to the entire surface of the sphere. It would be the boundary of the solid tetrahedron, but does not count as a boundary because the interior of the tetrahedron is not part of the simplicial complex.

Example: *The torus.* Consider the 2-torus T^2. We will see that $H_0(T^2) = \mathbb{R}$, $H_1(T^2) = \mathbb{R}^2 \equiv \mathbb{R} \oplus \mathbb{R}$ and $H_2(T^2) = \mathbb{R}$. A natural basis for the two-dimensional $H_1(T^2)$ consists of the 1-cycles α, β portrayed in Figure 13.7.
The cycle γ that, in Figure 13.2, winds once around the torus is homologous to $\alpha + \beta$. In terms of the second triangulation of the torus (Figure 13.3) we would have

$$\begin{aligned} \alpha &= P_1P_2 + P_2P_3 + P_3P_1, \\ \beta &= P_1P_7 + P_7P_4 + P_4P_1, \end{aligned} \tag{13.31}$$

and

$$\begin{aligned} \gamma &= P_1P_8 + P_8P_6 + P_6P_1 \\ &= \alpha + \beta + \partial(P_1P_8P_2 + P_8P_9P_2 + P_2P_9P_3 + \cdots). \end{aligned} \tag{13.32}$$

Example: *The projective plane.* The projective plane $\mathbb{R}P^2$ can be regarded as a rectangle with diametrically opposite points identified. Suppose we decompose $\mathbb{R}P^2$ into eight triangles, as in Figure 13.8.

Consider the "entire surface"

$$\sigma = P_1P_2P_5 + P_1P_5P_4 + \cdots \in C_2(\mathbb{R}P^2), \tag{13.33}$$

consisting of the sum of all eight 2-simplices with the orientation indicated in the figure. Let $\alpha = P_1P_2 + P_2P_3$ and $\beta = P_1P_4 + P_4P_3$ be the sides of the rectangle running along the bottom horizontal and left vertical sides of the figure, respectively. In each case they

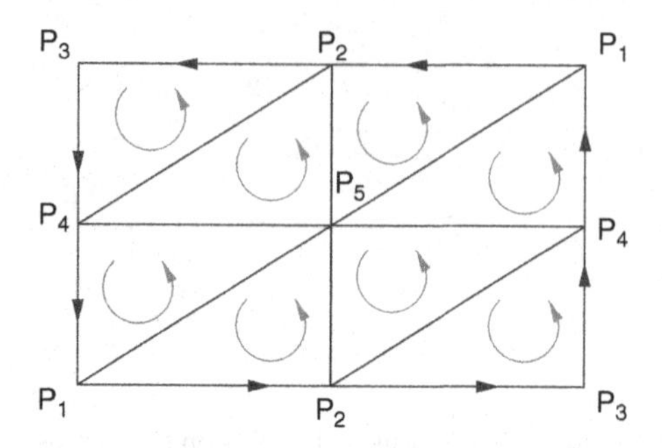

Figure 13.8 A triangulation of the projective plane.

run from P_1 to P_3. Then

$$\begin{aligned}\partial(\sigma) &= P_1P_2 + P_2P_3 + P_3P_4 + P_4P_1 + P_1P_2 + P_2P_3 + P_3P_4 + P_4P_1 \\ &= 2(\alpha - \beta) \neq 0. \end{aligned} \tag{13.34}$$

Although $\mathbb{R}P^2$ has no actual edge that we can fall off, from the homological viewpoint it does have a boundary! This represents the conflict between local orientation of each of the 2-simplices and the global non-orientability of $\mathbb{R}P^2$. The surface σ of $\mathbb{R}P^2$ is not a two-cycle, therefore. Indeed $Z_2(\mathbb{R}P^2)$, and *a fortiori* $H_2(\mathbb{R}P^2)$, contain only the zero vector. The only 1-cycle is $\alpha - \beta$ which runs from P_1 to P_1 via P_2, P_3 and P_4, but (13.34) shows that this is the boundary of $\frac{1}{2}\sigma$. Thus $H_2(\mathbb{R}P^2, \mathbb{R}) = \{\mathbf{0}\}$ and $H_1(\mathbb{R}P^2, \mathbb{R}) = \{\mathbf{0}\}$, while $H_0(\mathbb{R}P^2, \mathbb{R}) = \mathbb{R}$.

We can now see the advantage of restricting ourselves to integer coefficients. When we are not allowed fractions, the cycle $\gamma = (\alpha - \beta)$ is no longer a boundary, although $2(\alpha - \beta)$ is the boundary of σ. Thus, using the symbol $\mathbb{Z}_2$ to denote the additive group of the integers *modulo* 2, we can write $H_1(\mathbb{R}P^2, \mathbb{Z}) = \mathbb{Z}_2$. This homology space is a set with only two members $\{0\gamma, 1\gamma\}$. The finite group $H_1(\mathbb{R}P^2, \mathbb{Z}) = \mathbb{Z}_2$ is said to be the *torsion* part of the homology – a confusing terminology because this torsion has nothing to do with the torsion tensor of Riemannian geometry.

We introduced real-number homology first, because the theory of vector spaces is simpler than that of modules, and more familiar to physicists. The torsion is, however, invisible to the real-number homology. We were therefore buying a simplification at the expense of throwing away information.

The Euler character

The sum

$$\chi(M) \stackrel{\text{def}}{=} \sum_{p=0}^{d} (-1)^p \dim H_p(M, \mathbb{R}) \tag{13.35}$$

is called the *Euler character* of the manifold M. For example, the 2-sphere has $\chi(S^2) = 2$, the projective plane has $\chi(\mathbb{R}P^2) = 1$ and the n-torus has $\chi(T^n) = 0$. This number is manifestly a topological invariant because the individual $\dim H_p(M,\mathbb{R})$ are. We will show that the Euler character is also equal to $V - E + F - \cdots$ where V is the number of vertices, E is the number of edges and F is the number of faces in the simplicial dissection. The dots are for higher dimensional spaces, where the alternating sum continues with $(-1)^p$ times the number of p-simplices. In other words, we are claiming that

$$\chi(M) = \sum_{p=0}^{d} (-1)^p \dim C_p(M). \tag{13.36}$$

It is not so obvious that this new sum is a topological invariant. The individual dimensions of the spaces of p-chains depend on the details of how we dissect M into simplices. If our claim is to be correct, the dependence must somehow drop out when we take the alternating sum.

A useful tool for working with alternating sums of vector-space dimensions is provided by the notion of an *exact sequence*. We say that a set of vector spaces V_p with maps $f_p : V_p \to V_{p+1}$ is an exact sequence if $\mathrm{Ker}\,(f_p) = \mathrm{Im}\,(f_{p-1})$. For example, if all cycles were boundaries then the set of spaces C_p with the maps ∂_p taking us from C_p to C_{p-1} would constitute an exact sequence – albeit with p decreasing rather than increasing, but this is irrelevent. When the homology is non-zero, however, we only have $\mathrm{Im}\,(f_{p-1}) \subset \mathrm{Ker}\,(f_p)$, and the number $\dim H_p = \dim\,(\mathrm{Ker} f_p) - \dim\,(\mathrm{Im} f_{p-1})$ provides a measure of how far this set inclusion falls short of being an equality.

Suppose that

$$\{\mathbf{0}\} \xrightarrow{f_0} V_1 \xrightarrow{f_1} V_2 \xrightarrow{f_2} \cdots \xrightarrow{f_{n-1}} V_n \xrightarrow{f_n} \{\mathbf{0}\} \tag{13.37}$$

is a finite-length exact sequence. Here, $\{\mathbf{0}\}$ is the vector space containing only the zero vector. Being linear, f_0 maps $\mathbf{0}$ to $\mathbf{0}$. Also f_n maps everything in V_n to $\mathbf{0}$. Since this last map takes everything to zero, and what is mapped to zero is the image of the penultimate map, we have $V_n = \mathrm{Im} f_{n-1}$. Similarly, the fact that $\mathrm{Ker} f_1 = \mathrm{Im} f_0 = \{\mathbf{0}\}$ shows that $\mathrm{Im} f_1 \subseteq V_2$ is an isomorphic image of V_1. This situation is represented pictorially in Figure 13.9.

Now the range–null-space theorem tells us that

$$\begin{aligned}\dim V_p &= \dim\,(\mathrm{Im} f_p) + \dim\,(\mathrm{Ker} f_p)\\ &= \dim\,(\mathrm{Im} f_p) + \dim\,(\mathrm{Im} f_{p-1}).\end{aligned} \tag{13.38}$$

When we take the alternating sum of the dimensions, and use $\dim\,(\mathrm{Im} f_0) = 0$ and $\dim\,(\mathrm{Im} f_n) = 0$, we find that the sum telescopes to give

$$\sum_{p=0}^{n} (-1)^p \dim V_p = 0. \tag{13.39}$$

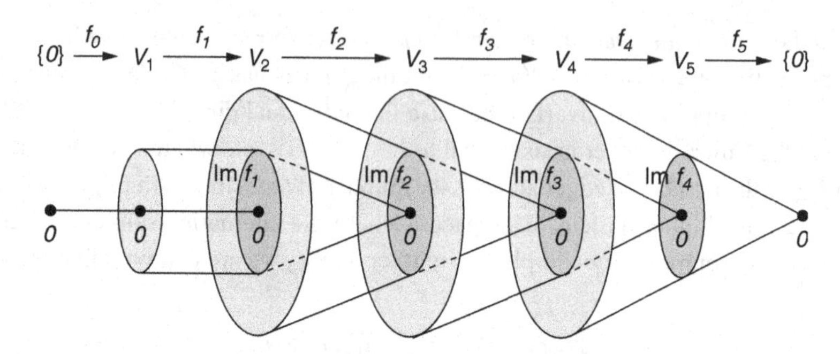

Figure 13.9 A schematic representation of an exact sequence.

The vanishing of this alternating sum is one of the principal properties of an exact sequence.

Now, for our sequence of spaces C_p with the maps $\partial_p : C_p \to C_{p-1}$, we have $\dim(\text{Ker}\,\partial_p) = \dim(\text{Im}\,\partial_{p+1}) + \dim H_p$. Using this and the range–null-space theorem in the same manner as above, shows that

$$\sum_{p=0}^{d}(-1)^p \dim C_p(M) = \sum_{p=0}^{d}(-1)^p \dim H_p(M). \tag{13.40}$$

This confirms our claim.

Exercise 13.1: Count the number of vertices, edges and faces in the triangulation we used to compute the homology groups of the real projective plane $\mathbb{R}P^2$. Verify that $V - E + F = 1$, and that this is the same number that we get by evaluating

$$\chi(\mathbb{R}P^2) = \dim H_0(\mathbb{R}P^2, \mathbb{R}) - \dim H_1(\mathbb{R}P^2, \mathbb{R}) + \dim H_2(\mathbb{R}P^2, \mathbb{R}).$$

Exercise 13.2: Show that the sequence

$$\{\mathbf{0}\} \to V \stackrel{\phi}{\to} W \to \{\mathbf{0}\}$$

of vector spaces being exact means that the map $\phi : V \to W$ is one-to-one and onto, and hence an isomorphism $V \cong W$.

Exercise 13.3: Show that a *short exact sequence*

$$\{\mathbf{0}\} \to A \stackrel{i}{\to} B \stackrel{\pi}{\to} C \to \{\mathbf{0}\}$$

of vector spaces is just a sophisticated way of asserting that $C \cong B/A$. More precisely, show that the map i is injective (one-to-one), so A can be considered to be a subspace of B. Then show that the map π is surjective (onto), and can be regarded as projecting B onto the equivalence classes B/A.

Exercise 13.4: Let $\alpha : A \to B$ be a linear map. Show that

$$\{\mathbf{0}\} \to \operatorname{Ker}\alpha \xrightarrow{i} A \xrightarrow{\alpha} B \xrightarrow{\pi} \operatorname{Coker}\alpha \to \{\mathbf{0}\}$$

is an exact sequence. (Recall that $\operatorname{Coker}\alpha \equiv B/\operatorname{Im}\alpha$.)

13.3.2 Relative homology

Mathematicians have invented powerful tools for computing homology. In this section we introduce one of them: the *exact sequence of a pair*. We describe this tool in detail because a homotopy analogue of this exact sequence is used in physics to classify defects such as dislocations, vortices and monopoles. Homotopy theory is, however, harder, and requires more technical apparatus than homology, so the ideas are easier to explain here.

We have seen that it is useful to think of complicated manifolds as being assembled out of simpler ones. We constructed the torus, for example, by gluing together edges of a rectangle. Another construction technique involves *shrinking* parts of a manifold to a point. Think, for example, of the unit 2-disc as being a circle of cloth with a drawstring sewn into its boundary. Now pull the string tight to form a spherical bag. The continuous functions on the resulting 2-sphere are those continuous functions on the disc that took the same value at all points on its boundary. Recall that we used this idea in Section 12.4.2, where we claimed that those spin textures in $\mathbb{R}^2$ that point in a fixed direction at infinity can be thought of as spin textures on the 2-sphere. We now extend this shrinking trick to homology.

Suppose that we have a chain complex consisting of spaces C_p and boundary operations ∂_p. We denote this chain complex by (C, ∂). Another set of spaces and boundary operations (C', ∂') is a *subcomplex* of (C, ∂) if each $C'_p \subseteq C_p$ and $\partial'_p(c) = \partial_p(c)$ for each $c \in C'_p$. This situation arises if we have a simplicial complex S and some subset S' that is itself a simplicial complex, and take $C'_p = C_p(S')$.

Since each C'_p is a subspace of C_p we can form the quotient spaces C_p/C'_p and make them into a chain complex by defining, for $c + C'_p \in C_p/C'_p$,

$$\overline{\partial}_p(c + C'_p) = \partial_p c + C'_{p-1}. \tag{13.41}$$

It easy to see that this operation is well defined (i.e. it gives the same output independent of the choice of representative in the equivalence class $c + C'_p$), that $\overline{\partial}_p : C_p \to C_{p-1}$ is a linear map, and that $\overline{\partial}_{p-1}\overline{\partial}_p = 0$. We have constructed a new chain complex $(C/C', \overline{\partial})$. We can therefore form its homology spaces in the usual way. The resulting vector space, or abelian group, $H_p(C/C')$ is the p-th *relative homology group of* C *modulo* C'. When C' and C arise from simplicial complexes $S' \subseteq S$, these spaces are what remains of the homology of S after every chain in S' has been shrunk to a point. In this case, it is customary to write $H_p(S, S')$ instead of $H_p(C/C')$, and similarly write the chain, cycle and boundary spaces as $C_p(S, S')$, $Z_p(S, S')$ and $B_p(S, S')$ respectively.

Example: *Constructing the 2-sphere S^2 from the 2-ball (or disc) B^2.* We regard B^2 to be the triangular simplex $P_1P_2P_3$, and its boundary, the 1-sphere or circle S^1, to be the simplicial complex containing the points P_1, P_2, P_3 and the sides P_1P_2, P_2P_3, P_3P_1, but not the interior of the triangle. We wish to contract this boundary complex to a point, and form the relative chain complexes and their homology spaces. Of the spaces we quotient by, $C_0(S^1)$ is spanned by the points P_1, P_2, P_3, the 1-chain space $C_1(S^1)$ is spanned by the sides P_1P_2, P_2P_3, P_3P_1, while $C_2(S^1) = \{\mathbf{0}\}$. The space of relative chains $C_2(B^1, S^1)$ consists of multiples of $P_1P_2P_3 + C_2(S^1)$, and the boundary

$$\overline{\partial}_2\Big(P_1P_2P_3 + C_2(S^1)\Big) = (P_2P_3 + P_3P_1 + P_1P_2) + C_1(S^1) \tag{13.42}$$

is equivalent to zero because $P_2P_3 + P_3P_1 + P_1P_2 \in C_1(S^1)$. Thus $P_1P_2P_3 + C_2(S^1)$ is a non-bounding cycle and spans $H_2(B^2, S^1)$, which is therefore one-dimensional. This space is isomorphic to the one-dimensional $H_2(S^2)$. Similarly $H_1(B^2, S^1)$ is zero dimensional, and so isomorphic to $H_1(S^2)$. This is because all chains in $C_1(B^2, S^1)$ are in $C_1(S^1)$ and therefore equivalent to zero.

A peculiarity, however, is that $H_0(B^2, S^1)$ is *not* isomorphic to $H_0(S^2) = \mathbb{R}$. Instead, we find that $H_0(B^2, S^1) = \{\mathbf{0}\}$ because all the points are equivalent to zero. This vanishing is characteristic of the zeroth relative homology space $H_0(S, S')$ for the simplicial triangulation of any connected manifold. It occurs because S being connected means that any point P in S can be reached by walking along edges from any other point, in particular from a point P' in S'. This makes P homologous to P', and so equivalent to zero in $H_0(S, S')$.

Exact homology sequence of a pair

Homological algebra is full of miracles. Here we describe one of them. From the ingredients we have at hand, we can construct a semi-infinite sequence of spaces and linear maps between them

$$\begin{aligned}
\dots \xrightarrow{\partial_{*p+1}} H_p(S') \xrightarrow{i_{*p}} \quad & H_p(S) \quad \xrightarrow{\pi_{*p}} H_p(S, S') \xrightarrow{\partial_{*p}} \\
H_{p-1}(S') \xrightarrow{i_{*p-1}} \quad & H_{p-1}(S) \quad \xrightarrow{\pi_{*p-1}} H_{p-1}(S, S') \xrightarrow{\partial_{*p-1}} \\
& \vdots \\
\xrightarrow{\partial_{*1}} H_0(S') \xrightarrow{i_{*0}} \quad & H_0(S) \quad \xrightarrow{\pi_{*0}} H_0(S, S') \xrightarrow{\partial_{*0}} \{\mathbf{0}\}.
\end{aligned} \tag{13.43}$$

The maps i_{*p} and π_{*p} are induced by the natural injection $i_p : C_p(S') \to C_p(S)$ and projection $\pi_p : C_p(S) \to C_p(S)/C_p(S')$. It is only necessary to check that

$$\begin{aligned}
\pi_{p-1}\partial_p &= \overline{\partial}_p \pi_p, \\
i_{p-1}\partial_p &= \partial_p i_p,
\end{aligned} \tag{13.44}$$

to see that they are compatible with the passage from the chain spaces to the homology spaces. More discussion is required of the *connection map* ∂_{*p} that takes us from one row to the next in the displayed form of (13.43).

The connection map is constructed as follows: let $h \in H_p(S, S')$. Then $h = z + B_p(S, S')$ for some cycle $z \in Z(S, S')$, and in turn $z = c + C_p(S')$ for some $c \in C_p(S)$. (So *two* choices of representative of equivalence class are being made here.) Now $\overline{\partial}_p z = 0$ which means that $\partial_p c \in C_{p-1}(S')$. This fact, when combined with $\partial_{p-1}\partial_p = 0$, tells us that $\partial_p c \in Z_{p-1}(S')$. We now define the ∂_{*p} image of h to be

$$\partial_{*p}(h) = \partial_p c + B_{p-1}(S'). \tag{13.45}$$

This sounds rather involved, but let's say it again in words: an element of $H_p(S, S')$ is a relative p-cycle *modulo* S'. This means that its boundary is not necessarily zero, but may be a non-zero element of $C_{p-1}(S')$. Since this element is the boundary of something its own boundary vanishes, so it is a $(p-1)$-cycle in $C_{p-1}(S')$ and hence a representative of a homology class in $H_{p-1}(S')$. This homology class is the output of the ∂_{*p} map.

The miracle is that the sequence of maps (13.43) is *exact*. It is an example of a standard homological algebra construction of a *long exact sequence* out of a family of short exact sequences, in this case out of the sequences

$$\{\mathbf{0}\} \to C_p(S') \to C_p(S) \to C_p(S, S') \to \{\mathbf{0}\}. \tag{13.46}$$

Proving that the long sequence is exact is straightforward. All one must do is check each map to see that it has the properties required. This exercise in what is called *diagram chasing* is left to the reader.

The long exact sequence that we have constructed is called the *exact homology sequence of a pair*. If we know that certain homology spaces are zero dimensional, it provides a powerful tool for computing other spaces in the sequence. As an illustration, consider the sequence of the pair B^{n+1} and S^n for $n > 0$:

$$\begin{array}{llll}
\cdots \xrightarrow{i_{*p}} & \underbrace{H_p(B^{n+1})}_{=\{\mathbf{0}\}} & \xrightarrow{\pi_{*p}} H_p(B^{n+1}, S^n) & \xrightarrow{\partial_{*p}} \quad H_{p-1}(S^n) \\
\xrightarrow{i_{*p-1}} & \underbrace{H_{p-1}(B^{n+1})}_{=\{\mathbf{0}\}} & \xrightarrow{\pi_{*p-1}} H_{p-1}(B^{n+1}, S^n) & \xrightarrow{\partial_{*p-1}} \quad H_{p-2}(S^n) \\
& & \vdots & \\
\xrightarrow{i_{*1}} & \underbrace{H_1(B^{n+1})}_{=\{\mathbf{0}\}} & \xrightarrow{\pi_{*1}} H_1(B^{n+1}, S^n) & \xrightarrow{\partial_{*1}} \quad \underbrace{H_0(S^n)}_{=\mathbb{R}} \\
\xrightarrow{i_{*0}} & \underbrace{H_0(B^{n+1})}_{=\mathbb{R}} & \xrightarrow{\pi_{*0}} H_0(B^{n+1}, S^n) & \xrightarrow{\partial_{*0}} \quad \{\mathbf{0}\}.
\end{array} \tag{13.47}$$

We have inserted here the easily established data that $H_p(B^{n+1}) = \{\mathbf{0}\}$ for $p > 0$ (which is a consequence of the $(n+1)$-ball being a contractible space), and that $H_0(B^{n+1})$ and $H_0(S^n)$ are one-dimensional because they consist of a single connected component. We read off, from the $\{\mathbf{0}\} \to A \to B \to \{\mathbf{0}\}$ exact subsequences, the isomorphisms

$$H_p(B^{n+1}, S^n) \cong H_{p-1}(S^n), \quad p > 1, \tag{13.48}$$

and from the exact sequence

$$\{\mathbf{0}\} \to H_1(B^{n+1}, S^1) \to \mathbb{R} \to \mathbb{R} \to H_0(B^{n+1}, S^n) \to \{\mathbf{0}\} \tag{13.49}$$

that $H_1(B^{n+1}, S^n) = \{\mathbf{0}\} = H_0(B^{n+1}, S^n)$. The first of these equalities holds because $H_1(B^{n+1}, S^n)$ is the kernel of the isomorphism $\mathbb{R} \to \mathbb{R}$, and the second because $H_0(B^{n+1}, S^n)$ is the range of a surjective null map.

In the case $n = 0$, we have to modify our last conclusion because $H_0(S^0) = \mathbb{R} \oplus \mathbb{R}$ is two-dimensional. (Remember that $H_0(M)$ counts the number of disconnected components of M, and the 0-sphere S^0 consists of the two disconnected points P_1, P_2 lying in the boundary of the interval $B^1 = P_1P_2$.) As a consequence, the last five maps in (13.47) become

$$\{\mathbf{0}\} \to H_1(B^1, S^0) \to \mathbb{R} \oplus \mathbb{R} \to \mathbb{R} \to H_0(B^1, S^0) \to \{\mathbf{0}\}. \tag{13.50}$$

This tells us that $H_1(B^1, S^0) = \mathbb{R}$ and $H_0(B^1, S^0) = \{\mathbf{0}\}$.

Exact homotopy sequence of a pair

The construction of a long exact sequence from a short exact sequence is a very powerful technique. It has become almost ubiquitous in advanced mathematics. Here we briefly describe an application to homotopy theory.

We have met the homotopy groups $\pi_n(M)$ in Section 12.4.4. As we saw there, homotopy groups can be used to classify defects or textures in physical systems in which some field takes values in a manifold M. Suppose that the local physical properties of a system are invariant under the action of a Lie group G – for example the high-temperature phase of a ferromagnet may be invariant under the rotation group SO(3). Now suppose that system undergoes spontaneous symmetry breaking and becomes invariant only under a subgroup H. Then manifold of inequivalent states is the coset G/H. For a ferromagnet the symmetry breaking will be from $G = \mathrm{SO}(3)$ to $H = \mathrm{SO}(2)$ where SO(2) is the group of rotations about the axis of magnetization. G/H is then the 2-sphere of the direction in which the magnetization can point.

The group $\pi_n(G)$ can be taken to be the set of continuous maps of an n-dimensional cube into the group G, with the surface of the cube mapping to the identity element

$e \in G$. We similarly define the relative homotopy group $\pi_n(G,H)$ of G *modulo* H to be the set of continuous maps of the cube into G, with all but one face of the cube mapping to e, but with the remaining face mapping to the subgroup H. It can then be shown that $\pi_n(G/H) \cong \pi_n(G,H)$ (the hard part is to show that any continuous map into G/H can be represented as the projection of some continuous map into G).

The short exact sequence

$$\{e\} \to H \xrightarrow{i} G \xrightarrow{\pi} G/H \to \{e\} \tag{13.51}$$

of group homomorphisms (where $\{e\}$ is the group consisting only of the identity element) then gives rise to the long exact sequence

$$\cdots \to \pi_n(H) \to \pi_n(G) \to \pi_n(G,H) \to \pi_{n-1}(H) \to \cdots. \tag{13.52}$$

The derivation and utility of this exact sequence is very well described in the review article by Mermin cited in Section 12.4.4. We have therefore contented ourselves with simply displaying the result so that the reader can see the similarity between the homology theorem and its homotopy-theory analogue.

13.4 De Rham's theorem

We still have not related homology to cohomology. The link is provided by integration.

The integral provides a natural pairing of a p-chain c and a p-form ω: if $c = a_1 s_1 + a_2 s_2 + \cdots + a_n s_n$, where the s_i are simplices, we set

$$(c,\omega) = \sum_i a_i \int_{s_i} \omega. \tag{13.53}$$

The perhaps mysterious notion of "adding" geometric simplices is thus given a concrete interpretation in terms of adding real numbers.

Stokes' theorem now reads

$$(\partial c, \omega) = (c, d\omega), \tag{13.54}$$

suggesting that d and ∂ should be regarded as adjoints of each other. From this observation follows the key fact that the pairing between chains and forms descends to a pairing between homology classes and cohomology classes. In other words,

$$(z + \partial c, \omega + d\chi) = (z, \omega), \tag{13.55}$$

so it does not matter which representatives of the two equivalence classes we take when we compute the integral. Let us see why this is so.

Suppose $z \in Z_p$ and $\omega_2 = \omega_1 + d\eta$. Then

$$\begin{aligned}(z, \omega_2) = \int_z \omega_2 &= \int_z \omega_1 + \int_z d\eta \\ &= \int_z \omega_1 + \int_{\partial z} \eta \\ &= \int_z \omega_1 \\ &= (z, \omega_1)\end{aligned} \tag{13.56}$$

because $\partial z = 0$. Thus, all elements of the cohomology class of ω return the same answer when integrated over a cycle.

Similarly, if $\omega \in Z^p$ and $c_2 = c_1 + \partial a$ then

$$\begin{aligned}(c_2, \omega) &= \int_{c_1} \omega + \int_{\partial a} \omega \\ &= \int_{c_1} \omega + \int_a d\omega \\ &= \int_{c_1} \omega \\ &= (c_1, \omega),\end{aligned}$$

since $d\omega = 0$.

All this means that we can consider the equivalence classes of closed forms composing $H^p_{\mathrm{dR}}(M)$ to be elements of $(H_p(M))^*$, the dual space of $H_p(M)$ – hence the "co" in cohomology. The existence of the pairing does not automatically mean that H^p_{dR} *is* the dual space to $H_p(M)$, however, because there might be elements of the dual space that are not in H^p_{dR}, and there might be distinct elements of H^p_{dR} that give identical answers when integrated over any cycle, and so correspond to the same element in $(H_p(M))^*$. This does not happen, however, when the manifold is *compact*: De Rham showed that, for compact manifolds, $(H_p(M, \mathbb{R}))^* = H^p_{\mathrm{dR}}(M, \mathbb{R})$. We will not try to prove this, but be satisfied with some examples.

The statement $(H_p(M))^* = H^p_{\mathrm{dR}}(M)$ neatly summarizes de Rham's results, but, in practice, the more explicit statements given below are more useful.

Theorem: *(de Rham) Suppose that M is a compact manifold.*

(1) A closed p-form ω is exact if and only if

$$\int_{z_i} \omega = 0 \tag{13.57}$$

for all cycles $z_i \in Z_p$. It suffices to check this for one representative of each homology class.

(2) *If* $z_i \in Z_p$, $i = 1, \dots, \dim H_p$, *is a basis for the p-th homology space, and* α_i *a set of numbers, one for each* z_i, *then there exists a closed p-form* ω *such that*

$$\int_{z_i} \omega = \alpha_i. \tag{13.58}$$

If ω^i *constitute a basis of the vector space* $H^p(M)$ *then the matrix of numbers*

$$\Omega_i^{\ j} = (z_i, \omega^j) = \int_{z_i} \omega^j \tag{13.59}$$

is called the period matrix, and the $\Omega_i^{\ j}$ *themselves are the periods.*

Example: $H_1(T^2) = \mathbb{R} \oplus \mathbb{R}$ is two-dimensional. Since a finite-dimensional vector space and its dual have the same dimension, de Rham tells us that $H^1_{\mathrm{dR}}(T^2)$ is also two-dimensional. If we take as coordinates on T^2 the angles θ and ϕ, then the basis elements, or *generators*, of the cohomology spaces are the forms "$d\theta$" and "$d\phi$". We have inserted the quotes to stress that these expressions are not the d of a function. The angles θ and ϕ are *not* functions on the torus, since they are not single-valued. The homology basis 1-cycles can be taken as z_θ running from $\theta = 0$ to $\theta = 2\pi$ along $\phi = \pi$, and z_ϕ running from $\phi = 0$ to $\phi = 2\pi$ along $\theta = \pi$. Clearly, $\omega = \alpha_\theta d\theta/2\pi + \alpha_\phi d\phi/2\pi$ returns $\int_{z_\theta} \omega = \alpha_\theta$ and $\int_{z_\phi} \omega = \alpha_\phi$ for any $\alpha_\theta, \alpha_\pi$, so $\{d\theta/2\pi, d\phi/2\pi\}$ and $\{z_\theta, z_\phi\}$ are dual bases.

Example: We have earlier computed the homology groups $H_2(\mathbb{R}P^2, \mathbb{R}) = \{\mathbf{0}\}$ and $H_1(\mathbb{R}P^2, \mathbb{R}) = \{\mathbf{0}\}$. De Rham therefore tells us that $H^2(\mathbb{R}P^2, \mathbb{R}) = \{\mathbf{0}\}$ and $H^1(\mathbb{R}P^2, \mathbb{R}) = \{\mathbf{0}\}$. From this we deduce that all closed 1- and 2-forms on the projective plane $\mathbb{R}P^2$ are exact.

Example: As an illustration of de Rham part (1), observe that it is easy to show that a closed 1-form ϕ can be written as df, provided that $\int_{z_i} \phi = 0$ for all cycles. We simply define $f = \int_{x0}^{x} \phi$, and observe that the proviso ensures that f is not multivalued.

Example: A more subtle problem is to show that, given a 2-form ω on S^2, with $\int_{S^2} \omega = 0$ there is a globally defined χ such that $\omega = d\chi$. We begin by covering S^2 by two open sets D_+ and D_- which have the form of caps such that D_+ includes all of S^2 except for a neighbourhood of the south pole, while D_- includes all of S^2 except a neighbourhood of the north pole, and the intersection, $D_+ \cap D_-$, has the topology of an annulus, or *cingulum*, encircling the equator (Figure 13.10).

Since both D_+ and D_- are contractible, there are one-forms χ_+ and χ_- such that $\omega = d\chi_+$ in D_+ and $\omega = d\chi_-$ in D_-. Thus,

$$d(\chi_+ - \chi_-) = 0, \quad \text{in} \quad D_+ \cap D_-. \tag{13.60}$$

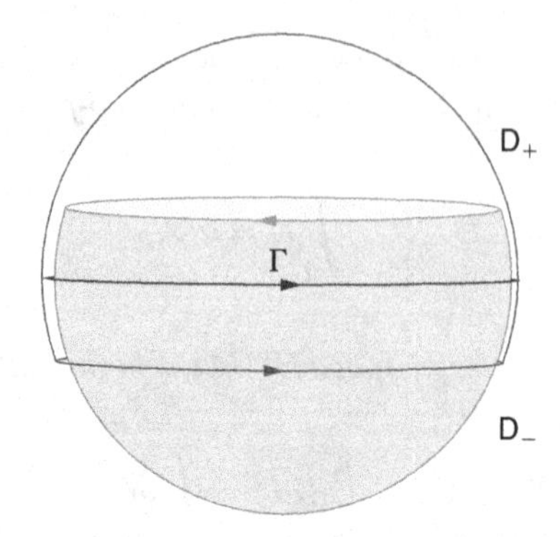

Figure 13.10 A covering of the 2-sphere by a pair of contractable caps.

Dividing the sphere into two disjoint sets with a common (but opposingly oriented) boundary $\Gamma \in D_+ \cap D_-$, we have

$$0 = \int_{S^2} \omega = \oint_\Gamma (\chi_+ - \chi_-), \tag{13.61}$$

and this is true for any such curve Γ. Thus, by the previous example,

$$\phi \equiv (\chi_+ - \chi_-) = df \tag{13.62}$$

for some smooth function f defined in $D_+ \cap D_-$. We now introduce a *partition of unity* subordinate to the cover of S^2 by D_+ and D_-. This partition is a pair of non-negative smooth functions, $\rho_\pm$, such that ρ_+ is non-zero only in D_+, ρ_- is non-zero only in D_- and $\rho_+ + \rho_- = 1$. Now

$$f = \rho_+ f - (-\rho_-) f, \tag{13.63}$$

and $f_- = \rho_+ f$ is a function defined everywhere on D_-. Similarly $f_+ = (-\rho_-) f$ is a function on D_+. Notice the interchange of $\pm$ labels! This is not a mistake. The function f is not defined outside $D_+ \cap D_-$, but we can define $\rho_- f$ everywhere on D_+ because f gets multiplied by zero wherever we have no specific value to assign to it. We now observe that

$$\chi_+ + df_+ = \chi_- + df_-, \quad \text{in} \quad D_+ \cap D_-. \tag{13.64}$$

Thus $\omega = d\chi$, where χ is defined everywhere by the rule

$$\chi = \begin{cases} \chi_+ + df_+, & \text{in } D_+, \\ \chi_- + df_-, & \text{in } D_-. \end{cases} \tag{13.65}$$

It does not matter which definition we take in the cingular region $D_+ \cap D_-$, because the two definitions coincide there.

The methods of this example can be extended to give a proof of de Rham's claims.

13.5 Poincaré duality

De Rham's theorem does not require that our manifold M be orientable. Our next results do, however, require orientability. We therefore assume throughout this section that M is a compact, orientable, D-dimensional manifold. We will also require that M is a *closed* manifold – meaning that it has no boundary.

We begin with the observation that if the forms ω_1 and ω_2 are closed then so is $\omega_1 \wedge \omega_2$. Furthermore, if one or both of ω_1, ω_2 is exact then the product $\omega_1 \wedge \omega_2$ is also exact. It follows that the cohomology class $[\omega_1 \wedge \omega_2]$ of $\omega_1 \wedge \omega_2$ depends only on the cohomology classes $[\omega_1]$ and $[\omega_2]$. The wedge product thus induces a map

$$H^p(M,\mathbb{R}) \times H^q(M,\mathbb{R}) \xrightarrow{\wedge} H^{p+q}(M,\mathbb{R}), \tag{13.66}$$

which is called the "cup product" of the cohomology classes. It is written as

$$[\omega_1 \wedge \omega_2] = [\omega_1] \cup [\omega_2], \tag{13.67}$$

and gives the cohomology the structure of a graded-commutative ring, denoted by $H^\bullet(M,R)$.

More significant for us than the ring structure is that, given $\omega \in H^D(M,\mathbb{R})$, we can obtain a real number by forming $\int_M \omega$. (This is the point at which we need orientability. We only know how to integrate over orientable chains, and so cannot even define $\int_M \omega$ when M is not orientable.) We can combine this integral with the cup product to make any cohomology class $[f] \in H^{D-p}(M,\mathbb{R})$ into an element F of $(H^p(M,\mathbb{R}))^*$. We do this by setting

$$F([g]) = \int_M f \wedge g \tag{13.68}$$

for each $[g] \in H^p(M,\mathbb{R})$. Furthermore, it is possible to show that we can get *any* element F of $(H^p(M,\mathbb{R}))^*$ in this way, and the corresponding $[f]$ is *unique*. But de Rham has already given us a way of identifying the elements of $(H^p(M,\mathbb{R}))^*$ with the cycles in $H_p(M,\mathbb{R})$! There is, therefore, a one-to-one onto map

$$H_p(M,\mathbb{R}) \leftrightarrow H^{D-p}(M,\mathbb{R}). \tag{13.69}$$

In particular the dimensions of these two spaces must coincide:

$$b_p(M) = b_{D-p}(M). \tag{13.70}$$

This equality of Betti numbers is called *Poincaré duality*. Poincaré originally conceived of it geometrically. His idea was to construct from each simplicial triangulation S of M a new "dual" triangulation S', where, in two dimensions for example, we place a new vertex at the centre of each triangle, and join the vertices by lines through each side of the old triangles to make new cells – each new cell containing one of the old vertices. If we are lucky, this process will have the effect of replacing each p-simplex by a $(D-p)$-simplex, and so set up a map between $C_p(S)$ and $C_{D-p}(S')$ that turns the homology "upside down". The new cells are not always simplices, however, and it is hard to make this construction systematic. Poincaré's original recipe was flawed.

Our present approach to Poincaré's result is asserting that for each basis p-cycle class $[z_i^p]$ there is a unique (up to cohomology) $(D-p)$-form ω_i^{D-p} such that

$$\int_{z_i^p} f = \int_M \omega_i^{D-p} \wedge f. \tag{13.71}$$

We can construct this ω_i^{D-p} "physically" by taking a representative cycle z_i^p in the homology class $[z_i^p]$ and thinking of it as a surface with a conserved unit $(d-p)$-form current flowing in its vicinity. An example would be the two-form topological current running along the one-dimensional world-line of a skyrmion. (See the discussion surrounding Equation (12.64).) The ω_i^{D-p} form a basis for $H^{D-p}(M,\mathbb{R})$. We can therefore expand $f \sim f^i \omega_i^{D-p}$, and similarly for the closed p-form g, to obtain

$$\int_M g \wedge f = f^i g^j I(i,j), \tag{13.72}$$

where the matrix

$$I(i,j) \stackrel{\text{def}}{=} I(z_i^p, z_j^{D-p}) = \int_M \omega_i^{D-p} \wedge \omega_j^p \tag{13.73}$$

is called the *intersection form*. From its definition we see that $I(i,j)$ satisfies the symmetry

$$I(i,j) = (-1)^{p(D-p)} I(j,i). \tag{13.74}$$

Less obvious is that $I(i,j)$ is an *integer* that reports the number of times (counted with orientation) that the cycles z_i^p and z_j^{D-p} intersect. This latter fact can be understood from our construction of the ω_i^p as unit currents localized near the z_i^{D-p} cycles. The integrand in (13.73) is non-zero only in the neighbourhood of the intersections of z_i^p with z_j^{D-p}, and at each intersection constitutes a D-form that integrates up to give ± 1.

This claim is illustrated in the left-hand part of Figure 13.11, which shows a region surrounding the intersection of the α and β 1-cycles on the 2-torus. The coordinate system has been chosen so that the α cycle runs along the x-axis and the β cycle along the y-axis. Each cycle is surrounded by the narrow shaded regions $-w < y < w$ and $-w < x < w$, respectively. To construct suitable forms ω_α and ω_β we select a smooth function $f(x)$ that vanishes for $|x| \geq w$ and such that $\int f\,dx = 1$. In the local chart we can then set

$$\begin{aligned} \omega_\alpha &= f(y)\,dy, \\ \omega_\beta &= -f(x)\,dx, \end{aligned}$$

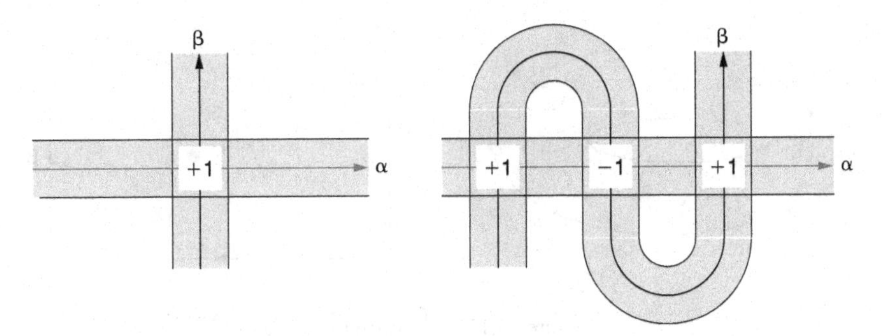

Figure 13.11 The intersection of two cycles: $I(\alpha,\beta) = 1 = 1 - 1 + 1$.

both these forms being closed. The intersection number is given by the integral

$$I(\alpha,\beta) = \int \omega_\alpha \wedge \omega_\beta = \iint f(x)f(y)\,dxdy = 1. \tag{13.75}$$

The right-hand part of Figure 13.11 illustrates why this intersection number depends only on the homology classes of the two 1-cycles, and not on their particular instantiation as curves.

We can more conveniently re-express (13.72) in terms of the *periods* of the forms

$$f_i \stackrel{\text{def}}{=} \int_{z_i^p} f = I(i,k)f^k, \quad g_j \stackrel{\text{def}}{=} \int_{z_j^{D-p}} g = I(j,l)g^l, \tag{13.76}$$

as

$$\int_M f \wedge g = \sum_{i,j} K(i,j) \int_{z_i^p} f \int_{z_j^{D-p}} g, \tag{13.77}$$

where

$$K(i,j) = I^{-1}(i,k)I^{-1}(j,l)I(k,l) = I^{-1}(j,i) \tag{13.78}$$

is the transpose of the inverse of the intersection-form matrix. The decomposition (13.77) of the integral of the product of a pair of closed forms into a bilinear form in their periods is one of the two principal results of this section, the other being (13.70).

In simple cases, we can obtain the decomposition (13.77) by more direct methods. Suppose, for example, that we label the cycles generating the homology group $H_1(T^2)$ of the 2-torus as α and β, and that a and b are closed ($da = db = 0$), but not necessarily exact, 1-forms. We will show that

$$\int_{T^2} a \wedge b = \int_\alpha a \int_\beta b - \int_\alpha b \int_\beta a. \tag{13.79}$$

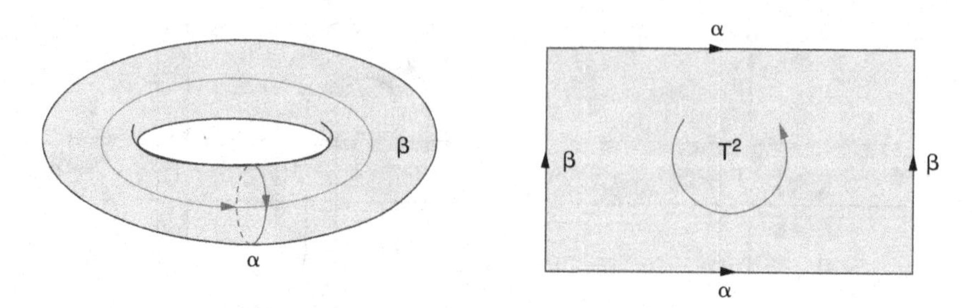

Figure 13.12 Cut-open torus.

To do this, we cut the torus along the cycles α and β and open it out into a rectangle with sides of length L_x and L_y (see Figure 13.12). The cycles α and β will form the sides of the rectangle, and we will take them as lying parallel to the x- and y-axes, respectively. Functions on the *torus* now become functions on the *rectangle*. Not all functions on the rectangle descend from functions on the torus, however. Only those functions that satisfy the periodic boundary conditions $f(0,y) = f(L_x,y)$ and $f(x,0) = f(x,L_y)$ can be considered (mathematicians would say "can be *lifted*") to be functions on the torus.

Since the rectangle (but not the torus) is retractable, we can write $a = df$ where f is a function on the rectangle – but not necessarily a function on the torus, i.e. f will not, in general, be periodic. Since $a \wedge b = d(fb)$, we can now use Stokes' theorem to evaluate

$$\int_{T^2} a \wedge b = \int_{T^2} d(fb) = \int_{\partial T^2} fb. \tag{13.80}$$

The two integrals on the two vertical sides of the rectangle can be combined to a single integral over the points of the 1-cycle β:

$$\int_{\text{vertical}} fb = \int_{\beta} [f(L_x,y) - f(0,y)]b. \tag{13.81}$$

We now observe that $[f(L_x,y) - f(0,y)]$ is a constant, and so can be taken out of the integral. It is a constant because all paths from the point $(0,y)$ to (L_x,y) are homologous to the one-cycle α, so the difference $f(L_x,y) - f(0,y)$ is equal to $\int_\alpha a$. Thus,

$$\int_{\beta} [f(L_x,y) - f(0,y)]b = \int_{\alpha} a \int_{\beta} b. \tag{13.82}$$

Similarly, the contribution of the two horizontal sides is

$$\int_{\alpha} [f(x,0) - f((x,L_y)]b = -\int_{\beta} a \int_{\alpha} b. \tag{13.83}$$

On putting the contributions of both pairs of sides together, the claimed result follows.

13.6 Characteristic classes

A supply of elements of $H^{2m}(M,\mathbb{R})$ and $H^{2m}(M,\mathbb{Z})$ is provided by the *characteristic classes* associated with connections on vector bundles over the manifold M.

Recall that connections appear in covariant derivatives

$$\nabla_\mu \stackrel{\text{def}}{=} \partial_\mu + A_\mu, \tag{13.84}$$

and are to be thought of as matrix-valued one-forms $A = A_\mu dx^\mu$. In the quantum mechanics of charged particles the covariant derivative that appears in the Schrödinger equation is

$$\nabla_\mu = \frac{\partial}{\partial x^\mu} - ieA_\mu^{\text{Maxwell}}. \tag{13.85}$$

Here, e is the charge of the particle on whose wavefunction the derivative acts, and A_μ^{Maxwell} is the usual electromagnetic vector potential. The matrix-valued connection 1-form is therefore

$$A = -ieA_\mu^{\text{Maxwell}} dx^\mu. \tag{13.86}$$

In this case the matrix is one-by-one.

In a non-abelian gauge theory with gauge group G the connection becomes

$$A = i\widehat{\lambda}_a A_\mu^a dx^\mu. \tag{13.87}$$

The $\widehat{\lambda}_a$ are hermitian matrices that have commutation relations $[\widehat{\lambda}_a, \widehat{\lambda}_b] = if_{ab}^c \widehat{\lambda}_c$, where the f_{ab}^c are the structure constants of the Lie algebra of the group G. The $\widehat{\lambda}_a$ therefore form a representation of the Lie algebra, and this representation plays the role of the "charge" of the non-abelian gauge particle.

For covariant derivatives acting on a tangent vector field $f^a \mathbf{e}_a$ on a Riemann n-manifold, where the $\mathbf{e}_a$ are an orthonormal vielbein frame, we have

$$A = \omega_{ab\mu} dx^\mu, \tag{13.88}$$

where, for each μ, the coefficients $\omega_{ab\mu} = -\omega_{ba\mu}$ can be thought of as the entries in a skew symmetric n-by-n matrix. These matrices are elements of the Lie algebra $\mathfrak{o}(n)$ of the orthogonal group $\mathrm{O}(n)$.

In all these cases we define the curvature two-form to be $F = dA + A^2$, where a combined matrix and wedge product is to be understood in A^2. In Exercises 11.19 and 11.20 you used the Bianchi identity to show that the gauge-invariant $2n$-forms $\mathrm{tr}\,(F^n)$ were closed. The integrals of these forms over cycles provide numbers that are topological invariants of the bundle. For example, in four-dimensional QCD, the integral

$$c_2 = -\frac{1}{8\pi^2}\int_\Omega \mathrm{tr}\,(F^2) \tag{13.89}$$

over a compactified four-dimensional manifold Ω is an integer that a mathematician would call the second Chern number of the non-abelian gauge bundle, and that a physicist would call the *instanton number* of the gauge field configuration. The closed forms themselves are called *characteristic classes*.

In the following section we will show that the integrals of characteristic classes are indeed topological invariants. We also explain something of what these invariants are measuring, and illustrate why, when suitably normalized, they are integer-valued.

13.6.1 Topological invariance

Suppose that we have been given a connection A and slightly deform it $A \to A + \delta A$. Then $F \to F + \delta F$ where

$$\delta F = d(\delta A) + \delta A\, A + A\, \delta A. \tag{13.90}$$

Using the Bianchi identity $dF = FA - AF$, we find that

$$\begin{aligned}\delta\,\mathrm{tr}(F^n) &= n\,\mathrm{tr}(\delta F\, F^{n-1})\\ &= n\,\mathrm{tr}(d(\delta A)F^{n-1}) + n\,\mathrm{tr}(\delta A\, AF^{n-1}) + n\,\mathrm{tr}(A\,\delta A F^{n-1})\\ &= n\,\mathrm{tr}(d(\delta A)F^{n-1}) + n\,\mathrm{tr}(\delta A\, AF^{n-1}) - n\,\mathrm{tr}(\delta A\, F^{n-1}A)\\ &= d\left\{n\,\mathrm{tr}(\delta A\, F^{n-1})\right\}.\end{aligned} \tag{13.91}$$

The last line of (13.91) is equal to the penultimate line because all but the first and last terms arising from the dF's in $d\left\{\mathrm{tr}(\delta A\, F^{n-1})\right\}$ cancel in pairs. A globally defined change in A therefore changes $\mathrm{tr}(F^n)$ by the d of something, and so does not change its cohomology class, or its integral over a cycle.

At first sight, this invariance under deformation suggests that all the $\mathrm{tr}(F^n)$ are exact forms – they can apparently all be written as $\mathrm{tr}(F^n) = d\omega_{2n-1}(A)$ for some $(2n-1)$-form $\omega_{2n-1}(A)$. To find $\omega_{2n-1}(A)$ all we have to do is deform the connection to zero by setting $A_t = tA$ and

$$F_t = dA_t + A_t^2 = tdA + t^2A^2. \tag{13.92}$$

Then $\delta A_t = A\delta t$, and

$$\frac{d}{dt}\mathrm{tr}(F_t^n) = d\left\{n\,\mathrm{tr}(AF_t^{n-1})\right\}. \tag{13.93}$$

Integrating up from $t = 0$, we find

$$\mathrm{tr}(F^n) = d\left\{n\int_0^1 \mathrm{tr}(AF_t^{n-1})\,dt\right\}. \tag{13.94}$$

For example

$$\begin{aligned}\mathrm{tr}(F^2) &= d\left\{2\int_0^1 \mathrm{tr}(A(tdA+t^2A^2)\,dt\right\}\\ &= d\left\{\mathrm{tr}\left(AdA+\frac{2}{3}A^3\right)\right\}. \end{aligned} \tag{13.95}$$

You should recognize here the $\omega_3(A) = \mathrm{tr}(AdA + \frac{2}{3}A^3)$ Chern–Simons form of Exercise 11.19. The naïve conclusion – that all the $\mathrm{tr}(F^n)$ are exact – is false, however. What the computation *actually* shows is that when $\int \mathrm{tr}(F^n) \neq 0$ we cannot find a globally defined 1-form A representing the connection or gauge field. With no global A, we cannot globally deform A to zero.

Consider, for example, an abelian U(1) gauge field on the 2-sphere S^2. When the first Chern number

$$c_1 = \frac{1}{2\pi i}\int_{S^2} F \tag{13.96}$$

is non-zero, there can be no globally defined 1-form A such that $F = dA$. Glance back, however, at Figure 13.10 on page 472. There we see that the retractability of the spherical caps $D_\pm$ guarantees that there are 1-forms $A_\pm$ defined on $D_\pm$ such that $F = dA_\pm$ in $D_\pm$. In the cingular region $D_+ \cap D_-$ where they are both defined, A_+ and A_- will be related by a gauge transformation. For a U(1) gauge field, the matrix g appearing in the general gauge transformation rule

$$A \to A^g \equiv g^{-1}Ag + g^{-1}dg \tag{13.97}$$

of Exercise 11.20 becomes the phase $e^{i\chi} \in \mathrm{U}(1)$. Consequently

$$A_+ = A_- + e^{-i\chi}de^{i\chi} = A_- + id\chi \quad \text{in} \quad D_+ \cap D_-. \tag{13.98}$$

The U(1) group element $e^{i\chi}$ is required to be single valued in $D_+ \cap D_-$, but the angle χ may be multivalued. We now write c_1 as the sum of integrals over the north and south hemispheres of S^2, and use Stokes' theorem to reduce this sum to a single integral over the hemispheres' common boundary, the equator Γ:

$$\begin{aligned} c_1 &= \frac{1}{2\pi i}\int_{\text{north}} F + \frac{1}{2\pi i}\int_{\text{south}} F\\ &= \frac{1}{2\pi i}\int_{\text{north}} dA_+ + \frac{1}{2\pi i}\int_{\text{south}} dA_-\\ &= \frac{1}{2\pi i}\int_\Gamma A_+ - \frac{1}{2\pi i}\int_\Gamma A_-\\ &= \frac{1}{2\pi}\int_\Gamma d\chi. \end{aligned} \tag{13.99}$$

We see that c_1 is an integer that counts the winding of χ as we circle Γ. A non-zero integer cannot be continuously reduced to zero, and if we attempt to deform $A \to tA \to 0$, we will violate the required single-valuedness of the U(1) group element $e^{i\chi}$.

Although the Chern–Simons forms $\omega_{2n-1}(A)$ cannot be defined globally, they are still very useful in physics. They occur as *Wess–Zumino terms* describing the low-energy properties of various quantum field theories, the prototype being the Skyrme–Witten model of hadrons.[3]

13.6.2 Chern characters and Chern classes

Any gauge-invariant polynomial (with exterior multiplication of forms understood) in F provides a closed, topologically invariant, differential form. Certain combinations, however, have additional desirable properties, and so have been given names.

The form

$$\mathrm{ch}_n(F) = \mathrm{tr}\left\{\frac{1}{n!}\left(\frac{i}{2\pi}F\right)^n\right\} \tag{13.100}$$

is called the n-th *Chern character*. It is convenient to think of this $2n$-form as being the n-th term in a generating-function expansion

$$\mathrm{ch}(F) \stackrel{\text{def}}{=} \mathrm{tr}\left\{\exp\left(\frac{i}{2\pi}F\right)\right\} = \mathrm{ch}_0(F) + \mathrm{ch}_1(F) + \mathrm{ch}_2(F) + \cdots, \tag{13.101}$$

where $\mathrm{ch}_0(F) \stackrel{\text{def}}{=} \mathrm{tr}\, I$ is the dimension of the space on which the $\widehat{\lambda}_a$ act. This formal sum of forms of different degree is called the *total Chern character*. The $n!$ normalization is chosen because it makes the Chern character behave nicely when we combine vector bundles – as we now do.

Given two vector bundles over the same manifold, having fibres U_x and V_x over the point x, we can make a new bundle with the direct sum $U_x \oplus V_x$ as fibre over x. This resulting bundle is called the *Whitney sum* of the bundles. Similarly we can make a tensor-product bundle whose fibre over x is $U_x \otimes V_x$.

Let us use the notation $\mathrm{ch}(U)$ to represent the Chern character of the bundle with fibres U_x, and $U \oplus V$ to denote the Whitney sum. Then we have

$$\mathrm{ch}(U \oplus V) = \mathrm{ch}(U) + \mathrm{ch}(V), \tag{13.102}$$

and

$$\mathrm{ch}(U \otimes V) = \mathrm{ch}(U) \wedge \mathrm{ch}(V). \tag{13.103}$$

The second of these formulæ comes about because if $\widehat{\lambda}_a^{(1)}$ is a Lie algebra element acting on $V^{(1)}$ and $\widehat{\lambda}_a^{(2)}$ the corresponding element acting on $V^{(2)}$, then they act on the tensor

[3] E. Witten, *Nucl. Phys.*, **B223** (1983) 422; *ibid.* **B223** (1983) 433.

product $V^{(1)} \otimes V^{(2)}$ as

$$\widehat{\lambda}_a^{(1\otimes 2)} = \widehat{\lambda}_a^{(1)} \otimes I + I \otimes \widehat{\lambda}_a^{(2)}, \tag{13.104}$$

where I is the identity operator on the appropriate space in the tensor product, and for matrices A and B we have

$$\mathrm{tr}\,\{\exp\,(A \otimes I + I \otimes B)\} = \mathrm{tr}\,\{\exp A \otimes \exp B\} = \mathrm{tr}\,\{\exp A\}\,\mathrm{tr}\,\{\exp B\}\,. \tag{13.105}$$

In terms of the individual $\mathrm{ch}_n(V)$, Equations (13.102) and (13.103) read

$$\mathrm{ch}_n(U \oplus V) = \mathrm{ch}_n(U) + \mathrm{ch}_n(V), \tag{13.106}$$

and

$$\mathrm{ch}_n(U \otimes V) = \sum_{m=0}^{n} \mathrm{ch}_{n-m}(U) \wedge \mathrm{ch}_m(V). \tag{13.107}$$

Related to the Chern characters are the *Chern classes*. These are wedge-product polynomials in the Chern characters, and are defined, *via* the matrix expansion

$$\det\,(I + A) = 1 + \mathrm{tr}\,A + \frac{1}{2}\Big((\mathrm{tr}\,A)^2 - \mathrm{tr}\,A^2\Big) + \ldots, \tag{13.108}$$

by the generating function for the total Chern class:

$$c(F) = \det\left(I + \frac{i}{2\pi}F\right) = 1 + c_1(F) + c_2(F) + \cdots. \tag{13.109}$$

Thus

$$c_1(F) = \mathrm{ch}_1(F), \quad c_2(F) = \frac{1}{2}\mathrm{ch}_1(F) \wedge \mathrm{ch}_1(F) - \mathrm{ch}_2(F), \tag{13.110}$$

and so on.

For matrices A and B we have $\det(A \oplus B) = \det(A)\det(B)$, and this leads to

$$c(U \oplus V) = c(U) \wedge c(V). \tag{13.111}$$

Although the Chern classes are more complicated in appearance than the Chern characters, they are introduced because their integrals over cycles turn out to be *integers*, and this property remains true of integer-coefficient sums of products of Chern classes. The cohomology classes $[c_n(F)]$ are therefore elements of the integer cohomology ring $H^\bullet(M, \mathbb{Z})$. This property does not hold for the Chern characters, whose integrals over cycles can be fractions. The cohomology classes $[\mathrm{ch}_n(F)]$ are therefore only elements of $H^\bullet(M, \mathbb{Q})$.

When we integrate products of Chern classes of total degree $2m$ over closed $2m$-dimensional orientable manifolds we get integer *Chern numbers*. These integers can be related to generalized winding numbers, and characterize the extent to which the gauge transformations that relate the connection fields in different patches serve to *twist* the vector bundle. Unfortunately it requires a considerable amount of combinatorial machinery (the Schubert calculus of complex Grassmannians) to explain these integers.

Pontryagin and Euler classes

When the fibres of a vector bundle are vector spaces over $\mathbb{R}$, the complex skew-hermitian matrices $\widehat{i\lambda_a}$ are replaced by real skew symmetric matrices. The Lie algebra of the n-by-n matrices $\widehat{i\lambda_a}$ was a subalgebra of $\mathfrak{u}(n)$. The Lie algebra of the n-by-n real, skew symmetric, matrices is a subalgebra of $\mathfrak{o}(n)$. Now, the trace of an odd power of any skew symmetric matrix is zero. As a consequence, Chern characters and Chern classes containing an odd number of F's all vanish. The remaining real $4n$-forms are known as *Pontryagin classes*. The precise definition is

$$p_k(V) \stackrel{\text{def}}{=} (-1)^k c_{2k}(V). \tag{13.112}$$

Pontryagin classes help to classify bundles whose gauge transformations are elements of $O(n)$. If we restrict ourselves to gauge transformations that lie in $SO(n)$, as we would when considering the tangent bundle of an *orientable* Riemann manifold, then we can make a gauge-invariant polynomial out of the skew-symmetric matrix-valued F by forming its *Pfaffian*.

Recall (or see Exercise A.18) that the Pfaffian of a skew symmetric $2n$-by-$2n$ matrix $\mathbf{A}$ with entries a_{ij} is

$$\mathrm{Pf}\,\mathbf{A} = \frac{1}{2^n n!}\epsilon_{i_1,\ldots i_{2n}} a_{i_1 i_2}\cdots a_{i_{2n-1}i_{2n}}. \tag{13.113}$$

The *Euler class* of the tangent bundle of a $2n$-dimensional orientable manifold is defined *via* its skew-symmetric Riemann-curvature form

$$\mathbf{R} = \frac{1}{2}R_{ab,\mu\nu}dx^\mu dx^\nu \tag{13.114}$$

to be

$$e(\mathbf{R}) = \mathrm{Pf}\left(\frac{1}{2\pi}\mathbf{R}\right). \tag{13.115}$$

In four dimensions, for example, this becomes the 4-form

$$e(\mathbf{R}) = \frac{1}{32\pi^2}\epsilon_{abcd}R_{ab}R_{cd}. \tag{13.116}$$

The generalized *Gauss–Bonnet theorem* asserts – for an oriented, even-dimensional, manifold without boundary – that the Euler character is given by

$$\chi(M) = \int_M e(\mathbf{R}). \tag{13.117}$$

We will not prove this theorem, but in Section 16.3.6 we will illustrate the strategy that leads to Chern's influential proof.

Exercise 13.5: Show that

$$c_3(F) = \frac{1}{6}\Big((\mathrm{ch}_1(F))^3 - 6\,\mathrm{ch}_1(F)\mathrm{ch}_2(F) + 12\,\mathrm{ch}_3(F)\Big).$$

13.7 Hodge theory and the Morse index

The Laplacian, when acting on a scalar function ϕ in $\mathbb{R}^3$, is simply div (grad ϕ), but when acting on a vector $\mathbf{v}$ it becomes

$$\nabla^2\mathbf{v} = \mathrm{grad}\,(\mathrm{div}\,\mathbf{v}) - \mathrm{curl}\,(\mathrm{curl}\,\mathbf{v}). \tag{13.118}$$

Why this weird expression? How should the Laplacian act on other types of fields?

For general curvilinear coordinates in $\mathbb{R}^n$, a reasonable definition for the Laplacian of a vector or tensor field $\mathbf{T}$ is $\nabla^2\mathbf{T} = g^{\mu\nu}\nabla_\mu\nabla_\nu\mathbf{T}$, where ∇_μ is the flat-space covariant derivative. This is the unique coordinate-independent object that reduces in cartesian coordinates to the ordinary Laplacian acting on the individual components of $\mathbf{T}$. The proof that the rather different-seeming (13.118) holds for vectors is that it too is constructed out of coordinate-independent operations, and in cartesian coordinates reduces to the ordinary Laplacian acting on the individual components of $\mathbf{v}$. It must therefore coincide with the covariant derivative definition. Why it should work out this way is not exactly obvious. Now, div, grad and curl can all be expressed in differential-form language, and therefore so can the scalar and vector Laplacian. Moreover, when we let the Laplacian act on any p-form the general pattern becomes clear. The differential-form definition of the Laplacian, and the exploration of its consequences, was the work of William Hodge in the 1930s. His theory has natural applications to the topology of manifolds.

13.7.1 The Laplacian on p-forms

Suppose that M is an oriented, compact, D-dimensional manifold without boundary. We can make the space $\Omega^p(M)$ of p-form fields on M into an L^2 Hilbert space by introducing the positive-definite inner product

$$\langle a, b\rangle_p = \langle b, a\rangle_p = \int_M a \star b = \frac{1}{p!}\int d^Dx\sqrt{g}\, a_{i_1 i_2 \ldots i_p} b^{i_1 i_2 \ldots i_p}. \tag{13.119}$$

Here, the subscript p denotes the order of the forms in the product, and should not be confused with the p we have elsewhere used to label the norm in L^p Banach spaces. The presence of the $\sqrt{g}$ and the Hodge $\star$ operator tells us that this inner product depends on both the metric on M and the global orientation.

We can use this new inner product to define a "hermitian adjoint" $\delta \equiv d^\dagger$ of the exterior differential operator d. The inverted commas "..." are because this hermitian adjoint is not quite an adjoint operator in the normal sense – d takes us from one vector space to another – but it is constructed in an analogous manner. We define δ by requiring that

$$\langle da, b\rangle_{p+1} = \langle a, \delta b\rangle_p, \tag{13.120}$$

where a is an arbitrary p-form and b an arbitrary $(p+1)$-form. Now recall that $\star$ takes p-forms to $(D-p)$ forms, and so $d \star b$ is a $(D-p)$ form. Acting twice on a $(D-p)$-form with $\star$ gives us back the original form multiplied by $(-1)^{p(D-p)}$. We use this to compute

$$\begin{aligned} d(a \star b) &= da \star b + (-1)^p a(d \star b) \\ &= da \star b + (-1)^p (-1)^{p(D-p)} a \star (\star d \star b) \\ &= da \star b - (-1)^{Dp+1} a \star (\star d \star b). \end{aligned} \tag{13.121}$$

In obtaining the last line we have observed that $p(p-1)$ is an even integer and so $(-1)^{p(1-p)} = 1$. Now, using Stokes' theorem, and the absence of a boundary to discard the integrated-out part, we conclude that

$$\int_M (da) \star b = (-1)^{Dp+1} \int_M a \star (\star d \star b), \tag{13.122}$$

or

$$\langle da, b\rangle_{p+1} = (-1)^{Dp+1} \langle a, (\star d \star) b\rangle_p \tag{13.123}$$

and so $\delta b = (-1)^{Dp+1}(\star d \star) b$. This was for δ acting on a $(p-1)$ form. Acting on a p form instead we have

$$\delta = (-1)^{Dp+D+1} \star d \star. \tag{13.124}$$

Observe how the sequence of maps in $\star d \star$ works:

$$\Omega^p(M) \xrightarrow{\star} \Omega^{D-p}(M) \xrightarrow{d} \Omega^{D-p+1}(M) \xrightarrow{\star} \Omega^{p-1}(M). \tag{13.125}$$

The net effect is that δ takes a p-form to a $(p-1)$-form. Observe also that $\delta^2 \propto \star d^2 \star = 0$.

We now define a second-order partial differential operator Δ_p to be the combination

$$\Delta_p = \delta d + d\delta, \tag{13.126}$$

acting on p-forms. This maps a p-form to a p-form. A slightly tedious calculation in cartesian coordinates will show that, for flat space,

$$\Delta_p = -\nabla^2 \tag{13.127}$$

on each component of a p-form. This Δ_p is therefore the natural definition for (minus) the Laplacian acting on differential forms. It is usually called the *Laplace–Beltrami* operator.

Using $\langle a, db\rangle = \langle \delta a, b\rangle$ we have

$$\langle (\delta d + d\delta)a, b\rangle_p = \langle \delta a, \delta b\rangle_{p-1} + \langle da, db\rangle_{p+1} = \langle a, (\delta d + d\delta)b\rangle_p, \tag{13.128}$$

and so we deduce that Δ_p is self-adjoint on $\Omega^p(M)$. The middle terms in (13.128) are both positive, so we also see that Δ_p is a positive operator – i.e. all its eigenvalues are positive or zero.

Suppose that $\Delta_p a = 0$. Then (13.128) for $a = b$ becomes

$$0 = \langle \delta a, \delta a\rangle_{p-1} + \langle da, da\rangle_{p+1}. \tag{13.129}$$

Because both of these inner products are positive or zero, the vanishing of their sum requires them to be individually zero. Thus $\Delta_p a = 0$ implies that $da = \delta a = 0$. By analogy with harmonic functions, we call a form that is annihilated by Δ_p a *harmonic form*. Recall that a form a is closed if $da = 0$. We correspondingly say that a is *co-closed* if $\delta a = 0$. A differential form is therefore harmonic if and only if it is both closed and co-closed.

When a self-adjoint operator A is Fredholm (i.e. the solutions of the equation $Ax = y$ are governed by the Fredholm alternative) the vector space on which A acts is decomposed into a direct sum of the kernel and range of the operator

$$V = \text{Ker}\,(A) \oplus \text{Im}\,(A). \tag{13.130}$$

It may be shown that our Laplace–Beltrami Δ_p is a Fredholm operator, and so for any p-form ω there is an η such that ω can be written as

$$\begin{aligned} \omega &= (d\delta + \delta d)\eta + \gamma \\ &= d\alpha + \delta\beta + \gamma, \end{aligned} \tag{13.131}$$

where $\alpha = \delta\eta$, $\beta = d\eta$ and γ is harmonic. This result is known as the *Hodge decomposition* of ω. It is a form-language generalization of the Hodge–Weyl and Helmholtz–Hodge decompositions of Chapter 6. It is easy to see that α, β and γ are uniquely determined by ω. If they were not, then we could find some α, β and γ such that

$$0 = d\alpha + \delta\beta + \gamma \tag{13.132}$$

with non-zero $d\alpha$, $\delta\beta$ and γ. To see that this is not possible, take the d of (13.132) and then the inner product of the result with β. Because $d(d\alpha) = d\gamma = 0$, we end up with

$$\begin{aligned} 0 &= \langle \beta, d\delta\beta \rangle \\ &= \langle \delta\beta, \delta\beta \rangle. \end{aligned} \tag{13.133}$$

Thus $\delta\beta = 0$. Now apply δ to the two remaining terms of (13.132) and take an inner product with α. Because $\delta\gamma = 0$, we find $\langle d\alpha, d\alpha \rangle = 0$, and so $d\alpha = 0$. What now remains of (13.132) asserts that $\gamma = 0$.

Suppose that ω is closed. Then our strategy of taking the d of the decomposition

$$\omega = d\alpha + \delta\beta + \gamma, \tag{13.134}$$

followed by an inner product with β, leads to $\delta\beta = 0$. A closed form can thus be decomposed as

$$\omega = d\alpha + \gamma, \tag{13.135}$$

with α and γ unique. Each cohomology class in $H^p(M)$ therefore contains a unique harmonic representative. Since any harmonic function is closed, and hence a representative of some cohomology class, we conclude that there is a one-to-one correspondence between p-form solutions of Laplace's equation and elements of $H^p(M)$. In particular

$$\dim(\mathrm{Ker}\, \Delta_\mathrm{p}) = \dim\left(H^p(M)\right) = b_p. \tag{13.136}$$

Here b_p is the p-th Betti number. From this we immediately deduce from the definition of the Euler character (13.35) that

$$\chi(M) = \sum_{p=0}^{D} (-1)^p \dim(\mathrm{Ker}\, \Delta_\mathrm{p}), \tag{13.137}$$

where $\chi(M)$ is the Euler character of the manifold M. There is therefore an intimate relationship between the null-spaces of the second-order partial differential operators Δ_p and the global topology of the manifold in which they live. This is an example of an *index theorem*.

Just as for the ordinary Laplace operator, Δ_p has a complete set of eigenfunctions with associated eigenvalues λ. Because the manifold is compact and hence has finite volume, the spectrum will be discrete. Remarkably, the topological influence we uncovered above is restricted to the zero-eigenvalue spaces of p-forms. To see this, suppose that we have a p-form eigenfunction u_λ for Δ_p:

$$\Delta_p u_\lambda = \lambda u_\lambda. \tag{13.138}$$

Then

$$\begin{aligned} \lambda\, du_\lambda &= d\,\Delta_p u_\lambda \\ &= d(d\delta + \delta d)u_\lambda \\ &= (d\delta) du_\lambda \\ &= (\delta d + d\delta) du_\lambda \\ &= \Delta_{p+1} du_\lambda. \end{aligned} \tag{13.139}$$

Thus, provided it is not identically zero, du_λ is a $(p+1)$-form eigenfunction of $\Delta_{(p+1)}$ with eigenvalue λ. Similarly, δu_λ is a $(p-1)$-form eigenfunction also with eigenvalue λ.

Can du_λ be zero? Yes! It will certainly be zero if u_λ itself is the d of something. What is less obvious is that it will be zero *only* if it is the d of something. To see this suppose that $du_\lambda = 0$ and $\lambda \neq 0$. Then

$$\lambda u_\lambda = (\delta d + d\delta)u_\lambda = d(\delta u_\lambda). \tag{13.140}$$

Thus $du_\lambda = 0$ implies that $u_\lambda = d\eta$, where $\eta = \delta u_\lambda/\lambda$. We see that for λ non-zero, the operators d and δ map the λ eigenspaces of Δ into one another, and the kernel of d acting on p-form eigenfunctions is precisely the image of d acting on $(p-1)$-form eigenfunctions. In other words, when restricted to positive λ eigenspaces of Δ, the cohomology is trivial.

The set of spaces V_p^λ together with the maps $d : V_p^\lambda \to V_{p+1}^\lambda$ therefore constitute an exact sequence when $\lambda \neq 0$, and so the alternating sum of their dimensions must be zero. We have therefore established that

$$\sum_p (-1)^p \dim V_p^\lambda = \begin{cases} \chi(M), & \lambda = 0, \\ 0, & \lambda \neq 0. \end{cases} \tag{13.141}$$

All the topological information resides in the null-spaces, therefore.

Exercise 13.6: Show that if ω is closed and co-closed then so is $\star\omega$. Deduce that for a compact orientable D-manifold we have $b_p = b_{D-p}$. This observation therefore gives another way of understanding Poincaré duality.

13.7.2 Morse theory

Suppose, as in the previous section, that M is a D-dimensional compact manifold without boundary and $V : M \to \mathbb{R}$ is a smooth function. The global topology of M imposes some constraints on the possible maxima, minima and saddle points of V. Suppose that P is a stationary point of V. Taking coordinates such that P is at $x^\mu = 0$, we can expand

$$V(x) = V(0) + \frac{1}{2} H_{\mu\nu} x^\mu x^\nu + \cdots . \tag{13.142}$$

Here, the matrix $H_{\mu\nu}$ is the *Hessian*

$$H_{\mu\nu} \stackrel{\text{def}}{=} \left. \frac{\partial^2 V}{\partial x^\mu \partial x^\nu} \right|_0 . \tag{13.143}$$

We can change coordinates so as to reduce the Hessian to a canonical form which is diagonal and has only $\pm 1, 0$ on its diagonal:

$$H_{\mu\nu} = \begin{pmatrix} -I_m & & \\ & I_n & \\ & & 0_{D-m-n} \end{pmatrix} . \tag{13.144}$$

If there are no zeros on the diagonal then the stationary point is said to be *non-degenerate*. The number m of downward-bending directions is then called the *index* of V at P. If P were a local maximum, then $m = D$, $n = 0$. If it were a local minimum then $m = 0$, $n = D$. When all its stationary points are non-degenerate, V is said to be a *Morse function*. This is the generic case. Degenerate stationary points can be regarded as arising from the merging of two or more non-degenerate points.

The *Morse index theorem* asserts that if V is a Morse function, and if we define N_0 to be the number of stationary points with index 0 (i.e. local minima), and N_1 to be the number of stationary points with index 1, etc. then

$$\sum_{m=0}^{D} (-1)^m N_m = \chi(M). \tag{13.145}$$

Here $\chi(M)$ is the Euler character of M. Thus, a function on the two-dimensional torus (which has $\chi = 0$) can have a local maximum, a local minimum and two saddle points, but cannot have only one local maximum, one local minimum and no saddle points. On a 2-sphere ($\chi = 2$), if V has one local maximum and one local minimum it can have no saddle points.

Closely related to the Morse index theorem is the *Poincaré–Hopf theorem*. This counts the isolated zeros of a tangent-vector field X on a compact D-manifold and, amongst other things, explains why we cannot comb a hairy ball. An *isolated zero* is a point z_n at which X becomes zero, and that has a neighbourhood in which there is no other zero. If X possesses only finitely many zeros then each of them will be isolated. For an isolated zero, we can define a *vector field index* at z_n by surrounding it with a small $(D-1)$-sphere on which X does not vanish. The direction of X at each point on this sphere then provides a map from the sphere to itself. The index $i(z_n)$ is defined to be the

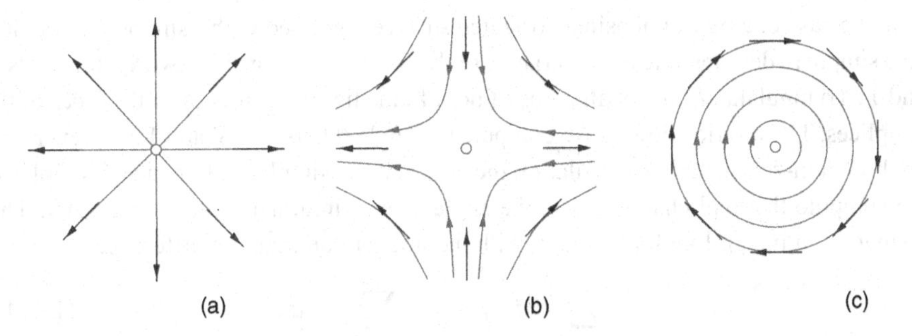

Figure 13.13 Two-dimensional vector fields and their streamlines near zeros with indices (a) $i(z_a) = +1$, (b) $i(z_b) = -1$, (c) $i(z_c) = +1$.

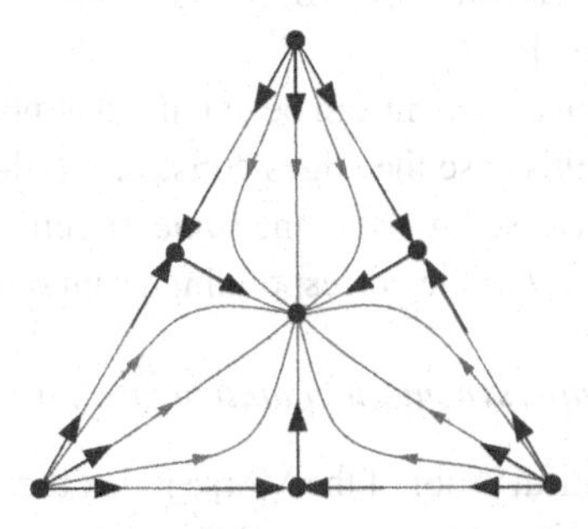

Figure 13.14 Gradient vector field and streamlines in a 2-simplex.

winding number (Brouwer degree) of this map (Figure 13.13). This index can be any integer, but in the special case that X is the gradient of a Morse function it takes the value $i(z_n) = (-1)^{m_n}$ where m is the Morse index at z_n.

The Poincaré–Hopf theorem states that, for a compact manifold without boundary, and for a tangent vector field with only finitely many zeros,

$$\sum_{\text{zeros } n} i(z_n) = \chi(M). \tag{13.146}$$

A tangent-vector field must therefore always have at least one zero unless $\chi(M) = 0$. For example, since the 2-sphere has $\chi = 2$, it cannot be combed.

If one is prepared to believe that $\sum_{\text{zeros}} i(z_n)$ is the same integer for all tangent vector fields X on M, it is simple to show that this integer must be equal to the Euler character of M. Consider, for ease of visualization, a 2-manifold. Triangulate M and take X to be the gradient field of a function with local minima at each vertex, saddle points on the edges and local maxima at the centre of each face (see Figure 13.14). It must be clear that this particular field X has

$$\sum_{\text{zeros } n} i(z_n) = V - E + F = \chi(M). \tag{13.147}$$

In the case of a two-dimensional oriented surface equipped with a smooth metric, it is also simple to demonstrate the invariance of the index sum. Consider two vector fields X and Y. Triangulate M so that all zeros of both fields lie in the interior of the faces of the simplices. The metric allows us to compute the angle θ between X and Y wherever they are both non-zero, and in particular on the edges of the simplices. For each 2-simplex σ we compute the total change $\Delta\theta$ in the angle as we circumnavigate its boundary. This change is an integral multiple of 2π, with the integer counting the difference

$$\sum_{\text{zeros of } X \in \sigma} i(z_n) - \sum_{\text{zeros of } Y \in \sigma} i(z_n) \tag{13.148}$$

of the indices of the zeros within σ. On summing over all triangles σ, each edge is traversed twice, once in each direction, so $\sum_\sigma \Delta\theta$ vanishes. The total index of X is therefore the same as that of Y.

This pairwise cancellation argument can be extended to non-orientable surfaces, such as the projective plane. In this case the edges constituting the homological "boundary" of the closed surface are traversed twice in the *same* direction, but the angle θ at a point on one edge is paired with $-\theta$ at the corresponding point of the other edge.

Supersymmetric quantum mechanics

Edward Witten gave a beautiful proof of the Morse index theorem for a closed orientable manifold M by re-interpreting the Laplace–Beltrami operator as the Hamiltonian of *supersymmetric quantum mechanics* on M. Witten's idea had a profound impact, and led to quantum physics serving as a rich source of inspiration and insight for mathematicians. We have seen most of the ingredients of this re-interpretation in previous chapters. Indeed you should have experienced a sense of *déjà vu* when you saw d and δ mapping eigenfunctions of one differential operator into eigenfunctions of a related operator.

We begin with a novel way to think of the calculus of differential forms. We introduce a set of fermion annihilation and creation operators ψ^μ and $\psi^{\dagger\mu}$ which anticommute, $\psi^\mu\psi^\nu = -\psi^\nu\psi^\mu$, and obey the anticommutation relation

$$\{\psi^{\dagger\mu}, \psi^\nu\} \equiv \psi^{\dagger\mu}\psi^\nu + \psi^\nu\psi^{\dagger\mu} = g^{\mu\nu}. \tag{13.149}$$

Here, $g^{\mu\nu}$ is the metric tensor, and the Greek indices μ and ν range from 1 to D. As is usual when we are given annihilation and creation operators, we also introduce a *vacuum state* $|0\rangle$ which is killed by all the annihilation operators: $\psi^\mu|0\rangle = 0$. The states

$$(\psi^{\dagger 1})^{p_1}(\psi^{\dagger 2})^{p_2}\ldots(\psi^{\dagger n})^{p_D}|0\rangle, \tag{13.150}$$

with each of the p_i taking the value 1 or 0, then constitute a basis for 2^D-dimensional Hilbert space. We call $p = \sum_i p_i$ the *fermion number* of the state. We assume that $\langle 0|0\rangle = 1$ and use the anticommutation relations to show that

$$\langle 0|\psi^{\mu_p}\ldots\psi^{\mu_2}\psi^{\mu_1}\ldots\psi^{\dagger\nu_1}\psi^{\dagger\nu_2}\ldots\psi^{\dagger\nu_q}|0\rangle$$

is zero unless $p = q$, in which case it is equal to

$$g^{\mu_1\nu_1}g^{\mu_2\nu_2}\dots g^{\mu_p\nu_p} \pm \text{(permutations)}.$$

We now make the correspondence

$$\frac{1}{p!}f_{\mu_1\mu_2\dots\mu_p}(x)\psi^{\dagger\mu_1}\psi^{\dagger\mu_2}\dots\psi^{\dagger\mu_p}|0\rangle \leftrightarrow \frac{1}{p!}f_{\mu_1\mu_2\dots\mu_p}(x)dx^{\mu_1}dx^{\mu_2}\cdots dx^{\mu_p}, \quad (13.151)$$

to identify p-fermion states with p-forms. We think of $f_{\mu_1\mu_2\dots\mu_p}(x)$ as being the wave-function of a particle moving on M, with the subscripts informing us there are fermions occupying the states μ_i. It is then natural to take the inner product of

$$|a\rangle = \frac{1}{p!}a_{\mu_1\mu_2\dots\mu_p}(x)\psi^{\dagger\mu_1}\psi^{\dagger\mu_2}\dots\psi^{\dagger\mu_p}|0\rangle \quad (13.152)$$

and

$$|b\rangle = \frac{1}{q!}b_{\mu_1\mu_2\dots\mu_q}(x)\psi^{\dagger\mu_1}\psi^{\dagger\mu_2}\dots\psi^{\dagger\mu_q}|0\rangle \quad (13.153)$$

to be

$$\begin{aligned}\langle a, b\rangle &= \int_M d^Dx\sqrt{g}\,\frac{1}{p!q!}\,a^*_{\mu_1\mu_2\dots\mu_p}(x)b_{\nu_1\nu_2\dots\nu_q}(x)\langle 0|\psi^{\mu_p}\dots\psi^{\mu_1}\psi^{\dagger\nu_1}\dots\psi^{\dagger\nu_q}|0\rangle\\ &= \delta_{pq}\int_M d^Dx\sqrt{g}\,\frac{1}{p!}\,a^*_{\mu_1\mu_2\dots\mu_p}(x)b^{\mu_1\mu_2\dots\mu_p}(x). \end{aligned} \quad (13.154)$$

This coincides with the Hodge inner product of the corresponding forms.

If we lower the index on ψ^μ by defining ψ_μ to be $g_{\mu\nu}\psi^\mu$ then the action of the annihilation operator $X^\mu\psi_\mu$ on a p-fermion state coincides with the action of the interior multiplication i_X on the corresponding p-form. All the other operations of the exterior calculus can also be expressed in terms of the ψ and $\psi^\dagger$'s. In particular, in cartesian coordinates where $g_{\mu\nu} = \delta_{\mu\nu}$, we can identify d with $\psi^{\dagger\mu}\partial_\mu$. To find the operator that corresponds to the Hodge δ, we compute

$$\delta = d^\dagger = (\psi^{\dagger\mu}\partial_\mu)^\dagger = \partial_\mu^\dagger\psi^\mu = -\partial_\mu\psi^\mu = -\psi^\mu\partial_\mu. \quad (13.155)$$

The hermitian adjoint of ∂_μ is here being taken with respect to the standard $L^2(\mathbb{R}^D)$ inner product. This computation becomes more complicated when when $g_{\mu\nu}$ becomes position dependent. The adjoint $\partial_\mu^\dagger$ then involves the derivative of $\sqrt{g}$, and ψ and ∂_μ no longer commute. For this reason, and because such complications are inessential for what follows, we will delay discussing this general case until the end of this section.

Having found a simple formula for δ, it is now automatic to compute

$$d\delta + \delta d = -\{\psi^{\dagger\mu}, \psi^\nu\}\,\partial_\mu\partial_\nu = -\delta^{\mu\nu}\partial_\mu\partial_\nu = -\nabla^2. \quad (13.156)$$

This is much easier than deriving the same result by using $\delta = (-1)^{Dp+D+1} \star d \star$.

Witten's fermionic formalism simplifies a number of computations involving δ, but his real innovation was to consider a *deformation* of the exterior calculus by introducing the operators

$$d_t = e^{-tV(x)} d\, e^{tV(x)}, \quad \delta_t = e^{tV(x)} \delta\, e^{-tV(x)}, \tag{13.157}$$

and the t-deformed

$$\Delta_t = d_t \delta_t + \delta_t d_t. \tag{13.158}$$

Here, $V(x)$ is the Morse function whose stationary points we seek to count.

It is easy to see that the deformed derivative continues to obey $d_t^2 = 0$. We also see that $d\omega = 0$ if and only if $d_t e^{-tV} \omega = 0$. Similarly, if $\omega = d\eta$ then $e^{-tV}\omega = d_t e^{-tV}\eta$. The cohomology of d is therefore transformed into the cohomology of d_t by multiplication by e^{-tV}. Since the exponential function is never zero, this correspondence is invertible and the mapping is an isomorphism. In particular the dimensions of the spaces $\mathrm{Ker}\,(d_t)_p/\mathrm{Im}\,(d_t)_{p-1}$ are t-independent and coincide with the $t = 0$ Betti numbers b_p. Furthermore, the t-deformed Laplace–Beltrami operator remains Fredholm with only positive or zero eigenvalues. We can therefore make a Hodge decomposition

$$\omega = d_t \alpha + \delta_t \beta + \gamma, \tag{13.159}$$

where $\Delta_t \gamma = 0$, and conclude that

$$\dim\,(\mathrm{Ker}\,(\Delta_t)_p) = b_p \tag{13.160}$$

as before. The non-zero eigenvalue spaces will also continue to form exact sequences. Nothing seems to have changed! Why do we introduce d_t then? The motivation is that when t becomes large we can use our knowledge of quantum mechanics to compute the Morse index.

To do this, we expand out

$$\begin{aligned} d_t &= \psi^{\dagger\mu}(\partial_\mu + t\partial_\mu V) \\ \delta_t &= -\psi^{\mu}(\partial_\mu - t\partial_\mu V) \end{aligned} \tag{13.161}$$

and find

$$d_t \delta_t + \delta_t d_t = -\nabla^2 + t^2 |\nabla V|^2 + t[\psi^{\dagger\mu}, \psi^\nu]\, \partial^2_{\mu\nu} V. \tag{13.162}$$

This can be thought of as a Schrödinger Hamiltonian on M containing a potential $t^2|\nabla V|^2$ and a fermionic term $t[\psi^{\dagger\mu}, \psi^\nu]\, \partial^2_{\mu\nu} V$. When t is large and positive the potential will be large and positive everywhere except near those points where $\nabla V = 0$. The wave-functions of all low-energy states, and in particular all zero-energy states, will therefore be concentrated at precisely the stationary points we are investigating. Let us focus on a particular stationary point, which we will take as the origin of our coordinate system, and see if any zero-energy state is localized there. We first rotate the coordinate system about the origin so that the Hessian matrix $\partial^2_{\mu\nu}V|_0$ becomes diagonal with eigenvalues λ_n. The Schrödinger problem can then be approximated by a sum of harmonic oscillator Hamiltonians

$$\Delta_{p,t} \approx \sum_{i=1}^{D}\left\{-\frac{\partial^2}{\partial x_i^2} + t^2\lambda_i^2 x_i^2 + t\lambda_i[\psi^{\dagger i}, \psi^i]\right\}. \tag{13.163}$$

The commutator $[\psi^{\dagger i}, \psi^i]$ takes the value $+1$ if the i-th fermion state is occupied, and -1 if it is not. The spectrum of the approximate Hamiltonian is therefore

$$t\sum_{i=1}^{D}\{|\lambda_i|(1+2n_i) \pm \lambda_i\}. \tag{13.164}$$

Here the n_i label the harmonic oscillator states. The lowest-energy states will have all the $n_i = 0$. To get a state with zero energy we must arrange for the $\pm$ sign to be negative (no fermion in state i) whenever λ_i is positive, and to be positive (fermion state i occupied) whenever λ_i is negative. The fermion number "p" of the zero-energy state is therefore equal to the number of negative λ_i – i.e. to the index of the critical point! We can, in this manner, find one zero-energy state for each critical point. All other states have energies proportional to t, and therefore large. Since the number of zero-energy states having fermion number p is the Betti number b_p, the harmonic oscillator approximation suggests that $b_p = N_p$.

If we could trust our computation of the energy spectrum, we would have established the Morse theorem

$$\sum_{p=0}^{D}(-1)^p N_p = \sum_{p=0}^{D}(-1)^p b_p = \chi(M), \tag{13.165}$$

by having the two sums agree term by term. Our computation is only approximate, however. While there can be no more zero-energy states than those we have found, some states that appear to be zero modes may instead have small positive energy. This might arise from tunnelling between the different potential minima, or from the higher-order corrections to the harmonic oscillator potentials, both effects we have neglected. We can therefore only be confident that

$$N_p \geq b_p. \tag{13.166}$$

The remarkable thing is that, for the Morse index, *this does not matter*! If one of our putative zero modes gains a small positive energy, it is now in the non-zero eigenvalue sector of the spectrum. The exact-sequence property therefore tells us that one of the other putative zero modes must also be a not-quite-zero mode state with exactly the same energy. This second state will have a fermion number that differs from the first by plus or minus one. An error in counting the zero energy states therefore cancels out when we take the alternating sum. Our unreliable estimate $b_p \approx N_p$ has thus provided us with an *exact* computation of the Morse index.

We have described Witten's argument as if the manifold M were flat. When the manifold M is not flat, however, the curvature will not affect our computations. Once the parameter t is large, the low-energy eigenfunctions will be so tightly localized about the critical points that they will be hard-pressed to detect the curvature. Even if the curvature can effect an infintesimal energy shift, the exact-sequence argument again shows that this does not affect the alternating sum.

The Weitzenböck formula

Although we were able to evade them when proving the Morse index theorem, it is interesting to uncover the workings of the nitty-gritty Riemann tensor index machinary that lie concealed behind the polished facade of Hodge's d, δ calculus.

Let us assume that our manifold M is equipped with a torsion-free connection $\Gamma^{\mu}{}_{\nu\lambda} = \Gamma^{\mu}{}_{\lambda\nu}$, and use this connection to define the action of an operator $\hat{\nabla}_\mu$ by specifying its commutators with c-number functions f, and with the ψ^μ and $\psi^{\dagger\mu}$'s:

$$\begin{aligned}
[\hat{\nabla}_\mu, f] &= \partial_\mu f, \\
[\hat{\nabla}_\mu, \psi^{\dagger\nu}] &= -\Gamma^{\nu}{}_{\mu\lambda}\psi^{\dagger\lambda}, \\
[\hat{\nabla}_\mu, \psi^{\nu}] &= -\Gamma^{\nu}{}_{\mu\lambda}\psi^{\lambda}.
\end{aligned} \tag{13.167}$$

We also set $\hat{\nabla}_\mu|0\rangle = 0$. These rules allow us to compute the action of $\hat{\nabla}_\mu$ on $f_{\mu_1\mu_2\ldots\mu_p}(x)\psi^{\dagger\mu_1}\ldots\psi^{\dagger\mu_p}|0\rangle$. For example

$$\begin{aligned}
\hat{\nabla}_\mu\left(f_\nu\psi^{\dagger\nu}|0\rangle\right) &= \left([\hat{\nabla}_\mu, f_\nu\psi^{\dagger\nu}] + f_\nu\psi^{\dagger\nu}\hat{\nabla}_\mu\right)|0\rangle \\
&= \left([\hat{\nabla}_\mu, f_\nu]\psi^{\dagger\nu} + f_\alpha[\hat{\nabla}_\mu, \psi^{\dagger\alpha}]\right)|0\rangle \\
&= (\partial_\mu f_\nu - f_\alpha\Gamma^{\alpha}{}_{\mu\nu})\psi^{\dagger\nu}|0\rangle \\
&= \left(\nabla_\mu f_\nu\right)\psi^{\dagger\nu}|0\rangle,
\end{aligned} \tag{13.168}$$

where

$$\nabla_\mu f_\nu = \partial_\mu f_\nu - \Gamma^{\alpha}{}_{\mu\nu} f_\alpha \tag{13.169}$$

is the usual covariant derivative acting on the components of a covariant vector.

The metric $g^{\mu\nu}$ counts as a c-number function, and so $[\hat{\nabla}_\alpha, g^{\mu\mu}]$ is not zero, but is instead $\partial_\alpha g^{\mu\nu}$. This might be disturbing – being able pass the metric through a covariant derivative is a basic compatibility condition in Riemann geometry – but all is not lost. $\hat{\nabla}_\mu$ (with a caret) is not quite the same beast as ∇_μ. We proceed as follows:

$$\begin{aligned}
\partial_\alpha g^{\mu\nu} &= [\hat{\nabla}_\alpha, g^{\mu\mu}] \\
&= [\hat{\nabla}_\alpha, \{\psi^{\dagger\mu}, \psi^\nu\}] \\
&= [\hat{\nabla}_\alpha, \psi^{\dagger\mu}\psi^\nu] + [\hat{\nabla}_\alpha, \psi^\nu\psi^{\dagger\mu}] \\
&= -\{\psi^{\dagger\mu}, \psi^\lambda\}\Gamma^\nu{}_{\alpha\lambda} - \{\psi^{\dagger\nu}, \psi^\lambda\}\Gamma^\mu{}_{\alpha\lambda} \\
&= -g^{\mu\lambda}\,\Gamma^\nu{}_{\alpha\lambda} - g^{\nu\lambda}\,\Gamma^\mu{}_{\alpha\lambda}.
\end{aligned} \tag{13.170}$$

Thus, we conclude that

$$\partial_\alpha g^{\mu\nu} + g^{\mu\lambda}\Gamma^\nu{}_{\alpha\lambda} + g^{\lambda\nu}\Gamma^\mu{}_{\alpha\lambda} \equiv \nabla_\alpha g^{\mu\nu} = 0. \tag{13.171}$$

Metric compatibility is therefore satisfied, and the connection is therefore the standard Riemannian

$$\Gamma^\alpha{}_{\mu\nu} = \frac{1}{2} g^{\alpha\lambda}\left(\partial_\mu g_{\lambda\nu} + \partial_\nu g_{\mu\lambda} - \partial_\lambda g_{\mu\nu}\right). \tag{13.172}$$

Knowing this, we can compute the adjoint of $\hat{\nabla}_\mu$:

$$\begin{aligned}
\left(\hat{\nabla}_\mu\right)^\dagger &= -\frac{1}{\sqrt{g}}\hat{\nabla}_\mu\sqrt{g} \\
&= -\hat{\nabla}_\mu - \partial_\mu \ln\sqrt{g} \\
&= -(\hat{\nabla}_\mu + \Gamma^\nu{}_{\nu\mu}).
\end{aligned} \tag{13.173}$$

That $\Gamma^\nu{}_{\nu\mu}$ is the logarithmic derivative of $\sqrt{g}$ is a standard identity for the Riemann connection (see Exercise 11.14). The resultant formula for $(\hat{\nabla}_\mu)^\dagger$ can be used to verify that the second and third equations in (13.167) are compatible with each other.

We can also compute $[[\hat{\nabla}_\mu, \hat{\nabla}_\nu], \psi^\alpha]$, and from it deduce that

$$[\hat{\nabla}_\mu, \hat{\nabla}_\nu] = R_{\sigma\lambda\mu\nu}\psi^{\dagger\sigma}\psi^\lambda, \tag{13.174}$$

where

$$R^\alpha{}_{\beta\mu\nu} = \partial_\mu\Gamma^\alpha{}_{\beta\nu} - \partial_\nu\Gamma^\alpha{}_{\beta\mu} + \Gamma^\alpha{}_{\lambda\mu}\Gamma^\lambda{}_{\beta\nu} - \Gamma^\alpha{}_{\lambda\nu}\Gamma^\lambda{}_{\beta\mu} \tag{13.175}$$

is the Riemann curvature tensor.

We now define d to be

$$d = \psi^{\dagger\mu}\,\hat{\nabla}_\mu. \tag{13.176}$$

Its action coincides with the usual d because the symmetry of the $\Gamma^{\alpha}_{\mu\nu}$'s ensures that their contributions cancel. From this we find that δ is

$$\begin{aligned}\delta &\equiv \left(\psi^{\dagger\mu}\,\hat{\nabla}_{\mu}\right)^{\dagger} \\ &= \hat{\nabla}^{\dagger}_{\mu}\,\psi^{\mu} \\ &= -(\hat{\nabla}_{\mu} + \Gamma^{\nu}{}_{\mu\nu})\psi^{\mu} \\ &= -\psi^{\mu}(\hat{\nabla}_{\mu} + \Gamma^{\nu}{}_{\mu\nu}) + \Gamma^{\mu}{}_{\mu\nu}\psi^{\nu} \\ &= -\psi^{\mu}\,\hat{\nabla}_{\mu}.\end{aligned} \tag{13.177}$$

The Laplace–Beltrami operator can now be worked out as

$$\begin{aligned}d\delta + \delta d &= -\left(\psi^{\dagger\mu}\hat{\nabla}_{\mu}\psi^{\nu}\hat{\nabla}_{\nu} + \psi^{\nu}\hat{\nabla}_{\nu}\psi^{\dagger\mu}\hat{\nabla}_{\mu}\right) \\ &= -\left(\{\psi^{\dagger\mu},\psi^{\nu}\}(\hat{\nabla}_{\mu}\hat{\nabla}_{\nu} - \Gamma^{\sigma}{}_{\mu\nu}\hat{\nabla}_{\sigma}) + \psi^{\nu}\psi^{\dagger\mu}[\hat{\nabla}_{\nu},\hat{\nabla}_{\mu}]\right) \\ &= -\left(g^{\mu\nu}(\hat{\nabla}_{\mu}\hat{\nabla}_{\nu} - \Gamma^{\alpha}{}_{\mu\nu}\hat{\nabla}_{\sigma}) + \psi^{\nu}\psi^{\dagger\mu}\psi^{\dagger\sigma}\psi^{\lambda}R_{\sigma\lambda\nu\mu}\right).\end{aligned} \tag{13.178}$$

By making use of the symmetries $R_{\sigma\lambda\nu\mu} = R_{\nu\mu\sigma\lambda}$ and $R_{\sigma\lambda\nu\mu} = -R_{\sigma\lambda\mu\nu}$ we can tidy up the curvature term to get

$$d\delta + \delta d = -g^{\mu\nu}(\hat{\nabla}_{\mu}\hat{\nabla}_{\nu} - \Gamma^{\sigma}{}_{\mu\nu}\hat{\nabla}_{\sigma}) - \psi^{\dagger\alpha}\psi^{\beta}\psi^{\dagger\mu}\psi^{\nu}R_{\alpha\beta\mu\nu}. \tag{13.179}$$

This result is called the *Weitzenböck formula*. An equivalent formula can be derived directly from (13.124), but only with a great deal more effort. The part without the curvature tensor is called the *Bochner Laplacian*. It is normally written as $B = -g^{\mu\nu}\nabla_{\mu}\nabla_{\nu}$ with ∇_{μ} being understood to be acting on the index ν, and therefore tacitly containing the extra $\Gamma^{\sigma}_{\mu\nu}$ that must be made explicit – as we have in (13.179) – when we define the action of $\hat{\nabla}_{\mu}$ *via* commutators. The Bochner Laplacian can also be written as

$$B = \hat{\nabla}^{\dagger}_{\mu}\,g^{\mu\nu}\,\hat{\nabla}_{\nu} \tag{13.180}$$

which shows that it is a positive operator.

13.8 Further exercises and problems

Exercise 13.7: Let

$$A = A_x\,dx + A_y\,dy + A_z\,dz$$

be a closed form in $\mathbb{R}^3$. Use the formula (13.6) of Section 13.2.1 to find a scalar $\varphi(x,y,z)$ such that $A = d\varphi$. Compute the exterior derivative from your expression for φ and verify that it reconstitutes A.

Exercise 13.8: By considering the example of the unit disc in two dimensions, show that the condition of being closed – in the sense of having no boundary – is a necessary condition in the statement of Poincaré duality. What goes wrong with our construction of the elements of $H^{D-p}(M)$ from cycles in $H_p(M)$ in this case?

Exercise 13.9: Use Poincaré duality to show that the Euler character of any odd-dimensional closed manifold is zero.

14

Groups and group representations

Groups usually appear in physics as symmetries of the system or model we are studying. Often the symmetry operation involves a linear transformation, and this naturally leads to the idea of finding sets of matrices having the same multiplication table as the group. These sets are called *representations* of the group. Given a group, we endeavour to find and classify all possible representations.

14.1 Basic ideas

We begin with a rapid review of basic group theory.

14.1.1 Group axioms

A *group* G is a set with a binary operation that assigns to each ordered pair (g_1, g_2) of elements a third element, g_3, usually written with multiplicative notation as $g_3 = g_1 g_2$. The binary operation, or *product*, obeys the following rules:

(i) Associativity: $g_1(g_2 g_3) = (g_1 g_2)g_3$.
(ii) Existence of an identity: there is an element[1] $e \in G$ such that $eg = g$ for all $g \in G$.
(iii) Existence of an inverse: for each $g \in G$ there is an element g^{-1} such that $g^{-1}g = e$.

From these axioms there follow some conclusions that are so basic that they are often included in the axioms themselves, but since they are not independent, we state them as corollaries.

Corollary: *(i)*: $gg^{-1} = e$.

Proof: Start from $g^{-1}g = e$, and multiply on the right by g^{-1} to get $g^{-1}gg^{-1} = eg^{-1} = g^{-1}$, where we have used the left identity property of e at the last step. Now multiply on the left by $(g^{-1})^{-1}$, and use associativity to get $gg^{-1} = e$.

Corollary: *(ii)*: $ge = g$.

Proof: Write $ge = g(g^{-1}g) = (gg^{-1})g = eg = g$.

Corollary: *(iii)*: The identity e is unique.

[1] The symbol "e" is often used for the identity element, from the German *Einheit*, meaning "unity".

Proof: Suppose there is another element e_1 such that $e_1 g = eg = g$. Multiply on the right by g^{-1} to get $e_1 e = e^2 = e$, but $e_1 e = e_1$, so $e_1 = e$.

Corollary: *(iv)*: The inverse of a given element g is unique.

Proof: Let $g_1 g = g_2 g = e$. Use the result of Corollary (i), that any left inverse is also a right inverse, to multiply on the right by g_1, and so find that $g_1 = g_2$.

Two elements g_1 and g_2 are said to *commute* if $g_1 g_2 = g_2 g_1$. If the group has the property that $g_1 g_2 = g_2 g_1$ for all $g_1, g_2 \in G$, it is said to be *abelian*, otherwise it is *non-abelian*.

If the set G contains only finitely many elements, the group G is said to be *finite*. The number of elements in the group, $|G|$, is called the *order* of the group.

Examples of groups

(1) The integers $\mathbb{Z}$ under addition. The binary operation is $(n, m) \mapsto n + m$, and "0" plays the role of the identity element. This is not a finite group.
(2) The integers modulo n under addition. $(m, m') \mapsto m + m'$, mod n. This group is denoted by $\mathbb{Z}_n$, and is finite.
(3) The non-zero integers modulo p (a prime) under *multiplication* $(m, m') \mapsto mm'$, mod p. Here "1" is the identity element. If the modulus is not a prime number, we do not get a group (why not?). This group is sometimes denoted by $(\mathbb{Z}_p)^\times$.
(4) The set of numbers $\{2, 4, 6, 8\}$ under multiplication modulo 10. Here, the number "6" plays the role of the identity!
(5) The set of functions

$$f_1(z) = z, \quad f_2(z) = \frac{1}{1-z}, \quad f_3(z) = \frac{z-1}{z},$$
$$f_4(z) = \frac{1}{z}, \quad f_5(z) = 1 - z, \quad f_6(z) = \frac{z}{z-1},$$

with $(f_i, f_j) \mapsto f_i \circ f_j$. Here, the "$\circ$" is a standard notation for composition of functions: $(f_i \circ f_j)(z) = f_i(f_j(z))$.
(6) The set of rotations in three dimensions, equivalently the set of 3-by-3 real matrices O, obeying $O^T O = I$ and $\det O = 1$. This is the group SO(3). SO(n) is defined analogously as the group of rotations in n dimensions. If we relax the condition on the determinant we get the *orthogonal group* O(n). Both SO(n) and O(n) are examples of *Lie groups*. A Lie group is a group that is also a manifold M, and whose multiplication law is a smooth function $M \times M \to M$.
(7) Groups are often specified by giving a list of *generators* and *relations*. For example the *cyclic group* of order n, denoted by C_n, is specified by giving the generator a and relation $a^n = e$. Similarly, the *dihedral group* D_n has two generators a, b and relations $a^n = e$, $b^2 = e$, $(ab)^2 = e$. This group has order $2n$.

14.1.2 Elementary properties

Here are the basic properties of groups that we need:

(i) *Subgroups*: If a subset of elements of a group forms a group, it is called a subgroup. For example, $\mathbb{Z}_{12}$ has a subgroup consisting of $\{0, 3, 6, 9\}$. Any group G possesses at least two subgroups: the entirety of G itself, and the subgroup containing only the identity element $\{e\}$. These are known as the *trivial* subgroups. Any other subgroups are called *proper* subgroups.

(ii) *Cosets*: Given a subgroup $H \subseteq G$, having elements $\{h_1, h_2, \ldots\}$, and an element $g \in G$, we form the (left) *coset* $gH = \{gh_1, gh_2, \ldots\}$. If two cosets g_1H and g_2H intersect, they coincide. (Proof: if $g_1h_1 = g_2h_2$, then $g_2 = g_1(h_1h_2^{-1})$ and so $g_1H = g_2H$.) If H is a finite group, each coset has the same number of distinct elements as H. (Proof: if $gh_1 = gh_2$ then left multiplication by g^{-1} shows that $h_1 = h_2$.) If the order of G is also finite, the group G is decomposed into an integer number of cosets,

$$G = g_1H + g_2H + \cdots , \tag{14.1}$$

where "+" denotes the union of disjoint sets. From this we see that the order of H must divide the order of G. This result is called *Lagrange's theorem*. The set whose elements are the cosets is denoted by G/H.

(iii) *Normal subgroups*: A subgroup $H = \{h_1, h_2, \ldots\}$ of G is said to be *normal*, or *invariant*, if $g^{-1}Hg = H$ for all $g \in G$. This notation means that the set of elements $g^{-1}Hg = \{g^{-1}h_1g, g^{-1}h_2g, \ldots\}$ coincides with H, or equivalently that the map $h \mapsto g^{-1}hg$ does not take $h \in H$ out of H, but simply scrambles the order of the elements of H.

(iv) *Quotient groups*: Given a normal subgroup H, we can define a multiplication rule on the set of cosets $G/H \equiv \{g_1H, g_2H, \ldots\}$ by taking a representative element from each of g_iH, and g_jH, taking the product of these elements, and defining $(g_iH)(g_jH)$ to be the coset in which this product lies. This coset is independent of the representative elements chosen (this would not be so were the subgroup not normal). The resulting group is called the *quotient group* of G by H, and is denoted by G/H. (Note that the symbol "G/H" is used to denote both the set of cosets, and, when it exists, the group whose elements are these cosets.)

(v) *Simple groups*: A group G with no normal subgroups is said to be *simple*. The finite simple groups have been classified. They fall into various infinite families (cyclic groups, alternating groups, 16 families of Lie type) together with 26 *sporadic groups*, the largest of which, the *Monster*, has order 808,017,424,794,512,875,886,459,904,961,710,757,005,754, 368,000,000, 000. The mysterious "Monstrous moonshine" links its representation theory to the elliptic modular function $J(\tau)$ and to string theory.

(vi) *Conjugacy and conjugacy classes*: Two group elements g_1, g_2 are said to be *conjugate* in G if there is an element $g \in G$ such that $g_2 = g^{-1}g_1g$. If g_1 is conjugate to g_2, we write $g_1 \sim g_2$. Conjugacy is an *equivalence relation*,[2] and, for finite groups, the resulting *conjugacy classes* have orders that divide the order of G. To see this, consider the conjugacy class containing the element g. Observe that the set H of elements $h \in G$ such that $h^{-1}gh = g$ forms a subgroup. The set of elements conjugate to g can be identified with the coset space G/H. The order of G divided by the order of the conjugacy class is therefore $|H|$.

Example: In the rotation group SO(3), the conjugacy classes are the sets of rotations through the same angle, but about different axes.

Example: In the group U(n), of n-by-n unitary matrices, the conjugacy classes are the set of matrices possessing the same eigenvalues.

Example: *Permutations.* The permutation group on n objects, S_n, has order $n!$. Suppose we consider permutations π_1, π_2 in S_8 such that π_1 maps

$$\pi_1 : \begin{pmatrix} 1 & 2 & 3 & 4 & 5 & 6 & 7 & 8 \\ \downarrow & \downarrow & \downarrow & \downarrow & \downarrow & \downarrow & \downarrow & \downarrow \\ 2 & 3 & 1 & 5 & 4 & 7 & 6 & 8 \end{pmatrix},$$

and π_2 maps

$$\pi_2 : \begin{pmatrix} 1 & 2 & 3 & 4 & 5 & 6 & 7 & 8 \\ \downarrow & \downarrow & \downarrow & \downarrow & \downarrow & \downarrow & \downarrow & \downarrow \\ 2 & 3 & 4 & 5 & 6 & 7 & 8 & 1 \end{pmatrix}.$$

The product $\pi_2 \circ \pi_1$ then takes

$$\pi_2 \circ \pi_1 : \begin{pmatrix} 1 & 2 & 3 & 4 & 5 & 6 & 7 & 8 \\ \downarrow & \downarrow & \downarrow & \downarrow & \downarrow & \downarrow & \downarrow & \downarrow \\ 3 & 4 & 2 & 6 & 5 & 8 & 7 & 1 \end{pmatrix}.$$

We can write these partitions out more compactly by using Paolo Ruffini's cycle notation:

$$\pi_1 = (123)(45)(67)(8), \quad \pi_2 = (12345678), \quad \pi_2 \circ \pi_1 = (132468)(5)(7).$$

In this notation, each number is mapped to the one immediately to its right, with the last number in each bracket, or *cycle*, wrapping round to map to the first. Thus $\pi_1(1) = 2$, $\pi_1(2) = 3$, $\pi_1(3) = 1$. The "8", being both first and last in its cycle, maps to itself: $\pi_1(8) = 8$. Any permutation with this cycle pattern, $(* * *)(**)(**)(*)$, is in the same

[2] An equivalence relation, $\sim$, is a binary relation that is
(i) *Reflexive*: $A \sim A$.
(ii) *Symmetric*: $A \sim B \iff B \sim A$.
(iii) *Transitive*: $A \sim B,\ B \sim C \implies A \sim C$.
Such a relation breaks a set up into disjoint *equivalence classes*.

conjugacy class as π_1. We say that π_1 possesses one 1-cycle, two 2-cycles and one 3-cycle. The class $(r_1, r_2, \dots, r_n)$ having r_1 1-cycles, r_2 2-cycles, etc., where $r_1 + 2r_2 + \cdots + nr_n = n$, contains

$$N_{(r_1,r_2,\dots)} = \frac{n!}{1^{r_1}(r_1!)\, 2^{r_2}(r_2!) \cdots n^{r_n}(r_n!)}$$

elements. The *sign* of the permutation,

$$\text{sgn}\, \pi = \epsilon_{\pi(1)\pi(2)\pi(3)\dots\pi,(n)}$$

is equal to

$$\text{sgn}\, \pi = (+1)^{r_1}(-1)^{r_2}(+1)^{r_3}(-1)^{r_4} \cdots .$$

We have, for any two permutations π_1, π_2,

$$\text{sgn}\,(\pi_1)\text{sgn}\,(\pi_2) = \text{sgn}\,(\pi_1 \circ \pi_2),$$

so the *even* ($\text{sgn}\, \pi = +1$) permutations form an invariant subgroup called the *alternating group*, A_n. The group A_n is simple for $n \geq 5$, and Ruffini (1801) showed that this simplicity prevents the solution of the general quintic by radicals. His work was ignored, however, and later independently rediscovered by Abel (1824) and Galois (1829).

If we write out the group elements in some order $\{e, g_1, g_2, \dots\}$, and then multiply on the left

$$g\{e, g_1, g_2, \dots\} = \{g, gg_1, gg_2, \dots\}$$

then the ordered list $\{g, gg_1, gg_2, \dots\}$ is a permutation of the original list. Any group G is therefore a subgroup of the permutation group $S_{|G|}$. This result is called *Cayley's theorem*. Cayley's theorem arguably held up the development of group theory for many years by its suggestion that permutations were the only groups worthy of study.

Exercise 14.1: Let H_1, H_2 be two subgroups of a group G. Show that $H_1 \cap H_2$ is also a subgroup.

Exercise 14.2: Let G be any group.

(a) The subset $Z(G)$ of G consisting of those $g \in G$ that commute with all other elements of the group is called the *centre* of the group. Show that $Z(G)$ is a subgroup of G.
(b) If g is an element of G, the set $C_G(g)$ of elements of G that commute with g is called the *centralizer* of g in G. Show that it is a subgroup of G.
(c) If H is a subgroup of G, the set of elements of G that commute with all elements of H is the *centralizer* $C_G(H)$ of H in G. Show that it is a subgroup of G.
(d) If H is a subgroup of G, the set $N_G(H) \subset G$ consisting of those g such that $g^{-1}Hg = H$ is called the *normalizer* of H in G. Show that $N_G(H)$ is a subgroup of G, and that H is a normal subgroup of $N_G(H)$.

Table 14.1 Multiplication table of $\mathcal{G}$. To find AB look in row A column B.

$\mathcal{G}$	I	A	B	C	D	E
I	I	A	B	C	D	E
A	A	B	I	E	C	D
B	B	I	A	D	E	C
C	C	D	E	I	A	B
D	D	E	C	B	I	A
E	E	C	D	A	B	I

Exercise 14.3: Show that the set of powers g_0^n of an element $g_0 \in G$ form a subgroup. Now, let p be a prime number. Recall that the set $\{1, 2, \dots p-1\}$ forms the group $(\mathbb{Z}_p)^\times$ under multiplication modulo p. By appealing to Lagrange's theorem, prove *Fermat's little theorem* that for any prime p, and positive integer a that is not divisible by p, we have $a^{p-1} = 1, \bmod p$. (Fermat actually used the binomial theorem to show that $a^p = a, \bmod p$ for any a – divisible by p or not.)

Exercise 14.4: Use Fermat's theorem from the previous exercise to establish the mathematical identity underlying the RSA algorithm for public-key cryptography: Let p, q be prime and $N = pq$. First, use Euclid's algorithm for the highest common factor (HCF) of two numbers to show that if the integer e is coprime to[3] $(p-1)(q-1)$, then there is an integer d such that

$$de = 1, \bmod (p-1)(q-1).$$

Then show that if

$$C = M^e, \bmod N \quad \text{(encryption)}$$

then

$$M = C^d, \bmod N \quad \text{(decryption)}.$$

The numbers e and N can be made known to the public, but it is hard to find the secret decoding key, d, unless the factors p and q of N are known.

Exercise 14.5: Consider the group $\mathcal{G}$ with multiplication table shown in Table 14.1.

This group has a proper subgroup $\mathcal{H} = \{I, A, B\}$, and the corresponding (left) cosets are $I\mathcal{H} = \{I, A, B\}$ and $C\mathcal{H} = \{C, D, E\}$.

(i) Construct the conjugacy classes of this group.
(ii) Show that $\{I, A, B\}$ and $\{C, D, E\}$ are indeed the left cosets of $\mathcal{H}$.

[3] Has no factors in common with.

(iii) Determine whether $\mathcal{H}$ is a normal subgroup.
(iv) If so, construct the group multiplication table for the corresponding quotient group.

Exercise 14.6: Let H and K be groups. Make the cartesian product $G = H \times K$ into a group by introducing a multiplication rule $*$ for elements of the cartesian product by setting:

$$(h_1, k_1) * (h_2, k_2) = (h_1 h_2, k_1 k_2).$$

Show that G, equipped with $*$ as its product, satisfies the group axioms. The resultant group is called the *direct product* of H and K.

Exercise 14.7: If F and G are groups, a map $\varphi : F \to G$ that preserves the group structure, i.e. if $\varphi(g_1)\varphi(g_2) = \varphi(g_1 g_2)$, is called a group homomorphism. If φ is such a homomorphism show that $\varphi(e_F) = e_G$, where e_F and e_G are the identity element in F, G respectively.

Exercise 14.8: If $\varphi : F \to G$ is a group homomorphism, and if we define $\text{Ker}(\varphi)$ as the set of elements $f \in F$ that map to e_G, show that $\text{Ker}(\varphi)$ is a normal subgroup of F.

14.1.3 Group actions on sets

Groups usually appear in physics as symmetries: they act on a physical object to change it in some way, perhaps while leaving some other property invariant.

Suppose X is a set. We call its elements "points". A *group action* on X is a map $g \in G : X \to X$ that takes a point $x \in X$ to a new point that we denote by $gx \in X$, and such that $g_2(g_1 x) = (g_2 g_1)x$, and $ex = x$. There is some standard vocabulary for group actions:

(i) Given a point $x \in X$ we define the *orbit* of x to be the set $Gx \stackrel{\text{def}}{=} \{gx : g \in G\} \subseteq X$.
(ii) The action of the group is *transitive* if any orbit is the whole of X.
(iii) The action is *effective*, or *faithful*, if the map $g : X \to X$ being the identity map implies that $g = e$. Another way of saying this is that the action is effective if the map $G \to \text{Map}\,(X \to X)$ is one-to-one. If the action of G is *not* faithful, the set of $g \in G$ that acts as the identity map forms an invariant subgroup H of G, and the quotient group G/H has a faithful action.
(iv) The action is *free* if the existence of an x such that $gx = x$ implies that $g = e$. In this case, we equivalently say that g acts without fixed points.

If the group acts freely and transitively then, having chosen a fiducial point x_0, we can uniquely label every point in X by the group element g such that $x = gx_0$. (If g_1 and g_2 both take $x_0 \to x$, then $g_1^{-1} g_2 x_0 = x_0$. By the free-action property we deduce that $g_1^{-1} g_2 = e$, and $g_1 = g_2$.) In this case we might, for some purposes, identify X with G.

Suppose the group acts transitively, but not freely. Let H be the set of elements that leaves x_0 fixed. This is clearly a subgroup of G, and if $g_1 x_0 = g_2 x_0$ we have $g_1^{-1} g_2 \in H$, or $g_1H = g_2H$. The space X can therefore be identified with the space of cosets G/H. Such sets are called *quotient spaces* or *homogeneous spaces*. Many spaces of significance in physics can be thought of as cosets in this way.

Example: The rotation group SO(3) acts transitively on the 2-sphere S^2. The SO(2) subgroup of rotations about the z-axis leaves the north pole of the sphere fixed. We can therefore identify $S^2 \simeq \mathrm{SO}(3)/\mathrm{SO}(2)$.

Many phase transitions are a result of *spontaneous symmetry breaking*. For example the water → ice transition results in the continuous translation invariance of the liquid water being broken down to the discrete translation invariance of the crystal lattice of the solid ice. When a system with symmetry group G spontaneously breaks the symmetry to a subgroup H, the set of inequivalent ground states can be identified with the homogeneous space G/H.

14.2 Representations

An n-dimensional *representation* of a group G is formally defined to be a homomorphism from G to a subgroup of $\mathrm{GL}(n, \mathbb{C})$, the group of invertible n-by-n matrices with complex entries. In effect, it is a set of n-by-n matrices that obey the group multiplication rules

$$D(g_1)D(g_2) = D(g_1g_2), \quad D(g^{-1}) = [D(g)]^{-1}. \tag{14.2}$$

Given such a representation, we can form another one $D'(g)$ by conjugation with any fixed invertible matrix C

$$D'(g) = C^{-1}D(g)C. \tag{14.3}$$

If $D'(g)$ is obtained from $D(g)$ in this way, we say that D and D' are *equivalent* representations and write $D \sim D'$. We can think of D and D' as being matrices representing the same linear map, but in different bases. Our task in the rest of this chapter is to find and classify all representations of a finite group G, up to equivalence.

Real and pseudo-real representations

We can form a new representation from $D(g)$ by setting

$$D'(g) = D^*(g),$$

where $D^*(g)$ denotes the matrix whose entries are the complex conjugates of those in $D(g)$. Suppose $D^* \sim D$. It may then be possible to find a basis in which the matrices have only real entries. In this case we say the representation is *real*. It may be, however, that $D^* \sim D$ but we cannot find a basis in which the matrices become real. In this case we say that D is *pseudo-real*.

Example: Consider the defining representation of SU(2) (the group of 2-by-2 unitary matrices with unit determinant). Such matrices are necessarily of the form

$$U = \begin{pmatrix} a & -b^* \\ b & a^* \end{pmatrix}, \tag{14.4}$$

where a and b are complex numbers with $|a|^2 + |b|^2 = 1$. They are therefore specified by *three* real parameters, and so the group manifold is three-dimensional. Now

$$\begin{aligned} \begin{pmatrix} a & -b^* \\ b & a^* \end{pmatrix}^* &= \begin{pmatrix} a^* & -b \\ b^* & a \end{pmatrix}, \\ &= \begin{pmatrix} 0 & 1 \\ -1 & 0 \end{pmatrix} \begin{pmatrix} a & -b^* \\ b & a^* \end{pmatrix} \begin{pmatrix} 0 & -1 \\ 1 & 0 \end{pmatrix}, \\ &= \begin{pmatrix} 0 & -1 \\ 1 & 0 \end{pmatrix}^{-1} \begin{pmatrix} a & -b^* \\ b & a^* \end{pmatrix} \begin{pmatrix} 0 & -1 \\ 1 & 0 \end{pmatrix}, \end{aligned} \tag{14.5}$$

and so $U \sim U^*$. It is not possible to find a basis in which all SU(2) matrices are simultaneously real, however. If such a basis existed then, in that basis, a and b would be real, and we could specify the matrices by only two real parameters – but we have seen that we need three real numbers to describe all possible SU(2) matrices.

Direct sum and direct product

We can obtain new representations from old by combining them.

Given two representations $D^{(1)}(g)$ and $D^{(2)}(g)$, we can form their *direct sum* $D^{(1)} \oplus D^{(2)}$ as the set of block-diagonal matrices

$$\begin{pmatrix} D^{(1)}(g) & 0 \\ 0 & D^{(2)}(g) \end{pmatrix}. \tag{14.6}$$

The dimension of this new representation is the sum of the dimensions of the two constituent representations. We are particularly interested in taking a representation and breaking it up as a direct sum of simpler representations.

Given two representations $D^{(1)}(g)$, $D^{(2)}(g)$, we can combine them in a different way by taking their *direct product* $D^{(1)} \otimes D^{(2)}$, which is the natural action of the group on the tensor product of the representation spaces. In other words, if $D^{(1)}(g)\mathbf{e}_j^{(1)} = \mathbf{e}_i^{(1)} D_{ij}^{(1)}(g)$ and $D^{(2)}(g)\mathbf{e}_j^{(2)} = \mathbf{e}_i^{(2)} D_{ij}^{(2)}(g)$, we define

$$[D^{(1)} \otimes D^{(2)}](g)(\mathbf{e}_i^{(1)} \otimes \mathbf{e}_j^{(2)}) = (\mathbf{e}_k^{(1)} \otimes \mathbf{e}_l^{(2)}) D_{ki}^{(1)}(g) D_{lj}^{(2)}(g). \tag{14.7}$$

We think of $D_{ki}^{(1)}(g) D_{lj}^{(2)}(g)$ being the entries in the direct-product matrix

$$[D^{(1)}(g) \otimes D^{(2)}(g)]_{kl,ij},$$

whose rows and columns are indexed by *pairs* of numbers. The dimension of the product representation is therefore the product of the dimensions of its factors.

Exercise 14.9: Show that if $D(g)$ is a representation, then so is

$$D'(g) = [D(g^{-1})]^{\mathrm{T}},$$

where the superscript T denotes the transposed matrix.

Exercise 14.10: Show that a map that assigns every element of a group G to the 1-by-1 identity matrix is a representation. It is, not unreasonably, called the *trivial* representation.

Exercise 14.11: A representation $D : G \to \mathrm{GL}(n, \mathbb{C})$ that assigns an element $g \in G$ to the n-by-n identity matrix I_n if and only if $g = e$ is said to be *faithful*. Let D be a non-trivial, but non-faithful, representation of G by n-by-n matrices. Let $H \subset G$ consist of those elements h such that $D(h) = I_n$. Show that H is a normal subgroup of G, and that D descends to a faithful representation of the quotient group G/H.

Exercise 14.12: Let A and B be linear maps from $U \to U$ and let C and D be linear maps from $V \to V$. Then the direct products $A \otimes C$ and $B \otimes D$ are linear maps from $U \otimes V \to U \otimes V$. Show that

$$(A \otimes C)(B \otimes D) = (AB) \otimes (CD).$$

Show also that

$$(A \oplus C)(B \oplus D) = (AB) \oplus (CD).$$

Exercise 14.13: Let A and B be m-by-m and n-by-n matrices, respectively, and let I_n denote the n-by-n unit matrix. Show that:

(i) $\mathrm{tr}(A \oplus B) = \mathrm{tr}(A) + \mathrm{tr}(B)$.
(ii) $\mathrm{tr}(A \otimes B) = \mathrm{tr}(A)\,\mathrm{tr}(B)$.
(iii) $\exp(A \oplus B) = \exp(A) \oplus \exp(B)$.
(iv) $\exp(A \otimes I_n + I_m \otimes B) = \exp(A) \otimes \exp(B)$.
(v) $\det(A \oplus B) = \det(A)\det(B)$.
(vi) $\det(A \otimes B) = (\det(A))^n(\det(B))^m$.

14.2.1 Reducibility and irreducibility

The "atoms" of representation theory are those representations that cannot, even by a clever choice of basis, be decomposed into, or *reduced* to, a direct sum of smaller representations. Such a representation is said to be *irreducible*. It is usually not easy to tell just by looking at a representation whether it is reducible or not. To do this, we need to develop some tools. We begin with a more powerful definition of irreducibility.

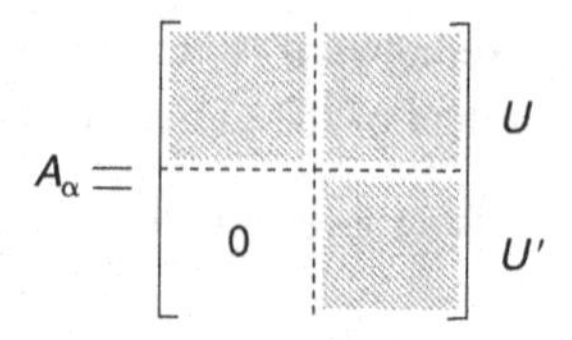

Figure 14.1 Block-partitioned reducible matrices.

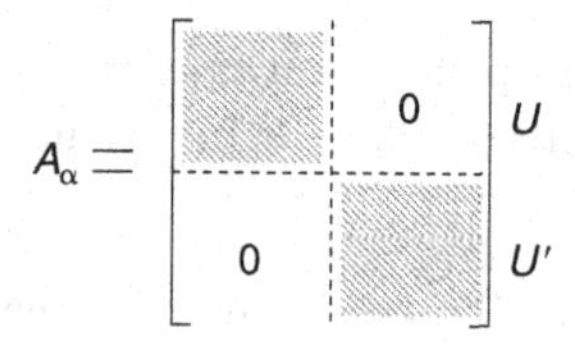

Figure 14.2 Completely reducible matrices.

We first introduce the notion of an *invariant subspace*. Suppose we have a set $\{A_\alpha\}$ of linear maps acting on a vector space V. A subspace $U \subseteq V$ is an invariant subspace for the set if $x \in U \Rightarrow A_\alpha x \in U$ for all A_α. The set $\{A_\alpha\}$ is *irreducible* if the only invariant subspaces are V itself and $\{0\}$. Conversely, if there is a non-trivial invariant subspace, then the set[4] of operators is *reducible*.

If the A_α's possess a non-trivial invariant subspace U, and we decompose $V = U \oplus U'$, where U' is a complementary subspace, then, in a basis adapted to this decomposition, the matrices A_α take the block-partitioned form of Figure 14.1.

If we can find a[5] complementary subspace U' that is also invariant, then we have the block partitioned form of Figure 14.2.

We say that such matrices are *completely reducible*. When our linear operators are unitary with respect to some inner product, we can take the complementary subspace to be the *orthogonal complement*. This, by unitarity, is automatically invariant. Thus, unitarity and reducibility implies complete reducibility.

Schur's lemma

The most useful results concerning irreducibility come from:

Schur's lemma: Suppose we have two sets of linear operators $A_\alpha : U \to U$, and $B_\alpha : V \to V$, that act irreducibly on their spaces, and an *intertwining operator* $\Lambda : U \to V$ such that

$$\Lambda A_\alpha = B_\alpha \Lambda, \tag{14.8}$$

for all α. Then *either*

[4] Irreducibility is a property of the set as a whole. Any individual matrix always has a non-trivial invariant subspace because it possesses at least one eigenvector.

[5] Remember that complementary subspaces are not unique.

(a) $\Lambda = 0$,

or

(b) Λ is one-to-one and onto (and hence invertible), in which case U and V have the same dimension and $A_\alpha = \Lambda^{-1} B_\alpha \Lambda$.

The proof is straightforward: The relation (14.8) shows that Ker $(\Lambda) \subseteq U$ and Im$(\Lambda) \subseteq V$ are invariant subspaces for the sets $\{A_\alpha\}$ and $\{B_\alpha\}$ respectively. Consequently, either $\Lambda = 0$, or Ker $(\Lambda) = \{0\}$ and Im$(\Lambda) = V$. In the latter case Λ is one-to-one and onto, and hence invertible.

Corollary: If $\{A_\alpha\}$ acts irreducibly on an n-dimensional vector space, and there is an operator Λ such that

$$\Lambda A_\alpha = A_\alpha \Lambda, \tag{14.9}$$

then either $\Lambda = 0$ or $\Lambda = \lambda I$.

To see this, observe that (14.9) remains true if Λ is replaced by $(\Lambda - xI)$. Now det $(\Lambda - xI)$ is a polynomial in x of degree n, and, by the fundamental theorem of algebra, has at least one root, $x = \lambda$. Since its determinant is zero, $(\Lambda - \lambda I)$ is not invertible, and so must vanish by Schur's lemma.

14.2.2 Characters and orthogonality

Unitary representations of finite groups

Let G be a finite group and let $g \mapsto D(g)$ be a representation of G by matrices acting on a vector space V. Let $(\mathbf{x}, \mathbf{y})$ denote a positive-definite, conjugate-symmetric, sesquilinear inner product of two vectors in V. From $(\ ,\ )$ we construct a new inner product $\langle\ ,\ \rangle$ by averaging over the group

$$\langle \mathbf{x}, \mathbf{y} \rangle = \frac{1}{|G|} \sum_{g \in G} (D(g)\mathbf{x}, D(g)\mathbf{y}). \tag{14.10}$$

It is easy to see that this new inner product remains positive definite, and in addition has the property that

$$\langle D(g)\mathbf{x}, D(g)\mathbf{y} \rangle = \langle \mathbf{x}, \mathbf{y} \rangle. \tag{14.11}$$

This means that the maps $D(g) : V \to V$ are unitary with respect to the new product. If we change basis to one that is orthonormal with respect to this new product, then the $D(g)$ become unitary matrices, with $D(g^{-1}) = D^{-1}(g) = D^\dagger(g)$, where $D^\dagger_{ij}(g) = [D_{ji}(g)]^*$ denotes the conjugate-transposed matrix.

We conclude that representations of finite groups can always be taken to be unitary. This leads to the important consequence that for such representations reducibility implies complete reducibility.

Warning: In this construction it is essential that the sum over the $g \in G$ converges. This is guaranteed for a finite group, but may not work for infinite groups. In particular, non-compact Lie groups, such as the Lorentz group, have no finite dimensional unitary representations.

Orthogonality of the matrix elements

Now let $D^J(g) : V_J \to V_J$ be the matrices of an irreducible representation or *irrep*. Here, J is a label that distinguishes inequivalent irreps from one another. We will use the symbol $\dim J$ to denote the dimension of the representation vector space V_J.

Let D^K be an irrep that is either identical to D^J or inequivalent to it, and let M_{ij} be a matrix possessing the appropriate number of rows and columns for the product $D^J M D^K$ to be defined, but otherwise arbitrary. The sum

$$\Lambda = \sum_{g \in G} D^J(g^{-1}) M D^K(g) \tag{14.12}$$

obeys $D^J(g)\Lambda = \Lambda D^K(g)$ for any g. Consequently, Schur's lemma tells us that

$$\Lambda_{il} = \sum_{g \in G} D^J_{ij}(g^{-1}) M_{jk} D^K_{kl}(g) = \lambda(M)\delta_{il}\delta^{JK}. \tag{14.13}$$

We are here summing over repeated indices, and have written $\lambda(M)$ to stress that the number λ depends on the chosen matrix M. Now take M to be zero everywhere except for one entry of unity in row j column k. Then we have

$$\sum_{g \in G} D^J_{ij}(g^{-1}) D^K_{kl}(g) = \lambda_{jk}\delta_{il}, \delta^{JK} \tag{14.14}$$

where we have relabelled λ to indicate its dependence on the location (j, k) of the non-zero entry in M. We can find the constants λ_{jk} by assuming that $K = J$, setting $i = l$ and summing over i. We find

$$|G|\delta_{jk} = \lambda_{jk} \dim J. \tag{14.15}$$

Putting these results together we find that

$$\frac{1}{|G|} \sum_{g \in G} D^J_{ij}(g^{-1}) D^K_{kl}(g) = \frac{1}{\dim J}\delta_{jk}\delta_{il}\delta^{JK}. \tag{14.16}$$

This matrix-element orthogonality theorem is often called the *grand orthogonality theorem* because of its utility.

When our matrices $D(g)$ are unitary, we can write the orthogonality theorem in a slightly prettier form:

$$\frac{1}{|G|} \sum_{g \in G} \left(D^J_{ij}(g)\right)^* D^K_{kl}(g) = \frac{1}{\dim J}\delta_{ik}\delta_{jl}\delta^{JK}. \tag{14.17}$$

If we consider complex-valued functions $G \to \mathbb{C}$ as forming a vector space, then the individual matrix entries D^J_{ij} are elements of this space and this form shows that they are mutually orthogonal with respect to the natural sesquilinear inner product.

There can be no more orthogonal functions on G than the dimension of the function space itself, which is $|G|$. We therefore have a constraint

$$\sum_J (\dim J)^2 \leq |G| \tag{14.18}$$

that places a limit on how many inequivalent representations can exist. In fact, as you will show later, the equality holds: the sum of the squares of the dimensions of the inequivalent irreducible representations is equal to the order of G, and consequently the matrix elements form a complete orthonormal set of functions on G.

Class functions and characters

Because

$$\operatorname{tr}(C^{-1}DC) = \operatorname{tr} D, \tag{14.19}$$

the trace of a representation matrix is the same for equivalent representations. Furthermore, because

$$\operatorname{tr} D(g_1^{-1}gg_1) = \operatorname{tr}\left(D^{-1}(g_1)D(g)D(g_1)\right) = \operatorname{tr} D(g), \tag{14.20}$$

the trace is the same for all group elements in a conjugacy class. The *character*,

$$\chi(g) \stackrel{\text{def}}{=} \operatorname{tr} D(g), \tag{14.21}$$

is therefore said to be a *class function*.

By taking the trace of the matrix-element orthogonality relation we see that the characters $\chi^J = \operatorname{tr} D^J$ of the irreducible representations obey

$$\frac{1}{|G|}\sum_{g\in G}\left(\chi^J(g)\right)^* \chi^K(g) = \frac{1}{|G|}\sum_i d_i \left(\chi_i^J\right)^* \chi_i^K = \delta^{JK}, \tag{14.22}$$

where d_i is the number of elements in the i-th conjugacy class.

The completeness of the matrix elements as functions on G implies that the characters form a complete orthonormal set of functions on the space of conjugacy classes equipped with the inner product

$$\langle \chi^1, \chi^2\rangle \stackrel{\text{def}}{=} \frac{1}{|G|}\sum_i d_i \left(\chi_i^1\right)^* \chi_i^2. \tag{14.23}$$

Table 14.2 Character table of S_4.

	Typical element and class size				
S_4	(1)	(12)	(123)	(1234)	(12)(34)
Irrep	1	6	8	6	3
A_1	1	1	1	1	1
A_2	1	-1	1	-1	1
E	2	0	-1	0	2
T_1	3	1	0	-1	-1
T_2	3	-1	0	1	-1

Consequently there are exactly as many inequivalent irreducible representations as there are conjugacy classes in the group.

Given a reducible representation, $D(g)$, we can find out exactly which irreps J it contains, and how many times, n_J, they occur. We do this by forming the *compound character*

$$\chi(g) = \operatorname{tr} D(g) \tag{14.24}$$

and observing that if we can find a basis in which

$$D(g) = \underbrace{(D^1(g) \oplus D^1(g) \oplus \cdots)}_{n_1 \text{ terms}} \oplus \underbrace{(D^2(g) \oplus D^2(g) \oplus \cdots)}_{n_2 \text{ terms}} \oplus \cdots, \tag{14.25}$$

then

$$\chi(g) = n_1 \chi^1(g) + n_2 \chi^2(g) + \cdots \tag{14.26}$$

From this we find that the multiplicities are given by

$$n_J = \langle \chi, \chi^J \rangle = \frac{1}{|G|} \sum_i d_i \, (\chi_i)^* \, \chi_i^J. \tag{14.27}$$

There are extensive tables of group characters. Table 14.2 shows, for example, the characters of the group S_4 of permutations on four objects.

Since $\chi^J(e) = \dim J$ we see that the irreps A_1 and A_2 are one-dimensional, that E is two-dimensional, and that $T_{1,2}$ are both three-dimensional. Also we confirm that the sum of the squares of the dimensions

$$1 + 1 + 2^2 + 3^2 + 3^2 = 24 = 4!$$

is equal to the order of the group.

As a further illustration of how to read Table 14.2, let us verify the orthonormality of the characters of the representations T_1 and T_2. We have

$$\langle \chi^{T_1}, \chi^{T_2} \rangle = \frac{1}{|G|} \sum_i d_i \left(\chi_i^{T_1}\right)^* \chi_i^{T_2} = \frac{1}{24}[1{\cdot}3{\cdot}3 - 6{\cdot}1{\cdot}1 + 8{\cdot}0{\cdot}0 - 6{\cdot}1{\cdot}1 + 3{\cdot}1 \cdot 1] = 0,$$

while

$$\langle \chi^{T_1}, \chi^{T_1} \rangle = \frac{1}{|G|} \sum_i d_i \left(\chi_i^{T_1}\right)^* \chi_i^{T_1} = \frac{1}{24}[1{\cdot}3{\cdot}3 + 6{\cdot}1{\cdot}1 + 8{\cdot}0{\cdot}0 + 6{\cdot}1{\cdot}1 + 3{\cdot}1{\cdot}1] = 1.$$

The sum giving $\langle \chi^{T_2}, \chi^{T_2} \rangle = 1$ is identical to this.

Exercise 14.14: Let D^1 and D^2 be representations with characters $\chi^1(g)$ and $\chi^2(g)$ respectively. Show that the character of the direct product representation $D^1 \otimes D^2$ is given by

$$\chi^{1\otimes 2}(g) = \chi^1(g)\,\chi^2(g).$$

14.2.3 The group algebra

Given a finite group G, we construct a vector space $\mathbb{C}(G)$ whose basis vectors are in one-to-one correspondence with the elements of the group. We denote the vector corresponding to the group element g by the boldface symbol $\mathbf{g}$. A general element of $\mathbb{C}(G)$ is therefore a formal sum

$$\mathbf{x} = x_1\mathbf{g}_1 + x_2\mathbf{g}_2 + \cdots + x_{|G|}\mathbf{g}_{|G|}. \tag{14.28}$$

We take products of these sums by using the group multiplication rule. If $g_1 g_2 = g_3$ we set $\mathbf{g}_1\mathbf{g}_2 = \mathbf{g}_3$, and require the product to be distributive with respect to vector-space addition. Thus

$$\mathbf{g}\mathbf{x} = x_1\mathbf{g}\mathbf{g}_1 + x_2\mathbf{g}\mathbf{g}_2 + \cdots + x_{|G|}\mathbf{g}\mathbf{g}_{|G|}. \tag{14.29}$$

The resulting mathematical structure is called the *group algebra*. It was introduced by Frobenius.

The group algebra, considered as a vector space, is automatically a representation. We define the natural action of G on $\mathbb{C}(G)$ by setting

$$D(g)\mathbf{g}_i = \mathbf{g}\,\mathbf{g}_i = \mathbf{g}_j D_{ji}(g). \tag{14.30}$$

The matrices $D_{ji}(g)$ make up the *regular representation*. Because the list $\mathbf{g}\,\mathbf{g}_1, \mathbf{g}\,\mathbf{g}_2, \ldots$ is a permutation of the list $\mathbf{g}_1, \mathbf{g}_2, \ldots$, their matrix entries consist of 1's and 0's, with exactly one non-zero entry in each row and each column.

Exercise 14.15: Show that the character of the regular representation has $\chi(e) = |G|$, and $\chi(g) = 0$, for $g \neq e$.

Exercise 14.16: Use the previous exercise to show that the number of times an n-dimensional irrep occurs in the regular representation is n. Deduce that $|G| = \sum_J (\dim J)^2$, and from this construct the completeness proof for the representations and characters.

Projection operators

A representation D^J of the group G automatically provides a representation of the group algebra. We simply set

$$D^J(x_1\mathbf{g}_1 + x_2\mathbf{g}_2 + \cdots) \stackrel{\text{def}}{=} x_1 D^J(g_1) + x_2 D^J(g_2) + \cdots . \tag{14.31}$$

Certain linear combinations of group elements turn out to be very useful because the corresponding matrices can be used to project out vectors possessing desirable symmetry properties.

Consider the elements

$$\mathbf{e}^J_{\alpha\beta} = \frac{\dim J}{|G|}\sum_{g\in G}\left[D^J_{\alpha\beta}(g)\right]^* \mathbf{g} \tag{14.32}$$

of the group algebra. These have the property that

$$\begin{aligned}
\mathbf{g}_1\mathbf{e}^J_{\alpha\beta} &= \frac{\dim J}{|G|}\sum_{g\in G}\left[D^J_{\alpha\beta}(g)\right]^* (\mathbf{g}_1\mathbf{g}) \\
&= \frac{\dim J}{|G|}\sum_{g\in G}\left[D^J_{\alpha\beta}(g_1^{-1}g)\right]^* \mathbf{g} \\
&= \left[D^J_{\alpha\gamma}(g_1^{-1})\right]^* \frac{\dim J}{|G|}\sum_{g\in G}\left[D^J_{\gamma\beta}(g)\right]^* \mathbf{g} \\
&= \mathbf{e}^J_{\gamma\beta} D^J_{\gamma\alpha}(g_1).
\end{aligned} \tag{14.33}$$

In going from the first to the second line we have changed summation variables from $g \to g_1^{-1}g$, and in going from the second to the third line we have used the representation property to write $D^J(g_1^{-1}g) = D^J(g_1^{-1})D^J(g)$.

From $\mathbf{g}_1\mathbf{e}^J_{\alpha\beta} = \mathbf{e}^J_{\gamma\beta}D^J_{\gamma\alpha}(g_1)$ and the matrix-element orthogonality, it follows that

$$
\begin{aligned}
\mathbf{e}^J_{\alpha\beta}\,\mathbf{e}^K_{\gamma\delta} &= \frac{\dim J}{|G|}\sum_{g\in G}\left[D^J_{\alpha\beta}(g)\right]^*\mathbf{g}\,\mathbf{e}^K_{\gamma\delta} \\
&= \frac{\dim J}{|G|}\sum_{g\in G}\left[D^J_{\alpha\beta}(g)\right]^* D^K_{\epsilon\gamma}(g)\mathbf{e}^K_{\epsilon\delta} \\
&= \delta^{JK}\delta_{\alpha\epsilon}\delta_{\beta\gamma}\,\mathbf{e}^K_{\epsilon\delta} \\
&= \delta^{JK}\delta_{\beta\gamma}\,\mathbf{e}^J_{\alpha\delta}.
\end{aligned}
\tag{14.34}
$$

For each J, this multiplication rule of the $\mathbf{e}^J_{\alpha\beta}$ is identical to that of matrices having zero entries everywhere except for the (α, β)-th, which is a "1". There are $(\dim J)^2$ of these $\mathbf{e}^J_{\alpha\beta}$ for each n-dimensional representation J, and they are linearly independent. Because $\sum_J(\dim J)^2 = |G|$, they form a basis for the algebra. In particular every element of G can be reconstructed as

$$
\mathbf{g} = \sum_J D^J_{ij}(g)\mathbf{e}^J_{ij}. \tag{14.35}
$$

We can also define the useful objects

$$
\mathbf{P}^J = \sum_i \mathbf{e}^J_{ii} = \frac{\dim J}{|G|}\sum_{g\in G}\left[\chi^J(g)\right]^*\mathbf{g}. \tag{14.36}
$$

They have the property

$$
\mathbf{P}^J\mathbf{P}^K = \delta^{JK}\mathbf{P}^K, \quad \sum_J \mathbf{P}^J = \mathbf{I}, \tag{14.37}
$$

where $\mathbf{I}$ is the identity element of $\mathbb{C}(G)$. The $\mathbf{P}^J$ are therefore projection operators composing a resolution of the identity. Their utility resides in the fact that when $D(g)$ is a reducible representation acting on a linear space

$$
V = \bigoplus_J V_J, \tag{14.38}
$$

then setting $\mathbf{g} \to D(g)$ in the formula for $\mathbf{P}^J$ results in a projection matrix from V onto the irreducible component V_J. To see how this comes about, let $\mathbf{v} \in V$ and, for any fixed p, set

$$
\mathbf{v}_i = \mathbf{e}^J_{ip}\mathbf{v}, \tag{14.39}
$$

where $\mathbf{e}^J_{ip}\mathbf{v}$ should be understood as shorthand for $D(\mathbf{e}^J_{ip})\mathbf{v}$. Then

$$
D(g)\mathbf{v}_i = \mathbf{g}\mathbf{e}^J_{ip}\mathbf{v} = \mathbf{e}^J_{jp}\mathbf{v}D^J_{ji}(g) = \mathbf{v}_jD^J_{ji}(g). \tag{14.40}
$$

We see that the $\mathbf{v}_i$, if not all zero, are basis vectors for V_J. Since $\mathbf{P}^J$ is a sum of the $\mathbf{e}^J_{ij}$, the vector $\mathbf{P}^J\mathbf{v}$ is a sum of such vectors, and therefore lies in V_J. The advantage of using $\mathbf{P}^J$ over any individual $\mathbf{e}^J_{ip}$ is that $\mathbf{P}^J$ can be computed from character tables, i.e. its construction does not require knowledge of the irreducible representation matrices.

The algebra of classes

If a conjugacy class C_i consists of the elements $\{g_1, g_2, \dots g_{d_i}\}$, we can define $\mathbf{C}_i$ to be the corresponding element of the group algebra:

$$\mathbf{C}_i = \frac{1}{d_i}(\mathbf{g}_1 + \mathbf{g}_2 + \cdots \mathbf{g}_{d_i}). \tag{14.41}$$

(The factor of $1/d_i$ is a conventional normalization.) Because conjugation merely permutes the elements of a conjugacy class, we have $\mathbf{g}^{-1}\mathbf{C}_i\mathbf{g} = \mathbf{C}_i$ for all $\mathbf{g} \in \mathbb{C}(G)$. The $\mathbf{C}_i$ therefore commute with every element of $\mathbb{C}(G)$. Conversely any element of $\mathbb{C}(G)$ that commutes with every element in $\mathbb{C}(G)$ must be a linear combination: $\mathbf{C} = c_1\mathbf{C}_1 + c_2\mathbf{C}_2 + \dots$ The subspace of $\mathbb{C}(G)$ consisting of sums of the classes is therefore the *centre* $Z[\mathbb{C}(G)]$ of the group algebra. Because the product $\mathbf{C}_i\mathbf{C}_j$ commutes with every element, it lies in $Z[\mathbb{C}(G)]$, and so there are constants $c_{ij}{}^k$ such that

$$\mathbf{C}_i\mathbf{C}_j = \sum_k c_{ij}{}^k\mathbf{C}_k. \tag{14.42}$$

We can regard the $\mathbf{C}_i$ as being linear maps from $Z[\mathbb{C}(G)]$ to itself, whose associated matrices have entries $(\mathbf{C}_i)^k{}_j = c_{ij}{}^k$. These matrices commute, and can be simultaneously diagonalized. We will leave it as an exercise for the reader to demonstrate that

$$\mathbf{C}_i\mathbf{P}^J = \left(\frac{\chi_i^J}{\chi_0^J}\right)\mathbf{P}^J. \tag{14.43}$$

Here $\chi_0^J \equiv \chi_{\{e\}}^J = \dim J$. The common eigenvectors of the $\mathbf{C}_i$ are therefore the projection operators $\mathbf{P}^J$, and the eigenvalues $\lambda_i^J = \chi_i^J/\chi_0^J$ are, up to normalization, the characters. Equation (14.43) provides a convenient method for computing the characters from knowledge only of the coefficients $c_{ij}{}^k$ appearing in the class multiplication table. Once we have found the eigenvalues λ_i^J, we recover the χ_i^J by noting that χ_0^J is real and positive, and that $\sum_i d_i|\chi_i^J|^2 = |G|$.

Exercise 14.17: Use Schur's lemma to show that for an irrep $D^J(g)$ we have

$$\frac{1}{d_i}\sum_{g\in C_i} D^J_{jk}(g) = \frac{1}{\dim J}\delta_{jk}\chi_i^J,$$

and hence establish (14.43).

14.3 Physics applications

14.3.1 Quantum mechanics

When a group $G = \{g_i\}$ acts on a mechanical system, then G will act as a set of linear operators $D(g)$ on the Hilbert space $\mathcal{H}$ of the corresponding quantum system. Thus $\mathcal{H}$ will be a representation[6] space for G. If the group is a symmetry of the system then the $D(g)$ will commute with the Hamiltonian $\hat{H}$. If this is so, and if we can decompose

$$\mathcal{H} = \bigoplus_{\text{irreps}\, J} \mathcal{H}_J \tag{14.44}$$

into $\hat{H}$-invariant irreps of G then Schur's lemma tells us that in each $\mathcal{H}_J$ the Hamiltonian $\hat{H}$ will act as a multiple of the identity operator. In other words every state in $\mathcal{H}_J$ will be an eigenstate of $\hat{H}$ with a common energy E_J.

This fact can greatly simplify the task of finding the energy levels. If an irrep J occurs only once in the decomposition of $\mathcal{H}$ then we can find the eigenstates directly by applying the projection operator $\mathbf{P}^J$ to vectors in $\mathcal{H}$. If the irrep occurs n_J times in the decomposition, then $\mathbf{P}^J$ will project to the reducible subspace

$$\underbrace{\mathcal{H}_J \oplus \mathcal{H}_J \oplus \cdots \mathcal{H}_J}_{n_J \text{ copies}} = \mathcal{M} \otimes \mathcal{H}_J.$$

Here $\mathcal{M}$ is an n_J-dimensional *multiplicity space*. The Hamiltonian $\hat{H}$ will act in $\mathcal{M}$ as an n_J-by-n_J matrix. In other words, if the vectors

$$|n, i\rangle \equiv |n\rangle \otimes |i\rangle \in \mathcal{M} \otimes \mathcal{H}_J \tag{14.45}$$

form a basis for $\mathcal{M} \otimes \mathcal{H}_J$, with n labelling which copy of $\mathcal{H}_J$ the vector $|n, i\rangle$ lies in, then

$$\begin{aligned} \hat{H}|n, i\rangle &= |m, i\rangle H^J_{mn}, \\ D(g)|n, i\rangle &= |n, j\rangle D^J_{ji}(g). \end{aligned} \tag{14.46}$$

Diagonalizing H^J_{nm} provides us with n_j $\hat{H}$-invariant copies of $\mathcal{H}_J$ and gives us the energy eigenstates.

Consider, for example, the molecule C_{60} (buckminsterfullerene) consisting of 60 carbon atoms in the form of a soccer ball. The chemically active electrons can be treated in a tight-binding approximation in which the Hilbert space has dimension 60 – one π-orbital basis state for each carbon atom. The geometric symmetry group of the molecule

[6] The rules of quantum mechanics only require that $D(g_1)D(g_2) = e^{i\phi(g_1,g_2)}D(g_1g_2)$. A set of matrices that obeys the group multiplication rule "up to a phase" is called a *projective* (or *ray*) representation. In many cases, however, we can choose the $D(g)$ so that ϕ is not needed. This is the case in all the examples we discuss.

Table 14.3 Character table for the group Y.

	Typical element and class size				
Y	e	C_5	C_5^2	C_2	C_3
Irrep	1	12	12	15	20
A	1	1	1	1	1
T_1	3	τ^{-1}	$-\tau$	-1	0
T_2	3	$-\tau$	τ^{-1}	-1	0
G	4	-1	-1	0	1
H	5	0	0	1	-1

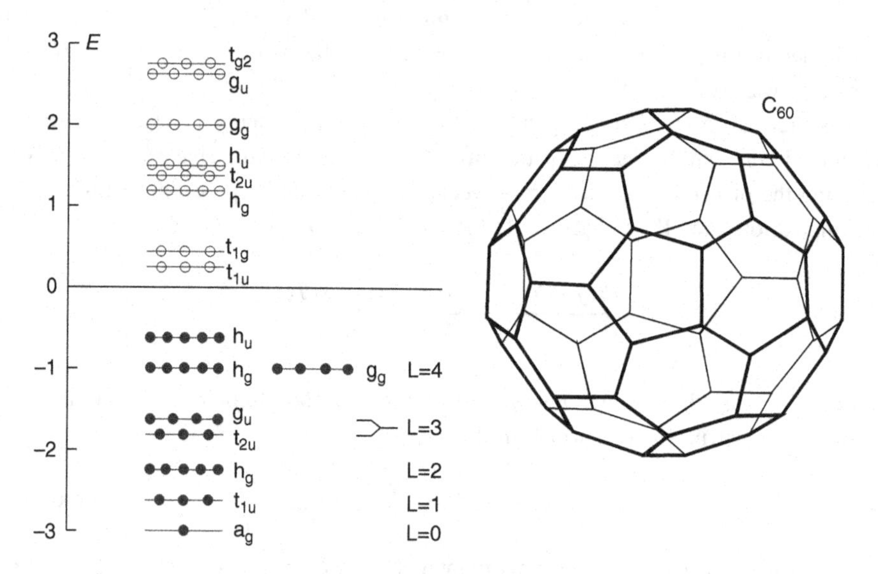

Figure 14.3 A sketch of the tight-binding electronic energy levels of C_{60}.

is $Y_h = Y \times \mathbb{Z}_2$, where Y is the rotational symmetry group of the icosahedron (a subgroup of SO(3)) and $\mathbb{Z}_2$ is the parity inversion $\sigma : \mathbf{r} \mapsto -\mathbf{r}$. The characters of Y are displayed in Table 14.3. In this table $\tau = \frac{1}{2}(\sqrt{5} - 1)$ denotes the golden mean. The class C_5 is the set of $2\pi/5$ rotations about an axis through the centres of a pair of antipodal pentagonal faces, the class C_3 is the set of of $2\pi/3$ rotations about an axis through the centres of a pair of antipodal hexagonal faces and C_2 is the set of π rotations through the midpoints of a pair of antipodal edges, each lying between two adjacent hexagonal faces. The geometric symmetry group acts on the 60-dimensional Hilbert space by permuting the basis states concurrently with their associated atoms. Figure 14.3 shows how the 60 states are disposed into energy levels.[7] Each level is labelled by a lower-case letter specifying the irrep of Y, and by a subscript g or u standing for *gerade* (German for

[7] After R. C. Haddon, L. E. Brus, K. Raghavachari, *Chem. Phys. Lett.*, **125** (1986) 459.

even) or *ungerade* (German for *odd*) that indicates whether the wavefunction is even or odd under the inversion $\sigma : \mathbf{r} \mapsto -\mathbf{r}$.

The buckyball is roughly spherical, and the lowest 25 states can be thought of as being derived from the $L = 0, 1, 2, 3, 4$ eigenstates, where L is the angular momentum quantum number that classifies the energy levels for an electron moving on a perfect sphere. In the many-electron ground-state, the 30 single-particle states with energy below $E < 0$ are each occupied by pairs of spin up/down electrons. The 30 states with $E > 0$ are empty.

To explain, for example, why three copies of T_1 appear, and why two of these are T_{1u} and one T_{1g}, we must investigate the manner in which the 60-dimensional Hilbert space decomposes into irreducible representations of the 120-element group Y_h. Problem 14.23 leads us through this computation, and shows that no irrep of Y_h occurs more than three times. In finding the energy levels, we therefore never have to diagonalize a matrix bigger than 3-by-3.

The equality of the energies of the h_g and g_g levels at $E = -1$ is an *accidental degeneracy*. It is not required by the symmetry, and will presumably disappear in a more sophisticated calculation. The appearance of many "accidental" degeneracies in an energy spectrum hints that there may be a *hidden symmetry* that arises from something beyond geometry. For example, in the Schrödinger spectrum of the hydrogen atom all states with the same principal quantum number n have the same energy although they correspond to different irreps $L = 1, \ldots, n-1$ of O(3). This degeneracy occurs because the classical Kepler-orbit problem has symmetry group O(4), rather than the naïvely expected O(3) rotational symmetry.

14.3.2 Vibrational spectrum of H_2O

The small vibrations of a mechanical system with n degrees of freedom are governed by a Lagrangian of the form

$$L = \frac{1}{2}\dot{\mathbf{x}}^T M \dot{\mathbf{x}} - \frac{1}{2}\mathbf{x}^T V \mathbf{x} \tag{14.47}$$

where M and V are symmetric n-by-n matrices, and with M being positive definite. This Lagrangian leads to the equations of motion

$$M\ddot{\mathbf{x}} = V\mathbf{x}. \tag{14.48}$$

We look for normal mode solutions $\mathbf{x}(t) \propto e^{i\omega_i t}\mathbf{x}_i$, where the vectors $\mathbf{x}_i$ obey

$$-\omega_i^2 M\mathbf{x}_i = V\mathbf{x}_i. \tag{14.49}$$

The normal-mode frequencies are solutions of the secular equation

$$\det(V - \omega^2 M) = 0, \tag{14.50}$$

and modes with distinct frequencies are orthogonal with respect to the inner product defined by M,

$$\langle \mathbf{x}, \mathbf{y} \rangle = \mathbf{x}^T M \mathbf{y}. \tag{14.51}$$

We are interested in solving this problem for vibrations about the equilibrium configuration of a molecule. Suppose this equilibrium configuration has a symmetry group G. This gives rise to an n-dimensional representation on the space of $\mathbf{x}$'s in which

$$g : \mathbf{x} \mapsto D(g)\mathbf{x} \tag{14.52}$$

leaves both the intertia matrix M and the potential matrix V unchanged:

$$[D(g)]^T M D(g) = M, \qquad [D(g)]^T V D(g) = V. \tag{14.53}$$

Consequently, if we have an eigenvector $\mathbf{x}_i$ with frequency ω_i,

$$-\omega_i^2 M \mathbf{x}_i = V \mathbf{x}_i \tag{14.54}$$

we see that $D(g)\mathbf{x}_i$ also satisfies this equation. The frequency eigenspaces are therefore left invariant by the action of $D(g)$ and, barring accidental degeneracy, there will be a one-to-one correspondence between the frequency eigenspaces and the irreducible representations occurring in $D(g)$.

Consider, for example, the vibrational modes of the water molecule H_2O (Figure 14.4). This familiar molecule has symmetry group C_{2v} which is generated by two elements: a rotation a through π about an axis through the oxygen atom, and a reflection b in the plane through the oxygen atom and bisecting the angle between the two hydrogens. The product ab is a reflection in the plane defined by the equilibrium position of the three atoms. The relations are $a^2 = b^2 = (ab)^2 = e$, and the characters are displayed in Table 14.4.

The group C_{2v} is abelian, so all the representations are one dimensional.

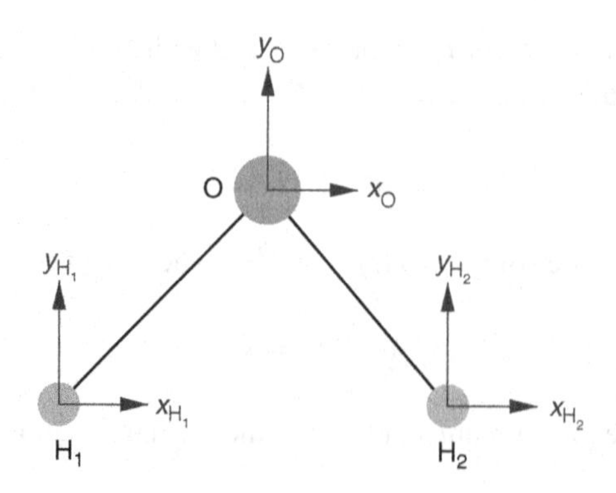

Figure 14.4 Water molecule.

Table 14.4 Character table of C_{2v}.

	Class and size			
C_{2v}	e	a	b	ab
Irrep	1	1	1	1
A_1	1	1	1	1
A_2	1	1	−1	−1
B_1	1	−1	1	−1
B_2	1	−1	−1	1

To find out what representations occur when C_{2v} acts, we need to find the character of its action $D(g)$ on the nine-dimensional vector

$$\mathbf{x} = (x_{\mathrm{O}}, y_{\mathrm{O}}, z_{\mathrm{O}}, x_{\mathrm{H}_1}, y_{\mathrm{H}_1}, z_{\mathrm{H}_1}, x_{\mathrm{H}_2}, y_{\mathrm{H}_2}, z_{\mathrm{H}_2}). \tag{14.55}$$

Here the coordinates $x_{\mathrm{H}_2}, y_{\mathrm{H}_2}, z_{\mathrm{H}_2}$ etc. denote the *displacements* of the labelled atom from its equilibrium position.

We take the molecule as lying in the xy-plane, with the z pointing towards us. The effect of the symmetry operations on the atomic displacements is

$$\begin{aligned}
D(a)\mathbf{x} &= (-x_{\mathrm{O}}, +y_{\mathrm{O}}, -z_{\mathrm{O}}, -x_{\mathrm{H}_2}, +y_{\mathrm{H}_2}, -z_{\mathrm{H}_2}, -x_{\mathrm{H}_1}, +y_{\mathrm{H}_1}, -z_{\mathrm{H}_1}),\\
D(b)\mathbf{x} &= (-x_{\mathrm{O}}, +y_{\mathrm{O}}, +z_{\mathrm{O}}, -x_{\mathrm{H}_2}, +y_{\mathrm{H}_2}, +z_{\mathrm{H}_2}, -x_{\mathrm{H}_1}, +y_{\mathrm{H}_1}, +z_{\mathrm{H}_1}),\\
D(ab)\mathbf{x} &= (+x_{\mathrm{O}}, +y_{\mathrm{O}}, -z_{\mathrm{O}}, +x_{\mathrm{H}_1}, +y_{\mathrm{H}_1}, -z_{\mathrm{H}_1}, +x_{\mathrm{H}_2}, +y_{\mathrm{H}_2}, -z_{\mathrm{H}_2}).
\end{aligned}$$

Notice how the transformations $D(a)$, $D(b)$ have interchanged the displacement coordinates of the two hydrogen atoms. In calculating the character of a transformation we need look only at the effect on atoms that are left fixed – those that are moved have matrix elements only in non-diagonal positions. Thus, when computing the compound characters for a b, we can focus on the oxygen atom. For ab we need to look at all three atoms. We find

$$\begin{aligned}
\chi^D(e) &= 9,\\
\chi^D(a) &= -1+1-1 = -1,\\
\chi^D(b) &= -1+1+1 = 1,\\
\chi^D(ab) &= 1+1-1+1+1-1+1+1-1 = 3.
\end{aligned}$$

By using the orthogonality relations, we find the decomposition

$$\begin{pmatrix} 9\\ -1\\ 1\\ 3\end{pmatrix} = 3\begin{pmatrix} 1\\ 1\\ 1\\ 1\end{pmatrix} + \begin{pmatrix} 1\\ 1\\ -1\\ -1\end{pmatrix} + 2\begin{pmatrix} 1\\ -1\\ 1\\ -1\end{pmatrix} + 3\begin{pmatrix} 1\\ -1\\ -1\\ 1\end{pmatrix} \tag{14.56}$$

or

$$\chi^D = 3\chi^{A_1} + \chi^{A_2} + 2\chi^{B_1} + 3\chi^{B_2}. \tag{14.57}$$

Thus, the nine-dimensional representation decomposes as

$$D = 3A_1 \oplus A_2 \oplus 2B_1 \oplus 3B_2. \tag{14.58}$$

How do we exploit this? First we cut out the junk. Out of the nine modes, six correspond to easily identified zero-frequency motions – three of translation and three rotations. A translation in the x-direction would have $x_{\rm O} = x_{\rm H_1} = x_{\rm H_2} = \xi$, all other entries being zero. This displacement vector changes sign under both a and b, but is left fixed by ab. This behaviour is characteristic of the representation B_2. Similarly we can identify A_1 as translation in y, and B_1 as translation in z. A rotation about the y-axis makes $z_{\rm H_1} = -z_{\rm H_2} = \phi$. This is left fixed by a, but changes sign under b and ab, so the y rotation mode is A_2. Similarly, rotations about the x- and z-axes correspond to B_1 and B_2, respectively. All that is left for genuine vibrational modes is $2A_1 \oplus B_2$.

We now apply the projection operator

$$P^{A_1} = \frac{1}{4}[(\chi^{A_1}(e))^* D(e) + (\chi^{A_1}(a))^* D(b) + (\chi^{A_1}(b))^* D(b) + (\chi^{A_1}(ab))^* D(ab)] \tag{14.59}$$

to $\mathbf{v}_{{\rm H}_1,x}$, a small displacement of H_1 in the x-direction. We find

$$\begin{aligned} P^{A_1}\mathbf{v}_{{\rm H}_1,x} &= \frac{1}{4}(\mathbf{v}_{{\rm H}_1,x} - \mathbf{v}_{{\rm H}_2,x} - \mathbf{v}_{{\rm H}_2,x} + \mathbf{v}_{{\rm H}_1,x}) \\ &= \frac{1}{2}(\mathbf{v}_{{\rm H}_1,x} - \mathbf{v}_{{\rm H}_2,x}). \end{aligned} \tag{14.60}$$

This mode is an eigenvector for the vibration problem.

If we apply P^{A_1} to $\mathbf{v}_{{\rm H}_1,y}$ and $\mathbf{v}_{{\rm O},y}$ we find

$$\begin{aligned} P^{A_1}\mathbf{v}_{{\rm H}_1,y} &= \frac{1}{2}(\mathbf{v}_{{\rm H}_1,y} + \mathbf{v}_{{\rm H}_2,y}), \\ P^{A_1}\mathbf{v}_{{\rm O},y} &= \mathbf{v}_{{\rm O},y}, \end{aligned} \tag{14.61}$$

but we are not quite done. These modes are contaminated by the y translation direction zero mode, which is also in an A_1 representation. After we make our modes orthogonal to this, there is only one left, and this has $y_{\rm H_1} = y_{\rm H_2} = -y_{\rm O} m_{\rm O}/(2m_{\rm H}) = a_1$, all other components vanishing.

We can similarly find vectors corresponding to B_2 as

$$P^{B_2}\mathbf{v}_{\mathrm{H}_1,x} = \frac{1}{2}(\mathbf{v}_{\mathrm{H}_1,x} + \mathbf{v}_{\mathrm{H}_2,x})$$
$$P^{B_2}\mathbf{v}_{\mathrm{H}_1,y} = \frac{1}{2}(\mathbf{v}_{\mathrm{H}_1,y} - \mathbf{v}_{\mathrm{H}_2,y})$$
$$P^{B_2}\mathbf{v}_{\mathrm{O},x} = \mathbf{v}_{\mathrm{O},x}$$

and these need to be cleared of both translations in the x-direction and rotations about the z-axis, both of which transform under B_2. Again there is only one mode left and it is

$$y_{\mathrm{H}_1} = -y_{\mathrm{H}_2} = \alpha x_{\mathrm{H}_1} = \alpha x_{\mathrm{H}_2} = \beta x_0 = a_2 \tag{14.62}$$

where α is chosen to ensure that there is no angular momentum about O, and β to make the total x linear momentum vanish. We have therefore found three true vibration eigenmodes, two transforming under A_1 and one under B_2 as advertised earlier. The eigenfrequencies, of course, depend on the details of the spring constants, but now that we have the eigenvectors we can just plug them in to find these.

14.3.3 Crystal field splittings

A quantum mechanical system has a symmetry G if the Hamiltonian $\hat{H}$ obeys

$$D^{-1}(g)\hat{H}D(g) = \hat{H}, \tag{14.63}$$

for some group action $D(g) : \mathcal{H} \to \mathcal{H}$ on the Hilbert space. If follows that the eigenspaces, $\mathcal{H}_\lambda$, of states with a common eigenvalue, λ, are invariant subspaces for the representation $D(g)$.

We often need to understand how a degeneracy is lifted by perturbations that break G down to a smaller subgroup H. An n-dimensional irreducible representation of G is automatically a representation of any subgroup of G, but in general it is no longer irreducible. Thus the n-fold degenerate level is split into multiplets, one for each of the irreducible representations of H contained in the original representation. The manner in which an originally irreducible representation decomposes under restriction to a subgroup is known as the *branching rule* for the representation.

A physically important case is given by the breaking of the full SO(3) rotation symmetry of an isolated atomic Hamiltonian by a crystal field. Suppose the crystal has octohedral symmetry. The characters of the octohedral group are displayed in Table 14.5.

The classes are labelled by the rotation angles, C_2 being a twofold rotation axis ($\theta = \pi$), C_3 a threefold axis ($\theta = 2\pi/3$), etc.

The character of the $J = l$ representation of $SO(3)$ is

$$\chi^l(\theta) = \frac{\sin(2l+1)\theta/2}{\sin\theta/2}, \tag{14.64}$$

Table 14.5 Character table of the octohedral group O.

O	e	Class (size)			
		$C_3(8)$	$C_4^2(3)$	$C_2(6)$	$C_4(6)$
A_1	1	1	1	1	1
A_2	1	1	1	−1	−1
E	2	−1	2	0	0
F_2	3	0	−1	1	−1
F_1	3	0	−1	−1	1

Table 14.6 Characters evaluated on rotation classes.

l	e	Class (size)			
		$C_3(8)$	$C_4^2(3)$	$C_2(6)$	$C_4(6)$
0	1	1	1	1	1
1	3	0	−1	−1	−1
2	5	−1	1	1	−1
3	7	1	−1	−1	−1
4	9	0	1	1	1

and the first few χ^l's evaluated on the rotation angles of the classes of O are dsiplayed in Table 14.6.

The ninefold degenerate $l = 4$ multiplet therefore decomposes as

$$\begin{pmatrix} 9 \\ 0 \\ 1 \\ 1 \\ 1 \end{pmatrix} = \begin{pmatrix} 1 \\ 1 \\ 1 \\ 1 \\ 1 \end{pmatrix} + \begin{pmatrix} 2 \\ -1 \\ 2 \\ 0 \\ 0 \end{pmatrix} + \begin{pmatrix} 3 \\ 0 \\ -1 \\ -1 \\ 1 \end{pmatrix} + \begin{pmatrix} 3 \\ 0 \\ -1 \\ 1 \\ -1 \end{pmatrix}, \tag{14.65}$$

or

$$\chi^4_{SO(3)} = \chi^{A_1} + \chi^E + \chi^{F_1} + \chi^{F_2}. \tag{14.66}$$

The octohedral crystal field splits the nine states into four multiplets with symmetries A_1, E, F_1, F_2 and degeneracies 1, 2, 3 and 3, respectively.

We have considered only the simplest case here, ignoring the complications introduced by reflection symmetries, and by two-valued spinor representations of the rotation group.

14.4 Further exercises and problems

We begin with some technologically important applications of group theory to cryptography and number theory.

Exercise 14.18: The set $\mathbb{Z}_n$ forms a group under multiplication only when n is a prime number. Show, however, that the subset $\mathrm{U}(\mathbb{Z}_n) \subset \mathbb{Z}_n$ of elements of $\mathbb{Z}_n$ that are co-prime to n is a group. It is the *group of units* of the ring $\mathbb{Z}_n$.

Exercise 14.19: *Cyclic groups*. A group G is said to be *cyclic* if its elements consist of powers a^n of an element a, called the *generator*. The group will be of finite order $|G| = m$ if $a^m = a^0 = e$ for some $m \in \mathbb{Z}^+$.

(a) Show that a group of prime order is necessarily cyclic, and that any element other than the identity can serve as its generator. (Hint: let a be any element other than e and consider the subgroup consisting of powers a^m.)
(b) Show that any subgroup of a cyclic group is itself cyclic.

Exercise 14.20: *Cyclic groups and cryptography*. In a large cyclic group G it can be relatively easy to compute a^x, but to recover x given $h = a^x$ one might have to compute a^y and compare it with h for every $1 < y < |G|$. If $|G|$ has several hundred digits, such a brute force search could take longer than the age of the Universe. Rather more efficient algorithms for this *discrete logarithm problem* exist, but the difficulty is still sufficient for it to be useful in cryptography.

(a) *Diffie–Hellman key exchange*. This algorithm allows Alice and Bob to establish a secret key that can be used with a conventional cypher without Eve, who is listening to their conversation, being able to reconstruct it. Alice chooses a random element $g \in G$ and an integer x between 1 and $|G|$ and computes g^x. She sends g and g^x to Bob, but keeps x to herself. Bob chooses an integer y and computes g^y and $g^{xy} = (g^x)^y$. He keeps y secret and sends g^y to Alice, who computes $g^{xy} = (g^y)^x$. Show that, although Eve knows g, g^y and g^x, she cannot obtain Alice and Bob's secret key g^{xy} without solving the discrete logarithm problem.
(b) *Elgamal public key encryption*. This algorithm, based on Diffie–Hellman, was invented by the Egyptian cryptographer Taher El Gamal. It is a component of PGP and other modern encryption packages. To use it, Alice first chooses a random integer x in the range 1 to $|G|$ and computes $h = a^x$. She publishes a description of G, together with the elements h and a, as her public key. She keeps the integer x secret. To send a message m to Alice, Bob chooses an integer y in the same range and computes $c_1 = a^y$, $c_2 = mh^y$. He transmits c_1 and c_2 to Alice, but keeps y secret. Alice can recover m from c_1, c_2 by computing $c_2(c_1^x)^{-1}$. Show that, although Eve knows Alice's public key and has overheard c_1 and c_2, she nonetheless cannot decrypt the message without solving the discrete logarithm problem.

Popular choices for G are subgroups of $(\mathbb{Z}_p)^\times$, for large prime p. $(\mathbb{Z}_p)^\times$ is itself cyclic (can you prove this?), but is unsuitable for technical reasons.

Exercise 14.21: *Modular arithmetic and number theory*. An integer a is said to be a *quadratic residue* mod p if there is an r such that $a = r^2 \pmod{p}$. Let p be an odd prime. Show that if $r_1^2 = r_2^2 \pmod{p}$ then $r_1 = \pm r_2 \pmod{p}$, and that $r \neq -r \pmod{p}$. Deduce that exactly *one half* of the $p - 1$ non-zero elements of $\mathbb{Z}_p$ are quadratic residues.

Now consider the *Legendre symbol*

$$\left(\frac{a}{p}\right) \stackrel{\text{def}}{=} \begin{cases} 0, & a = 0, \\ 1, & a \text{ a quadratic residue} \pmod{p}, \\ -1 & a \text{ not a quadratic residue} \pmod{p}. \end{cases}$$

Show that

$$\left(\frac{a}{p}\right)\left(\frac{b}{p}\right) = \left(\frac{ab}{p}\right),$$

and so the Legendre symbol forms a one-dimensional representation of the multiplicative group $(\mathbb{Z}_p)^\times$. Combine this fact with the character orthogonality theorem to give an alternative proof that precisely half the $p - 1$ elements of $(\mathbb{Z}_p)^\times$ are quadratic residues. (Hint: to show that the product of two non-residues is a residue, observe that the set of residues is a normal subgroup of $(\mathbb{Z}_p)^\times$, and consider the multiplication table of the resulting quotient group.)

Exercise 14.22: *More practice with modular arithmetic*. Again let p be an odd prime. Prove *Euler's theorem* that

$$a^{(p-1)/2} \pmod{p} = \left(\frac{a}{p}\right).$$

(Hint: begin by showing that the usual school-algebra proof that an equation of degree n can have no more than n solutions remains valid for arithmetic modulo a prime number, and so $a^{(p-1)/2} = 1 \pmod{p}$ can have no more than $(p - 1)/2$ roots. Cite Fermat's little theorem to show that these roots must be the quadratic residues. Cite Fermat again to show that the quadratic non-residues must then have $a^{(p-1)/2} = -1 \pmod{p}$.)

The harder-to-prove *law of quadratic reciprocity* asserts that for p, q odd primes, we have

$$(-1)^{(p-1)(q-1)/4}\left(\frac{p}{q}\right) = \left(\frac{q}{p}\right).$$

Problem 14.23: *Buckyball spectrum*. Consider the symmetry group of the C_{60} buckyball molecule of Figure 14.3.

(a) Starting from the character table of the orientation-preserving icosohedral group Y (Table 14.3), and using the fact that the $\mathbb{Z}_2$ parity inversion $\sigma : \mathbf{r} \to -\mathbf{r}$ combines with $g \in Y$ so that $D^{J_g}(\sigma g) = D^{J_g}(g)$, whilst $D^{J_u}(\sigma g) = -D^{J_u}(g)$, write down the character table of the extended group $Y_h = Y \times \mathbb{Z}_2$ that acts as a symmetry on the C_{60} molecule. There are now 10 conjugacy classes, and the 10 representations will be labelled A_g, A_u, etc. Verify that your character table has the expected row-orthogonality properties.

(b) By counting the number of atoms left fixed by each group operation, compute the compound character of the action of Y_h on the C_{60} molecule. (Hint: examine the pattern of panels on a regulation soccer ball, and deduce that four carbon atoms are left unmoved by operations in the class σC_2.)

(c) Use your compound character from part (b) to show that the 60-dimensional Hillbert space decomposes as

$$\mathcal{H}_{C_{60}} = A_g \oplus T_{1g} \oplus 2T_{1u} \oplus T_{2g} \oplus 2T_{2u} \oplus 2G_g \oplus 2G_u \oplus 3H_g \oplus 2H_u,$$

consistent with the energy levels sketched in Figure 14.3.

Problem 14.24: *The Frobenius–Schur indicator.* Recall that a real or pseudo-real representation is one such that $D(g) \sim D^*(g)$, and for unitary matrices D we have $D^*(g) = [D^T(g)]^{-1}$. In this unitary case $D(g)$ being real or pseudo-real is equivalent to the statement that there exists an invertible matrix F such that

$$FD(g)F^{-1} = [D^T(g)]^{-1}.$$

We can rewrite this statement as $D^T(g)FD(g) = F$, and so F can be interpreted as the matrix representing a G-invariant quadratic form.

(i) Use Schur's lemma to show that when D is irreducible the matrix F is unique up to an overall constant. In other words, $D^T(g)F_1D(g) = F_1$ and $D^T(g)F_2D(g) = F_2$ for all $g \in G$ implies that $F_2 = \lambda F_1$. Deduce that for irreducible D we have $F^T = \pm F$.

(ii) By reducing F to a suitable canonical form, show that F is symmetric ($F = F^T$) in the case that $D(g)$ is a real representation, and F is skew symmetric ($F = -F^T$) when $D(g)$ is a pseudo-real representation.

(iii) Now let G be a *finite* group. For any matrix U, the sum

$$F_U = \frac{1}{|G|} \sum_{g \in G} D^T(g)UD(g)$$

is a G-invariant matrix. Deduce that F_U is always zero when $D(g)$ is neither real nor pseudo-real, and, by specializing both U and the indices on F_U, show that in the real or pseudo-real case

$$\sum_{g \in G} \chi(g^2) = \pm \sum_{g \in G} \chi(g)\chi(g),$$

where $\chi(g) = \operatorname{tr} D(g)$ is the character of the irreducible representation $D(g)$. Deduce that the *Frobenius–Schur indicator*

$$\kappa \stackrel{\text{def}}{=} \frac{1}{|G|}\sum_{g\in G}\chi(g^2)$$

takes the value $+1$, -1 or 0 when $D(g)$ is, respectively, real, pseudo-real or not real.

(iv) Show that the identity representation occurs in the decomposition of the tensor product $D(g)\otimes D(g)$ of an irrep with itself if, and only if, $D(g)$ is real or pseudo-real. Given a basis $\mathbf{e}_i$ for the vector space V on which $D(g)$ acts, show that the matrix F can be used to construct the basis for the identity-representation subspace V^{id} in the decomposition

$$V\otimes V = \bigoplus_{\text{irreps } J} V^J.$$

Problem 14.25: *Induced representations*. Suppose we know a representation $D^W(h): W\to W$ for a subgroup $H\subset G$. From this representation we can construct an *induced representation* $\text{Ind}_{\text{H}}^{G}(D^W)$ for the larger group G. The construction cleverly combines the coset space G/H with the representation space W to make a (usually reducible) representation space $\text{Ind}_{\text{H}}^{G}(W)$ of dimension $|G/H|\times \dim W$.

Recall that there is a natural action of G on the coset space G/H. If $x=\{g_1,g_2,\ldots\}\in G/H$ then gx is the coset $\{gg_1,gg_2,\ldots\}$. We select from each coset $x\in G/H$ a representative element a_x, and observe that the product ga_x can be decomposed as $ga_x = a_{gx}h$, where a_{gx} is the selected representative from the coset gx and h is some element of H. Next we introduce a basis $|n,x\rangle$ for $\text{Ind}_{\text{H}}^{G}(W)$. We use the symbol "0" to label the coset $\{e\}$, and take $|n,0\rangle$ to be the basis vectors for W. For $h\in H$ we can therefore set

$$D(h)|n,0\rangle \stackrel{\text{def}}{=} |m,0\rangle D^W_{mn}(h).$$

We also define the result of the action of a_x on $|n,0\rangle$ to be the vector $|n,x\rangle$:

$$D(a_x)|n,0\rangle \stackrel{\text{def}}{=} |n,x\rangle.$$

We may now obtain the action of a general element of G on the vectors $|n,x\rangle$ by requiring $D(g)$ to be a representation, and so computing

$$\begin{aligned}
D(g)|n,x\rangle &= D(g)D(a_x)|n,0\rangle\\
&= D(ga_x)|n,0\rangle\\
&= D(a_{gx}h)|n,0\rangle\\
&= D(a_{gx})D(h)|n,0\rangle
\end{aligned}$$

$$= D(a_{gx})|m,0\rangle D^W_{mn}(h)$$
$$= |m,gx\rangle D^W_{mn}(h).$$

(i) Confirm that the action $D(g)|n,x\rangle = |m,gx\rangle D^W_{mn}(h)$, with h obtained from g and x *via* the decomposition $ga_x = a_{gx}h$, does indeed define a representation of G. Show also that if we set $|f\rangle = \sum_{n,x} f_n(x)|n,x\rangle$, then the action of g on the components takes

$$f_n(x) \mapsto D^W_{nm}(h) f_m(g^{-1}x).$$

(ii) Let $f(h)$ be a class function on H. Let us extend it to a function on G by setting $f(g) = 0$ if $g \notin H$, and define

$$\mathrm{Ind}^G_{\mathrm{H}}[f](s) = \frac{1}{|H|}\sum_{g\in G} f(g^{-1}sg).$$

Show that $\mathrm{Ind}^G_{\mathrm{H}}[f](s)$ is a class function on G, and further show that if χ_W is the character of the starting representation for H then $\mathrm{Ind}^G_{\mathrm{H}}[\chi_W]$ is the character of the induced representation of G. (Hint: only fixed points of the G-action on G/H contribute to the character, and $gx = x$ means that $ga_x = a_x h$. Thus $D^W(h) = D^W(a_x^{-1}ga_x)$.)

(iii) Given a representation $D^V(g) : V \to V$ of G we can trivially obtain a (generally reducible) representation $\mathrm{Res}^G_{\mathrm{H}}(V)$ of $H \subset G$ by restricting G to H. Define the usual inner product on the group functions by

$$\langle \phi_1, \phi_2\rangle_G = \frac{1}{|G|}\sum_{g\in G} \phi_1(g^{-1})\phi_2(g),$$

and show that if ψ is a class function on H and ϕ a class function on G then

$$\langle \psi, \mathrm{Res}^G_{\mathrm{H}}[\phi]\rangle_{\mathrm{H}} = \langle \mathrm{Ind}^G_{\mathrm{H}}[\psi], \phi\rangle_G.$$

Thus, $\mathrm{Ind}^G_{\mathrm{H}}$ and $\mathrm{Res}^G_{\mathrm{H}}$ are, in some sense, adjoint operations. Mathematicians would call them a pair of mutually *adjoint functors*.

(iv) By applying the result from part (iii) to the characters of the irreducible representations of G and H, deduce *Frobenius' reciprocity theorem*: the number of times an irrep $D^J(g)$ of G occurs in the representation induced from an irrep $D^K(h)$ of H is equal to the number of times that D^K occurs in the decomposition of D^J into irreps of H.

The representation of the Poincaré group (= the SO(1, 3) Lorentz group together with space-time translations) that classifies the states of a spin-J elementary particle are those induced from the spin-J representation of its SO(3) rotation subgroup. The quantum state of a mass m elementary particle is therefore of the form $|k,\sigma\rangle$ where k is the particle's four-momentum, which lies in the coset SO(1, 3)/SO(3), and σ is the label from the $|J,\sigma\rangle$ spin state.

15

Lie groups

A *Lie group* is a group which is also a smooth manifold G. The group operation of multiplication $(g_1, g_2) \mapsto g_3$ is required to be a smooth function, as is the operation of taking the inverse of a group element. Lie groups are named after the Norwegian mathematician Sophus Lie. The examples most commonly met in physics are the infinite families of *matrix groups* GL(n), SL(n), O(n), SO(n), U(n), SU(n), and Sp(n), all of which we shall describe in this chapter, togther with the family of five *exceptional* Lie groups: G_2, F_4, E_6, E_7 and E_8, which have applications in string theory.

One of the properties of a Lie group is that, considered as a manifold, the neighbourhood of any point looks exactly like that of any other. Accordingly, the group's dimension and much of its structure can be understood by examining the immediate vicinity of any chosen point, which we may as well take to be the identity element. The vectors lying in the tangent space at the identity element make up the *Lie algebra* of the group. Computations in the Lie algebra are often easier than those in the group, and provide much of the same information. This chapter will be devoted to studying the interplay between the Lie group itself and this Lie algebra of infinitesimal elements.

15.1 Matrix groups

The *Classical Groups* are described in a book with this title by Hermann Weyl. They are subgroups of the *general linear group*, $\mathrm{GL}(n, \mathbb{F})$, which consists of invertible n-by-n matrices over the field $\mathbb{F}$. We will mostly consider the cases $\mathbb{F} = \mathbb{C}$ or $\mathbb{F} = \mathbb{R}$.

A near-identity matrix in $\mathrm{GL}(n, \mathbb{R})$ can be written $g = I + \epsilon A$, where A is an arbitrary n-by-n real matrix. This matrix contains n^2 real entries, so we can move away from the identity in n^2 distinct directions. The tangent space at the identity, and hence the group manifold itself, is therefore n^2 dimensional. The manifold of $\mathrm{GL}(n, \mathbb{C})$ has n^2 *complex* dimensions, and this corresponds to $2n^2$ real dimensions.

If we restrict the determinant of a $\mathrm{GL}(n, \mathbb{F})$ matrix to be unity, we get the *special linear group*, $\mathrm{SL}(n, \mathbb{F})$. An element near the identity in this group can still be written as $g = I + \epsilon A$, but since

$$\det(I + \epsilon A) = 1 + \epsilon \operatorname{tr}(A) + O(\epsilon^2), \tag{15.1}$$

this requires $\operatorname{tr}(A) = 0$. The restriction on the trace means that $\mathrm{SL}(n, \mathbb{R})$ has dimension $n^2 - 1$.

15.1.1 The unitary and orthogonal groups

Perhaps the most important of the matrix groups are the unitary and orthogonal groups.

The unitary group

The unitary group U(n) comprises the set of n-by-n complex matrices U such that $U^\dagger = U^{-1}$. If we consider matrices near the identity

$$U = I + \epsilon A, \tag{15.2}$$

with ϵ real, then unitarity requires

$$\begin{aligned} I + O(\epsilon^2) &= (I + \epsilon A)(I + \epsilon A^\dagger) \\ &= I + \epsilon(A + A^\dagger) + O(\epsilon^2), \end{aligned} \tag{15.3}$$

so $A_{ij} = -(A_{ji})^*$ i.e. A is skew-hermitian. A complex skew-hermitian matrix contains

$$n + \left[2 \times \frac{1}{2} n(n-1)\right] = n^2$$

real parameters. In this counting, the first "n" is the number of entries on the diagonal, each of which must be of the form i times a real number. The $n(n-1)/2$ is the number of entries above the main diagonal, each of which can be an arbitrary complex number. The number of real dimensions in the group manifold is therefore n^2. As $U^\dagger U = I$, the rows or columns in the matrix U form an orthonormal set of vectors. Their entries are therefore bounded, $|U_{ij}| \le 1$, and this property leads to the n^2-dimensional group manifold of U(n) being a compact set.

When a group manifold is compact, we say that the group itself is a *compact group*. There is a natural notion of volume on a group manifold and compact Lie groups have finite total volume. Because of this, they have many properties in common with the finite groups we studied in the previous chapter.

Recall that a group is *simple* if it possesses no invariant subgroups. U(n) is not simple. Its centre Z is an invariant U(1) subgroup consisting of matrices of the form $U = e^{i\theta} I$. The *special unitary group* SU(n) consists of n-by-n unimodular (having determinant $+1$) unitary matrices. It is not strictly simple because its centre Z consists of the discrete subgroup of matrices $U_m = \omega^m I$ with ω an n-th root of unity, and this is an invariant subgroup. Because the centre, its only invariant subgroup, is not a *continuous* group, SU(n) is counted as being simple in Lie theory. With $U = I + \epsilon A$, as above, the unimodularity imposes the additional constraint on A that $\operatorname{tr} A = 0$, so the SU(n) group manifold is $(n^2 - 1)$-dimensional.

The orthogonal group

The orthogonal group O(n) consists of the set of real matrices O with the property that $O^{\mathrm{T}} = O^{-1}$. For a matrix in the neighbourhood of the identity, $O = I + \epsilon A$, this property requires that A be skew symmetric: $A_{ij} = -A_{ij}$. Skew-symmetric real matrices have $n(n-1)/2$ independent entries, and so the group manifold of $\mathrm{O}(n)$ is $n(n-1)/2$ dimensional. The condition $O^{\mathrm{T}}O = I$ means that the rows or columns of O, considered as row or column vectors, are orthonormal. All entries are bounded, i.e. $|O_{ij}| \le 1$, and again this leads to $\mathrm{O}(n)$ being a compact group.

The identity

$$1 = \det(O^{\mathrm{T}}O) = \det O^{\mathrm{T}} \det O = (\det O)^2 \tag{15.4}$$

tells us that $\det O = \pm 1$. The subset of orthogonal matrices with $\det O = +1$ constitute a subgroup of $\mathrm{O}(n)$ called the *special orthogonal group* $\mathrm{SO}(n)$. The unimodularity condition discards a disconnected part of the group manifold and does not reduce its dimension, which remains $n(n-1)/2$.

15.1.2 Symplectic groups

The symplectic groups (named from the Greek, meaning to "fold together") are perhaps less familiar than the other matrix groups.

We start with a non-degenerate skew-symmetric matrix ω. The symplectic group $\mathrm{Sp}(2n, \mathbb{F})$ is then defined by

$$\mathrm{Sp}(2n, \mathbb{F}) = \{S \in \mathrm{GL}(2n, \mathbb{F}) : S^{\mathrm{T}}\omega S = \omega\}. \tag{15.5}$$

Here $\mathbb{F}$ can be $\mathbb{R}$ or $\mathbb{C}$. When $\mathbb{F} = \mathbb{C}$, we still use the transpose "T", not the adjoint "$\dagger$", in this definition. Setting $S = I_{2n} + \epsilon A$ and demanding that $S^{\mathrm{T}}\omega S = \omega$ shows that $A^{\mathrm{T}}\omega + \omega A = 0$.

It does not matter what skew matrix ω we start from, because we can always find a basis in which ω takes its canonical form:

$$\omega = \begin{pmatrix} 0 & -I_n \\ I_n & 0 \end{pmatrix}. \tag{15.6}$$

In this basis we find, after a short computation, that the most general form for A is

$$A = \begin{pmatrix} a & b \\ c & -a^{\mathrm{T}} \end{pmatrix}. \tag{15.7}$$

Here, a is any n-by-n matrix, and b and c are symmetric (i.e. $b^{\mathrm{T}} = b$ and $c^{\mathrm{T}} = c$) n-by-n matrices. If the matrices are real, then counting the degrees of freedom gives the dimension of the *real symplectic group* as

$$\dim \mathrm{Sp}(2n, \mathbb{R}) = n^2 + \left[2 \times \frac{n}{2}(n+1)\right] = n(2n+1). \tag{15.8}$$

The entries in a, b, c can be arbitrarily large. $\mathrm{Sp}(2n, \mathbb{R})$ is not compact.

The determinant of any symplectic matrix is $+1$. To see this take the elements of ω to be ω_{ij}, and let

$$\omega(x,y) = \omega_{ij}x^i y^j \tag{15.9}$$

be the associated skew bilinear (*not* sesquilinear) form. Then Weyl's identity from Exercise A.19 shows that

$$\begin{aligned}&\mathrm{Pf}\,(\omega)\,(\det M)\det|x_1,\ldots,x_{2n}|\\ &= \frac{1}{2^n n!}\sum_{\pi\in S_{2n}} \mathrm{sgn}\,(\pi)\omega(Mx_{\pi(1)},Mx_{\pi(2)})\cdots\omega(Mx_{\pi(2n-1)},Mx_{\pi(2n)}),\end{aligned}$$

for any linear map M. If $\omega(x,y) = \omega(Mx,My)$, we conclude that $\det M = 1$ – but preserving ω is exactly the condition that M be an element of the symplectic group. Since the matrices in $\mathrm{Sp}(2n,\mathbb{F})$ are automatically unimodular there is no "special symplectic" group.

Unitary symplectic group

The intersection of two groups is also a group. We therefore define the *unitary symplectic group* as

$$\mathrm{Sp}(n) = \mathrm{Sp}(2n,\mathbb{C}) \cap \mathrm{U}(2n). \tag{15.10}$$

This group is compact – a property it inherits from the compactness of the $\mathrm{U}(n)$ in which it is embedded as a subgroup. We will see that its dimension is $n(2n+1)$, the same as the non-compact $\mathrm{Sp}(2n,\mathbb{R})$. $\mathrm{Sp}(n)$ may also be defined as $\mathrm{U}(n,\mathbb{H})$ where $\mathbb{H}$ denotes the skew field of quaternions.

Warning: Physics papers often make no distinction between $\mathrm{Sp}(n)$, which is a compact group, and $\mathrm{Sp}(2n,\mathbb{R})$ which is non-compact. To add to the confusion the compact $\mathrm{Sp}(n)$ is also sometimes called $\mathrm{Sp}(2n)$. You have to judge from the context what group the author has in mind.

Physics application: Kramers'degeneracy. Let $\widehat{\sigma}_i$ be the Pauli matrices, and $\mathbf{L}$ the orbital angular momentum operator. The matrix $C = i\widehat{\sigma}_2$ has the property that

$$C^{-1}\widehat{\sigma}_i C = -\widehat{\sigma}_i^*. \tag{15.11}$$

A time-reversal invariant Hamiltonian containing $\mathbf{L}\cdot\mathbf{S}$ spin–orbit interactions obeys

$$C^{-1}HC = H^*. \tag{15.12}$$

We regard the $2n$-by-$2n$ matrix H as being an n-by-n matrix whose entries H_{ij} are themselves 2-by-2 matrices. We therefore expand these entries as

$$H_{ij} = h^0_{ij} + i\sum_{n=1}^{3} h^n_{ij}\widehat{\sigma}_n.$$

The condition (15.12) now implies that the h^a_{ij} are real numbers. We therefore say that H is *real quaternionic*. This is because the Pauli sigma matrices are algebraically isomorphic to Hamilton's quaternions under the identification

$$\begin{aligned} i\widehat{\sigma}_1 &\leftrightarrow \mathbf{i}, \\ i\widehat{\sigma}_2 &\leftrightarrow \mathbf{j}, \\ i\widehat{\sigma}_3 &\leftrightarrow \mathbf{k}. \end{aligned} \tag{15.13}$$

The hermiticity of H requires that $H_{ji} = \overline{H}_{ij}$ where the overbar denotes quaternionic conjugation, i.e. the mapping

$$q^0 + iq^1\widehat{\sigma}_1 + iq^2\widehat{\sigma}_2 + iq^3\widehat{\sigma}_3 \to q^0 - iq^1\widehat{\sigma}_1 - iq^2\widehat{\sigma}_2 - iq^3\widehat{\sigma}_3. \tag{15.14}$$

If $H\psi = E\psi$, then $HC\psi^* = E\psi^*$. Since C is skew, ψ and $C\psi^*$ are necessarily orthogonal. Therefore all states are doubly degenerate. This is *Kramers'* degeneracy.

H may be diagonalized by a matrix in $\mathrm{U}(n, \mathbb{H})$, where $\mathrm{U}(n, \mathbb{H})$ consists of those elements of $\mathrm{U}(2n)$ that satisfy $C^{-1}UC = U^*$. We may rewrite this condition as

$$C^{-1}UC = U^* \Rightarrow UCU^{\mathrm{T}} = C,$$

so $\mathrm{U}(n, \mathbb{H})$ consists of the unitary matrices that preserve the skew symmetric matrix C. Thus $\mathrm{U}(n, \mathbb{H}) \subseteq \mathrm{Sp}(n)$. Further investigation shows that $\mathrm{U}(n, \mathbb{H}) = \mathrm{Sp}(n)$.

We can exploit the quaternionic viewpoint to count the dimensions. Let $U = I + \epsilon B$ be in $\mathrm{U}(n, \mathbb{H})$. Then $B_{ij} + \overline{B}_{ji} = 0$. The diagonal elements of B are thus pure "imaginary" quaternions having no part proportional to I. There are therefore three parameters for each diagonal element. The upper triangle has $n(n-1)/2$ independent elements, each with four parameters. Counting up, we find

$$\dim \mathrm{U}(n, \mathbb{H}) = \dim \mathrm{Sp}(n) = 3n + \left[4 \times \frac{n}{2}(n-1)\right] = n(2n+1). \tag{15.15}$$

Thus, as promised, we see that the compact group $\mathrm{Sp}(n)$ and the non-compact group $\mathrm{Sp}(2n, \mathbb{R})$ have the same dimension.

We can also count the dimension of $\mathrm{Sp}(n)$ by looking at our previous matrices

$$A = \begin{pmatrix} a & b \\ c & -a^{\mathrm{T}} \end{pmatrix}$$

where a, b and c are now allowed to be complex, but with the restriction that $S = I + \epsilon A$ be unitary. This requires A to be skew-hermitian, so $a = -a^\dagger$, and $c = -b^\dagger$, while b (and hence c) remains symmetric. There are n^2 free real parameters in a, and $n(n+1)$ in b, so

$$\dim \mathrm{Sp}(n) = (n^2) + n(n+1) = n(2n+1),$$

as before.

Exercise 15.1: Show that

$$\mathrm{SO}(2N) \cap \mathrm{Sp}(2N, \mathbb{R}) \cong \mathrm{U}(N).$$

Hint: group the $2N$ basis vectors on which $\mathrm{O}(2N)$ acts into pairs $\mathbf{x}_n$ and $\mathbf{y}_n$, $n = 1, \ldots, N$. Assemble these pairs into $\mathbf{z}_n = \mathbf{x}_n + i\mathbf{y}_n$ and $\bar{\mathbf{z}} = \mathbf{x}_n - i\mathbf{y}_n$. Let ω be the linear map that takes $\mathbf{x}_n \to \mathbf{y}_n$ and $\mathbf{y}_n \to -\mathbf{x}_n$. Show that the subset of $\mathrm{SO}(2N)$ that commutes with ω mixes $\mathbf{z}_i$'s only with $\mathbf{z}_i$'s and $\bar{\mathbf{z}}_i$'s only with $\bar{\mathbf{z}}_i$'s.

15.2 Geometry of SU(2)

To get a sense of Lie groups as geometric objects, we will study the simplest non-trivial case, SU(2), in some detail.

A general 2-by-2 complex matrix can be parametrized as

$$U = \begin{pmatrix} x^0 + ix^3 & ix^1 + x^2 \\ ix^1 - x^2 & x^0 - ix^3 \end{pmatrix}. \tag{15.16}$$

The determinant of this matrix is unity, provided

$$(x^0)^2 + (x^1)^2 + (x^2)^2 + (x^3)^2 = 1. \tag{15.17}$$

When this condition is met, and if in addition the x^i are real, the matrix is unitary: $U^\dagger = U^{-1}$. The group manifold of SU(2) can therefore be identified with the 3-sphere S^3. We will take as local coordinates x^1, x^2, x^3. When we desire to know x^0 we will find it from $x^0 = \sqrt{1 - (x^1)^2 - (x^2)^2 - (x^3)^2}$. This coordinate chart only labels the points in the half of the 3-sphere having $x^0 > 0$, but this is typical of any non-trivial manifold. A complete atlas of charts can be constructed if needed.

We can simplify our notation by using the Pauli sigma matrices

$$\widehat{\sigma}_1 = \begin{pmatrix} 0 & 1 \\ 1 & 0 \end{pmatrix}, \quad \widehat{\sigma}_2 = \begin{pmatrix} 0 & -i \\ i & 0 \end{pmatrix}, \quad \widehat{\sigma}_3 = \begin{pmatrix} 1 & 0 \\ 0 & -1 \end{pmatrix}. \tag{15.18}$$

These obey

$$[\widehat{\sigma}_i, \widehat{\sigma}_j] = 2i\epsilon_{ijk}\widehat{\sigma}_k, \quad \text{and} \quad \sigma_i\widehat{\sigma}_j + \widehat{\sigma}_j\widehat{\sigma}_i = 2\delta_{ij}I. \tag{15.19}$$

In terms of them, we can write the general element of SU(2) as

$$g = U = x^0 I + ix^1\widehat{\sigma}_1 + ix^2\widehat{\sigma}_2 + ix^3\widehat{\sigma}_3. \tag{15.20}$$

Elements of the group in the neighbourhood of the identity differ from $e \equiv I$ by real linear combinations of the $i\widehat{\sigma}_i$. The three-dimensional vector space spanned by these matrices is therefore the tangent space TG_e at the identity element. For any Lie group, this tangent space is called the *Lie algebra*, $\mathfrak{g} = \mathrm{Lie}\, G$ of the group. There will be a similar set of matrices $i\widehat{\lambda}_i$ for any matrix group. They are called the *generators* of the Lie algebra, and satisfy commutation relations of the form

$$[i\widehat{\lambda}_i, i\widehat{\lambda}_j] = -f_{ij}{}^k (i\widehat{\lambda}_k), \tag{15.21}$$

or equivalently

$$[\widehat{\lambda}_i, \widehat{\lambda}_j] = if_{ij}{}^k \widehat{\lambda}_k. \tag{15.22}$$

The $f_{ij}{}^k$ are called the *structure constants* of the algebra. The "i"'s associated with the $\widehat{\lambda}$'s in this expression are conventional in physics texts because for quantum mechanics application we usually desire the $\widehat{\lambda}_i$ to be hermitian. They are usually absent in books aimed at mathematicians.

Exercise 15.2: Let $\widehat{\lambda}_1$ and $\widehat{\lambda}_2$ be hermitian matrices. Show that if we define $\widehat{\lambda}_3$ by the relation $[\widehat{\lambda}_1, \widehat{\lambda}_2] = i\widehat{\lambda}_3$, then $\widehat{\lambda}_3$ is also a hermitian matrix.

Exercise 15.3: For the group O(n) the matrices "$i\widehat{\lambda}$" are real n-by-n skew symmetric matrices A. Show that if A_1 and A_2 are real skew symmetric matrices, then so is $[A_1, A_2]$.

Exercise 15.4: For the group Sp($2n, \mathbb{R}$) the $i\widehat{\lambda}$ matrices are of the form

$$A = \begin{pmatrix} a & b \\ c & -a^{\mathrm{T}} \end{pmatrix}$$

where a is any real n-by-n matrix and b and c are symmetric ($c^{\mathrm{T}} = c$ and $b^{\mathrm{T}} = b$) real n-by-n matrices. Show that the commutator of any two matrices of this form is also of this form.

15.2.1 Invariant vector fields

Consider a matrix group, and in it a group element $I + i\epsilon\widehat{\lambda}_i$ lying close to the identity $e \equiv I$. Draw an arrow connecting I to $I + i\epsilon\widehat{\lambda}_i$, and regard this arrow as a vector L_i lying in TG_e. Next, map the infinitesimal element $I + i\epsilon\widehat{\lambda}_i$ to the neighbourhood an arbitrary group element g by multiplying on the *left* to get $g(I + i\epsilon\widehat{\lambda}_i)$. By drawing an arrow from g to $g(I + i\epsilon\widehat{\lambda}_i)$, we obtain a vector $L_i(g)$ lying in TG_g. This vector at g is the push-forward of the vector at e by left multiplication by g. For example, consider SU(2) with infinitesimal element $I + i\epsilon\widehat{\sigma}_3$. We find

$$\begin{aligned} g(I + i\epsilon\widehat{\sigma}_3) &= (x^0 + ix^1\widehat{\sigma}_1 + ix^2\widehat{\sigma}_2 + ix^3\widehat{\sigma}_3)(I + i\epsilon\widehat{\sigma}_3) \\ &= (x^0 - \epsilon x^3) + i\widehat{\sigma}_1(x^1 - \epsilon x^2) + i\widehat{\sigma}_2(x^2 + \epsilon x^1) + i\widehat{\sigma}_3(x^3 + \epsilon x^0). \end{aligned} \tag{15.23}$$

This computation can also be interpreted as showing that the multiplication of $g \in \mathrm{SU}(2)$ on the *right* by $(I + i\epsilon\widehat{\sigma}_3)$ displaces the point g, changing its x^i parameters by an amount

$$\delta \begin{pmatrix} x^0 \\ x^1 \\ x^2 \\ x^3 \end{pmatrix} = \epsilon \begin{pmatrix} -x^3 \\ -x^2 \\ x^1 \\ x^0 \end{pmatrix}. \tag{15.24}$$

Knowing how the displacement looks in terms of the x^1, x^2, x^3 coordinate system let us read off the $\partial/\partial x^\mu$ components of a vector L_3 lying in TG_g:

$$L_3 = -x^2\partial_1 + x^1\partial_2 + x^0\partial_3. \tag{15.25}$$

Since g can be any point in the group, we have constructed a globally defined vector field L_3 that acts on a function $F(g)$ on the group manifold as

$$L_3 F(g) = \lim_{\epsilon \to 0} \left\{ \frac{1}{\epsilon} \left[F\left(g(I + i\epsilon\widehat{\sigma}_3)\right) - F(g) \right] \right\}. \tag{15.26}$$

Similarly, we obtain

$$\begin{aligned} L_1 &= x^0\partial_1 - x^3\partial_2 + x^2\partial_3 \\ L_2 &= x^3\partial_1 + x^0\partial_2 - x^1\partial_3. \end{aligned} \tag{15.27}$$

The vector fields L_i are said to be *left invariant* because the push-forward of the vector $L_i(g)$ lying in the tangent space at g by multiplication on the left by any g' produces a vector $g'_*[L_i(g)]$ lying in the tangent space at $g'g$, and this pushed-forward vector coincides with the $L_i(g'g)$ already there. We can express this statement tersely as $g_* L_i = L_i$.

Using $\partial_i x^0 = -x^i/x_0$, $i = 1, 2, 3$, we can compute the Lie brackets and find

$$[L_1, L_2] = -2L_3. \tag{15.28}$$

In general

$$[L_i, L_j] = -2\epsilon_{ijk}L_k, \tag{15.29}$$

which coincides with the matrix commutator of the $\widehat{i\sigma}_i$.

This construction works for all Lie groups. For each basis vector L_i in the tangent space at the identity e, we push it forward to the tangent space at g by left multiplication by g, and so construct the global left-invariant vector field L_i. The Lie bracket of these vector fields will be

$$[L_i, L_j] = -f_{ij}{}^k L_k, \tag{15.30}$$

where the coefficients $f_{ij}{}^k$ are guaranteed to be position independent because (see Exercise 12.5) the operation of taking the Lie bracket of two vector fields commutes with the operation of pushing-forward the vector fields. Consequently, the Lie bracket at any point is just the image of the Lie bracket calculated at the identity. When the group is a matrix group, this Lie bracket will coincide with the commutator of the $\widehat{i\lambda}_i$, that group's analogue of the $\widehat{i\sigma}_i$ matrices.

The exponential map

Recall that given a vector field $X \equiv X^\mu \partial_\mu$ we define associated *flow* by solving the equation

$$\frac{dx^\mu}{dt} = X^\mu(x(t)). \tag{15.31}$$

If we do this for the left-invariant vector field L, with initial condition $x(0) = e$, we obtain a t-dependent group element $g(x(t))$, which we denote by Exp (tL). The symbol "Exp" stands for the *exponential map* which takes elements of the Lie algebra to elements of the Lie group. The reason for the name and notation is that for matrix groups this operation corresponds to the usual exponentiation of matrices. Elements of the matrix Lie group are therefore exponentials of matrices in the Lie algebra. To see this, suppose that L_i is the left-invariant vector field derived from $\widehat{i\lambda}_i$. Then the matrix

$$g(t) = \exp(it\widehat{\lambda}_i) \equiv I + it\widehat{\lambda}_i - \frac{1}{2}t^2\widehat{\lambda}_i^2 - i\frac{1}{3!}t^3\widehat{\lambda}_i^3 + \cdots \tag{15.32}$$

is an element of the group, and

$$g(t+\epsilon) = \exp(it\widehat{\lambda}_i)\exp(i\epsilon\widehat{\lambda}_i) = g(t)\left(I + i\epsilon\widehat{\lambda}_i + O(\epsilon^2)\right). \tag{15.33}$$

From this we deduce that

$$\frac{d}{dt}g(t) = \lim_{\epsilon\to 0}\left\{\frac{1}{\epsilon}[g(t)(I + i\epsilon\widehat{\lambda}_i) - g(t)]\right\} = L_i g(t). \tag{15.34}$$

Since $\exp(it\widehat{\lambda}) = I$ when $t = 0$, we deduce that Exp $(tL_i) = \exp(it\widehat{\lambda}_i)$.

Right-invariant vector fields

We can also use multiplication on the *right* to push forward an infinitesimal group element. For example:

$$\begin{aligned}(I + i\epsilon\widehat{\sigma}_3)g &= (I + i\epsilon\widehat{\sigma}_3)(x^0 + ix^1\widehat{\sigma}_1 + ix^2\widehat{\sigma}_2 + ix^3\widehat{\sigma}_3)\\ &= (x^0 - \epsilon x^3) + i\widehat{\sigma}_1(x^1 + \epsilon x^2) + i\widehat{\sigma}_2(x^2 - \epsilon x^1) + i\widehat{\sigma}_3(x^3 + \epsilon x^0)\end{aligned} \tag{15.35}$$

This motion corresponds to the *right-invariant vector field*

$$R_3 = x^2\partial_1 - x^1\partial_2 + x^0\partial_3. \tag{15.36}$$

Similarly, we obtain

$$\begin{aligned}R_1 &= x^0\partial_1 + x^3\partial_2 - x^2\partial_3,\\ R_2 &= -x^3\partial_1 + x^0\partial_2 + x^1\partial_3,\end{aligned} \tag{15.37}$$

and find that

$$[R_1, R_2] = +2R_3. \tag{15.38}$$

In general,

$$[R_i, R_j] = +2\epsilon_{ijk}R_k. \tag{15.39}$$

For any Lie group, the Lie brackets of the right-invariant fields will be

$$[R_i, R_j] = +f_{ij}{}^k R_k \tag{15.40}$$

whenever

$$[L_i, L_j] = -f_{ij}{}^k L_k \tag{15.41}$$

are the Lie brackets of the left-invariant fields. The relative minus sign between the bracket algebra of the left- and right-invariant vector fields has the same origin as the relative sign between the commutators of space- and body-fixed rotations in classical mechanics. Because multiplication from the left does not interfere with multiplication from the right, the left and right invariant fields commute:

$$[L_i, R_j] = 0. \tag{15.42}$$

15.2.2 Maurer–Cartan forms

Suppose that g is an element of a group and dg denotes its exterior derivative. Then the combination $dg\, g^{-1}$ is a Lie-algebra-valued 1-form. For example, starting from the elements of SU(2)

$$\begin{aligned} g &= x^0 + ix^1\widehat{\sigma}_1 + ix^2\widehat{\sigma}_2 + ix^3\widehat{\sigma}_3 \\ g^{-1} = g^\dagger &= x^0 - ix^1\widehat{\sigma}_1 - ix^2\widehat{\sigma}_2 - ix^3\widehat{\sigma}_3 \end{aligned} \tag{15.43}$$

we compute

$$\begin{aligned} dg &= dx^0 + idx^1\widehat{\sigma}_1 + idx^2\widehat{\sigma}_2 + idx^3\widehat{\sigma}_3 \\ &= (x^0)^{-1}(-x^1dx^1 - x^2dx^2 - x^3dx^3) + idx^1\widehat{\sigma}_1 + idx^2\widehat{\sigma}_2 + idx^3\widehat{\sigma}_3. \end{aligned} \tag{15.44}$$

From this we find

$$\begin{aligned} dgg^{-1} &= i\widehat{\sigma}_1\left((x^0 + (x^1)^2/x^0)dx^1 + (x^3 + (x^1x^2)/x^0)dx^2 + (-x^2 + (x^1x^3)/x^0)dx^3\right) \\ &\quad + i\widehat{\sigma}_2\left((-x^3 + (x^2x^1)/x^0)dx^1 + (x^0 + (x^2)^2/x^0)dx^2 + (x^1 + (x^2x^3)/x^0)dx^3\right) \\ &\quad + i\widehat{\sigma}_3\left((x^2 + (x^3x^1)/x^0)dx^1 + (-x^1 + (x^3x^2)/x^0)dx^2 + (x^0 + (x^3)^2/x^0)dx^3\right). \end{aligned} \tag{15.45}$$

Observe that the part proportional to the identity matrix has cancelled. The result of inserting a vector $X^i\partial_i$ into $dg\, g^{-1}$ is therefore an element of the Lie algebra of SU(2). This is what we mean when we say that $dg\, g^{-1}$ is *Lie-algebra-valued*.

For a general group, we define the (right-invariant) Maurer–Cartan forms ω_R^i as being the coefficient of the Lie algebra generator $i\widehat{\lambda}_i$. Thus, for SU(2), we have

$$dgg^{-1} = \omega_R = (i\widehat{\sigma}_i)\omega_R^i. \tag{15.46}$$

If we evaluate the 1-form ω_R^1 on the right-invariant vector field R_1, we find

$$\begin{aligned} \omega_R^1(R_1) &= (x^0 + (x^1)^2/x^0)x^0 + (x^3 + (x^1x^2)/x^0)x^3 \\ &\quad + (-x^2 + (x^1x^3)/x^0)(-x^2) \\ &= (x^0)^2 + (x^1)^2 + (x^2)^2 + (x^3)^2 \\ &= 1. \end{aligned} \tag{15.47}$$

Working similarly, we find

$$\begin{aligned}\omega_R^1(R_2) &= (x^0 + (x^1)^2/x^0)(-x^3) + (x^3 + (x^1x^2)/x^0)x^0 \\ &\quad + (-x^2 + (x^1x^3)/x^0)x^1 \\ &= 0. \end{aligned} \tag{15.48}$$

In general, we discover that $\omega_R^i(R_j) = \delta_j^i$. The Maurer–Cartan forms therefore constitute the dual basis to the right-invariant vector fields.

We may similarly define the *left-invariant* Maurer–Cartan forms by

$$g^{-1}dg = \omega_L = (i\widehat{\sigma}_i)\omega_L^i. \tag{15.49}$$

These obey $\omega_L^i(L_j) = \delta_j^i$, showing that the ω_L^i are the dual basis to the left-invariant vector fields.

Acting with the exterior derivative d on $gg^{-1} = I$ tells us that $d(g^{-1}) = -g^{-1}dgg^{-1}$. By exploiting this fact, together with the anti-derivation property

$$d(a \wedge b) = da \wedge b + (-1)^p a \wedge db,$$

we may compute the exterior derivative of ω_R. We find that

$$d\omega_R = d(dgg^{-1}) = (dgg^{-1}) \wedge (dgg^{-1}) = \omega_R \wedge \omega_R. \tag{15.50}$$

A matrix product is implicit here. If it were not, the product of the two identical 1-forms on the right would automatically be zero. When we make this matrix structure explicit, we see that

$$\begin{aligned}\omega_R \wedge \omega_R &= \omega_R^i \wedge \omega_R^j (i\widehat{\sigma}_i)(i\widehat{\sigma}_j) \\ &= \frac{1}{2}\omega_R^i \wedge \omega_R^j\,[i\widehat{\sigma}_i, i\widehat{\sigma}_j] \\ &= -\frac{1}{2}f_{ij}{}^k\,(i\widehat{\sigma}_k)\,\omega_R^i \wedge \omega_R^j, \end{aligned} \tag{15.51}$$

so

$$d\omega_R^k = -\frac{1}{2}f_{ij}{}^k\,\omega_R^i \wedge \omega_R^j. \tag{15.52}$$

These equations are known as the *Maurer–Cartan relations* for the right-invariant forms.

For the left-invariant forms we have

$$d\omega_L = d(g^{-1}dg) = -(g^{-1}dg) \wedge (g^{-1}dg) = -\omega_L \wedge \omega_L, \tag{15.53}$$

or

$$d\omega_L^k = +\frac{1}{2}f_{ij}{}^k\,\omega_L^i \wedge \omega_L^j. \tag{15.54}$$

The Maurer–Cartan relations appear in the physics literature when we quantize gauge theories. They are one part of the BRST transformations of the Fadeev–Popov ghost fields. We will provide a further discussion of these transformations in the next chapter.

15.2.3 Euler angles

In physics it is common to use *Euler angles* to parametrize the group SU(2). We can write an arbitrary SU(2) matrix U as a product

$$
\begin{aligned}
U &= \exp\{-i\phi\widehat{\sigma}_3/2\}\exp\{-i\theta\widehat{\sigma}_2/2\}\exp\{-i\psi\widehat{\sigma}_3/2\},\\
&= \begin{pmatrix} e^{-i\phi/2} & 0 \\ 0 & e^{i\phi/2} \end{pmatrix}\begin{pmatrix} \cos\theta/2 & -\sin\theta/2 \\ \sin\theta/2 & \cos\theta/2 \end{pmatrix}\begin{pmatrix} e^{-i\psi/2} & 0 \\ 0 & e^{i\psi/2} \end{pmatrix},\\
&= \begin{pmatrix} e^{-i(\phi+\psi)/2}\cos\theta/2 & -e^{i(\psi-\phi)/2}\sin\theta/2 \\ e^{i(\phi-\psi)/2}\sin\theta/2 & e^{i(\psi+\phi)/2}\cos\theta/2 \end{pmatrix}.
\end{aligned} \tag{15.55}
$$

Comparing with the earlier expression for U in terms of the coordinates x^μ, we obtain the Euler-angle parametrization of the 3-sphere

$$
\begin{aligned}
x^0 &= \cos\theta/2\cos(\psi+\phi)/2,\\
x^1 &= \sin\theta/2\sin(\phi-\psi)/2,\\
x^2 &= -\sin\theta/2\cos(\phi-\psi)/2,\\
x^3 &= -\cos\theta/2\sin(\psi+\phi)/2.
\end{aligned} \tag{15.56}
$$

When the angles are taken in the range $0 \le \phi < 2\pi, 0 \le \theta < \pi, 0 \le \psi < 4\pi$ we cover the entire 3-sphere exactly once.

Exercise 15.5: Show that the Hopf map, defined in Chapter 3, Hopf : $S^3 \to S^2$ is the "forgetful" map $(\theta,\phi,\psi) \to (\theta,\phi)$, where θ and ϕ are spherical polar coordinates on the 2-sphere.

Exercise 15.6: Show that

$$
U^{-1}dU = -\frac{i}{2}\widehat{\sigma}_i\,\Omega^i_{\rm L},
$$

where

$$
\begin{aligned}
\Omega^1_{\rm L} &= \sin\psi\, d\theta - \sin\theta\cos\psi\, d\phi,\\
\Omega^2_{\rm L} &= \cos\psi\, d\theta + \sin\theta\sin\psi\, d\phi,\\
\Omega^3_{\rm L} &= d\psi + \cos\theta\, d\phi.
\end{aligned}
$$

Observe that these 1-forms are essentially the components

$$\begin{aligned}\omega_X &= \sin\psi\,\dot\theta - \sin\theta\cos\psi\,\dot\phi,\\ \omega_Y &= \cos\psi\,\dot\theta + \sin\theta\sin\psi\,\dot\phi,\\ \omega_Z &= \dot\psi + \cos\theta\,\dot\phi\end{aligned}$$

of the angular velocity $\boldsymbol{\omega}$ of a body with respect to the *body-fixed XYZ* axes in the Euler-angle conventions of Exercise 11.17.

Similarly, show that

$$dUU^{-1} = -\frac{i}{2}\widehat{\sigma}_i\,\Omega^i_{\mathrm{R}},$$

where

$$\begin{aligned}\Omega^1_{\mathrm{R}} &= -\sin\phi\,d\theta + \sin\theta\cos\psi\,d\psi,\\ \Omega^2_{\mathrm{R}} &= \quad\cos\phi\,d\theta + \sin\theta\sin\psi\,d\psi,\\ \Omega^3_{\mathrm{R}} &= \quad d\phi + \cos\theta\,d\psi.\end{aligned}$$

Compute the components ω_x, ω_y, ω_z of the same angular velocity vector $\boldsymbol{\omega}$, but now taken with respect to the *space-fixed xyz* frame. Compare your answer with the Ω^i_{R}.

15.2.4 Volume and metric

The manifold of any Lie group has a natural metric, which is obtained by transporting the Killing form (see Section 15.3.2) from the tangent space at the identity to any other point g by either left or right multiplication by g. For a compact group, the resultant left- and right-invariant metrics coincide. In the particular case of SU(2) this metric is the usual metric on the 3-sphere.

Using the Euler angle expression for the x^μ to compute the dx^μ, we can express the metric on the sphere as

$$\begin{aligned}ds^2 &= (dx^0)^2 + (dx^1)^2 + (dx^2)^2 + (dx^3)^2,\\ &= \frac{1}{4}\left(d\theta^2 + \cos^2\theta/2(d\psi + d\phi)^2 + \sin^2\theta/2(d\psi - d\phi)^2\right),\\ &= \frac{1}{4}\left(d\theta^2 + d\psi^2 + d\phi^2 + 2\cos\theta\,d\phi\,d\psi\right).\end{aligned} \tag{15.57}$$

Here, to save space, we have used the traditional physics way of writing a metric. In the more formal notation, where we think of the metric as being a bilinear function, we would write the last line as

$$g(\ ,\ ) = \frac{1}{4}\left(d\theta\otimes d\theta + d\psi\otimes d\psi + d\phi\otimes d\phi + \cos\theta(d\phi\otimes d\psi + d\psi\otimes d\phi)\right). \tag{15.58}$$

From (15.58) we find

$$g = \det(g_{\mu\nu}) = \frac{1}{4^3}\begin{vmatrix} 1 & 0 & 0 \\ 0 & 1 & \cos\theta \\ 0 & \cos\theta & 1 \end{vmatrix}$$
$$= \frac{1}{64}(1-\cos^2\theta) = \frac{1}{64}\sin^2\theta. \tag{15.59}$$

The volume element, $\sqrt{g}\, d\theta d\phi d\psi$, is therefore

$$d(\text{Volume}) = \frac{1}{8}\sin\theta d\theta d\phi d\psi, \tag{15.60}$$

and the total volume of the sphere is

$$\text{Vol}(S^3) = \frac{1}{8}\int_0^\pi \sin\theta d\theta \int_0^{2\pi} d\phi \int_0^{4\pi} d\psi = 2\pi^2. \tag{15.61}$$

This volume coincides, for $d = 4$, with the standard expression for the volume of S^{d-1}, the surface of the d-dimensional unit ball,

$$\text{Vol}(S^{d-1}) = \frac{2\pi^{d/2}}{\Gamma(\frac{d}{2})}. \tag{15.62}$$

Exercise 15.7: Evaluate the Maurer–Cartan form ω_L^3 in terms of the Euler angle parametrization, and hence show that

$$i\omega_L^3 = \frac{1}{2}\text{tr}\,(\widehat{\sigma}_3 U^{-1} dU) = -\frac{i}{2}(d\psi + \cos\theta\, d\phi).$$

Now recall that the Hopf map takes the point on the 3-sphere with Euler angle coordinates (θ, ϕ, ψ) to the point on the 2-sphere with spherical polar coordinates (θ, ϕ). Thus, if we set $A = -d\psi - \cos\theta\, d\phi$, then we find

$$F \equiv dA = \sin\theta\, d\theta\, d\phi = \text{Hopf}^*\left(d\left[\text{Area}\, S^2\right]\right).$$

Also observe that

$$A \wedge F = -\sin\theta\, d\theta\, d\phi\, d\psi.$$

From this, show that the Hopf index of the Hopf map itself is equal to

$$\frac{1}{16\pi^2}\int_{S^3} A \wedge F = -1.$$

Exercise 15.8: Show that for U, the defining 2-by-2 matrices of SU(2), we have

$$\int_{\mathrm{SU}(2)} \mathrm{tr}\left[(U^{-1}dU)^3\right] = 24\pi^2.$$

Suppose we have a map $g : \mathbb{R}^3 \to \mathrm{SU}(2)$ such that $g(x)$ goes to the identity element at infinity. Consider the integral

$$S[g] = \frac{1}{24\pi^2}\int_{\mathbb{R}^3} \mathrm{tr}\,(g^{-1}dg)^3,$$

where the 3-form $\mathrm{tr}\,(g^{-1}dg)^3$ is the pull-back to $\mathbb{R}^3$ of the form $\mathrm{tr}\,[(U^{-1}dU)^3]$ on SU(2). Show that if we make the variation $g \to g + \delta g$, then

$$\delta S[g] = \frac{1}{24\pi^2}\int_{\mathbb{R}^3} d\left\{3\,\mathrm{tr}\left((g^{-1}\delta g)(g^{-1}dg)^2\right)\right\} = 0,$$

and so $S[g]$ is a topological invariant of the map g. Conclude that the functional $S[g]$ is an integer, that integer being the Brouwer degree, or winding number, of the map $g : S^3 \to S^3$.

Exercise 15.9: Generalize the result of the previous problem to show, for any mapping $x \mapsto g(x)$ into a Lie group G, and for n an odd integer, that the n-form $\mathrm{tr}\,(g^{-1}dg)^n$ constructed from the Maurer–Cartan form is closed, and that

$$\delta\,\mathrm{tr}\,(g^{-1}dg)^n = d\left\{n\,\mathrm{tr}\left((g^{-1}\delta g)(g^{-1}dg)^{n-1}\right)\right\}.$$

(Note that for even n the trace of $(g^{-1}dg)^n$ vanishes identically.)

15.2.5 SO(3) $\simeq$ SU(2)/$\mathbb{Z}_2$

The groups SU(2) and SO(3) are *locally isomorphic*. They have the same Lie algebra, but differ in their global topology. Although rotations in space are elements of SO(3), electrons respond to these rotations by transforming under the two-dimensional defining representation of SU(2). As we shall see, this means that after a rotation through 2π the electron wavefunction comes back to minus itself. The resulting orientation entanglement is characteristic of the *spinor* representation of rotations and is intimately connected with the Fermi statistics of the electron. The spin representations were discovered by Élie Cartan in 1913, some years before they were needed in physics.

The simplest way to motivate the spin/rotation connection is via the Pauli sigma matrices. These matrices are hermitian, traceless and obey

$$\widehat{\sigma}_i\widehat{\sigma}_j + \widehat{\sigma}_j\widehat{\sigma}_i = 2\delta_{ij}I. \tag{15.63}$$

If, for any $U \in SU(2)$, we define

$$\widehat{\sigma}_i' = U\widehat{\sigma}_i U^{-1}, \tag{15.64}$$

then the $\widehat{\sigma}_i'$ are also hermitian, traceless and obey (15.63). Since the original $\widehat{\sigma}_i$ form a basis for the space of hermitian traceless matrices, we must have

$$\widehat{\sigma}_i' = \widehat{\sigma}_j R_{ji} \tag{15.65}$$

for some real 3-by-3 matrix having entries R_{ij}. From (15.63) we find that

$$\begin{aligned} 2\delta_{ij} &= \widehat{\sigma}_i'\widehat{\sigma}_j' + \widehat{\sigma}_j'\widehat{\sigma}_i' \\ &= (\widehat{\sigma}_l R_{li})(\widehat{\sigma}_m R_{mj}) + (\widehat{\sigma}_m R_{mj})(\widehat{\sigma}_l R_{li}) \\ &= (\widehat{\sigma}_l\widehat{\sigma}_m + \widehat{\sigma}_m\widehat{\sigma}_l) R_{li} R_{mj} \\ &= 2\delta_{lm} R_{li} R_{mj}. \end{aligned}$$

Thus

$$R_{mi} R_{mk} = \delta_{ik}. \tag{15.66}$$

In other words, $R^{\mathrm{T}} R = I$, and so R is an element of O(3). Now the determinant of any orthogonal matrix is ± 1, but the manifold of SU(2) is a connected set and $R = I$ when $U = I$. Since a continuous map from a connected set to the integers must be a constant, we conclude that $\det R = 1$ for all U. The R matrices are therefore in SO(3).

We now exploit the principle of the sextant to show that the correspondance goes both ways, i.e. we can find a $U(R)$ for any element $R \in SO(3)$. This familiar instrument is used to measure the altitude of the Sun above the horizon while standing on the unsteady deck of a ship at sea (Figure 15.1). A theodolite or similar device would be rendered useless by the ship's pitching and rolling. The sextant exploits the fact that successive reflection in two mirrors inclined at an angle θ to one another serves to rotate the image through an angle 2θ about the line of intersection of the mirror planes. This rotation is used to superimpose the image of the sun onto the image of the horizon, where it stays even if the instrument is rocked back and forth. Exactly the same trick is used in constructing the spinor representations of the rotation group.

Consider a vector $\mathbf{x}$ with components x^i and form the matrix $\widehat{\mathbf{x}} = x^i\widehat{\sigma}_i$. Now, if $\mathbf{n}$ is a unit vector with components n^i, then

$$(-\widehat{\sigma}_i n^i)\widehat{\mathbf{x}}(\widehat{\sigma}_k n^k) = \left(x^j - 2(\mathbf{n}\cdot\mathbf{x})(n^j)\right)\widehat{\sigma}_j = \widehat{\mathbf{x}} - 2(\mathbf{n}\cdot\mathbf{x})\widehat{\mathbf{n}}. \tag{15.67}$$

The vector $\mathbf{x} - 2(\mathbf{n}\cdot\mathbf{x})\mathbf{n}$ is the result of reflecting $\mathbf{x}$ in the plane perpendicular to $\mathbf{n}$. Consequently

$$-(\widehat{\sigma}_1\cos\theta/2 + \widehat{\sigma}_2\sin\theta/2)(-\widehat{\sigma}_1)\,\widehat{\mathbf{x}}\,(\widehat{\sigma}_1)(\widehat{\sigma}_1\cos\theta/2 + \widehat{\sigma}_2\sin\theta/2) \tag{15.68}$$

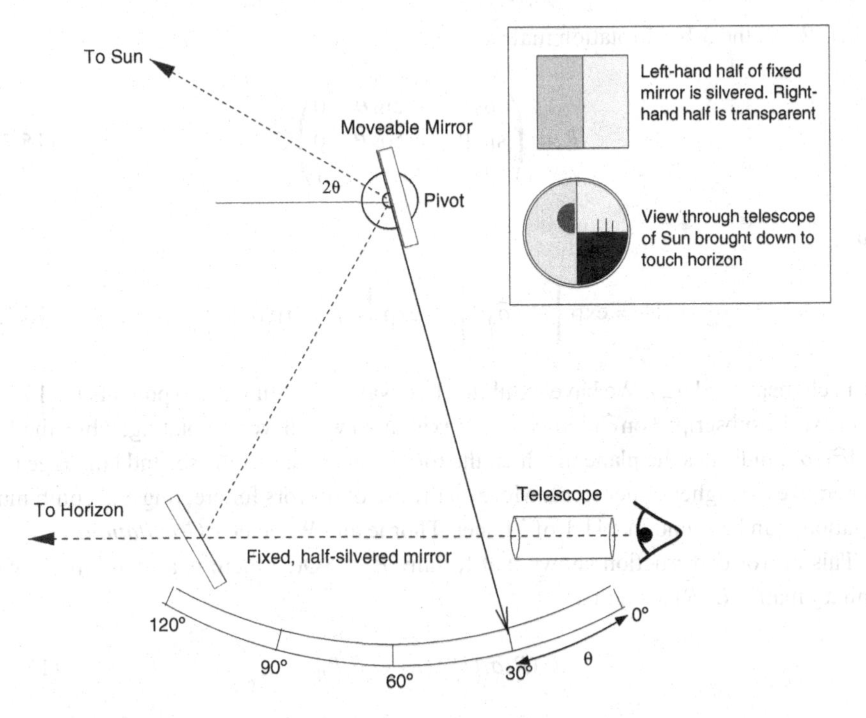

Figure 15.1 The sextant. The telescope and the half-silvered mirror are fixed to the frame of the instrument, which also holds the scale. The second mirror and attached pointer pivot so that the angle θ between the mirrors can be varied and accurately recorded. The scale is calibrated so as to display the altitude 2θ. For the configuration shown, $\theta = 15°$ while the pointer indicates that the sun is 30° above the horizon.

performs two successive reflections on $\mathbf{x}$. The first, a reflection in the "1" plane, is performed by the inner $\widehat{\sigma}_1$'s. The second reflection, in a plane at an angle $\theta/2$ to the "1" plane, is performed by the $(\widehat{\sigma}_1 \cos\theta/2 + \widehat{\sigma}_2 \sin\theta/2)$'s. Multiplying out the factors, and using the $\widehat{\sigma}_i$ algebra, we find

$$\begin{aligned}(\cos\theta/2 - \widehat{\sigma}_1\widehat{\sigma}_2 \sin\theta/2)\widehat{\mathbf{x}}(\cos\theta/2 + \widehat{\sigma}_1\widehat{\sigma}_2 \sin\theta/2) \\ = \widehat{\sigma}_1(\cos\theta\, x^1 - \sin\theta\, x^2) + \widehat{\sigma}_2(\sin\theta\, x^1 + \cos\theta\, x^2) + \widehat{\sigma}_3 x^3.\end{aligned} \tag{15.69}$$

The effect on $\mathbf{x}$ is a rotation

$$\begin{aligned} x^1 &\mapsto \cos\theta\, x^1 - \sin\theta\, x^2, \\ x^2 &\mapsto \sin\theta\, x^1 + \cos\theta\, x^2, \\ x^3 &\mapsto x^3, \end{aligned} \tag{15.70}$$

through the angle θ about the 3-axis. We can drop the x^i and re-express (15.69) as

$$U\widehat{\sigma}_i U^{-1} = \widehat{\sigma}_j R_{ji}, \tag{15.71}$$

where R_{ij} is the 3-by-3 rotation matrix

$$R = \begin{pmatrix} \cos\theta & -\sin\theta & 0 \\ \sin\theta & \cos\theta & 0 \\ 0 & 0 & 1 \end{pmatrix}, \tag{15.72}$$

and

$$U = \exp\left\{-\frac{i}{2}\widehat{\sigma}_3\theta\right\} = \exp\left\{-i\frac{1}{4i}[\widehat{\sigma}_1, \widehat{\sigma}_2]\theta\right\} \tag{15.73}$$

is an element of SU(2). We have exhibited two ways of writing the exponents in (15.73) because the subscript 3 on $\widehat{\sigma}_3$ indicates the axis about which we are rotating, while the 1, 2 in $[\widehat{\sigma}_1, \widehat{\sigma}_2]$ indicates the plane in which the rotation occurs. It is the second language that generalizes to higher dimensions. More on the use of mirrors for creating and combining rotations can be found in §41.1 of Misner, Thorne and Wheeler's *Gravitation.*

This mirror construction shows that for any $R \in$ SO(3) there is a two-dimensional unitary matrix $U(R)$ such that

$$U(R)\widehat{\sigma}_i U^{-1}(R) = \widehat{\sigma}_j R_{ji}. \tag{15.74}$$

This $U(R)$ is not unique, however. If $U \in$ SU(2) then so is $-U$. Furthermore

$$U(R)\widehat{\sigma}_i U^{-1}(R) = (-U(R))\widehat{\sigma}_i(-U(R))^{-1}, \tag{15.75}$$

and so $U(R)$ and $-U(R)$ implement exactly the same rotation R. Conversely, if two SU(2) matrices U, V obey

$$U\sigma_i U^{-1} = V\sigma_i V^{-1} \tag{15.76}$$

then $V^{-1}U$ commutes with all 2-by-2 matrices and, by Schur's lemma, must be a multiple of the identity. But if $\lambda I \in$ SU(2) then $\lambda = \pm 1$. Thus, $U = \pm V$. The mapping between SU(2) and SO(3) is therefore two-to-one. Since U and $-U$ correspond to the same R, the group manifold of SO(3) is the 3-sphere *with antipodal points identified*. Unlike the 2-sphere, where the identification of antipodal points gives the non-orientable projective plane, the SO(3) group manifold remains orientable. It is not, however, simply connected: a path on the 3-sphere from a point to its antipode forms a closed loop in SO(3), but one that is not contractible to a point. If we continue on from the antipode back to the original point, the complete path *is* contractible. This means that the first *homotopy group*, the group $\pi_1(\text{SO}(3))$ of based paths in SO(3) with composition given by concatenation, is isomorphic to $\mathbb{Z}_2$. This two-element group encodes the topology behind the Balinese Candle Dance, and keeps track of whether a sequence of rotations that eventually bring a spin-$\frac{1}{2}$ particle back to its original orientation should be counted as a 360° rotation ($U = -I$) or a 720° $\sim$ 0° rotation ($U = +I$).

Exercise 15.10: Verify that

$$U(R)\widehat{\sigma}_i U^{-1}(R) = \widehat{\sigma}_j R_{ji}$$

is consistent with $U(R_2)U(R_1) = \pm U(R_2 R_1)$.

Spinor representations of SO(N)

The mirror trick can be extended to perform rotations in N dimensions. We replace the three $\widehat{\sigma}_i$ matrices by a set of N *Dirac gamma matrices*, which obey the defining relations of a *Clifford algebra*:

$$\widehat{\gamma}_\mu \widehat{\gamma}_\nu + \widehat{\gamma}_\nu \widehat{\gamma}_\mu = 2\delta_{\mu\nu} I. \tag{15.77}$$

These relations are a generalization of the key algebraic property of the Pauli sigma matrices.

If $N\ (= 2n)$ is even, then we can find 2^n-by-2^n hermitian matrices $\widehat{\gamma}_\mu$ satisfying this algebra. If $N\ (= 2n+1)$ is odd, we append to the matrices for $N = 2n$ the hermitian matrix $\widehat{\gamma}_{2n+1} = -(i)^n \widehat{\gamma}_1 \widehat{\gamma}_2 \cdots \widehat{\gamma}_{2n}$ which obeys $\widehat{\gamma}_{2n+1}^2 = 1$ and anticommutes with all the other $\widehat{\gamma}_\mu$. The $\widehat{\gamma}$ matrices therefore act on a $2^{\lfloor N/2 \rfloor}$-dimensional space, where the symbol $\lfloor N/2 \rfloor$ denotes the *integer part* of $N/2$.

The $\widehat{\gamma}$'s do not form a Lie algebra as they stand, but a rotation through θ in the $\mu\nu$-plane is obtained from

$$\exp\left\{-i\frac{1}{4i}[\widehat{\gamma}_\mu, \widehat{\gamma}_\nu]\theta\right\} \widehat{\gamma}_i \exp\left\{i\frac{1}{4i}[\widehat{\gamma}_\mu, \widehat{\gamma}_\nu]\theta\right\} = \widehat{\gamma}_j R_{ji}, \tag{15.78}$$

and we find that the hermitian matrices $\widehat{\Gamma}_{\mu\nu} = \frac{1}{4i}[\widehat{\gamma}_\mu, \widehat{\gamma}_\nu]$ form a basis for the Lie algebra of SO(N). The $2^{\lfloor N/2 \rfloor}$-dimensional space on which they act is the Dirac spinor representation of SO(N). Although the matrices $\exp\{i\widehat{\Gamma}_{\mu\nu}\theta_{\mu\nu}\}$ are unitary, they are not the entirety of $\mathrm{U}(2^{\lfloor N/2 \rfloor})$, but instead constitute a subgroup called Spin(N).

If N is even then we can still construct the matrix $\widehat{\gamma}_{2n+1}$ that anti-commutes with all the other $\widehat{\gamma}_\mu$'s. It cannot be the identity matrix, therefore, but it commutes with all the $\Gamma_{\mu\nu}$. By Schur's lemma, this means that the SO($2n$) Dirac spinor representation space V is *reducible*. Now, $\widehat{\gamma}_{2n+1}^2 = I$, and so $\widehat{\gamma}_{2n+1}$ has eigenvalues ± 1. The two eigenspaces are invariant under the action of the group, and thus the Dirac spinor space decomposes into two irreducible *Weyl spinor* representations:

$$V = V_{\text{odd}} \oplus V_{\text{even}}. \tag{15.79}$$

Here V_{even} and V_{odd}, the plus and minus eigenspaces of $\widehat{\gamma}_{2n+1}$, are called the spaces of right and left *chirality*. When N is odd the spinor representation is irreducible.

Exercise 15.11: Starting from the defining relations of the Clifford algebra (15.77) show that, for $N = 2n$,

$$\begin{aligned}
\operatorname{tr}(\widehat{\gamma}_\mu) &= 0,\\
\operatorname{tr}(\widehat{\gamma}_{2n+1}) &= 0,\\
\operatorname{tr}(\widehat{\gamma}_\mu\widehat{\gamma}_\nu) &= \operatorname{tr}(I)\,\delta_{\mu\nu},\\
\operatorname{tr}(\widehat{\gamma}_\mu\widehat{\gamma}_\nu\widehat{\gamma}_\sigma) &= 0,\\
\operatorname{tr}(\widehat{\gamma}_\mu\widehat{\gamma}_\nu\widehat{\gamma}_\sigma\widehat{\gamma}_\tau) &= \operatorname{tr}(I)\,(\delta_{\mu\nu}\delta_{\sigma\tau} - \delta_{\mu\sigma}\delta_{\nu\tau} + \delta_{\mu\tau}\delta_{\nu\sigma}).
\end{aligned}$$

Exercise 15.12: Consider the space $\Omega(\mathbb{C}) = \bigoplus_p \Omega^p(\mathbb{C})$ of complex-valued skew symmetric tensors $A_{\mu_1\ldots\mu_p}$ for $0 \le p \le N = 2n$. Let

$$\psi_{\alpha\beta} = \sum_{p=0}^{N} \frac{1}{p!}\left(\widehat{\gamma}_{\mu_1}\cdots\widehat{\gamma}_{\mu_p}\right)_{\alpha\beta} A_{\mu_1\ldots\mu_p}$$

define a mapping from $\Omega(\mathbb{C})$ into the space of complex matrices of the same size as the $\widehat{\gamma}_\mu$. Show that this mapping is invertible, i.e. given $\psi_{\alpha\beta}$ we can recover the $A_{\mu_1\ldots\mu_p}$. By showing that the dimension of $\Omega(\mathbb{C})$ is 2^N, deduce that the $\widehat{\gamma}_\mu$ must be at least 2^n-by-2^n matrices.

Exercise 15.13: Show that the $\mathbb{R}^{2n}$ *Dirac operator* $D = \widehat{\gamma}_\mu\partial_\mu$ obeys $D^2 = \nabla^2$. Recall that the Hodge operator $d-\delta$ from Section 13.7.1 is also a "square root" of the Laplacian:

$$(d-\delta)^2 = -(d\delta + \delta d) = \nabla^2.$$

Show that

$$\psi_{\alpha\beta} \to (D\psi)_{\alpha\beta} = (\widehat{\gamma}_\mu)_{\alpha\alpha'}\partial_\mu\psi_{\alpha'\beta}$$

corresponds to the action of $d-\delta$ on the space $\Omega(\mathbb{R}^{2n},\mathbb{C})$ of differential forms

$$A = \frac{1}{p!}A_{\mu_1\ldots\mu_p}(x)dx^{\mu_1}\cdots dx^{\mu_p}.$$

The space of complex-valued differential forms has thus been made to look like a collection of 2^n Dirac spinor fields, one for each value of the "flavour index" β. These $\psi_{\alpha\beta}$ are called *Kähler–Dirac* fields. They are not really flavoured spinors because a rotation transforms both the α and β indices.

Exercise 15.14: That a set of $2n$ Dirac γ's has a 2^n-by-2^n matrix representation is most naturally established by using the tools of second quantization. To this end, let a_i, $a_i^\dagger$, $i = 1,\ldots,n$, be set of anticommuting annihilation and creation operators obeying

$$a_ia_j + a_ja_i = 0, \quad a_ia_j^\dagger + a_j^\dagger a_i = \delta_{ij}I,$$

and let $|0\rangle$ be the "no particle" state for which $a_i|0\rangle = 0, i = 1, \ldots, n$. Then the 2^n states

$$|m_1, \ldots, m_n\rangle = (a_1^\dagger)^{m_1} \cdots (a_n^\dagger)^{m_n}|0\rangle,$$

where the m_i take the value 0 or 1, constitute a basis for a space on which the a_i and $a_i^\dagger$ act irreducibly. Show that the $2n$ operators

$$\gamma_i = a_i + a_i^\dagger,$$

$$\gamma_{i+n} = i(a_i - a_i^\dagger),$$

obey

$$\gamma_\mu \gamma_\nu + \gamma_\nu \gamma_\mu = 2\delta_{\mu\nu} I,$$

and hence can be represented by 2^n-by-2^n matrices. Deduce further that spaces of left and right chirality are the spaces of odd or even "particle number".

The adjoint representation

We established the connection between SU(2) and SO(3) by means of a conjugation: $\widehat{\sigma}_i \mapsto U\widehat{\sigma}_i U^{-1}$. The idea of obtaining a representation by conjugation works for an arbitrary Lie group. It is easiest, however, to describe in the case of a matrix group where we can consider an infinitesimal element $I + i\epsilon\widehat{\lambda}_i$. The conjugate element $g(I + i\epsilon\widehat{\lambda}_i)g^{-1}$ will also be an infinitesimal element. Since $gIg^{-1} = I$, this means that $g(i\widehat{\lambda}_i)g^{-1}$ must be expressible as a linear combination of the $i\widehat{\lambda}_i$ matrices. Consequently, we can define a linear map acting on the element $X = \xi^i\widehat{\lambda}_i$ of the Lie algebra by setting

$$\mathrm{Ad}(g)\widehat{\lambda}_i \equiv g\widehat{\lambda}_i g^{-1} = \widehat{\lambda}_j [\mathrm{Ad}\,(g)]^j{}_i. \tag{15.80}$$

The matrices with entries $[\mathrm{Ad}\,(g)]^j{}_i$ form the *adjoint* representation of the group. The dimension of the adjoint representation coincides with that of the group manifold. The spinor construction shows that the defining representation of SO(3) is the adjoint representation of SU(2).

For a general Lie group, we make $\mathrm{Ad}(g)$ act on a vector in the tangent space at the identity by pushing the vector forward to TG_g by left multiplication by g, and then pushing it back from TG_g to TG_e by right multiplication by g^{-1}.

Exercise 15.15: Show that

$$[\mathrm{Ad}\,(g_1 g_2)]^j{}_i = [\mathrm{Ad}\,(g_1)]^j{}_k [\mathrm{Ad}\,(g_2)]^k{}_i,$$

thus confirming that $\mathrm{Ad}(g)$ is a representation.

15.2.6 Peter–Weyl theorem

The volume element constructed in Section 15.2.4 has the feature that it is *invariant*. In other words if we have a subset Ω of the group manifold with volume V, then the image set $g\Omega$ under left multiplication has exactly the same volume. We can also construct a volume element that is invariant under right multiplication by g, and in general these will be different. For a group whose manifold is a compact set, however, both left- and right-invariant volume elements coincide. The resulting measure on the group manifold is called the *Haar* measure.

For a *compact* group, therefore, we can replace the sums over the group elements that occur in the representation theory of finite groups, by convergent integrals over the group elements using the invariant Haar measure, which is usually denoted by $d[g]$. The invariance property is expressed by $d[g_1 g] = d[g]$ for any constant element g_1. This allows us to make a change-of-variables transformation, $g \to g_1 g$, identical to that which played such an important role in deriving the finite-group theorems. Consequently, all the results from finite groups, such as the existence of an invariant inner product and the orthogonality theorems, can be taken over by the simple replacement of a sum by an integral. In particular, if we normalize the measure so that the volume of the group manifold is unity, we have the orthogonality relation

$$\int d[g] \left(D^J_{ij}(g)\right)^* D^K_{lm}(g) = \frac{1}{\dim J} \delta^{JK} \delta_{il} \delta_{jm}. \tag{15.81}$$

The Peter–Weyl theorem asserts that the representation matrices $D^J_{mn}(g)$ form a complete set of orthogonal functions on the group manifold. In the case of SU(2) this tells us that the spin J representation matrices

$$\begin{aligned} D^J_{mn}(\theta, \phi, \psi) &= \langle J, m | e^{-iJ_3\phi} e^{-iJ_2\theta} e^{-iJ_3\psi} | J, n\rangle, \\ &= e^{-im\phi} d^J_{mn}(\theta) e^{-in\psi}, \end{aligned} \tag{15.82}$$

which you will likely have seen in quantum mechanics courses,[1] are a complete set of functions on the 3-sphere with orthogonality relation

$$\begin{aligned} &\frac{1}{16\pi^2} \int_0^\pi \sin\theta d\theta \int_0^{2\pi} d\phi \int_0^{4\pi} d\psi \left(D^J_{mn}(\theta, \phi, \psi)\right)^* D^{J'}_{m'n'}(\theta, \phi, \psi) \\ &\quad = \frac{1}{2J+1} \delta^{JJ'} \delta_{mm'} \delta_{nn'}. \end{aligned} \tag{15.83}$$

Since the D^L_{m0} (where L has to be an integer for $n = 0$ to be possible) are independent of the third Euler angle, ψ, we can do the trivial integral over ψ to obtain the special case

$$\frac{1}{4\pi} \int_0^\pi \sin\theta d\theta \int_0^{2\pi} d\phi \left(D^L_{m0}(\theta, \phi)\right)^* D^{L'}_{m'0}(\theta, \phi) = \frac{1}{2L+1} \delta^{LL'} \delta_{mm'}. \tag{15.84}$$

[1] See, for example, G. Baym, *Lectures on Quantum Mechanics*, Chapter 17.

Comparing with the definition of the spherical harmonics, we see that we can identify

$$Y^L_m(\theta,\phi) = \sqrt{\frac{2L+1}{4\pi}}\left(D^L_{m0}(\theta,\phi,\psi)\right)^*. \tag{15.85}$$

The complex conjugation is necessary here because $D^J_{mn}(\theta,\phi,\psi) \propto e^{-im\phi}$, while $Y^L_m(\theta,\phi) \propto e^{im\phi}$.

The character, $\chi^J(g) = \sum_n D^J_{nn}(g)$, will be a function only of the rotation angle θ and not the axis of rotation – all rotations through a common angle being conjugate to one another. Because of this, $\chi^J(\theta)$ can be found most simply by looking at rotations about the z-axis, since these give rise to easily computed diagonal matrices. Thus, we find

$$\begin{aligned}\chi(\theta) &= e^{iJ\theta} + e^{i(J-1)\theta} + \cdots + e^{-i(J-1)\theta} + e^{-iJ\theta}, \\ &= \frac{\sin(2J+1)\theta/2}{\sin\theta/2}.\end{aligned} \tag{15.86}$$

Warning: The angle θ in this formula and the next is *not* the Euler angle.

For integer J, corresponding to non-spinor rotations, a rotation through an angle θ about an axis $\mathbf{n}$ and a rotation though an angle $2\pi - \theta$ about $-\mathbf{n}$ are the same operation. The maximum rotation angle is therefore π. For spinor rotations this equivalence does not hold, and the rotation angle θ runs from 0 to 2π. The character orthogonality relation must therefore be

$$\frac{1}{\pi}\int_0^{2\pi} \chi^J(\theta)\chi^{J'}(\theta)\sin^2(\theta/2)d\theta = \delta^{JJ'}, \tag{15.87}$$

implying that the volume fraction of the rotation group containing rotations through angles between θ and $\theta + d\theta$ is $\sin^2(\theta/2)d\theta/\pi$.

Exercise 15.16: Prove this last statement about the volume of the equivalence classes by showing that the volume of the unit 3-sphere that lies between a rotation angle of θ and $\theta + d\theta$ is $2\pi\sin^2(\theta/2)d\theta$.

15.2.7 Lie brackets vs. commutators

There is an irritating minus-sign problem that needs to be acknowledged. The Lie bracket $[X, Y]$ of two vector fields is defined by first running along X, then Y and then back in the reverse order. If we do this for the action of matrices, $\widehat{X}$ and $\widehat{Y}$, on a vector space, then, since the sequence of matrix operations is to be read from right to left, we have

$$e^{-t_2\widehat{Y}}e^{-t_1\widehat{X}}e^{t_2\widehat{Y}}e^{t_1\widehat{X}} = I - t_1t_2[\widehat{X},\widehat{Y}] + \cdots, \tag{15.88}$$

which has the other sign. Consider, for example, rotations about the x, y, z axes, and look at the effect these have on the coordinates of a point:

$$L_x : \left.\begin{cases} \delta y = -z\,\delta\theta_x \\ \delta z = +y\,\delta\theta_x \end{cases}\right\} \Longrightarrow L_x = y\partial_z - z\partial_y, \quad \widehat{L}_x = \begin{pmatrix} 0 & 0 & 0 \\ 0 & 0 & -1 \\ 0 & 1 & 0 \end{pmatrix},$$

$$L_y : \left.\begin{cases} \delta z = -x\,\delta\theta_y \\ \delta x = +z\,\delta\theta_y \end{cases}\right\} \Longrightarrow L_y = z\partial_x - x\partial_z, \quad \widehat{L}_y = \begin{pmatrix} 0 & 0 & 1 \\ 0 & 0 & 0 \\ -1 & 0 & 0 \end{pmatrix},$$

$$L_z : \left.\begin{cases} \delta x = -y\,\delta\theta_z \\ \delta y = +x\,\delta\theta_z \end{cases}\right\} \Longrightarrow L_z = x\partial_y - y\partial_x, \quad \widehat{L}_z = \begin{pmatrix} 0 & -1 & 0 \\ 1 & 0 & 0 \\ 0 & 0 & 0 \end{pmatrix}.$$

From this we find

$$[L_x, L_y] = -L_z, \tag{15.89}$$

as a Lie bracket of vector fields, but

$$[\widehat{L}_x, \widehat{L}_y] = +\widehat{L}_z, \tag{15.90}$$

as a commutator of matrices. This is the reason why it is the *left*-invariant vector fields whose Lie bracket coincides with the commutator of the $i\widehat{\lambda}_i$ matrices.

Some insight into all this can be had by considering the action of the left-invariant fields on the representation matrices, $D^J_{mn}(g)$. For example,

$$\begin{aligned} L_i D^J_{mn}(g) &= \lim_{\epsilon\to 0}\left[\frac{1}{\epsilon}\left(D^J_{mn}(g(1+i\epsilon\widehat{\lambda}_i)) - D^J_{mn}(g)\right)\right] \\ &= \lim_{\epsilon\to 0}\left[\frac{1}{\epsilon}\left(D^J_{mn'}(g)D^J_{n'n}(1+i\epsilon\widehat{\lambda}_i) - D^J_{mn}(g)\right)\right] \\ &= \lim_{\epsilon\to 0}\left[\frac{1}{\epsilon}\left(D^J_{mn'}(g)(\delta_{n'n} + i\epsilon(\widehat{\Lambda}^J_i)_{n'n}) - D^J_{mn}(g)\right)\right] \\ &= D^J_{mn'}(g)(i\widehat{\Lambda}^J_i)_{n'n}, \end{aligned} \tag{15.91}$$

where $\widehat{\Lambda}^J_i$ is the matrix representing $\widehat{\lambda}_i$ in the representation J. Repeating this exercise we find that

$$L_i\left(L_j D^J_{mn}(g)\right) = D^J_{mn''}(g)(i\widehat{\Lambda}^J_i)_{n''n'}(i\widehat{\Lambda}^J_j)_{n'n}. \tag{15.92}$$

Thus

$$[L_i, L_j]D^J_{mn}(g) = D^J_{mn'}(g)[i\widehat{\Lambda}^J_i, i\widehat{\Lambda}^J_j]_{n'n}, \tag{15.93}$$

and we get the commutator of the representation matrices in the "correct" order only if we multiply the infinitesimal elements successively from the right.

There appears to be no escape from this sign problem. Many texts simply ignore it, a few define the Lie bracket of vector fields with the opposite sign, and a few simply point out the inconvenience and get on with the job. We will follow the last route.

15.3 Lie algebras

A Lie algebra $\mathfrak{g}$ is a (real or complex) finite-dimensional vector space with a non-associative binary operation $\mathfrak{g} \times \mathfrak{g} \to \mathfrak{g}$ that assigns to each ordered pair of elements, X_1, X_2, a third element called the Lie bracket, $[X_1, X_2]$. The bracket is:

(a) Skew symmetric: $[X, Y] = -[Y, X]$;
(b) Linear: $[\lambda X + \mu Y, Z] = \lambda[X, Z] + \mu[Y, Z]$;
and in place of associativity, obeys

(c) The Jacobi identity: $[[X, Y], Z] + [[Y, Z], X] + [[Z, X], Y] = 0$.

Example: Let $M(n)$ denote the algebra of real n-by-n matrices. As a vector space over $\mathbb{R}$, this algebra is n^2-dimensional. Setting $[A, B] = AB - BA$ makes $M(n)$ into a Lie algebra.

Example: Let $\mathfrak{b}^+$ denote the subset of $M(n)$ consisting of upper triangular matrices with any number (including zero) allowed on the diagonal. Then $\mathfrak{b}^+$ with the above bracket is a Lie algebra. (The "b" stands for the French mathematician and statesman Émile *Borel*.)

Example: Let $\mathfrak{n}^+$ denote the subset of $\mathfrak{b}^+$ consisting of strictly upper triangular matrices – those with zero on the diagonal. Then $\mathfrak{n}^+$ with the above bracket is a Lie algebra. (The "n" stands for *nilpotent*.)

Example: Let G be a Lie group, and L_i the left-invariant vector fields. We know that

$$[L_i, L_j] = f_{ij}{}^k L_k \tag{15.94}$$

where [,] is the Lie bracket of vector fields. The resulting Lie algebra, $\mathfrak{g} = \text{Lie}\, G$ is the Lie algebra of the group.

Example: The set N^+ of upper triangular matrices with 1's on the diagonal forms a Lie group and has $\mathfrak{n}^+$ as its Lie algebra. Similarly, the set B^+ consisting of upper triangular matrices, with any non-zero number allowed on the diagonal, is also a Lie group, and has $\mathfrak{b}^+$ as its Lie algebra.

Ideals and quotient algebras

As we saw in the examples, we can define subalgebras of a Lie algebra. If we want to define quotient algebras by analogy to quotient groups, we need a concept analogous to that of invariant subgroups. This is provided by the notion of an *ideal*. A ideal is a subalgebra $\mathfrak{i} \subseteq \mathfrak{g}$ with the property that

$$[\mathfrak{i}, \mathfrak{g}] \subseteq \mathfrak{i}. \tag{15.95}$$

In other words, taking the bracket of any element of $\mathfrak{g}$ with any element of $\mathfrak{i}$ gives an element in $\mathfrak{i}$. With this definition we can form $\mathfrak{g} - \mathfrak{i}$ by identifying $X \sim X + I$ for any $I \in \mathfrak{i}$. Then

$$[X + \mathfrak{i}, Y + \mathfrak{i}] = [X, Y] + \mathfrak{i}, \tag{15.96}$$

and the bracket of two equivalence classes is insensitive to the choice of representatives.

If a Lie group G has an invariant subgroup H that is also a Lie group, then the Lie algebra $\mathfrak{h}$ of the subgroup is an ideal in $\mathfrak{g} = \mathrm{Lie}\, G$, and the Lie algebra of the quotient group G/H is the quotient algebra $\mathfrak{g} - \mathfrak{h}$.

If the Lie algebra has no non-trivial ideals, then it is said to be *simple*. The Lie algebra of a simple Lie group will be simple.

Exercise 15.17: Let $\mathfrak{i}_1$ and $\mathfrak{i}_2$ be ideals in $\mathfrak{g}$. Show that $\mathfrak{i}_1 \cap \mathfrak{i}_2$ is also an ideal in $\mathfrak{g}$.

15.3.1 Adjoint representation

Given an element $X \in \mathfrak{g}$, let it act on the Lie algebra, considered as a vector space, by a linear map ad (x) defined by

$$\mathrm{ad}\,(X)Y = [X, Y]. \tag{15.97}$$

The Jacobi identity is then equivalent to the statement:

$$(\mathrm{ad}\,(X)\mathrm{ad}\,(Y) - \mathrm{ad}\,(Y)\mathrm{ad}\,(X))\,Z = \mathrm{ad}\,([X, Y])Z. \tag{15.98}$$

Thus

$$(\mathrm{ad}\,(X)\mathrm{ad}\,(Y) - \mathrm{ad}\,(Y)\mathrm{ad}\,(X)) = \mathrm{ad}\,([X, Y]), \tag{15.99}$$

or

$$[\mathrm{ad}\,(X), \mathrm{ad}\,(Y)] = \mathrm{ad}\,([X, Y]), \tag{15.100}$$

and the map $X \to \mathrm{ad}\,(X)$ is a representation of the algebra called the *adjoint representation*.

The linear map "ad (X)" exponentiates to give a map $\exp[\mathrm{ad}\,(tX)]$ defined by

$$\exp[\mathrm{ad}\,(tX)]Y = Y + t[X,Y] + \frac{1}{2}t^2[X,[X,Y]] + \cdots . \tag{15.101}$$

You probably know the matrix identity[2]

$$e^{tA}Be^{-tA} = B + t[A,B] + \frac{1}{2}t^2[A,[A,B]] + \cdots . \tag{15.102}$$

Now, earlier in the chapter, we defined the adjoint representation "Ad" of the *group* on the vector space of the Lie algebra. We did this setting $gXg^{-1} = \mathrm{Ad}\,(g)X$. Comparing the two previous equations we see that

$$\mathrm{Ad}\,(\mathrm{Exp}\,Y) = \exp(\mathrm{ad}\,(Y)). \tag{15.103}$$

15.3.2 The Killing form

Using "ad" we can define an inner product $\langle\ ,\ \rangle$ on a real Lie algebra by setting

$$\langle X,Y\rangle = \mathrm{tr}\,(\mathrm{ad}\,(X)\mathrm{ad}\,(Y)). \tag{15.104}$$

This inner product is called the *Killing form*, after Wilhelm Killing. Using the Jacobi identity and the cyclic property of the trace, we find that

$$\langle \mathrm{ad}\,(X)Y,Z\rangle + \langle Y,\mathrm{ad}\,(X)Z\rangle = 0, \tag{15.105}$$

or, equivalently,

$$\langle [X,Y],Z\rangle + \langle Y,[X,Z]\rangle = 0. \tag{15.106}$$

From this we deduce (by differentiating with respect to t) that

$$\langle \exp(\mathrm{ad}\,(tX))Y, \exp(\mathrm{ad}\,(tX))Z\rangle = \langle Y,Z\rangle, \tag{15.107}$$

so the Killing form is invariant under the action of the adjoint representation of the *group* on the algebra. When our group is simple, any other invariant inner product will be proportional to this Killing-form product.

[2] In case you do not, it is easily proved by setting $F(t) = e^{tA}Be^{-tA}$, noting that $\frac{d}{dt}F(t) = [A,F(t)]$, and observing that the RHS is the unique series solution to this equation satisfying the boundary condition $F(0) = B$.

Exercise 15.18: Let $\mathfrak{i}$ be an ideal in $\mathfrak{g}$. Show that for $I_1, I_2 \in \mathfrak{i}$

$$\langle I_1, I_2 \rangle_{\mathfrak{g}} = \langle I_1, I_2 \rangle_{\mathfrak{i}}$$

where $\langle\ ,\ \rangle_{\mathfrak{i}}$ is the Killing form on $\mathfrak{i}$ considered as a Lie algebra in its own right. (This equality of inner products is not true for subalgebras that are not ideals.)

Semisimplicity

Recall that a Lie algebra containing no non-trivial ideals is said to be *simple*. When the Killing form is non-degenerate, the Lie algebra is said to be *semisimple*. The reason for this name is that a semisimple algebra is *almost* simple, in that it can be decomposed into a direct sum of decoupled simple algebras:

$$\mathfrak{g} = \mathfrak{s}_1 \oplus \mathfrak{s}_2 \oplus \cdots \oplus \mathfrak{s}_n. \tag{15.108}$$

By "decoupled" we mean that the direct sum symbol "$\oplus$" implies not only a direct sum of vector spaces but also that $[\mathfrak{s}_i, \mathfrak{s}_j] = 0$ for $i \neq j$.

The Lie algebra of all the matrix groups O(n), Sp(n), SU(n), etc. are semisimple (indeed they are usually simple) but this is not true of the algebras $\mathfrak{n}^+$ and $\mathfrak{b}^+$.

Cartan showed that our Killing-form definition of semisimplicity is equivalent to his original definition of a Lie algebra being semisimple if the algebra contains no non-zero *abelian* ideal – i.e. no ideal with $[I_i, I_j] = 0$ for all $I_i \in \mathfrak{i}$. The following exercises establish the direct sum decomposition, and, *en passant*, the easy half of Cartan's result.

Exercise 15.19: Use the identity (15.106) to show that if $\mathfrak{i} \subset \mathfrak{g}$ is an ideal, then $\mathfrak{i}^\perp$, the set of elements orthogonal to $\mathfrak{i}$ with respect to the Killing form, is also an ideal.

Exercise 15.20: Show that if $\mathfrak{a}$ is an abelian ideal, then every element of $\mathfrak{a}$ is Killing perpendicular to the entire Lie algebra. (Thus, non-degeneracy implies no non-trivial abelian ideal. The null space of the Killing form is not necessarily an abelian ideal, though, so establishing the converse is harder.)

Exercise 15.21: Let $\mathfrak{g}$ be a semisimple Lie algebra and $\mathfrak{i} \subset \mathfrak{g}$ an ideal. We know from Exercise 15.17 that $\mathfrak{i} \cap \mathfrak{i}^\perp$ is an ideal. Use (15.106), coupled with the non-degeneracy of the Killing form, to show that it is an *abelian* ideal. Use the previous exercise to conclude that $\mathfrak{i} \cap \mathfrak{i}^\perp = \{0\}$, and from this that $[\mathfrak{i}, \mathfrak{i}^\perp] = 0$.

Exercise 15.22: Let $\langle\ ,\ \rangle$ be a non-degenerate inner product on a vector space V. Let $W \subseteq V$ be a subspace. Show that

$$\dim W + \dim W^\perp = \dim V.$$

(This is not as obvious as it looks. For a non-positive-definite inner product, W and $W^\perp$ can have a non-trivial intersection. Consider two-dimensional Minkowski space. If W

is the space of right-going, light-like, vectors then $W \equiv W^\perp$, but $\dim W + \dim W^\perp$ still equals two.)

Exercise 15.23: Put the two preceding exercises together to show that

$$\mathfrak{g} = \mathfrak{i} \oplus \mathfrak{i}^\perp.$$

Show that $\mathfrak{i}$ and $\mathfrak{i}^\perp$ are semisimple in their own right as Lie algebras. We can therefore continue to break up $\mathfrak{i}$ and $\mathfrak{i}^\perp$ until we end with $\mathfrak{g}$ decomposed into a direct sum of simple algebras.

Compactness

If the Killing form is negative definite, a real Lie algebra is said to be *compact*, and is the Lie algebra of a compact group. With the physicist's habit of writing iX_i for the generators of the Lie algebra, a compact group has Killing metric tensor

$$g_{ij} \stackrel{\text{def}}{=} \text{tr}\,\{\text{ad}\,(X_i)\text{ad}\,(X_j)\} \tag{15.109}$$

that is a *positive-definite* matrix. In a basis where $g_{ij} = \delta_{ij}$, the $\exp(\text{ad}\,X)$ matrices of the adjoint representations of a compact group G form a subgroup of the orthogonal group $\text{O}(N)$, where N is the dimension of G.

Totally anti-symmetric structure constants

Given a basis iX_i for the Lie-algebra vector space, we define the structure constants $f_{ij}{}^k$ through

$$[X_i, X_j] = if_{ij}{}^k X_k. \tag{15.110}$$

In terms of the $f_{ij}{}^k$, the skew symmetry of $\text{ad}\,(X_i)$, as expressed by Equation (15.105), becomes

$$\begin{aligned} 0 &= \langle \text{ad}\,(X_k)X_i, X_j\rangle + \langle X_i, \text{ad}\,(X_k)X_j\rangle \\ &\equiv \langle [X_k, X_i], X_j\rangle + \langle X_i, [X_k, X_j]\rangle \\ &= i(f_{ki}{}^l g_{lj} + g_{il} f_{kj}{}^l) \\ &= i(f_{kij} + f_{kji}). \end{aligned} \tag{15.111}$$

In the last line we have used the Killing metric to "lower" the index l and so define the symbol f_{ijk}. Thus, f_{ijk} is skew symmetric under the interchange of its second pair of indices. Since the skew symmetry of the Lie bracket ensures that f_{ijk} is skew symmetric under the interchange of the first pair of indices, it follows that f_{ijk} is skew symmetric under the interchange of *any* pair of its indices.

By comparing the definition of the structure constants with

$$[X_i, X_j] = \mathrm{ad}\,(X_i)X_j = X_k[\mathrm{ad}\,(X_i)]^k{}_j, \tag{15.112}$$

we read off that the matrix representing $\mathrm{ad}\,(X_i)$ has entries

$$[(\mathrm{ad}\,(X_i)]^k{}_j = if_{ij}{}^k. \tag{15.113}$$

Consequently

$$g_{ij} = \mathrm{tr}\,\{\mathrm{ad}\,(X_i)\mathrm{ad}\,(X_j)\} = -f_{ik}{}^l f_{jl}{}^k. \tag{15.114}$$

The quadratic Casimir

The only "product" that is defined in the abstract Lie algebra $\mathfrak{g}$ is the Lie bracket $[X, Y]$. Once we have found matrices forming a representation of the Lie algebra, however, we can form the ordinary matrix product of these. Suppose that we have a Lie algebra $\mathfrak{g}$ with basis X_i, and have found matrices $\widehat{X}_i$ with the same commutation relations as the X_i. Suppose, further, that the algebra is semisimple and so g^{ij}, the inverse of the Killing metric, exists. We can use g^{ij} to construct the matrix

$$\widehat{C}_2 = g^{ij}\widehat{X}_i\widehat{X}_j. \tag{15.115}$$

This matrix is called the *quadratic Casimir* operator, after Hendrik Casimir. Its chief property is that it commutes with all the $\widehat{X}_i$:

$$[\widehat{C}_2, \widehat{X}_i] = 0. \tag{15.116}$$

If our representation is irreducible then Shur's lemma tells us that

$$\widehat{C}_2 = c_2 I, \tag{15.117}$$

where the number c_2 is referred to as the "value" of the quadratic Casimir in that irrep.[3]

Exercise 15.24: Show that $[\widehat{C}_2, X_i] = 0$ is another consequence of the complete skew symmetry of the f_{ijk}.

[3] Mathematicians do sometimes consider formal products of Lie algebra elements $X, Y \in \mathfrak{g}$. When they do, they equip them with the rule that $XY - YX - [X, Y] = 0$, where XY and YX are formal products, and $[X, Y]$ is the Lie algebra product. These formal products are not elements of the Lie algebra, but instead live in an extended mathematical structure called the *Universal enveloping algebra* of $\mathfrak{g}$, and denoted by $U(\mathfrak{g})$. The quadratic Casimir operator can then be considered to be an element of this larger algebra.

15.3.3 Roots and weights

We now want to study the representation theory of Lie groups. It is, in fact, easier to study the representations of the corresponding Lie algebra and then exponentiate these to find the representations of the group. In other words, given an abstract Lie algebra with bracket

$$[X_i, X_j] = if_{ij}{}^k X_k, \tag{15.118}$$

we seek to find all matrices $\widehat{X}_i^J$ such that

$$[\widehat{X}_i^J, \widehat{X}_j^J] = if_{ij}{}^k \widehat{X}_k^J. \tag{15.119}$$

(Here, as with the representations of finite groups, we use the superscript J to distinguish one representation from another.) Then, given a representation $\widehat{X}_i^J$ of the Lie algebra, the matrices

$$D^J(g(\xi)) = \exp\left\{i\xi^i \widehat{X}_i^J\right\}, \tag{15.120}$$

where $g(\xi) = \mathrm{Exp}\{i\xi^i X_i\}$, will form a representation of the Lie *group*. To be more precise, they will form a representation of the part of the group that is connected to the identity element. The numbers ξ^i serve as coordinates for some neighbourhood of the identity. For compact groups there will be a restriction on the range of the ξ^i, because there must be ξ^i for which $\exp\{i\xi^i \widehat{X}_i^J\} = I$.

The Lie algebra of SU(2)

The quantum mechanical angular momentum algebra consists of the commutation relation

$$[J_1, J_2] = i\hbar J_3, \tag{15.121}$$

together with two similar equations related by cyclic permutations. This, once we set $\hbar = 1$, is the Lie algebra $\mathfrak{su}(2)$ of the group SU(2). The goal of representation theory is to find all possible sets of matrices that have the same commutation relations as these operators. Since the group SU(2) is compact, we can use the group-averaging trick from Section 14.2.2 to define an inner product with respect to which these representations are unitary, and the matrices J_i are hermitian.

Remember how this problem is solved in quantum mechanics courses, where we find a representation for each spin $j = \frac{1}{2}, 1, \frac{3}{2}$, etc. We begin by constructing "ladder" operators

$$J_+ \stackrel{\text{def}}{=} J_1 + iJ_2, \quad J_- \stackrel{\text{def}}{=} J_+^\dagger = J_1 - iJ_2, \tag{15.122}$$

which are eigenvectors of $\mathrm{ad}\,(J_3)$

$$\mathrm{ad}\,(J_3)J_\pm = [J_3, J_\pm] = \pm J_\pm. \tag{15.123}$$

From (15.123) we see that if $|j,m\rangle$ is an eigenstate of J_3 with eigenvalue m, then $J_\pm|j,m\rangle$ is an eigenstate of J_3 with eigenvalue $m\pm 1$.

Now, in any finite-dimensional representation there must be a *highest weight* state, $|j,j\rangle$, such that $J_3|j,j\rangle = j|j,j\rangle$ for some real number j, and such that $J_+|j,j\rangle = 0$. From $|j,j\rangle$ we work down by successive applications of J_- to find $|j,j-1\rangle$, $|j,j-2\rangle,\ldots$ We can find the normalization factors of the states $|j,m\rangle \propto (J_-)^{j-m}|j,j\rangle$ by repeated use of the identities

$$\begin{aligned}
J_+J_- &= (J_1^2+J_2^2+J_3^2)-(J_3^2-J_3),\\
J_-J_+ &= (J_1^2+J_2^2+J_3^2)-(J_3^2+J_3).
\end{aligned} \tag{15.124}$$

The combination $J^2 \equiv J_1^2+J_2^2+J_3^2$ is the quadratic Casimir of $\mathfrak{su}(2)$, and hence in any irrep is proportional to the identity matrix: $J^2 = c_2 I$. Because

$$\begin{aligned}
0 &= \|J_+|j,j\rangle\|^2\\
&= \langle j,j|J_+^\dagger J_+|j,j\rangle\\
&= \langle j,j|J_-J_+|j,j\rangle\\
&= \langle j,j|\left(J^2-J_3(J_3+1)\right)|j,j\rangle\\
&= [c_2 - j(j+1)]\langle j,j|j,j\rangle,
\end{aligned} \tag{15.125}$$

and $\langle j,j|j,j\rangle \equiv \||j,j\rangle\|^2$ is not zero, we must have $c_2 = j(j+1)$.

We now compute

$$\begin{aligned}
\|J_-|j,m\rangle\|^2 &= \langle j,m|J_-^\dagger J_-|j,m\rangle\\
&= \langle j,m|J_+J_-|j,m\rangle\\
&= \langle j,m|\left(J^2-J_3(J_3-1)\right)|j,m\rangle\\
&= [j(j+1)-m(m-1)]\langle j,m|j,m\rangle,
\end{aligned} \tag{15.126}$$

and deduce that the resulting set of normalized states $|j,m\rangle$ can be chosen to obey

$$\begin{aligned}
J_3|j,m\rangle &= m|j,m\rangle,\\
J_-|j,m\rangle &= \sqrt{j(j+1)-m(m-1)}\,|j,m-1\rangle,\\
J_+|j,m\rangle &= \sqrt{j(j+1)-m(m+1)}\,|j,m+1\rangle.
\end{aligned} \tag{15.127}$$

If we take j to be an integer or a half-integer, we will find that $J_-|j,-j\rangle = 0$. In this case we are able to construct a total of $2j+1$ states, one for each integer-spaced m in the range $-j\le m\le j$. If we select some other fractional value for j, then the set of states will not terminate, and we will find an infinity of states with $m < -j$. These will have $\|J_-|j,m\rangle\|^2 < 0$, so the resultant representation cannot be unitary.

SU(3)

The strategy of finding ladder operators works for any semisimple Lie algebra. Consider, for example, $\mathfrak{su}(3) = \mathrm{Lie}(\mathrm{SU}(3))$. The matrix Lie algebra $\mathfrak{su}(3)$ is spanned by the Gell-Mann λ-matrices

$$\widehat{\lambda}_1 = \begin{pmatrix} 0 & 1 & 0 \\ 1 & 0 & 0 \\ 0 & 0 & 0 \end{pmatrix}, \quad \widehat{\lambda}_2 = \begin{pmatrix} 0 & -i & 0 \\ i & 0 & 0 \\ 0 & 0 & 0 \end{pmatrix}, \quad \widehat{\lambda}_3 = \begin{pmatrix} 1 & 0 & 0 \\ 0 & -1 & 0 \\ 0 & 0 & 0 \end{pmatrix},$$

$$\widehat{\lambda}_4 = \begin{pmatrix} 0 & 0 & 1 \\ 0 & 0 & 0 \\ 1 & 0 & 0 \end{pmatrix}, \quad \widehat{\lambda}_5 = \begin{pmatrix} 0 & 0 & -i \\ 0 & 0 & 0 \\ i & 0 & 0 \end{pmatrix}, \quad \widehat{\lambda}_6 = \begin{pmatrix} 0 & 0 & 0 \\ 0 & 0 & 1 \\ 0 & 1 & 0 \end{pmatrix},$$

$$\widehat{\lambda}_7 = \begin{pmatrix} 0 & 0 & 0 \\ 0 & 0 & -i \\ 0 & i & 0 \end{pmatrix}, \quad \widehat{\lambda}_8 = \frac{1}{\sqrt{3}}\begin{pmatrix} 1 & 0 & 0 \\ 0 & 1 & 0 \\ 0 & 0 & -2 \end{pmatrix}, \tag{15.128}$$

which form a basis for the real vector space of 3-by-3 traceless, hermitian matrices. They have been chosen and normalized so that

$$\mathrm{tr}\,(\widehat{\lambda}_i\widehat{\lambda}_j) = 2\delta_{ij}, \tag{15.129}$$

by analogy with the properties of the Pauli matrices. Notice that $\widehat{\lambda}_3$ and $\widehat{\lambda}_8$ commute with each other, and that this will be true in any representation.

The matrices

$$\begin{aligned} t_\pm &= \frac{1}{2}(\widehat{\lambda}_1 \pm i\widehat{\lambda}_2), \\ v_\pm &= \frac{1}{2}(\widehat{\lambda}_4 \pm i\widehat{\lambda}_5), \\ u_\pm &= \frac{1}{2}(\widehat{\lambda}_6 \pm i\widehat{\lambda}_7) \end{aligned} \tag{15.130}$$

have unit entries, rather like the step-up and step-down matrices $\widehat{\sigma}_\pm = \frac{1}{2}(\widehat{\sigma}_1 \pm i\widehat{\sigma}_2)$.

Let us define Λ_i to be abstract operators with the same commutation relations as $\widehat{\lambda}_i$, and define

$$\begin{aligned} T_\pm &= \frac{1}{2}(\Lambda_1 \pm i\Lambda_2), \\ V_\pm &= \frac{1}{2}(\Lambda_4 \pm i\Lambda_5), \\ U_\pm &= \frac{1}{2}(\Lambda_6 \pm i\Lambda_7). \end{aligned} \tag{15.131}$$

These are simultaneous eigenvectors of the commuting pair of operators ad (Λ_3) and ad (Λ_8):

$$\begin{aligned}
\text{ad}\,(\Lambda_3)T_\pm &= [\Lambda_3, T_\pm] = \pm 2T_\pm, \\
\text{ad}\,(\Lambda_3)V_\pm &= [\Lambda_3, V_\pm] = \pm V_\pm, \\
\text{ad}\,(\Lambda_3)U_\pm &= [\Lambda_3, U_\pm] = \mp U_\pm, \\
\text{ad}\,(\Lambda_8)T_\pm &= [\Lambda_8, T_\pm] = 0, \\
\text{ad}\,(\Lambda_8)V_\pm &= [\Lambda_8, V_\pm] = \pm\sqrt{3}V_\pm, \\
\text{ad}\,(\Lambda_8)U_\pm &= [\Lambda_8, U_\pm] = \pm\sqrt{3}U_\pm. \qquad (15.132)
\end{aligned}$$

Thus, in any representation, the $T_\pm$, $U_\pm$, $V_\pm$ act as ladder operators, changing the simultaneous eigenvalues of the commuting pair Λ_3, Λ_8. Their eigenvalues, λ_3, λ_8, are called the *weights*, and there will be a set of such weights for each possible representation. By using the ladder operators one can go from any weight in a representation to any other, but one cannot get outside this set. The amount by which the ladder operators change the weights are called the *roots* or *root vectors*, and the root diagram characterizes the Lie algebra (Figure 15.2).

In a finite-dimensional representation there must be a highest-weight state $|\lambda_3, \lambda_8\rangle$ that is killed by all three of U_+, T_+ and V_+. We can then obtain all other states in the representation by repeatedly acting on the highest-weight state with U_-, T_- or V_- and their products. Since there is usually more than one route by which we can step down from the highest weight to another weight, the weight spaces may be *degenerate* – i.e. there may be more than one linearly independent state with the same eigenvalues of Λ_3 and Λ_8. Exactly what states are obtained, and with what multiplicity, is not immediately

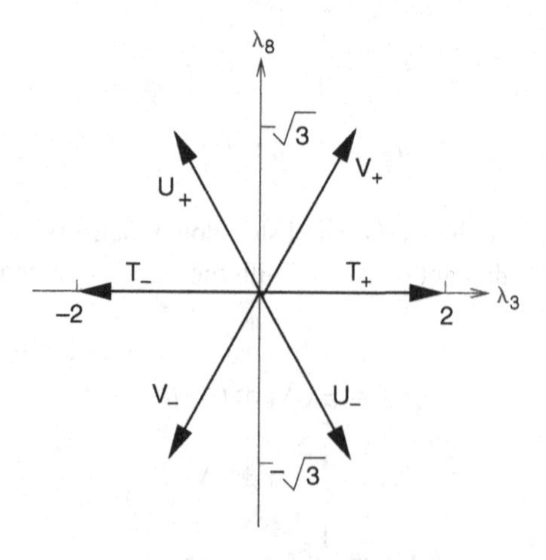

Figure 15.2 The root vectors of $\mathfrak{su}(3)$.

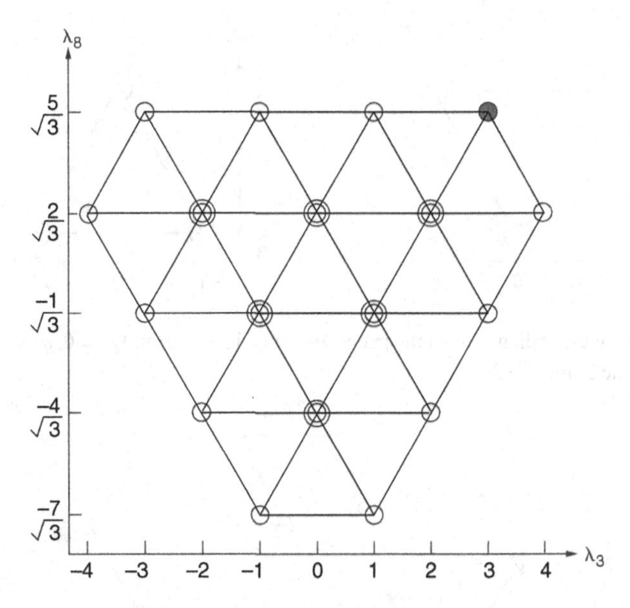

Figure 15.3 The weight diagram of the 24-dimensional irrep with $p = 3$, $q = 1$. The highest weight is shaded.

obvious. We will therefore restrict ourselves to describing the outcome of this procedure without giving proofs.

What we find is that the weights in a finite-dimensional representation of $\mathfrak{su}(3)$ form a hexagonally symmetric "crystal" lying on a triangular lattice, and the representations may be labelled by pairs of integers (zero allowed) p, q which give the length of the sides of the crystal. These representations have dimension $d = \frac{1}{2}(p+1)(q+1)(p+q+2)$.

Figure 15.3 shows the set of weights occurring in the representation of SU(3) with $p = 3$ and $q = 1$. Each circle represents a state, whose weight (λ_3, λ_8) may be read off from the displayed axes. A double circle indicates that there are two linearly independent vectors with the same weight. A count confirms that the number of independent weights, and hence the dimension of the representation, is 24. For SU(3) representations the degeneracy – i.e. the number of states with a given weight – increases by unity at each "layer" until we reach a triangular inner core, all of whose weights have the same degeneracy.

In particle physics applications, representations are often labelled by their dimension. The defining representation of SU(3) and its complex conjugate are denoted by 3 and $\bar{3}$ (see Figure 15.4), while the weight diagrams of the adjoint representation 8 and the representation 10 have shapes shown in Figure 15.5.

Cartan algebras: Roots and co-roots

For a general simple Lie algebra we may play the same game. We first find a maximal linearly independent set of commuting generators h_i. These h_i form a basis for the *Cartan*

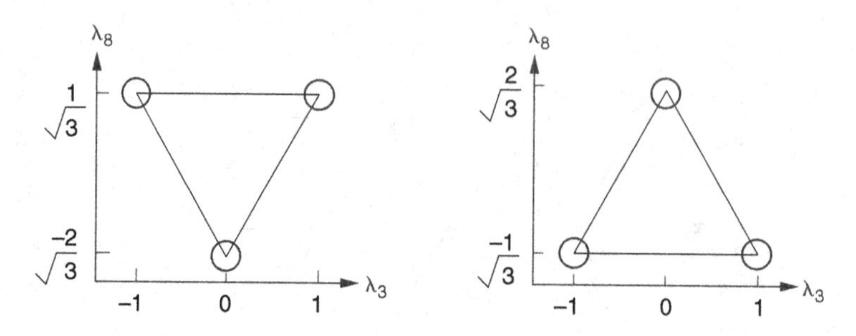

Figure 15.4 The weight diagrams of the irreps with $p = 1, q = 0$, and $p = 0, q = 1$, also known, respectively, as the 3 and the $\overline{3}$.

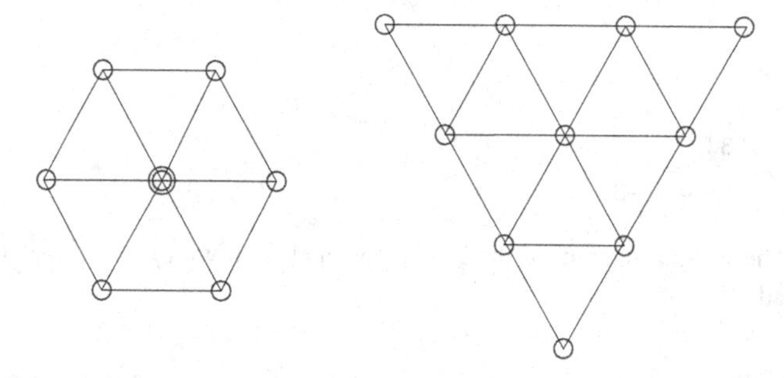

Figure 15.5 The irreps 8 (the adjoint) and 10.

algebra $\mathfrak{h}$, whose dimension is the *rank* of the Lie algebra. We next find ladder operators by diagonalizing the "ad" action of the h_i on the rest of the algebra:

$$\text{ad}\,(h_i)e_\alpha = [h_i, e_\alpha] = \alpha_i e_\alpha. \tag{15.133}$$

The simultaneous eigenvectors e_α are the ladder operators that change the eigenvalues of the h_i. The corresponding eigenvalues α, thought of as vectors with components α_i, are the *roots*, or root vectors. The roots are therefore the weights of the adjoint representation. It is possible to put factors of "i" in appropriate places so that the α_i are real, and we will assume that this has been done. For example, in $\mathfrak{su}(3)$ we have already seen that $\alpha_T = (2, 0)$, $\alpha_V = (1, \sqrt{3})$, $\alpha_U = (-1, \sqrt{3})$.

Here are the basic properties and ideas that emerge from this process:

(i) Since $\alpha_i \langle e_\alpha, h_j\rangle = \langle \text{ad}\,(h_i)e_\alpha, h_j\rangle = -\langle e_\alpha, [h_i, h_j]\rangle = 0$, we see that $\langle h_i, e_\alpha\rangle = 0$.

(ii) Similarly, we see that $(\alpha_i + \beta_i)\langle e_\alpha, e_\beta\rangle = 0$, so the e_α are orthogonal to one another unless $\alpha + \beta = 0$. Since our Lie algebra is semisimple, and consequently the Killing form non-degenerate, we deduce that if α is a root, so is $-\alpha$.

(iii) Since the Killing form is non-degenerate, yet the h_i are orthogonal to all the e_α, it must also be non-degenerate when restricted to the Cartan algebra. Thus, the metric tensor, $g_{ij} = \langle h_i, h_j\rangle$, must be invertible with inverse g^{ij}. We will use the notation $\alpha \cdot \beta$ to represent $\alpha_i \beta_j g^{ij}$.

(iv) If α, β are roots, then the Jacobi identity shows that

$$[h_i, [e_\alpha, e_\beta]] = (\alpha_i + \beta_i)[e_\alpha, e_\beta],$$

so if $[e_\alpha, e_\beta]$ is non-zero then $\alpha + \beta$ is also a root, and $[e_\alpha, e_\beta] \propto e_{\alpha+\beta}$.

(v) It follows from (iv) that $[e_\alpha, e_{-\alpha}]$ commutes with all the h_i, and since $\mathfrak{h}$ was assumed maximal, it must either be zero or a linear combination of the h_i. A short calculation shows that

$$\langle h_i, [e_\alpha, e_{-\alpha}]\rangle = \alpha_i \langle e_\alpha, e_{-\alpha}\rangle,$$

and, since $\langle e_\alpha, e_{-\alpha}\rangle$ does not vanish, $[e_\alpha, e_{-\alpha}]$ is non-zero. We can therefore choose to normalize the $e_{\pm\alpha}$ so that

$$[e_\alpha, e_{-\alpha}] = \frac{2\alpha^i}{\alpha^2} h_i \stackrel{\text{def}}{=} h_\alpha,$$

where $\alpha^i = g^{ij}\alpha_j$, and h_α obeys

$$[h_\alpha, e_{\pm\alpha}] = \pm 2 e_{\pm\alpha}.$$

The h_α are called the *co-roots*.

(vi) The importance of the co-roots stems from the observation that the triad h_α, $e_{\pm\alpha}$ obeys the same commutation relations as $\widehat{\sigma}_3$ and $\sigma_\pm$, and so forms an $\mathfrak{su}(2)$ subalgebra of $\mathfrak{g}$. In particular h_α (being the analogue of $2J_3$) has only *integer* eigenvalues. For example, in $\mathfrak{su}(3)$

$$[T_+, T_-] = h_T = \Lambda_3,$$
$$[V_+, V_-] = h_V = \frac{1}{2}\Lambda_3 + \frac{\sqrt{3}}{2}\Lambda_8,$$
$$[U_+, U_-] = h_U = -\frac{1}{2}\Lambda_3 + \frac{\sqrt{3}}{2}\Lambda_8,$$

and in the defining representation

$$h_T = \begin{pmatrix} 1 & 0 & 0 \\ 0 & -1 & 0 \\ 0 & 0 & 0 \end{pmatrix}$$
$$h_V = \begin{pmatrix} 1 & 0 & 0 \\ 0 & 0 & 0 \\ 0 & 0 & -1 \end{pmatrix}$$
$$h_U = \begin{pmatrix} 0 & 0 & 0 \\ 0 & 1 & 0 \\ 0 & 0 & -1 \end{pmatrix},$$

have eigenvalues ± 1.

(vii) Since

$$\mathrm{ad}\,(h_\alpha)e_\beta = [h_\alpha, e_\beta] = \frac{2\alpha\cdot\beta}{\alpha^2}e_\beta,$$

we conclude that $2\alpha\cdot\beta/\alpha^2$ must be an integer for any pair of roots α, β.

(viii) Finally, there can only be one e_α for each root α. If not, and there were an independent e'_α, we could take linear combinations so that $e_{-\alpha}$ and e'_α are Killing orthogonal, and hence $[e_{-\alpha}, e'_\alpha] = \alpha^i h_i \langle e_{-\alpha}, e'_\alpha\rangle = 0$. Thus $\mathrm{ad}\,(e_{-\alpha})e'_\alpha = 0$, and e'_α is killed by the step-down operator. It would therefore be the lowest weight in some $\mathfrak{su}(2)$ representation. At the same time, however, $\mathrm{ad}\,(h_\alpha)e'_\alpha = 2e'_\alpha$, and we know that the lowest weight in any spin J representation cannot have positive eigenvalue.

The conditions that

$$\frac{2\alpha\cdot\beta}{\alpha^2} \in \mathbb{Z}$$

for any pair of roots tightly constrains the possible root systems, and is the key to Cartan and Killing's classification of the semisimple Lie algebras. For example the angle θ between any pair of roots obeys $\cos^2\theta = n/4$ so θ can take only the values $0°, 30°, 45°, 60°, 90°, 120°, 135°, 150°$ or $180°$.

These constraints lead to a complete classification of possible root systems into the following infinite families:

$$\begin{array}{lll} A_n, & n = 1, 2, \cdots. & \mathfrak{sl}(n+1, \mathbb{C}), \\ B_n, & n = 2, 3, \cdots. & \mathfrak{so}(2n+1, \mathbb{C}), \\ C_n, & n = 3, 4, \cdots. & \mathfrak{sp}(2n, \mathbb{C}), \\ D_n, & n = 4, 5, \cdots. & \mathfrak{so}(2n, \mathbb{C}), \end{array}$$

together with the root systems G_2, F_4, E_6, E_7 and E_8 of the exceptional algebras. The latter do not correspond to any of the classical matrix groups. For example, G_2 is the root system of $\mathfrak{g}_2$, the Lie algebra of the group G_2 of automorphisms of the *octonions*. This group is also the subgroup of SL(7) preserving the general totally antisymmetric trilinear form.

The restrictions on the starting values of n in these families are to avoid repeats arising from "accidental" isomorphisms. If we allow $n = 1, 2, 3$, in each series, then $C_1 = D_1 = A_1$. This corresponds to $\mathfrak{sp}(2, \mathbb{C}) \cong \mathfrak{so}(3, \mathbb{C}) \cong \mathfrak{sl}(2, \mathbb{C})$. Similarly, $D_2 = A_1 + A_1$, corresponding to the isomorphism $\mathrm{SO}(4) \cong \mathrm{SU}(2) \times \mathrm{SU}(2)/\mathbb{Z}_2$, while $C_2 = B_2$ implies that, locally, the compact $\mathrm{Sp}(2) \cong \mathrm{SO}(5)$. Finally, $D_3 = A_3$ implies that $\mathrm{SU}(4)/\mathbb{Z}_2 \cong \mathrm{SO}(6)$.

15.3.4 Product representations

Given two representations $\Lambda_i^{(1)}$ and $\Lambda_i^{(2)}$ of $\mathfrak{g}$, we can form a new representation that exponentiates to the tensor product of the corresponding representations of the group G. Motivated by the result of Exercise 14.13

$$\exp(A \otimes I_n + I_m \otimes B) = \exp(A) \otimes \exp(B), \tag{15.134}$$

we take the representation matrices to act on the tensor product space as

$$\Lambda_i^{(1\otimes 2)} = \Lambda_i^{(1)} \otimes I^{(2)} + I^{(1)} \otimes \Lambda_i^{(2)}. \tag{15.135}$$

With this definition

$$\begin{aligned}
[\Lambda_i^{(1\otimes 2)}, \Lambda_j^{(1\otimes 2)}] &= ([\Lambda_i^{(1)} \otimes I^{(2)} + I^{(1)} \otimes \Lambda_i^{(2)}), (\Lambda_j^{(1)} \otimes I^{(2)} + I^{(1)} \otimes \Lambda_j^{(2)})] \\
&= [\Lambda_i^{(1)}, \Lambda_j^{(1)}] \otimes I^{(2)} + [\Lambda_i^{(1)}, I^{(1)}] \otimes \Lambda_j^{(2)} \\
&\quad + \Lambda_i^{(1)} \otimes [I^{(2)}, \Lambda_j^{(2)}] + I^{(1)} \otimes [\Lambda_i^{(2)}, \Lambda_j^{(2)}] \\
&= [\Lambda_i^{(1)}, \Lambda_j^{(1)}] \otimes I^{(2)} + I^{(1)} \otimes [\Lambda_i^{(2)}, \Lambda_j^{(2)}],
\end{aligned} \tag{15.136}$$

showing that the $\Lambda_i^{(1\otimes 2)}$ obey the Lie algebra as required.

This process of combining representations is analogous to the addition of angular momentum in quantum mechanics. Perhaps more precisely, the addition of angular momentum is an example of this general construction. If representation $\Lambda_i^{(1)}$ has weights $m_i^{(1)}$, i.e. $h_i^{(1)}|m^{(1)}\rangle = m_i^{(1)}|m^{(1)}\rangle$, and $\Lambda_i^{(2)}$ has weights $m_i^{(2)}$, then, writing $|m^{(1)}, m^{(2)}\rangle$ for $|m^{(1)}\rangle \otimes |m^{(2)}\rangle$, we have

$$\begin{aligned}
h_i^{(1\otimes 2)}|m^{(1)}, m^{(2)}\rangle &= (h_i^{(1)} \otimes 1 + 1 \otimes h_i^{(2)})|m^{(1)}, m^{(2)}\rangle \\
&= (m_i^{(1)} + m_i^{(2)})|m^{(1)}, m^{(2)}\rangle
\end{aligned} \tag{15.137}$$

so the weights appearing in the representation $\Lambda_i^{(1\otimes 2)}$ are $m_i^{(1)} + m_i^{(2)}$.

The new representation is usually decomposable. We are familiar with this decomposition for angular momentum where, if $j \geq j'$,

$$j \otimes j' = (j + j') \oplus (j + j' - 1) \oplus \cdots \oplus (j - j'). \tag{15.138}$$

This can be understood from adding weights. For example consider adding the weights of $j = 1/2$, which are $m = \pm 1/2$ to those of $j = 1$, which are $m = -1, 0, 1$. We get $m = -3/2, -1/2$ (twice), $+1/2$ (twice) and $m = +3/2$. These decompose as shown in Figure 15.6.

The rules for decomposing products in other groups are more complicated than for SU(2), but can be obtained from weight diagrams in the same manner. In SU(3), we

Figure 15.6 The weights for $1/2 \otimes 1 = 3/2 \oplus 1/2$.

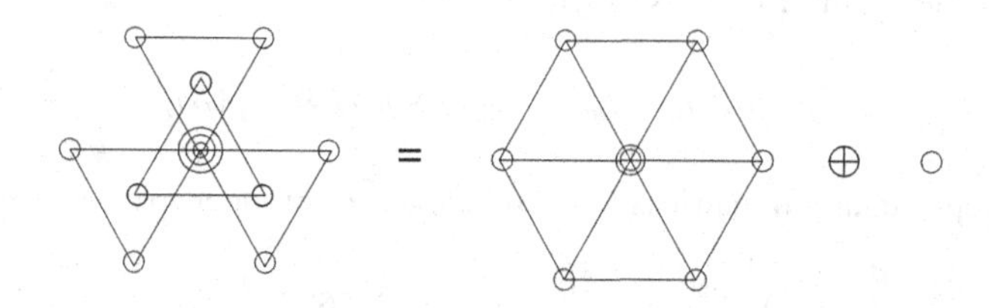

Figure 15.7 Adding the weights of 3 and $\bar{3}$.

have, for example

$$
\begin{aligned}
3 \otimes \bar{3} &= 1 \oplus 8, \\
3 \otimes 8 &= 3 \oplus \bar{6} \oplus 15, \\
8 \otimes 8 &= 1 \oplus 8 \oplus 8 \oplus 10 \oplus \overline{10} \oplus 27.
\end{aligned} \tag{15.139}
$$

To illustrate the first of these we show, in Figure 15.7, the addition of the weights in $\bar{3}$ to each of the weights in the 3. The resultant weights decompose (uniquely) into the weight diagrams for the 8 together with a singlet.

15.3.5 Subalgebras and branching rules

As with finite groups, a representation that is irreducible under the full Lie group or algebra will in general become reducible when restricted to a subgroup or subalgebra. The pattern of the decomposition is again called a *branching rule*. Here, we provide some examples to illustrate the ideas.

The three operators $V_\pm$ and $h_V = \frac{1}{2}\Lambda_3 + \frac{\sqrt{3}}{2}\Lambda_8$ of $\mathfrak{su}(3)$ form a Lie subalgebra that is isomorphic to $\mathfrak{su}(2)$ under the map that takes them to $\sigma_\pm$ and σ_3, respectively. When restricted to this subalgebra, the eight-dimensional representation of $\mathfrak{su}(3)$ becomes reducible, decomposing as

$$
8 = 3 \oplus 2 \oplus 2 \oplus 1, \tag{15.140}
$$

where the 3, 2 and 1 are the $j = 1, \frac{1}{2}$ and 0 representations of $\mathfrak{su}(2)$.

We can visualize this decomposition as coming about by first projecting the (λ_3, λ_8) weights to the "m" of the $|j, m\rangle$ labelling of $\mathfrak{su}(2)$ as

$$
m = \frac{1}{4}\lambda_3 + \frac{\sqrt{3}}{4}\lambda_8, \tag{15.141}
$$

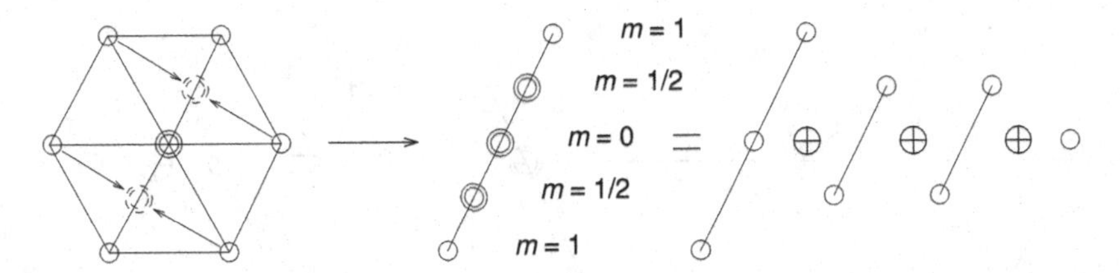

Figure 15.8 Projection of the $\mathfrak{su}(3)$ weights on to $\mathfrak{su}(2)$, and the decomposition $8 = 3 \oplus 2 \oplus 2 \oplus 1$.

and then stripping off the $\mathfrak{su}(2)$ irreps as we did when decomposing product represention (see Figure 15.8).

This branching pattern occurs in the strong interactions, where the mass of the strange quark s being much larger than that of the light quarks u and d causes the octet of pseudo-scalar mesons, which would all have the same mass if SU(3) flavour symmetry were exact, to decompose into the triplet of pions π^+, π^0 and π^-, the pair K^+ and K^0, their antiparticles K^- and $\bar{K}^0$, and the singlet η.

There are obviously other $\mathfrak{su}(2)$ subalgebras consisting of $\{T_\pm, h_T\}$ and $\{U_\pm, h_U\}$, each giving rise to similar decompositions. These subalgebras, and a continuous infinity of related ones, are obtained from the $\{V_\pm, h_V\}$ algebra by conjugation by elements of SU(3).

Another, unrelated, $\mathfrak{su}(2)$ subalgebra consists of

$$\begin{aligned}\sigma_+ &\simeq \sqrt{2}(U_+ + T_+),\\ \sigma_- &\simeq \sqrt{2}(U_- + T_-),\\ \sigma_3 &\simeq 2h_V = (\Lambda_3 + \sqrt{3}\Lambda_8).\end{aligned} \tag{15.142}$$

The factor of two between the assignment $\sigma_3 \simeq h_V$ of our previous example and the present assignment $\sigma_3 \simeq 2h_V$ has a non-trivial effect on the branching rules. Under restriction to this new subalgebra, the 8 of $\mathfrak{su}(3)$ decomposes as

$$8 = 5 \oplus 3, \tag{15.143}$$

where the 5 and 3 are the $j = 2$ and $j = 1$ representations of $\mathfrak{su}(2)$; see Figure 15.9. A clue to the origin and significance of this subalgebra is found by noting that the 3 and $\bar{3}$ representations of $\mathfrak{su}(3)$ both remain irreducible, but project to the same $j = 1$

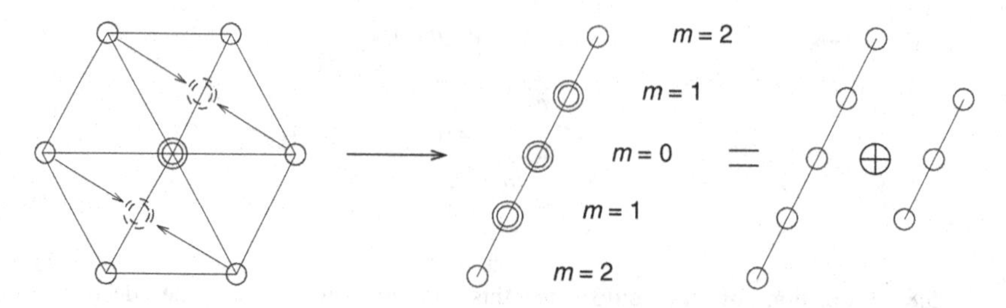

Figure 15.9 The projection and decomposition for $8 = 5 \oplus 3$.

representation of $\mathfrak{su}(2)$. Interpreting this $j = 1$ representation as the defining vector representation of $\mathfrak{so}(3)$ suggests (correctly) that our new $\mathfrak{su}(2)$ subalgebra is the Lie algebra of the SO(3) subgroup of SU(3) consisting of SU(3) matrices with real entries.

15.4 Further exercises and problems

Exercise 15.25: A Lie group manifold G has the property that it is *parallelizable*. This term means that we can find a globally smooth basis for the tangent spaces. We can, for example, take the basis vectors to be the left-invariant fields L_i. The existence of a positive-definite Killing metric also makes a compact Lie group into a Riemann manifold. In the basis formed from the L_i, the metric tensor $g_{ij} = \langle L_i, L_j \rangle$ is then numerically constant.

We may use the globally defined L_i basis to define a connection and covariant derivative by setting $\nabla_{L_i} L_j = 0$. When we do this, the connection components $\omega^k{}_{ij}$ are all zero, as are all components of the Riemann curvature tensor. The connection is therefore *flat*. The individual vectors composing a vector field with position-independent components are therefore, by definition, parallel to each other.

(a) Show that this flat connection is compatible with the metric, but is *not* torsion free.
(b) Define a new connection and covariant derivative by setting $\nabla_{L_i} L_j = \frac{1}{2}[L_i, L_j]$. Show that this new connection remains compatible with the metric but is now torsion free. It is therefore the Riemann connection. Compute components $\omega^k{}_{ij}$ of the new connection in terms of the structure constants defined by $[L_i, L_j] = -f_{ij}{}^k L_k$. Similarly compute the components of the Riemann curvature tensor.
(c) Show that, for any constants α^i, the parametrized curves $g(t) = \mathrm{Exp}(t\alpha^i L_i)g(0)$ are geodesics of the Riemann metric.

Exercise 15.26: *Campbell–Baker–Hausdorff formulæ*. Here are some useful formula for working with exponentials of matrices that do not commute with each other.

(a) Let X and Y be matrices. Show that

$$e^{tX} Y e^{-tX} = Y + t[X, Y] + \frac{1}{2} t^2 [X, [X, Y]] + \cdots,$$

the terms on the right being the series expansion of $\exp[\mathrm{ad}(tX)]Y$.

(b) Let X and δX be matrices. Show that

$$\begin{aligned} e^{-X}e^{X+\delta X} &= 1 + \int_0^1 e^{-tX}\delta X e^{tX}\,dt + O\left[(\delta X)^2\right] \\ &= 1 + \delta X - \frac{1}{2}[X,\delta X] + \frac{1}{3!}[X,[X,\delta X]] + \cdots + O\left[(\delta X)^2\right] \\ &= 1 + \left(\frac{1-e^{-\mathrm{ad}(X)}}{\mathrm{ad}(X)}\right)\delta X + O\left[(\delta X)^2\right]. \end{aligned} \tag{15.144}$$

(c) By expanding out the exponentials, show that

$$e^X e^Y = e^{X+Y+\frac{1}{2}[X,Y]+\text{higher}},$$

where "higher" means terms of higher order in X, Y. The next two terms are, in fact, $\frac{1}{12}[X,[X,Y]] + \frac{1}{12}[Y,[Y,X]]$. You will find the general formula in part (d).

(d) By using the formula from part (b), show that that $e^X e^Y$ can be written as e^Z, where

$$Z = X + \int_0^1 g(e^{\mathrm{ad}(X)}e^{\mathrm{ad}(tY)})Y\,dt.$$

Here,

$$g(z) \equiv \frac{\ln z}{1-1/z}$$

has a power-series expansion

$$g(z) = 1 + \frac{1}{2}(z-1) + \frac{1}{6}(z-1)^2 + \frac{1}{12}(z-1)^3 + \cdots,$$

which is convergent for $|z| < 1$. Show that $g(e^{\mathrm{ad}(X)}e^{\mathrm{ad}(tY)})$ can be expanded as a double power series in $\mathrm{ad}(X)$ and $\mathrm{ad}(tY)$, provided X and Y are small enough. This $\mathrm{ad}(X)$, $\mathrm{ad}(tY)$ expansion allows us to evaluate the product of two matrix exponentials as a third matrix exponential provided we know their commutator algebra.

Exercise 15.27: SU(2) *disentangling theorems*: Almost any 2×2 matrix can be factored (a Gaussian decomposition) as

$$\begin{pmatrix} a & b \\ c & d \end{pmatrix} = \begin{pmatrix} 1 & \alpha \\ 0 & 1 \end{pmatrix}\begin{pmatrix} \lambda & 0 \\ 0 & \mu \end{pmatrix}\begin{pmatrix} 1 & 0 \\ \beta & 1 \end{pmatrix}.$$

Use this trick to work the following problems:

(a) Show that

$$\exp\left\{\frac{\theta}{2}(e^{i\phi}\widehat{\sigma}_+ - e^{-i\phi}\widehat{\sigma}_-)\right\} = \exp(\alpha\widehat{\sigma}_+)\exp(\lambda\widehat{\sigma}_3)\exp(\beta\widehat{\sigma}_-),$$

where $\widehat{\sigma}_\pm = (\widehat{\sigma}_1 \pm i\widehat{\sigma}_2)/2$, and

$$\begin{aligned}\alpha &= e^{i\phi}\tan\theta/2,\\ \lambda &= -\ln\cos\theta/2,\\ \beta &= -e^{-i\phi}\tan\theta/2.\end{aligned}$$

(b) Use the fact that the spin-$\frac{1}{2}$ representation of $SU(2)$ is faithful, to show that

$$\exp\left\{\frac{\theta}{2}(e^{i\phi}\widehat{J}_+ - e^{-i\phi}\widehat{J}_-)\right\} = \exp(\alpha\widehat{J}_+)\exp(2\lambda\widehat{J}_3)\exp(\beta\widehat{J}_-),$$

where $\widehat{J}_\pm = \widehat{J}_1 \pm i\widehat{J}_2$. Take care, the reasoning here is subtle! Notice that the series expansion of exponentials of $\widehat{\sigma}_\pm$ truncates after the second term, but the same is *not* true of the expansion of exponentials of the $\widehat{J}_\pm$. You need to explain why the formula continues to hold in the absence of this truncation.

Exercise 15.28: Recall that the Lie algebra $\mathfrak{so}(N)$ of the group SO(N) consists of the skew-symmetric N-by-N matrices A with entries $A_{\mu\nu} = -A_{\nu\mu}$. Let $\widehat{\gamma}_\mu$, $\mu = 1,\dots,N$ be the Dirac gamma matrices, and define $\widehat{\Gamma}_{\mu\nu}$ to be the hermitian matrix $\frac{1}{4i}[\widehat{\gamma}_\mu, \widehat{\gamma}_\nu]$. Construct the skew-hermitian matrix $\Gamma(A)$ from A by setting

$$\Gamma(A) = \frac{i}{2}\sum_{\mu\nu} A_{\mu\nu}\widehat{\Gamma}_{\mu\nu},$$

and similarly construct $\Gamma(B)$ and $\Gamma([A,B])$ from the skew-symmetric matrices B and $[A,B]$. Show that

$$[\Gamma(A), \Gamma(B)] = \Gamma([A,B]).$$

Conclude that the map $A \to \Gamma(A)$ is a representation of $\mathfrak{so}(N)$.

Exercise 15.29: *Invariant tensors for* SU(3). Let $\widehat{\lambda}_i$ be the Gell-Mann lambda matrices. The totally anti-symmetric structure constants, f_{ijk}, and a set of totally symmetric constants, d_{ijk}, are defined by

$$f_{ijk} = \frac{1}{2}\mathrm{tr}\,(\widehat{\lambda}_i[\widehat{\lambda}_j, \widehat{\lambda}_k]), \qquad d_{ijk} = \frac{1}{2}\mathrm{tr}\,(\widehat{\lambda}_i\{\widehat{\lambda}_j, \widehat{\lambda}_k\}).$$

In the second expression, the braces denote an anticommutator:

$$\{x, y\} \stackrel{\text{def}}{=} xy + yx.$$

Let $D^8_{ij}(g)$ be the matrices representing SU(3) in "8" – the eight-dimensional adjoint representation.

(a) Show that

$$f_{ijk} = D^8_{il}(g)D^8_{jm}(g)D^8_{kn}(g)f_{lmn},$$
$$d_{ijk} = D^8_{il}(g)D^8_{jm}(g)D^8_{kn}(g)d_{lmn},$$

and so f_{ijk} and d_{ijk} are *invariant tensors* in the same sense that δ_{ij} and $\epsilon_{i_1 \ldots i_n}$ are invariant tensors for SO(n).

(b) Let $w_i = f_{ijk}u_j v_k$. Show that if $u_i \to D^8_{ij}(g)u_j$ and $v_i \to D^8_{ij}(g)v_j$, then $w_i \to D^8_{ij}(g)w_j$. Similarly for $w_i = d_{ijk}u_j v_k$. (Hint: show first that the D^8 matrices are real and orthogonal.) Deduce that f_{ijk} and d_{ijk} are *Clebsh–Gordan coefficients* for the $8 \oplus 8$ part of the decomposition

$$8 \otimes 8 = 1 \oplus 8 \oplus 8 \oplus 10 \oplus \overline{10} \oplus 27.$$

(c) Similarly show that $\delta_{\alpha\beta}$ and the entries in the lambda matrices $(\widehat{\lambda_i})_{\alpha\beta}$ can be regarded as Clebsch–Gordan coefficients for the decomposition

$$\bar{3} \otimes 3 = 1 \oplus 8.$$

(d) Use the graphical method of plotting weights and peeling off irreps to obtain the tensor product decomposition in part (b).

16

The geometry of fibre bundles

In earlier chapters we have used the language of bundles and connections, but in a relatively casual manner. We deferred proper mathematical definitions until now, because, for the applications we meet in physics, it helps to first have acquired an understanding of the geometry of Lie groups.

16.1 Fibre bundles

We begin with a formal definition of a bundle and then illustrate the definition with examples from quantum mechanics. These allow us to appreciate the physics that the definition is designed to capture.

16.1.1 Definitions

A smooth *bundle* comprises three ingredients: E, π and M, where E and M are manifolds, and $\pi : E \to M$ is a smooth surjective (onto) map. The manifold E is the *total space*, M is the *base space* and π is the *projection map*. The inverse image $\pi^{-1}(x)$ of a point in M (i.e. the set of points in E that map to x in M) is the *fibre* over x.

We usually require that all fibres be diffeomorphic to some fixed manifold F. The bundle is then a *fibre bundle*, and F is "the fibre" of the bundle. In a similar vein, we sometimes also refer to the total space E as "the bundle". Examples of possible fibres are vector spaces (in which case we have a *vector bundle*), spheres (in which case we have a sphere bundle) and Lie groups. When the fibre is a Lie group we speak of a *principal bundle*. A principal bundle can be thought of as the parent of various *associated bundles*, which are constructed by allowing the Lie group to act on a fibre. A bundle whose fibre is a one-dimensional vector space is called a *line bundle*.

The simplest example of a fibre bundle consists of setting E equal to the cartesian product $M \times F$ of the base space and the fibre. In this case the projection just "forgets" the point $f \in F$, and so $\pi : (x,f) \mapsto x$.

A more interesting example can be constructed by taking M to be the circle S^1 equipped with coordinate θ, and F as the one-dimensional interval $I = [-1, 1]$. We can assemble these ingredients to make E into a *Möbius strip*. We do this by gluing the copy of I over $\theta = 2\pi$ to that over $\theta = 0$ with a half-twist so that the end $-1 \in [-1, 1]$ is attached to $+1$, and *vice versa*.

A bundle that is a cartesian product $E = M \times F$ is said to be *trivial*. The Möbius strip is not a cartesian product, and is said to be a *twisted* bundle. The Möbius strip is,

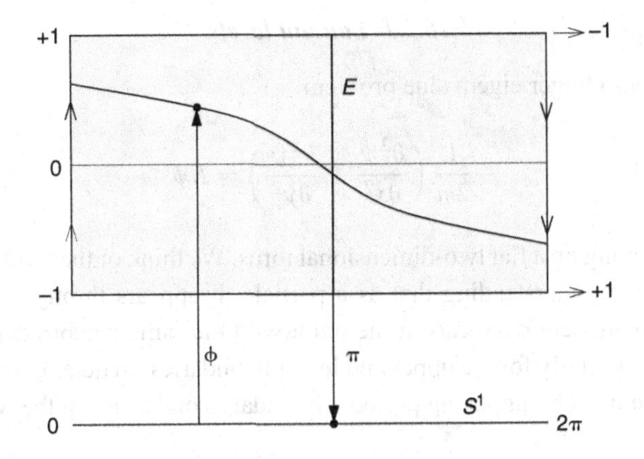

Figure 16.1 Möbius strip bundle, together with a section ϕ.

however, *locally trivial* in that for each $x \in M$ there is an open retractable neighbourhood $U \subset M$ of x in which E looks like a product $U \times F$. We will assume that all our bundles are locally trivial in this sense. If $\{U_i\}$ is a cover of M (i.e. if $M = \bigcup U_i$) by such retractable neighbourhoods, and F is a fixed fibre, then a bundle can be assembled out of the collection of $U_i \times F$ product bundles by giving gluing rules that identify points on the fibre over $x \in U_i$ in the product $U_i \times F$ with points in the fibre over $x \in U_j$ in $U_j \times F$ for each $x \in U_i \cap U_j$. These identifications are made by means of invertible maps $\varphi_{U_iU_j}(x) : F \to F$ that are defined for each x in the overlap $U_i \cap U_j$. The $\varphi_{U_iU_j}$ are known as *transition functions*. They must satisfy the consistency conditions

$$\begin{aligned} \varphi_{U_iU_i}(x) &= \text{Identity}, \\ \varphi_{U_iU_j}(x) &= \phi^{-1}_{U_jU_i}(x), \\ \varphi_{U_iU_j}(x)\varphi_{U_jU_k}(x)\varphi_{U_kU_i} &= \text{Identity}, \quad x \in U_i \cap U_j \cap U_k \neq \emptyset. \end{aligned} \tag{16.1}$$

A *section* of a fibre bundle (E, π, M) is a smooth map $\phi : M \to E$ such that $\phi(x)$ lies in the fibre $\pi^{-1}(x)$ over x. Thus $\pi \circ \phi =$ Identity. When the total space E is a product $M \times F$ this ϕ is simply a function $\phi : M \to F$. When the bundle is twisted, as is the Möbius strip (see Figure 16.1), then the section is no longer a function as it takes no unique value at the points x above which the fibres are being glued together. Observe that in the Möbius strip the half-twist forces the section $\phi(x)$ to pass through $0 \in [-1, 1]$. The Möbius bundle therefore has no nowhere-zero globally defined sections. Many twisted bundles have no globally defined sections at all.

16.2 Physics examples

We now provide three applications where the bundle concept appears in quantum mechanics. The first two illustrations are re-expressions of well-known physics. The third, the geometric approach to quantization, is perhaps less familiar.

16.2.1 Landau levels

Consider the Schrödinger eigenvalue problem

$$-\frac{1}{2m}\left(\frac{\partial^2\psi}{\partial x^2}+\frac{\partial^2\psi}{\partial y^2}\right)=E\psi \tag{16.2}$$

for a particle moving on a flat two-dimensional torus. We think of the torus as an $L_x \times L_y$ rectangle with the understanding that as a particle disappears through the right-hand boundary it immediately reappears at the point with the same y coordinate on the left-hand boundary; similarly for the upper and lower boundaries. In quantum mechanics we implement these rules by imposing periodic boundary conditions on the wavefunction:

$$\psi(0,y)=\psi(L_x,y), \qquad \psi(x,0)=\psi(x,L_y). \tag{16.3}$$

These conditions make the wavefunction a well-defined and continuous function on the torus, in the sense that after pasting the edges of the rectangle together to make a real toroidal surface the function has no jumps, and each point on the surface assigns a unique value to ψ. The wavefunction is a section of an untwisted line bundle with the torus as its base-space, the fibre over (x,y) being the one-dimensional complex vector space $\mathbb{C}$ in which $\psi(x,y)$ takes its value.

Now try to carry out the same programme for a particle of charge e moving in a uniform magnetic field B perpendicular to the xy-plane. The Schrödinger equation becomes

$$-\frac{1}{2m}\left(\frac{\partial}{\partial x}-ieA_x\right)^2\psi-\frac{1}{2m}\left(\frac{\partial}{\partial y}-ieA_y\right)^2\psi=E\psi, \tag{16.4}$$

where (A_x, A_y) is the vector potential. We at once meet a problem. Although the magnetic field is constant, the vector potential cannot be chosen to be constant – or even periodic. In the *Landau gauge*, for example, where we set $A_x = 0$, the remaining component becomes $A_y = Bx$. This means that as the particle moves out of the right-hand edge of the rectangle representing the torus we must perform a gauge transformation that prepares it for motion in the (A_x, A_y) field it will encounter when it reappears at the left. If Equation (16.4) holds, then it continues to hold after the simultaneous change

$$\begin{aligned}\psi(x,y) &\to e^{-ieBL_x y}\psi(x,y)\\ -ieA_y &\to -ieA_y + e^{-iBL_x y}\frac{\partial}{\partial y}e^{+ieBL_x y} = -ie(A_y - BL_x).\end{aligned} \tag{16.5}$$

At the right-hand boundary $x = L_x$ this gauge transformation resets the vector potential A_y back to its value at the left-hand boundary. Accordingly, we modify the boundary conditions to

$$\psi(0,y)=e^{-ieBL_x y}\psi(L_x,y), \qquad \psi(x,0)=\psi(x,L_y). \tag{16.6}$$

The new boundary conditions make the wavefunction into a section[1] of a *twisted* line bundle over the torus. The fibre is again the one-dimensional complex vector space $\mathbb{C}$.

We have already met the language in which the gauge field $-ieA_\mu$ is called a *connection* on the bundle, and the associated ieB field is the *curvature*. We will explain how connections fit into the formal bundle language in Section 16.3.

The twisting of the boundary conditions by the gauge transformation seems innocent, but within it lurks an important constraint related to the consistency conditions in (16.1). We can find the value of $\psi(L_x, L_y)$ from that of $\psi(0,0)$ by using the relations in (16.6) in the order $\psi(0,0) \to \psi(0, L_y) \to \psi(L_x, L_y)$, or in the order $\psi(0,0) \to \psi(L_x, 0) \to \psi(L_x, L_y)$. Since we must obtain the same $\psi(L_x, L_y)$ whichever route we use, we need to satisfy the condition

$$e^{ieBL_xL_y} = 1. \tag{16.7}$$

This tells us that the Schrödinger problem makes sense only when the magnetic flux BL_xL_y through the torus obeys

$$eBL_xL_y = 2\pi N \tag{16.8}$$

for some integer N. We cannot continuously vary the flux through a finite torus. This means that if we introduce torus boundary conditions as a mathematical convenience in a calculation, then physical effects may depend discontinuously on the field.

The integer N counts the number of times the phase of the wavefunction is twisted as we travel from $(x,y) = (L_x, 0)$ to $(x,y) = (L_x, L_y)$ gluing the right-hand edge wavefunction back to the left-hand edge wavefunction. This twisting number is a topological invariant. We have met this invariant before, in Section 13.6. It is the first *Chern number* of the wavefunction bundle. If we permit B to become position dependent without altering the total twist N, then quantities such as energies and expectation values can change smoothly with B. If N is allowed to change, however, these quantities may jump discontinuously.

The energy $E = E_n$ solutions to (16.4) with boundary conditions (16.6) are given by

$$\Psi_{n,k}(x,y) = \sum_{p=-\infty}^{\infty} \psi_n\left(x - \frac{k}{B} - pL_x\right) e^{i(eBpL_x+k)y}. \tag{16.9}$$

Here, $\psi_n(x)$ is a harmonic oscillator wavefunction obeying

$$-\frac{1}{2m}\frac{d^2\psi_n}{dx^2} + \frac{1}{2}m\omega^2\psi_n = E_n\psi_n, \tag{16.10}$$

with $\omega = eB/m$ the classical cyclotron frequency, and $E_n = \omega(n + 1/2)$. The parameter k takes the values $2\pi q/L_y$ for q an integer. At each energy E_n we obtain N independent

[1] That the wave "function" is no longer a function should not be disturbing. Schrödinger's ψ is never really a *function* of space-time. Seen from a frame moving at velocity v, $\psi(x,t)$ acquires factor of $\exp(-imvx - mv^2t/2)$, and this is no way for a self-respecting function of x and t to behave.

eigenfunctions as q runs from 1 to $eBL_xL_y/2\pi$. These N-fold degenerate states are the *Landau levels*. The degeneracy, being of necessity an integer, provides yet another explanation for why the flux must be quantized.

16.2.2 The Berry connection

Suppose we are in possession of a quantum-mechanical Hamiltonian $\widehat{H}(\xi)$ depending on some parameters $\xi = (\xi^1, \xi^2, \dots) \in M$, and know the eigenstates $|n;\xi\rangle$ that obey

$$\widehat{H}(\xi)|n;\xi\rangle = E_n(\xi)|n;\xi\rangle. \tag{16.11}$$

If, for fixed n, we can find a smooth family of eigenstates $|n;\xi\rangle$, one for every ξ in the parameter space M, we have a vector bundle over the space M. The fibre above ξ is the one-dimensional vector space spanned by $|n;\xi\rangle$. This bundle is a *sub-bundle* of the product bundle $M \times \mathcal{H}$ where $\mathcal{H}$ is the Hilbert space on which $\widehat{H}$ acts. Although the larger bundle is not twisted, the sub-bundle may be. It may also not exist: if the state $|n;\xi\rangle$ becomes degenerate with another state $|m;\xi\rangle$ at some value of ξ, then both states can vary discontinuously with the parameters, and we wish to exclude this possibility.

In the previous paragraph we considered the evolution of the eigenstates of a time-independent Hamiltonian as we varied its parameters. Another, more physical, evolution is given by solving the *time-dependent* Schrödinger equation

$$i\partial_t|\psi(t)\rangle = \widehat{H}(\xi(t))|\psi(t)\rangle \tag{16.12}$$

so as to follow the evolution of a state $|\psi(t)\rangle$ as the parameters are slowly varied. If the initial state $|\psi(0)\rangle$ coincides with the eigenstate $|0,\xi(0)\rangle$, and if the time evolution of the parameters is slow enough, then $|\psi\rangle$ is expected to remain close to the corresponding eigenstate $|0;\xi(t)\rangle$ of the time-independent Schrödinger equation for the Hamiltonian $\widehat{H}(\xi(t))$. To determine exactly how "close" it stays, insert the expansion

$$|\psi(t)\rangle = \sum_n a_n(t)|n;\xi(t)\rangle \exp\left\{-i\int_0^t E_0(\xi(t))\,dt\right\} \tag{16.13}$$

into (16.12) and take the inner product with $|m;\xi\rangle$. For $m \neq 0$, we expect that the overlap $\langle m;\xi|\psi(t)\rangle$ will be small and of order $O(\partial\xi/\partial t)$. Assuming that this is so, we read off that

$$\dot{a}_0 + a_0\langle 0;\xi|\partial_\mu|0;\xi\rangle\frac{\partial\xi^\mu}{\partial t} = 0, \quad (m = 0) \tag{16.14}$$

$$a_m = ia_0\frac{\langle m;\xi|\partial_\mu|0;\xi\rangle}{E_m - E_0}\frac{\partial\xi^\mu}{\partial t}, \quad (m \neq 0) \tag{16.15}$$

up to first-order accuracy in the time-derivatives of the $|n;\xi(t)\rangle$. Hence,

$$|\psi(t)\rangle = e^{i\gamma_{\text{Berry}}(t)}\left\{|0;\xi\rangle + i\sum_{m\neq 0}\frac{|m;\xi\rangle\langle m;\xi|\partial_\mu|0;\xi\rangle}{E_m - E_0}\frac{\partial\xi^\mu}{\partial t} + \cdots\right\}e^{-i\int_0^t E_0(t)dt}, \tag{16.16}$$

where the dots refer to terms of higher order in time-derivatives.

Equation (16.16) constitutes the first two terms in a systematic *adiabatic series expansion*. The factor $a_0(t) = \exp\{i\gamma_{\text{Berry}}(t)\}$ is the solution of the differential equation (16.14). The angle γ_{Berry} is known as *Berry's phase*, after the British mathematical physicist Michael Berry. It is needed to take up the slack between the arbitrary ξ-dependent phase-choice at our disposal when defining the $|0;\xi\rangle$, and the specific phase selected by the Schrödinger equation as it evolves the state $|\psi(t)\rangle$. Berry's phase is also called the *geometric phase* because it depends only on the Hilbert-space geometry of the family of states $|0;\xi\rangle$, and not on their energies. We can write

$$\gamma_{\text{Berry}}(t) = i\int_0^t \langle 0;\xi|\partial_\mu|0;\xi\rangle\frac{\partial\xi^\mu}{\partial t}\,dt, \tag{16.17}$$

and regard the 1-form

$$A_{\text{Berry}} \stackrel{\text{def}}{=} \langle 0;\xi|\partial_\mu|0;\xi\rangle d\xi^\mu = \langle 0;\xi|d|0;\xi\rangle \tag{16.18}$$

as a connection on the bundle of states over the space of parameters. The equation

$$\dot{\xi}^\mu\left(\frac{\partial}{\partial\xi^\mu} + A_{\text{Berry},\mu}\right)|\psi\rangle = 0 \tag{16.19}$$

then identifies the Schrödinger time evolution with parallel transport. It seems reasonable to refer to this particular form of parallel transport as "Berry transport".

In order for corrections to the approximation $|\psi(t)\rangle \approx (\text{phase})|0;\xi(t)\rangle$ to remain small, we need the denominator $(E_m - E_0)$ to remain large when compared to its numerator. The state that we are following must therefore never become degenerate with any other state.

Monopole bundle

Consider, for example, a spin-1/2 particle in a magnetic field. If the field points in direction $\mathbf{n}$, the Hamiltonian is

$$\widehat{H}(\mathbf{n}) = \mu|B|\widehat{\boldsymbol{\sigma}}\cdot\mathbf{n}. \tag{16.20}$$

There are are two eigenstates, with energy $E_\pm = \pm\mu|B|$. Let us focus on the eigenstate $|\psi_+\rangle$, corresponding to E_+. For each $\mathbf{n}$ we can obtain an E_+ eigenstate by applying the projection operator

$$\widehat{P} = \frac{1}{2}(\mathbf{I} + \mathbf{n} \cdot \widehat{\boldsymbol{\sigma}}) = \frac{1}{2}\begin{pmatrix} 1+n_z & n_x - in_y \\ n_x + in_y & 1 - n_z \end{pmatrix} \tag{16.21}$$

to almost any vector, and then multiplying by a real normalization constant $\mathcal{N}$. Applying $\widehat{P}$ to a "spin-up" state, for example, gives

$$\mathcal{N}\frac{1}{2}(\mathbf{I} + \mathbf{n} \cdot \widehat{\boldsymbol{\sigma}})\begin{pmatrix} 1 \\ 0 \end{pmatrix} = \begin{pmatrix} \cos\theta/2 \\ e^{i\phi}\sin\theta/2 \end{pmatrix}. \tag{16.22}$$

Here, θ and ϕ are spherical polar angles on S^2 that specify the direction of $\mathbf{n}$.

Although the bundle of $E = E_+$ eigenstates is globally defined, the family of states $|\psi_+^{(1)}(\mathbf{n})\rangle$ that we have obtained, and would like to use as a base for the fibre over $\mathbf{n}$, becomes singular when $\mathbf{n}$ is in the vicinity of the south pole $\theta = \pi$. This is because the factor $e^{i\phi}$ is multivalued at the south pole. There is no problem at the north pole because the ambiguous phase $e^{i\phi}$ multiples $\sin\theta/2$, which is zero there.

Near the south pole, however, we can project from a "spin-down" state to find

$$|\psi_+^{(2)}(\mathbf{n})\rangle = \mathcal{N}\frac{1}{2}(\mathbf{I} + \mathbf{n} \cdot \widehat{\boldsymbol{\sigma}})\begin{pmatrix} 0 \\ 1 \end{pmatrix} = \begin{pmatrix} e^{-i\phi}\cos\theta/2 \\ \sin\theta/2 \end{pmatrix}. \tag{16.23}$$

This family of eigenstates is smooth near the south pole, but is ill-defined at the north pole. As in Section 13.6, we are compelled to cover the sphere S^2 by two caps, D_+ and D_-, and use $|\psi_+^{(1)}\rangle$ in D_+ and $|\psi_+^{(2)}\rangle$ in D_-. The two families are related by

$$|\psi_+^{(1)}(\mathbf{n})\rangle = e^{i\phi}|\psi_+^{(2)}(\mathbf{n})\rangle \tag{16.24}$$

in the cingular overlap region $D_+ \cap D_-$. Here, $e^{i\phi}$ is the transition function that glues the two families of eigenstates together.

The Berry connections are

$$\begin{aligned} A_+^{(1)} &= \langle\psi_+^{(1)}|d|\psi_+^{(1)}\rangle = \frac{i}{2}(\cos\theta - 1)d\phi \\ A_+^{(2)} &= \langle\psi_+^{(2)}|d|\psi_+^{(2)}\rangle = \frac{i}{2}(\cos\theta + 1)d\phi. \end{aligned} \tag{16.25}$$

In their common domain of definition, they are related by a gauge transformation:

$$A_+^{(2)} = A_+^{(1)} + id\phi. \tag{16.26}$$

The curvature of either connection is

$$dA = -\frac{i}{2}\sin\theta d\theta d\phi = -\frac{i}{2}d(\text{Area}). \tag{16.27}$$

Being the area 2-form, the curvature tells us that when we slowly change the direction of B and bring it back to its original orientation the spin state will, in addition to the *dynamical phase* $\exp\{-iE_+t\}$, have accumulated a phase equal to (minus) one-half of the area enclosed by the trajectory of $\mathbf{n}$ on S^2. The 2-form field dA can be thought of as the flux of a magnetic monopole residing at the centre of the sphere. The corresponding bundle of one-dimensional vector spaces, spanned by $|\psi_+(\mathbf{n})\rangle$, over $\mathbf{n} \in S^2$ is therefore called the *monopole bundle*.

16.2.3 Quantization

In this section we provide a short introduction to *geometric quantization*. This idea, due largely to Kirilov, Kostant and Souriau, extends the familiar technique of canonical quantization to phase spaces with more structure than that of the harmonic oscillator. We illustrate the formalism by quantizing spin, and show how the resulting Hilbert space provides an example of the Borel–Weil–Bott construction of the representations of a semisimple Lie group as spaces of sections of holomorphic line bundles.

Prequantization

The passage from classical mechanics to quantum mechanics involves replacing the classical variables by operators in such a way that the classical Poisson-bracket algebra is mirrored by the operator commutator algebra. In general, this process of *quantization* is not possible without making some compromises. It is, however, usually possible to *prequantize* a phase space and its associated Poisson algebra.

Let M be a $2n$-dimensional classical phase space with its closed symplectic form ω. Classically a function $f : M \to \mathbb{R}$ gives rise to a Hamiltonian vector field v_f *via* Hamilton's equations

$$df = -i_{v_f}\omega. \tag{16.28}$$

We saw in Section 11.4.2 that the closure condition $d\omega = 0$ ensures that the Poisson bracket

$$\{f, g\} = v_f g = \omega(v_f, v_g) \tag{16.29}$$

obeys

$$[v_f, v_g] = v_{\{f,g,\}}. \tag{16.30}$$

Now suppose that the cohomology class of $(2\pi\hbar)^{-1}\omega$ in $H^2(M, \mathbb{R})$ has the property that its integrals over cycles in $H_2(M, \mathbb{Z})$ are integers. Then (it can be shown) there exists a line bundle L over M with curvature $F = -i\hbar^{-1}\omega$. If we locally write $\omega = d\eta$, where $\eta = \eta_\mu dx^\mu$, then the connection 1-form is $A = -i\hbar^{-1}\eta$ and the covariant derivative

$$\nabla_v \stackrel{\text{def}}{=} v^\mu(\partial_\mu - i\hbar^{-1}\eta_\mu), \tag{16.31}$$

acts on sections of the line bundle. The corresponding curvature is

$$F(u,v)=[\nabla_u,\nabla_v]-\nabla_{[u,v]}=-i\hbar^{-1}\omega(u,v). \tag{16.32}$$

We define a prequantized operator $\widehat{\rho}(f)$ that, when acting on sections $\Psi(x)$ of the line bundle, corresponds to the classical function f:

$$\widehat{\rho}(f)\stackrel{\text{def}}{=}-i\hbar\nabla_{v_f}+f. \tag{16.33}$$

For Hamiltonian vector fields v_f and v_g we have

$$\begin{aligned}[\hbar\nabla_{v_f}+if,\nabla_{v_g}]&=\hbar\nabla_{[v_f,v_g]}-i\omega(v_f,v_g)+i[f,\nabla_{v_g}]\\&=\hbar\nabla_{[v_f,v_g]}-i(i_{v_f}\omega+df)(v_g)\\&=\hbar\nabla_{[v_f,v_g]},\end{aligned} \tag{16.34}$$

and so

$$\begin{aligned}[-i\hbar\nabla_{v_f}+f,-i\hbar\nabla_{v_g}+g]&=-\hbar^2\nabla_{[v_f,v_g]}-i\hbar\,v_fg\\&=-i\hbar(-i\hbar\nabla_{[v_f,v_g]}+\{f,g\})\\&=-i\hbar(-i\hbar\nabla_{v_{\{f,g\}}}+\{f,g\}).\end{aligned} \tag{16.35}$$

Equation (16.35) is Dirac's quantization rule:

$$i\,[\widehat{\rho}(f),\widehat{\rho}(g)]=\hbar\,\widehat{\rho}(\{f,g\}). \tag{16.36}$$

The process of quantization is completed, when possible, by defining a *polarization*. This is a restriction on the variables that we allow the wavefunctions to depend on. For example, if there is a global set of Darboux coordinates p,q we might demand that the wavefunction depend only on q, only on p, or only on the combination $p+iq$. Such a restriction is necessary so that the representation $f\mapsto\widehat{\rho}(f)$ be *irreducible*. As globally defined Darboux coordinates do not usually exist, this step is the hard part of quantization.

The general definition of a polarized section is rather complicated. We sketch it here, but give a concrete example in the next section. We begin by observing that, at each point $x\in M$, the symplectic form defines a skew bilinear form. We seek a *Lagrangian subspace* of $V_x\subset TM_x$ for this form. A Lagrangian subspace is one such that $V_x=V_x^{\perp}$. For example, if

$$\begin{aligned}\omega&=dp_1\wedge dq_1+dp_2\wedge dq_2\\&=\frac{1}{2i}\{d(p_1-iq_1)\wedge d(p_1+iq_1)+d(p_2-iq_2)\wedge d(p_2+iq_2)\}\end{aligned} \tag{16.37}$$

then the space spanned by the ∂_q's is Lagrangian, as is the space spanned by the ∂_p's, and the space spanned by the ∂_{p+iq}'s. In the last case, we have allowed the coefficients of the vectors in V_x to be complex numbers. Now we let x vary and consider the distribution defined by the vector fields spanning the V_x's. We require this distribution to be globally integrable so that the V_x are the tangent spaces to a global foliation of M. With these ingredients at hand, we declare a section Ψ of the line bundle to be *polarized* if $\nabla_{\bar{\xi}}\Psi = 0$ for all $\xi \in V_x$. Here, $\bar{\xi}$ is the vector field whose components are the complex conjugates of those in ξ.

We define an inner product on the space of polarized sections by using the Liouville measure $\omega^n/n!$ on the phase space. The quantum Hilbert space then consists of finite-norm polarized sections of the line bundle. Only classical functions that give rise to polarization-compatible vector fields will have their Poisson-bracket algebra coincide with the quantum commutator algebra.

Quantizing spin

To illustrate these ideas, we quantize spin. The classical mechanics of spin was discussed in Section 11.4.2. There we showed that the appropriate phase space is the 2-sphere equipped with a symplectic form proportional to the area form. Here we must be specific about the constant of proportionality. We choose units in which $\hbar = 1$, and take $\omega = j\,d(\text{Area})$. The integrality of $\omega/2\pi$ requires that j be an integer or half-integer. We will assume that j is positive.

We parametrize the 2-sphere using complex stereographic coordinates $z, \bar{z}$ which are constructed similarly to those in Section 12.4.3. This choice will allow us to impose a natural complex polarization on the wavefunctions. In contrast to Section 12.4.3, however, it is here convenient to make the point $z = 0$ correspond to the *south* pole, so the polar coordinates θ, ϕ, on the sphere are related to $z, \bar{z}$ *via*

$$
\begin{aligned}
\cos\theta &= \frac{|z|^2 - 1}{|z|^2 + 1}, \\
e^{i\phi}\sin\theta &= \frac{2z}{|z|^2 + 1}, \\
e^{-i\phi}\sin\theta &= \frac{2\bar{z}}{|z|^2 + 1}.
\end{aligned}
\tag{16.38}
$$

In terms of the $z, \bar{z}$ coordinates the symplectic form is given by

$$
\omega = \frac{2ij}{(1 + |z|^2)^2}\, dz \wedge d\bar{z}. \tag{16.39}
$$

As long as we avoid the north pole, where $z = \infty$, we can write

$$
\omega = d\left\{ij \frac{z\,d\bar{z} - \bar{z}\,dz}{1 + |z|^2}\right\} = d\eta, \tag{16.40}
$$

and so the local connection form has components proportional to

$$\eta_z = -ij\frac{\bar{z}}{|z|^2+1}, \qquad \eta_{\bar{z}} = ij\frac{z}{|z|^2+1}. \tag{16.41}$$

The covariant derivatives are therefore

$$\nabla_z = \frac{\partial}{\partial z} - j\frac{\bar{z}}{|z|^2+1}, \qquad \nabla_{\bar{z}} = \frac{\partial}{\partial \bar{z}} + j\frac{z}{|z|^2+1}. \tag{16.42}$$

We impose the polarization condition that $\nabla_{\bar{z}}\Psi = 0$. This condition requires the allowed sections to be of the form

$$\Psi(z,\bar{z}) = (1+|z|^2)^{-j}\psi(z), \tag{16.43}$$

where ψ depends only on z, and not on $\bar{z}$. It is natural to combine the $(1+|z|^2)^{-j}$ prefactor with the Liouville measure so that the inner product becomes

$$\langle\psi|\chi\rangle = \frac{2j+1}{2\pi i}\int_{\mathbb{C}} \frac{d\bar{z}\wedge dz}{(1+|z|^2)^{2j+2}}\,\overline{\psi(z)}\chi(z). \tag{16.44}$$

The normalizable wavefunctions are then polynomials in z of degree less than or equal to $2j$, and a complete orthonormal set is given by

$$\psi_m(z) = \sqrt{\frac{2j!}{(j-m)!(j+m)!}}\,z^{j+m}, \qquad -j \le m \le j. \tag{16.45}$$

We desire to find the quantum operators $\widehat{\rho}(J_i)$ corresponding to the components

$$J_1 = j\sin\theta\cos\phi, \quad J_2 = j\sin\theta\sin\phi, \quad J_3 = j\cos\theta, \tag{16.46}$$

of a classical spin $\mathbf{J}$ of magnitude j, and also to the ladder-operator components $J_\pm = J_1 \pm iJ_2$. In our complex coordinates, these functions become

$$\begin{aligned} J_3 &= j\frac{|z|^2-1}{|z|^2+1}, \\ J_+ &= j\frac{2z}{|z|^2+1}, \\ J_- &= j\frac{2\bar{z}}{|z|^2+1}. \end{aligned} \tag{16.47}$$

Also in these coordinates, Hamilton's equations $dH = -\omega(v_H, \ )$ take the form

$$\begin{aligned} \dot{z} &= \ i\frac{(1+|z|^2)^2}{2j}\frac{\partial H}{\partial \bar{z}}, \\ \dot{\bar{z}} &= -i\frac{(1+|z|^2)^2}{2j}\frac{\partial H}{\partial z}, \end{aligned} \tag{16.48}$$

and the Hamiltonian vector fields corresponding to the classical phase space functions J_3, J_+ and J_- are

$$\begin{aligned} v_{J_3} &= \quad iz\partial_z - i\bar{z}\partial_{\bar{z}}, \\ v_{J_+} &= -iz^2\partial_z - i\partial_{\bar{z}}, \\ v_{J_-} &= \quad i\partial_z + i\bar{z}^2\partial_{\bar{z}}. \end{aligned} \tag{16.49}$$

Using the recipe (16.33) for $\widehat{\rho}(H)$ from the previous section, together with the fact that $\nabla_{\bar{z}}\Psi = 0$, we find, for example, that

$$\begin{aligned} &\widehat{\rho}(J_+)(1+|z|^2)^{-j}\psi(z) \\ &\quad = \left[-z^2\left(\frac{\partial}{\partial z} - \frac{j\bar{z}}{(1+|z|^2)}\right) + \frac{2jz}{(1+|z|^2)}\right](1+|z|^2)^{-j}\psi(z), \\ &\quad = (1+|z|^2)^{-j}\left[-z^2\frac{\partial}{\partial z} + 2jz\right]\psi. \end{aligned} \tag{16.50}$$

It is natural to define operators

$$\widehat{J}_i \stackrel{\text{def}}{=} (1+|z|^2)^j\widehat{\rho}(J_i)(1+|z|^2)^{-j} \tag{16.51}$$

that act only on the z-polynomial part $\psi(z)$ of the section $\Psi(z,\bar{z})$. We then have

$$\widehat{J}_+ = -z^2\frac{\partial}{\partial z} + 2jz. \tag{16.52}$$

Similarly, we find that

$$\widehat{J}_- = \frac{\partial}{\partial z}, \tag{16.53}$$

$$\widehat{J}_3 = z\frac{\partial}{\partial z} - j. \tag{16.54}$$

These operators obey the $\mathfrak{su}(2)$ Lie-algebra relations

$$\begin{aligned} [\widehat{J}_3, \widehat{J}_\pm] &= \pm\widehat{J}_\pm, \\ [\widehat{J}_+, \widehat{J}_-] &= 2\widehat{J}_3, \end{aligned} \tag{16.55}$$

and act on the $\psi_m(z)$ monomials as

$$\begin{aligned} \widehat{J}_3\psi_m(z) &= m\,\psi_m(z), \\ \widehat{J}_\pm\psi_m(z) &= \sqrt{j(j+1) - m(m\pm 1)}\,\psi_{m\pm 1}(z). \end{aligned} \tag{16.56}$$

This is the familiar action of the $\mathfrak{su}(2)$ generators on $|j,m\rangle$ basis states.

Exercise 16.1: Show that with respect to the inner product (16.44) we have

$$\widehat{J}_3^\dagger = \widehat{J}_3, \quad \widehat{J}_+^\dagger = \widehat{J}_-.$$

Coherent states and the Borel–Weil–Bott theorem

We now explain how the spin wavefunctions $\psi_m(z)$ can be understood as sections of a *holomorphic* line bundle.

Suppose that we have a compact Lie group G and a unitary irreducible representation $g \in G \mapsto D^J(g)$. Let $|0\rangle$ be the normalized highest (or lowest) weight state in the representation space. Consider the states

$$|g\rangle = D^J(g)|0\rangle, \quad \langle g| = \langle 0| \left[D^J(g)\right]^\dagger. \tag{16.57}$$

The $|g\rangle$ compose a family of *generalized coherent states*.[2] There is a continuous infinity of the $|g\rangle$, and so they cannot constitute an orthonormal set on the finite-dimensional representation space. The matrix-element orthogonality property (15.81), however, provides us with a useful *over-completeness relation*:

$$\mathbf{I} = \frac{\dim(J)}{\mathrm{Vol}\, G} \int_G |g\rangle\langle g|. \tag{16.58}$$

The integral is over all of G, but many points in G give the same contribution. The *maximal torus*, denoted by T, is the abelian subgroup of G obtained by exponentiating elements of the Cartan algebra. Because any weight vector is a common eigenvector of the Cartan algebra, elements of T leave $|0\rangle$ fixed up to a phase. The set of distinct $|g\rangle$ in the integral can therefore be identified with G/T. This coset space is always an even-dimensional manifold, and thus a candidate phase space.

Consider, in particular, the spin-j representation of SU(2). The coset space G/T is then $\mathrm{SU}(2)/U(1) \simeq S^2$. We can write a general element of SU(2) as

$$U = \exp(\bar{z}J_+)\exp(\theta J_3)\exp(\gamma J_-) \tag{16.59}$$

for some complex parameters $\bar{z}$, θ and γ which are functions of the three real coordinates that parameterize SU(2). We let U act on the lowest-weight state $|j,-j\rangle$. The rightmost factor has no effect on the lowest weight state, and the middle factor only multiplies it by a constant. We therefore restrict our attention to the states

$$|\bar{z}\rangle = \exp(\bar{z}J_+)|j,-j\rangle, \quad \langle z| = \langle j,-j|\exp(zJ_-) = (|\bar{z}\rangle)^\dagger. \tag{16.60}$$

These states are not normalized, but have the advantage that the $\langle z|$ are holomorphic in the parameter z – i.e. they depend on z but not on $\bar{z}$.

The set of distinct $|\bar{z}\rangle$ can still be identified with the 2-sphere, and $z, \bar{z}$ are its complex stereographic coordinates. This identification is an example of a general property of compact Lie groups:

$$G/T \cong G_{\mathbb{C}}/B_+. \tag{16.61}$$

[2] A. Perelomov, *Generalized Coherent States and their Applications* (Springer-Verlag, 1986).

Here, $G_{\mathbb{C}}$ is the *complexification* of G – the group G, but with its parameters allowed to be complex – and B_+ is the Borel group whose Lie algebra consists of the Cartan algebra together with the step-up ladder operators.

The inner product of two $|\bar{z}\rangle$ states is

$$\langle z'|\bar{z}\rangle = (1+\bar{z}z')^{2j}, \tag{16.62}$$

and the eigenstates $|j,m\rangle$ of J^2 and J_3 possess *coherent state wavefunctions*:

$$\psi_m^{(1)}(z) \equiv \langle z|j,m\rangle = \sqrt{\frac{2j!}{(j-m)!(j+m)!}}\, z^{j+m}. \tag{16.63}$$

We recognize these as our spin wavefunctions from the previous section.

The over-completeness relation can be written as

$$\mathbf{I} = \frac{2j+1}{2\pi i}\int \frac{d\bar{z}\wedge dz}{(1+\bar{z}z)^{2j+2}}\,|\bar{z}\rangle\langle z|, \tag{16.64}$$

and provides the inner product for the coherent-state wavefunctions. If $\psi(z) = \langle z|\psi\rangle$ and $\chi(z) = \langle z|\chi\rangle$ then

$$\begin{aligned}\langle\psi|\chi\rangle &= \frac{2j+1}{2\pi i}\int \frac{d\bar{z}\wedge dz}{(1+\bar{z}z)^{2j+2}}\,\langle\psi|\bar{z}\rangle\langle z|\chi\rangle\\ &= \frac{2j+1}{2\pi i}\int \frac{d\bar{z}\wedge dz}{(1+\bar{z}z)^{2j+2}}\,\overline{\psi(z)}\chi(z),\end{aligned} \tag{16.65}$$

which coincides with (16.44).

The wavefunctions $\psi_m^{(1)}(z)$ are singular at the north pole, where $z=\infty$. Indeed, there is no actual state $\langle\infty|$ because the phase of this putative limiting state would depend on the direction from which we approach the point at infinity. We may, however, define a second family of coherent states:

$$|\bar{\zeta}\rangle_2 = \exp(\bar{\zeta}J_-)|j,j\rangle, \quad {}_2\langle\zeta| = \langle j,j|\exp(\zeta J_+), \tag{16.66}$$

and form the wavefunctions

$$\psi_m^{(2)}(\zeta) = {}_2\langle\zeta|j,m\rangle. \tag{16.67}$$

These new states and wavefunctions are well defined in the vicinity of the north pole, but singular near the south pole.

To find the relation between $\psi^{(2)}(\zeta)$ and $\psi^{(1)}(z)$ we note that the matrix identity

$$\begin{bmatrix}0 & -1\\ 1 & 0\end{bmatrix}\begin{bmatrix}1 & 0\\ z & 1\end{bmatrix} = \begin{bmatrix}1 & 0\\ -z^{-1} & 1\end{bmatrix}\begin{bmatrix}-z & 0\\ 0 & -z^{-1}\end{bmatrix}\begin{bmatrix}1 & z^{-1}\\ 0 & 1\end{bmatrix}, \tag{16.68}$$

coupled with the faithfulness of the spin-$\frac{1}{2}$ representation of SU(2), implies the relation

$$\widehat{w}\exp(zJ_+) = \exp(-z^{-1}J_-)(-z)^{2J_3}\exp(z^{-1}J_+), \tag{16.69}$$

where $\widehat{w} = \exp(-i\pi J_2)$. We also note that

$$\langle j,j|\widehat{w} = (-1)^{2j}\langle j,-j|, \quad \langle j,-j|\widehat{w} = \langle j,j|. \tag{16.70}$$

Thus,

$$\begin{aligned}
\psi_m^{(1)}(z) &= \langle j,-j|e^{zJ_-}|j,m\rangle \\
&= (-1)^{2j}\langle j,j|\widehat{w}\,e^{zJ_-}|j,m\rangle \\
&= (-1)^{2j}\langle j,j|e^{-z^{-1}J_-}(-z)^{2J_3}e^{z^{-1}J_+}|j,m\rangle \\
&= (-1)^{2j}(-z)^{2j}\langle j,j|e^{z^{-1}J_+}|j,m\rangle \\
&= z^{2j}\psi_m^{(2)}(z^{-1}).
\end{aligned} \tag{16.71}$$

The transition function z^{2j} that relates $\psi_m^{(1)}(z)$ to $\psi_m^{(2)}(\zeta \equiv 1/z)$ depends only on z. We therefore say that the wavefunctions $\psi_m^{(1)}(z)$ and $\psi_m^{(2)}(\zeta)$ are the local components of a global section $\psi_m \leftrightarrow |j,m\rangle$ of a *holomorphic line bundle*. The requirement that the transition function and its inverse be holomorphic and single valued in the overlap of the z and ζ coordinate patches forces $2j$ to be an integer. The ψ_m form a basis for the space of global holomorphic sections of this bundle.

Borel, Weil and Bott showed that any finite-dimensional representation of a semisimple Lie group G can be realized as the space of global holomorphic sections of a line bundle over $G_{\mathbb{C}}/B_+$. This bundle is constructed from the highest (or lowest) weight vectors in the representation by a natural generalization of the method we have used for spin. This idea has been extended by Witten and others to infinite-dimensional Lie groups, where it can be used, for example, to quantize two-dimensional gravity.

Exercise 16.2: Normalize the states $|\bar{z}\rangle$, $\langle z|$, by multiplying them by $N = (1+|z|^2)^{-j}$. Show that

$$\begin{aligned}
N^2\langle z|J_3|\bar{z}\rangle &= j\frac{|z|^2-1}{|z|^2+1}, \\
N^2\langle z|J_+|\bar{z}\rangle &= j\frac{2z}{|z|^2+1}, \\
N^2\langle z|J_-|\bar{z}\rangle &= j\frac{2\bar{z}}{|z|^2+1},
\end{aligned}$$

thus confirming the identification of $z, \bar{z}$ with the complex stereographic coordinates on the sphere.

16.3 Working in the total space

We have mostly considered a bundle to be a collection of mathematical objects and a base-space to which they are attached, rather than treating the bundle as a geometric object in its own right. In this section we demonstrate the advantages to be gained from the latter viewpoint.

16.3.1 Principal bundles and associated bundles

The fibre bundles that arise in a gauge theory with Lie group G are called *principal G-bundles*, and the fields and wavefunctions are sections of *associated* bundles. A principal G-bundle comprises the total space, which we here call P, together with the projection π to the base space M. The fibre can be regarded as a copy of G, i.e.

$$\pi : P \to M, \qquad \pi^{-1}(x) \cong G. \tag{16.72}$$

Strictly speaking, the fibre is only required to be a homogeneous space on which G acts freely and transitively on the *right*; $x \to xg$. Such a set can be identified with G after we have selected a fiducial point $f_0 \in F$ to be the group identity. There is no canonical choice for f_0 and, if the bundle is twisted, there can be no globally smooth choice. This is because a smooth choice for f_0 in the fibres above an open subset $U \subseteq M$ makes P locally into a product $U \times G$. Being able to extend U to the entirety of M means that P is trivial. We will, however, make use of local assignments $f_0 \mapsto e$ to introduce bundle coordinate charts in which P is locally a product, and therefore parametrized by ordered pairs (x, g) with $x \in U$ and $g \in G$.

To understand the bundles *associated* with P, it is simplest to define the sections of the associated bundle. Let $\varphi_i(x, g)$ be a function on the total space P with a set of indices i carrying some representation $g \mapsto D(g)$ of G. We say that $\varphi_i(x, g)$ is a section of an associated bundle if it varies in a particular way as we run up and down the fibres by acting on them from the *right* with elements of G; we require that

$$\varphi_i(x, gh) = D_{ij}(h^{-1})\varphi_j(x, g). \tag{16.73}$$

These sections can be thought of as wavefunctions for a particle moving in a gauge field on the base-space. The choice of representation D plays the role of "charge", and (16.73) are the gauge transformations. Note that we must take h^{-1} as the argument of D in order for the transformation to be consistent under group multiplication:

$$\begin{aligned}\varphi_i(x, gh_1h_2) &= D_{ij}(h_2^{-1})\varphi_j(x, gh_1)\\ &= D_{ij}(h_2^{-1})D_{jk}(h_1^{-1})\varphi_k(x, g)\\ &= D_{ik}(h_2^{-1}h_1^{-1})\varphi_k(x, g)\\ &= D_{ik}((h_1h_2)^{-1})\varphi_k(x, g).\end{aligned} \tag{16.74}$$

The construction of the associated bundle itself requires rather more abstraction. Suppose that the matrices $D(g)$ act on the vector space V. Then the total space P_V of the associated bundle consists of equivalence classes of $P \times V$ under the relation $((x,g),\mathbf{v}) \sim ((x,gh),D(h^{-1})\mathbf{v})$ for all $\mathbf{v} \in V$, $(x,g) \in P$ and $h \in G$. The set of G-action equivalence classes in a cartesian product $A \times B$ is usually denoted by $A \times_G B$. Our total space is therefore

$$P_V = P \times_G V. \tag{16.75}$$

We find it conceptually easier to work with the sections as defined above, rather than with these equivalence classes.

16.3.2 Connections

A gauge field is a connection on a principal bundle. The formal definition of a connection is a decomposition of the tangent space TP_p of P at $p \in P$ into a *horizontal subspace* $H_p(P)$ and a *vertical subspace* $V_p(P)$. We require that $V_p(P)$ be the tangent space to the fibres and $H_p(P)$ to be a complementary subspace, i.e. the direct sum should be the whole tangent space

$$TP_p = H_p(P) \oplus V_p(P). \tag{16.76}$$

The horizontal subspaces must also be invariant under the push-forward induced from the action on the fibres from the *right* of a fixed element of G. More formally, if $R[g] : P \to P$ acts to take $p \to pg$, i.e. by $R[g](x,g') = (x,g'g)$, we require that

$$R[g]_* H_p(P) = H_{pg}(P). \tag{16.77}$$

Thus, we get to choose one horizontal subspace in each fibre, the rest being determined by the right-invariance condition.

We now show how this geometric definition of a connection leads to parallel-transport. We begin with a curve $x(t)$ in the base-space. By solving the equation

$$\dot{g} + \frac{\partial x^\mu}{\partial t}\mathcal{A}_\mu(x) g = 0, \tag{16.78}$$

we can *lift* the curve $x(t)$ to a new curve $(x(t), g(t))$ in the total space, whose tangent is everywhere horizontal. This lifting operation corresponds to parallel-transporting the initial value $g(0)$ along the curve $x(t)$ to get $g(t)$. The $\mathcal{A}_\mu = \widehat{i\lambda_a} A^a_\mu$ are a set of Lie-algebra-valued functions that are determined by our choice of horizontal subspace. They are defined so that the vector $(\delta x, -\mathcal{A}_\mu \delta x^\mu g)$ is horizontal for each small displacement δx^μ in the tangent space of M. Here, $-\mathcal{A}_\mu \delta x^\mu g$ is to be understood as the displacement that takes $g \to (1 - \mathcal{A}_\mu \delta x^\mu)g$. Because we are multiplying $\mathcal{A}$ in from the *left*, the lifted curve can be slid rigidly up and down the fibres by the right action of any fixed group element. The right-invariance condition is therefore automatically satisfied.

The directional derivative along the lifted curve is

$$\dot{x}^\mu \mathcal{D}_\mu = \dot{x}^\mu \left(\left(\frac{\partial}{\partial x^\mu} \right)_g - \mathcal{A}^a_\mu R_a \right), \tag{16.79}$$

where R_a is a right-invariant vector field on G, i.e. a differential operator on functions defined on the fibres. The $\mathcal{D}_\mu$ are a set of vector fields in TP. These *covariant derivatives* span the horizontal subspace at each point $p \in P$, and have Lie brackets

$$[\mathcal{D}_\mu, \mathcal{D}_\nu] = -\mathcal{F}^a_{\mu\nu} R_a. \tag{16.80}$$

Here, $\mathcal{F}_{\mu\nu}$ is given in terms of the structure constants appearing in the Lie brackets $[R_a, R_b] = f^c_{ab} R_c$ by

$$\mathcal{F}^c_{\mu\nu} = \partial_\mu \mathcal{A}^c_\nu - \partial_\nu \mathcal{A}^c_\mu - f^c_{ab} \mathcal{A}^a_\mu \mathcal{A}^b_\nu. \tag{16.81}$$

We can also write

$$\mathcal{F}_{\mu\nu} = \partial_\mu \mathcal{A}_\nu - \partial_\nu \mathcal{A}_\mu + [\mathcal{A}_\mu, \mathcal{A}_\nu] \tag{16.82}$$

where $\mathcal{F}_{\mu\nu} = i\widehat{\lambda}_a \mathcal{F}^a_{\mu\nu}$ and $[\widehat{\lambda}_a, \widehat{\lambda}_b] = if^c_{ab} \widehat{\lambda}_c$.

Because the Lie bracket of the $\mathcal{D}_\mu$ is a linear combination of the R_a, it lies entirely in the vertical subspace. Consequently, when $\mathcal{F}_{\mu\nu} \neq 0$ the $\mathcal{D}_\mu$ are not in involution, so Frobenius' theorem tells us that the horizontal subspaces cannot fit together to form the tangent spaces to a smooth foliation of P.

We now make contact with the more familiar definition of a covariant derivative. We begin by recalling that *right*-invariant vector fields are derivatives that involve infinitesimal multiplication from the *left*. Their definition is

$$R_a \varphi_i(x, g) = \lim_{\epsilon \to 0} \frac{1}{\epsilon} \left(\varphi_i(x, (1 + i\epsilon \widehat{\lambda}_a) g) - \varphi_i(x, g) \right), \tag{16.83}$$

where $[\widehat{\lambda}_a, \widehat{\lambda}_b] = if^c_{ab} \widehat{\lambda}_c$.

As $\varphi_i(x, g)$ is a section of the associated bundle, we know how it varies when we multiply group elements in on the right. We therefore write

$$(1 + i\epsilon \widehat{\lambda}_a) g = g\, g^{-1} (1 + i\epsilon \widehat{\lambda}_a) g, \tag{16.84}$$

and from this (and writing g for $D(g)$ where it makes for compact notation) we find

$$\begin{aligned} R_a \varphi_i(x, g) &= \lim_{\epsilon \to 0} \left(D_{ij}(g^{-1}(1 - i\epsilon \widehat{\lambda}_a) g) \varphi_j(x, g) - \varphi_i(x, g) \right) / \epsilon \\ &= -D_{ij}(g^{-1})(i\widehat{\lambda}_a)_{jk} D_{kl}(g) \varphi_l(x, g) \\ &= -i(g^{-1} \widehat{\lambda}_a g)_{ij} \varphi_j. \end{aligned} \tag{16.85}$$

Here, $i(\widehat{\lambda_a})_{ij}$ is the matrix representing the Lie algebra generator $i\widehat{\lambda_a}$ in the representation $g \mapsto D(g)$. Acting on sections, we therefore have

$$\mathcal{D}_\mu \varphi = \left(\partial_\mu \varphi\right)_g + (g^{-1} \mathcal{A}_\mu g)\varphi. \tag{16.86}$$

This still does not look too familiar, because the derivatives with respect to x_μ are being taken at *fixed* g. We normally *fix a gauge* by making a choice of $g = \sigma(x)$ for each x_μ. The conventional wavefunction $\varphi(x)$ is then $\varphi(x, \sigma(x))$. We can use $\varphi(x, \sigma(x)) = \sigma^{-1}(x)\varphi(x, e)$, to obtain

$$\partial_\mu \varphi = \left(\partial_\mu \varphi\right)_\sigma + \left(\partial_\mu \sigma^{-1}\right)\sigma\varphi = \left(\partial_\mu \varphi\right)_\sigma - \left(\sigma^{-1}\partial_\mu \sigma\right)\varphi. \tag{16.87}$$

From this, we get a derivative

$$\nabla_\mu \stackrel{\text{def}}{=} \partial_\mu + (\sigma^{-1}\mathcal{A}_\mu \sigma + \sigma^{-1}\partial_\mu \sigma) = \partial_\mu + A_\mu \tag{16.88}$$

on functions $\varphi(x) \stackrel{\text{def}}{=} \varphi(x, \sigma(x))$ defined (locally) on the base-space M. This is the conventional covariant derivative, now containing gauge fields

$$A_\mu(x) = \sigma^{-1}\mathcal{A}_\mu \sigma + \sigma^{-1}\partial_\mu \sigma \tag{16.89}$$

that are gauge transformations of our g-independent $\mathcal{A}_\mu$. The derivative has been constructed so that

$$\nabla_\mu \varphi(x) = \mathcal{D}_\mu \varphi(x, g)\big|_{g=\sigma(x)}, \tag{16.90}$$

and has commutator

$$[\nabla_\mu, \nabla_\nu] = \sigma^{-1}\mathcal{F}_{\mu\nu}\sigma = F_{\mu\nu}. \tag{16.91}$$

Note the sign change *vis-à-vis* Equation (16.80).

It is the curvature tensor $F_{\mu\nu}$ that we have met previously. Recall that it provides a Lie-algebra-valued 2-form

$$F = \frac{1}{2}F_{\mu\nu}dx^\mu dx^\nu = dA + A^2 \tag{16.92}$$

on the base-space. The connection $A = A_\mu dx^\mu$ is a 1-form on the base space, and both F and A have been defined only in the region $U \subset M$ in which the smooth gauge-choice section $\sigma(x)$ has been selected.

16.3.3 Monopole harmonics

The total-space operations and definitions in these sections may seem rather abstract. We therefore demonstrate their power by solving the Schrödinger problem for a charged particle confined to a unit sphere surrounding a magnetic monopole. The conventional approach to this problem involves first selecting a gauge for the vector potential A, which, because of the monopole, is necessarily singular at a Dirac string located somewhere on the sphere, and then delving into properties of Gegenbauer polynomials. Eventually we find the gauge-dependent wavefunction. By working with the total space, however, we can solve the problem in all gauges *at once*, and the problem becomes a simple exercise in Lie-group geometry.

Recall that the SU(2) representation matrices $D^J_{mn}(\theta,\phi,\psi)$ form a complete orthonormal set of functions on the group manifold S^3. There will be a similar complete orthonormal set of representation matrices on the manifold of any compact Lie group G. Given a subgroup $H \subset G$, we will use these matrices to construct bundles associated to a principal H-bundle that has G as its total space and the coset space G/H as its base-space. The fibres will be copies of H, and the projection π the usual projection $G \to G/H$.

The functions $D^J(g)$ are not in general functions on the coset space G/H as they depend on the choice of representative. Instead, because of the representation property, they vary with the choice of representative in a well-defined way:

$$D^J_{mn}(gh) = D^J_{mn'}(g)D^J_{n'n}(h). \tag{16.93}$$

Since we are dealing with compact groups, the representations can be taken to be unitary and therefore

$$[D^J_{mn}(gh)]^* = [D^J_{mn'}(g)]^*[D^J_{n'n}(h)]^* \tag{16.94}$$

$$= D^J_{nn'}(h^{-1})[D^J_{mn'}(g)]^*. \tag{16.95}$$

This is the correct variation under the right action of the group H for the set of functions $[D^J_{mn}(gh)]^*$ to be sections of a bundle associated with the principal fibre bundle $G \to G/H$. The representation $h \mapsto D(h)$ of H is not necessarily that defined by the label J because irreducible representations of G may be reducible under H; D depends on what representation of H the index n belongs to. If D is the identity representation, then the functions are functions on G/H in the ordinary sense. For $G = \mathrm{SU}(2)$ and H being the $U(1)$ subgroup generated by J_3, the quotient space is just S^2, and the projection is the Hopf map: $S^3 \to S^2$. The resulting bundle can be called the Hopf bundle. It is not really a new object, however, because it is a generalization of the monopole bundle of the preceding section. Parametrizing SU(2) with Euler angles, so that

$$D^J_{mn}(\theta,\phi,\psi) = \langle J,m|e^{-i\phi J_3}e^{-i\theta J_2}e^{-i\psi J_3}|J,n\rangle, \tag{16.96}$$

shows that the Hopf map consists of simply forgetting about ψ, so

$$\text{Hopf} : [(\theta,\phi,\psi) \in S^3] \mapsto [(\theta,\phi) \in S^2]. \tag{16.97}$$

The bundle is twisted because S^3 is not a product $S^2 \times S^1$. Taking $n = 0$ gives us functions independent of ψ, and we obtain the well-known identification of the spherical harmonics with representation matrices

$$Y^L_m(\theta,\phi) = \sqrt{\frac{2L+1}{4\pi}}[D^{(L)}_{m0}(\theta,\phi,0)]^*. \tag{16.98}$$

For $n = \Lambda \neq 0$ we get sections of a bundle whose Chern number is 2Λ. These sections are the *monopole harmonics*:

$$\mathcal{Y}^J_{m;\Lambda}(\theta,\phi,\psi) = \sqrt{\frac{2J+1}{4\pi}}[D^J_{m\Lambda}(\theta,\phi,\psi)]^* \tag{16.99}$$

for a monopole of flux $\int eB\, d(\text{Area}) = 4\pi\Lambda$. The integrality of the Chern number tells us that the flux $4\pi\Lambda$ must be an integer multiple of 2π. This gives us a geometric reason for why the eigenvalues m of J_3 can only be an integer or half-integer.

The monopole harmonics have a non-trivial dependence, $\propto e^{i\psi\Lambda}$, on the choice we make for ψ at each point on S^2, and we cannot make a globally smooth choice; we always encounter at least one point where there is a singularity. Considered as functions on the base-space, the sections of the twisted bundle have to be constructed in patches and glued together using transition functions. As functions on the total space of the principal bundle, however, they are globally smooth.

We now show that the monopole harmonics are eigenfunctions of the Schrödinger operator $-\nabla^2$ containing the gauge field connection, just as the spherical harmonics are eigenfunctions of the Laplacian on the sphere. This is a simple geometrical exercise. Because they are irreducible representations, the $D^J(g)$ are automatically eigenfunctions of the quadratic Casimir operator

$$(J_1^2 + J_2^2 + J_3^2)D^J(g) = J(J+1)D^J(g). \tag{16.100}$$

The J_i can be either right- or left-invariant vector fields on G; the quadratic Casimir is the same second-order differential operator in either case, and it is a good guess that it is proportional to the Laplacian on the group manifold. Taking a locally geodesic coordinate system (in which the connection vanishes) confirms this: $J^2 = -\nabla^2$ on the three-sphere. The operator in (16.100) is not the Laplacian we want, however. What we need is the ∇^2 on the 2-sphere $S^2 = G/H$, including the connection. This ∇^2 operator differs from the one on the total space since it must contain only differential operators lying in the horizontal subspaces. There is a natural notion of orthogonality in the Lie group, deriving from the Killing form, and it is natural to choose the horizontal subspaces to be orthogonal to the fibres of G/H. Because multiplication on the right by the subgroup generated by J_3 moves one up and down the fibres, the orthogonal displacements are obtained by multiplication on the right by infinitesimal elements made by exponentiating J_1 and J_2. The desired ∇^2 is thus made out of the left-invariant vector fields (which act by multiplication on the right), J_1 and J_2 only. The wave operator must therefore be

$$-\nabla^2 = J_1^2 + J_2^2 = J^2 - J_3^2. \tag{16.101}$$

Applying this to the $\mathcal{Y}^J_{m;\Lambda}$, we see that they are eigenfunctions of $-\nabla^2$ on S^2 with eigenvalues $J(J+1) - \Lambda^2$. The Laplace eigenvalues for our flux $= 4\pi\Lambda$ monopole problem are therefore

$$E_{J,m} = (J(J+1) - \Lambda^2), \quad J \geq |\Lambda|, \quad -J \leq m \leq J. \tag{16.102}$$

The utility of the monopole harmonics is not restricted to exotic monopole physics. They occur in molecular and nuclear physics as the wavefunctions for the rotational degrees of freedom of diatomic molecules and uniaxially deformed nuclei that possess angular momentum Λ about their axis of symmetry.[3]

Exercise 16.3: Compare these energy levels for a particle on a sphere with those of the Landau level problem on the plane. Show that for any fixed flux the low-lying energies remain close to $E = (eB/m_{\text{particle}})(n + 1/2)$, with $n = 0, 1, \ldots$, but their degeneracy is equal to the number of flux units penetrating the sphere *plus one*.

16.3.4 Bundle connection and curvature forms

Recall that in Section 16.3.2 we introduced the Lie-algebra-valued functions $\mathcal{A}_\mu(x)$. We now use these functions to introduce the *bundle connection form* $\mathbb{A}$ that lives in T^*P. We set

$$\mathcal{A} = \mathcal{A}_\mu \, dx^\mu \tag{16.103}$$

and

$$\mathbb{A} \stackrel{\text{def}}{=} g^{-1}\left(\mathcal{A} + \delta g\, g^{-1}\right) g. \tag{16.104}$$

In these definitions, x and g are the local coordinates in which points in the total space are labelled as (x, g), and d acts on functions of x, and the "δ" is used to denote the exterior derivative acting on the fibre.[4] We have, then, that $\delta x^\mu = 0$ and $dg = 0$. The combinations $\delta g\, g^{-1}$ and $g^{-1}\delta g$ are respectively the right- and left-invariant Maurer–Cartan forms on the group.

[3] This is explained, with chararacteristic terseness, in a footnote on page 317 of L. D. Landau and E. M. Lifshitz, *Quantum Mechanics* (third edition).

[4] It is *not* therefore to be confused with the Hodge $\delta = d^\dagger$ operator.

The complete exterior derivative in the total space requires us to differentiate both with respect to g and with respect to x, and is given by $d_{\text{tot}} = d + \delta$. Because d^2, δ^2 and $(d+\delta)^2 = d^2 + \delta^2 + d\delta + \delta d$ are all zero, we must have

$$\delta d + d\delta = 0. \tag{16.105}$$

We now define the *bundle curvature form* in terms of $\mathbb{A}$ to be

$$\mathbb{F} \stackrel{\text{def}}{=} d_{\text{tot}}\mathbb{A} + \mathbb{A}^2. \tag{16.106}$$

To compute $\mathbb{F}$ in terms of $\mathcal{A}(x)$ and g we need the ingredients

$$d\mathbb{A} = g^{-1}(d\mathcal{A})g, \tag{16.107}$$

and

$$\delta\mathbb{A} = -(g^{-1}\delta g)\mathbb{A} - \mathbb{A}(g^{-1}\delta g) - (g^{-1}\delta g)^2. \tag{16.108}$$

We find that

$$\begin{aligned} \mathbb{F} = (d+\delta)\mathbb{A} + \mathbb{A}^2 &= g^{-1}\left(d\mathcal{A} + \mathcal{A}^2\right)g \\ &= g^{-1}\mathcal{F}g, \end{aligned} \tag{16.109}$$

where

$$\mathcal{F} = \frac{1}{2}\mathcal{F}_{\mu\nu}dx^\mu dx^\nu, \tag{16.110}$$

and

$$\mathcal{F}_{\mu\nu} = \partial_\mu \mathcal{A}_\nu - \partial_\nu \mathcal{A}_\mu + [\mathcal{A}_\mu, \mathcal{A}_\nu]. \tag{16.111}$$

Although we have defined the connection form $\mathbb{A}$ in terms of the local bundle coordinates (x, g), it is, in fact, an intrinsic quantity, i.e. it has a global existence, independent of the choice of these coordinates. $\mathbb{A}$ has been constructed so that

(i) A vector is annihilated by $\mathbb{A}$ if and only if it is horizontal. In particular, $\mathbb{A}(\mathcal{D}_\mu) = 0$ for all covariant derivatives $\mathcal{D}_\mu$.
(ii) The connection form is constant on *left*-invariant vector fields on the fibres. In particular, $\mathbb{A}(L_a) = \widehat{i\lambda_a}$.

Between them, the globally defined fields $\mathcal{D}_\mu \in H_p(P)$ and $L_a \in V_p(P)$ span the tangent space TP_p. Consequently the two properties listed above tell us how to evaluate $\mathbb{A}$ on any vector, and so define it uniquely and globally.

From the globally defined and gauge invariant $\mathbb{A}$ and its associated curvature $\mathbb{F}$, and for any local gauge-choice section $\sigma : (U \subset M) \to P$, we can recover the gauge-dependent base-space forms A and F as the pull-backs

$$A = \sigma^* \mathbb{A}, \quad F = \sigma^* \mathbb{F}, \tag{16.112}$$

to $U \subset M$ of the total-space forms. The resulting forms are

$$A = \left(\sigma^{-1} \mathcal{A}_\mu \sigma + \sigma^{-1} \partial_\mu \sigma\right) dx^\mu, \quad F = \frac{1}{2}\left(\sigma^{-1} \mathcal{F}_{\mu\nu} \sigma\right) dx^\mu dx^\nu, \tag{16.113}$$

and coincide with the equations connecting A_μ with $\mathcal{A}_\mu$ and $F_{\mu\nu}$ with $\mathcal{F}_{\mu\nu}$ that we obtained in Section 16.3.2. We should take care to note that the dx^μ that appear in A and F are differential forms on M, while the dx^μ that appear in $\mathcal{A}$ and $\mathcal{F}$ are differential forms on P. Now the projection π is a left inverse of the gauge-choice section σ, i.e. $\pi \circ \sigma =$ identity. The associated pull-backs are also inverses, but with the order reversed: $\sigma^* \circ \pi^* =$ identity. These maps relate the two sets of "dx^μ" by

$$dx^\mu|_M = \sigma^* \left(dx^\mu|_P\right), \quad \text{or} \quad dx^\mu|_P = \pi^* \left(dx^\mu|_M\right). \tag{16.114}$$

We now explain the advantage of knowing the total space connection and curvature forms. Consider the Chern character $\propto \operatorname{tr} F^2$ on the base-space M. We can use the bundle projection π to pull this form back to the total space. From

$$\mathbb{F}_{\mu\nu} = (g\sigma^{-1})^{-1} F_{\mu\nu} (g\sigma^{-1}), \tag{16.115}$$

we find that

$$\pi^* \left(\operatorname{tr} F^2\right) = \operatorname{tr} \mathbb{F}^2. \tag{16.116}$$

Now $\mathbb{A}$, $\mathbb{F}$ and d_{tot} have the same calculus properties as A , F and d. The manipulations that give

$$\operatorname{tr} F^2 = d \operatorname{tr}\left(A\, dA + \frac{2}{3} A^3\right)$$

also show, therefore, that

$$\operatorname{tr} \mathbb{F}^2 = d_{\text{tot}} \operatorname{tr}\left(\mathbb{A}\, d_{\text{tot}} \mathbb{A} + \frac{2}{3} \mathbb{A}^3\right). \tag{16.117}$$

There is a big difference in the significance of the computation, however. The bundle connection $\mathbb{A}$ is globally defined; consequently, the form

$$\omega_3(\mathbb{A}) \equiv \operatorname{tr}\left(\mathbb{A}\, d_{\text{tot}} \mathbb{A} + \frac{2}{3} \mathbb{A}^3\right) \tag{16.118}$$

is also globally defined. The pull-back to the total space of the Chern character is d_{tot}-exact! This miracle works for all characteristic classes: but on the base-space they are exact only when the bundle is trivial; on the total space they are always exact.

We have seen this phenomenon before, for example in Exercise 15.7. The area form $d[\text{Area}] = \sin\theta\, d\theta d\phi$ is closed but not exact on S^2. When pulled back to S^3 by the Hopf map, the area form becomes exact:

$$\text{Hopf}^* d[\text{Area}] = \sin\theta\, d\theta d\phi = d(-\cos\theta d\phi + d\psi). \tag{16.119}$$

16.3.5 Characteristic classes as obstructions

The generalized Gauss–Bonnet theorem states that, for a compact orientable even-dimensional manifold M, the integral of the Euler class over M is equal to the Euler character $\chi(M)$. Shiing-Shen Chern used the exactness of the pull-back of the Euler class to give an elegant intrinsic proof[5] of this theorem. He showed that the integral of the Euler class over M was equal to the sum of the Poincaré–Hopf indices of any tangent vector field on M, a sum we independently know to equal the Euler character $\chi(M)$. We illustrate his strategy by showing how a non-zero $\text{ch}_2(F)$ provides a similar index sum for the singularities of any section of an SU(2)-bundle over a four-dimensional base-space. This result provides an interpretation of characteristic classes as obstructions to the existence of global sections.

Let $\sigma : M \to P$ be a section of an SU(2) principal bundle P over a four-dimensional compact orientable manifold M without boundary. For any SU(n) group we have $\text{ch}_1(F) \equiv 0$, but

$$\int_M \text{ch}_2(F) = -\frac{1}{8\pi^2}\int_M \text{tr}\,(F^2) = n \tag{16.120}$$

can be non-zero.

The section σ will, in general, have points x_i where it becomes singular. We punch infinitesimal holes in M surrounding the singular points. The manifold $M' = (M\setminus\text{holes})$ will have as its boundary $\partial M'$ a disjoint union of small 3-spheres. We denote by Σ the image of M' under the map $\sigma : M' \to P$. This Σ will be a submanifold of P, whose boundary will be equal in homology to a linear combination of the boundary components of M' with integer coefficients. We show that the Chern number n is equal to the sum of these coefficients.

We begin by using the projection π to pull back $\text{ch}_2(F)$ to the bundle, where we know that

$$\pi^*\text{ch}_2(F) = -\frac{1}{8\pi^2} d_{\text{tot}}\, \omega_3(\mathbb{A}). \tag{16.121}$$

[5] S.-J. Chern, *Ann. Math.*, **47** (1946) 85. This paper is a readable classic.

Now we can decompose $\omega_3(\mathbb{A})$ into terms of different bi-degree, i.e. into terms that are p-forms in d and q-forms in δ:

$$\omega_3(\mathbb{A}) = \omega_3^0 + \omega_2^1 + \omega_1^2 + \omega_0^3. \tag{16.122}$$

Here the superscript counts the form-degree in δ, and the subscript the form-degree in d. The only term we need to know explicitly is ω_0^3. This comes from the $g^{-1}\delta g$ part of $\mathbb{A}$, and is

$$\begin{aligned}\omega_0^3 &= \mathrm{tr}\left((g^{-1}\delta g)\,\delta(g^{-1}\delta g) + \frac{2}{3}(g^{-1}\delta g)^3\right)\\ &= \mathrm{tr}\left(-(g^{-1}\delta g)^3 + \frac{2}{3}(g^{-1}\delta g)^3\right)\\ &= -\frac{1}{3}(g^{-1}\delta g)^3.\end{aligned} \tag{16.123}$$

We next use the map $\sigma : M' \to P$ to pull the right-hand side of (16.121) back from P to M'. We recall that acting on forms on M' we have $\sigma^* \circ \pi^* =$ identity. Thus

$$\begin{aligned}\int_M \mathrm{ch}_2(F) = \int_{M'} \mathrm{ch}_2(F) &= \int_{M'} \sigma^* \circ \pi^* \mathrm{ch}_2(F)\\ &= -\frac{1}{8\pi^2}\int_{M'} \sigma^* d_{\mathrm{tot}}\,\omega_3(\mathbb{A})\\ &= -\frac{1}{8\pi^2}\int_{\Sigma} d_{\mathrm{tot}}\,\omega_3(\mathbb{A})\\ &= -\frac{1}{8\pi^2}\int_{\partial\Sigma} \omega_3(\mathbb{A})\\ &= \frac{1}{24\pi^2}\int_{\partial\Sigma} (g^{-1}\delta g)^3.\end{aligned} \tag{16.124}$$

At the first step we have observed that the omitted spheres make a negligible contribution to the integral over M, and at the last step we have used the fact that the boundary of Σ has significant extent only along the fibres, so all contributions to the integral over $\partial\Sigma$ come from the purely vertical component of $\omega_3(\mathbb{A})$, which is $\omega_0^3 = -\frac{1}{3}(g^{-1}dg)$.

We know (see Exercise 15.8) that for maps $g \mapsto U \in \mathrm{SU}(2)$ we have

$$\int \mathrm{tr}\,(g^{-1}dg)^3 = 24\pi^2 \times \text{winding number}.$$

We conclude that

$$\int_M \mathrm{ch}_2(F) = \frac{1}{24\pi^2}\int_{\partial\Sigma} (g^{-1}\delta g)^3 = \sum_{\text{singularities}\, x_i} N_i \tag{16.125}$$

where N_i is the Brouwer degree of the map $\sigma : S^3 \to \mathrm{SU}(2) \cong S^3$ on the small sphere surrounding x_i.

It turns out that for any SU(n) the integral of $\mathrm{tr}\,(g^{-1}\delta g)^3$ is $24\pi^2$ times an integer winding number of g about homology spheres. The second Chern number of a SU(n)-bundle is therefore also equal to the sum of the winding-number indices of the section about its singularities. Chern's strategy can be used to relate other characteristic classes to obstructions to the existence of global sections of appropriate bundles.

16.3.6 Stora–Zumino descent equations

In the previous sections we met the forms

$$\mathbb{A} = g^{-1}\mathcal{A}g + g^{-1}\delta g \tag{16.126}$$

and

$$A = \sigma^{-1}\mathcal{A}\sigma + \sigma^{-1}d\sigma. \tag{16.127}$$

The group element g labelled points on the fibres and was independent of x, while $\sigma(x)$ was the gauge-choice section of the bundle and depended on x. The two quantities $\mathbb{A}$ and A look similar, but are not identical. A third superficially similar but distinct object is met with in the BRST (Becchi–Rouet–Stora–Tyutin) approach to quantizing gauge theories, and also in the geometric theory of anomalies. We describe it here to alert the reader to the potential for confusion.

Rather than attempting to define this new differential form rigorously, we will first explain how to calculate with it, and only then indicate what it is. We begin by considering a fixed connection form A on M, and its orbit under the action of the group $\mathcal{G}$ of gauge transformations. These elements of this infinite dimensional group are maps $g : M \to G$ equipped with pointwise product $g_1g_2(x) = g_1(x)g_2(x)$. This $g(x)$ is neither the fibre coordinate g, nor the gauge choice section $\sigma(x)$. The gauge transformation $g(x)$ acts on A to give A^g where

$$A^g = g^{-1}Ag + g^{-1}dg. \tag{16.128}$$

We now introduce an object

$$v(x) = g^{-1}\delta g, \tag{16.129}$$

and consider

$$\mathfrak{A} = A^g + v = g^{-1}Ag + g^{-1}dg + g^{-1}\delta g. \tag{16.130}$$

This 1-form appears to be a hybrid of the earlier quantities, but we will see that it has to be considered as something new. The essential difference from what has gone before is that we want v to behave like $g^{-1}\delta g$, in that $\delta v = -v^2$, and yet to depend on x. In particular we want δ to behave as an exterior derivative that implements an infinitesimal gauge transformation that takes $g \to g + \delta g$. Thus,

$$\begin{aligned}\delta(g^{-1}dg) &= -(g^{-1}\delta g)(g^{-1}dg) + g^{-1}\delta dg \\ &= -(g^{-1}\delta g)(g^{-1}dg) - (g^{-1}dg)(g^{-1}\delta g) + (g^{-1}dg)(g^{-1}\delta g) - g^{-1}d\delta g \\ &= -v(g^{-1}dg) - (g^{-1}dg)v - dv, \end{aligned} \tag{16.131}$$

and hence

$$\delta A^g = -vA^g - A^g v - dv. \tag{16.132}$$

Previously $g^{-1}dg \equiv 0$, and so there was no "dv" in δ(gauge field).

We can define a curvature associated with $\mathfrak{A}$

$$\mathfrak{F} \stackrel{\text{def}}{=} d_{\text{tot}}\mathfrak{A} + \mathfrak{A}^2, \tag{16.133}$$

and compute

$$\begin{aligned}\mathfrak{F} &= (d+\delta)(A^g + v) + (A^g + v)^2 \\ &= dA^g + dv + \delta A^g + \delta v + (A^g)^2 + A^g v + vA^g + v^2 \\ &= dA^g + (A^g)^2 \\ &= g^{-1}Fg. \end{aligned} \tag{16.134}$$

Stora calls (16.134) the *Russian formula*.

Because $\mathfrak{F}$ is yet another gauge transform of F, we have

$$\operatorname{tr} F^2 = \operatorname{tr} \mathfrak{F}^2 = (d+\delta) \operatorname{tr}\left(\mathfrak{A}(d+\delta)\mathfrak{A} + \frac{2}{3}\mathfrak{A}^3\right) \tag{16.135}$$

and can decompose the right-hand side into terms that are simultaneously p-foms in d and q-forms in δ.

The left-hand side, $\operatorname{tr} \mathfrak{F}^2 = \operatorname{tr} F^2$, of (16.135) is independent of v. The right-hand side of (16.135) contains $\omega_3(\mathfrak{A})$ which we expand as

$$\omega_3(A^g + v) = \omega_3^0(A^g) + \omega_2^1(v, A^g) + \omega_1^2(v, A^g) + \omega_0^3(v). \tag{16.136}$$

As in the previous section, the superscript counts the form-degree in δ, and the subscript the form-degree in d. Explicit computation shows that

$$\begin{aligned}\omega_3^0(A^g) &= \operatorname{tr}\left(A^g\, dA^g + \tfrac{2}{3}(A^g)^3\right), \\ \omega_2^1(v, A^g) &= \operatorname{tr}(v\, dA^g), \\ \omega_1^2(v, A^g) &= -\operatorname{tr}(A^g v^2), \\ \omega_0^3(v) &= -\tfrac{1}{3}v^3. \end{aligned} \tag{16.137}$$

For example,

$$\omega_0^3(v) = \mathrm{tr}\left(v\,\delta v + \frac{2}{3}v^3\right) = \mathrm{tr}\left(v(-v^2) + \frac{2}{3}v^3\right) = -\frac{1}{3}v^3. \tag{16.138}$$

With this decomposition, (16.117) falls apart into the chain of *descent equations*

$$\begin{aligned} \mathrm{tr}\,F^2 &= \ d\omega_3^0(A^g), \\ \delta\omega_3^0(A^g) &= -d\,\omega_2^1(v, A^g), \\ \delta\omega_2^1(v, A^g) &= -d\,\omega_1^2(v, A^g), \\ \delta\omega_1^2(v, A^g) &= -d\,\omega_0^3(v), \\ \delta\omega_0^3(v) &= \ 0. \end{aligned} \tag{16.139}$$

Let us verify, for example, the penultimate equation $\delta\omega_1^2(v, A^g) = -d\,\omega_0^3(v)$. The left-hand side is

$$-\delta\,\mathrm{tr}\,(A^g v^2) = -\mathrm{tr}\,(-Av^3 - vA^g v^2 - dv\,v^2) = \mathrm{tr}\,(dv\,v^2), \tag{16.140}$$

the terms involving A^g having cancelled *via* the cyclic property of the trace and the fact that A^g anticommutes with v. The right-hand side is

$$-d\left(-\tfrac{1}{3}\mathrm{tr}\,v^3\right) = \mathrm{tr}\,(dv\,v^2) \tag{16.141}$$

as required.

The descent equations were introduced by Raymond Stora and Bruno Zumino as a tool for obtaining and systematizing information about *anomalies* in the quantum field theory of fermions interacting with the gauge field A^g. The $\omega_p^q(v, A^g)$ are p-forms in the dx^μ, and before use they are integrated over p-cycles in M. This process is understood to produce local functionals of A^g that remain q-forms in δg. For example, in $2n$ space-time dimensions, the integral

$$I[g^{-1}\delta g, A^g] = \int_M \omega_{2n}^1(g^{-1}\delta g, A^g) \tag{16.142}$$

has the properties required for it to be a candidate for the anomalous variation $\delta S[A^g]$ of the fermion effective action due to an infinitesimal gauge transformation $g \to g + \delta g$. In particular, when $\partial M = \emptyset$, we have

$$\delta I[g^{-1}\delta g, A^g] = \int_M \delta\omega_{2n}^1(v, A^g) = -\int_M d\omega_{2n-1}^2(v, A^g) = 0. \tag{16.143}$$

This is the *Wess–Zumino consistency condition* that $\delta(\delta S)$ must obey as a consequence of $\delta^2 = 0$.

In addition to producing a convenient solution of the Wess–Zumino condition, the descent equations provide a compact derivation of the gauge transformation properties of useful differential forms. We will not seek to explain further the physical meaning of these forms, leaving this to a quantum field theory course.

The similarity between $\mathbb{A}$ and $\mathfrak{A}$ led various authors to attempt to identify them, and in particular to identify $v(x)$ with the $g^{-1}\delta g$ Maurer–Cartan form appearing in $\mathbb{A}$. However the physical meaning of expressions such as $d(g^{-1}\delta g)$ precludes such a simple interpretation. In evaluating $dv \sim d(g^{-1}\delta g)$ on a vector field $\xi^a(x)L_a$ representing an infinitesimal gauge transformation, we first insert the field into $v \sim g^{-1}\delta g$ to obtain the x-dependent Lie algebra element $i\xi^a(x)\widehat{\lambda}_a$, and only then take the exterior derivative to obtain $i\widehat{\lambda}_a\partial_\mu\xi^a\, dx^\mu$. The result therefore involves derivatives of the components $\xi^a(x)$. The evaluation of an ordinary differential form on a vector field never produces derivatives of the vector components.

To understand what the Stora–Zumino forms are, imagine that we equip a two-dimensional fibre bundle $E = M \times F$ with base-space coordinate x and fibre coordinate y. A $p = 1$, $q = 1$ form on E will then be $F = f(x,y)\, dx\, \delta y$ for some function $f(x,y)$. There is only one object δy, and there is no meaning to integrating F over x to leave a 1-form in δy on E. The space of forms introduced by Stora and Zumino, on the other hand, would contain elements such as

$$J = \int_M j(x,y)\, dx\, \delta y_x \tag{16.144}$$

where there is a distinct δy_x for each $x \in M$. If we take, for example, $j(x,y) = \delta'(x-a)$, we evaluate J on the vector field $Y(x,y)\partial_y$ as

$$J[Y(x,y)\partial_y] = \int \delta'(x-a)Y(x,y)\, dx = -Y'(a,y). \tag{16.145}$$

The conclusion is that the 1-form field $v(x) \sim g^{-1}\delta g$ must be considered as the left-invariant Maurer–Cartan form on the infinite dimensional Lie group $\mathcal{G}$, rather than a Maurer–Cartan form on the finite dimensional Lie group G. The $\int_M \omega_{2n}^q(v, A^g)$ are therefore elements of the cohomology group $H^q(A^{\mathcal{G}})$ of the $\mathcal{G}$ orbit of A, a rather complicated object. For a thorough discussion see: J. A. de Azcárraga, J. M. Izquierdo, *Lie groups, Lie Algebras, Cohomology and some Applications in Physics* (Cambridge University Press).

17

Complex analysis

Although this chapter is called complex *analysis*, we will try to develop the subject as complex *calculus* – meaning that we shall follow the calculus course tradition of telling you how to do things, and explaining why theorems are true, with arguments that would not pass for rigorous proofs in a course on real analysis. We try, however, to tell no lies.

This chapter will focus on the basic ideas that need to be understood before we apply complex methods to evaluating integrals, analysing data and solving differential equations.

17.1 Cauchy–Riemann equations

We focus on functions, $f(z)$, of a single complex variable, z, where $z = x + iy$. We can think of these as being complex valued functions of two real variables, x and y. For example

$$\begin{aligned} f(z) = \sin z \equiv \sin(x+iy) &= \sin x \cos iy + \cos x \sin iy \\ &= \sin x \cosh y + i \cos x \sinh y. \end{aligned} \tag{17.1}$$

Here, we have used

$$\begin{aligned} \sin x &= \frac{1}{2i}\left(e^{ix} - e^{-ix}\right), \quad \sinh x = \frac{1}{2}\left(e^{x} - e^{-x}\right), \\ \cos x &= \frac{1}{2}\left(e^{ix} + e^{-ix}\right), \quad \cosh x = \frac{1}{2}\left(e^{x} + e^{-x}\right), \end{aligned}$$

to make the connection between the circular and hyperbolic functions. We shall often write $f(z) = u + iv$, where u and v are real functions of x and y. In the present example, $u = \sin x \cosh y$ and $v = \cos x \sinh y$.

If all four partial derivatives

$$\frac{\partial u}{\partial x}, \quad \frac{\partial v}{\partial y}, \quad \frac{\partial v}{\partial x}, \quad \frac{\partial u}{\partial y}, \tag{17.2}$$

exist and are continuous, then $f = u + iv$ is differentiable as a complex-valued function of two real variables. This means that we can approximate the variation in f as

$$\delta f = \frac{\partial f}{\partial x}\delta x + \frac{\partial f}{\partial y}\delta y + \cdots, \tag{17.3}$$

where the dots represent a remainder that goes to zero faster than linearly as δx, δy go to zero. We now regroup the terms, setting $\delta z = \delta x + i\delta y$, $\delta \bar{z} = \delta x - i\delta y$, so that

$$\delta f = \frac{\partial f}{\partial z}\delta z + \frac{\partial f}{\partial \bar{z}}\delta \bar{z} + \cdots, \tag{17.4}$$

where we have defined

$$\begin{aligned}\frac{\partial f}{\partial z} &= \frac{1}{2}\left(\frac{\partial f}{\partial x} - i\frac{\partial f}{\partial y}\right),\\ \frac{\partial f}{\partial \bar{z}} &= \frac{1}{2}\left(\frac{\partial f}{\partial x} + i\frac{\partial f}{\partial y}\right).\end{aligned} \tag{17.5}$$

Now our function $f(z)$ does not depend on $\bar{z}$, and so it must satisfy

$$\frac{\partial f}{\partial \bar{z}} = 0. \tag{17.6}$$

Thus, with $f = u + iv$,

$$\frac{1}{2}\left(\frac{\partial}{\partial x} + i\frac{\partial}{\partial y}\right)(u + iv) = 0 \tag{17.7}$$

i.e.

$$\left(\frac{\partial u}{\partial x} - \frac{\partial v}{\partial y}\right) + i\left(\frac{\partial v}{\partial x} + \frac{\partial u}{\partial y}\right) = 0. \tag{17.8}$$

Since the vanishing of a complex number requires the real and imaginary parts to be separately zero, this implies that

$$\begin{aligned}\frac{\partial u}{\partial x} &= +\frac{\partial v}{\partial y},\\ \frac{\partial v}{\partial x} &= -\frac{\partial u}{\partial y}.\end{aligned} \tag{17.9}$$

These two relations between u and v are known as the *Cauchy–Riemann equations*, although they were probably discovered by Gauss. If our continuous partial derivatives satisfy the Cauchy–Riemann equations at $z_0 = x_0 + iy_0$ then we say that the function is *complex differentiable* (or just differentiable) at that point. By taking $\delta z = z - z_0$, we have

$$\delta f \stackrel{\text{def}}{=} f(z) - f(z_0) = \frac{\partial f}{\partial z}(z - z_0) + \cdots, \tag{17.10}$$

where the remainder, represented by the dots, tends to zero faster than $|z-z_0|$ as $z \to z_0$. The validity of this linear approximation to the variation in $f(z)$ is equivalent to the statement that the ratio

$$\frac{f(z)-f(z_0)}{z-z_0} \tag{17.11}$$

tends to a definite limit as $z \to z_0$ from any direction. It is the direction-independence of this limit that provides a proper meaning to the phrase "does not depend on $\bar{z}$". Since we are not allowing dependence on $\bar{z}$, it is natural to drop the partial derivative signs and write the limit as an ordinary derivative

$$\lim_{z\to z_0} \frac{f(z)-f(z_0)}{z-z_0} = \frac{df}{dz}. \tag{17.12}$$

We will also use Newton's fluxion notation

$$\frac{df}{dz} = f'(z). \tag{17.13}$$

The complex derivative obeys exactly the same calculus rules as ordinary real derivatives:

$$\begin{aligned} \frac{d}{dz} z^n &= nz^{n-1}, \\ \frac{d}{dz} \sin z &= \cos z, \\ \frac{d}{dz}(fg) &= \frac{df}{dz} g + f\frac{dg}{dz}, \quad \text{etc.} \end{aligned} \tag{17.14}$$

If the function is differentiable at all points in an arcwise-connected[1] open set, or *domain*, D, the function is said to be *analytic* there. The words *regular* or *holomorphic* are also used.

17.1.1 Conjugate pairs

The functions u and v comprising the real and imaginary parts of an analytic function are said to form a pair of *harmonic conjugate functions*. Such pairs have many properties that are useful for solving physical problems.

From the Cauchy–Riemann equations we deduce that

$$\begin{aligned} \left(\frac{\partial^2}{\partial x^2} + \frac{\partial^2}{\partial y^2}\right) u &= 0, \\ \left(\frac{\partial^2}{\partial x^2} + \frac{\partial^2}{\partial y^2}\right) v &= 0 \end{aligned} \tag{17.15}$$

[1] *Arcwise-connected* means that any two points in D can be joined by a continuous path that lies wholly within D.

and so both the real and imaginary parts of $f(z)$ are automatically *harmonic* functions of x, y.

Further, from the Cauchy–Riemann conditions, we deduce that

$$\frac{\partial u}{\partial x}\frac{\partial v}{\partial x}+\frac{\partial u}{\partial y}\frac{\partial v}{\partial y}=0. \tag{17.16}$$

This means that $\nabla u\cdot\nabla v=0$. We conclude that, provided that neither of these gradients vanishes, the pair of curves $u=$ const. and $v=$ const. intersect at right angles. If we regard u as the potential ϕ solving some electrostatics problem $\nabla^2\phi=0$, then the curves $v=$ const. are the associated field lines.

Another application is to fluid mechanics. If $\mathbf{v}$ is the velocity field of an irrotational (curl $\mathbf{v}=\mathbf{0}$) flow, then we can (perhaps only locally) write the flow field as a gradient

$$\begin{aligned} v_x &= \partial_x\phi, \\ v_y &= \partial_y\phi, \end{aligned} \tag{17.17}$$

where ϕ is a *velocity potential*. If the flow is incompressible (div $\mathbf{v}=0$), then we can (locally) write it as a curl

$$\begin{aligned} v_x &= \partial_y\chi, \\ v_y &= -\partial_x\chi, \end{aligned} \tag{17.18}$$

where χ is a *stream function*. The curves $\chi=$ const. are the flow streamlines. If the flow is both irrotational and incompressible, then we may use either ϕ or χ to represent the flow, and, since the two representations must agree, we have

$$\begin{aligned} \partial_x\phi &= +\partial_y\chi, \\ \partial_y\phi &= -\partial_x\chi. \end{aligned} \tag{17.19}$$

Thus ϕ and χ are harmonic conjugates, and so the complex combination $\Phi=\phi+i\chi$ is an analytic function called the *complex stream function*.

A conjugate v exists (at least locally) for any harmonic function u. To see why, assume first that we have a (u,v) pair obeying the Cauchy–Riemann equations. Then we can write

$$\begin{aligned} dv &= \frac{\partial v}{\partial x}dx+\frac{\partial v}{\partial y}dy \\ &= -\frac{\partial u}{\partial y}dx+\frac{\partial u}{\partial x}dy. \end{aligned} \tag{17.20}$$

This observation suggests that if we are given a harmonic function u in some simply connected domain D, we can *define* a v by setting

$$v(z)=\int_{z_0}^{z}\left(-\frac{\partial u}{\partial y}dx+\frac{\partial u}{\partial x}dy\right)+v(z_0), \tag{17.21}$$

for some real constant $v(z_0)$ and point z_0. The integral does not depend on choice of path from z_0 to z, and so $v(z)$ is well defined. The path independence comes about because the curl

$$\frac{\partial}{\partial y}\left(-\frac{\partial u}{\partial y}\right)-\frac{\partial}{\partial x}\left(\frac{\partial u}{\partial x}\right)=-\nabla^2 u \tag{17.22}$$

vanishes, and because in a simply connected domain all paths connecting the same endpoints are homologous.

We now verify that this candidate $v(z)$ satisfies the Cauchy–Riemann relations. The path independence allows us to make our final approach to $z = x + iy$ along a straight line segment lying on either the x- or y-axis. If we approach along the x-axis, we have

$$v(z)=\int^{x}\left(-\frac{\partial u}{\partial y}\right)dx' + \text{ rest of integral,} \tag{17.23}$$

and may use

$$\frac{d}{dx}\int^{x} f(x',y)\,dx' = f(x,y) \tag{17.24}$$

to see that

$$\frac{\partial v}{\partial x}=-\frac{\partial u}{\partial y} \tag{17.25}$$

at (x,y). If, instead, we approach along the y-axis, we may similarly compute

$$\frac{\partial v}{\partial y}=\frac{\partial u}{\partial x}. \tag{17.26}$$

Thus $v(z)$ does indeed obey the Cauchy–Riemann equations.

Because of the utility of the harmonic conjugate it is worth giving a practical recipe for finding it, and so obtaining $f(z)$ when given only its real part $u(x,y)$. The method we give below is one we learned from John d'Angelo. It is more efficient than those given in most textbooks. We first observe that if f is a function of z only, then $\overline{f(z)}$ depends only on $\overline{z}$. We can therefore define a function $\overline{f}$ of $\overline{z}$ by setting $\overline{f(z)}=\overline{f}(\overline{z})$. Now

$$\frac{1}{2}\left(f(z)+\overline{f(z)}\right)=u(x,y). \tag{17.27}$$

Set

$$x=\frac{1}{2}(z+\overline{z}),\quad y=\frac{1}{2i}(z-\overline{z}), \tag{17.28}$$

so

$$u\left(\frac{1}{2}(z+\overline{z}),\frac{1}{2i}(z-\overline{z})\right)=\frac{1}{2}\left(f(z)+\overline{f}(\overline{z})\right). \tag{17.29}$$

Now set $\bar{z} = 0$, while keeping z fixed! Thus

$$f(z) + \bar{f}(0) = 2u\left(\frac{z}{2}, \frac{z}{2i}\right). \tag{17.30}$$

The function f is not completely determined of course, because we can always add a constant to v, and so we have the result

$$f(z) = 2u\left(\frac{z}{2}, \frac{z}{2i}\right) + iC, \quad C \in \mathbb{R}. \tag{17.31}$$

For example, let $u = x^2 - y^2$. We find

$$f(z) + \bar{f}(0) = 2\left(\frac{z}{2}\right)^2 - 2\left(\frac{z}{2i}\right)^2 = z^2, \tag{17.32}$$

or

$$f(z) = z^2 + iC, \quad C \in \mathbb{R}. \tag{17.33}$$

The business of setting $\bar{z} = 0$, while keeping z fixed, may feel like a dirty trick, but it can be justified by the (as yet to be proved) fact that f has a convergent expansion as a power series in $z = x + iy$. In this expansion it is meaningful to let x and y themselves be complex, and so allow z and $\bar{z}$ to become two independent complex variables. Anyway, you can always check *ex post facto* that your answer is correct.

17.1.2 Conformal mapping

An analytic function $w = f(z)$ maps subsets of its domain of definition in the "z" plane on to subsets in the "w" plane. These maps are often useful for solving problems in two-dimensional electrostatics or fluid flow. Their simplest property is geometrical: such maps are *conformal* (Figure 17.1).

Suppose that the derivative of $f(z)$ at a point z_0 is non-zero. Then, for z near z_0 we have

$$f(z) - f(z_0) \approx A(z - z_0), \tag{17.34}$$

where

$$A = \left.\frac{df}{dz}\right|_{z_0}. \tag{17.35}$$

If you think about the geometric interpretation of complex multiplication (multiply the magnitudes, add the arguments) you will see that the "f" image of a small neighbourhood of z_0 is stretched by a factor $|A|$, and rotated through an angle $\arg A$ – but relative angles are not altered. The map $z \mapsto f(z) = w$ is therefore *isogonal*. Our map also preserves

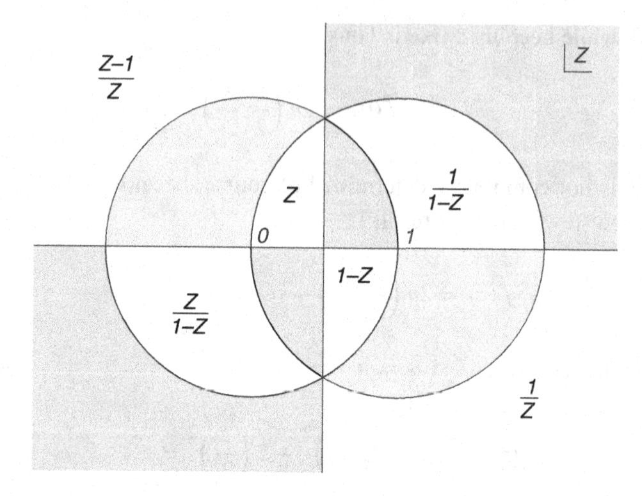

Figure 17.1 An illustration of conformal mapping. The unshaded "triangle" marked z is mapped into the other five unshaded regions by the functions labelling them. Observe that although the regions are distorted, the angles of the "triangle" are preserved by the maps (with the exception of those corners that get mapped to infinity).

orientation (the sense of rotation of the relative angle) and these two properties, isogonality and orientation-preservation, are what make the map conformal.[2] The conformal property fails at points where the derivative vanishes or becomes infinite.

If we can find a conformal map $z\ (\equiv x+iy) \mapsto w\ (\equiv u+iv)$ of some domain D to another D' then a function $f(z)$ that solves a potential theory problem (a Dirichlet boundary value problem, for example) in D will lead to $f(z(w))$ solving an analogous problem in D'.

Consider, for example, the map $z \mapsto w = z + e^z$. This map takes the strip $-\infty < x < \infty$, $-\pi \le y \le \pi$ to the entire complex plane with cuts from $-\infty + i\pi$ to $-1 + i\pi$ and from $-\infty - i\pi$ to $-1 - i\pi$. The cuts occur because the images of the lines $y = \pm\pi$ get folded back on themselves at $w = -1 \pm i\pi$, where the derivative of $w(z)$ vanishes (see Figure 17.2).

In this case, the imaginary part of the function $f(z) = x + iy$ trivially solves the Dirichlet problem $\nabla^2_{x,y} y = 0$ in the infinite strip, with $y = \pi$ on the upper boundary and $y = -\pi$ on the lower boundary. The function $y(u, v)$, now quite non-trivially, solves $\nabla^2_{u,v} y = 0$ in the entire w plane, with $y = \pi$ on the half-line running from $-\infty + i\pi$ to $-1 + i\pi$, and $y = -\pi$ on the half-line running from $-\infty - i\pi$ to $-1 - i\pi$. We may regard the images of the lines $y = \text{const.}$ (solid curves) as being the streamlines of an irrotational and incompressible flow out of the end of a tube into an infinite region, or as the equipotentials near the edge of a pair of capacitor plates. In the latter case, the images of the lines $x = \text{const.}$ (dotted curves) are the corresponding field-lines

[2] If f were a function of $\bar{z}$ only, then the map would still be isogonal, but would reverse the orientation. We call such maps *antiholomorphic* or *anticonformal*.

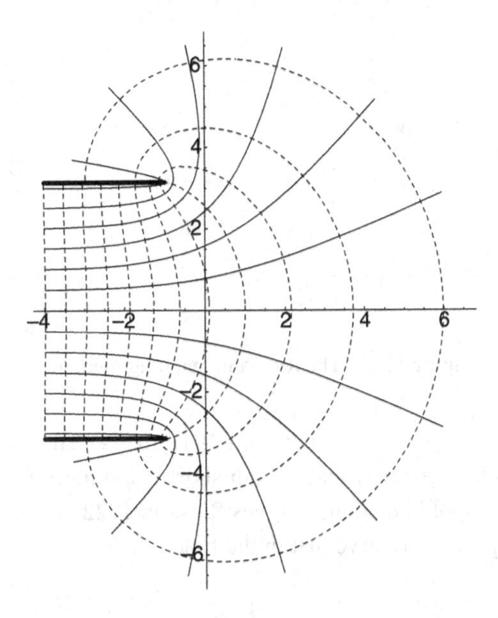

Figure 17.2 Image of part of the strip $-\pi \leq y \leq \pi$, $-\infty < x < \infty$ under the map $z \mapsto w = z + e^z$.

Example: *The Joukowski map*. This map is famous in the history of aeronautics because it can be used to map the exterior of a circle to the exterior of an aerofoil-shaped region. We can use the *Milne–Thomson circle theorem* (see Section 17.3.2) to find the streamlines for the flow past a circle in the z plane, and then use Joukowski's transformation,

$$w = f(z) = \frac{1}{2}\left(z + \frac{1}{z}\right), \tag{17.36}$$

to map this simple flow to the flow past the aerofoil. To produce an aerofoil shape, the circle must go through the point $z = 1$, where the derivative of f vanishes, and the image of this point becomes the sharp trailing edge of the aerofoil.

The Riemann mapping theorem

There are tables of conformal maps for D, D' pairs, but an underlying principle is provided by the Riemann mapping theorem (Figure 17.3):

Theorem: *The interior of any simply connected domain D in $\mathbb{C}$ whose boundary consists of more that one point can be mapped conformally one-to-one and onto the interior of the unit circle. It is possible to choose an arbitrary interior point w_0 of D and map it to the origin, and to take an arbitrary direction through w_0 and make it the direction of the real axis. With these two choices the mapping is* unique.

This theorem was first stated in Riemann's PhD thesis in 1851. He regarded it as "obvious" for the reason that we will give as a physical "proof". Riemann's argument

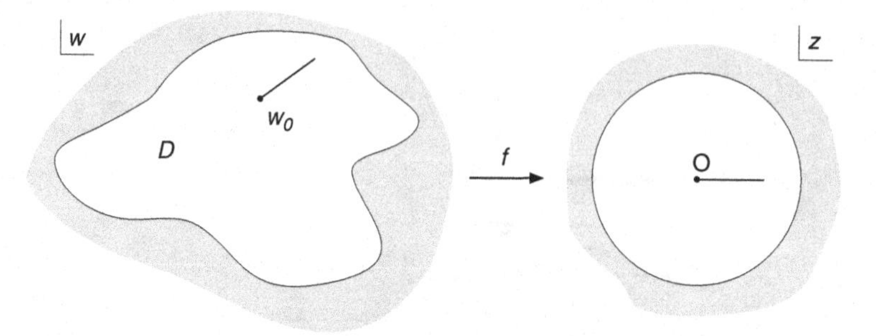

Figure 17.3 The Riemann mapping theorem.

is not rigorous, however, and it was not until 1912 that a real proof was obtained by Constantin Carathéodory. A proof that is both shorter and more in spirit of Riemann's ideas was given by Leopold Fejér and Frigyes Riesz in 1922.

For the physical "proof", observe that in the function

$$-\frac{1}{2\pi}\ln z = -\frac{1}{2\pi}\{\ln|z| + i\theta\}, \tag{17.37}$$

the real part $\phi = -\frac{1}{2\pi}\ln|z|$ is the potential of a unit charge at the origin, and with the additive constant chosen so that $\phi = 0$ on the circle $|z| = 1$. Now imagine that we have solved the two-dimensional electrostatics problem of finding the potential for a unit charge located at $w_0 \in D$, also with the boundary of D being held at zero potential. We have

$$\nabla^2\phi_1 = -\delta^2(w - w_0), \quad \phi_1 = 0 \quad \text{on} \quad \partial D. \tag{17.38}$$

Now find the ϕ_2 that is harmonically conjugate to ϕ_1. Set

$$\phi_1 + i\phi_2 = \Phi(w) = -\frac{1}{2\pi}\ln(ze^{i\alpha}) \tag{17.39}$$

where α is a real constant. We see that the transformation $w \mapsto z$, or

$$z = e^{-i\alpha}e^{-2\pi\Phi(w)}, \tag{17.40}$$

does the job of mapping the interior of D into the interior of the unit circle, and the boundary of D to the boundary of the unit circle. Note how our freedom to choose the constant α is what allows us to "take an arbitrary direction through w_0 and make it the direction of the real axis".

Example: To find the map that takes the upper half-plane into the unit circle, with the point $z = i$ mapping to the origin, we use the method of images to solve for the complex potential of a unit charge at $w = i$:

$$\phi_1 + i\phi_2 = -\frac{1}{2\pi}(\ln(w - i) - \ln(w + i))$$
$$= -\frac{1}{2\pi}\ln(e^{i\alpha}z).$$

Therefore

$$z = e^{-i\alpha}\frac{w - i}{w + i}. \tag{17.41}$$

We immediately verify that this works: we have $|z| = 1$ when w is real, and $z = 0$ at $w = i$.

The difficulty with the physical argument is that it is not clear that a solution to the point-charge electrostatics problem exists. In three dimensions, for example, there is no solution when the boundary has a sharp inward directed spike. (We cannot physically realize such a situation either: the electric field becomes unboundedly large near the tip of a spike, and boundary charge will leak off and neutralize the point charge.) There might well be analogous difficulties in two dimensions if the boundary of D is pathological. However, the fact that there *is* a proof of the Riemann mapping theorem shows that the two-dimensional electrostatics problem does always have a solution, at least in the *interior* of D – even if the boundary is an infinite-length fractal. However, unless ∂D is reasonably smooth the resulting Riemann map cannot be continuously extended to the boundary. When the boundary of D *is* a smooth closed curve, then the boundary of D *will* map one-to-one and continuously onto the boundary of the unit circle.

Exercise 17.1: *Van der Pauw's theorem.*[3] This problem explains a practical method of determining the conductivity σ of a material, given a sample in the form of a wafer of uniform thickness d, but of irregular shape (Figure 17.4). In practice at the Phillips company in Eindhoven, this was a wafer of semiconductor cut from an unmachined boule.

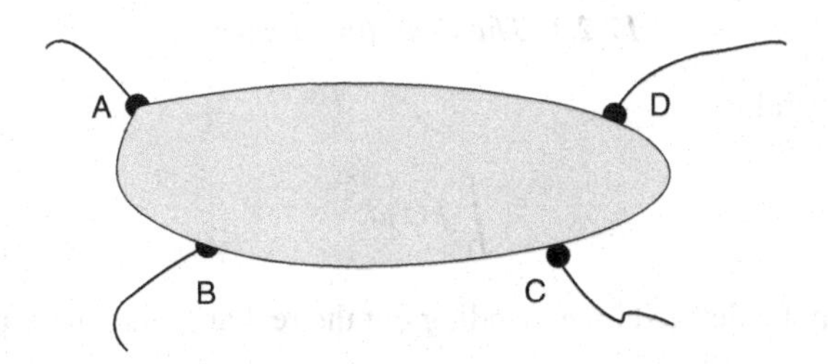

Figure 17.4 A thin semiconductor wafer with attached leads.

[3] L. J. Van der Pauw, *Phillips Research Reps.*, **13** (1958) 1. See also A. M. Thompson, D. G. Lampard, *Nature*, **177** (1956) 888, and D. G. Lampard. *Proc. Inst. Elec. Eng. C.*, **104** (1957) 271, for the "calculable capacitor".

We attach leads to point contacts A, B, C, D, taken in anticlockwise order, on the periphery of the wafer and drive a current I_{AB} from A to B. We record the potential difference $V_D - V_C$ and so find $R_{AB,DC} = (V_D - V_C)/I_{AB}$. Similarly we measure $R_{BC,AD}$. The current flow in the wafer is assumed to be two dimensional, and to obey

$$\mathbf{J} = -(\sigma d)\nabla V, \qquad \nabla \cdot \mathbf{J} = 0,$$

and $\mathbf{n} \cdot \mathbf{J} = 0$ at the boundary (except at the current source and drain). The potential V is therefore harmonic, with Neumann boundary conditions.

Van der Pauw claims that

$$\exp\{-\pi\sigma d R_{AB,DC}\} + \exp\{-\pi\sigma d R_{BC,AD}\} = 1.$$

From this σd can be found numerically.

(a) First show that Van der Pauw's claim is true if the wafer were the entire upper half-plane with A, B, C, D on the real axis with $x_A < x_B < x_C < x_D$.

(b) Next, taking care to consider the transformation of the current source terms and the Neumann boundary conditions, show that the claim is invariant under conformal maps, and, by mapping the wafer to the upper half-plane, show that it is true in general.

17.2 Complex integration: Cauchy and Stokes

In this section we will define the integral of an analytic function, and make contact with the exterior calculus from Chapters 11–13. The most obvious difference between the real and complex integral is that in evaluating the definite integral of a function in the complex plane we must specify the path along which we integrate. When this path of integration is the boundary of a region, it is often called a *contour* from the use of the word in the graphic arts to describe the outline of something. The integrals themselves are then called *contour integrals*.

17.2.1 The complex integral

The complex integral

$$\int_\Gamma f(z)dz \tag{17.42}$$

over a path Γ may be defined by expanding out the real and imaginary parts

$$\int_\Gamma f(z)dz \stackrel{\text{def}}{=} \int_\Gamma (u+iv)(dx+idy) = \int_\Gamma (udx - vdy) + i\int_\Gamma (vdx + udy) \tag{17.43}$$

and treating the two integrals on the right-hand side as standard vector-calculus line integrals of the form $\int \mathbf{v} \cdot d\mathbf{r}$, one with $\mathbf{v} \to (u, -v)$ and one with $\mathbf{v} \to (v, u)$.

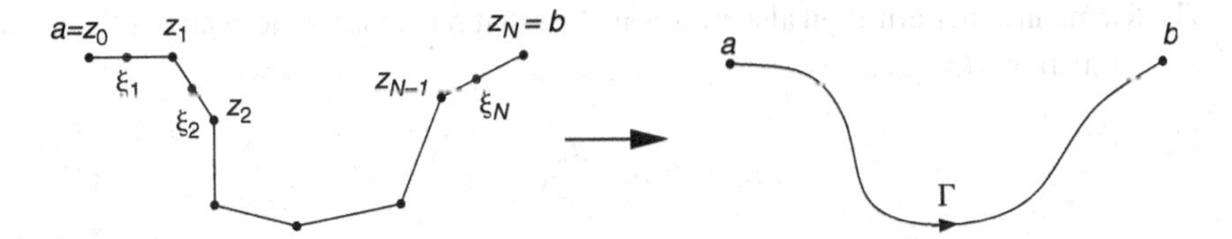

Figure 17.5 A chain approximation to the curve Γ.

The complex integral can also be constructed as the limit of a Riemann sum in a manner parallel to the definition of the real-variable Riemann integral of elementary calculus. Replace the path Γ with a chain composed of N line segments z_0-to-z_1, z_1-to-z_2, all the way to z_{N-1}-to-z_N (Figure 17.5). Now let ξ_m lie on the line segment joining z_{m-1} and z_m. Then the integral $\int_\Gamma f(z)dz$ is the limit of the (Riemann) sum

$$S = \sum_{m=1}^{N} f(\xi_m)(z_m - z_{m-1}) \tag{17.44}$$

as N gets large and all the $|z_m - z_{m-1}| \to 0$. For this definition to make sense and be useful, the limit must be independent of both how we chop up the curve and how we select the points ξ_m. This will be the case when the integration path is smooth and the function being integrated is continuous.

The Riemann sum definition of the integral leads to a useful inequality: combining the triangle inequality $|a+b| \leq |a| + |b|$ with $|ab| = |a|\,|b|$ we deduce that

$$\begin{aligned}\left|\sum_{m=1}^{N} f(\xi_m)(z_m - z_{m-1})\right| &\leq \sum_{m=1}^{N} |f(\xi_m)(z_m - z_{m-1})| \\ &= \sum_{m=1}^{N} |f(\xi_m)|\,|(z_m - z_{m-1})|. \end{aligned} \tag{17.45}$$

For sufficiently smooth curves the last sum converges to the real integral $\int_\Gamma |f(z)|\,|dz|$, and we deduce that

$$\left|\int_\Gamma f(z)\,dz\right| \leq \int_\Gamma |f(z)|\,|dz|. \tag{17.46}$$

For curves Γ that are smooth enough to have a well-defined length $|\Gamma|$, we will have $\int_\Gamma |dz| = |\Gamma|$. From this identification we conclude that if $|f| \leq M$ on Γ, then we have the *Darboux inequality*

$$\left|\int_\Gamma f(z)\,dz\right| \leq M|\Gamma|. \tag{17.47}$$

We shall find many uses for this inequality.

The Riemann sum definition also makes it clear that if $f(z)$ is the derivative of another analytic function $g(z)$, i.e.

$$f(z) = \frac{dg}{dz}, \tag{17.48}$$

then, for Γ a smooth path from $z = a$ to $z = b$, we have

$$\int_\Gamma f(z)dz = g(b) - g(a). \tag{17.49}$$

This claim is established by approximating $f(\xi_m) \approx (g(z_m) - g(z_{m-1}))/(z_m - z_{m-1})$, and observing that the resulting Riemann sum

$$\sum_{m=1}^{N} \Big(g(z_m) - g(z_{m-1}) \Big) \tag{17.50}$$

telescopes. The approximation to the derivative will become accurate in the limit $|z_m - z_{m-1}| \to 0$. Thus, when $f(z)$ is the derivative of another function, the integral is independent of the route that Γ takes from a to b.

We shall see that any analytic function is (at least locally) the derivative of another analytic function, and so this path independence holds generally – provided that we do not try to move the integration contour over a place where f ceases to be differentiable. This is the essence of what is known as *Cauchy's theorem* – although, as with much of complex analysis, the result was known to Gauss.

17.2.2 Cauchy's theorem

Before we state and prove Cauchy's theorem, we must introduce an orientation convention and some traditional notation. Recall that a p-chain is a finite formal sum of p-dimensional oriented surfaces or curves, and that a p-cycle is a p-chain Γ whose boundary vanishes: $\partial\Gamma = 0$. A 1-cycle that consists of only a single connected component is a closed curve. We will mostly consider integrals over *simple* closed curves – these being curves that do not self-intersect – or 1-cycles consisting of finite formal sums of such curves. The orientation of a simple closed curve can be described by the sense, clockwise or anticlockwise, in which we traverse it. We will adopt the convention that a positively oriented curve is one such that the integration is performed in an *anticlockwise* direction. The integral over a chain Γ of oriented simple closed curves will be denoted by the symbol $\oint_\Gamma f\,dz$.

We now establish Cauchy's theorem by relating it to our previous work with exterior derivatives: suppose that f is analytic with domain D, so that $\partial_{\bar z} f = 0$ within D. We therefore have that the exterior derivative of f is

$$df = \partial_z f\, dz + \partial_{\bar z} f\, d\bar z = \partial_z f\, dz. \tag{17.51}$$

Now suppose that the simple closed curve Γ is the boundary of a region $\Omega \subset D$. We can exploit Stokes' theorem to deduce that

$$\oint_{\Gamma=\partial\Omega} f(z)dz = \int_{\Omega} d(f(z)dz) = \int_{\Omega} (\partial_{\bar{z}} f)\, d\bar{z} \wedge dz = 0. \tag{17.52}$$

The last integral is zero because $dz \wedge dz = 0$. We may state our result as:

Theorem (Cauchy, in modern language): The integral of an analytic function over a 1-cycle that is homologous to zero vanishes.

The zero result is only guaranteed if the function f is analytic throughout the region Ω. For example, if Γ is the unit circle $z = e^{i\theta}$ then

$$\oint_{\Gamma} \left(\frac{1}{z}\right) dz = \int_0^{2\pi} e^{-i\theta}\, d\left(e^{i\theta}\right) = i\int_0^{2\pi} d\theta = 2\pi i. \tag{17.53}$$

Cauchy's theorem is not applicable because $1/z$ is *singular*, i.e. not differentiable, at $z = 0$. The formula (17.53) will hold for Γ any contour homologous to the unit circle in $\mathbb{C} \setminus 0$, the complex plane punctured by the removal of the point $z = 0$. Thus

$$\oint_{\Gamma} \left(\frac{1}{z}\right) dz = 2\pi i \tag{17.54}$$

for any contour Γ that encloses the origin. We can deduce a rather remarkable formula from (17.54): writing $\Gamma = \partial\Omega$ with anticlockwise orientation, we use Stokes' theorem to obtain

$$\oint_{\partial\Omega} \left(\frac{1}{z}\right) dz = \int_{\Omega} \partial_{\bar{z}} \left(\frac{1}{z}\right) d\bar{z} \wedge dz = \begin{cases} 2\pi i, & 0 \in \Omega, \\ 0, & 0 \notin \Omega. \end{cases} \tag{17.55}$$

Since $d\bar{z} \wedge dz = 2i dx \wedge dy$, we have established that

$$\partial_{\bar{z}} \left(\frac{1}{z}\right) = \pi \delta(x)\delta(y). \tag{17.56}$$

This rather cryptic formula encodes one of the most useful results in mathematics.

Perhaps perversely, functions that are more singular than $1/z$ have vanishing integrals about their singularities. With Γ again the unit circle, we have

$$\oint_{\Gamma} \left(\frac{1}{z^2}\right) dz = \int_0^{2\pi} e^{-2i\theta}\, d\left(e^{i\theta}\right) = i\int_0^{2\pi} e^{-i\theta}\, d\theta = 0. \tag{17.57}$$

The same is true for all higher integer powers:

$$\oint_{\Gamma} \left(\frac{1}{z^n}\right) dz = 0, \quad n \geq 2. \tag{17.58}$$

We can understand this vanishing in another way, by evaluating the integral as

$$\oint_\Gamma \left(\frac{1}{z^n}\right) dz = \oint_\Gamma \frac{d}{dz}\left(-\frac{1}{n-1}\frac{1}{z^{n-1}}\right) dz = \left[-\frac{1}{n-1}\frac{1}{z^{n-1}}\right]_\Gamma = 0, \quad n \neq 1. \tag{17.59}$$

Here, the notation $[A]_\Gamma$ means the difference in the value of A at two ends of the integration path Γ. For a closed curve the difference is zero because the two ends are at the same point. This approach reinforces the fact that the complex integral can be computed from the "anti-derivative" in the same way as the real-variable integral. We also see why $1/z$ is special. It is the derivative of $\ln z = \ln |z| + i \arg z$, and $\ln z$ is not really a function, as it is multivalued. In evaluating $[\ln z]_\Gamma$ we must follow the continuous evolution of $\arg z$ as we traverse the contour. As thc origin is within the contour, this angle increases by 2π, and so

$$[\ln z]_\Gamma = [i \arg z]_\Gamma = i\left(\arg e^{2\pi i} - \arg e^{0i}\right) = 2\pi i. \tag{17.60}$$

Exercise 17.2: Suppose $f(z)$ is analytic in a simply connected domain D, and $z_0 \in D$. Set $g(z) = \int_{z_0}^{z} f(z)\, dz$ along some path in D from z_0 to z. Use the path-independence of the integral to compute the derivative of $g(z)$ and show that

$$f(z) = \frac{dg}{dz}.$$

This confirms our earlier claim that any analytic function is the derivative of some other analytic function.

Exercise 17.3: *The "D-bar" problem*: Suppose we are given a simply connected domain Ω, and a function $f(z, \bar{z})$ defined on it, and wish to find a function $F(z, \bar{z})$ such that

$$\frac{\partial F(z, \bar{z})}{\partial \bar{z}} = f(z, \bar{z}), \quad (z, \bar{z}) \in \Omega.$$

Use (17.56) to argue formally that the general solution is

$$F(\zeta, \bar{\zeta}) = -\frac{1}{\pi}\int_\Omega \frac{f(z, \bar{z})}{z - \zeta} dx \wedge dy + g(\zeta),$$

where $g(\zeta)$ is an arbitrary analytic function. This result can be shown to be correct by more rigorous reasoning.

17.2.3 The residue theorem

The essential tool for computations with complex integrals is provided by the *residue theorem*. With the aid of this theorem, the evaluation of contour integrals becomes easy. All one has to do is identify points at which the function being integrated blows up, and examine just how it blows up.

If, near the point z_i, the function can be written

$$f(z) = \left\{ \frac{a_N^{(i)}}{(z - z_i)^N} + \cdots + \frac{a_2^{(i)}}{(z - z_i)^2} + \frac{a_1^{(i)}}{(z - z_i)} \right\} g^{(i)}(z), \tag{17.61}$$

where $g^{(i)}(z)$ is analytic and non-zero at z_i, then $f(z)$ has a *pole* of order N at z_i. If $N = 1$ then $f(z)$ is said to have a *simple pole* at z_i. We can normalize $g^{(i)}(z)$ so that $g^{(i)}(z_i) = 1$, and then the coefficient, $a_1^{(i)}$, of $1/(z - z_i)$ is called the *residue* of the pole at z_i. The coefficients of the more singular terms do not influence the result of the integral, but N must be finite for the singularity to be called a pole.

Theorem: Let the function $f(z)$ be analytic within and on the boundary $\Gamma = \partial D$ of a simply connected domain D, with the exception of a finite number of points at which $f(z)$ has poles. Then

$$\oint_\Gamma f(z)\, dz = \sum_{\text{poles} \in D} 2\pi i \,(\text{residue at pole}), \tag{17.62}$$

the integral being traversed in the positive (anticlockwise) sense.

We prove the residue theorem by drawing small circles C_i about each singular point z_i in D (Figure 17.6).

We now assert that

$$\oint_\Gamma f(z)\, dz = \sum_i \oint_{C_i} f(z)\, dz, \tag{17.63}$$

because the 1-cycle

$$C \equiv \Gamma - \sum_i C_i = \partial\Omega \tag{17.64}$$

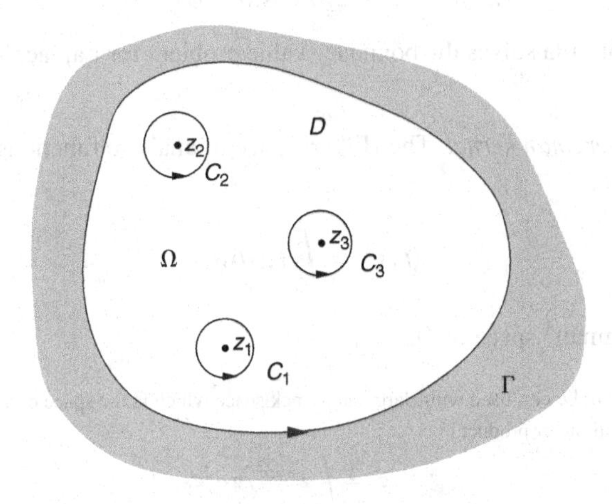

Figure 17.6 Circles for the residue theorem.

is the boundary of a region Ω in which f is analytic, and hence C is homologous to zero. If we make the radius R_i of the circle C_i sufficiently small, we may replace each $g^{(i)}(z)$ by its limit $g^{(i)}(z_i) = 1$, and so take

$$f(z) \to \left\{ \frac{a_1^{(i)}}{(z - z_i)} + \frac{a_2^{(i)}}{(z - z_i)^2} + \cdots + \frac{a_N^{(i)}}{(z - z_i)^N} \right\} g^{(i)}(z_i)$$
$$= \frac{a_1^{(i)}}{(z - z_i)} + \frac{a_2^{(i)}}{(z - z_i)^2} + \cdots + \frac{a_N^{(i)}}{(z - z_i)^N}, \tag{17.65}$$

on C_i. We then evaluate the integral over C_i by using our previous results to get

$$\oint_{C_i} f(z)\, dz = 2\pi i a_1^{(i)}. \tag{17.66}$$

The integral around Γ is therefore equal to $2\pi i \sum_i a_1^{(i)}$.

The restriction to contours containing only finitely many poles arises for two reasons: firstly, with infinitely many poles, the sum over i might not converge; secondly, there may be a point whose every neighbourhood contains infinitely many of the poles, and there our construction of drawing circles around each individual pole would not be possible.

Exercise 17.4: *Poisson's formula.* The function $f(z)$ is analytic in $|z| < R'$. Prove that if $|a| < R < R'$,

$$f(a) = \frac{1}{2\pi i} \oint_{|z|=R} \frac{R^2 - \bar{a}a}{(z - a)(R^2 - \bar{a}z)} f(z) dz.$$

Deduce that, for $0 < r < R$,

$$f(re^{i\theta}) = \frac{1}{2\pi} \int_0^{2\pi} \frac{R^2 - r^2}{R^2 - 2Rr\cos(\theta - \phi) + r^2} f(Re^{i\phi}) d\phi.$$

Show that this formula solves the boundary value problem for Laplace's equation in the disc $|z| < R$.

Exercise 17.5: *Bergman kernel.* The Hilbert space of analytic functions on a domain D with inner product

$$\langle f, g \rangle = \int_D \bar{f} g \, dxdy$$

is called the Bergman[4] space of D.

[4] This space should not be confused with Bargmann–Fock space which is the space of analytic functions on the entirety of $\mathbb{C}$ with inner product

$$\langle f, g \rangle = \int_{\mathbb{C}} e^{-|z|^2} \bar{f} g \, d^2 z.$$

Stefan Bergman and Valentine Bargmann are two different people.

(a) Suppose that $\varphi_n(z)$, $n = 0, 1, 2, \ldots$, are a complete set of orthonormal functions on the Bergman space. Show that

$$K(\zeta, z) = \sum_{m=0}^{\infty} \varphi_m(\zeta)\overline{\varphi_m(z)}$$

has the property that

$$g(\zeta) = \iint_D K(\zeta, z) g(z)\, dxdy$$

for any function g analytic in D. Thus $K(\zeta, z)$ plays the role of the delta function on the space of analytic functions on D. This object is called the *reproducing* or *Bergman kernel*. By taking $g(z) = \varphi_n(z)$, show that it is the unique integral kernel with the reproducing property.

(b) Consider the case of D being the unit circle. Use the Gramm–Schmidt procedure to construct an orthonormal set from the functions z^n, $n = 0, 1, 2, \ldots$ Use the result of part (a) to conjecture (because we have not proved that the set is complete) that, for the unit circle,

$$K(\zeta, z) = \frac{1}{\pi} \frac{1}{(1 - \zeta\bar{z})^2}.$$

(c) For any smooth, complex valued, function g defined on a domain D and its boundary, use Stokes' theorem to show that

$$\iint_D \partial_{\bar{z}} g(z, \bar{z}) dxdy = \frac{1}{2i} \oint_{\partial D} g(z, \bar{z}) dz.$$

Use this to verify that the $K(\zeta, z)$ you constructed in part (b) is indeed a (and hence "the") reproducing kernel.

(d) Now suppose that D is a simply connected domain whose boundary ∂D is a smooth curve. We know from the Riemann mapping theorem that there exists an analytic function $f(z) = f(z; \zeta)$ that maps D onto the interior of the unit circle in such a way that $f(\zeta) = 0$ and $f'(\zeta)$ is real and non-zero. Show that if we set $K(\zeta, z) = \overline{f'(z)} f'(\zeta)/\pi$, then, by using part (c) together with the residue theorem to evaluate the integral over the boundary, we have

$$g(\zeta) = \iint_D K(\zeta, z) g(z)\, dxdy.$$

This $K(\zeta, z)$ must therefore be the reproducing kernel. We see that if we know K we can recover the map f from

$$f'(z; \zeta) = \sqrt{\frac{\pi}{K(\zeta, \zeta)}} K(z, \zeta).$$

(e) Apply the formula from part (d) to the unit circle, and so deduce that

$$f(z;\zeta) = \frac{z-\zeta}{1-\bar{\zeta}z}$$

is the unique function that maps the unit circle onto itself with the point ζ mapping to the origin and with the horizontal direction through ζ remaining horizontal.

17.3 Applications

We now know enough about complex variables to work through some interesting applications, including the mechanism by which an aeroplane flies.

17.3.1 Two-dimensional vector calculus

It is often convenient to use complex coordinates for vectors and tensors. In these coordinates the standard metric on $\mathbb{R}^2$ becomes

$$\begin{aligned} \text{“}ds^2\text{”} &= dx \otimes dx + dy \otimes dy \\ &= d\bar{z} \otimes dz \\ &= g_{zz} dz \otimes dz + g_{\bar{z}z} d\bar{z} \otimes dz + g_{z\bar{z}} dz \otimes d\bar{z} + g_{\bar{z}\bar{z}} d\bar{z} \otimes d\bar{z}, \end{aligned} \tag{17.67}$$

so the complex coordinate components of the metric tensor are $g_{zz} = g_{\bar{z}\bar{z}} = 0$, $g_{z\bar{z}} = g_{\bar{z}z} = \frac{1}{2}$. The inverse metric tensor is $g^{z\bar{z}} = g^{\bar{z}z} = 2$, $g^{zz} = g^{\bar{z}\bar{z}} = 0$.

In these coordinates the Laplacian is

$$\nabla^2 = g^{ij}\partial^2_{ij} = 2(\partial_z\partial_{\bar{z}} + \partial_{\bar{z}}\partial_z). \tag{17.68}$$

When f has singularities, it is not safe to assume that $\partial_z\partial_{\bar{z}}f = \partial_{\bar{z}}\partial_z f$. For example, from

$$\partial_{\bar{z}}\left(\frac{1}{z}\right) = \pi\delta^2(x,y), \tag{17.69}$$

we deduce that

$$\partial_{\bar{z}}\partial_z \ln z = \pi\delta^2(x,y). \tag{17.70}$$

When we evaluate the derivatives in the opposite order, however, we have

$$\partial_z\partial_{\bar{z}} \ln z = 0. \tag{17.71}$$

To understand the source of the non-commutativity, take real and imaginary parts of these last two equations. Write $\ln z = \ln |z| + i\theta$, where $\theta = \arg z$, and add and subtract. We find

$$\begin{aligned} \nabla^2 \ln |z| &= 2\pi \delta^2(x,y), \\ (\partial_x \partial_y - \partial_y \partial_x)\theta &= 2\pi \delta^2(x,y). \end{aligned} \tag{17.72}$$

The first of these shows that $\frac{1}{2\pi} \ln |z|$ is the Green function for the Laplace operator, and the second reveals that the vector field $\nabla\theta$ is singular, having a delta function "curl" at the origin.

If we have a vector field $\mathbf{v}$ with contravariant components (v^x, v^y) and (numerically equal) covariant components (v_x, v_y) then the covariant components in the complex coordinate system are $v_z = \frac{1}{2}(v_x - iv_y)$ and $v_{\bar{z}} = \frac{1}{2}(v_x + iv_y)$. This can be obtained by using the change of coordinates rule, but a quicker route is to observe that

$$\mathbf{v} \cdot d\mathbf{r} = v_x dx + v_y dy = v_z dz + v_{\bar{z}} d\bar{z}. \tag{17.73}$$

Now

$$\partial_{\bar{z}} v_z = \frac{1}{4}(\partial_x v_x + \partial_y v_y) + i\frac{1}{4}(\partial_y v_x - \partial_x v_y). \tag{17.74}$$

Thus the statement that $\partial_{\bar{z}} v_z = 0$ is equivalent to the vector field $\mathbf{v}$ being both solenoidal (incompressible) and irrotational. This can also be expressed in form language by setting $\eta = v_z\, dz$ and saying that $d\eta = 0$ means that the corresponding vector field is both solenoidal and irrotational.

17.3.2 Milne–Thomson circle theorem

As we mentioned earlier, we can describe an irrotational and incompressible fluid motion either by a velocity potential

$$v_x = \partial_x \phi, \quad v_y = \partial_y \phi, \tag{17.75}$$

where $\mathbf{v}$ is automatically irrotational but incompressibility requires $\nabla^2 \phi = 0$, or by a stream function

$$v_x = \partial_y \chi, \quad v_y = -\partial_x \chi, \tag{17.76}$$

where $\mathbf{v}$ is automatically incompressible but irrotationality requires $\nabla^2 \chi = 0$. We can combine these into a single *complex stream function* $\Phi = \phi + i\chi$ which, for an irrotational incompressible flow, satisfies the Cauchy–Riemann equations and is therefore an analytic function of z. We see that

$$2v_z = \frac{d\Phi}{dz}, \tag{17.77}$$

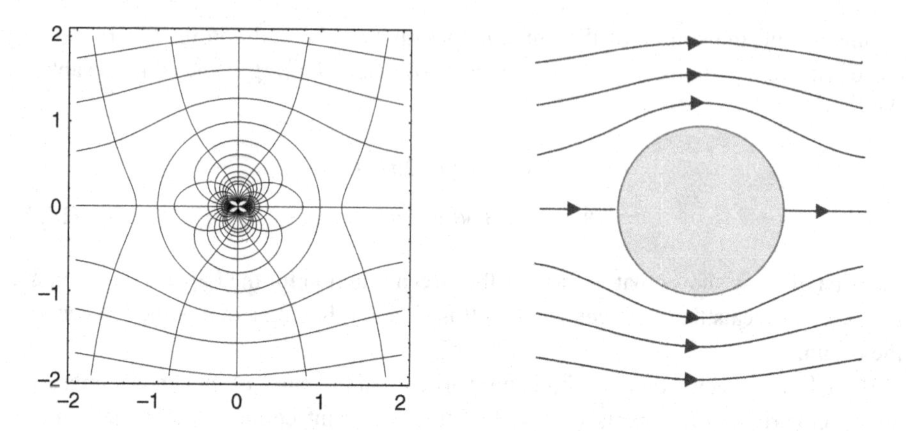

Figure 17.7 The real and imaginary parts of the function $z+z^{-1}$ provide the velocity potentials and streamlines for irrotational incompressible flow past a cylinder of unit radius.

ϕ and χ making equal contributions.

The Milne–Thomson theorem says that if Φ is the complex stream function for a flow in unobstructed space, then

$$\widetilde{\Phi} = \Phi(z) + \overline{\Phi}\left(\frac{a^2}{z}\right) \tag{17.78}$$

is the stream function after the cylindrical obstacle $|z| = a$ is inserted into the flow. Here $\overline{\Phi}(z)$ denotes the analytic function defined by $\overline{\Phi}(z) = \overline{\Phi(\overline{z})}$. To see that this works, observe that $a^2/z = \overline{z}$ on the curve $|z| = a$, and so on this curve $\operatorname{Im}\widetilde{\Phi} = \chi = 0$. The surface of the cylinder has therefore become a streamline, and so the flow does not penetrate into the cylinder. If the original flow is created by sources and sinks exterior to $|z| = a$, which will be singularities of Φ, the additional term has singularities that lie only within $|z| = a$. These will be the "images" of the sources and sinks in the sense of the "method of images".

Example: A uniform flow with speed U in the x-direction has $\Phi(z) = Uz$. Inserting a cylinder makes this (see Figure 17.7)

$$\tilde{\Phi}(z) = U\left(z + \frac{a^2}{z}\right). \tag{17.79}$$

Because v_z is the derivative of this, we see that the perturbing effect of the obstacle on the velocity field falls off as the square of the distance from the cylinder. This is a general result for obstructed flows.

17.3.3 Blasius and Kutta–Joukowski theorems

We now derive the celebrated result, discovered independently by Martin Wilhelm Kutta (1902) and Nikolai Egorovich Joukowski (1906), that the lift per unit span of an aircraft

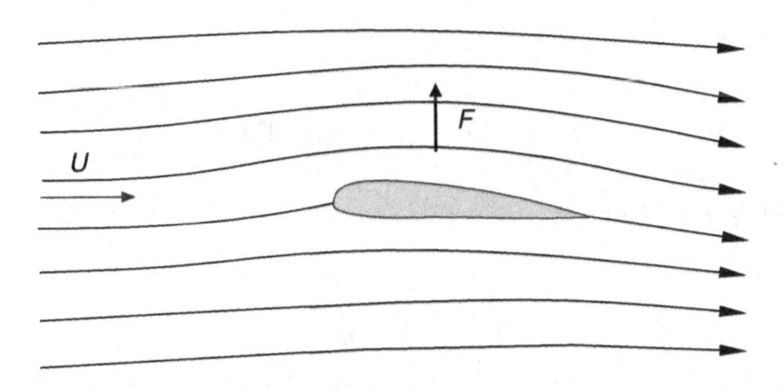

Figure 17.8 Flow past an aerofoil.

wing is equal to the product of the density of the air ρ, the circulation $\kappa \equiv \oint \mathbf{v} \cdot d\mathbf{r}$ about the wing and the forward velocity U of the wing through the air (Figure 17.8). Their theory treats the air as being incompressible – a good approximation unless the flow-velocities approach the speed of sound – and assumes that the wing is long enough that the flow can be regarded as being two dimensional.

Begin by recalling how the momentum flux tensor

$$T_{ij} = \rho v_i v_j + g_{ij} P \tag{17.80}$$

enters fluid mechanics. In cartesian coordinates, and in the presence of an external body force f_i acting on the fluid, the Euler equation of motion for the fluid is

$$\rho(\partial_t v_i + v^j \partial_j v_i) = -\partial_i P + f_i. \tag{17.81}$$

Here P is the pressure and we are distinguishing between co- and contravariant components, although at the moment $g_{ij} \equiv \delta_{ij}$. We can combine Euler's equation with the law of mass conservation,

$$\partial_t \rho + \partial^i(\rho v_i) = 0, \tag{17.82}$$

to obtain

$$\partial_t(\rho v_i) + \partial^j(\rho v_j v_i + g_{ij} P) = f_i. \tag{17.83}$$

This momemtum-tracking equation shows that the external force acts as a source of momentum, and that for steady flow f_i is equal to the divergence of the momentum flux tensor:

$$f_i = \partial^l T_{li} = g^{kl} \partial_k T_{li}. \tag{17.84}$$

As we are interested in steady, irrotational motion with uniform density we may use Bernoulli's theorem, $P + \frac{1}{2}\rho|v|^2 = \text{const.}$, to substitute $-\frac{1}{2}\rho|v|^2$ in place of P. (The constant will not affect the momentum flux.) With this substitution T_{ij} becomes a traceless symmetric tensor:

$$T_{ij} = \rho(v_i v_j - \frac{1}{2} g_{ij} |v|^2). \tag{17.85}$$

Using $v_z = \frac{1}{2}(v_x - iv_y)$ and

$$T_{zz} = \frac{\partial x^i}{\partial z} \frac{\partial x^j}{\partial z} T_{ij}, \tag{17.86}$$

together with

$$x \equiv x^1 = \frac{1}{2}(z + \bar{z}), \qquad y \equiv x^2 = \frac{1}{2i}(z - \bar{z}) \tag{17.87}$$

we find

$$T \equiv T_{zz} = \frac{1}{4}(T_{xx} - T_{yy} - 2iT_{xy}) = \rho(v_z)^2. \tag{17.88}$$

This is the only component of T_{ij} that we will need to consider. $T_{\bar{z}\bar{z}}$ is simply $\overline{T}$, whereas $T_{z\bar{z}} = 0 = T_{\bar{z}z}$ because T_{ij} is traceless.

In our complex coordinates, the equation

$$f_i = g^{kl} \partial_k T_{li} \tag{17.89}$$

reads

$$f_z = g^{\bar{z}z} \partial_{\bar{z}} T_{zz} + g^{z\bar{z}} \partial_z T_{\bar{z}z} = 2\partial_{\bar{z}} T. \tag{17.90}$$

We see that in steady flow the net momentum flux $\dot{P}_i$ out of a region Ω is given by

$$\dot{P}_z = \int_\Omega f_z \, dxdy = \frac{1}{2i} \int_\Omega f_z \, d\bar{z}dz = \frac{1}{i} \int_\Omega \partial_{\bar{z}} T \, d\bar{z}dz = \frac{1}{i} \oint_{\partial\Omega} T \, dz. \tag{17.91}$$

We have used Stokes' theorem at the last step. In regions where there is no external force, T is analytic, $\partial_{\bar{z}} T = 0$, and the integral will be independent of the choice of contour $\partial\Omega$. We can subsititute $T = \rho v_z^2$ to get

$$\dot{P}_z = -i\rho \oint_{\partial\Omega} v_z^2 \, dz. \tag{17.92}$$

To apply this result to our aerofoil we can take $\partial\Omega$ to be its boundary. Then $\dot{P}_z$ is the total force exerted on the fluid by the wing, and, by Newton's third law, this is minus the force exerted by the fluid on the wing. The total force on the aerofoil is therefore

$$F_z = i\rho \oint_{\partial\Omega} v_z^2 \, dz. \tag{17.93}$$

The result (17.93) is often called *Blasius' theorem*.

Evaluating the integral in (17.93) is not immediately possible because the velocity $\mathbf{v}$ on the boundary will be a complicated function of the shape of the body. We can, however, exploit the contour independence of the integral and evaluate it over a path encircling the aerofoil at large distance where the flow field takes the asymptotic form

$$v_z = U_z + \frac{\kappa}{4\pi i}\frac{1}{z} + O\left(\frac{1}{z^2}\right). \tag{17.94}$$

The $O(1/z^2)$ term is the velocity perturbation due to the air having to flow round the wing, as with the cylinder in a free flow. To confirm that this flow has the correct circulation we compute

$$\oint \mathbf{v}\cdot d\mathbf{r} = \oint v_z dz + \oint v_{\bar{z}}\, d\bar{z} = \kappa. \tag{17.95}$$

Substituting v_z in (17.93) we find that the $O(1/z^2)$ term cannot contribute as it cannot affect the residue of any pole. The only part that does contribute is the cross-term that arises from multiplying U_z by $\kappa/(4\pi iz)$. This gives

$$F_z = i\rho\left(\frac{U_z\kappa}{2\pi i}\right)\oint\frac{dz}{z} = i\rho\kappa U_z \tag{17.96}$$

so that

$$\frac{1}{2}(F_x - iF_y) = i\rho\kappa\frac{1}{2}(U_x - iU_y). \tag{17.97}$$

Thus, in conventional coordinates, the reaction force on the body is

$$\begin{aligned} F_x &= \rho\kappa U_y, \\ F_y &= -\rho\kappa U_x. \end{aligned} \tag{17.98}$$

The fluid therefore provides a lift force proportional to the product of the circulation with the asymptotic velocity. The force is at right angles to the incident airstream, so there is no *drag*.

The circulation around the wing is determined by the *Kutta condition* that the velocity of the flow at the sharp trailing edge of the wing be finite. If the wing starts moving into the air and the requisite circulation is not yet established then the flow under the wing does not leave the trailing edge smoothly but tries to whip round to the topside. The velocity gradients become very large and viscous forces become important and prevent the air from making the sharp turn. Instead, a *starting vortex* is shed from the trailing edge. Kelvin's theorem on the conservation of vorticity shows that this causes a circulation of equal and opposite strength to be induced about the wing.

For finite wings, the path independence of $\oint \mathbf{v} \cdot d\mathbf{r}$ means that the wings must leave a pair of trailing *wingtip vortices* of strength κ that connect back to the starting vortex to form a closed loop. The velocity field induced by the trailing vortices causes the airstream incident on the aerofoil to come from a slightly different direction than the asymptotic flow. Consequently, the lift is not quite perpendicular to the motion of the wing. For finite-length wings, therefore, lift comes at the expense of an inevitable *induced drag* force. The work that has to be done against this drag force in driving the wing forwards provides the kinetic energy in the trailing vortices.

17.4 Applications of Cauchy's theorem

Cauchy's theorem provides the Royal Road to complex analysis. It is possible to develop the theory without it, but the path is harder going.

17.4.1 Cauchy's integral formula

If $f(z)$ is analytic within and on the boundary of a simply connected domain Ω, with $\partial\Omega = \Gamma$, and if ζ is a point in Ω, then, noting that the integrand has a simple pole at $z = \zeta$ and applying the residue formula, we have *Cauchy's integral formula*

$$f(\zeta) = \frac{1}{2\pi i}\oint_\Gamma \frac{f(z)}{z-\zeta}\,dz, \quad \zeta \in \Omega. \tag{17.99}$$

This formula holds only if ζ lies within Ω (Figure 17.9). If it lies outside, then the integrand is analytic everywhere inside Ω, and so the integral gives zero.

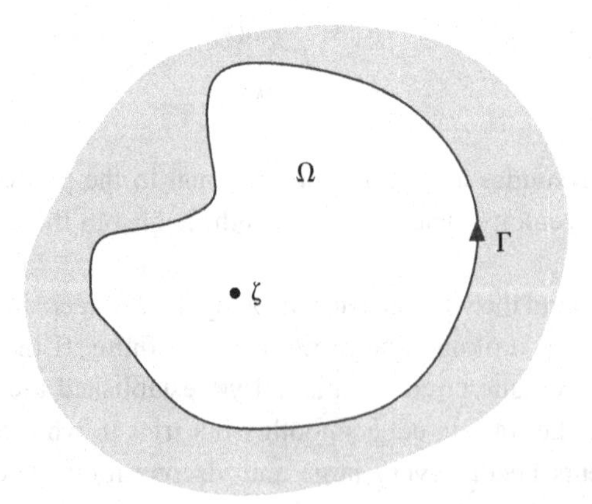

Figure 17.9 Cauchy contour.

We may show that it is legitimate to differentiate under the integral sign in Cauchy's formula. If we do so n times, we have the useful corollary that

$$f^{(n)}(\zeta) = \frac{n!}{2\pi i} \oint_\Gamma \frac{f(z)}{(z-\zeta)^{n+1}}\, dz. \tag{17.100}$$

This shows that being *once* differentiable (analytic) in a region automatically implies that $f(z)$ is differentiable *arbitrarily many times*!

Exercise 17.6: *The generalized Cauchy formula.* Suppose that we have solved a D-bar problem (see Exercise 17.3), and so found an $F(z,\bar{z})$ with $\partial_{\bar{z}} F = f(z,\bar{z})$ in a region Ω. Compute the exterior derivative of

$$\frac{F(z,\bar{z})}{z-\zeta}$$

using (17.56). Now, manipulating formally with delta functions, apply Stokes' theorem to show that, for $(\zeta,\bar{\zeta})$ in the interior of Ω, we have

$$F(\zeta,\bar{\zeta}) = \frac{1}{2\pi i}\oint_{\partial\Omega} \frac{F(z,\bar{z})}{z-\zeta}\, dz - \frac{1}{\pi}\int_\Omega \frac{f(z,\bar{z})}{z-\zeta}\, dx\, dy.$$

This is called the *generalized Cauchy formula*. Note that the first term on the right, unlike the second, is a function only of ζ, and so is analytic.

Liouville's theorem

A dramatic corollary of Cauchy's integral formula is provided by

Liouville's theorem: *If $f(z)$ is analytic in all of $\mathbb{C}$, and is bounded there, meaning that there is a positive real number K such that $|f(z)| < K$, then $f(z)$ is a constant.*

This result provides a powerful strategy for proving that two formulæ, $f_1(z)$ and $f_2(z)$, represent the same analytic function. If we can show that the difference $f_1 - f_2$ is analytic and tends to zero at infinity then Liouville's theorem tells us that $f_1 = f_2$.

Because the result is perhaps unintuitive, and because the methods are typical, we will spell out in detail how Liouville's theorem works. We select any two points, z_1 and z_2, and use Cauchy's formula to write

$$f(z_1) - f(z_2) = \frac{1}{2\pi i}\oint_\Gamma \left(\frac{1}{z-z_1} - \frac{1}{z-z_2}\right) f(z)\, dz. \tag{17.101}$$

We take the contour Γ to be a circle of radius ρ centred on z_1 (Figure 17.10). We make $\rho > 2|z_1 - z_2|$, so that when z is on Γ we are sure that $|z - z_2| > \rho/2$. Then, using

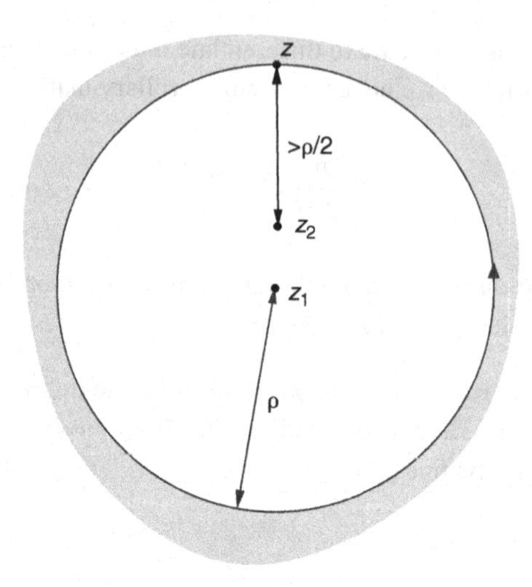

Figure 17.10 Contour for Liouville's theorem.

$|\int f(z)dz| \le \int |f(z)||dz|$, we have

$$\begin{aligned} |f(z_1) - f(z_2)| &= \frac{1}{2\pi}\left|\oint_\Gamma \frac{(z_1 - z_2)}{(z - z_1)(z - z_2)} f(z)\, dz\right| \\ &\le \frac{1}{2\pi}\int_0^{2\pi} \frac{|z_1 - z_2|K}{\rho/2}\, d\theta = \frac{2|z_1 - z_2|K}{\rho}. \end{aligned} \tag{17.102}$$

The right-hand side can be made arbitrarily small by taking ρ large enough, so we must have $f(z_1) = f(z_2)$. As z_1 and z_2 were any pair of points, we deduce that $f(z)$ takes the same value everywhere.

Exercise 17.7: Let $a_1, \ldots, a_N$ be N distinct complex numbers. Use Liouville's theorem to prove that

$$\sum_{k\neq j}\sum_{j=1}^{N} \frac{1}{(z - a_j)}\frac{1}{(z - a_k)^2} = \sum_{k\neq j}\sum_{j=1}^{N} \frac{1}{(a_k - a_j)}\frac{1}{(z - a_k)^2}.$$

17.4.2 Taylor and Laurent series

We have defined a function to be analytic in a domain D if it is (once) complex differentiable at all points in D. It turned out that this apparently mild requirement automatically implied that the function is differentiable *arbitrarily many times* in D. In this section we shall see that knowledge of all derivatives of $f(z)$ at any single point in D is enough to completely determine the function at any other point in D. Compare this with functions of a real variable, for which it is easy to construct examples that are once but not twice differentiable, and where complete knowledge of function at a point, or in even in a neighbourhood of a point, tells us absolutely nothing of the behaviour of the function away from the point or neighbourhood.

The key ingredient in these almost magical properties of complex analytic functions is that any analytic function has a Taylor series expansion that actually converges to the function. Indeed an alternative definition of analyticity is that $f(z)$ be representable by a convergent power series. For real variables this is the definition of a *real analytic* function.

To appreciate the utility of power series representations we do need to discuss some basic properties of power series. Most of these results are extensions to the complex plane of what we hope are familiar notions from real analysis.

Consider the power series

$$\sum_{n=0}^{\infty} a_n(z-z_0)^n \equiv \lim_{N\to\infty} S_N, \tag{17.103}$$

where S_N are the *partial sums*

$$S_N = \sum_{n=0}^{N} a_n(z-z_0)^n. \tag{17.104}$$

Suppose that this limit exists (i.e. the series is convergent) for some $z = \zeta$; then it turns out that the series is *absolutely convergent*[5] for any $|z-z_0| < |\zeta - z_0|$.

To establish this absolute convergence we may assume, without loss of generality, that $z_0 = 0$. Then, convergence of the sum $\sum a_n\zeta^n$ requires that $|a_n\zeta^n| \to 0$, and thus $|a_n\zeta^n|$ is bounded. In other words, there is a B such that $|a_n\zeta^n| < B$ for any n. We now write

$$|a_n z^n| = |a_n\zeta^n| \left|\frac{z}{\zeta}\right|^n < B\left|\frac{z}{\zeta}\right|^n. \tag{17.105}$$

The sum $\sum |a_n z^n|$ therefore converges for $|z/\zeta| < 1$, by comparison with a geometric progression.

This result, that if a power series in $(z-z_0)$ converges at a point then it converges at all points closer to z_0, shows that a power series possesses some *radius of convergence* R. The series converges for all $|z-z_0| < R$, and diverges for all $|z-z_0| > R$. What happens *on* the circle $|z-z_0| = R$ is usually delicate, and harder to establish. A useful result, however, is *Abel's theorem*, which we will not try to prove. Abel's theorem says

[5] Recall that absolute convergence of $\sum a_n$ means that $\sum |a_n|$ converges. Absolute convergence implies convergence, and also allows us to rearrange the order of terms in the series without changing the value of the sum. Compare this with *conditional convergence*, where $\sum a_n$ converges, but $\sum |a_n|$ does not. You may remember that Riemann showed that the terms of a conditionally convergent series can be rearranged so as to *get any answer whatsoever*!

that if the sum $\sum a_n$ is convergent, and if $A(z) = \sum_{n=0}^{\infty} a_n z^n$ for $|z| < 1$, then

$$\lim_{z\to 1_-} A(z) = \sum_{n=0}^{\infty} a_n. \tag{17.106}$$

The converse is not true: if $A(z)$ has a finite limit as we approach the circle of convergence, the corresponding sum need not converge

By comparison with a geometric progression, we may establish the following useful formulæ giving R for the series $\sum a_n z^n$:

$$\begin{aligned} R &= \lim_{n\to\infty} \frac{|a_{n-1}|}{|a_n|} \\ &= \lim_{n\to\infty} |a_n|^{1/n}. \end{aligned} \tag{17.107}$$

The proof of these formulæ is identical to the real-variable version.

We soon show that the radius of convergence of a power series is the distance from z_0 to the nearest singularity of the function that it represents.

When we differentiate the terms in a power series, and thus take $a_n z^n \to n a_n z^{n-1}$, this does not alter R. This observation suggests that it is legitimate to evaluate the derivative of the function represented by the power series by differentiating term-by-term. As a step on the way to justifying this, observe that if the series converges at $z = \zeta$ and D_r is the domain $|z| < r < |\zeta|$ then, using the same bound as in the proof of absolute convergence, we have

$$|a_n z^n| < B\frac{|z^n|}{|\zeta|^n} < B\frac{r^n}{|\zeta|^n} = M_n \tag{17.108}$$

where $\sum M_n$ is convergent. As a consequence $\sum a_n z^n$ is *uniformly convergent* in D_r by the Weierstrass "M" test. You probably know that uniform convergence allows us to interchange the order of sums and *integrals*: $\int(\sum f_n(x))dx = \sum \int f_n(x)dx$. For real variables, uniform convergence is *not* a strong enough condition for us to safely interchange order of sums and *derivatives*: $(\sum f_n(x))'$ is not necessarily equal to $\sum f_n'(x)$. For complex analytic functions, however, Cauchy's integral formula reduces the operation of differentiation to that of integration, and so this interchange *is* permitted. In particular we have that if

$$f(z) = \sum_{n=0}^{\infty} a_n z^n, \tag{17.109}$$

and R is defined by $R = |\zeta|$ for any ζ for which the series converges, then $f(z)$ is analytic in $|z| < R$ and

$$f'(z) = \sum_{n=0}^{\infty} n a_n z^{n-1}, \tag{17.110}$$

is also analytic in $|z| < R$.

Morera's theorem

There is a partial converse of Cauchy's theorem:

Theorem (Morera): If $f(z)$ is defined and continuous in a domain D, and if $\oint_\Gamma f(z)\,dz = 0$ for all closed contours, then $f(z)$ is analytic in D.

To prove this we set $F(z) = \int_P^z f(\zeta)\,d\zeta$. The integral is path-independent by the hypothesis of the theorem, and because $f(z)$ is continuous we can differentiate with respect to the integration limit to find that $F'(z) = f(z)$. Thus $F(z)$ is complex differentiable, and so analytic. Then, by Cauchy's formula for higher derivatives, $F''(z) = f'(z)$ exists, and so $f(z)$ itself is analytic.

A corollary of Morera's theorem is that if $f_n(z) \to f(z)$ uniformly in D, with all the f_n analytic, then

(i) $f(z)$ is analytic in D, and
(ii) $f_n'(z) \to f'(z)$ uniformly.

We use Morera's theorem to prove (i) (appealing to the uniform convergence to justify the interchange of the order of summation and integration), and use Cauchy's theorem to prove (ii).

Taylor's theorem for analytic functions

Theorem: Let Γ be a circle of radius ρ centred on the point a. Suppose that $f(z)$ is analytic within and on Γ, and that the point $z = \zeta$ is within Γ. Then $f(\zeta)$ can be expanded as a Taylor series

$$f(\zeta) = f(a) + \sum_{n=1}^{\infty} \frac{(\zeta - a)^n}{n!} f^{(n)}(a), \tag{17.111}$$

meaning that this series converges to $f(\zeta)$ for all ζ such that $|\zeta - a| < \rho$.

To prove this theorem we use the identity

$$\frac{1}{z-\zeta} = \frac{1}{z-a} + \frac{(\zeta - a)}{(z-a)^2} + \cdots + \frac{(\zeta-a)^{N-1}}{(z-a)^N} + \frac{(\zeta-a)^N}{(z-a)^N}\frac{1}{z-\zeta} \tag{17.112}$$

and Cauchy's integral, to write

$$\begin{aligned} f(\zeta) &= \frac{1}{2\pi i}\oint_\Gamma \frac{f(z)}{(z-\zeta)}\,dz \\ &= \sum_{n=0}^{N-1} \frac{(\zeta-a)^n}{2\pi i} \oint \frac{f(z)}{(z-a)^{n+1}}\,dz + \frac{(\zeta-a)^N}{2\pi i} \oint \frac{f(z)}{(z-a)^N (z-\zeta)}\,dz \\ &= \sum_{n=0}^{N-1} \frac{(\zeta-a)^n}{n!} f^{(n)}(a) + R_N, \end{aligned} \tag{17.113}$$

where

$$R_N \stackrel{\text{def}}{=} \frac{(\zeta - a)^N}{2\pi i} \oint_\Gamma \frac{f(z)}{(z-a)^N (z-\zeta)}\, dz. \tag{17.114}$$

This is Taylor's theorem with remainder. For real variables this is as far as we can go. Even if a real function is differentiable infinitely many times, there is no reason for the remainder to become small. For analytic functions, however, we can show that $R_N \to 0$ as $N \to \infty$. This means that the complex-variable Taylor series is convergent, and its limit is actually equal to $f(z)$. To show that $R_N \to 0$, recall that Γ is a circle of radius ρ centred on $z = a$. Let $r = |\zeta - a| < \rho$, and let M be an upper bound for $f(z)$ on Γ. (This exists because f is continuous and Γ is a compact subset of $\mathbb{C}$.) Then, estimating the integral using methods similar to those invoked in our proof of Liouville's theorem, we find that

$$R_N < \frac{r^N}{2\pi} \left(\frac{2\pi \rho M}{\rho^N (\rho - r)} \right). \tag{17.115}$$

As $r < \rho$, this tends to zero as $N \to \infty$.

We can take ρ as large as we like provided there are no singularities of f that end up within, or on, the circle. This confirms the claim made earlier: the radius of convergence of the power series representation of an analytic function is the distance to the nearest singularity.

Laurent series

Theorem (Laurent): Let Γ_1 and Γ_2 be two anticlockwise circular paths with centre a, radii ρ_1 and ρ_2 and with $\rho_2 < \rho_1$ (see Figure 17.11). If $f(z)$ is analytic on the circles

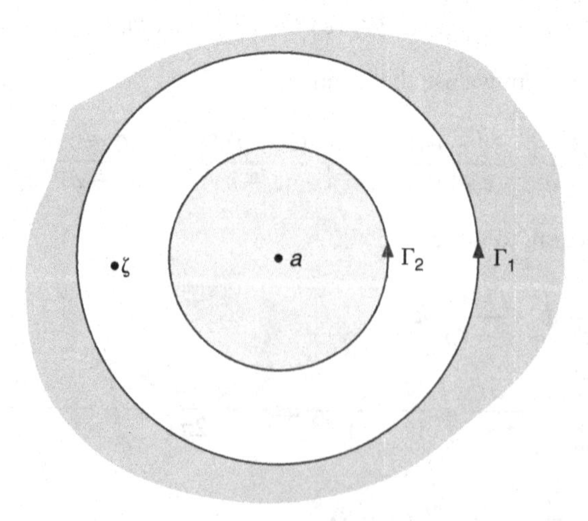

Figure 17.11 Contours for Laurent's theorem.

and within the annulus between them, then, for ζ in the annulus:

$$f(\zeta) = \sum_{n=0}^{\infty} a_n(\zeta - a)^n + \sum_{n=1}^{\infty} b_n(\zeta - a)^{-n}. \tag{17.116}$$

The coefficients a_n and b_n are given by

$$a_n = \frac{1}{2\pi i}\oint_{\Gamma_1} \frac{f(z)}{(z-a)^{n+1}}\,dz, \quad b_n = \frac{1}{2\pi i}\oint_{\Gamma_2} f(z)(z-a)^{n-1}\,dz. \tag{17.117}$$

Laurent's theorem is proved by observing that

$$f(\zeta) = \frac{1}{2\pi i}\oint_{\Gamma_1} \frac{f(z)}{(z-\zeta)}\,dz - \frac{1}{2\pi i}\oint_{\Gamma_2} \frac{f(z)}{(z-\zeta)}\,dz, \tag{17.118}$$

and using the identities

$$\frac{1}{z-\zeta} = \frac{1}{z-a} + \frac{(\zeta-a)}{(z-a)^2} + \cdots + \frac{(\zeta-a)^{N-1}}{(z-a)^N} + \frac{(\zeta-a)^N}{(z-a)^N}\frac{1}{z-\zeta}, \tag{17.119}$$

and

$$-\frac{1}{z-\zeta} = \frac{1}{\zeta-a} + \frac{(z-a)}{(\zeta-a)^2} + \cdots + \frac{(z-a)^{N-1}}{(\zeta-a)^N} + \frac{(z-a)^N}{(\zeta-a)^N}\frac{1}{\zeta-z}. \tag{17.120}$$

Once again we can show that the remainder terms tend to zero.

Warning: Although the coefficients a_n are given by the same integrals as in Taylor's theorem, they are not interpretable as derivatives of f unless $f(z)$ is analytic within the inner circle, in which case all the b_n are zero.

17.4.3 Zeros and singularities

This section is something of a *nosology* – a classification of diseases – but you should study it carefully as there is some tight reasoning here, and the conclusions are the essential foundations for the rest of subject.

First a review and some definitions:

(a) If $f(z)$ is analytic with a domain D, we have seen that f may be expanded in a Taylor series about any point $z_0 \in D$:

$$f(z) = \sum_{n=0}^{\infty} a_n(z - z_0)^n. \tag{17.121}$$

If $a_0 = a_1 = \cdots = a_{n-1} = 0$, and $a_n \neq 0$, so that the first non-zero term in the series is $a_n(z-z_0)^n$, we say that $f(z)$ has a *zero* of order n at z_0.

(b) A *singularity* of $f(z)$ is a point at which $f(z)$ ceases to be differentiable. If $f(z)$ has no singularities at finite z (for example, $f(z) = \sin z$) then it is said to be an *entire* function.

(c) If $f(z)$ is analytic in D except at $z = a$, an *isolated singularity*, then we may draw two concentric circles of centre a, both within D, and in the annulus between them we have the Laurent expansion

$$f(z) = \sum_{n=0}^{\infty} a_n(z-a)^n + \sum_{n=1}^{\infty} b_n(z-a)^{-n}. \tag{17.122}$$

The second term, consisting of negative powers, is called the *principal part* of $f(z)$ at $z = a$. It may happen that $b_m \neq 0$ but $b_n = 0$, $n > m$. Such a singularity is called a pole of order m at $z = a$. The coefficient b_1, which may be 0, is called the residue of f at the pole $z = a$. If the series of negative powers does not terminate, the singularity is called an *isolated essential singularity*.

Now some observations:

(i) Suppose $f(z)$ is analytic in a domain D containing the point $z = a$. Then we can expand: $f(z) = \sum a_n(z-a)^n$. If $f(z)$ is zero at $z = 0$, then there are exactly two possibilities: (a) all the a_n vanish, and then $f(z)$ is identically zero; (b) there is a first non-zero coefficient, a_m say, and so $f(z) = z^m\varphi(z)$, where $\varphi(a) \neq 0$. In the second case f is said to possess a *zero of order m* at $z = a$.

(ii) If $z = a$ is a zero of order m of $f(z)$ then the zero is *isolated* – i.e. there is a neighbourhood of a which contains no other zero. To see this observe that $f(z) = (z-a)^m\varphi(z)$ where $\varphi(z)$ is analytic and $\varphi(a) \neq 0$. Analyticity implies continuity, and by continuity there is a neighbourhood of a in which $\varphi(z)$ does not vanish.

(iii) Limit points of zeros I: Suppose that we know that $f(z)$ is analytic in D and we know that it vanishes at a sequence of points $a_1, a_2, a_3, \ldots \in D$. If these points have a limit point[6] that is interior to D then $f(z)$ must, by continuity, be zero there. But this would be a non-isolated zero, in contradiction to item (ii), unless $f(z)$ actually vanishes identically in D. This, then, is the only option.

(iv) From the definition of poles, they too are isolated.

(v) If $f(z)$ has a pole at $z = a$ then $f(z) \to \infty$ as $z \to a$ in any manner.

(vi) Limit points of zeros II: Suppose we know that f is analytic in D, except possibly at $z = a$ which is a limit point of zeros as in (iii), but we also know that f is not identically zero. Then $z = a$ must be a singularity of f – but not a pole (because f would tend to infinity and could not have arbitrarily close zeros) – so a must be an isolated essential singularity. For example, $\sin 1/z$ has an isolated essential singularity at $z = 0$, this being a limit point of the zeros at $z = 1/n\pi$.

(vii) A limit point of poles or other singularities would be a *non-isolated essential singularity*.

[6] A point z_0 is a limit point of a set S if for every $\epsilon > 0$ there is some $a \in S$, other than z_0 itself, such that $|a - z_0| \leq \epsilon$. A sequence need not have a limit for it to possess one or more limit points.

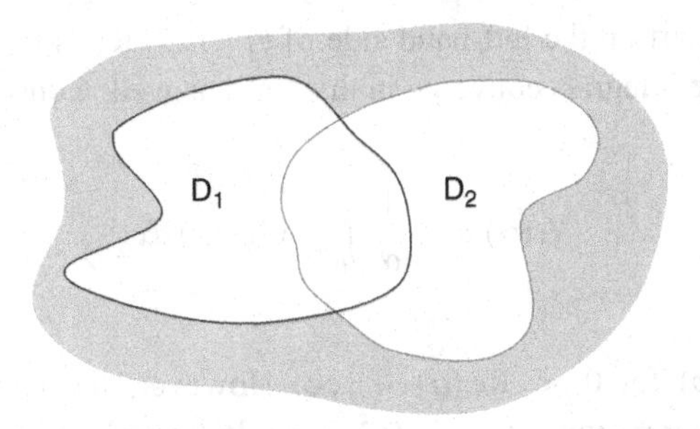

Figure 17.12 Intersecting domains.

17.4.4 Analytic continuation

Suppose that $f_1(z)$ is analytic in the (open, arcwise-connected) domain D_1, and $f_2(z)$ is analytic in D_2, with $D_1 \cap D_2 \neq \emptyset$ (Figure 17.12). Suppose further that $f_1(z) = f_2(z)$ in $D_1 \cap D_2$. Then we say that f_2 is an analytic continuation of f_1 to D_2. Such analytic continuations are *unique*: if f_3 is also analytic in D_2, and $f_3 = f_1$ in $D_1 \cap D_2$, then $f_2 - f_3 = 0$ in $D_1 \cap D_2$. Because the intersection of two open sets is also open, $f_1 - f_2$ vanishes on an open set and so by observation (iii) of the previous section, it vanishes everywhere in D_2.

We can use this uniqueness result, coupled with the circular domains of convergence of the Taylor series, to extend the definition of analytic functions beyond the domain of their initial definition.

The distribution $x_+^{\alpha-1}$

An interesting and useful example of analytic continuation is provided by the distribution $x_+^{\alpha-1}$, which, for real positive α, is defined by its evaluation on a test function $\varphi(x)$ as

$$(x_+^{\alpha-1}, \varphi) = \int_0^\infty x^{\alpha-1}\varphi(x)\,dx. \tag{17.123}$$

The pairing $(x_+^{\alpha-1}, \varphi)$ extends to a complex analytic function of α provided the integral converges. Test functions are required to decrease at infinity faster than any power of x, and so the integral always converges at the upper limit. It will converge at the lower limit provided $\mathrm{Re}(\alpha) > 0$. Assume that this is so, and integrate by parts using

$$\frac{d}{dx}\left(\frac{x^\alpha}{\alpha}\varphi(x)\right) = x^{\alpha-1}\varphi(x) + \frac{x^\alpha}{\alpha}\varphi'(x). \tag{17.124}$$

We find that, for $\epsilon > 0$,

$$\left[\frac{x^\alpha}{\alpha}\varphi(x)\right]_\epsilon^\infty = \int_\epsilon^\infty x^{\alpha-1}\varphi(x)\,dx + \int_\epsilon^\infty \frac{x^\alpha}{\alpha}\varphi'(x)\,dx. \tag{17.125}$$

The integrated-out part on the left-hand side of (17.125) tends to zero as we take ϵ to zero, and both of the integrals converge in this limit as well. Consequently

$$I_1(\alpha) \equiv -\frac{1}{\alpha}\int_0^\infty x^\alpha \varphi'(x)\,dx \tag{17.126}$$

is equal to $(x_+^{\alpha-1}, \varphi)$ for $0 < \mathrm{Re}\,(\alpha) < \infty$. However, the integral defining $I_1(\alpha)$ converges in the larger region $-1 < \mathrm{Re}\,(\alpha) < \infty$. It therefore provides an analytic continuation to this larger domain. The factor of $1/\alpha$ reveals that the analytically continued function possesses a pole at $\alpha = 0$, with residue

$$-\int_0^\infty \varphi'(x)\,dx = \varphi(0). \tag{17.127}$$

We can repeat the integration by parts, and find that

$$I_2(\alpha) \equiv \frac{1}{\alpha(\alpha+1)}\int_0^\infty x^{\alpha+1}\varphi''(x)\,dx \tag{17.128}$$

provides an analytic continuation to the region $-2 < \mathrm{Re}\,(\alpha) < \infty$. By proceeding in this manner, we can continue $(x_+^{\alpha-1}, \varphi)$ to a function analytic in the entire complex α plane with the exception of zero and the negative integers, at which it has simple poles. The residue of the pole at $\alpha = -n$ is $\varphi^{(n)}(0)/n!$.

There is another, much more revealing, way of expressing these analytic continuations. To obtain this, suppose that $\phi \in C^\infty[0,\infty]$ and $\phi \to 0$ at infinity as least as fast as $1/x$. (Our test function φ decreases much more rapidly than this, but $1/x$ is all we need for what follows.) Now the function

$$I(\alpha) \equiv \int_0^\infty x^{\alpha-1}\phi(x)\,dx \tag{17.129}$$

is convergent and analytic in the strip $0 < \mathrm{Re}\,(\alpha) < 1$. By the same reasoning as above, $I(\alpha)$ is there equal to

$$-\int_0^\infty \frac{x^\alpha}{\alpha}\phi'(x)\,dx. \tag{17.130}$$

Again this new integral provides an analytic continuation to the larger strip $-1 < \mathrm{Re}\,(\alpha) < 1$. But in the left-hand half of this strip, where $-1 < \mathrm{Re}\,(\alpha) < 0$, we can write

$$-\int_0^\infty \frac{x^\alpha}{\alpha}\phi'(x)\,dx = \lim_{\epsilon\to 0}\left\{\int_\epsilon^\infty x^{\alpha-1}\phi(x)\,dx - \left[\frac{x^\alpha}{\alpha}\phi(x)\right]_\epsilon^\infty\right\}$$
$$= \lim_{\epsilon\to 0}\left\{\int_\epsilon^\infty x^{\alpha-1}\phi(x)\,dx + \phi(\epsilon)\frac{\epsilon^\alpha}{\alpha}\right\}$$
$$= \lim_{\epsilon\to 0}\left\{\int_\epsilon^\infty x^{\alpha-1}[\phi(x)-\phi(\epsilon)]\,dx\right\},$$
$$= \int_0^\infty x^{\alpha-1}[\phi(x)-\phi(0)]\,dx. \tag{17.131}$$

Observe how the integrated out part, which tends to zero in $0 < \mathrm{Re}\,(\alpha) < 1$, becomes divergent in the strip $-1 < \mathrm{Re}\,(\alpha) < 0$. This divergence is there craftily combined with the integral to cancel *its* divergence, leaving a finite remainder. As a consequence, for $-1 < \mathrm{Re}\,(\alpha) < 0$, the analytic continuation is given by

$$I(\alpha) = \int_0^\infty x^{\alpha-1}[\phi(x)-\phi(0)]\,dx. \tag{17.132}$$

Next we observe that $\chi(x) = [\phi(x)-\phi(0)]/x$ tends to zero as $1/x$ for large x, and at $x = 0$ can be defined by its limit as $\chi(0) = \phi'(0)$. This $\chi(x)$ then satisfies the same hypotheses as $\phi(x)$. With $I(\alpha)$ denoting the analytic continuation of the original I, we therefore have

$$I(\alpha) = \int_0^\infty x^{\alpha-1}[\phi(x)-\phi(0)]\,dx, \qquad -1 < \mathrm{Re}\,(\alpha) < 0$$
$$= \int_0^\infty x^{\beta-1}\left[\frac{\phi(x)-\phi(0)}{x}\right]dx, \quad \text{where } \beta = \alpha+1$$
$$\to \int_0^\infty x^{\beta-1}\left[\frac{\phi(x)-\phi(0)}{x} - \phi'(0)\right]dx, \quad -1 < \mathrm{Re}\,(\beta) < 0$$
$$= \int_0^\infty x^{\alpha-1}[\phi(x)-\phi(0)-x\phi'(0)]\,dx, \quad -2 < \mathrm{Re}\,(\alpha) < -1, \tag{17.133}$$

the arrow denoting the same analytic continuation process that we used with ϕ.

We can now apply this machinery to our original $\varphi(x)$, and so deduce that the analytically continued distribution is given by

$$(x_+^{\alpha-1},\varphi) = \begin{cases} \displaystyle\int_0^\infty x^{\alpha-1}\varphi(x)\,dx, & 0 < \mathrm{Re}\,(\alpha) < \infty \\ \displaystyle\int_0^\infty x^{\alpha-1}[\varphi(x)-\varphi(0)]\,dx, & -1 < \mathrm{Re}\,(\alpha) < 0 \\ \displaystyle\int_0^\infty x^{\alpha-1}[\varphi(x)-\varphi(0)-x\varphi'(0)]\,dx, & -2 < \mathrm{Re}\,(\alpha) < -1, \end{cases} \tag{17.134}$$

and so on. The analytic continuation automatically subtracts more and more terms of the Taylor series of $\varphi(x)$ the deeper we penetrate into the left-hand half-plane. This property, that analytic continuation covertly subtracts the minimal number of Taylor-series terms required to ensure convergence, lies behind a number of physics applications, most notably the method of *dimensional regularization* in quantum field theory.

The following exercise illustrates some standard techniques of reasoning *via* analytic continuation.

Exercise 17.8: Define the *dilogarithm* function by the series

$$\mathrm{Li}_2(z) = \frac{z}{1^2} + \frac{z^2}{2^2} + \frac{z^3}{3^2} + \cdots .$$

The radius of convergence of this series is unity, but the domain of $\mathrm{Li}_2(z)$ can be extended to $|z| > 1$ by analytic continuation.

(a) Observe that the series converges at $z = \pm 1$, and at $z = 1$ is

$$\mathrm{Li}_2(1) = 1 + \frac{1}{2^2} + \frac{1}{3^2} + \cdots = \frac{\pi^2}{6}.$$

Rearrange the series to show that

$$\mathrm{Li}_2(-1) = -\frac{\pi^2}{12}.$$

(b) Identify the derivative of the power series for $\mathrm{Li}_2(z)$ with that of an elementary function. Exploit your identification to extend the definition of $[\mathrm{Li}_2(z)]'$ outside $|z| < 1$. Use the properties of this derivative function, together with part (a), to prove that the extended function obeys

$$\mathrm{Li}_2(-z) + \mathrm{Li}_2\left(-\frac{1}{z}\right) = -\frac{1}{2}(\ln z)^2 - \frac{\pi^2}{6}.$$

This formula allows us to calculate values of the dilogarithm for $|z| > 1$ in terms of those with $|z| < 1$.

Many weird identities involving dilogarithms exist. Some, such as

$$\mathrm{Li}_2\left(-\frac{1}{2}\right) + \frac{1}{6}\mathrm{Li}_2\left(\frac{1}{9}\right) = -\frac{1}{18}\pi^2 + \ln 2 \ln 3 - \frac{1}{2}(\ln 2)^2 - \frac{1}{3}(\ln 3)^2,$$

were found by Ramanujan. Others, originally discovered by sophisticated numerical methods, have been given proofs based on techniques from quantum mechanics. *Polylogarithms*, defined by

$$\mathrm{Li}_k(z) = \frac{z}{1^k} + \frac{z^2}{2^k} + \frac{z^3}{3^k} + \cdots ,$$

occur frequently when evaluating Feynman diagrams.

17.4.5 Removable singularities and the Weierstrass–Casorati theorem

Sometimes we are given a definition that makes a function analytic in a region with the exception of a single point. Can we extend the definition to make the function analytic in the entire region? Provided that the function is well enough behaved near the point, the answer is yes, and the extension is unique. Curiously, the proof that this is so gives us insight into the wild behaviour of functions near essential singularities.

Removable singularities

Suppose that $f(z)$ is analytic in $D \setminus a$, but that $\lim_{z\to a}(z-a)f(z) = 0$. Then f may be extended to a function analytic in all of D, i.e. $z = a$ is a *removable singularity*. To see this, let ζ lie between two simple closed contours Γ_1 and Γ_2, with a within the smaller, Γ_2. We use Cauchy to write

$$f(\zeta) = \frac{1}{2\pi i}\oint_{\Gamma_1} \frac{f(z)}{z-\zeta}\,dz - \frac{1}{2\pi i}\oint_{\Gamma_2} \frac{f(z)}{z-\zeta}\,dz. \tag{17.135}$$

Now we can shrink Γ_2 down to be very close to a, and because of the condition on $f(z)$ near $z = a$, we see that the second integral vanishes. We can also arrange for Γ_1 to enclose any chosen point in D. Thus, if we set

$$\tilde{f}(\zeta) = \frac{1}{2\pi i}\oint_{\Gamma_1} \frac{f(z)}{z-\zeta}\,dz \tag{17.136}$$

within Γ_1, we see that $\tilde{f} = f$ in $D \setminus a$, and is analytic in all of D. The extension is unique because any two analytic functions that agree everywhere except for a single point, must also agree at that point.

Weierstrass–Casorati theorem

We apply the idea of removable singularities to show just how pathological a beast is an isolated essential singularity:

Theorem (Weierstrass–Casorati): Let $z = a$ be an isolated essential singularity of $f(z)$. Then in any neighbourhood of a, the function $f(z)$ comes arbitrarily close to any assigned values in $\mathbb{C}$.

To prove this, define $N_\delta(a) = \{z \in \mathbb{C} : |z-a| < \delta\}$, and $N_\epsilon(\zeta) = \{z \in \mathbb{C} : |z-\zeta| < \epsilon\}$. The claim is then that there is an $z \in N_\delta(a)$ such that $f(z) \in N_\epsilon(\zeta)$. Suppose that the claim is *not* true. Then we have $|f(z) - \zeta| > \epsilon$ for all $z \in N_\delta(a)$. Therefore

$$\left|\frac{1}{f(z)-\zeta}\right| < \frac{1}{\epsilon} \tag{17.137}$$

in $N_\delta(a)$, while $1/(f(z)-\zeta)$ is analytic in $N_\delta(a) \setminus a$. Therefore $z = a$ is a removable singularity of $1/(f(z)-\zeta)$, and there is an analytic $g(z)$ which coincides with $1/(f(z)-\zeta)$

at all points except a. Therefore

$$f(z) = \zeta + \frac{1}{g(z)} \tag{17.138}$$

except at a. Now $g(z)$, being analytic, may have a zero at $z = a$ giving a pole in f, but it cannot give rise to an essential singularity. The claim is true, therefore.

Picard's theorems

The Weierstrass–Casorati theorem is elementary. There are much stronger results:

Theorem (Picard's little theorem): Every non-constant entire function attains every complex value with at most one exception.

Theorem (Picard's big theorem): In any neighbourhood of an isolated essential singularity, $f(z)$ takes every complex value with at most one exception.

The proofs of these theorems are hard.

As an illustration of Picard's little theorem, observe that the function $\exp z$ is entire, and takes all values except 0. For the big theorem observe that the function $f(z) = \exp(1/z)$ has an essential singularity at $z = 0$, and takes all values, with the exception of 0, in any neighbourhood of $z = 0$.

17.5 Meromorphic functions and the winding number

A function whose only singularities in D are poles is said to be *meromorphic* there. These functions have a number of properties that are essentially topological in character.

17.5.1 Principle of the argument

If $f(z)$ is meromorphic in D with $\partial D = \Gamma$, and $f(z) \neq 0$ on Γ, then

$$\frac{1}{2\pi i} \oint_\Gamma \frac{f'(z)}{f(z)}\, dz = N - P \tag{17.139}$$

where N is the number of zeros in D and P is the number of poles. To show this, we note that if $f(z) = (z - a)^m \varphi(z)$ where φ is analytic and non-zero near a, then

$$\frac{f'(z)}{f(z)} = \frac{m}{z - a} + \frac{\varphi'(z)}{\varphi(z)} \tag{17.140}$$

so f'/f has a simple pole at a with residue m. Here m can be either positive or negative. The term $\varphi'(z)/\varphi(z)$ is analytic at $z = a$, so collecting all the residues from each zero or pole gives the result.

Since $f'/f = \frac{d}{dz}\ln f$ the integral may be written

$$\oint_\Gamma \frac{f'(z)}{f(z)}\,dz = \Delta_\Gamma \ln f(z) = i\Delta_\Gamma \arg f(z), \tag{17.141}$$

the symbol Δ_Γ denoting the total change in the quantity after we traverse Γ. Thus

$$N - P = \frac{1}{2\pi}\Delta_\Gamma \arg f(z). \tag{17.142}$$

This result is known as the principle of the argument.

Local mapping theorem

Suppose the function $w = f(z)$ maps a region Ω holomorphically onto a region Ω', and a simple closed curve $\gamma \subset \Omega$ onto another closed curve $\Gamma \subset \Omega'$, which will in general have self intersections. Given a point $a \in \Omega'$, we can ask ourselves how many points within the simple closed curve γ map to a. The answer is given by the *winding number* of the image curve Γ about a (Figure 17.13).

To see that this is so, we appeal to the principle of the argument as

$$\begin{aligned}\#\text{ of zeros of } (f-a) \text{ within } \gamma &= \frac{1}{2\pi i}\oint_\gamma \frac{f'(z)}{f(z)-a}\,dz,\\ &= \frac{1}{2\pi i}\oint_\Gamma \frac{dw}{w-a},\\ &= n(\Gamma, a),\end{aligned} \tag{17.143}$$

where $n(\Gamma, a)$ is called the winding number of the image curve Γ about a. It is equal to

$$n(\Gamma, a) = \frac{1}{2\pi}\Delta_\gamma \arg(w-a), \tag{17.144}$$

and is the number of times the image point w encircles a as z traverses the original curve γ.

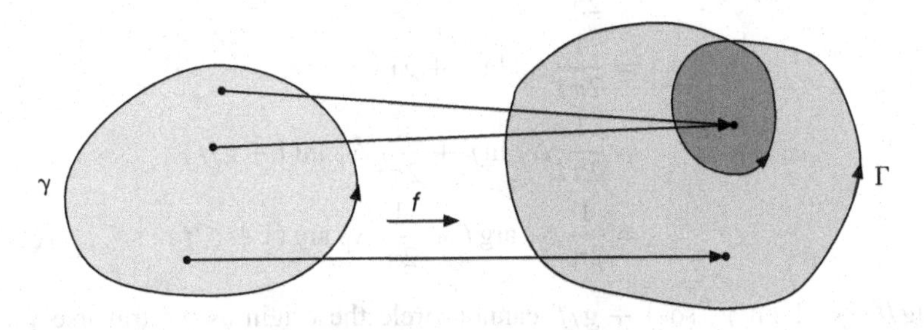

Figure 17.13 An analytic map is one-to-one where the winding number is unity, but two-to-one at points where the image curve winds twice.

Since the number of pre-image points cannot be negative, these winding numbers must be positive. This means that the holomorphic image of a curve winding in the anticlockwise direction is also a curve winding anticlockwise.

For mathematicians, another important consequence of this result is that a holomorphic map is *open* – i.e. the holomorphic image of an open set is itself an open set. The local mapping theorem is therefore sometimes called the *open mapping theorem*.

17.5.2 Rouché's theorem

Here we provide an effective tool for locating zeros of functions.

Theorem (Rouché): Let $f(z)$ and $g(z)$ be analytic within and on a simple closed contour γ. Suppose further that $|g(z)| < |f(z)|$ everywhere on γ. Then $f(z)$ and $f(z)+g(z)$ have the same number of zeros within γ.

Before giving the proof, we illustrate Rouché's theorem by giving its most important corollary: the algebraic completeness of the complex numbers, a result otherwise known as the *fundamental theorem of algebra*. This asserts that, if R is sufficiently large, a polynomial $P(z) = a_n z^n + a_{n-1} z^{n-1} + \cdots + a_0$ has exactly n zeros, when counted with their multiplicity, lying within the circle $|z| = R$. To prove this note that we can take R sufficiently big that

$$\begin{aligned} |a_n z^n| &= |a_n| R^n \\ &> |a_{n-1}| R^{n-1} + |a_{n-2}| R^{n-2} \cdots + |a_0| \\ &> |a_{n-a} z^{n-1} + a_{n-2} z^{n-2} \cdots + a_0|, \end{aligned} \tag{17.145}$$

on the circle $|z| = R$. We can therefore take $f(z) = a_n z^n$ and $g(z) = a_{n-a} z^{n-1} + a_{n-2} z^{n-2} \cdots + a_0$ in Rouché. Since $a_n z^n$ has exactly n zeros, all lying at $z = 0$, within $|z| = R$, we conclude that so does $P(z)$.

The proof of Rouché is a corollary of the principle of the argument. We observe that

$$\begin{aligned} \#\text{ of zeros of } f + g &= n(\Gamma, 0) \\ &= \frac{1}{2\pi} \Delta_\gamma \arg(f + g) \\ &= \frac{1}{2\pi i} \Delta_\gamma \ln(f + g) \\ &= \frac{1}{2\pi i} \Delta_\gamma \ln f + \frac{1}{2\pi i} \Delta_\gamma \ln(1 + g/f) \\ &= \frac{1}{2\pi} \Delta_\gamma \arg f + \frac{1}{2\pi} \Delta_\gamma \arg(1 + g/f). \end{aligned} \tag{17.146}$$

Now $|g/f| < 1$ on γ, so $1 + g/f$ cannot circle the origin as we traverse γ. As a consequence $\Delta_\gamma \arg(1 + g/f) = 0$. Thus the number of zeros of $f + g$ inside γ is the same as that of f alone. (Naturally, they are not usually in the same places.)

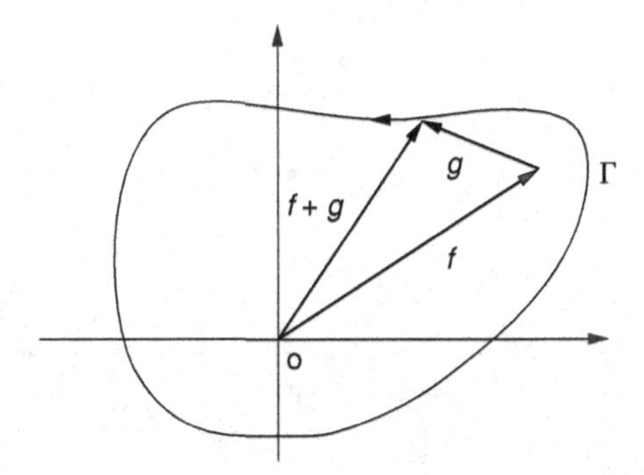

Figure 17.14 The curve Γ is the image of γ under the map $f+g$. If $|g| < |f|$, then, as z traverses γ, $f+g$ winds about the origin the same number of times that f does.

The geometric part of this argument is often illustrated by a dog on a lead. If the lead has length L, and the dog's owner stays a distance $R > L$ away from a lamp post, then the dog cannot run round the lamp post unless the owner does the same (Figure 17.14).

Exercise 17.9: *Jacobi Theta function.* The function $\theta(z|\tau)$ is defined for $\mathrm{Im}\,\tau > 0$ by the sum

$$\theta(z|\tau) = \sum_{n=-\infty}^{\infty} e^{i\pi\tau n^2} e^{2\pi i n z}.$$

Show that $\theta(z+1|\tau) = \theta(z|\tau)$, and $\theta(z+\tau|\tau) = e^{-i\pi\tau-2\pi i z}\theta(z|\tau)$. Use this information and the principle of the argument to show that $\theta(z|\tau)$ has exactly one zero in each unit cell of the Bravais lattice comprising the points $z = m + n\tau$; $m, n \in \mathbb{Z}$. Show that these zeros are located at $z = (m + 1/2) + (n + 1/2)\tau$.

Exercise 17.10: Use Rouché's theorem to find the number of roots of the equation $z^5 + 15z + 1 = 0$ lying within the circles, (i) $|z| = 2$, ii) $|z| = 3/2$.

17.6 Analytic functions and topology

17.6.1 The point at infinity

Some functions, $f(z) = 1/z$ for example, tend to a fixed limit (here 0) as z becomes large, independently of in which direction we set off towards infinity. Others, such as $f(z) = \exp z$, behave quite differently depending on what direction we take as $|z|$ becomes large.

To accommodate the former type of function, and to be able to legitimately write $f(\infty) = 0$ for $f(z) = 1/z$, it is convenient to add "∞" to the set of complex numbers. Technically, we are constructing the *one-point compactification* of the locally compact space $\mathbb{C}$. We often portray this extended complex plane as a sphere S^2 (the Riemann

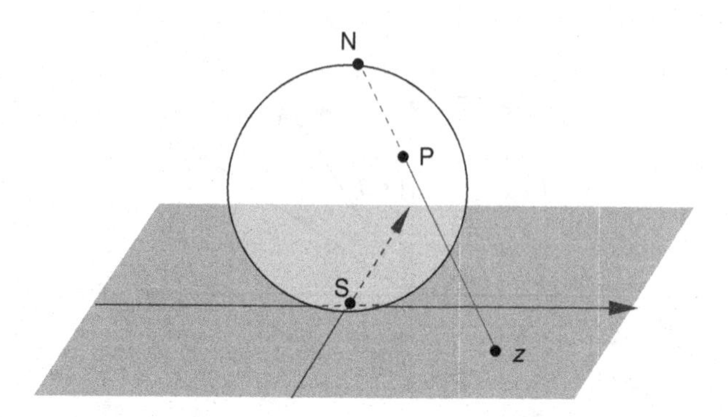

Figure 17.15 Stereographic mapping of the complex plane to the 2-sphere.

sphere), using stereographic projection (see Figure 17.15) to locate infinity at the north pole, and 0 at the south pole.

By the phrase an open *neighbourhood* of z, we mean an open set containing z. We use the stereographic map to define an open *neighbourhood of infinity* as the stereographic image of an open neighbourhood of the north pole. With this definition, the extended complex plane $\mathbb{C} \cup \{\infty\}$ becomes topologically a sphere and, in particular, becomes a compact set.

If we wish to study the behaviour of a function "at infinity", we use the map $z \mapsto \zeta = 1/z$ to bring ∞ to the origin, and study the behaviour of the function there. Thus the polynomial

$$f(z) = a_0 + a_1 z + \cdots + a_N z^N \tag{17.147}$$

becomes

$$f(\zeta) = a_0 + a_1 \zeta^{-1} + \cdots + a_N \zeta^{-N}, \tag{17.148}$$

and so has a pole of order N at infinity. Similarly, the function $f(z) = z^{-3}$ has a zero of order three at infinity, and $\sin z$ has an isolated essential singularity there.

We must be careful about defining *residues* at infinity. The residue is more a property of the 1-form $f(z)\,dz$ than of the function $f(z)$ alone, and to find the residue we need to transform the dz as well as $f(z)$. For example, if we set $z = 1/\zeta$ in dz/z we have

$$\frac{dz}{z} = \zeta\, d\left(\frac{1}{\zeta}\right) = -\frac{d\zeta}{\zeta}, \tag{17.149}$$

so the 1-form $(1/z)\,dz$ has a pole at $z = 0$ with residue 1, and has a pole with residue -1 at infinity, even though the *function* $1/z$ has no pole there. This 1-form viewpoint is required for compatability with the residue theorem: the integral of $1/z$ around the positively oriented unit circle is simultaneously minus the integral of $1/z$ about the oppositely oriented unit circle, now regarded as a positively oriented circle enclosing the point at infinity. Thus if $f(z)$ has of pole of order N at infinity, and

$$\begin{aligned} f(z) &= \cdots + a_{-2}z^{-2} + a_{-1}z^{-1} + a_0 + a_1 z + a_2 z^2 + \cdots + A_N z^N \\ &= \cdots + a_{-2}\zeta^2 + a_{-1}\zeta + a_0 + a_1\zeta^{-1} + a_2\zeta^{-2} + \cdots + A_N\zeta^{-N} \end{aligned} \tag{17.150}$$

near infinity, then the residue at infinity must be defined to be $-a_{-1}$, and not a_1 as one might naïvely have thought.

Once we have allowed ∞ as a point in the set we map *from*, it is only natural to add it to the set we map *to* – in other words to allow ∞ as a possible value for $f(z)$. We will set $f(a) = \infty$, if $|f(z)|$ becomes unboundedly large as $z \to a$ in any manner. Thus, if $f(z) = 1/z$ we have $f(0) = \infty$.

The map

$$w = \left(\frac{z - z_0}{z - z_\infty}\right)\left(\frac{z_1 - z_\infty}{z_1 - z_0}\right) \tag{17.151}$$

takes

$$\begin{aligned} z_0 &\to 0, \\ z_1 &\to 1, \\ z_\infty &\to \infty, \end{aligned} \tag{17.152}$$

for example. Using this language, the Möbius maps

$$w = \frac{az + b}{cz + d} \tag{17.153}$$

become one-to-one maps of $S^2 \to S^2$. They are the only such globally conformal one-to-one maps. When the matrix

$$\begin{pmatrix} a & b \\ c & d \end{pmatrix} \tag{17.154}$$

is an element of SU(2), the resulting one-to-one map is a rigid rotation of the Riemann sphere. Stereographic projection is thus revealed to be the geometric origin of the spinor representations of the rotation group.

If an analytic function $f(z)$ has no essential singularities anywhere on the Riemann sphere then f is *rational*, meaning that it can be written as $f(z) = P(z)/Q(z)$ for some polynomials P, Q.

We begin the proof of this fact by observing that $f(z)$ can have only a finite number of poles. If, to the contrary, f had an infinite number of poles then the compactness of S^2 would ensure that the poles would have a limit point somewhere. This would be a non-isolated singularity of f, and hence an essential singularity. Now suppose we have poles at $z_1, z_2, \ldots, z_N$ with principal parts

$$\sum_{m=1}^{m_n} \frac{b_{n,m}}{(z-z_n)^m}.$$

If one of the z_n is ∞, we first use a Möbius map to move it to some finite point. Then

$$F(z) = f(z) - \sum_{n=1}^{N}\sum_{m=1}^{m_n} \frac{b_{n,m}}{(z-z_n)^m} \tag{17.155}$$

is everywhere analytic, and therefore continuous, on S^2. But S^2 being compact and $F(z)$ being continuous implies that F is bounded. Therefore, by Liouville's theorem, it is a constant. Thus

$$f(z) = \sum_{n=1}^{N}\sum_{m=1}^{m_n} \frac{b_{n,m}}{(z-z_n)^m} + C, \tag{17.156}$$

and this is a rational function. If we made use of a Möbius map to move a pole at infinity, we use the inverse map to restore the original variables. This manoeuvre does not affect the claimed result because Möbius maps take rational functions to rational functions.

The map $z \mapsto f(z)$ given by the rational function

$$f(z) = \frac{P(z)}{Q(z)} = \frac{a_n z^n + a_{n-1}z^{n-1} + \cdots a_0}{b_n z^n + b_{n-1}z^{n-1} + \cdots b_0} \tag{17.157}$$

wraps the Riemann sphere n times around the target S^2. In other words, it is an n-to-one map.

17.6.2 Logarithms and branch cuts

The function $y = \ln z$ is defined to be the solution to $z = \exp y$. Unfortunately, since $\exp 2\pi i = 1$, the solution is not unique: if y is a solution, so is $y + 2\pi i$. Another way of looking at this is that if $z = \rho \exp i\theta$, with ρ real, then $y = \ln \rho + i\theta$, and the angle θ has the same $2\pi i$ ambiguity. Now there is no such thing as a "many valued function". By definition, a function is a machine into which we plug something and get a unique output. To make $\ln z$ into a legitimate function we must select a unique $\theta = \arg z$ for each z. This can be achieved by cutting the z plane along a curve extending from the *branch point* at $z = 0$ all the way to infinity. Exactly where we put this *branch cut* is not important; what *is* important is that it serves as an impenetrable fence preventing us from following the continuous evolution of the function along a path that winds around the origin.

Similar branch cuts serve to make fractional powers single-valued. We define the power z^α for non-integral α by setting

$$z^\alpha = \exp\{\alpha \ln z\} = |z|^\alpha e^{i\alpha\theta}, \tag{17.158}$$

where $z = |z|e^{i\theta}$. For the square root $z^{1/2}$ we get

$$z^{1/2} = \sqrt{|z|}e^{i\theta/2}, \tag{17.159}$$

where $\sqrt{|z|}$ represents the *positive* square root of $|z|$. We can therefore make this single-valued by a cut from 0 to ∞. To make $\sqrt{(z-a)(z-b)}$ single valued we only need to cut from a to b. (Why? – think this through!)

We can get away without cuts if we imagine the functions being maps *from* some set other than the complex plane. The new set is called a *Riemann surface*. It consists of a number of copies of the complex plane, one for each possible value of our "multivalued function". The map from this new surface is then single-valued, because each possible value of the function is the value of the function evaluated at a point on a different copy. The copies of the complex plane are called *sheets*, and are connected to each other in a manner dictated by the function. The cut plane may now be thought of as a drawing of one level of the multilayered Riemann surface. Think of an architect's floor plan of a spiral-floored multi-storey car park: if the architect starts drawing at one parking spot and works her way round the central core, at some point she will find that the floor has become the ceiling of the part already drawn. The rest of the structure will therefore have to be plotted on the plan of the next floor up – but exactly where she draws the division between one floor and the one above is rather arbitrary. The spiral carpark is a good model for the Riemann surface of the $\ln z$ function. See Figure 17.16.

To see what happens for a square root, follow $z^{1/2}$ along a curve circling the branch point singularity at $z = 0$. We come back to our starting point with the function having changed sign; a second trip along the same path would bring us back to the original value. The square root thus has only two sheets, and they are cross-connected as shown in Figure 17.17.

In Figures 17.16 and 17.17, we have shown the cross-connections being made rather abruptly along the cuts. This is not necessary – there is no singularity in the function at the cut – but it is often a convenient way to think about the structure of the surface. For example, the surface for $\sqrt{(z-a)(z-b)}$ also consists of two sheets. If we include the point at infinity, this surface can be thought of as two spheres, one inside the other, and cross-connected along the cut from a to b.

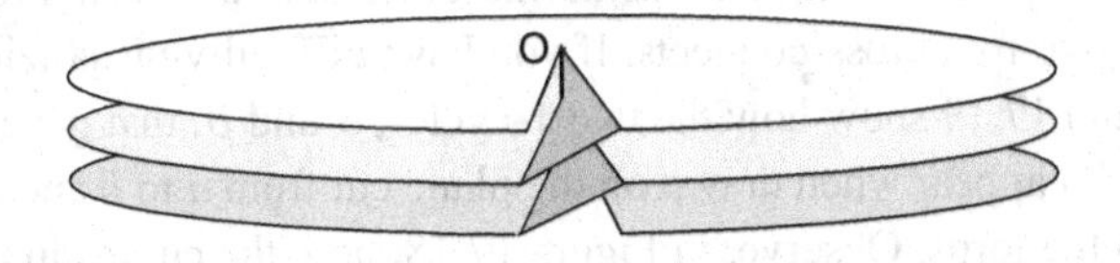

Figure 17.16 Part of the Riemann surface for $\ln Z$. Each time we circle the origin, we go up one level.

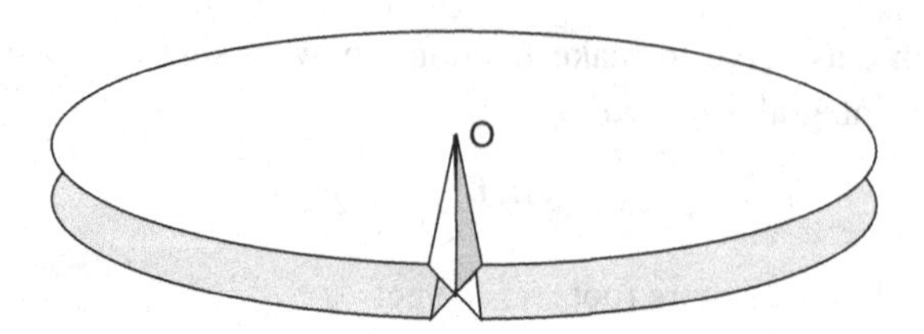

Figure 17.17 Part of the Riemann surface for $\sqrt{z}$. Two copies of $\mathbb{C}$ are cross-connected. Circling the origin once takes you to the lower level. A second circuit brings you back to the upper level.

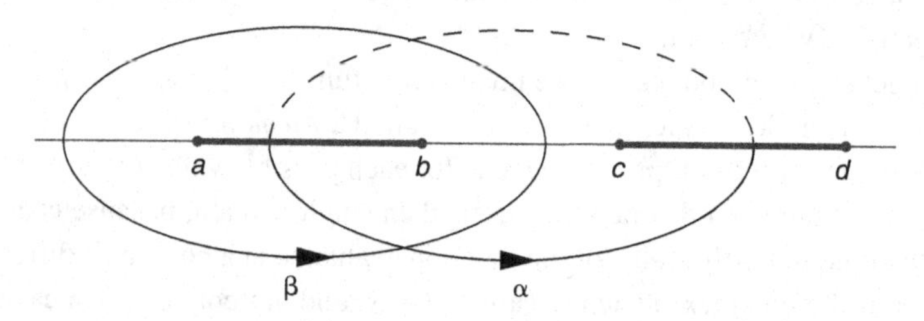

Figure 17.18 The 1-cycles α and β on the plane with two square-root branch cuts. The dashed part of α lies hidden on the second sheet of the Riemann surface.

17.6.3 Topology of Riemann surfaces

Riemann surfaces often have interesting topology. Indeed much of modern algebraic topology emerged from the need to develop tools to understand multiply connected Riemann surfaces. As we have seen, the complex numbers, with the point at infinity included, have the topology of a sphere. The $\sqrt{(z-a)(z-b)}$ surface is still topologically a sphere. To see this imagine continuously deforming the Riemann sphere by pinching it at the equator down to a narrow waist. Now squeeze the front and back of the waist together and (imagining that the surface can pass freely through itself) fold the upper half of the sphere inside the lower. The result is precisely the two-sheeted $\sqrt{(z-a)(z-b)}$ surface described above. The Riemann surface of the function $\sqrt{(z-a)(z-b)(z-c)(z-d)}$, which can be thought of as two spheres, one inside the other and connected along two cuts, one from a to b and one from c to d, is, however, a *torus*. Think of the torus as a bicycle inner tube. Imagine using the fingers of your left hand to pinch the front and back of the tube together and the fingers of your right hand to do the same on the diametrically opposite part of the tube. Now fold the tube about the pinch lines through itself so that one half of the tube is inside the other, and connected to the outer half through two square-root cross-connects. If you have difficulty visualizing this process, Figures 17.18 and 17.19 show how the two 1-cycles, α and β, that generate the homology group $H_1(T^2)$ appear when drawn on the plane cut from a to b and c to d, and then when drawn on the torus. Observe, in Figure 17.18, how the curves in the two-sheeted plane manage to intersect in only one point, just as they do when drawn on the torus in Figure 17.19.

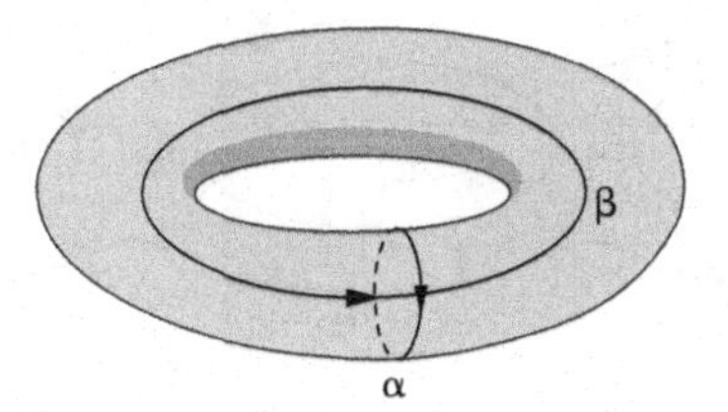

Figure 17.19 The 1-cycles α and β on the torus.

That the topology of the twice-cut plane is that of a torus has important consequences. This is because the *elliptic integral*

$$w = I^{-1}(z) = \int_{z_0}^{z} \frac{dt}{\sqrt{(t-a)(t-b)(t-c)(t-d)}} \tag{17.160}$$

maps the twice-cut z-plane one-to-one onto the torus, the latter being considered as the complex w plane with the points w and $w + n\omega_1 + m\omega_2$ identified. The two numbers $\omega_{1,2}$ are given by

$$\omega_1 = \oint_\alpha \frac{dt}{\sqrt{(t-a)(t-b)(t-c)(t-d)}},$$
$$\omega_2 = \oint_\beta \frac{dt}{\sqrt{(t-a)(t-b)(t-c)(t-d)}}, \tag{17.161}$$

and are called the *periods* of the *elliptic function* $z = I(w)$. The map $w \mapsto z = I(w)$ is a genuine function because the original z is uniquely determined by w. It is *doubly periodic* because

$$I(w + n\omega_1 + m\omega_2) = I(w), \quad n, m \in \mathbb{Z}. \tag{17.162}$$

The inverse "function" $w = I^{-1}(z)$ is not a genuine function of z, however, because w increases by ω_1 or ω_2 each time z goes around a curve deformable into α or β, respectively. The periods are complicated functions of a, b, c, d.

If you recall our discussion of de Rham's theorem from Chapter 4, you will see that the ω_i are the results of pairing the closed holomorphic 1-form.

$$\text{“}dw\text{”} = \frac{dz}{\sqrt{(z-a)(z-b)(z-c)(z-d)}} \in H^1(T^2) \tag{17.163}$$

with the two generators of $H_1(T^2)$. The quotation marks about dw are there to remind us that dw is not an exact form, i.e. it is not the exterior derivative of a single-valued function w. This cohomological interpretation of the periods of the elliptic function is the origin of the use of the word "period" in the context of de Rham's theorem. (See Section 19.5 for more information on elliptic functions.)

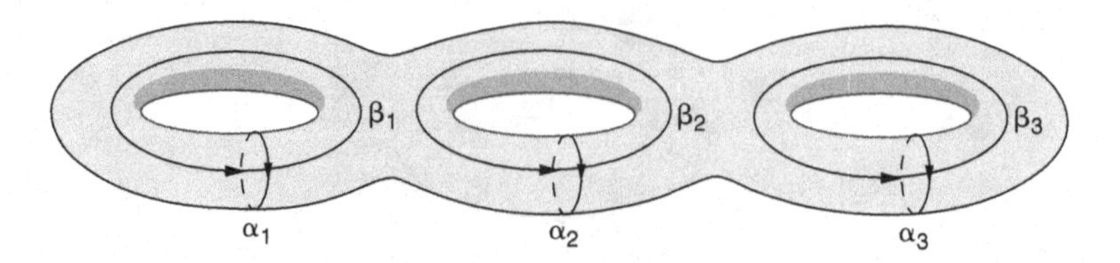

Figure 17.20 A surface M of genus 3. The non-bounding 1-cycles α_i and β_i form a basis of $H_1(M)$. The entire surface forms the single 2-cycle that spans $H_2(M)$.

More general Riemann surfaces are oriented 2-manifolds that can be thought of as the surfaces of doughnuts with g holes. The number g is called the *genus* of the surface. The sphere has $g = 0$ and the torus has $g = 1$. The Euler character of the Riemann surface of genus g is $\chi = 2(1 - g)$. For example, Figure 17.20 shows a surface of genus 3. The surface is in one piece, so $\dim H_0(M) = 1$. The other Betti numbers are $\dim H_1(M) = 6$ and $\dim H_2(M) = 1$, so

$$\chi = \sum_{p=0}^{2} (-1)^p \dim H_p(M) = 1 - 6 + 1 = -4, \tag{17.164}$$

in agreement with $\chi = 2(1 - 3) = -4$. For complicated functions, the genus may be infinite.

If we have two complex variables z and w then a polynomial relation $P(z, w) = 0$ defines a *complex algebraic curve*. Except for degenerate cases, this one (complex) dimensional curve is simultaneously a two (real) dimensional Riemann surface. With

$$P(z, w) = z^3 + 3w^2 z + w + 3 = 0, \tag{17.165}$$

for example, we can think of $z(w)$ as being a three-sheeted function of w defined by solving this cubic. Alternatively we can consider $w(z)$ to be the two-sheeted function of z obtained by solving the quadratic equation

$$w^2 + \frac{1}{3z} w + \frac{(3 + z^3)}{3z} = 0. \tag{17.166}$$

In each case the branch points will be located where two or more roots coincide. The roots of (17.166), for example, coincide when

$$1 - 12z(3 + z^3) = 0. \tag{17.167}$$

This quartic equation has four solutions, so there are four square-root branch points. Although constructed differently, the Riemann surface for $w(z)$ and the Riemann surface for $z(w)$ will have the same genus (in this case $g = 1$) because they are really one and the same object – the algebraic curve defined by the original polynomial equation.

In order to capture all its points at infinity, we often consider a complex algebraic curve as being a subset of $\mathbb{C}P^2$. To do this we make the defining equation homogeneous by introducing a third coordinate. For example, for (17.165) we make

$$P(z,w) = z^3 + 3w^2z + w + 3 \to P(z,w,v) = z^3 + 3w^2z + wv^2 + 3v^3. \quad (17.168)$$

The points where $P(z,w,v) = 0$ define[7] a *projective curve* lying in $\mathbb{C}P^2$. Places on this curve where the coordinate v is zero are the added points at infinity. Places where v is non-zero (and where we may as well set $v = 1$) constitute the original *affine curve*.

A generic (non-singular) curve

$$P(z,w) = \sum_{r,s} a_{rs} z^r w^s = 0, \quad (17.169)$$

with its points at infinity included, has genus

$$g = \frac{1}{2}(d-1)(d-2). \quad (17.170)$$

Here $d = \max(r+s)$ is the *degree* of the curve. This *degree–genus* relation is due to Plücker. It is not, however, trivial to prove. Also not easy to prove is Riemann's theorem of 1852 that *any* finite genus Riemann surface is the complex algebraic curve associated with some two-variable polynomial.

The two assertions in the previous paragraph seem to contradict each other. "Any" finite genus must surely include $g = 2$, but how can a genus 2 surface be a complex algebraic curve? There is no integer value of d such that $(d-1)(d-2)/2 = 2$. This is where the "non-singular" caveat becomes important. An affine curve $P(z,w) = 0$ is said to be *singular* at $\mathrm{P} = (z_0, w_0)$ if all of

$$P(z,w), \quad \frac{\partial P}{\partial z}, \quad \frac{\partial P}{\partial w},$$

vanish at P. A projective curve is singular at $\mathrm{P} \in \mathbb{C}P^2$ if all of

$$P(z,w,v), \quad \frac{\partial P}{\partial z}, \quad \frac{\partial P}{\partial w}, \quad \frac{\partial P}{\partial v}$$

are zero there. If the curve has a singular point then it degenerates and ceases to be a manifold. Now Riemann's construction does not guarantee an *embedding* of the surface into $\mathbb{C}P^2$, only an *immersion*. The distinction between these two concepts is that an immersed surface is allowed to self-intersect, while an embedded one is not. Being a double root of the defining equation $P(z,w) = 0$, a point of self-intersection is necessarily a singular point.

[7] A homogeneous polynomial $P(z,w,v)$ of degree n does not provide a map from $\mathbb{C}P^2 \to \mathbb{C}$ because $P(\lambda z, \lambda w, \lambda v) = \lambda^n P(z,w,v)$ usually depends on λ, while the coordinates $(\lambda z, \lambda w, \lambda v)$ and (z,w,v) correspond to the same point in $\mathbb{C}P^2$. The *zero set* where $P = 0$ is, however, well defined in $\mathbb{C}P^2$.

As an illustration of a singular curve, consider our earlier example of the curve

$$w^2 = (z-a)(z-b)(z-c)(z-d) \tag{17.171}$$

whose Riemann surface we know to be a torus once some points are added at infinity, and when a, b, c, d are all distinct. The degree–genus formula applied to this degree-four curve gives, however, $g = 3$ instead of the expected $g = 1$. This is because the corresponding projective curve

$$w^2v^2 = (z-av)(z-bv)(z-cv)(z-dv) \tag{17.172}$$

has a *tacnode* singularity at the point $(z, w, v) = (0, 1, 0)$. Rather than investigate this rather complicated singularity at infinity, we will consider the simpler case of what happens if we allow b to coincide with c. When b and c merge, the finite point $\mathrm{P} = (w_0, z_0) = (0, b)$ becomes singular. Near the singularity, the equation defining our curve looks like

$$0 = w^2 - ad\,(z-b)^2, \tag{17.173}$$

which is the equation of two lines, $w = \sqrt{ad}\,(z-b)$ and $w = -\sqrt{ad}\,(z-b)$, that intersect at the point $(w, z) = (0, b)$. To understand what is happening topologically it is first necessary to realize that a *complex* line is a copy of $\mathbb{C}$ and hence, after the point at infinity is included, is topologically a sphere. A pair of intersecting complex lines is therefore topologically a pair of spheres sharing a common point. Our degenerate curve only looks like a pair of lines near the point of intersection however. To see the larger picture, look back at the figure of the twice-cut plane where we see that as b approaches c we have an α cycle of zero total length. A zero length cycle means that the circumference of the torus becomes zero at P, so that it looks like a bent sausage with its two ends sharing the common point P. Instead of two separate spheres, our sausage is equivalent to a single two-sphere with two points identified.

As it stands, such a set is no longer a manifold because any neighbourhood of P will contain bits of both ends of the sausage, and therefore cannot be given coordinates that

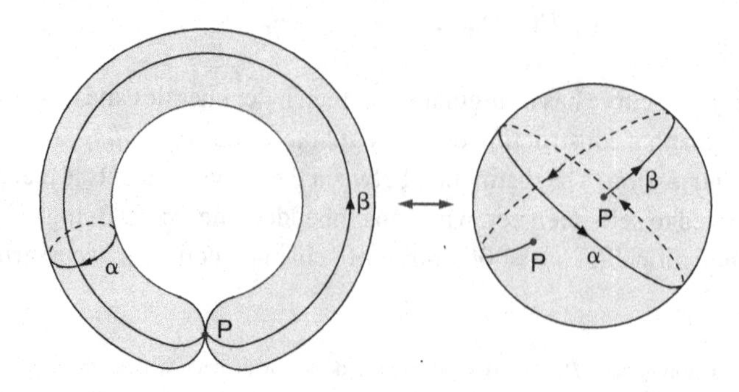

Figure 17.21 A degenerate torus is topologically the same as a sphere with two points identified.

make it look like a region in $\mathbb{R}^2$. We can, however, simply agree to delete the common point, and then plug the resulting holes in the sausage ends with two distinct points. The new set is again a manifold, and topologically a sphere. From the viewpoint of the pair of intersecting lines, this construction means that we stay on one line, and ignore the other as it passes through.

A similar *resolution of singularities* allows us to regard immersed surfaces as non-singular manifolds, and it is in this sense that Riemann's theorem is to be understood. When n such self-intersection double points are deleted and replaced by pairs of distinct points the degree–genus formula becomes

$$g = \frac{1}{2}(d-1)(d-2) - n, \tag{17.174}$$

and this can take any integer value.

17.6.4 Conformal geometry of Riemann surfaces

In this section we recall Hodge's theory of harmonic forms from Section 13.7.1, and see how it looks from a complex-variable perspective. This viewpoint reveals a relationship between Riemann surfaces and Riemann manifolds that forms an important ingredient in string and conformal field theory.

Isothermal coordinates and complex structure

Suppose we have a two-dimensional orientable Riemann manifold M with metric

$$ds^2 = g_{ij}\, dx^i dx^j. \tag{17.175}$$

In two dimensions g_{ij} has three independent components. When we make a coordinate transformation we have two arbitrary functions at our disposal, and so we can use this freedom to select local coordinates in which only one independent component remains. The most useful choice is *isothermal* (also called *conformal*) coordinates x, y in which the metric tensor is diagonal, $g_{ij} = e^\sigma \delta_{ij}$, and so

$$ds^2 = e^\sigma (dx^2 + dy^2). \tag{17.176}$$

The e^σ is called the *scale factor* or *conformal factor*. If we set $z = x + iy$ and $\bar{z} = x - iy$ the metric becomes

$$ds^2 = e^{\sigma(z,\bar{z})} d\bar{z}\, dz. \tag{17.177}$$

We can construct isothermal coordinates for some open neighbourhood of any point in M. If in an overlapping isothermal coordinate patch the metric is

$$ds^2 = e^{\tau(\zeta,\bar{\zeta})} d\bar{\zeta}\, d\zeta, \tag{17.178}$$

and if the coordinates have the same orientation, then in the overlap region ζ must be a function only of z and $\overline{\zeta}$ a function only of $\overline{z}$. This is so that

$$e^{\tau(\zeta,\overline{\zeta})}d\overline{\zeta}\,d\zeta = e^{\sigma(z,\overline{z})}\left|\frac{dz}{d\zeta}\right|^2 d\overline{\zeta}\,d\zeta \tag{17.179}$$

without any $d\zeta^2$ or $d\overline{\zeta}^2$ terms appearing. A manifold with an atlas of complex charts whose change-of-coordinate formulae are holomorphic in this way is said to be a *complex manifold*, and the coordinates endow it with a *complex structure*. The existence of a global complex structure allows us to define the notion of meromorphic and rational functions on M. Our Riemann *manifold* is therefore also a Riemann *surface*.

While any compact, orientable, two-dimensional Riemann manifold has a complex structure that is determined by the metric, the mapping: *metric* $\to$ *complex structure* is not one-to-one. Two metrics g_{ij}, $\tilde{g}_{ij}$ that are related by a conformal scale factor

$$g_{ij} = \lambda(x^1, x^2)\tilde{g}_{ij} \tag{17.180}$$

give rise to the same complex structure. Conversely, a pair of two-dimensional Riemann manifolds having the same complex structure have metrics that are related by a scale factor.

The use of isothermal coordinates simplifies many computations. Firstly, observe that $g^{ij}/\sqrt{g} = \delta_{ij}$, the conformal factor having cancelled. If you look back at its definition, you will see that this means that when the Hodge "$\star$" map acts on 1-forms, the result is independent of the metric. If ω is a 1-form

$$\omega = p\,dx + q\,dy, \tag{17.181}$$

then

$$\star\omega = -q\,dx + p\,dy. \tag{17.182}$$

Note that, on 1-forms,

$$\star\star = -1. \tag{17.183}$$

With $z = x + iy$, $\overline{z} = x - iy$, we have

$$\omega = \frac{1}{2}(p - iq)\,dz + \frac{1}{2}(p + iq)\,d\overline{z}. \tag{17.184}$$

Let us focus on the dz part:

$$A = \frac{1}{2}(p - iq)\,dz = \frac{1}{2}(p - iq)(dx + idy). \tag{17.185}$$

Then

$$\star A = \frac{1}{2}(p - iq)(dy - idx) = -iA. \tag{17.186}$$

Similarly, with

$$B = \frac{1}{2}(p + iq)\, d\bar{z} \tag{17.187}$$

we have

$$\star B = iB. \tag{17.188}$$

Thus the dz and $d\bar{z}$ parts of the original form are separately eigenvectors of $\star$ with different eigenvalues. We use this observation to construct a resolution of the identity Id into the sum of two projection operators

$$\begin{aligned} Id &= \frac{1}{2}(1 + i\star) + \frac{1}{2}(1 - i\star), \\ &= P + \overline{P}, \end{aligned} \tag{17.189}$$

where P projects on the dz part and $\overline{P}$ onto the $d\bar{z}$ part of the form.

The original form is harmonic if it is both closed $d\omega = 0$, and co-closed $d\star\omega = 0$. Thus, in two dimensions, the notion of being harmonic (i.e. a solution of Laplace's equation) is independent of what metric we are given. If ω is a harmonic form, then $(p - iq)dz$ and $(p + iq)d\bar{z}$ are separately closed. Observe that $(p - iq)dz$ being closed means that $\partial_{\bar{z}}(p - iq) = 0$, and so $p - iq$ is a holomorphic (and hence harmonic) function. Since both $(p - iq)$ and dz depend only on z, we will call $(p - iq)dz$ a holomorphic 1-form. The complex conjugate form

$$\overline{(p - iq)dz} = (p + iq)d\bar{z} \tag{17.190}$$

then depends only on $\bar{z}$ and is antiholomorphic.

Riemann bilinear relations

As an illustration of the interplay of harmonic forms and two-dimensional topology, we derive some famous formulæ due to Riemann. These formulæ have applications in string theory and in conformal field theory.

Suppose that M is a Riemann surface of genus g, with α_i, β_i, $i = 1, \ldots, g$, the representative generators of $H_1(M)$ that intersect as shown in Figure 17.20. By applying Hodge–de Rham to this surface, we know that we can select a set of $2g$ independent, real, harmonic, 1-forms as a basis of $H^1(M, \mathbb{R})$. With the aid of the projector P we can assemble these into g holomorphic closed 1-forms ω_i, together with g antiholomorphic closed 1-forms $\overline{\omega}_i$, the original $2g$ real forms being recovered from these as $\omega_i + \overline{\omega}_i$

and $\star(\omega_i + \overline{\omega}_i) = i(\overline{\omega}_i - \omega_i)$. A physical interpretation of these forms is as the z and $\overline{z}$ components of irrotational and incompressible fluid flows on the surface M. It is not surprising that such flows form a $2g$ real dimensional, or g complex dimensional, vector space because we can independently specify the circulation $\oint \mathbf{v}\cdot d\mathbf{r}$ around each of the $2g$ generators of $H_1(M)$. If the flow field has (covariant) components v_x, v_y, then $\omega = v_z dz$ where $v_z = (v_x - iv_y)/2$, and $\overline{\omega} = v_{\overline{z}} d\overline{z}$ where $v_{\overline{z}} = (v_x + iv_y)/2$.

Suppose now that a and b are closed 1-forms on M. Then, either by exploiting the powerful and general intersection-form formula (13.77) or by cutting open the surface along the curves α_i, β_i and using the more direct strategy that gave us (13.79), we find that

$$\int_M a \wedge b = \sum_{i=1}^{g} \left\{ \int_{\alpha_i} a \int_{\beta_i} b - \int_{\beta_i} a \int_{\alpha_i} b \right\}. \tag{17.191}$$

We use this formula to derive two *bilinear relations* associated with a closed holomorphic 1-form ω. Firstly we compute its Hodge inner-product norm

$$\begin{aligned} \|\omega\|^2 \equiv \int_M \omega \wedge \star\overline{\omega} &= \sum_{i=1}^{g} \left\{ \int_{\alpha_i} \omega \int_{\beta_i} \star\overline{\omega} - \int_{\beta_i} \omega \int_{\alpha_i} \star\overline{\omega} \right\} \\ &= i \sum_{i=1}^{g} \left\{ \int_{\alpha_i} \omega \int_{\beta_i} \overline{\omega} - \int_{\beta_i} \omega \int_{\alpha_i} \overline{\omega} \right\} \\ &= i \sum_{i=1}^{g} \left\{ A_i \overline{B}_i - B_i \overline{A}_i \right\}, \end{aligned} \tag{17.192}$$

where $A_i = \int_{\alpha_i} \omega$ and $B_i = \int_{\beta_i} \omega$. We have used the fact that $\overline{\omega}$ is an antiholomorphic 1 form and thus an eigenvector of $\star$ with eigenvalue i. It follows, therefore, that if all the A_i are zero then $\|\omega\| = 0$ and so $\omega = 0$.

Let $A_{ij} = \int_{\alpha_i} \omega_j$. The determinant of the matrix A_{ij} is non-zero: if it were zero, then there would be numbers λ_i, not all zero, such that

$$0 = A_{ij}\lambda_j = \int_{\alpha_i} (\omega_j \lambda_j), \tag{17.193}$$

but, by (17.192), this implies that $\|\omega_j\lambda_j\| = 0$ and hence $\omega_j\lambda_j = 0$, contrary to the linear independence of the ω_i. We can therefore solve the equations

$$A_{ij}\lambda_{jk} = \delta_{ik} \tag{17.194}$$

for the numbers λ_{jk} and use these to replace each of the ω_i by the linear combination $\omega_j\lambda_{ji}$. The new ω_i then obey $\int_{\alpha_i} \omega_j = \delta_{ij}$. From now on we suppose that this has been done.

Define $\tau_{ij} = \int_{\beta_i} \omega_j$. Observe that $dz \wedge dz = 0$ forces $\omega_i \wedge \omega_j = 0$, and therefore we have a second relation

$$\begin{aligned} 0 = \int_M \omega_m \wedge \omega_n &= \sum_{i=1}^{g} \left\{ \int_{\alpha_i} \omega_m \int_{\beta_i} \omega_n - \int_{\beta_i} \omega_m \int_{\alpha_i} \omega_n \right\} \\ &= \sum_{i=1}^{g} \{\delta_{im}\tau_{in} - \tau_{im}\delta_{in}\} \\ &= \tau_{mn} - \tau_{nm}. \end{aligned} \tag{17.195}$$

The matrix τ_{ij} is therefore symmetric. A similar compuation shows that

$$\|\lambda_i \omega_i\|^2 = 2\overline{\lambda}_i (\text{Im}\, \tau_{ij}) \lambda_j \tag{17.196}$$

so the matrix $(\text{Im}\, \tau_{ij})$ is positive definite. The set of such symmetric matrices whose imaginary part is positive definite is called the *Siegel upper half-plane*. Not every such matrix corresponds to a Riemann surface, but when it does it encodes all information about the shape of the Riemann manifold M that is left invariant under conformal rescaling.

17.7 Further exercises and problems

Exercise 17.11: *Harmonic partners.* Show that the function

$$u = \sin x \cosh y + 2 \cos x \sinh y$$

is harmonic. Determine the corresponding analytic function $u + iv$.

Exercise 17.12: *Möbius maps.* The map

$$z \mapsto w = \frac{az + b}{cz + d}$$

is called a Möbius transformation. These maps are important because they are the only one-to-one conformal maps of the Riemann sphere onto itself.

(a) Show that two successive Möbius transformations

$$z' = \frac{az + b}{cz + d}, \qquad z'' = \frac{Az' + B}{Cz' + D}$$

give rise to another Möbius transformation, and show that the rule for combining them is equivalent to matrix multiplication.

(b) Let z_1, z_2, z_3, z_4 be complex numbers. Show that a necessary and sufficient condition for the four points to be concyclic is that their *cross-ratio*

$$\{z_1, z_2, z_3, z_4\} \stackrel{\text{def}}{=} \frac{(z_1 - z_4)(z_3 - z_2)}{(z_1 - z_2)(z_3 - z_4)}$$

be real. (Hint: use a well-known property of opposite angles of a cyclic quadrilateral.) Show that Möbius transformations leave the cross-ratio invariant, and thus take circles into circles.

Exercise 17.13: *Hyperbolic geometry*. The Riemann metric for the Poincaré-disc model of Lobachevski's hyperbolic plane (see Exercises 1.7 and 12.13) can be taken to be

$$ds^2 = \frac{4|dz|^2}{(1-|z|^2)^2}, \quad |z|^2 < 1.$$

(a) Show that the Möbius transformation

$$z \mapsto w = e^{i\lambda}\frac{z-a}{\bar{a}z-1}, \qquad |a| < 1, \quad \lambda \in \mathbb{R}$$

provides a one-to-one map of the interior of the unit disc onto itself. Show that these maps form a group.

(b) Show that the hyperbolic-plane metric is left invariant under the group of maps in part (a). Deduce that such maps are orientation-preserving *isometries* of the hyperbolic plane.

(c) Use the circle-preserving property of the Möbius maps to deduce that circles in hyperbolic geometry are represented in the Poincaré disc by Euclidean circles that lie entirely within the disc.

The conformal maps of part (a) are in fact the *only* orientation-preserving isometries of the hyperbolic plane. With the exception of circles centred at $z = 0$, the center of the hyperbolic circle does not coincide with the centre of its representative Euclidean circle. Euclidean circles that are internally tangent to the boundary of the unit disc have infinite hyperbolic radius and their hyperbolic centres lie on the boundary of the unit disc and hence at hyperbolic infinity. They are known as *horocycles*.

Exercise 17.14: *Rectangle to ellipse.* Consider the map $w \mapsto z = \sin w$. Draw a picture of the image, in the z plane, of the interior of the rectangle with corners $u = \pm\pi/2$, $v = \pm\lambda$ ($w = u + iv$). Show which points correspond to the corners of the rectangle, and verify that the vertex angles remain $\pi/2$. At what points does the isogonal property fail?

Exercise 17.15: The part of the negative real axis where $x < -1$ is occupied by a conductor held at potential $-V_0$. The positive real axis for $x > +1$ is similarly occupied by a conductor held at potential $+V_0$. The conductors extend to infinity in both directions perpendicular to the x–y plane, and so the potential V satisfies the two-dimensional Laplace equation.

(a) Find the image in the ζ plane of the cut z plane where the cuts run from -1 to $-\infty$ and from $+1$ to $+\infty$ under the map $z \mapsto \zeta = \sin^{-1} z$.

(b) Use your answer from part (a) to solve the electrostatic problem and show that the field lines and equipotentials are conic sections of the form $ax^2 + by^2 = 1$. Find expressions for a and b for both the field lines and the equipotentials and draw a labelled sketch to illustrate your results.

Exercise 17.16: Draw the image under the map $z \mapsto w = e^{\pi z/a}$ of the infinite strip S, consisting of those points $z = x + iy \in \mathbb{C}$ for which $0 < y < a$. Label enough points to show which point in the w plane corresponds to which in the z plane. Hence or otherwise show that the Dirichlet Green function $G(x,y;x_0,y_0)$ that obeys

$$\nabla^2 G = \delta(x - x_0)\delta(y - y_0)$$

in S, and $G(x,y;x_0,y_0) = 0$ for (x,y) on the boundary of S, can be written as

$$G(x,y;x_0,y_0) = \frac{1}{2\pi} \ln |\sinh(\pi(z - z_0)/2a)| + \dots$$

The dots indicate the presence of a second function, similar to the first, that you should find. Assume that $(x_0,y_0) \in S$.

Exercise 17.17: State Laurent's theorem for functions analytic in an annulus. Include formulae for the coefficients of the expansion. Show that, suitably interpreted, this theorem reduces to a form of Fourier's theorem for functions analytic in a neighbourhood of the unit circle.

Exercise 17.18: *Laurent paradox.* Show that in the annulus $1 < |z| < 2$ the function

$$f(z) = \frac{1}{(z-1)(2-z)}$$

has a Laurent expansion in powers of z. Find the coefficients. The part of the series with negative powers of z does not terminate. Does this mean that $f(z)$ has an essential singularity at $z = 0$?

Exercise 17.19: Assuming the following series

$$\frac{1}{\sinh z} = \frac{1}{z} - \frac{1}{6}z + \frac{7}{16}z^3 + \dots,$$

evaluate the integral

$$I = \oint_{|z|=1} \frac{1}{z^2 \sinh z} dz.$$

Now evaluate the integral

$$I = \oint_{|z|=4} \frac{1}{z^2 \sinh z} dz.$$

(Hint: the zeros of $\sinh z$ lie at $z = n\pi i$.)

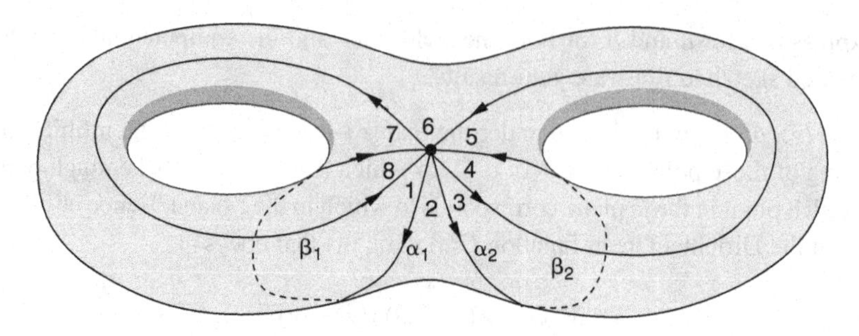

Figure 17.22 Concurrent 1-cycles on a genus-2 surface.

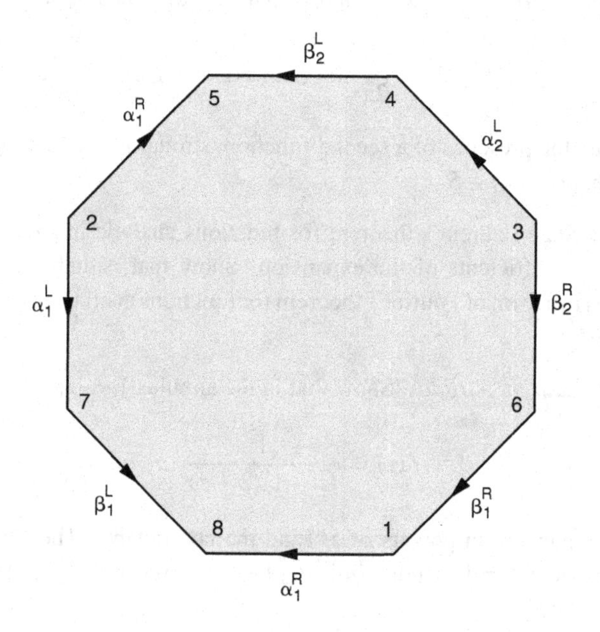

Figure 17.23 The cut-open genus-2 surface. The superscripts L and R denote respectively the left and right sides of each 1-cycle, viewed from the direction of the arrow orienting the cycle.

Exercise 17.20: State the theorem relating the difference between the number of poles and zeros of $f(z)$ in a region to the winding number of the argument of $f(z)$. Hence, or otherwise, evaluate the integral

$$I = \oint_C \frac{5z^4 + 1}{z^5 + z + 1} dz$$

where C is the circle $|z| = 2$. Prove, including a statement of any relevent theorem, any assertions you make about the locations of the zeros of $z^5 + z + 1$.

Exercise 17.21: *Arcsine branch cuts.* Let $w = \sin^{-1} z$. Show that

$$w = n\pi \pm i \ln\{iz + \sqrt{1 - z^2}\}$$

with the $\pm$ being selected depending on whether n is odd or even. Where would you put cuts to ensure that w is a single-valued function?

Problem 17.22: *Cutting open a genus-2 surface.* The Riemann surface for the function

$$y = \sqrt{(z-a_1)(z-a_2)(z-a_3)(z-a_4)(z-a_5)(z-a_6)}$$

has genus $g = 2$. Such a surface M is sketched in Figure 17.22, where the four independent 1-cycles $\alpha_{1,2}$ and $\beta_{1,2}$ that generate $H_1(M)$ have been drawn so that they share a common vertex.

(a) Realize the genus-2 surface as two copies of $\mathbb{C} \cup \{\infty\}$ cross-connected by three square-root branch cuts. Sketch how the 1-cycles α_i and β_i, $i = 1, 2$ of Figure 17.22 appear when drawn on your thrice-cut plane.
(b) Cut the surface open along the four 1-cycles, and convince yourself that the resulting surface is homeomorphic to the octagonal region appearing in Figure 17.23.
(c) Apply the direct method that gave us (13.79) to the octagonal region of part (b). Hence show that for closed 1-forms a, b, on the surface we have

$$\int_M a \wedge b = \sum_{i=1}^{2} \left\{ \int_{\alpha_i} a \int_{\beta_i} b - \int_{\beta_i} a \int_{\alpha_i} b \right\}.$$

18

Applications of complex variables

In this chapter we will find uses for what we have learned of complex variables. The applications will range from the elementary to the sophisticated.

18.1 Contour integration technology

The goal of contour integration technology is to evaluate ordinary, real-variable, definite integrals. We have already met the basic tool, the *residue theorem*:

Theorem: *Let $f(z)$ be analytic within and on the boundary $\Gamma = \partial D$ of a simply connected domain D, with the exception of a finite number of points at which the function has poles. Then*

$$\oint_\Gamma f(z)\, dz = \sum_{\text{poles} \in D} 2\pi i\, (\text{residue at pole}).$$

18.1.1 Tricks of the trade

The effective application of the residue theorem is something of an art, but there are useful classes of integrals which we can learn to recognize.

Rational trigonometric expressions

Integrals of the form

$$\int_0^{2\pi} F(\cos\theta, \sin\theta)\, d\theta \tag{18.1}$$

are dealt with by writing $\cos\theta = \frac{1}{2}(z + \bar{z})$, $\sin\theta = \frac{1}{2i}(z - \bar{z})$ and integrating around the unit circle. For example, let a, b be real and $b < a$. Then

$$I = \int_0^{2\pi} \frac{d\theta}{a + b\cos\theta} = \frac{2}{i}\oint_{|z|=1} \frac{dz}{bz^2 + 2az + b} = \frac{2}{ib}\oint \frac{dz}{(z-\alpha)(z-\beta)}. \tag{18.2}$$

Since $\alpha\beta = 1$, only one pole is within the contour. This is at

$$\alpha = (-a + \sqrt{a^2 - b^2})/b. \tag{18.3}$$

The residue is

$$\frac{2}{ib}\frac{1}{\alpha-\beta}=\frac{1}{i}\frac{1}{\sqrt{a^2-b^2}}. \tag{18.4}$$

Therefore, the integral is given by

$$I=\frac{2\pi}{\sqrt{a^2-b^2}}. \tag{18.5}$$

These integrals are, of course, also do-able by the "t" substitution $t=\tan(\theta/2)$, whence

$$\sin\theta=\frac{2t}{1+t^2},\quad \cos\theta=\frac{1-t^2}{1+t^2},\quad d\theta=\frac{2dt}{1+t^2}, \tag{18.6}$$

followed by a partial fraction decomposition. The labour is perhaps slightly less using the contour method.

Rational functions

Integrals of the form

$$\int_{-\infty}^{\infty} R(x)\,dx, \tag{18.7}$$

where $R(x)$ is a rational function of x with the degree of the denominator exceeding the degree of the numerator by two or more, may be evaluated by integrating around a rectangle from $-A$ to $+A$, A to $A+iB$, $A+iB$ to $-A+iB$, and back down to $-A$. Because the integrand decreases at least as fast as $1/|z|^2$ as z becomes large, we see that if we let $A, B\to\infty$, the contributions from the unwanted parts of the contour become negligible. Thus

$$I=2\pi i\left(\sum \text{Residues of poles in upper half-plane}\right). \tag{18.8}$$

We could also use a rectangle in the lower half-plane with the result

$$I=-2\pi i\left(\sum \text{Residues of poles in lower half-plane}\right). \tag{18.9}$$

This must give the same answer.

For example, let n be a positive integer and consider

$$I=\int_{-\infty}^{\infty}\frac{dx}{(1+x^2)^n}. \tag{18.10}$$

The integrand has an n-th order pole at $z = \pm i$. Suppose we close the contour in the upper half-plane. The new contour encloses the pole at $z = +i$ and we therefore need to compute its residue. We set $z - i = \zeta$ and expand

$$\begin{aligned}\frac{1}{(1+z^2)^n} &= \frac{1}{[(i+\zeta)^2+1]^n} = \frac{1}{(2i\zeta)^n}\left(1-\frac{i\zeta}{2}\right)^{-n} \\ &= \frac{1}{(2i\zeta)^n}\left(1+n\left(\frac{i\zeta}{2}\right)+\frac{n(n+1)}{2!}\left(\frac{i\zeta}{2}\right)^2+\cdots\right). \end{aligned} \tag{18.11}$$

The coefficient of ζ^{-1} is

$$\frac{1}{(2i)^n}\frac{n(n+1)\cdots(2n-2)}{(n-1)!}\left(\frac{i}{2}\right)^{n-1} = \frac{1}{2^{2n-1}i}\frac{(2n-2)!}{((n-1)!)^2}. \tag{18.12}$$

The integral is therefore

$$I = \frac{\pi}{2^{2n-2}}\frac{(2n-2)!}{((n-1)!)^2}. \tag{18.13}$$

These integrals can also be done by partial fractions.

18.1.2 Branch-cut integrals

Integrals of the form

$$I = \int_0^\infty x^{\alpha-1}R(x)dx, \tag{18.14}$$

where $R(x)$ is rational, can be evaluated by integration round a slotted circle (or "key-hole") contour (Figure 18.1). A little more work is required to extract the answer, though.

For example, consider

$$I = \int_0^\infty \frac{x^{\alpha-1}}{1+x}dx, \quad 0 < \mathrm{Re}\,\alpha < 1. \tag{18.15}$$

The restrictions on the range of α are necessary for the integral to converge at its upper and lower limits.

We take Γ to be a circle of radius Λ centred at $z = 0$, with a slot indentation designed to exclude the positive real axis, which we take as the branch cut of $z^{\alpha-1}$, and a small circle of radius ϵ about the origin. The branch of the fractional power is defined by setting

$$z^{\alpha-1} = \exp[(\alpha-1)(\ln|z|+i\theta)], \tag{18.16}$$

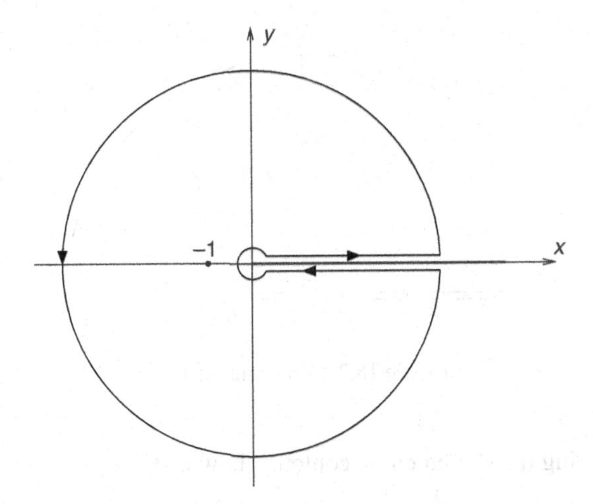

Figure 18.1 A slotted circle contour Γ of outer radius Λ and inner radius ϵ.

where we will take θ to be zero immediately above the real axis, and 2π immediately below it. With this definition the residue at the pole at $z = -1$ is $e^{i\pi(\alpha-1)}$. The residue theorem therefore tells us that

$$\oint_\Gamma \frac{z^{\alpha-1}}{1+z}dz = 2\pi i e^{\pi i(\alpha-1)}. \tag{18.17}$$

The integral decomposes as

$$\oint_\Gamma \frac{z^{\alpha-1}}{1+z}dz = \oint_{|z|=\Lambda} \frac{z^{\alpha-1}}{1+z}dz + (1 - e^{2\pi i(\alpha-1)}) \int_\epsilon^\Lambda \frac{x^{\alpha-1}}{1+x}dx - \oint_{|z|=\epsilon} \frac{z^{\alpha-1}}{1+z}dz. \tag{18.18}$$

As we send Λ off to infinity we can ignore the "1" in the denominator compared to the z, and so estimate

$$\left|\oint_{|z|=\Lambda} \frac{z^{\alpha-1}}{1+z}dz\right| \to \left|\oint_{|z|=\Lambda} z^{\alpha-2}dz\right| \le 2\pi\Lambda \times \Lambda^{\mathrm{Re}\,(\alpha)-2}. \tag{18.19}$$

This tends to zero provided that $\mathrm{Re}\,\alpha < 1$. Similarly, provided $0 < \mathrm{Re}\,\alpha$, the integral around the small circle about the origin tends to zero with ϵ. Thus

$$-e^{\pi i\alpha} 2\pi i = \left(1 - e^{2\pi i(\alpha-1)}\right) I. \tag{18.20}$$

We conclude that

$$I = \frac{2\pi i}{(e^{\pi i\alpha} - e^{-\pi i\alpha})} = \frac{\pi}{\sin \pi\alpha}. \tag{18.21}$$

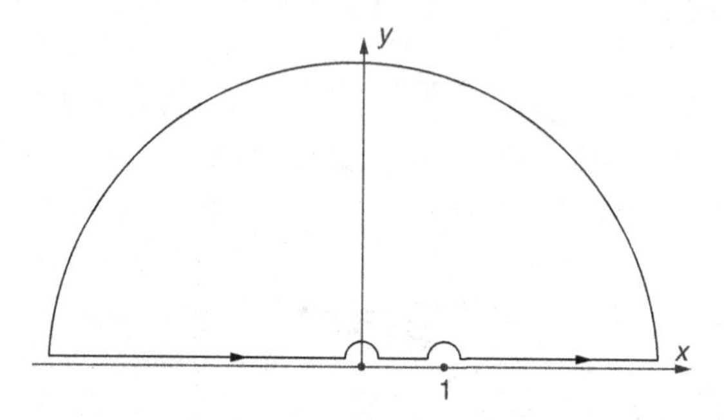

Figure 18.2 The contour Γ_1.

Exercise 18.1: Using the slotted circle contour, show that

$$I = \int_0^\infty \frac{x^{p-1}}{1+x^2}dx = \frac{\pi}{2\sin(\pi p/2)} = \frac{\pi}{2}\operatorname{cosec}(\pi p/2), \quad 0 < p < 2.$$

Exercise 18.2: Integrate $z^{a-1}/(z-1)$ around a contour Γ_1 consisting of a semicircle in the upper half-plane together with the real axis indented at $z = 0$ and $z = 1$ (see Figure 18.2) to get

$$0 = \oint_\Gamma \frac{z^{a-1}}{z-1}dz = P\int_0^\infty \frac{x^{a-1}}{x-1}dx - i\pi + (\cos\pi a + i\sin\pi a)\int_0^\infty \frac{x^{a-1}}{x+1}dx.$$

As usual, the symbol P in front of the integral sign denotes a *principal-part* integral, meaning that we must omit an infinitesimal segment of the contour symmetrically disposed about the pole at $z = 1$. The term $-i\pi$ comes from integrating around the small semicircle about this point. We get $-1/2$ of the residue because we have only a half-circle, and that traversed in the "wrong" direction. **Warning**: this fractional residue result is only true when we indent to avoid a *simple pole* – i.e. one that is of order one.

Now take real and imaginary parts and deduce that

$$\int_0^\infty \frac{x^{a-1}}{1+x}dx = \frac{\pi}{\sin\pi\alpha}, \quad 0 < \operatorname{Re} a < 1,$$

and

$$P\int_0^\infty \frac{x^{a-1}}{1-x}dx = \pi\cot\pi a, \quad 0 < \operatorname{Re} a < 1.$$

18.1.3 Jordan's lemma

We often need to evaluate Fourier integrals

$$I(k) = \int_{-\infty}^{\infty} e^{ikx} R(x)\, dx \tag{18.22}$$

with $R(x)$ a rational function. For example, the Green function for the operator $-\partial_x^2 + m^2$ is given by

$$G(x) = \int_{-\infty}^{\infty} \frac{dk}{2\pi} \frac{e^{ikx}}{k^2 + m^2}. \tag{18.23}$$

Suppose $x \in \mathbb{R}$ and $x > 0$. Then, in contrast to the analogous integral without the exponential function, we have no flexibility in closing the contour in the upper or lower half-plane. The function e^{ikx} grows without limit as we head south in the lower half-plane, but decays rapidly in the upper half-plane. This means that we may close the contour without changing the value of the integral by adding a large upper-half-plane semicircle (Figure 18.3).

The modified contour encloses a pole at $k = im$, and this has residue $i/(2m)e^{-mx}$. Thus

$$G(x) = \frac{1}{2m} e^{-mx}, \quad x > 0. \tag{18.24}$$

For $x < 0$, the situation is reversed, and we must close in the lower half-plane. The residue of the pole at $k = -im$ is $-i/(2m)e^{mx}$, but the minus sign is cancelled because the contour goes the "wrong way" (clockwise). Thus

$$G(x) = \frac{1}{2m} e^{+mx}, \quad x < 0. \tag{18.25}$$

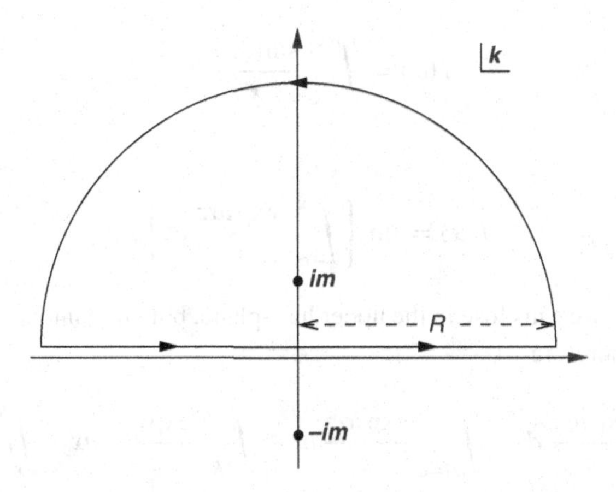

Figure 18.3 Closing the contour in the upper half-plane.

We can combine the two results as

$$G(x) = \frac{1}{2m} e^{-m|x|}. \tag{18.26}$$

The formal proof that the added semicircles make no contribution to the integral when their radius becomes large is known as *Jordan's lemma*:

Lemma: *Let Γ be a semicircle, centred at the origin, and of radius R. Suppose that*

(i) $f(z)$ is meromorphic in the upper half-plane;
(ii) $f(z)$ tends uniformly to zero as $|z| \to \infty$ for $0 < \arg z < \pi$;
(iii) the number λ is real and positive.

Then

$$\int_\Gamma e^{i\lambda z} f(z)\, dz \to 0, \quad \text{as} \quad R \to \infty. \tag{18.27}$$

To establish this, we assume that R is large enough that $|f| < \epsilon$ on the contour, and make a simple estimate

$$\begin{aligned}\left| \int_\Gamma e^{i\lambda z} f(z)\, dz \right| &< 2R\epsilon \int_0^{\pi/2} e^{-\lambda R \sin\theta}\, d\theta \\ &< 2R\epsilon \int_0^{\pi/2} e^{-2\lambda R\theta/\pi}\, d\theta \\ &= \frac{\pi\epsilon}{\lambda}(1 - e^{-\lambda R}) < \frac{\pi\epsilon}{\lambda}. \end{aligned} \tag{18.28}$$

In the second inequality we have used the fact that $(\sin\theta)/\theta \geq 2/\pi$ for angles in the range $0 < \theta < \pi/2$. Since ϵ can be made as small as we like, the lemma follows.

Example: Evaluate

$$I(\alpha) = \int_{-\infty}^{\infty} \frac{\sin(\alpha x)}{x}\, dx. \tag{18.29}$$

We have

$$I(\alpha) = \text{Im}\left\{ \int_{-\infty}^{\infty} \frac{\exp i\alpha z}{z}\, dz \right\}. \tag{18.30}$$

If we take $\alpha > 0$, we can close in the upper half-plane, but our contour must exclude the pole at $z = 0$. Therefore

$$0 = \int_{|z|=R} \frac{\exp i\alpha z}{z}\, dz - \int_{|z|=\epsilon} \frac{\exp i\alpha z}{z}\, dz + \int_{-R}^{-\epsilon} \frac{\exp i\alpha x}{x}\, dx + \int_{\epsilon}^{R} \frac{\exp i\alpha x}{x}\, dx. \tag{18.31}$$

As $R \to \infty$, we can ignore the big semicircle, the rest, after letting $\epsilon \to 0$, gives

$$0 = -i\pi + P\int_{-\infty}^{\infty} \frac{e^{i\alpha x}}{x}\,dx. \tag{18.32}$$

Again, the symbol P denotes a principal-part integral. The $-i\pi$ comes from the small semicircle. We get $-1/2$ of the residue because we have only a half circle, and that traversed in the "wrong" direction. (Remember that this fractional residue result is only true when we indent to avoid a *simple pole* – i.e one that is of order one.)

Reading off the real and imaginary parts, we conclude that

$$\int_{-\infty}^{\infty} \frac{\sin\alpha x}{x}\,dx = \pi, \quad P\int_{-\infty}^{\infty} \frac{\cos\alpha x}{x}\,dx = 0, \qquad \alpha > 0. \tag{18.33}$$

No "P" is needed in the sine integral, as the integrand is finite at $x = 0$.

If we relax the condition that $\alpha > 0$ and take into account that sine is an odd function of its argument, we have

$$\int_{-\infty}^{\infty} \frac{\sin\alpha x}{x}\,dx = \pi\,\mathrm{sgn}\,\alpha. \tag{18.34}$$

This identity is called *Dirichlet's discontinuous integral*.

We can interpret Dirichlet's integral as giving the Fourier transform of the principal-part distribution $P(1/x)$ as

$$P\int_{-\infty}^{\infty} \frac{e^{i\omega x}}{x}\,dx = i\pi\,\mathrm{sgn}\,\omega. \tag{18.35}$$

This will be of use later in the chapter.

Example: We will evaluate the integral

$$\oint_C e^{iz} z^{a-1}\,dz \tag{18.36}$$

about the first-quadrant contour shown in Figure 18.4. Observe that when $0 < a < 1$ neither the large nor the small arc makes a contribution, and that there are no poles. Hence, we deduce that

$$0 = \int_0^{\infty} e^{ix} x^{a-1}\,dx - i\int_0^{\infty} e^{-y} y^{a-1} e^{(a-1)\frac{\pi}{2}i}\,dy, \quad 0 < a < 1. \tag{18.37}$$

Taking real and imaginary parts, we find

$$\begin{aligned}
\int_0^{\infty} x^{a-1}\cos x\,dx &= \Gamma(a)\cos\left(\frac{\pi}{2}a\right), \quad 0 < a < 1,\\
\int_0^{\infty} x^{a-1}\sin x\,dx &= \Gamma(a)\sin\left(\frac{\pi}{2}a\right), \quad 0 < a < 1,
\end{aligned} \tag{18.38}$$

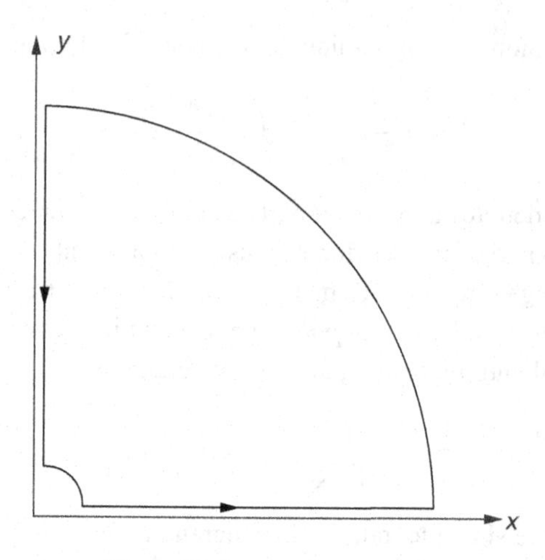

Figure 18.4 Quadrant contour.

where

$$\Gamma(a) = \int_0^\infty y^{a-1} e^{-y}\, dy \tag{18.39}$$

is the Euler Gamma function.

Example: *Fresnel integrals*. Integrals of the form

$$C(t) = \int_0^t \cos(\pi x^2/2)\, dx, \tag{18.40}$$

$$S(t) = \int_0^t \sin(\pi x^2/2)\, dx, \tag{18.41}$$

occur in the theory of diffraction and are called *Fresnel integrals* after Augustin Fresnel. They are naturally combined as

$$C(t) + iS(t) = \int_0^t e^{i\pi x^2/2}\, dx. \tag{18.42}$$

The limit as $t \to \infty$ exists and is finite. Even though the integrand does not tend to zero at infinity, its rapid oscillation for large x is just sufficient to ensure convergence.[1]

[1] We can exhibit this convergence by setting $x^2 = s$ and then integrating by parts to get

$$\int_0^t e^{i\pi x^2/2}\, dx = \frac{1}{2}\int_0^1 e^{i\pi s/2} \frac{ds}{s^{1/2}} + \left[\frac{e^{i\pi s/2}}{\pi i s^{1/2}}\right]_1^{t^2} + \frac{1}{2\pi i}\int_1^{t^2} e^{i\pi s/2} \frac{ds}{s^{3/2}}.$$

The right-hand side is now manifestly convergent as $t \to \infty$.

As t varies, the complex function $C(t) + iS(t)$ traces out the *Cornu spiral*, named after Marie Alfred Cornu, a nineteenth-century French optical physicist.

We can evaluate the limiting value

$$C(\infty) + iS(\infty) = \int_0^\infty e^{i\pi x^2/2}\, dx \tag{18.43}$$

by deforming the contour off the real axis and onto a line of length L running into the first quadrant at 45°, this being the direction of most rapid decrease of the integrand (Figure 18.6).

A circular arc returns the contour to the axis whence it continues to ∞, but an estimate similar to that in Jordan's lemma shows that the arc and the subsequent segment on the real axis make a negligible contribution when L is large. To evaluate the integral on the radial line we set $z = e^{i\pi/4}s$, and so

$$\int_0^{e^{i\pi/4}\infty} e^{i\pi z^2/2}\, dz = e^{i\pi/4}\int_0^\infty e^{-\pi s^2/2}\, ds = \frac{1}{\sqrt{2}}e^{i\pi/4} = \frac{1}{2}(1+i). \tag{18.44}$$

Figure 18.5 shows how $C(t)+iS(t)$ orbits the limiting point $0.5+0.5i$ and slowly spirals in towards it. Taking real and imaginary parts we have

$$\int_0^\infty \cos\left(\frac{\pi x^2}{2}\right) dx = \int_0^\infty \sin\left(\frac{\pi x^2}{2}\right) dx = \frac{1}{2}. \tag{18.45}$$

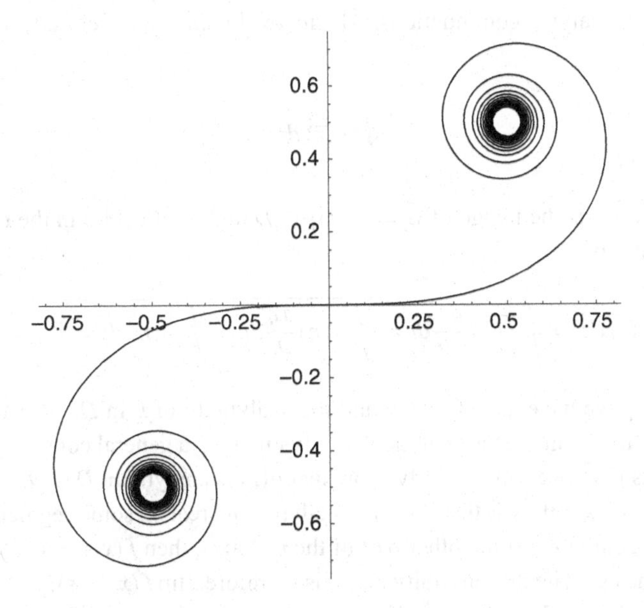

Figure 18.5 The Cornu spiral $C(t) + iS(t)$ for t in the range $-8 < t < 8$. The spiral in the first quadrant corresponds to positive values of t.

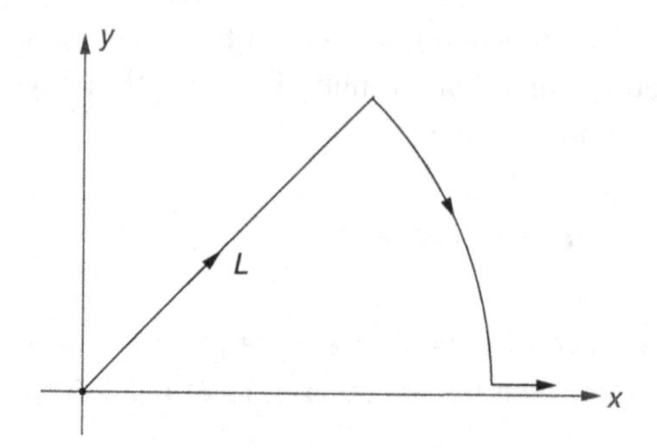

Figure 18.6 Fresnel contour.

18.2 The Schwarz reflection principle

Theorem (Schwarz): Let $f(z)$ be analytic in a domain D where ∂D includes a segment of the real axis. Assume that $f(z)$ is real when z is real. Then there is a unique *analytic continuation of f into the region $\overline{D}$ (the mirror image of D in the real axis; see Figure 18.7) given by*

$$g(z) = \begin{cases} f(z), & z \in D, \\ \overline{f(\overline{z})}, & z \in \overline{D}, \\ \text{either}, & z \in \mathbb{R}. \end{cases} \tag{18.46}$$

The proof invokes Morera's theorem to show analyticity, and then appeals to the uniqueness of analytic continuations. Begin by looking at a closed contour lying only in $\overline{D}$:

$$\oint_C \overline{f(\overline{z})}\, dz, \tag{18.47}$$

where $C = \{\overline{\eta(t)}\}$ is the image of $\overline{C} = \{\eta(t)\} \subset D$ under reflection in the real axis. We can rewrite this as

$$\oint_C \overline{f(\overline{z})}\, dz = \oint \overline{f(\eta)}\, \frac{d\overline{\eta}}{dt} dt = \overline{\oint f(\eta) \frac{d\eta}{dt} dt} = \overline{\oint_{\overline{C}} f(\eta)\, dz} = 0. \tag{18.48}$$

At the last step we have used Cauchy and the analyticity of f in D. Morera's theorem therefore confirms that $g(z)$ is analytic in $\overline{D}$. By breaking a general contour up into parts in D and parts in $\overline{D}$, we can similarly show that $g(z)$ is analytic in $D \cup \overline{D}$.

The important corollary is that if $f(z)$ is analytic, and real on some segment of the real axis, but has a cut along some other part of the real axis, then $f(x + i\epsilon) = \overline{f(x - i\epsilon)}$ as we go over the cut. The discontinuity disc f is therefore $2\mathrm{Im} f(x + i\epsilon)$.

Suppose $f(z)$ is real on the negative real axis, and goes to zero as $|z| \to \infty$. Then applying Cauchy to the contour Γ depicted in Figure 18.8.

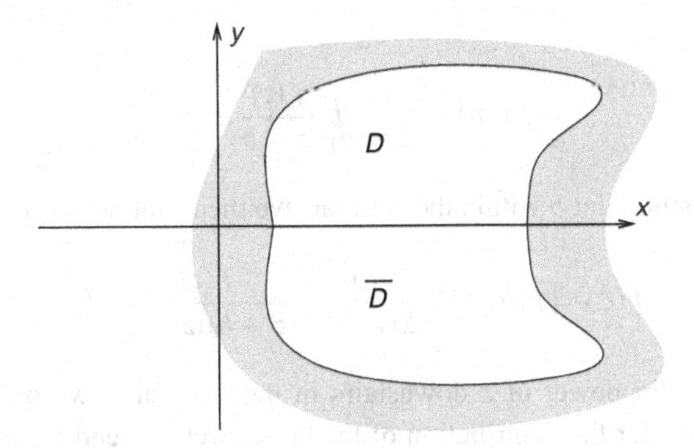

Figure 18.7 The domain D and its mirror image $\overline{D}$.

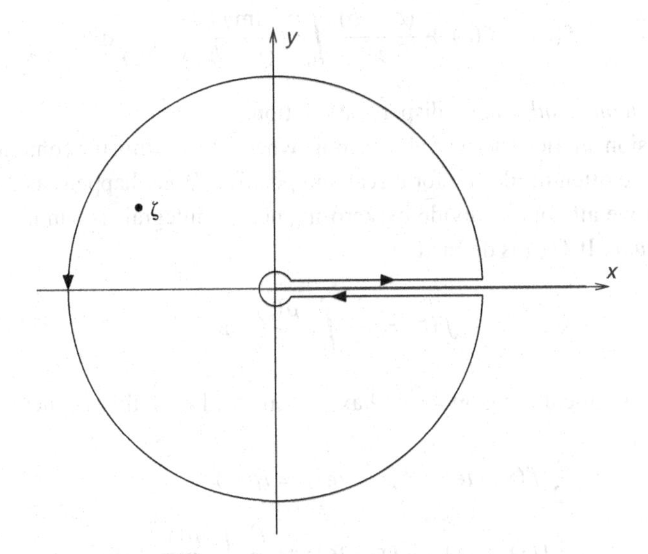

Figure 18.8 The contour Γ for the dispersion relation.

we find

$$f(\zeta) = \frac{1}{\pi} \int_0^{\infty} \frac{\operatorname{Im} f(x + i\epsilon)}{x - \zeta} \, dx, \tag{18.49}$$

for ζ within the contour. This is an example of a *dispersion relation*. The name comes from the prototypical application of this technology to optical dispersion, i.e. the variation of the refractive index with frequency.

If $f(z)$ does not tend to zero at infinity then we cannot ignore the contribution to Cauchy's formula from the large circle. We can, however, still write

$$f(\zeta) = \frac{1}{2\pi i} \oint_{\Gamma} \frac{f(z)}{z - \zeta} \, dz, \tag{18.50}$$

and

$$f(b) = \frac{1}{2\pi i}\oint_\Gamma \frac{f(z)}{z-b}\,dz, \tag{18.51}$$

for some convenient point b within the contour. We then subtract to get

$$f(\zeta) = f(b) + \frac{(\zeta-b)}{2\pi i}\int_\Gamma \frac{f(z)}{(z-b)(z-\zeta)}\,dz. \tag{18.52}$$

Because of the extra power of z downstairs in the integrand, we only need f to be bounded at infinity for the contribution of the large circle to tend to zero. If this is the case, we have

$$f(\zeta) = f(b) + \frac{(\zeta-b)}{\pi}\int_0^\infty \frac{\mathrm{Im} f(x+i\epsilon)}{(x-b)(x-\zeta)}\,dx. \tag{18.53}$$

This is called a *once-subtracted* dispersion relation.

The dispersion relations derived above apply when ζ lies within the contour. In physics applications we often need $f(\zeta)$ for ζ real and positive. What happens as ζ approaches the axis, and we attempt to divide by zero in such an integral, is summarized by the *Plemelj formulæ*: If $f(\zeta)$ is defined by

$$f(\zeta) = \frac{1}{\pi}\int_\Gamma \frac{\rho(z)}{z-\zeta}\,dz, \tag{18.54}$$

where Γ has a segment lying on the real axis, then, if x lies in this segment,

$$\begin{aligned}\frac{1}{2}(f(x+i\epsilon) - f(x-i\epsilon)) &= i\rho(x)\\ \frac{1}{2}(f(x+i\epsilon) + f(x-i\epsilon)) &= \frac{P}{\pi}\int_\Gamma \frac{\rho(x')}{x'-x}\,dx'.\end{aligned} \tag{18.55}$$

As always, the "P" means that we are to delete an infinitesimal segment of the contour lying symmetrically about the pole.

The Plemelj formulæ hold under relatively mild conditions on the function $\rho(x)$. We won't try to give a general proof, but in the case that ρ is analytic the result is easy to understand: we can push the contour out of the way and let $\zeta \to x$ on the real axis from either above or below. In that case the drawing in Figure 18.9 shows how the sum of these two limits gives the principal-part integral and how their difference gives an integral round a small circle, and hence the residue $\rho(x)$.

The Plemelj equations usually appear in physics papers as the "$i\epsilon$" cabala (see Figure 18.10)

$$\frac{1}{x'-x\pm i\epsilon} = P\left(\frac{1}{x'-x}\right) \mp i\pi\delta(x'-x). \tag{18.56}$$

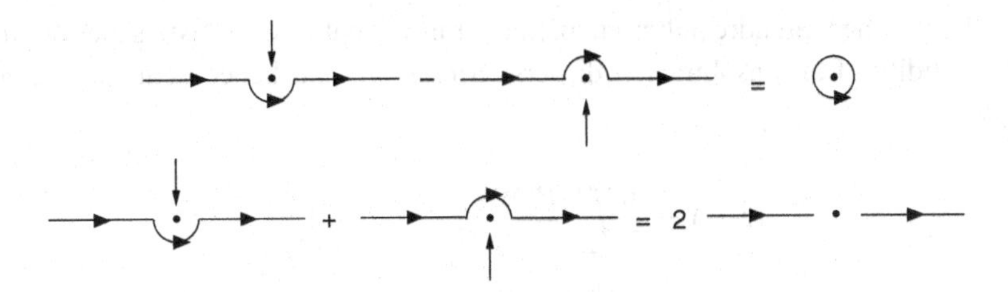

Figure 18.9 Origin of the Plemelj formulae.

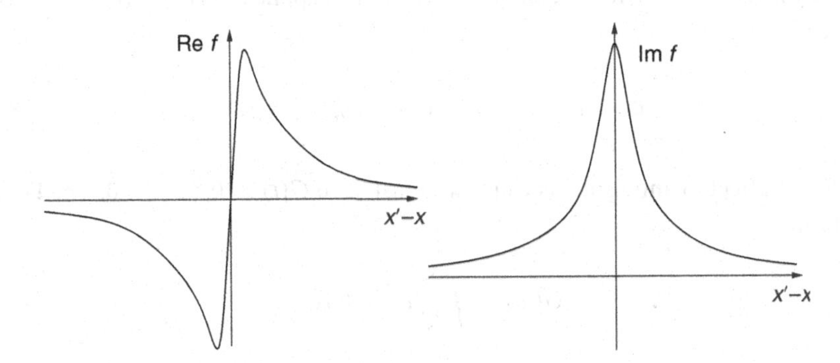

Figure 18.10 Sketch of the real and imaginary parts of $f(x') = 1/(x' - x - i\epsilon)$.

A limit $\epsilon \to 0$ is always to be understood in this formula.

We can also appreciate the origin of the $i\epsilon$ rule by examining the following identity:

$$\frac{1}{x' - (x \pm i\epsilon)} = \frac{x - x'}{(x' - x)^2 + \epsilon^2} \pm \frac{i\epsilon}{(x' - x)^2 + \epsilon^2}. \tag{18.57}$$

The first term is a symmetrically cut-off version of $1/(x' - x)$ and provides the principal-part integral. The second term sharpens and tends to the delta function $\pm i\pi\delta(x' - x)$ as $\epsilon \to 0$.

Exercise 18.3: The Legendre function of the second kind, $Q_n(z)$, may be defined for positive integer n by the integral

$$Q_n(z) = \frac{1}{2}\int_{-1}^{1} \frac{(1-t^2)^n}{2^n(z-t)^{n+1}}\,dt, \quad z \notin [-1, 1].$$

Use Rodriguez' formula to show that for $x \in [-1, 1]$ we have

$$Q_n(x + i\epsilon) - Q_n(x - i\epsilon) = -i\pi P_n(x),$$

where $P_n(x)$ is the Legendre polynomial. Show further that $Q_n(z)$ satisfies the conditions for the validity of an unsubtracted dispersion relation, and hence deduce *Neumann's formula*:

$$Q_n(z) = \frac{1}{2}\int_{-1}^{1} \frac{P_n(t)}{z-t}\,dt, \quad z \notin [-1,1].$$

18.2.1 Kramers–Kronig relations

Causality is the usual source of analyticity in physical applications. If $G(t)$ is a response function

$$\phi_{\text{response}}(t) = \int_{-\infty}^{\infty} G(t-t') f_{\text{cause}}(t')\,dt' \tag{18.58}$$

then for no effect to anticipate its cause we must have $G(t) = 0$ for $t < 0$. The Fourier transform

$$G(\omega) = \int_{-\infty}^{\infty} e^{i\omega t} G(t)\,dt \tag{18.59}$$

is then automatically analytic everywhere in the upper half-plane. Suppose, for example, we look at a forced, damped, harmonic oscillator whose displacement $x(t)$ obeys

$$\ddot{x} + 2\gamma\dot{x} + (\Omega^2+\gamma^2)x = F(t), \tag{18.60}$$

where the friction coefficient γ is positive. As we saw earlier, the solution is of the form

$$x(t) = \int_{-\infty}^{\infty} G(t,t')F(t')dt',$$

where the Green function $G(t,t') = 0$ if $t < t'$. In this case

$$G(t,t') = \begin{cases} \Omega^{-1} e^{-\gamma(t-t')} \sin\Omega(t-t') & t > t' \\ 0, & t < t' \end{cases} \tag{18.61}$$

and so

$$x(t) = \frac{1}{\Omega}\int_{-\infty}^{t} e^{-\gamma(t-t')} \sin\Omega(t-t')\,F(t')\,dt'. \tag{18.62}$$

Because the integral extends only from 0 to $+\infty$, the Fourier transform of $G(t,0)$,

$$\tilde{G}(\omega) \equiv \frac{1}{\Omega}\int_{0}^{\infty} e^{i\omega t} e^{-\gamma t} \sin\Omega t\,dt, \tag{18.63}$$

is nicely convergent when $\operatorname{Im}\omega > 0$, as evidenced by

$$\tilde{G}(\omega) = -\frac{1}{(\omega + i\gamma)^2 - \Omega^2} \tag{18.64}$$

having no singularities in the upper half-plane.[2]

Another example of such a causal function is provided by the complex, frequency-dependent, *refractive index* of a material $n(\omega)$. This is defined so that a travelling wave takes the form

$$\varphi(\mathbf{x}, t) = e^{in(\omega)\mathbf{k}\cdot\mathbf{x} - i\omega t}. \tag{18.65}$$

We can decompose n into its real and imaginary parts

$$\begin{aligned} n(\omega) &= n_R(\omega) + in_I(\omega) \\ &= n_R(\omega) + \frac{i}{2|k|}\gamma(\omega) \end{aligned} \tag{18.66}$$

where γ is the extinction coefficient, defined so that the intensity falls off as $I \propto \exp(-\gamma \mathbf{n}\cdot\mathbf{x})$, where $\mathbf{n} = \mathbf{k}/|k|$ is the direction of propagation. A non-zero γ can arise from either energy absorption or scattering out of the forward direction.

Being a causal response, the refractive index extends to a function analytic in the upper half-plane and $n(\omega)$ for real ω is the boundary value

$$n(\omega)_{\text{physical}} = \lim_{\epsilon\to 0} n(\omega + i\epsilon) \tag{18.67}$$

of this analytic function. Because a real ($\mathbf{E} = \mathbf{E}^*$) incident wave must give rise to a real wave in the material, and because the wave must decay in the direction in which it is propagating, we have the reality conditions

$$\begin{aligned} \gamma(-\omega + i\epsilon) &= -\gamma(\omega + i\epsilon), \\ n_R(-\omega + i\epsilon) &= +n_R(\omega + i\epsilon) \end{aligned} \tag{18.68}$$

with γ positive for positive frequency.

Many materials have a frequency range $|\omega| < |\omega_{\min}|$ where $\gamma = 0$, so the material is transparent. For any such material $n(\omega)$ obeys the Schwarz reflection principle and so there is an analytic continuation into the lower half-plane. At frequencies ω where the material is not perfectly transparent, the refractive index has an imaginary part even when ω is real. By Schwarz, n must be discontinuous across the real axis at these frequencies:

[2] If a pole in a response function manages to sneak into the upper half-plane, then the system will be unstable to exponentially growing oscillations. This may happen, for example, when we design an electronic circuit containing a feedback loop. Such poles, and the resultant instabilities, can be detected by applying the principle of the argument from the last chapter. This method leads to the *Nyquist stability criterion*.

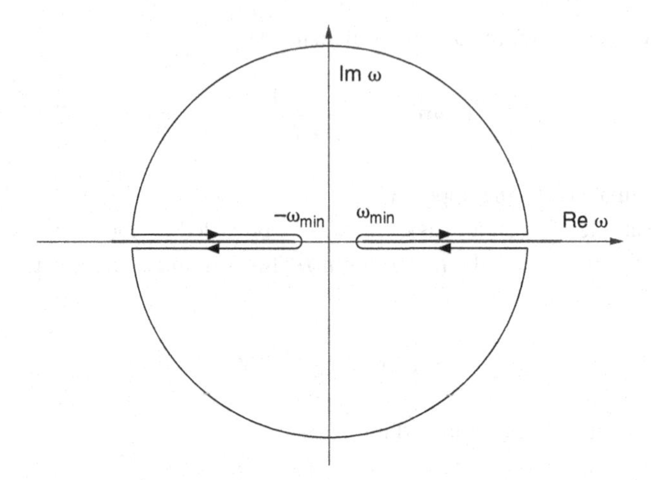

Figure 18.11 Contour for the $n - 1$ dispersion relation.

$n(\omega + i\epsilon) = n_R + in_I \neq n(\omega - i\epsilon) = n_R - in_I$. These discontinuities of $2in_I$ usually correspond to branch cuts.

No substance is able to respond to infinitely high frequency disturbances, so $n \to 1$ as $|\omega| \to \infty$, and we can apply our dispersion relation technology to the function $n - 1$. We will need the contour shown in Figure 18.11, which has cuts for both positive and negative frequencies.

By applying the dispersion-relation strategy, we find

$$n(\omega) = 1 + \frac{1}{\pi}\int_{-\infty}^{\omega_{\min}} \frac{n_I(\omega')}{\omega' - \omega}\,d\omega' + \frac{1}{\pi}\int_{\omega_{\min}}^{\infty} \frac{n_I(\omega')}{\omega' - \omega}\,d\omega' \tag{18.69}$$

for ω within the contour. Using Plemelj we can now take ω onto the real axis to get

$$\begin{aligned} n_R(\omega) &= 1 + \frac{P}{\pi}\int_{-\infty}^{\omega_{\min}} \frac{n_I(\omega')}{\omega' - \omega}\,d\omega' + \frac{P}{\pi}\int_{\omega_{\min}}^{\infty} \frac{n_I(\omega')}{\omega' - \omega}\,d\omega' \\ &= 1 + \frac{P}{\pi}\int_{\omega_{\min}^2}^{\infty} \frac{n_I(\omega')}{\omega'^2 - \omega^2}\,d\omega'^2, \\ &= 1 + \frac{c}{\pi}P\int_{\omega_{\min}}^{\infty} \frac{\gamma(\omega')}{\omega'^2 - \omega^2}\,d\omega'. \end{aligned} \tag{18.70}$$

In the second line we have used the anti-symmetry of $n_I(\omega)$ to combine the positive and negative frequency range integrals. In the last line we have used the relation $\omega/k = c$ to make connection with the way this equation is written in R. G. Newton's authoritative *Scattering Theory of Waves and Particles*. This relation, between the real and absorptive parts of the refractive index, is called a *Kramers–Kronig* dispersion relation, after the original authors.[3]

[3] H. A. Kramers, *Nature*, **117** (1926) 775; R. de L. Kronig, *J. Opt. Soc. Am.*, **12** (1926) 547.

If $n \to 1$ fast enough that $\omega^2(n-1) \to 0$ as $|\omega| \to \infty$, we can take the f in the dispersion relation to be $\omega^2(n-1)$ and deduce that

$$n_R = 1 + \frac{c}{\pi} P \int_{\omega^2_{min}}^{\infty} \left(\frac{\omega'^2}{\omega^2}\right) \frac{\gamma(\omega')}{\omega'^2 - \omega^2} \, d\omega', \tag{18.71}$$

another popular form of Kramers–Kronig. This second relation implies the first, but not *vice versa*, because the second demands more restrictive behaviour for $n(\omega)$.

Similar equations can be derived for other causal functions. A quantity closely related to the refractive index is the frequency-dependent dielectric "constant"

$$\epsilon(\omega) = \epsilon_1 + i\epsilon_2. \tag{18.72}$$

Again $\epsilon \to 1$ as $|\omega| \to \infty$, and, proceeding as before, we deduce that

$$\epsilon_1(\omega) = 1 + \frac{P}{\pi} \int_{\omega^2_{\text{min}}}^{\infty} \frac{\epsilon_2(\omega')}{\omega'^2 - \omega^2} \, d\omega'^2. \tag{18.73}$$

18.2.2 Hilbert transforms

Suppose that $f(x)$ is the boundary value on the real axis of a function everywhere analytic in the upper half-plane, and suppose further that $f(z) \to 0$ as $|z| \to \infty$ there. Then we have

$$f(z) = \frac{1}{2\pi i} \int_{-\infty}^{\infty} \frac{f(x)}{x - z} \, dx \tag{18.74}$$

for z in the upper half-plane. This is because we may close the contour with an upper semicircle without changing the value of the integral. For the same reason the integral must give zero when z is taken in the lower half-plane. Using the Plemelj formulæ we deduce that on the real axis,

$$f(x) = \frac{P}{\pi i} \int_{-\infty}^{\infty} \frac{f(x')}{x' - x} \, dx'. \tag{18.75}$$

We can use this strategy to derive the Kramers–Kronig relations even if n_I never vanishes, and so we cannot use the Schwarz reflection principle.

The relation (18.75) suggests the definition of the *Hilbert transform*, $\mathcal{H}\psi$, of a function $\psi(x)$, as

$$(\mathcal{H}\psi)(x) = \frac{P}{\pi} \int_{-\infty}^{\infty} \frac{\psi(x')}{x - x'} \, dx'. \tag{18.76}$$

Note the interchange of x, x' in the denominator of (18.76) when compared with (18.75). This switch is to make the Hilbert transform into a convolution integral. Equation (18.75) shows that a function that is the boundary value of a function analytic and tending to zero at infinity in the upper half-plane is automatically an eigenvector of $\mathcal{H}$ with eigenvalue $-i$. Similarly a function that is the boundary value of a function analytic and tending to zero at infinity in the lower half-plane will be an eigenvector with eigenvalue $+i$. (A function analytic in the *entire* complex plane and tending to zero at infinity must vanish identically by Liouville's theorem.)

Returning now to our original f, which had eigenvalue $-i$, and decomposing it as $f(x) = f_R(x) + if_I(x)$, we find that (18.75) becomes

$$\begin{aligned} f_I(x) &= (\mathcal{H}f_R)(x), \\ f_R(x) &= -(\mathcal{H}f_I)(x). \end{aligned} \tag{18.77}$$

Conversely, if we are given a real function $u(x)$ and set $v(x) = (\mathcal{H}u)(x)$, then, under some mild restrictions on u (that it lie in some $L^p(\mathbb{R}), p > 1$, for example, in which case $v(x)$ is also in $L^p(\mathbb{R})$) the function

$$f(z) = \frac{1}{2\pi i}\int_{-\infty}^{\infty} \frac{u(x) + iv(x)}{x - z}\, dx \tag{18.78}$$

will be analytic in the upper half-plane, tend to zero at infinity there and have $u(x)+iv(x)$ as its boundary value as z approaches the real axis from above. The last line of (18.77) therefore shows that we may recover $u(x)$ from $v(x)$ as $u(x) = -(\mathcal{H}v)(x)$. The Hilbert transform $\mathcal{H} : L^p(\mathbb{R}) \to L^p(\mathbb{R})$ is therefore invertible, and its inverse is given by $\mathcal{H}^{-1} = -\mathcal{H}$. (Note that the Hilbert transform of a constant is zero, but the $L^p(\mathbb{R})$ condition excludes constants from the domain of $\mathcal{H}$, and so this fact does not conflict with invertibility.)

Hilbert transforms are useful in signal processing. Given a real signal $X_R(t)$ we can take its Hilbert transform so as to find the corresponding imaginary part, $X_I(t)$, which serves to make the sum

$$Z(t) = X_R(t) + iX_I(t) = A(t)e^{i\phi(t)} \tag{18.79}$$

analytic in the upper half-plane. This complex function is the *analytic signal*.[4] The real quantity $A(t)$ is then known as the *instantaneous amplitude*, or *envelope*, while $\phi(t)$ is the *instantaneous phase* and

$$\omega_{\rm IF}(t) = \dot{\phi}(t) \tag{18.80}$$

is called the *instantaneous frequency* (IF). These quantities are used, for example, in narrow-band FM radio, in NMR, in geophysics and in image processing.

[4] D. Gabor, *J. Inst. Elec. Eng. (Part 3)*, **93** (1946) 429.

Exercise 18.4: Let $\widetilde{f}(\omega) = \int_{-\infty}^{\infty} e^{i\omega t} f(t)\, dt$ denote the Fourier transform of $f(t)$. Use the formula (18.35) for the Fourier transform of $P(1/t)$, combined with the convolution theorem for Fourier transforms, to show that the Fourier transform of the Hilbert transform of $f(t)$ is

$$\widetilde{(\mathcal{H}f)}(\omega) = i\,\mathrm{sgn}(\omega)\widetilde{f}(\omega).$$

Deduce that the analytic signal is derived from the original real signal by suppressing all positive frequency components (those proportional to $e^{-i\omega t}$ with $\omega > 0$) and multiplying the remaining negative-frequency amplitudes by 2.

Exercise 18.5: Suppose that $\varphi_1(x)$ and $\varphi_2(x)$ are real functions with finite $L^2(\mathbb{R})$ norms.

(a) Use the Fourier transform result from the previous exercise to show that

$$\langle \varphi_1, \varphi_2 \rangle = \langle \mathcal{H}\varphi_1, \mathcal{H}\varphi_2 \rangle.$$

Thus, $\mathcal{H}$ is a unitary transformation from $L^2(\mathbb{R}) \to L^2(\mathbb{R})$.

(b) Use the fact that $\mathcal{H}^2 = -I$ to deduce that

$$\langle \mathcal{H}\varphi_1, \varphi_2 \rangle = -\langle \varphi_1, \mathcal{H}\varphi_2 \rangle$$

and so $\mathcal{H}^\dagger = -\mathcal{H}$.

(c) Conclude from part (b) that

$$\int_{-\infty}^{\infty} \varphi_1(x) \left(P \int_{-\infty}^{\infty} \frac{\varphi_2(y)}{x-y}\, dy \right) dx = \int_{-\infty}^{\infty} \varphi_2(y) \left(P \int_{-\infty}^{\infty} \frac{\varphi_1(x)}{x-y}\, dx \right) dy,$$

i.e. for $L^2(\mathbb{R})$ functions, it is legitimate to interchange the order of "P" integration with ordinary integration.

d) By replacing $\varphi_1(x)$ by a constant, and $\varphi_2(x)$ by the Hilbert transform of a function f with $\int f\, dx \neq 0$, show that it is not *always* safe to interchange the order of "P" integration with ordinary integration.

Exercise 18.6: Suppose that we are given real functions $u_1(x)$ and $u_2(x)$ and substitute their Hilbert transforms $v_1 = \mathcal{H}u_1$, $v_2 = \mathcal{H}u_2$ into (18.78) to construct analytic functions $f_1(z)$ and $f_2(z)$ that are analytic in the upper half-plane and tend to zero at infinity there. Then, as we approach the real axis from above, the product $f_1(z)f_2(z) = F(z)$ has boundary value

$$F_R(x+i\epsilon) + iF_I(x+i\epsilon) = (u_1u_2 - v_1v_2) + i(u_1v_2 + u_2v_1).$$

By assuming that $F(z)$ satisfies the conditions for (18.77) to be applicable to this boundary value, deduce that

$$\mathcal{H}((\mathcal{H}u_1)u_2) + \mathcal{H}((\mathcal{H}u_2)u_1) - (\mathcal{H}u_1)(\mathcal{H}u_2) = -u_1u_2. \qquad (\star)$$

This result[5] sometimes appears in the physics literature[6] in the guise of the distributional identity

$$\frac{P}{x-y}\frac{P}{y-z}+\frac{P}{y-z}\frac{P}{z-x}+\frac{P}{z-x}\frac{P}{x-y}=-\pi^2\delta(x-y)\delta(x-z),$$

where $P/(x-y)$ denotes the principal-part distribution $P\Big(1/(x-y)\Big)$. This attractively symmetric form conceals the fact that x is being kept fixed, while y and z are being integrated over in specific orders. As the next exercise shows, were we to freely rearrange the integration order we could use the identity

$$\frac{1}{x-y}\frac{1}{y-z}+\frac{1}{y-z}\frac{1}{z-x}+\frac{1}{z-x}\frac{1}{x-y}=0 \quad x,y,z \text{ distinct}$$

to wrongly conclude that the right-hand side of (⋆) is zero.

Problem 18.7: Show that the identity (⋆) from Exercise 18.6 can be written as

$$\int_{-\infty}^{\infty}\left(\int_{-\infty}^{\infty}\frac{u_1(y)u_2(z)}{(z-y)(y-x)}dz\right)dy$$
$$=\int_{-\infty}^{\infty}\left(\int_{-\infty}^{\infty}\frac{u_1(y)u_2(z)}{(z-y)(y-x)}dy\right)dz-\pi^2u_1(x)u_2(x),$$

principal-part integrals being understood where necessary. This is a special case of a more general change-of-integration-order formula

$$\int_{-\infty}^{\infty}\left(\int_{-\infty}^{\infty}\frac{f(x,y,z)}{(z-y)(y-x)}dz\right)dy$$
$$=\int_{-\infty}^{\infty}\left(\int_{-\infty}^{\infty}\frac{f(x,y,z)}{(z-y)(y-x)}dy\right)dz-\pi^2f(x,x,x),$$

which was first obtained by G. H. Hardy in 1908. Hardy's result is often referred to as the *Poincaré–Bertrand theorem*.

Verify Hardy's formula for the particular case where x is zero and $f(0,y,z)$ is unity when both y and z lie within the interval $[-a,a]$ but zero elsewhere. You will need to show that

$$\int_0^{\infty}\ln\left(\frac{a-x}{a+x}\right)^2\frac{dx}{x}=-\pi^2\text{sgn}(a).$$

(Hint: observe that the integrand is singular at $x=|a|$. Explain why it is legitimate to evaluate the improper integral by expanding the logarithms as a power series in x or x^{-1}, and then integrating term-by-term.)

[5] A sufficient condition for its validity is that $u_1\in L^{p_1}(\mathbb{R})$, $u_2\in L^{p_2}(\mathbb{R})$, where p_1 and p_2 are greater than unity and $1/p_1+1/p_2<1$. See F. G. Tricomi, *Quart. J. Math. (Oxford)*, series 2, **2** (1951) 199.

[6] For example, in R. Jackiw, A. Strominger, *Phys. Lett.*, **99B** (1981) 133.

Exercise 18.8: Use the licit interchange of "P" integration with ordinary integration to show that

$$\int_{-\infty}^{\infty} \varphi(x) \left(P \int_{-\infty}^{\infty} \frac{\varphi(y)}{x-y}\, dy \right)^2 dx = \frac{\pi^2}{3} \int_{-\infty}^{\infty} \varphi^3(x)\, dx.$$

Exercise 18.9: Let $f(z)$ be analytic within the unit circle, and let $u(\theta)$ and $v(\theta)$ be the boundary values of its real and imaginary parts, respectively, at $z = e^{i\theta}$. Use Plemelj to show that

$$u(\theta) = -\frac{1}{2\pi} P \int_0^{2\pi} v(\theta') \cot\left(\frac{\theta - \theta'}{2}\right) d\theta' + \frac{1}{2\pi} \int_0^{2\pi} u(\theta')\, d\theta',$$

$$v(\theta) = \frac{1}{2\pi} P \int_0^{2\pi} u(\theta') \cot\left(\frac{\theta - \theta'}{2}\right) d\theta' + \frac{1}{2\pi} \int_0^{2\pi} v(\theta')\, d\theta'.$$

18.3 Partial-fraction and product expansions

In this section we will study other useful representations of functions which devolve from their analyticity properties.

18.3.1 Mittag-Leffler partial-fraction expansion

Let $f(z)$ be a meromorphic function with poles (perhaps infinitely many) at $z = z_j$, $(j = 1, 2, 3, \ldots)$, where $|z_1| < |z_2| < \ldots$ Let Γ_n be a contour enclosing the first n poles. Suppose further (for ease of description) that the poles are simple and have residue r_n. Then, for z inside Γ_n, we have

$$\frac{1}{2\pi i} \oint_{\Gamma_n} \frac{f(z')}{z' - z}\, dz' = f(z) + \sum_{j=1}^{n} \frac{r_j}{z_j - z}. \tag{18.81}$$

We often want to apply this formula to trigonometric functions whose periodicity means that they do not tend to zero at infinity. We therefore employ the same strategy that we used for dispersion relations: we subtract $f(0)$ from $f(z)$ to find

$$f(z) - f(0) = \frac{z}{2\pi i} \oint_{\Gamma_n} \frac{f(z')}{z'(z' - z)}\, dz' + \sum_{j=1}^{n} r_j \left(\frac{1}{z - z_j} + \frac{1}{z_j} \right). \tag{18.82}$$

If we now assume that $f(z)$ is uniformly bounded on the Γ_n – this meaning that $|f(z)| < A$ on Γ_n, with the same constant A working for all n – then the integral tends to zero as n becomes large, yielding the partial-fraction, or *Mittag-Leffler*, decomposition

$$f(z) = f(0) + \sum_{j=1}^{\infty} r_j \left(\frac{1}{z - z_j} + \frac{1}{z_j} \right). \tag{18.83}$$

Example: Consider cosec z. The residues of $1/(\sin z)$ at its poles at $z = n\pi$ are $r_n = (-1)^n$. We can take the Γ_n to be squares with corners $(n+1/2)(\pm 1 \pm i)\pi$. A bit of effort shows that cosec is uniformly bounded on them. To use the formula as given, we first need to subtract the pole at $z = 0$, then

$$\operatorname{cosec} z - \frac{1}{z} = \sum_{n=-\infty}^{\infty}{}' (-1)^n \left(\frac{1}{z - n\pi} + \frac{1}{n\pi} \right). \tag{18.84}$$

The prime on the summation symbol indicates that we are to omit the $n = 0$ term. The positive and negative n series converge separately, so we can add them, and write the more compact expression

$$\operatorname{cosec} z = \frac{1}{z} + 2z \sum_{n=1}^{\infty} (-1)^n \frac{1}{z^2 - n^2\pi^2}. \tag{18.85}$$

Example: A similar method applied to $\cot z$ yields

$$\cot z = \frac{1}{z} + \sum_{n=-\infty}^{\infty}{}' \left(\frac{1}{z - n\pi} + \frac{1}{n\pi} \right). \tag{18.86}$$

We can pair terms together to write this as

$$\begin{aligned} \cot z &= \frac{1}{z} + \sum_{n=1}^{\infty} \left(\frac{1}{z - n\pi} + \frac{1}{z + n\pi} \right), \\ &= \frac{1}{z} + \sum_{n=1}^{\infty} \frac{2z}{z^2 - n^2\pi^2} \end{aligned} \tag{18.87}$$

or

$$\cot z = \lim_{N\to\infty} \sum_{n=-N}^{N} \frac{1}{z - n\pi}. \tag{18.88}$$

In the last formula it is important that the upper and lower limits of summation be the same. Neither the sum over positive n nor the sum over negative n converges separately. By taking asymmetric upper and lower limits we could therefore obtain any desired number as the limit of the sum.

Exercise 18.10: Use Mittag-Leffler to show that

$$\operatorname{cosec}^2 z = \sum_{n=-\infty}^{\infty} \frac{1}{(z + n\pi)^2}.$$

Now use this infinite series to give a one-line proof of the trigonometric identity

$$\sum_{m=0}^{N-1} \operatorname{cosec}^2\left(z + \frac{m\pi}{N}\right) = N^2 \operatorname{cosec}^2(Nz).$$

(Is there a comparably easy *elementary* derivation of this finite sum?) Take a limit to conclude that

$$\sum_{m=1}^{N-1} \operatorname{cosec}^2\left(\frac{m\pi}{N}\right) = \frac{1}{3}(N^2 - 1).$$

Exercise 18.11: From the partial fraction expansion for $\cot z$, deduce that

$$\frac{d}{dz}\ln[(\sin z)/z] = \frac{d}{dz}\sum_{n=1}^{\infty}\ln(z^2 - n^2\pi^2).$$

Integrate this along a suitable path from $z = 0$, and so conclude that

$$\sin z = z\prod_{n=1}^{\infty}\left(1 - \frac{z^2}{n^2\pi^2}\right).$$

Exercise 18.12: By differentiating the partial fraction expansion for $\cot z$, show that, for k an integer ≥ 1, and $\operatorname{Im} z > 0$, we have

$$\sum_{n=-\infty}^{\infty}\frac{1}{(z+n)^{k+1}} = \frac{(-2\pi i)^{k+1}}{k!}\sum_{n=1}^{\infty} n^k e^{2\pi i n z}.$$

This is called *Lipshitz' formula*.

Exercise 18.13: The *Bernoulli numbers* are defined by

$$\frac{x}{e^x - 1} = 1 + B_1 x + \sum_{k=1}^{\infty} B_{2k}\frac{x^{2k}}{(2k)!}.$$

The first few are $B_1 = -1/2$, $B_2 = 1/6$, $B_4 = -1/30$. Except for B_1, the B_n are zero for n odd. Show that

$$x\cot x = ix + \frac{2ix}{e^{2ix} - 1} = 1 - \sum_{k=1}^{\infty}(-1)^{k+1}B_{2k}\frac{2^{2k}x^{2k}}{(2k)!}.$$

By expanding $1/(x^2 - n^2\pi^2)$ as a power series in x and comparing coefficients, deduce that, for positive integer k,

$$\sum_{n=1}^{\infty}\frac{1}{n^{2k}} = (-1)^{k+1}\pi^{2k}\frac{2^{2k-1}}{(2k)!}B_{2k}.$$

Exercise 18.14: *Euler–Maclaurin sum formula*. Let $f(x)$ be a real analytic function. Use the formal expansion

$$\frac{D}{e^D - 1} = \sum_k B_k \frac{D^k}{k!} = 1 - \frac{1}{2}D + \frac{1}{6}\frac{D^2}{2!} - \frac{1}{30}\frac{D^4}{4!} + \cdots,$$

with D interpreted as d/dx, to obtain

$$(-f'(x) - f'(x+1) - f'(x+2) + \cdots) = f(x) - \frac{1}{2}f'(x) + \frac{1}{6}\frac{f''(x)}{2!} - \frac{1}{30}\frac{f^{(4)}}{4!} + \cdots.$$

By integrating this formula from 0 to M, where M is an integer, motivate the Euler–Maclaurin sum formula:

$$\frac{1}{2}f(0) + f(1) + \cdots + f(M-1) + \frac{1}{2}f(M)$$
$$= \int_0^M f(x)\,dx + \sum_{k=1}^{\infty} \frac{B_{2k}}{(2k)!}(f^{(2k-1)}(M) - f^{(2k-1)}(0)).$$

The left-hand side is the *trapezium rule* approximation to the integral on the right-hand side. This derivation gives no insight into whether the infinite-sum correction to the trapezium rule converges (usually it does not), or what the error will be if we truncate the sum after a finite number of terms. When $f(x)$ is a polynomial, however, only a finite number of derivatives $f^{(2k-1)}$ are non-zero, and the result is exact.

18.3.2 Infinite product expansions

We can play a variant of the Mittag-Leffler game with suitable entire functions $g(z)$ and derive for them a representation as an infinite product. Suppose that $g(z)$ has simple zeros at z_i. Then $(\ln g)' = g'(z)/g(z)$ is meromorphic with poles at z_i, all with unit residues. Assuming that it satisfies the uniform boundedness condition, we now use Mittag-Leffler to write

$$\frac{d}{dz}\ln g(z) = \left.\frac{g'(z)}{g(z)}\right|_{z=0} + \sum_{j=1}^{\infty}\left(\frac{1}{z - z_j} + \frac{1}{z_j}\right). \tag{18.89}$$

Integrating up we have

$$\ln g(z) = \ln g(0) + cz + \sum_{j=1}^{\infty}\left(\ln(1 - z/z_j) + \frac{z}{z_j}\right), \tag{18.90}$$

where $c = g'(0)/g(0)$. We now re-exponentiate to get

$$g(z) = g(0)e^{cz}\prod_{j=1}^{\infty}\left(1 - \frac{z}{z_j}\right)e^{z/z_j}. \tag{18.91}$$

Example: Let $g(z) = \sin z/z$. Then $g(0) = 1$, while the constant c, which is the logarithmic derivative of g at $z = 0$, is zero, and

$$\frac{\sin z}{z} = \prod_{n=1}^{\infty} \left(1 - \frac{z}{n\pi}\right) e^{z/n\pi} \left(1 + \frac{z}{n\pi}\right) e^{-z/n\pi}. \tag{18.92}$$

Thus

$$\sin z = z \prod_{n=1}^{\infty} \left(1 - \frac{z^2}{n^2\pi^2}\right). \tag{18.93}$$

Convergence of infinite products

We have derived several infinite problem formulæ without discussing the issue of their convergence. For products of terms of the form $(1 + a_n)$ with positive a_n we can reduce the question of convergence to that of $\sum_{n=1}^{\infty} a_n$.

To see why this is so, let

$$p_N = \prod_{n=1}^{N} (1 + a_n), \quad a_n > 0. \tag{18.94}$$

Then we have the inequalities

$$1 + \sum_{n=1}^{N} a_n < p_N < \exp\left\{\sum_{n=1}^{N} a_n\right\}. \tag{18.95}$$

The infinite sum and product therefore converge or diverge together. If

$$P = \prod_{n=1}^{\infty} (1 + |a_n|) \tag{18.96}$$

converges, we say that

$$p = \prod_{n=1}^{\infty} (1 + a_n) \tag{18.97}$$

converges absolutely. As with infinite sums, absolute convergence implies convergence, but not *vice versa*. Unlike infinite sums, however, an infinite product containing negative a_n can diverge to *zero*. If $(1 + a_n) > 0$ then $\prod(1 + a_n)$ converges if $\sum \ln(1 + a_n)$ does, and we will say that $\prod(1 + a_n)$ diverges to zero if $\sum \ln(1 + a_n)$ diverges to $-\infty$.

Exercise 18.15: Show that

$$\prod_{n=1}^{N}\left(1+\frac{1}{n}\right)=N+1,$$

$$\prod_{n=2}^{N}\left(1-\frac{1}{n}\right)=\frac{1}{N}.$$

From these deduce that

$$\prod_{n=2}^{\infty}\left(1-\frac{1}{n^2}\right)=\frac{1}{2}.$$

Exercise 18.16: For $|z|<1$, show that

$$\prod_{n=0}^{\infty}\left(1+z^{2^n}\right)=\frac{1}{1-z}.$$

(Hint: think binary.)

Exercise 18.17: For $|z|<1$, show that

$$\prod_{n=1}^{\infty}(1+z^n)=\prod_{n=1}^{\infty}\frac{1}{1-z^{2n-1}}.$$

(Hint: $1-x^{2n}=(1-x^n)(1+x^n)$.)

18.4 Wiener–Hopf equations II

The theory of Hilbert transforms has shown us some of the consequences of functions being analytic in the upper or lower half-plane. Another application of these ideas is to *Wiener–Hopf equations*. Although we have discussed Wiener–Hopf integral equations in Chapter 9, it is only now that we possess the tools to appreciate the general theory. We begin, however, with the slightly simpler Wiener–Hopf *sum* equations, which are their discrete analogue. Here, analyticity in the upper or lower half-plane is replaced by analyticity within or without the unit circle.

18.4.1 Wiener–Hopf sum equations

Consider the infinite system of equations

$$y_n=\sum_{m=-\infty}^{\infty}a_{n-m}x_m,\qquad -\infty<n<\infty \tag{18.98}$$

where we are given the y_n and are seeking the x_n.

If the a_n, y_n are the Fourier coefficients of smooth complex-valued functions

$$A(\theta) = \sum_{n=-\infty}^{\infty} a_n e^{in\theta},$$

$$Y(\theta) = \sum_{n=-\infty}^{\infty} y_n e^{in\theta}, \tag{18.99}$$

then the system of equations is, in principle at least, easy to solve. We introduce the function

$$X(\theta) = \sum_{n=-\infty}^{\infty} x_n e^{in\theta}, \tag{18.100}$$

and (18.98) becomes

$$Y(\theta) = A(\theta)X(\theta). \tag{18.101}$$

From this, the desired x_n may be read off as the Fourier expansion coefficients of $Y(\theta)/A(\theta)$. We see that $A(\theta)$ must be nowhere zero or else the operator A represented by the infinite matrix a_{n-m} will not be invertible. This technique is a discrete version of the Fourier transform method for solving the integral equation

$$y(s) = \int_{-\infty}^{\infty} A(s-t)x(t)\,dt, \quad -\infty < s < \infty. \tag{18.102}$$

The connection with complex analysis is made by regarding $A(\theta), X(\theta), Y(\theta)$ as being functions on the unit circle in the z plane. If they are smooth enough we can extend their definition to an annulus about the unit circle, so that

$$A(z) = \sum_{n=-\infty}^{\infty} a_n z^n,$$

$$X(z) = \sum_{n=-\infty}^{\infty} x_n z^n,$$

$$Y(z) = \sum_{n=-\infty}^{\infty} y_n z^n. \tag{18.103}$$

The x_n may now be read off as the Laurent expansion coefficients of $Y(z)/A(z)$.

The discrete analogue of the *Wiener–Hopf integral equation*

$$y(s) = \int_{0}^{\infty} A(s-t)x(t)\,dt, \quad 0 \leq s < \infty \tag{18.104}$$

is the *Wiener–Hopf sum equation*

$$y_n = \sum_{m=0}^{\infty} a_{n-m} x_m, \quad 0 \le n < \infty. \tag{18.105}$$

This requires a more sophisticated approach. If you look back at our earlier discussion of Wiener–Hopf integral equations in Chapter 9, you will see that the trick for solving them is to extend the definition $y(s)$ to negative s (analogously, the y_n to negative n) and find these values at the same time as we find $x(s)$ for positive s (analogously, the x_n for positive n).

We proceed by introducing the same functions $A(z), X(z), Y(z)$ as before, but now keep careful track of whether their power-series expansions contain positive or negative powers of z. In doing so, we will discover that the Fredholm alternative governing the existence and uniqueness of the solutions will depend on the winding number $N = n(\Gamma, 0)$ where Γ is the image of the unit circle under the map $z \mapsto A(z)$ – in other words, on how many times $A(z)$ wraps around the origin as z goes once round the unit circle.

Suppose that $A(z)$ is smooth enough that it is analytic in an annulus including the unit circle, and that we can factorize $A(z)$ so that

$$A(z) = \lambda q_+(z) z^N [q_-(z)]^{-1}, \tag{18.106}$$

where

$$\begin{aligned} q_+(z) &= 1 + \sum_{n=1}^{\infty} q_n^+ z^n, \\ q_-(z) &= 1 + \sum_{n=1}^{\infty} q_{-n}^- z^{-n}. \end{aligned} \tag{18.107}$$

Here we demand that $q_+(z)$ be analytic and non-zero for $|z| < 1 + \epsilon$, and that $q_-(z)$ be analytic and non-zero for $|1/z| < 1 + \epsilon$. These no-pole, no-zero, conditions ensure, *via* the principle of the argument, that the winding numbers of $q_\pm(z)$ about the origin are zero, and so all the winding of $A(z)$ is accounted for by the N-fold winding of the z^N factor. The non-zero condition also ensures that the reciprocals $[q_\pm(z)]^{-1}$ have the same class of expansions (i.e. in positive or negative powers of z only) as the direct functions.

We now introduce the notation $[F(z)]_+$ and $[F(z)]_-$, meaning that we expand $F(z)$ as a Laurent series and retain only the positive powers of z (including z^0), or only the negative powers (starting from z^{-1}), respectively. Thus $F(z) = [F(z)]_+ + [F(z)]_-$. We will write $Y_\pm(z) = [Y(z)]_\pm$, and similarly for $X(z)$. We can therefore rewrite (18.105) in the form

$$\lambda z^N q_+(z) X_+ = [Y_+(z) + Y_-(z)] q_-(z). \tag{18.108}$$

If $N \geq 0$, and we break this equation into its positive and negative powers, we find

$$\begin{aligned} [Y_+ q_-]_+ &= \lambda z^N q_+(z) X_+, \\ [Y_+ q_-]_- &= -Y_- q_-(z). \end{aligned} \tag{18.109}$$

From the first of these equations we can read off the desired x_n as the positive-power Laurent coefficients of

$$X_+(z) = [Y_+ q_-]_+ (\lambda z^N q_+(z))^{-1}. \tag{18.110}$$

As a byproduct, the second allows us to find the coefficient y_{-n} of $Y_-(z)$. Observe that there is a condition on Y_+ for this to work: the power series expansion of $\lambda z^N q_+(z) X_+$ starts with z^N, and so for a solution to exist the first N terms of $(Y_+ q_-)_+$ as a power series in z must be zero. The given vector y_n must therefore satisfy N consistency conditions. A formal way of expressing this constraint begins by observing that it means that the range of the operator A represented by the matrix a_{n-m} falls short, by N dimensions, of being the entire space of possible y_n. This is exactly the situation that the notion of a "co-kernel" is intended to capture. Recall that if $A : V \to V$, then $\mathrm{Coker}\, A = V/\mathrm{Im}\, A$. We therefore have

$$\dim\, [\mathrm{Coker}\, A] = N.$$

When $N < 0$, on the other hand, we have

$$\begin{aligned} [Y_+(z) q_-(z)]_+ &= \ [\lambda z^{-|N|} q_+(z) X_+(z)]_+ \\ [Y_+(z) q_-(z)]_- &= -Y_-(z) q_-(z) + [\lambda z^{-|N|} q_+(z) X_+(z)]_-. \end{aligned} \tag{18.111}$$

Here the last term in the second equation contains no more than N terms. Because of the $z^{-|N|}$, we can add to X_+ any multiple of $Z_+(x) = z^n [q_+(z)]^{-1}$ for $n = 0, \ldots, N-1$, and still have a solution. Thus the solution is not unique. Instead, we have $\dim\, [\mathrm{Ker}\, (A)] = |N|$.

We have therefore shown that

$$\boxed{\mathrm{Index}\, (A) \stackrel{\mathrm{def}}{=} \dim\, (\mathrm{Ker}\, A) - \dim\, (\mathrm{Coker}\, A) = -N.}$$

This connection between a topological quantity – in the present case the winding number – and the difference in dimension of the kernel and co-kernel is an example of an index theorem.

We now need to show that we can indeed factorize $A(z)$ in the desired manner. When $A(z)$ is a rational function, the factorization is straightforward: if

$$A(z) = C \frac{\prod_n (z - a_n)}{\prod_m (z - b_m)}, \tag{18.112}$$

we simply take

$$q_+(z) = \frac{\prod_{|a_n|>0}(1 - z/a_n)}{\prod_{|b_m|>0}(1 - z/b_m)}, \tag{18.113}$$

where the products are over the linear factors corresponding to poles and zeros outside the unit circle, and

$$q_-(z) = \frac{\prod_{|b_m|<0}(1 - b_m/z)}{\prod_{|a_n|<0}(1 - a_n/z)}, \tag{18.114}$$

containing the linear factors corresponding to poles and zeros inside the unit circle. The constant λ and the power z^N in Equation (18.106) are the factors that we have extracted from the right-hand sides of (18.113) and (18.114), respectively, in order to leave 1's as the first term in each linear factor.

More generally, we take the logarithm of

$$z^{-N}A(z) = \lambda q_+(z)(q_-(z))^{-1} \tag{18.115}$$

to get

$$\ln[z^{-N}A(z)] = \ln[\lambda q_+(z)] - \ln[q_-(z)], \tag{18.116}$$

where we desire $\ln[\lambda q_+(z)]$ to be the boundary value of a function analytic within the unit circle, and $\ln[q_-(z)]$ the boundary value of a function analytic outside the unit circle and with $q_-(z)$ tending to unity as $|z| \to \infty$. The factor of z^{-N} in the logarithm serves to undo the winding of the argument of $A(z)$, and results in a single-valued logarithm on the unit circle. Plemelj now shows that

$$Q(z) = \frac{1}{2\pi i} \oint_{|z|=1} \frac{\ln[\zeta^{-N}A(\zeta)]}{\zeta - z} \, d\zeta \tag{18.117}$$

provides us with the desired factorization. This function $Q(z)$ is everywhere analytic except for a branch cut along the unit circle, and its branches, Q_+ within and Q_- without the circle, differ by $\ln[z^{-N}A(z)]$. We therefore have

$$\begin{aligned} \lambda q_+(z) &= e^{Q_+(z)}, \\ q_-(z) &= e^{Q_-(z)}. \end{aligned} \tag{18.118}$$

The expression for Q as an integral shows that $Q(z) \sim$ const.$/z$ as $|z|$ goes to infinity and so guarantees that $q_-(z)$ has the desired limit of unity there.

The task of finding this factorization is known as the *scalar Riemann–Hilbert problem*. In effect, we are decomposing the infinite matrix

$$\mathbf{A} = \begin{pmatrix} \ddots & \vdots & \vdots & \vdots & \\ \cdots & a_0 & a_1 & a_2 & \cdots \\ \cdots & a_{-1} & a_0 & a_1 & \cdots \\ \cdots & a_{-2} & a_{-1} & a_0 & \cdots \\ & \vdots & \vdots & \vdots & \ddots \end{pmatrix} \tag{18.119}$$

into the product of an upper triangular matrix

$$\mathbf{U} = \lambda \begin{pmatrix} \ddots & \vdots & \vdots & \vdots & \\ \cdots & 1 & q_1^+ & q_2^+ & \cdots \\ \cdots & 0 & 1 & q_1^+ & \cdots \\ \cdots & 0 & 0 & 1 & \cdots \\ & \vdots & \vdots & \vdots & \ddots \end{pmatrix}, \tag{18.120}$$

a lower triangular matrix $\mathbf{L}$, where

$$\mathbf{L}^{-1} = \begin{pmatrix} \ddots & \vdots & \vdots & \vdots & \\ \cdots & 1 & 0 & 0 & \cdots \\ \cdots & q_{-1}^- & 1 & 0 & \cdots \\ \cdots & q_{-2}^- & q_{-1}^- & 1 & \cdots \\ & \vdots & \vdots & \vdots & \ddots \end{pmatrix} \tag{18.121}$$

has 1's on the diagonal, and a matrix $\mathbf{\Lambda}^N$ which is zero everywhere except for a line of 1's located N steps above the main diagonal. The set of triangular matrices with unit diagonal form a group, so the inversion required to obtain $\mathbf{L}$ results in a matrix of the same form. The resulting *Birkhoff factorization*

$$\mathbf{A} = \mathbf{L}\mathbf{\Lambda}^N\mathbf{U}, \tag{18.122}$$

is an infinite-dimensional extension of the Gauss–Bruhat (or generalized LU) decomposition of a matrix. The finite-dimensional Gauss–Bruhat decomposition provides a factorization of a matrix $\mathbf{A} \in \mathrm{GL}(n)$ as

$$\mathbf{A} = \mathbf{L}\mathbf{\Pi}\mathbf{U}, \tag{18.123}$$

where $\mathbf{L}$ is a lower triangular matrix with 1's on the diagonal, $\mathbf{U}$ is an upper triangular matrix with no zeros on the diagonal and $\mathbf{\Pi}$ is a permutation matrix, i.e. a matrix that permutes the basis vectors by having one entry of 1 in each row and in each column, and all other entries zero. Our present $\mathbf{\Lambda}^N$ is playing the role of such a matrix. The matrix $\mathbf{\Pi}$ is uniquely determined by $\mathbf{A}$. The $\mathbf{L}$ and $\mathbf{U}$ matrices become unique if $\mathbf{L}$ is chosen so that $\mathbf{\Pi}^T\mathbf{L}\mathbf{\Pi}$ is also lower triangular.

18.4.2 Wiener–Hopf integral equations

We now carry over our insights from the simpler sum equations to Weiner–Hopf integral equations

$$\int_0^\infty K(x-y)\phi(y)\,dy = f(x), \quad x>0, \tag{18.124}$$

by imagining replacing the unit circle by a circle of radius R, and then taking $R\to\infty$ in such a way that the sums go over to integrals. In this way many features are retained: the problem is still solved by factorizing the Fourier transform

$$\widetilde{K}(k) = \int_{-\infty}^{\infty} K(x)e^{ikx}\,dx \tag{18.125}$$

of the kernel, and there remains an index theorem

$$\dim(\operatorname{Ker} K) - \dim(\operatorname{Coker} K) = -N, \tag{18.126}$$

but N now counts the winding of the phase of $\widetilde{K}(k)$ as k ranges over the real axis:

$$N = \frac{1}{2\pi}\arg\widetilde{K}\Big|_{k=-\infty}^{k=+\infty}. \tag{18.127}$$

One restriction arises though: we will require K to be of the form

$$K(x-y) = \delta(x-y) + g(x-y) \tag{18.128}$$

for some continuous function $g(x)$. Our discussion is therefore being restricted to Wiener–Hopf integral equations of the *second kind.*

The restriction comes about because we will seek to obtain a factorization of $\widetilde{K}$ as

$$\tau(\kappa)\widetilde{K}(k) = \exp\{Q_+(k) - Q_-(k)\} = q_+(k)(q_-(k))^{-1} \tag{18.129}$$

where $q_+(k) \equiv \exp\{Q_+(k)\}$ is analytic and non-zero in the upper half k-plane and $q_-(k) \equiv \exp\{Q_-(k)\}$ analytic and non-zero in the lower half-plane. The factor $\tau(\kappa)$ is a phase such as

$$\tau(k) = \left(\frac{k+i}{k-i}\right)^N, \tag{18.130}$$

which winds $-N$ times and serves to undo the $+N$ phase winding in $\widetilde{K}$. The $Q_\pm(k)$ will be the boundary values from above and below the real axis, respectively, of

$$Q(k) = \frac{1}{2\pi i}\int_{-\infty}^{\infty} \frac{\ln[\tau(\kappa)\widetilde{K}(\kappa)]}{\kappa - k}\,d\kappa. \tag{18.131}$$

The convergence of this infinite integral requires that $\ln[\tau(\kappa)\widetilde{K}(k)]$ go to zero at infinity, or, in other words,

$$\lim_{k\to\infty} \widetilde{K}(k) = 1. \tag{18.132}$$

This, in turn, requires that the original $K(x)$ contain a delta function.

Example: We will solve the problem

$$\phi(x) - \lambda \int_0^\infty e^{-|x-y|-\alpha(x-y)}\phi(y)\,dy = f(x), \quad x > 0. \tag{18.133}$$

We require that $0 < \alpha < 1$. The upper bound on α is necessary for the integral kernel to be bounded. We will also assume for simplicity that $\lambda < 1/2$. Following the same strategy as in the sum case, we extend the integral equation to the entire range of x by writing

$$\phi(x) - \lambda \int_0^\infty e^{-|x-y|-\alpha(x-y)}\phi(y)\,dy = f(x) + g(x), \tag{18.134}$$

where $f(x)$ is non-zero only for $x > 0$ and $g(x)$ is non-zero only for $x < 0$. The Fourier transform of this equation is

$$\left(\frac{(k+i\alpha)^2 + a^2}{(k+i\alpha)^2 + 1}\right)\widetilde{\phi}_+(k) = \widetilde{f}_+(k) + \widetilde{g}_-(k), \tag{18.135}$$

where $a^2 = 1 - 2\lambda$ and the $\pm$ subscripts are to remind us that $\widetilde{\phi}(k)$ and $\widetilde{f}(k)$ are analytic in the upper half-plane, and $\widetilde{g}(k)$ in the lower. We will use the notation H_+ for the space of functions analytic in the upper half plane, and H_- for functions analytic in the lower half-plane, and so

$$\widetilde{\phi}_+(k),\ \widetilde{f}(_+k) \in H_+, \quad \widetilde{g}_-(k) \in H_-. \tag{18.136}$$

We can factorize

$$\widetilde{K}(k) = \frac{(k+i\alpha)^2 + a^2}{(k+i\alpha)^2 + 1} = \frac{[k + i(\alpha - a)]\,[k + i(\alpha + a)]}{[k + i(\alpha - 1)]\,[k + i(\alpha + 1)]}. \tag{18.137}$$

Now suppose that a is small enough that $\alpha \pm a > 0$ and so the numerator has two zeros in the lower half-plane, and the numerator has one zero in each of the upper and lower half-planes. The change of phase in $\widetilde{K}(k)$ as we go from minus to plus infinity is therefore -2π, and so the index is $N = -1$. We should therefore multiply $\widetilde{K}$ by

$$\tau(k) = \left(\frac{k+i}{k-i}\right)^{-1} \tag{18.138}$$

before seeking to break it into its $q_\pm$ factors. We can however equally well take

$$\tau(k)=\left(\frac{k+i(\alpha-1)}{k+i(\alpha-a)}\right) \tag{18.139}$$

as this also undoes the winding and allows us to factorize with

$$q_-(k)=1, \quad q_+(k)=\left(\frac{k+i(\alpha+a)}{k+i(\alpha+1)}\right). \tag{18.140}$$

The resultant equation analogous to (18.108) is therefore

$$\begin{aligned}\left(\frac{k+i(\alpha+a)}{k+i(\alpha+1)}\right)\widetilde{\phi}_+ &= \left(\frac{k+i(\alpha-1)}{k+i(\alpha-a)}\right)\widetilde{f}_+ + \left(\frac{k+i(\alpha-1)}{k+i(\alpha-a)}\right)\widetilde{g}_- \\ q_+\widetilde{\phi}_+ &= \quad (\tau q_-)\widetilde{f}_+ \quad + \quad \tau q_-\widetilde{g}_-.\end{aligned} \tag{18.141}$$

The second line of this equation shows the interpretation of the first line in terms of the objects in the general theory. The left-hand side is in H_+ – i.e. analytic in the upper half-plane. The first term on the right is also in H_+. (We are lucky. More generally it would have to be decomposed into its $H_\pm$ parts.) If it were not for the $\tau(\kappa)$, the last term would be in H_-, but it has a potential pole at $k=-i(\alpha-a)$. We therefore remove this pole by subtracting a term

$$-\frac{\beta}{k+i(\alpha-a)}$$

(an element of H_+) from each side of the equation before projecting onto the $H^\pm$ parts. After projecting, we find that

$$\begin{aligned} H_+ &: \left(\frac{k+i(\alpha+a)}{k+i(\alpha+1)}\right)\widetilde{\phi}_+ - \left(\frac{k+i(\alpha-1)}{k+i(\alpha-a)}\right)\widetilde{f}_+ - \frac{\beta}{k+i(\alpha-a)} = 0, \\ H_- &: \left(\frac{k+i(\alpha-1)}{k+i(\alpha-a)}\right)\widetilde{g}_- - \frac{\beta}{k+i(\alpha-a)} = 0.\end{aligned} \tag{18.142}$$

We solve for $\widetilde{\phi}_+(k)$ and $\widetilde{g}_-(k)$:

$$\begin{aligned}\widetilde{\phi}_+(k) &= \left(\frac{(k+i\alpha)^2+1}{(k+i\alpha)^2+a^2}\right)\widetilde{f}_- - \beta\left(\frac{k+i(\alpha+1)}{(k+i\alpha)^2+a^2}\right) \\ \widetilde{g}_-(k) &= \frac{\beta}{k+i(\alpha-1)}.\end{aligned} \tag{18.143}$$

Observe that $g_-(k)$ is always in H_- because its only singularity is in the upper half-plane for any β. The constant β is therefore arbitrary. Finally, we invert the Fourier transform, using

$$\mathcal{F}\left(\theta(x)e^{-\alpha x}\sinh ax\right) = -\frac{a}{(k+i\alpha)^2+a^2}, \quad (\alpha\pm a)>0, \tag{18.144}$$

to find that

$$\phi(x) = f(x) - \frac{2\lambda}{a}\int_0^x e^{-\alpha(x-y)} \sinh a(x-y) f(y)\,dy \\ + \beta'\left\{(a-1)e^{-(\alpha+a)x} + (a+1)e^{-(\alpha-a)x}\right\}, \tag{18.145}$$

where β' (proportional to β) is an arbitrary constant.

By taking α in the range $-1 < \alpha < 0$ with $(\alpha \pm a) < 0$, we make the index to be $N = +1$. We will then find there is condition on $f(x)$ for the solution to exist. This condition is, of course, that $f(x)$ be orthogonal to the solution

$$\phi_0(x) = \left\{(a-1)e^{-(\alpha+a)x} + (a+1)e^{-(\alpha-a)x}\right\} \tag{18.146}$$

of the homogeneous adjoint problem, this being the $f(x) = 0$ case of the $\alpha > 0$ problem that we have just solved.

18.5 Further exercises and problems

Exercise 18.18: *Contour integration*: Use the calculus of residues to evaluate the following integrals:

$$I_1 = \int_0^{2\pi} \frac{d\theta}{(a + b\cos\theta)^2}, \qquad 0 < b < a$$

$$I_2 = \int_0^{2\pi} \frac{\cos^2 3\theta}{1 - 2a\cos 2\theta + a^2}\,d\theta, \qquad 0 < a < 1$$

$$I_3 = \int_0^{\infty} \frac{x^\alpha}{(1+x^2)^2}\,dx, \qquad -1 < \alpha < 2.$$

These are not meant to be easy! You will have to dig for the residues.

Answers:

$$I_1 = \frac{2\pi a}{(a^2 - b^2)^{3/2}}$$

$$I_2 = \frac{\pi(a^3+1)}{a^2-1} = \frac{\pi(1-a+a^2)}{a-1}$$

$$I_3 = \frac{\pi(1-\alpha)}{4\cos(\pi\alpha/2)}.$$

Exercise 18.19: By considering the integral of

$$f(z) = \ln(1 - e^{2iz}) = \ln(-2ie^{iz}\sin z)$$

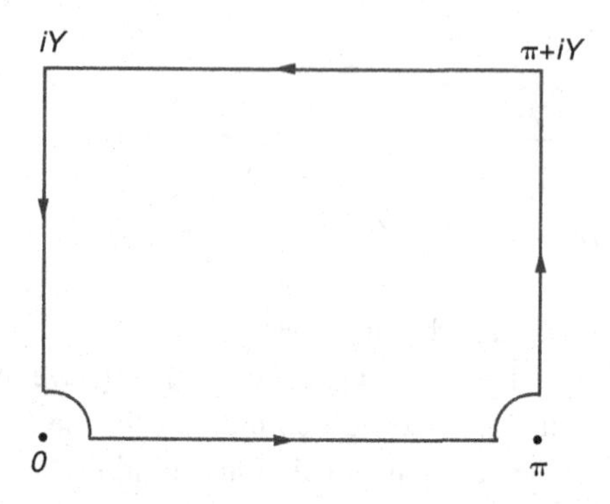

Figure 18.12 Indented rectangle.

around the indented rectangle in Figure 18.12, with vertices $0, \pi, \pi + iY, iY$, and letting Y become large, evaluate the integral

$$I = \int_0^{\pi} \ln(\sin x)\, dx.$$

Explain how the fact that $\epsilon \ln \epsilon \to 0$ as $\epsilon \to 0$ allows us to ignore contributions from the small indentations. You should also provide justification for any other discarded contributions. Take care to make consistent choices of the branch of the logarithm, especially if expanding $\ln(-2ie^{ix} \sin x) = ix + \ln 2 + \ln(\sin x) + \ln(-i)$.
(Ans: $-\pi \ln 2$.)

Exercise 18.20: By integrating a suitable function around the quadrant containing the point $z_0 = e^{i\pi/4}$, evaluate the integral

$$I(\alpha) = \int_0^{\infty} \frac{x^{\alpha-1}}{1+x^4} dx \qquad 0 < \alpha < 4.$$

It should only be necessary to consider the residue at z_0.
(Ans: $(\pi/4)\text{cosec}\,(\pi\alpha/4)$.)

Exercise 18.21: In Section 5.5.1 we considered the causal Green function for the damped harmonic oscillator

$$G(t) = \begin{cases} \frac{1}{\Omega} e^{-\gamma t} \sin(\Omega t), & t > 0, \\ 0, & t < 0, \end{cases}$$

and showed that its Fourier transform

$$\int_{-\infty}^{\infty} e^{i\omega t} G(t)\, dt = \frac{1}{\Omega^2 - (\omega + i\gamma)^2} \tag{18.147}$$

had no singularities in the upper half-plane. Use Jordan's lemma to compute the inverse Fourier transform

$$\frac{1}{2\pi}\int_{-\infty}^{\infty}\frac{e^{-i\omega t}}{\Omega^2-(\omega+i\gamma)^2}\,d\omega,$$

and verify that it reproduces $G(t)$.

Problem 18.22: *Jordan's lemma and one-dimensional scattering theory*. In Problem 4.13 we considered the one-dimensional scattering problem solutions

$$\begin{aligned}\psi_k(x) &= \begin{cases} e^{ikx}+r_L(k)e^{-ikx}, & x\in L,\\ t_L(k)e^{ikx}, & x\in R,\end{cases} \quad k>0\\ &= \begin{cases} t_R(k)e^{ikx}, & x\in L,\\ e^{ikx}+r_R(k)e^{-ikx}, & x\in R,\end{cases} \quad k<0\end{aligned}$$

and claimed that the bound-state contributions to the completeness relation were given in terms of the reflection and transmission coefficients as

$$\begin{aligned}\sum_{\text{bound}}\psi_n^*(x)\psi_n(x') &= -\int_{-\infty}^{\infty}\frac{dk}{2\pi}r_L(k)e^{-ik(x+x')}, && x,x'\in L,\\ &= -\int_{-\infty}^{\infty}\frac{dk}{2\pi}t_L(k)e^{-ik(x-x')}, && x\in L,\ x'\in R,\\ &= -\int_{-\infty}^{\infty}\frac{dk}{2\pi}t_R(k)e^{-ik(x-x')}, && x\in R,\ x'\in L,\\ &= -\int_{-\infty}^{\infty}\frac{dk}{2\pi}r_R(k)e^{-ik(x+x')}, && x,x'\in R.\end{aligned}$$

The eigenfunctions

$$\psi_k^{(+)}(x) = \begin{cases} e^{ikx}+r_L(k)e^{-ikx}, & x\in L,\\ t_L(k)e^{ikx}, & x\in,\end{cases}$$

and

$$\psi_k^{(-)}(x) = \begin{cases} t_R(k)e^{ikx}, & x\in L,\\ e^{ikx}+r_R(k)e^{-ikx}, & x\in R,\end{cases}$$

are initially refined for k real and positive ($\psi_k^{(+)}$) or for k real and negative ($\psi_k^{(-)}$), but they separately have analytic continuations to all of $k\in\mathbb{C}$. The reflection and transmission coefficients $r_{L,R}(k)$ and $t_{L,R}(k)$ are also analytic functions of k, and obey $r_{L,R}(k)=r^*_{L,R}(-k^*)$, $t_{L,R}(k)=t^*_{L,R}(-k^*)$.

(a) By inspecting the formulæ for $\psi_k^{(+)}(x)$, show that the bound states $\psi_n(x)$, with $E_n = -\kappa_n^2$, are proportional to $\psi_k^{(+)}(x)$ evaluated at points $k = i\kappa_n$ on the positive imaginary axis at which $r_L(k)$ and $t_L(k)$ simultaneously have poles. Similarly show that these same bound states are proportional to $\psi_k^{(-)}(x)$ evaluated at points $-i\kappa_n$ on the *negative* imaginary axis at which $r_R(k)$ and $t_R(k)$ have poles. (All these functions $\psi_k^{(\pm)}(x)$, $r_{R,L}(k)$, $t_{R,L}(k)$, may have branch points and other singularities in the half-plane on the opposite side of the real axis from the bound-state poles.)

(b) Use Jordan's lemma to evaluate the Fourier transforms given above in terms of the position and residues of the bound-state poles. Confirm that your answers are of the form

$$\sum_n A_n^*[\mathrm{sgn}(x)]e^{-\kappa_n|x|}A_n[\mathrm{sgn}(x')]e^{-\kappa_n|x'|},$$

as you would expect for the bound-state contribution to the completeness relation.

Exercise 18.23: *Lattice Matsubara sums.* Let $\omega_n = \exp\{i\pi(2n+1)/N\}$, for $n = 0, \ldots, N-1$, be the N-th roots of (-1). Show that, for suitable analytic functions $f(z)$, the sum

$$S = \frac{1}{N}\sum_{n=0}^{N-1} f(\omega_n)$$

can be written as an integral

$$S = \frac{1}{2\pi i}\int_C \frac{dz}{z}\frac{z^N}{z^N+1}f(z).$$

Here C consists of a pair of oppositely oriented concentric circles. The annulus formed by the circles should include all the roots of (-1), but exclude all singularites of f. Use this result to show that, for N even,

$$\frac{1}{N}\sum_{n=0}^{N-1}\frac{\sinh E}{\sinh^2 E + \sin^2\frac{(2n+1)\pi}{N}} = \frac{1}{\cosh E}\tanh\frac{NE}{2}.$$

Let $N \to \infty$ while scaling $E \to 0$ in some suitable manner, and hence show that

$$\sum_{n=-\infty}^{\infty}\frac{a}{a^2 + [(2n+1)\pi]^2} = \frac{1}{2}\tanh\frac{a}{2}.$$

(Hint: if you are careless, you will end up differing by a factor of two from this last formula. There are *two* regions in the finite sum that tend to the infinite sum in the large N limit.)

Problem 18.24: If we define $\chi(h) = e^{\alpha x}\phi(x)$, and $F(x) = e^{\alpha x}f(x)$, then the Wiener–Hopf equation

$$\phi(x) - \lambda \int_0^\infty e^{-|x-y|-\alpha(x-y)}\phi(y)\, dy = f(x), \quad x > 0$$

becomes

$$\chi(x) - \lambda \int_0^\infty e^{-|x-y|}\chi(y)\, dy = F(x), \quad x > 0,$$

all mention of α having disappeared! Why then does our answer, worked out in such detail, in Section 18.4.2, depend on the parameter α? Show that if α is small enough that $\alpha + a$ is positive and $\alpha - a$ is negative, then $\phi(x)$ really is independent of α. (Hint: what tacit assumptions about function spaces does our use of Fourier transforms entail? How does the inverse Fourier transform of $[(k + i\alpha)^2 + a^2]^{-1}$ vary with α?)

19

Special functions and complex variables

In this chapter we will apply complex analytic methods so as to obtain a wider view of some of the special functions of mathematical physics than can be obtained from the real axis. The standard text in this field remains the venerable *Course of Modern Analysis* of E. T. Whittaker and G. N. Watson.

19.1 The Gamma function

We begin by examining how Euler's "Gamma function" $\Gamma(z)$ behaves when z is allowed to become complex. You probably have some acquaintance with this creature. The usual definition is

$$\Gamma(z) = \int_0^\infty t^{z-1} e^{-t}\, dt, \quad \operatorname{Re} z > 0 \quad \text{(definition A)}. \tag{19.1}$$

The restriction on the real part of z is necessary to make the integral converge. We can, however, analytically continue $\Gamma(z)$ to a meromorphic function on all of $\mathbb{C}$. An integration by parts, based on

$$\frac{d}{dt}\left(t^z e^{-t}\right) = z t^{z-1} e^{-t} - t^z e^{-t}, \tag{19.2}$$

shows that

$$\left[t^z e^{-t}\right]_0^\infty = z\int_0^\infty t^{z-1} e^{-t}\, dt - \int_0^\infty t^z e^{-t}\, dt. \tag{19.3}$$

The integrated out part vanishes at both limits, provided the real part of z is greater than zero. Thus we obtain the recurrence relation

$$\Gamma(z+1) = z\Gamma(z). \tag{19.4}$$

Since $\Gamma(1) = 1$, we deduce that

$$\Gamma(n) = (n-1)!, \quad n = 1, 2, 3, \ldots \tag{19.5}$$

We can use the recurrence relation (19.4) to extend the definition of $\Gamma(z)$ to the left half-plane, where the real part of z is negative. Choosing an integer n such that the real part of $z+n$ is positive, we write

$$\Gamma(z) = \frac{\Gamma(z+n)}{z(z+1)\cdots(z+n-1)}. \tag{19.6}$$

We see that the extended $\Gamma(z)$ has poles at zero, and at the negative integers. The residue of the pole at $z=-n$ is $(-1)^n/n!$.

We can also view the analytic continuation as an example of Taylor series subtraction. Let us recall how this works. Suppose that $-1 < \mathrm{Re}\, x < 0$. Then, from

$$\frac{d}{dt}(t^x e^{-t}) = xt^{x-1}e^{-t} - t^x e^{-t} \tag{19.7}$$

we have

$$\left[t^x e^{-t}\right]_\epsilon^\infty = x\int_\epsilon^\infty dt\, t^{x-1}e^{-t} - \int_\epsilon^\infty dt\, t^x e^{-t}. \tag{19.8}$$

Here we have cut off the integral at the lower limit so as to avoid the divergence near $t=0$. Evaluating the left-hand side and dividing by x we find

$$-\frac{1}{x}\epsilon^x = \int_\epsilon^\infty dt\, t^{x-1}e^{-t} - \frac{1}{x}\int_\epsilon^\infty dt\, t^x e^{-t}. \tag{19.9}$$

Since, for this range of x,

$$-\frac{1}{x}\epsilon^x = \int_\epsilon^\infty dt\, t^{x-1}, \tag{19.10}$$

we can rewrite (19.9) as

$$\frac{1}{x}\int_\epsilon^\infty dt\, t^x e^{-t} = \int_\epsilon^\infty dt\, t^{x-1}\left(e^{-t}-1\right). \tag{19.11}$$

The integral on the right-hand side of this last expression is convergent as $\epsilon \to 0$, so we may safely take the limit and find

$$\frac{1}{x}\Gamma(x+1) = \int_0^\infty dt\, t^{x-1}\left(e^{-t}-1\right). \tag{19.12}$$

Since the left-hand side is equal to $\Gamma(x)$, we have shown that

$$\Gamma(x) = \int_0^\infty dt\, t^{x-1}\left(e^{-t}-1\right), \qquad -1 < \mathrm{Re}\, x < 0. \tag{19.13}$$

Similarly, if $-2 < \mathrm{Re}\, x < -1$, we can show that

$$\Gamma(x) = \int_0^\infty dt\, t^{x-1}\left(e^{-t} - 1 + t\right). \tag{19.14}$$

Thus the analytic continuation of the original integral is given by a new integral in which we have subtracted exactly as many terms from the Taylor expansion of e^{-t} as are needed to just make the integral convergent at the lower limit.

Other useful identities, usually proved by elementary real-variable methods, include Euler's "Beta function" identity,

$$B(a,b) \stackrel{\text{def}}{=} \frac{\Gamma(a)\Gamma(b)}{\Gamma(a+b)} = \int_0^1 (1-t)^{a-1} t^{b-1}\, dt \tag{19.15}$$

(which, as the *Veneziano formula*, was the original inspiration for string theory) and

$$\Gamma(z)\Gamma(1-z) = \pi \operatorname{cosec} \pi z. \tag{19.16}$$

The proofs of both formulæ begin in the same way: set $t = y^2, x^2$, so that

$$\begin{aligned}
\Gamma(a)\Gamma(b) &= 4\int_0^\infty y^{2a-1} e^{-y^2}\, dy \int_0^\infty x^{2b-1} e^{-x^2}\, dx \\
&= 4\int_0^\infty\int_0^\infty e^{-(x^2+y^2)} x^{2b-1} y^{2a-1}\, dx\, dy \\
&= 2\int_0^\infty e^{-r^2} (r^2)^{a+b-1}\, d(r^2) \int_0^{\pi/2} \sin^{2a-1}\theta \cos^{2b-1}\theta\, d\theta.
\end{aligned}$$

We have appealed to Fubini's theorem twice: once to turn a product of integrals into a double integral, and once (after setting $x = r\cos\theta$, $y = r\sin\theta$) to turn the double integral back into a product of decoupled integrals. In the second factor of the third line we can now change variables to $t = \sin^2\theta$ and obtain the Beta function identity. If, on the other hand, we put $a = 1 - z$, $b = z$, we have

$$\Gamma(z)\Gamma(1-z) = 2\int_0^\infty e^{-r^2}\, d(r^2) \int_0^{\pi/2} \cot^{2z-1}\theta\, d\theta = 2\int_0^{\pi/2} \cot^{2z-1}\theta\, d\theta. \tag{19.17}$$

Now set $\cot\theta = \zeta$. The last integral then becomes (see Exercise 18.1):

$$2\int_0^\infty \frac{\zeta^{2z-1}}{\zeta^2+1}\, d\zeta = \pi \operatorname{cosec} \pi z, \quad 0 < z < 1, \tag{19.18}$$

establishing the claimed result. Although this last integral has a restriction on the range of z (19.16) holds for all z by analytic continuation. If we put $z = 1/2$, we find that $(\Gamma(1/2))^2 = \pi$. Because the definition-A integral for $\Gamma(1/2)$ is manifestly positive, the positive square root is the correct one, and

$$\Gamma(1/2) = \sqrt{\pi}. \tag{19.19}$$

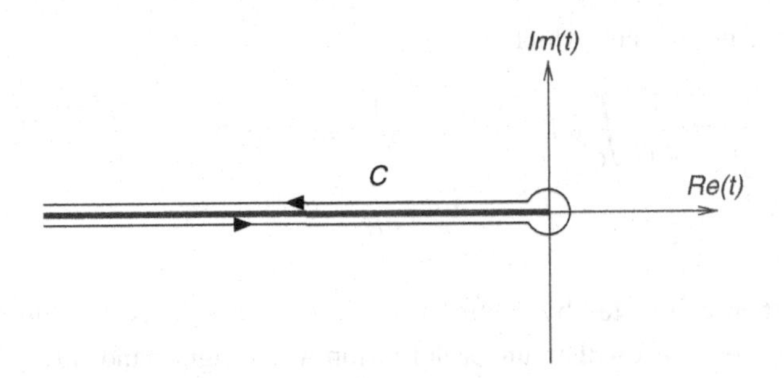

Figure 19.1 Definition B contour for $\Gamma(z)$.

The integral in definition A for $\Gamma(z)$ is only convergent for $\mathrm{Re}\, z > 0$. A more powerful definition, involving an integral that converges for all z, is (see Figure 19.1)

$$\frac{1}{\Gamma(z)} = \frac{1}{2\pi i}\int_C \frac{e^t}{t^z}\, dt \quad \text{(definition B)}. \tag{19.20}$$

Here, C is a contour originating at $z = -\infty - i\epsilon$, below the negative real axis (on which a cut serves to make t^{-z} single valued) rounding the origin, and then heading back to $z = -\infty + i\epsilon$, this time staying above the cut. We take $\arg t$ to be $+\pi$ immediately above the cut, and $-\pi$ immediately below it. This new definition is due to Hankel.

For z an integer, the cut is unnecessary and we can replace the contour by a circle about $z = 0$ and so find

$$\frac{1}{\Gamma(0)} = 0; \qquad \frac{1}{\Gamma(n)} = \frac{1}{(n-1)!}, \qquad n > 0. \tag{19.21}$$

Thus, definitions A and B agree on the integers. It is less obvious that they agree for all z. A hint that this is true stems from integrating by parts

$$\frac{1}{\Gamma(z)} = \frac{1}{2\pi i}\left[\frac{e^t}{(z-1)t^{z-1}}\right]_{-\infty-i\epsilon}^{-\infty+i\epsilon} + \frac{1}{(z-1)2\pi i}\int_C \frac{e^t}{t^{z-1}}\, dt = \frac{1}{(z-1)\Gamma(z-1)}. \tag{19.22}$$

The integrated-out part vanishes because e^t is zero at $-\infty$. Thus the "new" Gamma function obeys the same functional relation as the "old" one.

To show that the equivalence holds for non-integer z we will examine the definition-B expression for $\Gamma(1-z)$:

$$\frac{1}{\Gamma(1-z)} = \frac{1}{2\pi i}\int_C e^t t^{z-1}\, dt. \tag{19.23}$$

We will assume initially that $\mathrm{Re}\, z > 0$, so that there is no contribution to the integral from the small circle about the origin. We can therefore focus on the contribution from the discontinuity across the cut, which is

$$\frac{1}{\Gamma(1-z)} = \frac{1}{2\pi i}\int_C e^t t^{z-1}\,dt = -\frac{1}{2\pi i}(2i\sin\pi(z-1))\int_0^\infty t^{z-1}e^{-t}\,dt$$
$$= \frac{1}{\pi}\sin\pi z\int_0^\infty t^{z-1}e^{-t}\,dt. \tag{19.24}$$

The proof is then completed by using $\Gamma(z)\Gamma(1-z) = \pi\operatorname{cosec}\pi z$, which we proved using definition A, to show that, under definition A, the right-hand side is indeed equal to $1/\Gamma(1-z)$. We now use the uniqueness of analytic continuation, noting that if two analytic functions agree on the region $\operatorname{Re} z > 0$, then they agree everywhere.

Infinite product for $\Gamma(z)$

The function $\Gamma(z)$ has poles at $z = 0, -1, -2, \ldots$, and therefore $(z\Gamma(z))^{-1} = (\Gamma(z+1))^{-1}$ has zeros as $z = -1, -2, \ldots$ Furthermore, the integral in definition B converges for all z, and so $1/\Gamma(z)$ has no singularities in the finite z plane, i.e. it is an entire function. This means that we can use the infinite product formula from Section 18.3.2:

$$g(z) = g(0)e^{cz}\prod_{j=1}^{\infty}\left\{\left(1-\frac{z}{z_j}\right)e^{z/z_j}\right\}. \tag{19.25}$$

We need to recall the definition of the Euler–Mascheroni constant $\gamma = -\Gamma'(1) = 0.5772157\ldots$, and that $\Gamma(1) = 1$. Then

$$\frac{1}{\Gamma(z)} = ze^{\gamma z}\prod_{n=1}^{\infty}\left\{\left(1+\frac{z}{n}\right)e^{-z/n}\right\}. \tag{19.26}$$

We can use this formula to compute

$$\frac{1}{\Gamma(z)\Gamma(1-z)} = \frac{1}{(-z)\Gamma(z)\Gamma(-z)} = z\prod_{n=1}^{\infty}\left\{\left(1+\frac{z}{n}\right)e^{-z/n}\left(1-\frac{z}{n}\right)e^{z/n}\right\}$$
$$= z\prod_{n=1}^{\infty}\left(1-\frac{z^2}{n^2}\right)$$
$$= \frac{1}{\pi}\sin\pi z,$$

and so obtain another (but not really independent) demonstration that $\Gamma(z)\Gamma(1-z) = \pi\operatorname{cosec}\pi z$.

Exercise 19.1: Starting from the infinite product formula for $\Gamma(z)$, show that

$$\frac{d^2}{dz^2}\ln\Gamma(z) = \sum_{n=0}^{\infty}\frac{1}{(z+n)^2}.$$

(Compare this "half series" with the expansion

$$\pi^2 \operatorname{cosec}^2 \pi z = \sum_{n=-\infty}^{\infty} \frac{1}{(z+n)^2}.)$$

19.2 Linear differential equations

When linear differential equations have coefficients that are meromorphic functions, their solutions can be extended off the real line and into the complex plane. The broader horizon then allows us to see much more of their structure.

19.2.1 Monodromy

Consider the linear differential equation

$$Ly \equiv y'' + p(z)y' + q(z)y = 0, \tag{19.27}$$

where p and q are meromorphic. Recall that the point $z = a$ is a *regular singular point* of the equation if p or q is singular there but

$$(z-a)p(z), \quad (z-a)^2 q(z) \tag{19.28}$$

are both analytic at $z = a$. We know, from the explicit construction of power series solutions, that near a regular singular point y is a sum of functions of the form $y = (z-a)^\alpha \varphi(z)$ or $y = (z-a)^\alpha (\ln(z-a)\varphi(z) + \chi(z))$, where both $\varphi(z)$ and $\chi(z)$ are analytic near $z = a$. We now examine this fact from a more topological perspective.

Suppose that y_1 and y_2 are linearly independent solutions of $Ly = 0$. Start from some ordinary (non-singular) point of the equation and analytically continue the solutions round the singularity at $z = a$ and back to the starting point. The continued functions $\tilde{y}_1$ and $\tilde{y}_2$ will not in general coincide with the original solutions but, still being solutions of the equation, they must be linear combinations of them. Therefore

$$\begin{pmatrix} \tilde{y}_1 \\ \tilde{y}_2 \end{pmatrix} = \begin{pmatrix} a_{11} & a_{12} \\ a_{21} & a_{22} \end{pmatrix} \begin{pmatrix} y_1 \\ y_2 \end{pmatrix}, \tag{19.29}$$

for some constants a_{ij}. By a suitable redefinition of $y_{1,2}$ we may either diagonalize this *monodromy* matrix to find

$$\begin{pmatrix} \tilde{y}_1 \\ \tilde{y}_2 \end{pmatrix} = \begin{pmatrix} \lambda_1 & 0 \\ 0 & \lambda_2 \end{pmatrix} \begin{pmatrix} y_1 \\ y_2 \end{pmatrix} \tag{19.30}$$

or, if the eigenvalues coincide and the matrix is not diagonalizable, reduce it to a Jordan form

$$\begin{pmatrix} \tilde{y}_1 \\ \tilde{y}_2 \end{pmatrix} = \begin{pmatrix} \lambda & 1 \\ 0 & \lambda \end{pmatrix} \begin{pmatrix} y_1 \\ y_2 \end{pmatrix}. \tag{19.31}$$

These matrix equations are satisfied, in the diagonalizable case, by functions of the form

$$y_1 = (z-a)^{\alpha_1}\varphi_1(z), \quad y_2 = (z-a)^{\alpha_2}\varphi_2(z), \tag{19.32}$$

where $\lambda_k = e^{2\pi i\alpha_k}$, and $\varphi_k(z)$ is single valued near $z = a$. In the Jordan-form case we must have

$$y_1 = (z-a)^{\alpha}\left[\varphi_1(z) + \frac{1}{2\pi i\lambda}\ln(z-a)\varphi_2(z)\right], \quad y_2 = (z-a)^{\alpha}\varphi_2(z), \tag{19.33}$$

where, again, the $\varphi_k(z)$ are single-valued. Notice that coincidence of the monodromy eigenvalues λ_1 and λ_2 does not require the exponents α_1 and α_2 to be the same, only that they differ by an integer. This is the same Frobenius condition that signals the presence of a logarithm in the traditional series solution.

The occurrence of fractional powers and logarithms in solutions near a regular singular point is therefore quite natural.

19.2.2 Hypergeometric functions

Most of the special functions of mathematical physics are special cases of the hypergeometric function $F(a, b; c; z)$, which may be defined by the series

$$\begin{aligned} F(a,b;c;z) &= 1 + \frac{a.b}{1.c}z + \frac{a(a+1)b(b+1)}{2!c(c+1)}z^2 + \\ &\quad + \frac{a(a+1)(a+2)b(b+1)(b+2)}{3!c(c+1)(c+2)}z^3 + \cdots \\ &= \frac{\Gamma(c)}{\Gamma(a)\Gamma(b)}\sum_{n=0}^{\infty}\frac{\Gamma(a+n)\Gamma(b+n)}{\Gamma(c+n)\Gamma(1+n)}z^n. \end{aligned} \tag{19.34}$$

For general values of a, b, c, this series converges for $|z| < 1$, the singularity restricting the convergence being a branch point at $z = 1$.

Examples:

$$(1+z)^n = F(-n, b; b; -z), \tag{19.35}$$

$$\ln(1+z) = zF(1, 1; 2; -z), \tag{19.36}$$

$$z^{-1}\sin^{-1}z = F\left(\frac{1}{2}, \frac{1}{2}; \frac{3}{2}; z^2\right), \tag{19.37}$$

$$e^z = \lim_{b\to\infty} F(1, b; 1/b; z/b), \tag{19.38}$$

$$P_n(z) = F\left(-n, n+1; 1; \frac{1-z}{2}\right), \tag{19.39}$$

where in the last line P_n is the Legendre polynomial.

For future reference, note that expanding the integrand on the right-hand side as a power series in z and integrating term by term shows that $F(a,b;c;z)$ has the integral representation

$$F(a,b;c;z) = \frac{\Gamma(c)}{\Gamma(b)\Gamma(c-b)}\int_0^1 (1-tz)^{-a}t^{b-1}(1-t)^{c-b-1}dt. \tag{19.40}$$

If $\mathrm{Re}\, c > \mathrm{Re}\,(a+b)$, we may set $z=1$ in this integral to get

$$F(a,b;c;1) = \frac{\Gamma(c)\Gamma(c-a-b)}{\Gamma(c-a)\Gamma(c-b)}. \tag{19.41}$$

The hypergeometric function is a solution of the second-order differential equation

$$z(1-z)y'' + [c-(a+b+1)z]y' - aby = 0. \tag{19.42}$$

This equation has regular singular points at $z = 0, 1, \infty$. Provided that $1-c$ is not an integer, the general solution is

$$y = AF(a,b;c;z) + Bz^{1-c}F(b-c+1, a-c+1; 2-c; z). \tag{19.43}$$

A differential equation possessing only regular singular points is known as a *Fuchsian equation.* The hypergeometric equation is a particular case of the general Fuchsian equation with three[1] regular singularities at $z = z_1, z_2, z_3$. This equation is

$$y'' + P(z)y' + Q(z)y = 0, \tag{19.44}$$

[1] The Fuchsian equation with *two* regular singularities is

$$y'' + p(z)y' + q(z)y = 0$$

with

$$p(z) = \left(\frac{1-\alpha-\alpha'}{z-z_1} + \frac{1+\alpha+\alpha'}{z-z_2}\right),$$

$$q(z) = \frac{\alpha\alpha'(z_1-z_2)^2}{(z-z_1)^2(z-z_2)^2}.$$

Its general solution is

$$y = A\left(\frac{z-z_1}{z-z_2}\right)^{\alpha} + B\left(\frac{z-z_1}{z-z_2}\right)^{\alpha'}.$$

where

$$
\begin{aligned}
P(z) &= \left(\frac{1-\alpha-\alpha'}{z-z_1}+\frac{1-\beta-\beta'}{z-z_2}+\frac{1-\gamma-\gamma'}{z-z_3}\right)\\
Q(z) &= \frac{1}{(z-z_1)(z-z_2)(z-z_3)}\\
&\quad\times\left(\frac{(z_1-z_2)(z_1-z_3)\alpha\alpha'}{z-z_1}+\frac{(z_2-z_3)(z_2-z_1)\beta\beta'}{z-z_2}\right.\\
&\quad\left.+\frac{(z_3-z_1)(z_3-z_2)\gamma\gamma'}{z-z_3}\right).
\end{aligned}
\tag{19.45}
$$

The parameters are subject to the constraint $\alpha+\beta+\gamma+\alpha'+\beta'+\gamma'=1$, which ensures that $z=\infty$ is not a singular point of the equation. This equation is sometimes called *Riemann's P-equation*. The P probably stands for Papperitz, who discovered it.

The indicial equation relative to the regular singular point at z_1 is

$$
r(r-1)+(1-\alpha-\alpha')r+\alpha\alpha'=0,
\tag{19.46}
$$

and has roots $r=\alpha,\alpha'$. From this, we deduce that Riemann's equation has solutions that behave like $(z-z_1)^{\alpha}$ and $(z-z_1)^{\alpha'}$ near z_1. Similarly, there are solutions that behave like $(z-z_2)^{\beta}$ and $(z-z_2)^{\beta'}$ near z_2, and like $(z-z_3)^{\gamma}$ and $(z-z_3)^{\gamma'}$ near z_3. The solution space of Riemann's equation is traditionally denoted by the Riemann "P" symbol

$$
P\left\{\begin{matrix} z_1 & z_2 & z_3 & \\ \alpha & \beta & \gamma & z \\ \alpha' & \beta' & \gamma' & \end{matrix}\right\}
\tag{19.47}
$$

where the six quantities $\alpha,\beta,\gamma,\alpha',\beta',\gamma'$ are called the *exponents* of the solution. A particular solution is

$$
\begin{aligned}
y &= \left(\frac{z-z_1}{z-z_2}\right)^{\alpha}\left(\frac{z-z_3}{z-z_2}\right)^{\gamma}\\
&\quad\times F\left(\alpha+\beta+\gamma,\alpha+\beta'+\gamma;1+\alpha-\alpha';\frac{(z-z_1)(z_3-z_2)}{(z-z_2)(z_3-z_1)}\right).
\end{aligned}
\tag{19.48}
$$

By permuting the triples (z_1,α,α'), (z_2,β,β'), (z_3,γ,γ'), and within them interchanging the pairs $\alpha\leftrightarrow\alpha'$, $\gamma\leftrightarrow\gamma'$, we may find a total[2] of $6\times 4=24$ solutions of this form. They are called the *Kummer* solutions. Only two of them can be linearly independent, and a large part of the theory of special functions is devoted to obtaining the linear relations between them.

[2] The interchange $\beta\leftrightarrow\beta'$ leaves the hypergeometric function invariant, and so does not give a new solution.

It is straightforward, but a trifle tedious, to show that

$$(z-z_1)^r(z-z_2)^s(z-z_3)^t P\begin{Bmatrix} z_1 & z_2 & z_3 & \\ \alpha & \beta & \gamma & z \\ \alpha' & \beta' & \gamma' & \end{Bmatrix}$$

$$= P\begin{Bmatrix} z_1 & z_2 & z_3 & \\ \alpha+r & \beta+s & \gamma+t & z \\ \alpha'+r & \beta'+s & \gamma'+t & \end{Bmatrix}, \tag{19.49}$$

provided $r+s+t=0$. Riemann's equation retains its form under Möbius maps, only the location of the singular points changing. We therefore deduce that

$$P\begin{Bmatrix} z_1 & z_2 & z_3 & \\ \alpha & \beta & \gamma & z \\ \alpha' & \beta' & \gamma' & \end{Bmatrix} = P\begin{Bmatrix} z_1' & z_2' & z_3' & \\ \alpha & \beta & \gamma & z' \\ \alpha' & \beta' & \gamma' & \end{Bmatrix}, \tag{19.50}$$

where

$$z' = \frac{az+b}{cz+d}, \quad z_1' = \frac{az_1+b}{cz_1+d}, \quad z_2' = \frac{az_2+b}{cz_2+d}, \quad z_3' = \frac{az_3+b}{cz_3+d}. \tag{19.51}$$

By using the Möbius map that takes $(z_1, z_2, z_3) \to (0, 1, \infty)$, and by extracting powers to shift the exponents, we can reduce the general eight-parameter Riemann equation to the three-parameter hypergeometric equation.

The P symbol for the hypergeometric equation is

$$P\begin{Bmatrix} 0 & \infty & 1 & \\ 0 & a & 0 & z \\ 1-c & b & c-a-b & \end{Bmatrix}. \tag{19.52}$$

By using this observation and a suitable Möbius map we see that

$$F(a, b; a+b-c; 1-z)$$

and

$$(1-z)^{c-a-b}F(c-b, c-a; c-a-b+1; 1-z)$$

are also solutions of the hypergeometric equation, each having pure (as opposed to a linear combination of) power-law behaviours near $z=1$. (The previous solutions had pure power-law behaviours near $z=0$.) These new solutions must be linear combinations of the old ones, and we may use

$$F(a, b; c; 1) = \frac{\Gamma(c)\Gamma(c-a-b)}{\Gamma(c-a)\Gamma(c-b)}, \quad \mathrm{Re}\,(c-a-b) > 0, \tag{19.53}$$

together with the trick of substituting $z = 0$ and $z = 1$, to determine the coefficients and show that

$$\begin{aligned} F(a,b;c;z) \\ &= \frac{\Gamma(c)\Gamma(c-a-b)}{\Gamma(c-a)\Gamma(c-b)} F(a,b;a+b-c;1-z) \\ &+ \frac{\Gamma(c)\Gamma(a+b-c)}{\Gamma(a)\Gamma(b)} (1-z)^{c-a-b} F(c-b,c-a;c-a-b+1;1-z). \end{aligned} \tag{19.54}$$

This last equation holds for all values of a, b, c such that the Gamma functions make sense.

A complete set of pure-power solutions to the hypergeometric equation can be taken to be

$$\begin{aligned} \phi_0^{(0)}(z) &= F(a,b;c;z), \\ \phi_0^{(1)}(z) &= z^{1-c}F(a+1-c,b+1-c;2-c;z), \\ \phi_1^{(0)}(z) &= F(a,b;1-c+a+b;1-z), \\ \phi_1^{(1)}(z) &= (1-z)^{c-a-b}F(c-a,c-b;1+c-a-b;1-z), \\ \phi_\infty^{(0)}(z) &= z^{-a}F(a,a+1-c;1+a-b;z^{-1}), \\ \phi_\infty^{(1)}(z) &= z^{-b}F(a,b+1-c;1-a+b;z^{-1}). \end{aligned} \tag{19.55}$$

Here the suffix denotes the point at which the solution has pure-power behaviour. The connection coefficients relating the solutions to one another are then

$$\begin{aligned} \phi_0^{(0)} &= \frac{\Gamma(c)\Gamma(c-a-b)}{\Gamma(c-a)\Gamma(c-b)}\phi_1^{(0)} + \frac{\Gamma(c)\Gamma(a+b-c)}{\Gamma(a)\Gamma(b)}\phi_1^{(1)}, \\ \phi_0^{(1)} &= \frac{\Gamma(2-c)\Gamma(c-a-b)}{\Gamma(1-a)\Gamma(1-b)}\phi_1^{(0)} + \frac{\Gamma(2-c)\Gamma(a+b-c)}{\Gamma(a+1-c)\Gamma(b+1-c)}\phi_1^{(1)}, \end{aligned} \tag{19.56}$$

and

$$\begin{aligned} \phi_0^{(0)} &= e^{-i\pi a}\frac{\Gamma(c)\Gamma(b-a)}{\Gamma(c-a)\Gamma(b)}\phi_\infty^{(0)} + e^{-i\pi b}\frac{\Gamma(2-c)\Gamma(a-b)}{\Gamma(a+1-c)\Gamma(1-b)}\phi_\infty^{(1)}, \\ \phi_0^{(1)} &= e^{-i\pi(a+1-c)}\frac{\Gamma(2-c)\Gamma(b-a)}{\Gamma(b+1-c)\Gamma(1-a)}\phi_\infty^{(0)} \\ &\quad + e^{-i\pi(b+1-c)}\frac{\Gamma(2-c)\Gamma(a-b)}{\Gamma(a+1-c)\Gamma(1-b)}\phi_\infty^{(1)}. \end{aligned} \tag{19.57}$$

These relations assume that $\operatorname{Im} z > 0$. The signs in the exponential factors must be reversed when $\operatorname{Im} z < 0$.

Example: *The Pöschel–Teller problem for general positive l*. A substitution $z = (1 + e^{2x})^{-1}$ shows that the Pöschel–Teller–Schrödinger equation

$$\left(-\frac{d^2}{dx^2} - l(l+1)\mathrm{sech}^2 x\right)\psi = E\psi \tag{19.58}$$

has solution

$$\psi(x) = (1+e^{2x})^{-\kappa/2}(1+e^{-2x})^{-\kappa/2}F\left(\kappa+l+1, \kappa-l; \kappa+1; \frac{1}{1+e^{2x}}\right), \tag{19.59}$$

where $E = -\kappa^2$. This solution behaves near $x = \infty$ as

$$\psi \sim e^{-\kappa x}F(\kappa+l+1, \kappa-l; \kappa+; 0) = e^{-\kappa x}. \tag{19.60}$$

We use the connection formula (19.54) to see that it behaves in the vicinity of $x = -\infty$ as

$$\begin{aligned}\psi &\sim e^{\kappa x}F(\kappa+l+1, \kappa-l; \kappa+1; 1-e^{2x}) \\ &\to e^{\kappa x}\frac{\Gamma(\kappa+1)\Gamma(-\kappa)}{\Gamma(-l)\Gamma(1+l)} + e^{-\kappa x}\frac{\Gamma(\kappa+1)\Gamma(\kappa)}{\Gamma(\kappa+l+1)\Gamma(\kappa-l)}.\end{aligned} \tag{19.61}$$

To find the bound-state spectrum, assume that κ is positive. Then $E = -\kappa^2$ will be an eigenvalue provided that the coefficient of $e^{-\kappa x}$ near $x = -\infty$ vanishes. In other words, the spectrum follows from the condition

$$\frac{\Gamma(\kappa+1)\Gamma(\kappa)}{\Gamma(\kappa+l+1)\Gamma(\kappa-l)} = 0. \tag{19.62}$$

This condition is satisfied for a finite set κ_n, $n = 1, \ldots, \lfloor l \rfloor$ (where $\lfloor l \rfloor$ denotes the integer part of l) at which κ is positive but $\kappa - l$ is zero or a negative integer.

By setting $\kappa = -ik$, for real k we find the scattering solution

$$\psi(x) = \begin{cases} e^{ikx} + r(k)e^{-ikx}, & x \ll 0 \\ t(k)e^{ikx}, & x \gg 0 \end{cases} \tag{19.63}$$

where

$$\begin{aligned}r(k) &= \frac{\Gamma(l+1-ik)\Gamma(-ik-l)\Gamma(ik)}{\Gamma(-l)\Gamma(1+l)\Gamma(ik)} \\ &= -\frac{\sin \pi l}{\pi}\frac{\Gamma(l+1-ik)\Gamma(-ik-l)\Gamma(ik)}{\Gamma(-ik)},\end{aligned} \tag{19.64}$$

and

$$t(k) = \frac{\Gamma(l+1-ik)\Gamma(-ik-l)}{\Gamma(1-ik)\Gamma(-ik)}. \tag{19.65}$$

Whenever l is a (positive) integer, the divergent factor of $\Gamma(-l)$ in the denominator of $r(k)$ causes the reflected wave to vanish. This is something we had observed in earlier chapters. In this reflectionless case, the transmission coefficient $t(k)$ reduces to a phase

$$t(k) = \frac{(-ik+1)(-ik+2)\cdots(-ik+l)}{(-ik-1)(-ik-2)\cdots(-ik-l)}. \tag{19.66}$$

19.3 Solving ODEs via contour integrals

Our task in this section is to understand the origin of contour integral solutions, such as the expression

$$F(a,b;c;z) = \frac{\Gamma(c)}{\Gamma(b)\Gamma(c-b)}\int_0^1 (1-tz)^{-a}t^{b-1}(1-t)^{c-b-1}dt, \tag{19.67}$$

which we have previously seen for the hypergeometric equation.

Suppose that we are given a differential operator

$$L_z = \partial^2_{zz} + p(z)\partial_z + q(z) \tag{19.68}$$

and seek a solution of $L_z u = 0$ as an integral

$$u(z) = \int_\Gamma F(z,t)\, dt \tag{19.69}$$

over some contour Γ. If we can find a kernel F such that

$$L_z F = \frac{\partial Q}{\partial t}, \tag{19.70}$$

for some function $Q(z,t)$, then

$$L_z u = \int_\Gamma L_z F(z,t)\, dt = \int_\Gamma \left(\frac{\partial Q}{\partial t}\right) dt = [Q]_\Gamma\, . \tag{19.71}$$

Thus, if Q vanishes at both ends of the contour, if it takes the same value at the two ends, or if the contour is closed and thus has no ends, then we have succeeded in our quest.

Example: Consider Legendre's equation,

$$L_z u \equiv (1-z^2)\frac{d^2u}{dz^2} - 2z\frac{du}{dz} + \nu(\nu+1)u = 0. \tag{19.72}$$

The identity

$$L_z\left\{\frac{(t^2-1)^\nu}{(t-z)^{\nu+1}}\right\} = (\nu+1)\frac{d}{dt}\left\{\frac{(t^2-1)^{\nu+1}}{(t-z)^{\nu+2}}\right\} \tag{19.73}$$

shows that

$$P_\nu(z) = \frac{1}{2\pi i}\int_\Gamma \left\{\frac{(t^2-1)^\nu}{2^\nu(t-z)^{\nu+1}}\right\} dt \tag{19.74}$$

will be a solution of Legendre's equation, provided that

$$[Q]_\Gamma \equiv \left[\frac{(t^2-1)^{\nu+1}}{(t-z)^{\nu+2}}\right]_\Gamma = 0. \tag{19.75}$$

We could, for example, take a contour that encircles the points $t = z$ and $t = 1$, but excludes the point $t = -1$. On going round this contour, the numerator acquires a phase of $e^{2\pi i(\nu+1)}$, while the denominator of $[Q]_\Gamma$ acquires a phase of $e^{2\pi i(\nu+2)}$. The net phase change is therefore $e^{-2\pi i} = 1$. The function in the integrated-out part is therefore single-valued, and so the integrated-out part vanishes. When ν is an integer, Cauchy's formula shows that (19.74) reduces to

$$P_n(z) = \frac{1}{2^n n!}\frac{d^n}{dz^n}(z^2-1)^n, \tag{19.76}$$

which is Rodriguez' formula for the Legendre polynomials.

The figure-of-eight contour shown in Figure 19.2 gives us a second solution

$$Q_\nu(z) = \frac{1}{4i\sin\pi\nu}\int_\Gamma \left\{\frac{(t^2-1)^\nu}{2^\nu(z-t)^{\nu+1}}\right\} dt, \quad \nu \notin \mathbb{Z}. \tag{19.77}$$

Here we define $\arg(t-1)$ and $\arg(t+1)$ to be zero for $t > 1$ and $t > -1$ respectively. The integrated-out part vanishes because the phase gained by the $(t^2-1)^{\nu+1}$ in the numerator of $[Q]_\Gamma$ during the clockwise winding about $t = 1$ is undone during the anticlockwise winding about $t = -1$, and, provided that z lies outside the contour, there is no phase change in the $(z-t)^{(\nu+2)}$ in the denominator.

When ν is real and positive the contributions from the circular arcs surrounding $t = \pm 1$ become negligeable as we shrink this new contour down onto the real axis. We observe that, with the arguments of $(t \pm 1)$ as specified above,

$$(t^2-1)^\nu \to (1-t^2)^\nu e^{-i\pi\nu}$$

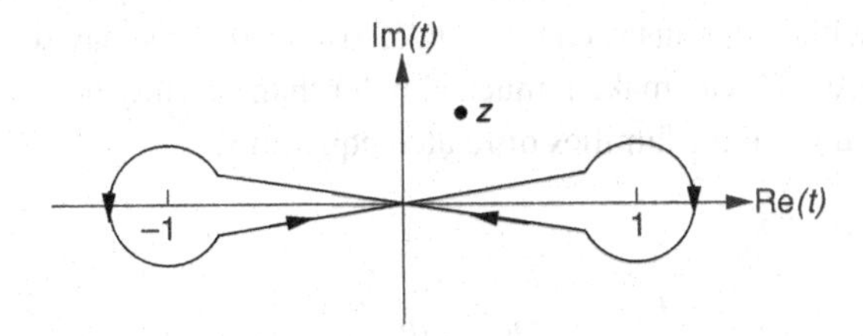

Figure 19.2 Figure-of-eight contour for $Q_\nu(z)$.

for the left-going part of the contour on the real axis between $t = +1$ and $t = -1$, and

$$(t^2 - 1)^\nu \to (1 - t^2)^\nu e^{-i\pi\nu} e^{2\pi i\nu} = (1 - t^2)e^{+i\pi\nu}$$

after we have rounded the branch point at $t = -1$ and are returning along the real axis to $t = +1$. Thus, after the shrinking manœuvre, the integral (19.77) becomes

$$Q_\nu(z) = \frac{1}{2}\int_{-1}^{1}\left\{\frac{(1-t^2)^\nu}{2^\nu(z-t)^{\nu+1}}\right\} dt, \quad \nu > 0. \tag{19.78}$$

In contrast to (19.77), this last formula continues to make sense when ν is a positive integer. It then provides a convenient definition of $Q_n(z)$, the Legendre function of the second kind (see Exercise 18.3).

It is usually hard to find a suitable $F(z, t)$ in one fell swoop. (The identity (19.73) exploited in the example is not exactly obvious!) An easier strategy is to seek a solution in the form of an integral operator with kernel K acting on the function $v(t)$. Thus we try

$$u(z) = \int_\Gamma K(z, t)v(t)\, dt. \tag{19.79}$$

Suppose that $L_z K(z, t) = M_t K(z, t)$, where M_t is a differential operator in t that does not involve z. The operator M_t will have a formal adjoint $M_t^\dagger$ such that

$$\int_\Gamma v(M_t K)\, dt - \int_\Gamma K(M_t^\dagger v)\, dt = [Q(K, v)]_\Gamma\, . \tag{19.80}$$

(This is Lagrange's identity.) Now

$$\begin{aligned} L_z u &= \int_\Gamma L_z K(z, t) v\, dt \\ &= \int_\Gamma (M_t K(z, t)) v\, dt \\ &= \int_\Gamma K(z, t)(M_t^\dagger v)\, dt + [Q(K, v)]_\Gamma\, . \end{aligned}$$

We can therefore solve the original equation, $L_z u = 0$, by finding a v such that $(M_t^\dagger v) = 0$, and a contour with endpoints such that $[Q(K, v)]_\Gamma = 0$. This may sound complicated, but an artful choice of K can make it much simpler than solving the original problem. A single K will often work for families of related equations.

Example: We will solve

$$L_z u = \frac{d^2 u}{dz^2} - z\frac{du}{dz} + \nu u = 0, \tag{19.81}$$

by using the kernel $K(z,t) = e^{-zt}$. It is easy to check that $L_z K(z,t) = M_t K(z,t)$ where

$$M_t = t^2 - t\frac{\partial}{\partial t} + \nu, \tag{19.82}$$

and so

$$M_t^\dagger = t^2 + \frac{\partial}{\partial t}t + \nu = t^2 + (\nu+1) + t\frac{\partial}{\partial t}. \tag{19.83}$$

The equation $M_t^\dagger v = 0$ has a solution

$$v(t) = t^{-(\nu+1)} e^{-\frac{1}{2}t^2}, \tag{19.84}$$

and so

$$u = \int_\Gamma t^{-(1+\nu)} e^{-(zt+\frac{1}{2}t^2)}\, dt, \tag{19.85}$$

for some suitable Γ.

19.3.1 Bessel functions

As an illustration of the general method we will explore the theory of Bessel functions. Bessel functions are members of the family of *confluent hypergeometric functions*, obtained by letting the two regular singular points z_2, z_3 of the Riemann–Papperitz equation coalesce at infinity. The resulting singular point is no longer regular, and confluent hypergeometric functions have an essential singularity at infinity. The confluent hypergeometric equation is

$$zy'' + (c - z)y' - ay = 0, \tag{19.86}$$

with solution

$$\Phi(a,c;z) = \frac{\Gamma(c)}{\Gamma(a)} \sum_{n=0}^{\infty} \frac{\Gamma(a+n)}{\Gamma(c+n)\Gamma(n+1)} z^n. \tag{19.87}$$

Observe that

$$\Phi(a,c;z) = \lim_{b\to\infty} F(a,b;c;z/b). \tag{19.88}$$

The second solution, provided that c is not an integer, is

$$z^{1-c}\Phi(a-c+1, 2-c; z). \tag{19.89}$$

Other functions of this family are the *parabolic cylinder functions*, which in special cases reduce to $e^{-z^2/4}$ times the *Hermite polynomials*, the *error function*,

$$\text{erf}\,(z) = \int_0^z e^{-t^2}\, dt = z\Phi\left(\frac{1}{2}, \frac{3}{2}; -z^2\right) \tag{19.90}$$

and the *Laguerre polynomials*,

$$L_n^m = \frac{\Gamma(n+m+1)}{\Gamma(n+1)\Gamma(m+1)} \Phi(-n, m+1; z). \tag{19.91}$$

Bessel's equation involves the operator

$$L_z = \partial_{zz}^2 + \frac{1}{z}\partial_z + \left(1 - \frac{\nu^2}{z^2}\right). \tag{19.92}$$

Experience shows that for Bessel functions a useful kernel is

$$K(z,t) = \left(\frac{z}{2}\right)^\nu \exp\left(t - \frac{z^2}{4t}\right). \tag{19.93}$$

Then

$$L_z K(z,t) = \left(\partial_t - \frac{\nu+1}{t}\right) K(z,t) \tag{19.94}$$

so, again, M is a first-order operator, which is simpler to deal with than the original second-order L_z. In this case the adjoint is

$$M^\dagger = \left(-\partial_t - \frac{\nu+1}{t}\right) \tag{19.95}$$

and we need a v such that

$$M^\dagger v = -\left(\partial_t + \frac{\nu+1}{t}\right) v = 0. \tag{19.96}$$

Clearly, $v = t^{-\nu-1}$ will work. The integrated-out part is

$$[Q(K,v)]_a^b = \left[t^{-\nu-1} \exp\left(t - \frac{z^2}{4t}\right)\right]_a^b, \tag{19.97}$$

and we see that

$$J_\nu(z) = \frac{1}{2\pi i}\left(\frac{z}{2}\right)^\nu \int_\Gamma t^{-\nu-1} e^{\left(t - \frac{z^2}{4t}\right)}\, dt \tag{19.98}$$

solves Bessel's equation provided we use a suitable contour.

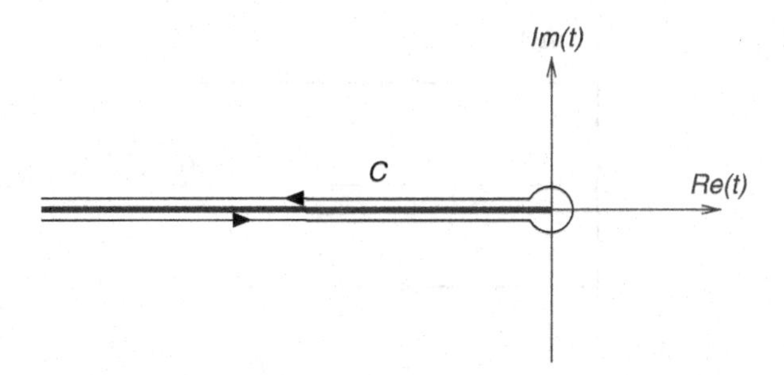

Figure 19.3 Contour for solving the Bessel equation.

We can take for Γ a curve C starting at $-\infty - i\epsilon$ and ending at $-\infty + i\epsilon$, and surrounding the branch cut of $t^{-\nu-1}$, which we take as the negative t axis (Figure 19.3). This contour works because Q is zero at both ends of the contour.

A cosmetic rewrite $t = uz/2$ gives

$$J_\nu(z) = \frac{1}{2\pi i} \int_C u^{-\nu-1} e^{\frac{z}{2}\left(u-\frac{1}{u}\right)}\, du. \tag{19.99}$$

For ν an integer, there is no discontinuity across the cut, so we can ignore it and take C to be the unit circle. Then, recognizing the resulting

$$J_n(z) = \frac{1}{2\pi i} \int_{|z|=1} u^{-n-1} e^{\frac{z}{2}\left(u-\frac{1}{u}\right)}\, du \tag{19.100}$$

to be a Laurent coefficient, we obtain the familiar Bessel-function generating function

$$e^{\frac{z}{2}\left(u-\frac{1}{u}\right)} = \sum_{n=-\infty}^{\infty} J_n(z) u^n. \tag{19.101}$$

When ν is not an integer, we see why we need a branch cut integral.

If we set $u = e^w$ we get

$$J_\nu(z) = \frac{1}{2\pi i} \int_{C'} dw\, e^{z \sinh w - \nu w}, \tag{19.102}$$

where C' goes from $(\infty - i\pi)$ to $-i\pi$ to $+i\pi$ to $(\infty + i\pi)$; see Figure 19.4.

If we set $w = t \pm i\pi$ on the horizontals and $w = i\theta$ on the vertical part, we can rewrite this as

$$J_\nu(z) = \frac{1}{\pi} \int_0^\pi \cos(\nu\theta - z \sin\theta)\, d\theta - \frac{\sin \nu\pi}{\pi} \int_0^\infty e^{-\nu t - z \sinh t}\, dt. \tag{19.103}$$

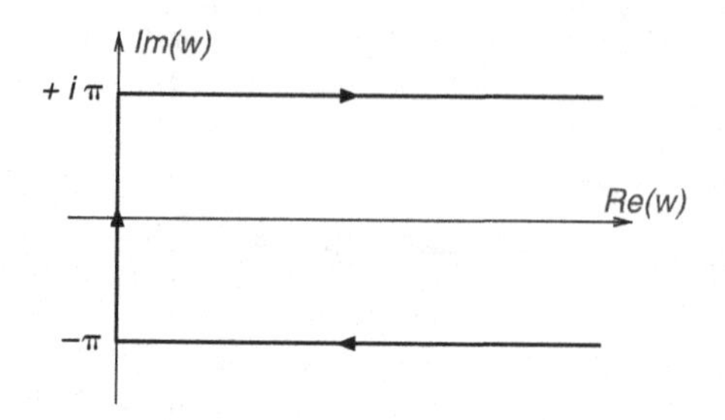

Figure 19.4 Bessel contour after the changes of variables.

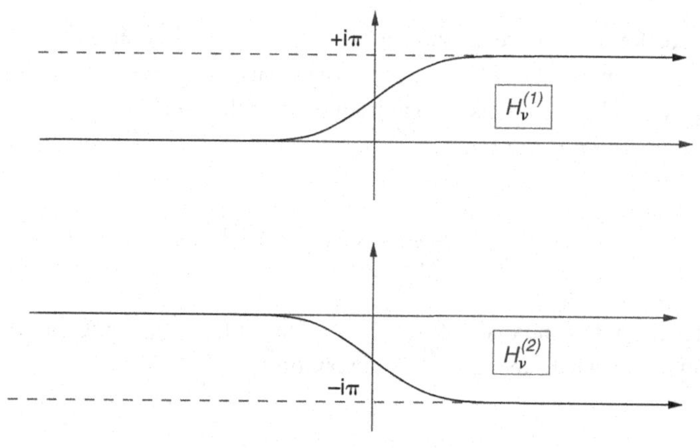

Figure 19.5 Contours defining $H_\nu^{(1)}(z)$ and $H_\nu^{(2)}(z)$.

All these are standard formulæ for the Bessel function but their origins would be hard to understand without the contour solutions trick.

When ν becomes an integer, the functions $J_\nu(z)$ and $J_{-\nu}(z)$ are no longer independent. In order to have a Bessel-equation solution that retains its independence from $J_\nu(z)$, even as ν becomes a whole number, we define the Neumann function by

$$\begin{aligned} N_\nu(z) &\stackrel{\text{def}}{=} \frac{J_\nu(z)\cos\nu\pi - J_{-\nu}(z)}{\sin\nu\pi} \\ &= \frac{\cot\nu\pi}{\pi}\int_0^\pi \cos(\nu\theta - z\sin\theta)\,d\theta - \operatorname{cosec}\nu\pi\pi\int_0^\pi \cos(\nu\theta + z\sin\theta)\,d\theta \\ &- \frac{\cos\nu\pi}{\pi}\int_0^\infty e^{-\nu t - z\sinh t}\,dt - \frac{1}{\pi}\int_0^\infty e^{\nu t - z\sinh t}\,dt. \end{aligned} \tag{19.104}$$

Both the Bessel and Neumann functions are real for positive real x. As x becomes large they oscillate as slowly decaying sines and cosines. It is sometimes convenient to decompose these real functions into solutions that oscillate as $e^{\pm ix}$. We therefore define the *Hankel functions* by (Figure 19.5)

$$H_\nu^{(1)}(z) = \frac{1}{i\pi}\int_{-\infty}^{\infty+i\pi} e^{z\sinh w-\nu w}\,dw, \qquad |\arg z| < \pi/2$$

$$H_\nu^{(2)}(z) = -\frac{1}{i\pi}\int_{-\infty}^{\infty-i\pi} e^{z\sinh w-\nu w}\,dw, \qquad |\arg z| < \pi/2. \tag{19.105}$$

Then

$$\frac{1}{2}(H_\nu^{(1)}(z) + H_\nu^{(2)}(z)) = J_\nu(z),$$

$$\frac{1}{2}(H_\nu^{(1)}(z) - H_\nu^{(2)}(z)) = N_\nu(z). \tag{19.106}$$

19.4 Asymptotic expansions

We often need to understand the behaviour of solutions of differential equations and functions, such as $J_\nu(x)$, when x takes values that are very large, or very small. This is the subject of *asymptotics*.

As an introduction to this art, consider the function

$$Z(\lambda) = \int_{-\infty}^{\infty} e^{-x^2-\lambda x^4}\,dx. \tag{19.107}$$

Those of you who have taken a course in quantum field theory based on path integrals will recognize that this is a "toy", 0-dimensional, version of the path integral for the $\lambda\varphi^4$ model of a self-interacting scalar field. Suppose we wish to obtain the perturbation expansion for $Z(\lambda)$ as a power series in λ. We naturally proceed as follows

$$\begin{aligned} Z(\lambda) &= \int_{-\infty}^{\infty} e^{-x^2-\lambda x^4}\,dx \\ &= \int_{-\infty}^{\infty} e^{-x^2}\sum_{n=0}^{\infty}(-1)^n\frac{\lambda^n x^{4n}}{n!}\,dx \\ &\stackrel{?}{=} \sum_{n=0}^{\infty}(-1)^n\frac{\lambda^n}{n!}\int_{-\infty}^{\infty} e^{-x^2}x^{4n}\,dx \\ &= \sum_{n=0}^{\infty}(-1)^n\frac{\lambda^n}{n!}\Gamma(2n+1/2). \end{aligned} \tag{19.108}$$

Something has clearly gone wrong here! The Gamma function $\Gamma(2n+1/2) \sim (2n)! \sim 4^n(n!)^2$ overwhelms the $n!$ in the denominator, so the radius of convergence of the final power series is zero.

The invalid, but popular, manœuvre is the interchange of the order of performing the integral and the sum. This interchange cannot be justified because the sum inside the integral does not converge uniformly on the domain of integration. Does this mean that the series is useless? It had better not! All quantum field theory (and most quantum mechanics) perturbation theory relies on versions of this manœuvre.

We are saved to some (often adequate) degree because, while the interchange of integral and sum does not lead to a convergent series, it does lead to a valid *asymptotic expansion*. We write

$$Z(\lambda) \sim \sum_{n=0}^{\infty} (-1)^n \frac{\lambda^n}{n!} \Gamma(2n + 1/2) \tag{19.109}$$

where

$$Z(\lambda) \sim \sum_{n=0}^{\infty} a_n \lambda^n \tag{19.110}$$

is shorthand for the more explicit

$$Z(\lambda) = \sum_{n=0}^{N} a_n \lambda^n + O\left(\lambda^{N+1}\right), \quad N = 1, 2, 3, \ldots \tag{19.111}$$

The "big O" notation,

$$Z(\lambda) - \sum_{n=0}^{N} a_n \lambda^n = O(\lambda^{N+1}) \tag{19.112}$$

as $\lambda \to 0$, means that

$$\lim_{\lambda \to 0} \left\{ \frac{|Z(\lambda) - \sum_0^N a_n \lambda^n|}{|\lambda^{N+1}|} \right\} = K < \infty. \tag{19.113}$$

The basic idea is that, given a *convergent* power series $\sum_n a_n \lambda^n$ for the function $f(\lambda)$, we fix the value of λ and take more and more terms. The sum then gets closer to $f(\lambda)$. Given an asymptotic series, on the other hand, we select a *fixed number of terms* in the series and then make λ smaller and smaller. The graph of $f(\lambda)$ and the graph of our polynomial approximation then approach each other. The more terms we take, the sooner they get close, but for any non-zero λ we can never get exacty $f(\lambda)$ – no matter how many terms we take.

We often consider asymptotic expansions where the independent variable becomes *large*. Here we have expansions in inverse powers of x:

$$F(x) = \sum_{n=0}^{N} b_n x^{-n} + O\left(x^{-N-1}\right), \quad N = 1, 2, 3, \ldots \tag{19.114}$$

In this case

$$F(x) - \sum_{n=0}^{N} b_n x^{-n} = O\left(x^{-N-1}\right) \tag{19.115}$$

means that

$$\lim_{x\to\infty} \left\{ \frac{|F(x) - \sum_0^N b_n x^{-n}|}{|x^{-N-1}|} \right\} = K < \infty. \tag{19.116}$$

Again we take a fixed number of terms, and as x becomes large the function and its approximation get closer.

Observations:

(i) Knowledge of the asymptotic expansion gives us useful knowledge about the function, but does not give us everything. In particular, two distinct functions may have the *same* asymptotic expansion. For example, for small positive λ, the functions $F(\lambda)$ and $F(\lambda) + ae^{-b/\lambda}$ have exactly the same asymptotic expansions as series in positive powers of λ. This is because $e^{-b/\lambda}$ goes to zero faster than any power of λ, and so its asymptotic expansion $\sum_n a_n \lambda^n$ has every coefficient a_n being zero. Physicists commonly say that $e^{-b/\lambda}$ is a *non-perturbative* function, meaning that it will not be visible to a perturbation expansion in powers of λ.

(ii) An asymptotic expansion is usually valid only in a sector $a < \arg z < b$ in the complex plane. Different sectors have different expansions. This is called the *Stokes phenomenon*.

The most useful methods for obtaining asymptotic expansions require that the function to be expanded be given in terms of an integral. This is the reason why we have stressed the contour-integral method of solving differential equations. If the integral can be approximated by a Gaussian, we are led to the *method of steepest descents*. This technique is best explained by means of examples.

19.4.1 Stirling's approximation for $n!$

We start from the integral representation of the Gamma function:

$$\Gamma(x+1) = \int_0^\infty e^{-t} t^x \, dt. \tag{19.117}$$

Set $t = x\zeta$, so

$$\Gamma(x+1) = x^{x+1} \int_0^\infty e^{xf(\zeta)} \, d\zeta, \tag{19.118}$$

where

$$f(\zeta) = \ln\zeta - \zeta. \tag{19.119}$$

We are going to be interested in evaluating this integral in the limit that $x \to \infty$ and finding the first term in the asymptotic expansion of $\Gamma(x+1)$ in powers of $1/x$. In this limit, the exponential will be dominated by the part of the integration region near the absolute maximum of $f(\zeta)$. Now, $f(\zeta)$ is a maximum at $\zeta = 1$ and

$$f(\zeta) = -1 - \frac{1}{2}(\zeta - 1)^2 + \cdots \tag{19.120}$$

so

$$\begin{aligned}
\Gamma(x+1) &= x^{x+1}e^{-x}\int_0^\infty e^{-\frac{x}{2}(\zeta-1)^2+\cdots}\,d\zeta \\
&\approx x^{x+1}e^{-x}\int_{-\infty}^\infty e^{-\frac{x}{2}(\zeta-1)^2}\,d\zeta \\
&= x^{x+1}e^{-x}\sqrt{\frac{2\pi}{x}} \\
&= \sqrt{2\pi}\,x^{x+1/2}e^{-x}.
\end{aligned} \tag{19.121}$$

By keeping more of the terms represented by the dots, and expanding them as

$$e^{-\frac{x}{2}(\zeta-1)^2+\cdots} = e^{-\frac{x}{2}(\zeta-1)^2}\left[1 + a_1(\zeta-1) + a_2(\zeta-1)^2 + \cdots\right], \tag{19.122}$$

we would find, on doing the integral, that

$$\begin{aligned}
\Gamma(z+1) \approx \sqrt{2\pi}x^{x+1/2}e^{-x}\Bigg[1 &+ \frac{1}{12x} + \frac{1}{288x^2} - \frac{139}{51840x^3} \\
&- \frac{571}{2488320x^4} + O\left(\frac{1}{x^5}\right)\Bigg].
\end{aligned} \tag{19.123}$$

Since $\Gamma(n+1) = n!$ we have the useful result

$$n! \approx \sqrt{2\pi}n^{n+1/2}e^{-n}\left[1 + \frac{1}{12n} + \cdots\right]. \tag{19.124}$$

We make contact with our discusion of asymptotic series by rewriting the expansion as

$$\frac{\Gamma(x+1)}{\sqrt{2\pi}x^{x+1/2}e^{-x}} \sim 1 + \frac{1}{12x} + \frac{1}{288x^2} - \frac{139}{51840x^3} - \frac{571}{2488320x^4} + \ldots \tag{19.125}$$

This is typical. We usually have to pull out a leading factor from the function whose asymptotic behaviour we are studying, before we are left with a plain asymptotic power series.

19.4.2 Airy functions

The Airy functions $\mathrm{Ai}(x)$ and $\mathrm{Bi}(x)$ are closely related to Bessel functions, and are named after the mathematician and astronomer George Biddell Airy. They occur widely in physics. We will investigate the behaviour of $\mathrm{Ai}(x)$ for large values of $|x|$. A more sophisticated treatment is needed for this problem, and we will meet with Stokes' phenomenon. Airy's differential equation is

$$\frac{d^2y}{dz^2} - zy = 0. \tag{19.126}$$

On the real axis Airy's equation becomes

$$-\frac{d^2y}{dx^2} + xy = 0, \tag{19.127}$$

and we can think of this as the Schrödinger equation for a particle running up a linear potential. A classical particle incident from the left with total energy $E = 0$ will come to rest at $x = 0$, and then retrace its path. The point $x = 0$ is therefore called a *classical turning point*. The corresponding quantum wavefunction, $\mathrm{Ai}\,(x)$, contains a travelling wave incident from the left and becoming evanescent as it tunnels into the classically forbidden region, $x > 0$, together with a reflected wave returning to $-\infty$ (see Figure 19.6). The sum of the incident and reflected waves is a real-valued standing wave.

We will look for contour integral solutions to Airy's equation of the form

$$y(x) = \int_C e^{xt} f(t)\, dt. \tag{19.128}$$

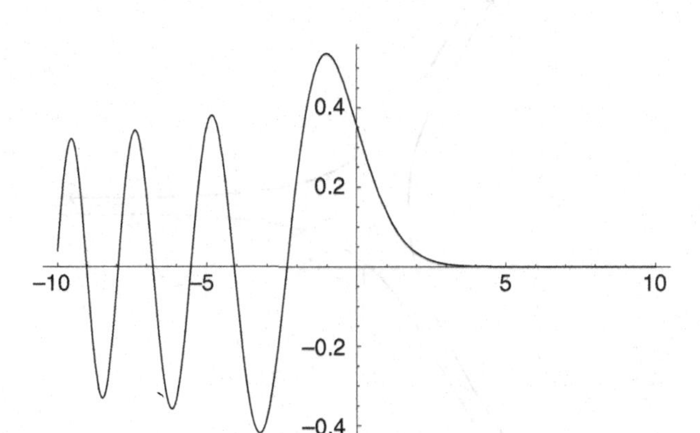

Figure 19.6 The Airy function $\mathrm{Ai}\,(x)$.

Denoting the Airy differential operator by $L_x \equiv \partial_x^2 - x$, we have

$$L_x y = \int_C (t^2 - x)e^{xt} f(t)\, dt = \int_a^b f(t) \left\{ t^2 - \frac{d}{dt} \right\} e^{xt}\, dt.$$

$$= \left[-e^{xt} f(t)\right]_C + \int_C \left(\left\{ t^2 + \frac{d}{dt} \right\} f(t) \right) e^{xt}\, dt. \tag{19.129}$$

Thus $f(t) = e^{-\frac{1}{3}t^3}$ and hence

$$y(x) = \int_C e^{xt - \frac{1}{3}t^3}\, dt, \tag{19.130}$$

provided the contour ends at points where the integrated-out term, $\left[e^{xt-\frac{1}{3}t^3}\right]_C$, vanishes. There are therefore three possible contours, shown in Figure 19.7, which end at any two of

$$+\infty, \quad \infty\, e^{2\pi i/3}, \quad \infty\, e^{-2\pi i/3}.$$

Since the integrand is an entire function, the sum $y_{C_1} + y_{C_2} + y_{C_3}$ is zero, so only two of the three solutions are linearly independent. The Airy function itself is defined by

$$\mathrm{Ai}\,(x) = \frac{1}{2\pi i} \int_{C_1} e^{xt - \frac{1}{3}t^3}\, dt = \frac{1}{\pi} \int_0^\infty \cos\left(xs + \frac{1}{3}s^3 \right) ds. \tag{19.131}$$

In obtaining the last equality, we have deformed the contour of integration, C_1, which ran from $\infty\, e^{-2\pi i/3}$ to $\infty\, e^{2\pi i/3}$, so that it lies on the imaginary axis, and there we have written $t = is$. You may check (by extending Jordan's lemma) that this deformation does not alter the value of the integral.

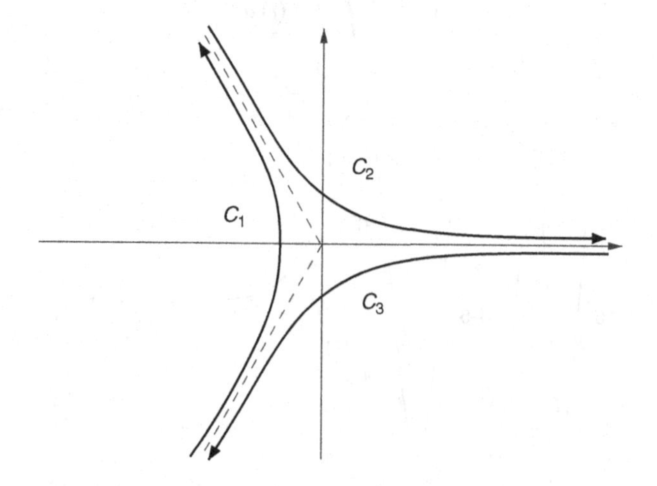

Figure 19.7 Contours providing solutions of Airy's equation.

To study the asymptotics of this function we need to examine separately two cases $x \gg 0$ and $x \ll 0$. For both ranges of x, the principal contribution to the integral will come from the neighbourhood of the stationary points of $f(t) = xt - t^3/3$. In the complex plane, stationary points are never pure maxima or minima of the real part of f (the real part alone determines the magnitude of the integrand) but are always *saddle points*. We must deform the contour so that on the integration path the stationary point is the highest point in a mountain pass. We must also ensure that everywhere on the contour the difference between f and its maximum value stays *real*. Because of the orthogonality of the real and imaginary part contours, this means that we must take a path of *steepest descent* from the pass – hence the name of the method. If we stray from the steepest descent path, the phase of the exponent will be changing. This means that the integrand will oscillate and we can no longer be sure that the result is dominated by the contributions near the saddle point.

(i) $x \gg 0$: The stationary points are at $t = \pm\sqrt{x}$. Writing $t = \xi - \sqrt{x}$ we have

$$f(\xi) = -\frac{2}{3}x^{3/2} + \xi^2\sqrt{x} - \frac{1}{3}\xi^3, \tag{19.132}$$

while near $t = +\sqrt{x}$ we write $t = \zeta + \sqrt{x}$ and find

$$f(\zeta) = +\frac{2}{3}x^{3/2} - \zeta^2\sqrt{x} - \frac{1}{3}\zeta^3. \tag{19.133}$$

We see that the saddle point near $-\sqrt{x}$ is a local maximum when we route the contour vertically, while the saddle point near $+\sqrt{x}$ is a local maximum as we go down the real axis. Since the contour in $\mathrm{Ai}\,(x)$ is aimed vertically we can distort it to pass through the saddle point near $-\sqrt{x}$, but cannot find a route through the point at $+\sqrt{x}$ without the integrand oscillating wildly. At the saddle point the exponent, $xt - t^3/3$, is real. If we write $t = u + iv$ we have

$$\mathrm{Im}\,(xt - t^3/3) = v(x - u^2 + v^2/3), \tag{19.134}$$

so the exact steepest descent path, on which the imaginary part remains zero, is given by the union of the real axis ($v = 0$) and the curve

$$u^2 - \frac{1}{3}v^2 = x. \tag{19.135}$$

This is a hyperbola, and the branch passing through the saddle point at $-\sqrt{x}$ is plotted in Figure 19.8(a).

Now setting $\xi = is$, we find

$$\mathrm{Ai}\,(x) = \frac{1}{2\pi}e^{-\frac{2}{3}x^{3/2}} \int_{-\infty}^{\infty} e^{-\sqrt{x}s^2+\cdots}\,ds \sim \frac{1}{2\sqrt{\pi}}x^{-1/4}e^{-\frac{2}{3}x^{3/2}}. \tag{19.136}$$

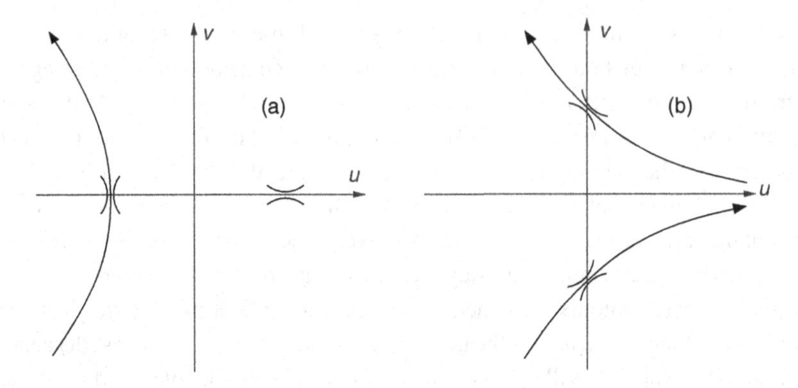

Figure 19.8 Steepest descent contours, location of the stationary points and orientation of the saddle passes for (a) $x \gg 0$, (b) $x \ll 0$.

(ii) $x \ll 0$: The stationary points are now at $\pm i\sqrt{|x|}$. Setting $t = \xi \pm i\sqrt{|x|}$ we find that

$$f(x) = \mp i\frac{2}{3}|x|^{3/2} \mp i\xi^2\sqrt{|x|}. \tag{19.137}$$

The exponent is no longer real, but the imaginary part will be constant and the integrand non-oscillatory provided we deform the contour so that it becomes the disconnected pair of curves shown in Figure 19.8(b). The new contour passes through both saddle points and we must sum their contributions. Near $t = i\sqrt{|x|}$ we set $\xi = e^{3\pi i/4}s$ and get

$$\begin{aligned}\frac{1}{2\pi i}e^{3\pi i/4}e^{-i\frac{2}{3}|x|^{3/2}}\int_{-\infty}^{\infty} e^{-\sqrt{|x|}s^2}\,ds &= \frac{1}{2i\sqrt{\pi}}e^{3\pi i/4}|x|^{-1/4}e^{-i\frac{2}{3}|x|^{3/2}}\\ &= -\frac{1}{2i\sqrt{\pi}}e^{-i\pi/4}|x|^{-1/4}e^{-i\frac{2}{3}|x|^{3/2}}.\end{aligned} \tag{19.138}$$

Near $t = -i\sqrt{|x|}$ we set $\xi = e^{\pi i/4}s$ and get

$$\frac{1}{2\pi i}e^{\pi i/4}e^{i\frac{2}{3}|x|^{3/2}}\int_{-\infty}^{\infty} e^{-\sqrt{|x|}s^2}\,ds = \frac{1}{2i\sqrt{\pi}}e^{\pi i/4}|x|^{-1/4}e^{i\frac{2}{3}|x|^{3/2}}. \tag{19.139}$$

The sum of these two contributions is

$$\mathrm{Ai}\,(x) \sim \frac{1}{\sqrt{\pi}|x|^{1/4}}\sin\left(\frac{2}{3}|x|^{3/2} + \frac{\pi}{4}\right). \tag{19.140}$$

The fruit of our labours is therefore

$$\begin{aligned}\mathrm{Ai}\,(x) &\sim \frac{1}{2\sqrt{\pi}}x^{-1/4}e^{-\frac{2}{3}x^{3/2}}\left[1+O\left(\frac{1}{x}\right)\right], \quad x>0,\\ &\sim \frac{1}{\sqrt{\pi}|x|^{1/4}}\sin\left(\frac{2}{3}|x|^{3/2}+\frac{\pi}{4}\right)\left[1+O\left(\frac{1}{x}\right)\right], \quad x<0. \qquad (19.141)\end{aligned}$$

Suppose that we allow x to become complex $x \to z = |z|e^{i\theta}$, with $-\pi < \theta < \pi$. Then Figure 19.9 shows how the steepest contour evolves and leads to two quite different

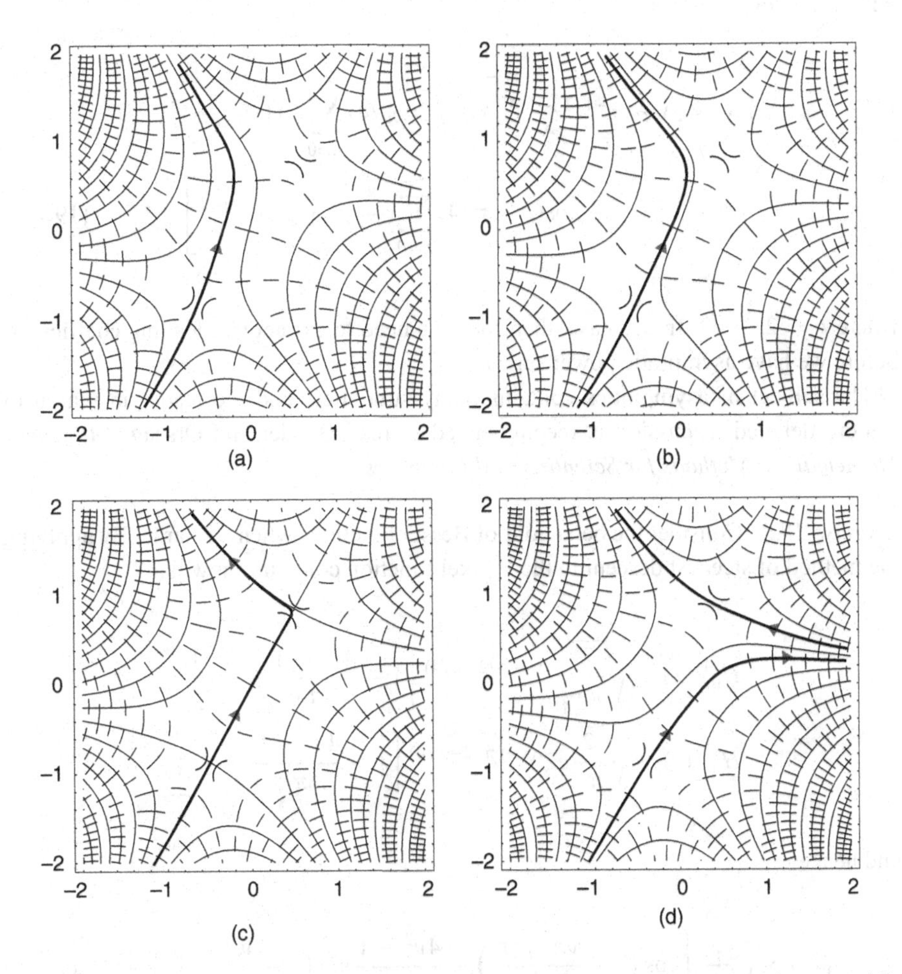

Figure 19.9 Evolution of the steepest-descent contour from passing through only one saddle point to passing through both. The dashed and solid lines are contours of the real and imaginary parts, respectively, of $(zt - t^3/3)$. $\theta = \operatorname{Arg} z$ takes the values (a) $7\pi/12$, (b) $5\pi/8$, (c) $2\pi/3$, (d) $3\pi/4$.

expansions for positive and negative x. We see that for $0 < \theta < 2\pi/3$ the steepest descent path continues to be routed through the single stationary point at $-\sqrt{|z|}e^{i\theta/2}$. Once θ reaches $2\pi/3$, though, it passes through both stationary points. The contribution to the integral from the newly acquired stationary point is, however, exponentially smaller as $|z| \to \infty$ than that of $t = -\sqrt{|z|}e^{i\theta/2}$. The new term is therefore said to be *subdominant*, and makes an insignificant contribution to the asymptotic behaviour of Ai (z). The two saddle points only make contributions of the same magnitude when θ reaches π. If we analytically continue beyond $\theta = \pi$, the new saddle point will now dominate over the old, and only its contribution is significant at large $|z|$. The *Stokes line*, at which we must change the form of the asymptotic expansion, is therefore at $\theta = \pi$.

If we try to systematically keep higher order terms we will find, for the oscillating Ai $(-z)$, a double series

$$\begin{aligned} \mathrm{Ai}\,(-z) \sim \pi^{-1/2} z^{-1/4} \Bigg[& \sin(\rho + \pi/4) \sum_{n=0}^{\infty} (-1)^n c_{2n} \rho^{-2n} \\ & - \cos(\rho + \pi/4) \sum_{n=0}^{\infty} (-1)^n c_{2n+1} \rho^{-2n-1} \Bigg] \end{aligned} \tag{19.142}$$

where $\rho = 2z^{3/2}/3$. In this case, therefore, we need to extract two leading coefficients before we have asymptotic power series.

The subject of asymptotics contains many subtleties, and the reader in search of a more detailed discussion is recommended to read Bender and Orszag's *Advanced Mathematical Methods for Scientists and Engineers*.

Exercise 19.2: Consider the behaviour of Bessel functions when x is large. By applying the method of steepest descent to the Hankel function contours show that

$$\begin{aligned} H_\nu^{(1)}(x) &\sim \sqrt{\frac{2}{\pi x}} e^{i(x-\nu\pi/2-\pi/4)} \left[1 - \frac{4\nu^2 - 1}{8\pi x} + \cdots \right] \\ H_\nu^{(2)}(x) &\sim \sqrt{\frac{2}{\pi x}} e^{-i(x-\nu\pi/2-\pi/4)} \left[1 + \frac{4\nu^2 - 1}{8\pi x} + \cdots \right], \end{aligned}$$

and hence

$$\begin{aligned} J_\nu(x) &\sim \sqrt{\frac{2}{\pi x}} \left[\cos\left(x - \frac{\nu\pi}{2} - \frac{\pi}{4} \right) - \frac{4\nu^2 - 1}{8x} \sin\left(x - \frac{\nu\pi}{2} - \frac{\pi}{4} \right) + \cdots \right], \\ N_\nu(x) &\sim \sqrt{\frac{2}{\pi x}} \left[\sin\left(x - \frac{\nu\pi}{2} - \frac{\pi}{4} \right) + \frac{4\nu^2 - 1}{8x} \cos\left(x - \frac{\nu\pi}{2} - \frac{\pi}{4} \right) + \cdots \right]. \end{aligned}$$

19.5 Elliptic functions

The subject of elliptic functions goes back to the remarkable identities of Guilio Fagnano (1750) and Leonhard Euler (1761). Euler's formula is

$$\int_0^u \frac{dx}{\sqrt{1-x^4}} + \int_0^v \frac{dy}{\sqrt{1-y^4}} = \int_0^r \frac{dz}{\sqrt{1-z^4}}, \tag{19.143}$$

where $0 \le u, v \le 1$, and

$$r = \frac{u\sqrt{1-v^4} + v\sqrt{1-u^4}}{1+u^2v^2}. \tag{19.144}$$

This looks mysterious, but perhaps so does

$$\int_0^u \frac{dx}{\sqrt{1-x^2}} + \int_0^v \frac{dy}{\sqrt{1-y^2}} = \int_0^r \frac{dz}{\sqrt{1-z^2}}, \tag{19.145}$$

where

$$r = u\sqrt{1-v^2} + v\sqrt{1-u^2}, \tag{19.146}$$

until you realize that the latter formula (19.146) is merely

$$\sin(a+b) = \sin a \cos b + \cos a \sin b \tag{19.147}$$

in disguise. To see this set

$$u = \sin a, \quad v = \sin b \tag{19.148}$$

and remember the integral formula for the inverse trigonometric sine function

$$a = \sin^{-1} u = \int_0^u \frac{dx}{\sqrt{1-x^2}}. \tag{19.149}$$

The Fagnano–Euler formula is a similarly disguised addition formula for an *elliptic function*. Just as we use the substitution $x = \sin y$ in the $1/\sqrt{1-x^2}$ integral, we can use an elliptic-function substitution to evaluate *elliptic integrals* which involve square-roots of quartic or cubic polynomials. Examples are

$$I_4 = \int_0^x \frac{dt}{\sqrt{(t-a_1)(t-a_2)(t-a_3)(t-a_4)}}, \tag{19.150}$$

$$I_3 = \int_0^x \frac{dt}{\sqrt{(t-a_1)(t-a_2)(t-a_3)}}. \tag{19.151}$$

Note that I_4 can be reduced to an integral of the form I_3 by using a Möbius-map substitution

$$t = \frac{at' + b}{ct' + d}, \quad dt = (ad - bc)\frac{dt'}{(ct' + d)^2} \tag{19.152}$$

to send a_4 to infinity. Indeed, we can use a suitable Möbius map to send any three of the four points a_n to $0, 1, \infty$.

The idea of elliptic functions (as opposed to elliptic *integrals*, which are their functional inverses) was known to Gauss, but Abel and Jacobi were the first to publish (1827). For developing the general theory, the simplest elliptic function is the Weierstrass $\wp$. This is really a family of functions that is parametrized by a pair of linearly independent complex numbers or *periods* ω_1, ω_2. For a given pair of periods, the $\wp$ function is defined by the double sum

$$\wp(z) = \frac{1}{z^2} + \sum_{(m,n)\neq(0,0)} \left\{ \frac{1}{(z - m\omega_1 - n\omega_2)^2} - \frac{1}{(m\omega_1 + n\omega_2)^2} \right\}. \tag{19.153}$$

Helped by the counter-term, the sum is absolutely convergent, so we can rearrange the terms to prove double periodicity

$$\wp(z + m\omega_1 + n\omega_2) = \wp(z), \quad m, n \in \mathbb{Z}. \tag{19.154}$$

The function is thus determined everywhere by its values in the period parallelogram $P = \{\lambda\omega_1 + \mu\omega_2 : 0 \leq \lambda, \mu < 1\}$. Double periodicity is the defining characteristic of elliptic functions (Figure 19.10).

Any non-constant meromorphic function $f(z)$ that is doubly periodic has four basic properties:

(a) The function must have at least one pole in its unit cell, otherwise it would be holomorphic and bounded, and therefore a constant by Liouville.
(b) The sum of the residues at the poles must add to zero. This follows from integrating $f(z)$ around the boundary of the period parallelogram and observing that the contributions from opposite edges cancel.
(c) The number of poles in each unit cell must equal the number of zeros. This follows from integrating f'/f around the boundary of the period parallelogram.
(d) If f has zeros at the N points z_i and poles at the N points p_i then

$$\sum_{i=1}^{N} z_i - \sum_{i=1}^{N} p_i = n\omega_1 + m\omega_2$$

where m, n are integers. This follows from integrating zf'/f around the boundary of the period parallelogram.

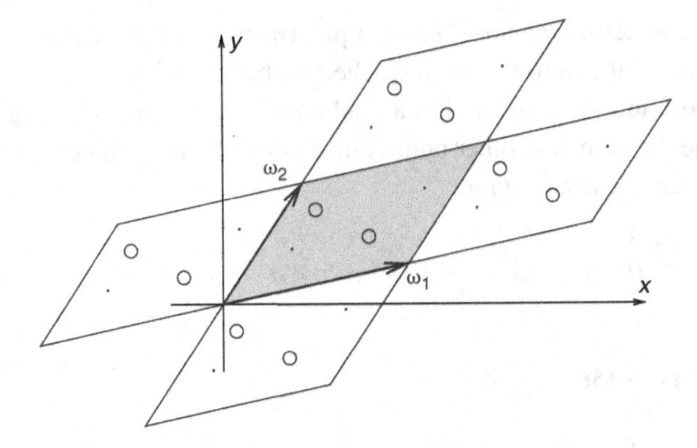

Figure 19.10 Unit cell and double periodicity.

The Weierstrass $\wp$ has a second-order pole at the origin. It also obeys

$$\lim_{|z|\to 0}\left(\wp(z)-\frac{1}{z^2}\right)=0,$$

$$\wp(z)=\wp(-z),$$

$$\wp'(z)=-\wp'(-z). \tag{19.155}$$

The property that makes $\wp(z)$ useful for evaluating integrals is

$$\left(\wp'(z)\right)^2=4\wp^3(z)-g_2\wp(z)-g_3, \tag{19.156}$$

where

$$g_2=60\sum_{(m,n)\neq(0,0)}\frac{1}{(m\omega_1+n\omega_2)^4},\quad g_3=140\sum_{(m,n)\neq(0,0)}\frac{1}{(m\omega_1+n\omega_2)^6}. \tag{19.157}$$

Equation (19.156) is proved by examining the first few terms in the Laurent expansion in z of the difference of the left-hand and right-hand sides. All negative powers cancel, as does the constant term. The difference is zero at $z=0$, has no poles or other singularities, and, being continuous and periodic, is automatically bounded. It is therefore identically zero by Liouville's theorem.

From the symmetry and periodicity of $\wp$ we see that $\wp'(z)=0$ at $\omega_1/2$, $\omega_2/2$ and $(\omega_1+\omega_2)/2$ where $\wp(z)$ takes the values $e_1=\wp(\omega_1/2)$, $e_2=\wp(\omega_2/2)$ and $e_3=\wp((\omega_1+\omega_2)/2)$. Now $\wp'$ must have exactly three zeros since it has a pole of order three at the origin and, by property (c), the number of zeros in the unit cell is equal to the number of poles. We therefore know the location of all three zeros, and can factorize:

$$4\wp^3(z)-g_2\wp(z)-g_3=4(\wp-e_1)(\wp-e_2)(\wp-e_3). \tag{19.158}$$

We note that the coefficient of $\wp^2$ in the polynomial on the left side is zero, implying that $e_1+e_2+e_3=0$.

The roots e_i can never coincide. For example, $(\wp(z) - e_1)$ has a double zero at $\omega_1/2$, but two zeros is all it is allowed because the number of poles per unit cell equals the number of zeros, and $(\wp(z) - e_1)$ has a double pole at 0 as its only singularity. Thus, $(\wp - e_1)$ cannot be zero at another point, but it would be if e_1 coincided with e_2 or e_3. As a consequence, the *discriminant*

$$\Delta \stackrel{\text{def}}{=} 16(e_1 - e_2)^2(e_2 - e_3)^2(e_1 - e_3)^2 = g_2^3 - 27g_3^2 \tag{19.159}$$

is never zero.

We use $\wp$ and (19.156) to write

$$z = \wp^{-1}(u) = \int_\infty^u \frac{dt}{2\sqrt{(t-e_1)(t-e_2)(t-e_3)}} = \int_\infty^u \frac{dt}{\sqrt{4t^3 - g_2 t - g_3}}. \tag{19.160}$$

This maps the u plane, with cuts that we can take from e_1 to e_2 and e_3 to ∞, one-to-one onto the 2-torus, regarded as the unit cell of the $\omega_{n,m} = n\omega_1 + m\omega_2$ lattice.

As z sweeps over the torus, the points $x = \wp(z), y = \wp'(z)$ move on the *elliptic curve*

$$y^2 = 4x^3 - g_2 x - g_3 \tag{19.161}$$

which should be thought of as a set in $\mathbb{C}P^2$. These curves, and the finite fields of rational points that lie on them, are exploited in modern cryptography.

The magic that leads to addition formulæ, such as the Euler–Fagnano relation (19.144) with which we began this section, lies in the (not immediately obvious) fact that any elliptic function having the same periods as $\wp(z)$ can be expressed as a rational function of $\wp(z)$ and $\wp'(z)$. From this it follows (after some thought) that any two such elliptic functions, $f_1(z)$ and $f_2(z)$, obey a relation $F(f_1, f_2) = 0$, where

$$F(x,y) = \sum_{m,n} a_{n,m} x^n y^m \tag{19.162}$$

is a polynomial in x and y. We can eliminate $\wp'(z)$ in these relations by writing $\wp'(z) = \sqrt{4\wp^3(z) - g_2\wp(z) - g_3}$.

Modular invariance

If ω_1 and ω_2 are periods and define a unit cell, so are

$$\omega_1' = a\omega_1 + b\omega_2$$
$$\omega_2' = c\omega_1 + d\omega_2$$

where a, b, c, d are integers with $ad - bc = \pm 1$. This condition on the determinant ensures that the matrix inverse also has integer entries, and so the ω_i can be expressed in terms of the ω_i' with integer coefficients. Consequently the set of integer linear combinations of the ω'_i generate the same lattice as the integer linear combinations of the original ω_i. This notion of redefining the unit cell should be familiar to you from solid state physics. If we wish to preserve the orientation of the basis vectors, we must restrict ourselves to maps whose determinant $ad - bc$ is unity. The set of such transforms constitutes the *modular* group $\mathrm{SL}(2,\mathbb{Z})$. Clearly $\wp$ is invariant under this group, as are g_2 and g_3 and Δ. Now define $\omega_2/\omega_1 = \tau$, and write

$$g_2(\omega_1,\omega_2) = \frac{1}{\omega_1^4}\tilde{g}_2(\tau), \quad g_3(\omega_1,\omega_2) = \frac{1}{\omega_1^6}\tilde{g}_3(\tau). \quad \Delta(\omega_1,\omega_2) = \frac{1}{\omega_1^{12}}\tilde{\Delta}(\tau), \tag{19.163}$$

and also

$$J(\tau) = \frac{\tilde{g}_2^3}{\tilde{g}_2^3 - 27\tilde{g}_3^2} = \frac{\tilde{g}_2^3}{\tilde{\Delta}}. \tag{19.164}$$

Because the denominator is never zero when $\mathrm{Im}\,\tau > 0$, the function $J(\tau)$ is holomorphic in the upper half-plane – but not on the real axis. The function $J(\tau)$ is called the *elliptic modular function*.

Except for the prefactors ω_1^n, the functions $\tilde{g}_i(\tau)$, $\tilde{\Delta}(\tau)$ and $J(\tau)$ are invariant under the Möbius transformation

$$\tau \to \frac{a\tau + b}{c\tau + d}, \tag{19.165}$$

with

$$\begin{pmatrix} a & b \\ c & d \end{pmatrix} \in \mathrm{SL}(2,\mathbb{Z}). \tag{19.166}$$

This Möbius transformation does not change if the entries in the matrix are multiplied by a common factor of ± 1, and so the transformation is an element of the modular group $\mathrm{PSL}(2,\mathbb{Z}) \equiv \mathrm{SL}(2,\mathbb{Z})/\{I,-I\}$.

Taking into account the change in the ω_1^α prefactors we have

$$\begin{aligned}
\tilde{g}_2\left(\frac{a\tau+b}{c\tau+d}\right) &= (c\tau+d)^4\tilde{g}_3(\tau), \\
\tilde{g}_3\left(\frac{a\tau+b}{c\tau+d}\right) &= (c\tau+d)^6\tilde{g}_3(\tau), \\
\tilde{\Delta}\left(\frac{a\tau+b}{c\tau+d}\right) &= (c\tau+d)^{12}\tilde{\Delta}(\tau).
\end{aligned} \tag{19.167}$$

Because $c = 0$ and $d = 1$ for the special case $\tau \to \tau + 1$, these three functions obey $f(\tau + 1) = f(\tau)$ and so depend on τ only via the combination $q^2 = e^{2\pi i\tau}$. For example, it is not hard to prove that

$$\tilde{\Delta}(\tau) = (2\pi)^{12} q^2 \prod_{n=1}^{\infty} \left(1 - q^{2n}\right)^{24}. \tag{19.168}$$

We can also expand these functions as power series in q^2 – and here things get interesting because the coefficients have number-theoretic properties. For example

$$\tilde{g}_2(\tau) = (2\pi)^4 \left[\frac{1}{12} + 20 \sum_{n=1}^{\infty} \sigma_3(n) q^{2n}\right],$$

$$\tilde{g}_3(\tau) = (2\pi)^6 \left[\frac{1}{216} - \frac{7}{3} \sum_{n=1}^{\infty} \sigma_5(n) q^{2n}\right]. \tag{19.169}$$

The symbol $\sigma_k(n)$ is defined by $\sigma_k(n) = \sum d^k$, where d runs over all positive divisors of the number n.

In the case of the function $J(\tau)$, the prefactors cancel and

$$J\left(\frac{a\tau + b}{c\tau + d}\right) = J(\tau), \tag{19.170}$$

so $J(\tau)$ is a *modular invariant*. One can show that if $J(\tau_1) = J(\tau_2)$, then

$$\tau_2 = \frac{a\tau_1 + b}{c\tau_1 + d} \tag{19.171}$$

for some modular transformation with integer a, b, c, d, where $ad - bc = 1$, and, further, that any modular-invariant function is a rational function of $J(\tau)$. It seems clear that $J(\tau)$ is rather a special object.

This $J(\tau)$ is the function referred to on page 500 in connection with the Monster group. As with the $\tilde{g}_i$, $J(\tau)$ depends on τ only through q^2. The first few terms in the power series expansion of $J(\tau)$ in terms of q^2 turn out to be

$$1728 J(\tau) = q^{-2} + 744 + 196884 q^2 + 21493760 q^4 + 864299970 q^6 + \cdots \tag{19.172}$$

Since $AJ(\tau) + B$ has all the same modular invariance properties as $J(\tau)$, the numbers 1728 $(= 12^3)$ and 744 are just conventional normalizations. Once we set the coefficient of q^{-2} to unity, however, the remaining integer coefficients are completely determined by the modular properties. A number-theory interpretation of these integers seemed lacking until John McKay and others observed that

$$\begin{aligned} 1 &= 1 \\ 196884 &= 1 + 196883 \\ 21493760 &= 1 + 196883 + 21296786 \\ 864299970 &= 2 \times 1 + 2 \times 196883 + 21296786 + 842609326, \end{aligned} \tag{19.173}$$

where "1" and the large integers on the right-hand side are the dimensions of the smallest irreducible representations of the Monster. This "Monstrous Moonshine" was originally mysterious and almost unbelievable ("moonshine" = "fantastic nonsense") but it was explained by Richard Borcherds by the use of techniques borrowed from string theory.[3] Borcherds received the 1998 Fields Medal for this work.

19.6 Further exercises and problems

Exercise 19.3: Show that the binomial series expansion of $(1+x)^{-\nu}$ can be written as

$$(1+x)^{-\nu} = \sum_{m=0}^{\infty}(-x)^m \frac{\Gamma(m+\nu)}{\Gamma(\nu)\, m!}, \qquad |x| < 1.$$

Exercise 19.4: *A Mellin transform and its inverse*. Combine the Beta-function identity (19.15) with a suitable change of variables to evaluate the Mellin transform

$$\int_0^{\infty} x^{s-1}(1+x)^{-\nu}\, dx, \quad \nu > 0,$$

of $(1+x)^{-\nu}$ as a product of Gamma functions. Now consider the *Bromwich contour* integral

$$\frac{1}{2\pi i\, \Gamma(\nu)} \int_{c-i\infty}^{c+i\infty} x^{-s}\Gamma(\nu - s)\Gamma(s)\, ds.$$

Here $\mathrm{Re}\, c \in (0, \nu)$. The Bromwich contour therefore runs parallel to the imaginary axis with the poles of $\Gamma(s)$ to its left and the poles of $\Gamma(\nu - s)$ to its right. Use the identity

$$\Gamma(s)\Gamma(1-s) = \pi \operatorname{cosec} \pi s$$

to show that when $|x| < 1$ the contour can be closed by a large semicircle lying to the left of the imaginary axis. By using the preceding exercise to sum the contributions from the enclosed poles at $s = -n$, evaluate the integral.

Exercise 19.5: *Mellin–Barnes integral*. Use the technique developed in the preceding exercise to show that

$$F(a,b,c;-x) = \frac{\Gamma(c)}{2\pi i\, \Gamma(a)\Gamma(b)} \int_{c-i\infty}^{c+i\infty} x^{-s} \frac{\Gamma(a-s)\Gamma(b-s)\Gamma(s)}{\Gamma(c-s)}\, ds,$$

[3] "I was in Kashmir. I had been traveling around northern India, and there was one really long tiresome bus journey, which lasted about 24 hours. Then the bus had to stop because there was a landslide and we couldn't go any further. It was all pretty darn unpleasant. Anyway, I was just toying with some calculations on this bus journey and finally I found an idea which made everything work" – Richard Borcherds (Interview in *The Guardian*, August 1998).

for a suitable range of x. This integral representation of the hypergeometric function is due to the English mathematician Ernest Barnes (1908), later a controversial Bishop of Birmingham.

Exercise 19.6: Let

$$Y = \begin{pmatrix} y_1 \\ y_2 \end{pmatrix}.$$

Show that the matrix differential equation

$$\frac{d}{dx}Y = \frac{A}{z}Y + \frac{B}{1-z}Y,$$

where

$$A = \begin{pmatrix} 0 & a \\ 0 & 1-c \end{pmatrix}, \quad B = \begin{pmatrix} 0 & 0 \\ b & a+b-c+1 \end{pmatrix},$$

has a solution

$$Y(z) = F(a,b;c;z)\begin{pmatrix} 1 \\ 0 \end{pmatrix} + \frac{z}{a}F'(a,b;c;z)\begin{pmatrix} 0 \\ 1 \end{pmatrix}.$$

Exercise 19.7: *Kniznik–Zamolodchikov equation.* The monodromy properties of solutions of differential equations play an important role in conformal field theory. The Fuchsian equations studied in this exercise are obeyed by the correlation functions in the level-k Wess–Zumino–Witten model.

Let $V^{(a)}$, $a = 1, \dots n$, be spin-j_a representation spaces for the group SU(2). Let $W(z_1, \dots, z_n)$ be a function taking values in $V^{(1)} \otimes V^{(2)} \otimes \cdots \otimes V^{(n)}$. (In other words W is a function $W_{i_1,\dots,i_n}(z_1, \dots, z_n)$ where the index i_a labels states in the spin-j_a factor.) Suppose that W obeys the *Kniznik–Zamolodchikov (K–Z) equations*

$$(k+2)\frac{\partial}{\partial z_a}W = \sum_{b, b\neq a} \frac{\mathbf{J}^{(a)} \cdot \mathbf{J}^{(b)}}{z_a - z_b}W, \quad a = 1, \dots, n,$$

where

$$\mathbf{J}^{(a)} \cdot \mathbf{J}^{(b)} \equiv J_1^{(a)}J_1^{(b)} + J_2^{(a)}J_2^{(b)} + J_3^{(a)}J_3^{(b)},$$

and $J_i^{(a)}$ indicates the $\mathfrak{su}(2)$ generator J_i acting on the $V^{(a)}$ factor in the tensor product. If we set $z_1 = z$, for example, and fix the position of $z_2, \dots z_n$, then the differential equation in z has regular singular points at the $n-1$ remaining z_b.

(a) By diagonalizing the operator $\mathbf{J}^{(a)} \cdot \mathbf{J}^{(b)}$ show that there are solutions $W(z)$ that behave for z_a close to z_b as

$$W(z) \sim (z_a - z_b)^{\Delta_j - \Delta_{j_a} - \Delta_{j_b}},$$

where

$$\Delta_j = \frac{j(j+1)}{k+2}, \quad \Delta_{j_a} = \frac{j_a(j_a+1)}{k+2},$$

and j is one of the spins $|j_a - j_b| \le j \le j_1 + j_a$ occuring in the decomposition of $j_a \otimes j_b$.

(b) Define covariant derivatives

$$\nabla_a = \frac{\partial}{\partial z_a} - \sum_{b, b \neq a} \frac{\mathbf{J}^{(a)} \cdot \mathbf{J}^{(b)}}{z_a - z_b}$$

and show that $[\nabla_a, \nabla_b] = 0$. Conclude that the effect of parallel transport of the solutions of the K–Z equations provides a representation of the braid group of the world-lines of the z_a.

Appendix A
Linear algebra review

In physics we often have to work with infinite-dimensional vector spaces. Navigating these vasty deeps is much easier if you have a sound grasp of the theory of finite-dimensional spaces. Most physics students have studied this as undergraduates, but not always in a systematic way. In this appendix we gather together and review those parts of linear algebra that we will find useful in the main text.

A.1 Vector space

A.1.1 Axioms

A *vector space* V over a field $\mathbb{F}$ is a set equipped with two operations: a binary operation called *vector addition* which assigns to each pair of elements $\mathbf{x}, \mathbf{y} \in V$ a third element denoted by $\mathbf{x}+\mathbf{y}$, and *scalar multiplication* which assigns to an element $\mathbf{x} \in V$ and $\lambda \in \mathbb{F}$ a new element $\lambda\mathbf{x} \in V$. There is also a distinguished element $\mathbf{0} \in V$ such that the following axioms are obeyed:[1]

(1) Vector addition is commutative: $\mathbf{x}+\mathbf{y}=\mathbf{y}+\mathbf{x}$.
(2) Vector addition is associative: $(\mathbf{x}+\mathbf{y})+\mathbf{z}=\mathbf{x}+(\mathbf{y}+\mathbf{z})$.
(3) Additive identity: $\mathbf{0}+\mathbf{x}=\mathbf{x}$.
(4) Existence of an additive inverse: for any $\mathbf{x} \in V$, there is an element $(-\mathbf{x}) \in V$, such that $\mathbf{x}+(-\mathbf{x})=\mathbf{0}$.
(5) Scalar distributive law (i) $\lambda(\mathbf{x}+\mathbf{y})=\lambda\mathbf{x}+\lambda\mathbf{y}$.
(6) Scalar distributive law (ii) $(\lambda+\mu)\mathbf{x}=\lambda\mathbf{x}+\mu\mathbf{x}$.
(7) Scalar multiplication is associative: $(\lambda\mu)\mathbf{x}=\lambda(\mu\mathbf{x})$.
(8) Multiplicative identity: $1\mathbf{x}=\mathbf{x}$.

The elements of V are called *vectors*. We will only consider vector spaces over the field of the real numbers, $\mathbb{F}=\mathbb{R}$, or the complex numbers, $\mathbb{F}=\mathbb{C}$.

You have no doubt been working with vectors for years, and are saying to yourself "I know this stuff". Perhaps so, but to see if you really understand these axioms try the following exercise. Its value lies not so much in the solution of its parts, which are easy, as in appreciating that these commonly used properties both can and need to be proved from the axioms. (Hint: work the problems in the order given; the later parts depend on the earlier.)

[1] In this list $1, \lambda, \mu, \in \mathbb{F}$ and $\mathbf{x}, \mathbf{y}, \mathbf{0} \in V$.

Exercise A.1: Use the axioms to show that:

(i) If $\mathbf{x} + \tilde{\mathbf{0}} = \mathbf{x}$, then $\tilde{\mathbf{0}} = \mathbf{0}$.
(ii) We have $0\mathbf{x} = \mathbf{0}$ for any $\mathbf{x} \in V$. Here 0 is the additive identity in $\mathbb{F}$.
(iii) If $\mathbf{x} + \mathbf{y} = \mathbf{0}$, then $\mathbf{y} = -\mathbf{x}$. Thus the additive inverse is unique.
(iv) Given $\mathbf{x}, \mathbf{y}$ in V, there is a unique $\mathbf{z}$ such that $\mathbf{x} + \mathbf{z} = \mathbf{y}$, to whit $\mathbf{z} = \mathbf{x} - \mathbf{y}$.
(v) $\lambda\mathbf{0} = \mathbf{0}$ for any $\lambda \in \mathbb{F}$.
(vi) If $\lambda\mathbf{x} = \mathbf{0}$, then either $\mathbf{x} = \mathbf{0}$ or $\lambda = 0$.
(vii) $(-1)\mathbf{x} = -\mathbf{x}$.

A.1.2 Bases and components

Let V be a vector space over $\mathbb{F}$. For the moment, this space has no additional structure beyond that of the previous section – no inner product and so no notion of what it means for two vectors to be orthogonal. There is still much that can be done, though. Here are the most basic concepts and properties that need to be understood:

(i) A set of vectors $\{\mathbf{e}_1, \mathbf{e}_2, \dots, \mathbf{e}_n\}$ is *linearly dependent* if there exist $\lambda^\mu \in \mathbb{F}$, not all zero, such that

$$\lambda^1\mathbf{e}_1 + \lambda^2\mathbf{e}_2 + \cdots + \lambda^n\mathbf{e}_n = \mathbf{0}. \tag{A.1}$$

(ii) If it is not linearly dependent, a set of vectors $\{\mathbf{e}_1, \mathbf{e}_2, \dots, \mathbf{e}_n\}$ is *linearly independent*. For a linearly independent set, a relation

$$\lambda^1\mathbf{e}_1 + \lambda^2\mathbf{e}_2 + \cdots + \lambda^n\mathbf{e}_n = \mathbf{0} \tag{A.2}$$

can hold only if all the λ^μ are zero.

(iii) A set of vectors $\{\mathbf{e}_1, \mathbf{e}_2, \dots, \mathbf{e}_n\}$ is said to *span* V if for any $\mathbf{x} \in V$ there are numbers x^μ such that $\mathbf{x}$ can be written (not necessarily uniquely) as

$$\mathbf{x} = x^1\mathbf{e}_1 + x^2\mathbf{e}_2 + \cdots + x^n\mathbf{e}_n. \tag{A.3}$$

A vector space is *finite dimensional* if a finite spanning set exists.

(iv) A set of vectors $\{\mathbf{e}_1, \mathbf{e}_2, \dots, \mathbf{e}_n\}$ is a *basis* if it is a *maximal linearly independent set* (i.e. introducing any additional vector makes the set linearly dependent). An alternative definition declares a basis to be a *minimal spanning set* (i.e. deleting any of the $\mathbf{e}_i$ destroys the spanning property). *Exercise*: Show that these two definitions are equivalent.

(v) If $\{\mathbf{e}_1, \mathbf{e}_2, \dots, \mathbf{e}_n\}$ is a basis then any $\mathbf{x} \in V$ can be written

$$\mathbf{x} = x^1\mathbf{e}_1 + x^2\mathbf{e}_2 + \dots x^n\mathbf{e}_n, \tag{A.4}$$

where the x^μ, the *components* of the vector with respect to this basis, are unique in that two vectors coincide if and only if they have the same components.

(vi) **Fundamental theorem**: If the sets $\{\mathbf{e}_1, \mathbf{e}_2, \ldots, \mathbf{e}_n\}$ and $\{\mathbf{f}_1, \mathbf{f}_2, \ldots, \mathbf{f}_m\}$ are both bases for the space V then $m = n$. This invariant integer is the *dimension*, $\dim(V)$, of the space. For a proof (not difficult) see a mathematics text such as Birkhoff and McLane's *Survey of Modern Algebra,* or Halmos' *Finite Dimensional Vector Spaces*.

Suppose that $\{\mathbf{e}_1, \mathbf{e}_2, \ldots, \mathbf{e}_n\}$ and $\{\mathbf{e}'_1, \mathbf{e}'_2, \ldots, \mathbf{e}'_n\}$ are both bases, and that

$$\mathbf{e}_\nu = a^\mu_\nu \mathbf{e}'_\mu. \tag{A.5}$$

Since $\{\mathbf{e}_1, \mathbf{e}_2, \ldots, \mathbf{e}_n\}$ is a basis, the $\mathbf{e}'_\nu$ can also be uniquely expressed in terms of the $\mathbf{e}_\mu$, and so the numbers a^μ_ν constitute an invertible matrix. (Note that we are, as usual, using the Einstein summation convention that repeated indices are to be summed over.) The components x'^μ of $\mathbf{x}$ in the new basis are then found by comparing the coefficients of $\mathbf{e}'_\mu$ in

$$x'^\mu \mathbf{e}'_\mu = \mathbf{x} = x^\nu \mathbf{e}_\nu = x^\nu \left(a^\mu_\nu \mathbf{e}'_\mu\right) = \left(x^\nu a^\mu_\nu\right) \mathbf{e}'_\mu \tag{A.6}$$

to be $x'^\mu = a^\mu_\nu x^\nu$, or equivalently, $x^\nu = (a^{-1})^\nu_{\ \mu} x'^\mu$. Note how the $\mathbf{e}_\mu$ and the x^μ transform in "opposite" directions. The components x^μ are therefore said to transform *contravariantly*.

A.2 Linear maps

Let V and W be vector spaces having dimensions n and m respectively. A *linear map*, or *linear operator*, A is a function $A : V \to W$ with the property that

$$A(\lambda \mathbf{x} + \mu \mathbf{y}) = \lambda A(\mathbf{x}) + \mu A(\mathbf{y}). \tag{A.7}$$

A.2.1 Matrices

The linear map A is an object that exists independently of any basis. Given bases $\{\mathbf{e}_\mu\}$ for V and $\{\mathbf{f}_\nu\}$ for W, however, the map may be represented by an *m-by-n matrix*. We obtain this matrix

$$\mathbf{A} = \begin{pmatrix} a^1{}_1 & a^1{}_2 & \ldots & a^1{}_n \\ a^2{}_1 & a^2{}_2 & \ldots & a^2{}_n \\ \vdots & \vdots & \ddots & \vdots \\ a^m{}_1 & a^m{}_2 & \ldots & a^m{}_n \end{pmatrix}, \tag{A.8}$$

having entries $a^\nu{}_\mu$, by looking at the action of A on the basis elements:

$$A(\mathbf{e}_\mu) = \mathbf{f}_\nu a^\nu{}_\mu. \tag{A.9}$$

To make the right-hand side of (A.9) look like a matrix product, where we sum over adjacent indices, the array $a^\nu{}_\mu$ has been written to the *right* of the basis vector.[2] The map $\mathbf{y} = A(\mathbf{x})$ is therefore

$$\mathbf{y} \equiv y^\nu \mathbf{f}_\nu = A(\mathbf{x}) = A(x^\mu \mathbf{e}_\mu) = x^\mu A(\mathbf{e}_\mu) = x^\mu (\mathbf{f}_\nu a^\nu{}_\mu) = (a^\nu{}_\mu x^\mu)\mathbf{f}_\nu, \tag{A.10}$$

whence, comparing coefficients of $\mathbf{f}_\nu$, we have

$$y^\nu = a^\nu{}_\mu x^\mu. \tag{A.11}$$

The action of the linear map on *components* is therefore given by the usual matrix multiplication from the *left*: $\mathbf{y} = \mathbf{A}\mathbf{x}$, or more explicitly

$$\begin{pmatrix} y^1 \\ y^2 \\ \vdots \\ y^m \end{pmatrix} = \begin{pmatrix} a^1{}_1 & a^1{}_2 & \dots & a^1{}_n \\ a^2{}_1 & a^2{}_2 & \dots & a^2{}_n \\ \vdots & \vdots & \ddots & \vdots \\ a^m{}_1 & a^m{}_2 & \dots & a^m{}_n \end{pmatrix} \begin{pmatrix} x^1 \\ x^2 \\ \vdots \\ x^n \end{pmatrix}. \tag{A.12}$$

The *identity map* $I : V \to V$ is represented by the n-by-n matrix

$$\mathbf{I}_n = \begin{pmatrix} 1 & 0 & 0 & \dots & 0 \\ 0 & 1 & 0 & \dots & 0 \\ 0 & 0 & 1 & \dots & 0 \\ \vdots & \vdots & \vdots & \ddots & \vdots \\ 0 & 0 & 0 & \dots & 1 \end{pmatrix}, \tag{A.13}$$

which has the same entries in any basis.

Exercise A.2: Let U, V, W be vector spaces, and $A : V \to W$, $B : U \to V$ linear maps which are represented by the matrices $\mathbf{A}$ with entries $a^\mu{}_\nu$ and $\mathbf{B}$ with entries $b^\mu{}_\nu$, respectively. Use the action of the maps on basis elements to show that the map $AB : U \to W$ is represented by the matrix product $\mathbf{AB}$ whose entries are $a^\mu{}_\lambda b^\lambda{}_\nu$.

A.2.2 Range–null-space theorem

Given a linear map $A : V \to W$, we can define two important subspaces:

(i) The *kernel* or *null-space* is defined by

$$\operatorname{Ker} A = \{\mathbf{x} \in V : A(\mathbf{x}) = \mathbf{0}\}. \tag{A.14}$$

It is a subspace of V.

[2] You have probably seen this "backward" action before in quantum mechanics. If we use Dirac notation $|n\rangle$ for an orthonormal basis, and insert a complete set of states, $|m\rangle\langle m|$, then $A|n\rangle = |m\rangle\langle m|A|n\rangle$. The matrix $\langle m|A|n\rangle$ representing the operator A operating on a vector from the *left* thus automatically appears to the *right* of the basis vectors used to expand the result.

(ii) The *range* or *image* space is defined by

$$\operatorname{Im} A = \{\mathbf{y} \in W : \mathbf{y} = A(\mathbf{x}), \mathbf{x} \in V\}. \tag{A.15}$$

It is a subspace of the *target space* W.

The key result linking these spaces is the *range–null-space theorem* which states that

$$\boxed{\dim(\operatorname{Ker} A) + \dim(\operatorname{Im} A) = \dim V.}$$

It is proved by taking a basis $\mathbf{n}_\mu$ for $\operatorname{Ker} A$ and extending it to a basis for the whole of V by appending $(\dim V - \dim(\operatorname{Ker} A))$ extra vectors $\mathbf{e}_\nu$. It is easy to see that the vectors $A(\mathbf{e}_\nu)$ are linearly independent and span $\operatorname{Im} A \subseteq W$. Note that this result is meaningless unless V is finite dimensional.

The number $\dim(\operatorname{Im} A)$ is the number of linearly independent columns in the matrix, and is often called the (column) rank of the matrix.

A.2.3 The dual space

Associated with the vector space V is its *dual space*, V^*, which is the set of linear maps $f : V \to \mathbb{F}$, in other words the set of linear functions $f(\ )$ that take in a vector and return a number. These functions are often also called *covectors*. (Mathematicians place the prefix *co-* in front of the name of a mathematical object to indicate a dual class of objects, consisting of the set of structure-preserving maps of the original objects into the field over which they are defined.)

Using linearity we have

$$f(\mathbf{x}) = f(x^\mu \mathbf{e}_\mu) = x^\mu f(\mathbf{e}_\mu) = x^\mu f_\mu. \tag{A.16}$$

The set of numbers $f_\mu = f(\mathbf{e}_\mu)$ are the components of the covector $f \in V^*$. If we change basis $\mathbf{e}_\nu = a_\nu^\mu \mathbf{e}'_\mu$ then

$$f_\nu = f(\mathbf{e}_\nu) = f(a_\nu^\mu \mathbf{e}'_\mu) = a_\nu^\mu f(\mathbf{e}'_\mu) = a_\nu^\mu f'_\mu. \tag{A.17}$$

Thus $f_\nu = a_\nu^\mu f'_\mu$ and the f_μ components transform in the same manner as the basis. They are therefore said to transform *covariantly*.

Given a basis $\mathbf{e}_\mu$ of V, we can define a *dual basis* for V^* as the set of covectors $\mathbf{e}^{*\mu} \in V^*$ such that

$$\mathbf{e}^{*\mu}(\mathbf{e}_\nu) = \delta_\nu^\mu. \tag{A.18}$$

It should be clear that this is a basis for V^*, and that f can be expanded

$$f = f_\mu \mathbf{e}^{*\mu}. \tag{A.19}$$

Although the spaces V and V^* have the same dimension, and are therefore isomorphic, there is no natural map between them. The assignment $\mathbf{e}_\mu \mapsto \mathbf{e}^{*\mu}$ is *unnatural* because it depends on the choice of basis.

One way of driving home the distinction between V and V^* is to consider the space V of fruit orders at a grocers. Assume that the grocer stocks only apples, oranges and pears. The elements of V are then vectors such as

$$\mathbf{x} = 3\text{kg}\,\mathbf{apples} + 4.5\text{kg}\,\mathbf{oranges} + 2\text{kg}\,\mathbf{pears}. \tag{A.20}$$

Take V^* to be the space of possible price lists, an example element being

$$f = (\pounds 3.00/\text{kg})\,\mathbf{apples}^* + (\pounds 2.00/\text{kg})\,\mathbf{oranges}^* + (\pounds 1.50/\text{kg})\,\mathbf{pears}^*. \tag{A.21}$$

The evaluation of f on $\mathbf{x}$,

$$f(\mathbf{x}) = 3 \times \pounds 3.00 + 4.5 \times \pounds 2.00 + 2 \times \pounds 1.50 = \pounds 21.00, \tag{A.22}$$

then returns the total cost of the order. You should have no difficulty in distinguishing between a price list and box of fruit!

We may consider the original vector space V to be the dual space of V^* since, given vectors in $\mathbf{x} \in V$ and $f \in V^*$, we naturally define $\mathbf{x}(f)$ to be $f(\mathbf{x})$. Thus $(V^*)^* = V$. Instead of giving one space priority as being the set of linear functions on the other, we can treat V and V^* on an equal footing. We then speak of the *pairing* of $\mathbf{x} \in V$ with $f \in V^*$ to get a number in the field. It is then common to use the notation $(f, \mathbf{x})$ to mean either of $f(\mathbf{x})$ or $\mathbf{x}(f)$.

Warning: despite the similarity of the notation, do not fall into the trap of thinking of the pairing $(f, \mathbf{x})$ as an inner product (see next section) of f with $\mathbf{x}$. The two objects being paired live in different spaces. In an inner product, the vectors being multiplied live in the same space.

A.3 Inner-product spaces

Some vector spaces V come equipped with an inner (or scalar) product. This additional structure allows us to relate V and V^*.

A.3.1 Inner products

We will use the symbol $\langle \mathbf{x}, \mathbf{y} \rangle$ to denote an *inner product*. An inner (or *scalar*) product is a conjugate-symmetric, sesquilinear, non-degenerate map $V \times V \to \mathbb{F}$. In this string of jargon, the phrase *conjugate symmetric* means that

$$\langle \mathbf{x}, \mathbf{y} \rangle = \langle \mathbf{y}, \mathbf{x} \rangle^*, \tag{A.23}$$

where the "$*$" denotes complex conjugation, and *sesquilinear*[3] means

$$\langle \mathbf{x}, \lambda\mathbf{y} + \mu\mathbf{z}\rangle = \lambda\langle \mathbf{x}, \mathbf{y}\rangle + \mu\langle \mathbf{x}, \mathbf{z}\rangle, \tag{A.24}$$

$$\langle \lambda\mathbf{x} + \mu\mathbf{y}, \mathbf{z}\rangle = \lambda^*\langle \mathbf{x}, \mathbf{z}\rangle + \mu^*\langle \mathbf{y}, \mathbf{z}\rangle. \tag{A.25}$$

The product is therefore linear in the second slot, but *anti-linear* in the first. When our field is the real numbers $\mathbb{R}$ then the complex conjugation is redundant and the product will be symmetric

$$\langle \mathbf{x}, \mathbf{y}\rangle = \langle \mathbf{y}, \mathbf{x}\rangle, \tag{A.26}$$

and bilinear

$$\langle \mathbf{x}, \lambda\mathbf{y} + \mu\mathbf{z}\rangle = \lambda\langle \mathbf{x}, \mathbf{y}\rangle) + \mu\langle \mathbf{x}, \mathbf{z}\rangle, \tag{A.27}$$

$$\langle \lambda\mathbf{x} + \mu\mathbf{y}, \mathbf{z}\rangle = \lambda\langle \mathbf{x}, \mathbf{z}\rangle + \mu\langle \mathbf{y}, \mathbf{z}\rangle. \tag{A.28}$$

The term *non-degenerate* means that if $\langle \mathbf{x}, \mathbf{y}\rangle = 0$ for all $\mathbf{y}$, then $\mathbf{x} = \mathbf{0}$. Many inner products satisfy the stronger condition of being *positive definite*. This means that $\langle \mathbf{x}, \mathbf{x}\rangle > 0$ unless $\mathbf{x} = \mathbf{0}$, in which case $\langle \mathbf{x}, \mathbf{x}\rangle = 0$. Positive definiteness implies non-degeneracy, but not *vice versa*.

Given a basis $\mathbf{e}_\mu$, we can form the pairwise products

$$\langle \mathbf{e}_\mu, \mathbf{e}_\nu\rangle = g_{\mu\nu}. \tag{A.29}$$

If the array of numbers $g_{\mu\nu}$ constituting the components of the *metric tensor* turns out to be $g_{\mu\nu} = \delta_{\mu\nu}$, then we say that the basis is *orthonormal* with respect to the inner product. We will not assume orthonormality without specifically saying so. The non-degeneracy of the inner product guarantees the existence of a matrix $g^{\mu\nu}$ which is the inverse of $g_{\mu\nu}$, i.e. $g_{\mu\nu}g^{\nu\lambda} = \delta^\lambda_\mu$.

If we take our field to be the real numbers $\mathbb{R}$ then the additional structure provided by a non-degenerate inner product allows us to identify V with V^*. For any $f \in V^*$ we can find a vector $\mathbf{f} \in V$ such that

$$f(\mathbf{x}) = \langle \mathbf{f}, \mathbf{x}\rangle. \tag{A.30}$$

In components, we solve the equation

$$f_\mu = g_{\mu\nu}f^\nu \tag{A.31}$$

for f^ν. We find $f^\nu = g^{\nu\mu}f_\mu$. Usually, we simply identify f with $\mathbf{f}$, and hence V with V^*. We say that the *covariant components* f_μ are related to the *contravariant components* f^μ by *raising*,

$$f^\mu = g^{\mu\nu}f_\nu, \tag{A.32}$$

[3] *Sesqui* is a Latin prefix meaning "one-and-a-half".

or *lowering*,

$$f_\mu = g_{\mu\nu} f^\nu, \tag{A.33}$$

the indices using the metric tensor. Obviously, this identification depends crucially on the inner product; a different inner product would, in general, identify an $f \in V^*$ with a completely different $\mathbf{f} \in V$.

A.3.2 Euclidean vectors

Consider $\mathbb{R}^n$ equipped with its Euclidean metric and associated "dot" inner product. Given a vector $\mathbf{x}$ and a basis $\mathbf{e}_\mu$ with $g_{\mu\nu} = \mathbf{e}_\mu \cdot \mathbf{e}_\nu$, we can define two sets of components for the same vector: firstly the coefficients x^μ appearing in the basis expansion

$$\mathbf{x} = x^\mu \mathbf{e}_\mu,$$

and secondly the "components"

$$x_\mu = \mathbf{x} \cdot \mathbf{e}_\mu = g_{\mu\nu} x^\nu$$

of $\mathbf{x}$ along the basis vectors. The x_μ are obtained from the x^μ by the same "lowering" operation as before, and so x^μ and x_μ are naturally referred to as the contravariant and covariant components, respectively, of the vector $\mathbf{x}$. When the $\mathbf{e}_\mu$ constitute an orthonormal basis, then $g_{\mu\nu} = \delta_{\mu\nu}$ and the two sets of components are numerically coincident.

A.3.3 Bra and ket vectors

When our vector space is over the field of complex numbers, the anti-linearity of the first slot of the inner product means we can no longer make a simple identification of V with V^*. Instead there is an *anti-linear* corresponence between the two spaces. The vector $\mathbf{x} \in V$ is mapped to $\langle \mathbf{x}, \ \rangle$ which, since it returns a number when a vector is inserted into its vacant slot, is an element of V^*. This mapping is anti-linear because

$$\lambda \mathbf{x} + \mu \mathbf{y} \mapsto \langle \lambda \mathbf{x} + \mu \mathbf{y}, \ \rangle = \lambda^* \langle \mathbf{x}, \ \rangle + \mu^* \langle \mathbf{y}, \ \rangle. \tag{A.34}$$

This anti-linear map is probably familiar to you from quantum mechanics, where V is the space of Dirac's "ket" vectors $|\psi\rangle$ and V^* the space of "bra" vectors $\langle\psi|$. The symbol, here ψ, in each of these objects is a label distinguishing one state-vector from another. We often use the eigenvalues of some complete set of commuting operators. To each vector $|\psi\rangle$ we use the $(\ldots)^\dagger$ map to assign it a dual vector

$$|\psi\rangle \mapsto |\psi\rangle^\dagger \equiv \langle\psi|$$

having the same labels. The dagger map is defined to be anti-linear

$$(\lambda|\psi\rangle + \mu|\chi\rangle)^\dagger = \lambda^*\langle\psi| + \mu^*\langle\chi|, \tag{A.35}$$

and Dirac denoted the number resulting from the pairing of the covector $\langle\psi|$ with the vector $|\chi\rangle$ by the "bra-c-ket" symbol $\langle\psi|\chi\rangle$:

$$\langle\psi|\chi\rangle \stackrel{\text{def}}{=} (\langle\psi|, |\chi\rangle). \tag{A.36}$$

We can regard the dagger map as either determining the inner-product on V *via*

$$\langle|\psi\rangle, |\chi\rangle\rangle \stackrel{\text{def}}{=} (|\psi\rangle^\dagger, |\chi\rangle) = (\langle\psi|, |\chi\rangle) \equiv \langle\psi|\chi\rangle, \tag{A.37}$$

or being determined by it as

$$|\psi\rangle^\dagger \stackrel{\text{def}}{=} \langle|\psi\rangle, \ \rangle \equiv \langle\psi|. \tag{A.38}$$

When we represent our vectors by their components with respect to an orthonormal basis, the dagger map is the familiar operation of taking the conjugate transpose,

$$\begin{pmatrix} x_1 \\ x_2 \\ \vdots \\ x_n \end{pmatrix} \mapsto \begin{pmatrix} x_1 \\ x_2 \\ \vdots \\ x_n \end{pmatrix}^\dagger = (x_1^*, x_2^*, \dots, x_n^*) \tag{A.39}$$

but this is not true in general. In a non-orthogonal basis the column vector with components x^μ is mapped to the row vector with components $(x^\dagger)_\mu = (x^\nu)^* g_{\nu\mu}$.

Much of Dirac notation tacitly assumes an orthonormal basis. For example, in the expansion

$$|\psi\rangle = \sum_n |n\rangle\langle n|\psi\rangle \tag{A.40}$$

the expansion coefficients $\langle n|\psi\rangle$ should be the *contravariant* components of $|\psi\rangle$, but the $\langle n|\psi\rangle$ have been obtained from the inner product, and so are in fact its *covariant* components. The expansion (A.40) is therefore valid only when the $|n\rangle$ constitute an orthonormal basis. This will always be the case when the labels on the states show them to be the eigenvectors of a complete commuting set of observables, but sometimes, for example, we may use the integer "n" to refer to an orbital centred on a particular atom in a crystal, and then $\langle n|m\rangle \neq \delta_{mn}$. When using such a non-orthonormal basis it is safer not to use Dirac notation.

Conjugate operator

A linear map $A : V \to W$ automatically induces a map $A^* : W^* \to V^*$. Given $f \in W^*$ we can evaluate $f(A(\mathbf{x}))$ for any $\mathbf{x}$ in V, and so $f(A(\ ))$ is an element of V^* that we may denote by $A^*(f)$. Thus,

$$A^*(f)(\mathbf{x}) = f(A(\mathbf{x})). \tag{A.41}$$

Functional analysts (people who spend their working day in Banach space) call A^* the *conjugate* of A. The word "conjugate" and the symbol A^* is rather unfortunate as it has the potential for generating confusion[4] – not least because the $(\ldots)^*$ map is *linear*. No complex conjugation is involved. Thus

$$(\lambda A + \mu B)^* = \lambda A^* + \mu B^*. \tag{A.42}$$

Dirac deftly sidesteps this notational problem by writing $\langle\psi|A$ for the action of the conjugate of the operator $A : V \to V$ on the bra vector $\langle\psi| \in V^*$. After setting $f \to \langle\psi|$ and $\mathbf{x} \to |\chi\rangle$, equation (A.41) therefore reads

$$(\langle\psi|A)\,|\chi\rangle = \langle\psi|\,(A|\chi\rangle)\,. \tag{A.43}$$

This shows that it does not matter where we place the parentheses, so Dirac simply drops them and uses one symbol $\langle\psi|A|\chi\rangle$ to represent both sides of (A.43). Dirac notation thus avoids the non-complex-conjugating "$*$" by suppressing the distinction between an operator and its conjugate. If, therefore, for some reason we need to make the distinction, we cannot use Dirac notation.

Exercise A.3: If $A : V \to V$ and $B : V \to V$ show that $(AB)^* = B^*A^*$.

Exercise A.4: How does the reversal of the operator order in the previous exercise manifest itself in Dirac notation?

Exercise A.5: Show that if the linear operator A is, in a basis $\mathbf{e}_\mu$, represented by the matrix $\mathbf{A}$, then the conjugate operator A^* is represented in the dual basis $\mathbf{e}^{*\mu}$ by the transposed matrix $\mathbf{A}^T$.

A.3.4 Adjoint operator

The "conjugate" operator of the previous section does not require an inner product for its definition, and is a map from V^* to V^*. When we do have an inner product, however, we can use it to define a different operator "conjugate" to A that, like A itself, is a map from V to V. This new conjugate is called the *adjoint* or the *hermitian conjugate* of A. To construct it, we first remind ourselves that for any linear map $f : V \to \mathbb{C}$, there is

[4] The terms *dual*, *transpose* or *adjoint* are sometimes used in place of "conjugate". Each of these words brings its own capacity for confusion.

a vector $\mathbf{f} \in V$ such that $f(\mathbf{x}) = \langle \mathbf{f}, \mathbf{x} \rangle$. (To find it we simply solve $f_\nu = (f^\mu)^* g_{\mu\nu}$ for f^μ.) We next observe that $\mathbf{x} \mapsto \langle \mathbf{y}, A\mathbf{x} \rangle$ is such a linear map, and so there is a $\mathbf{z}$ such that $\langle \mathbf{y}, A\mathbf{x} \rangle = \langle \mathbf{z}, \mathbf{x} \rangle$. It should be clear that $\mathbf{z}$ depends linearly on $\mathbf{y}$, so we may define the adjoint linear map, $A^\dagger$, by setting $A^\dagger \mathbf{y} = \mathbf{z}$. This gives us the identity

$$\boxed{\langle \mathbf{y}, A\mathbf{x} \rangle = \langle A^\dagger \mathbf{y}, \mathbf{x} \rangle.}$$

The correspondence $A \mapsto A^\dagger$ is anti-linear

$$(\lambda A + \mu B)^\dagger = \lambda^* A^\dagger + \mu^* B^\dagger. \tag{A.44}$$

The adjoint of A depends on the inner product being used to define it. Different inner products give different $A^\dagger$'s.

In the particular case that our chosen basis $\mathbf{e}_\mu$ is orthonormal with respect to the inner product, i.e.

$$\langle \mathbf{e}_\mu, \mathbf{e}_\nu \rangle = \delta_{mu\nu}, \tag{A.45}$$

then the hermitian conjugate $A^\dagger$ of the operator A is represented by the hermitian conjugate matrix $\mathbf{A}^\dagger$ which is obtained from the matrix $\mathbf{A}$ by interchanging rows and columns and complex conjugating the entries.

Exercise A.6: Show that $(AB)^\dagger = B^\dagger A^\dagger$.

Exercise A.7: When the basis is not orthonormal, show that

$$(A^\dagger)^\rho{}_\sigma = \left(g_{\sigma\mu} A^\mu{}_\nu g^{\nu\rho}\right)^*. \tag{A.46}$$

A.4 Sums and differences of vector spaces

A.4.1 Direct sums

Suppose that U and V are vector spaces. We define their *direct sum* $U \oplus V$ to be the vector space of ordered pairs $(\mathbf{u}, \mathbf{v})$ with

$$\lambda(\mathbf{u}_1, \mathbf{v}_1) + \mu(\mathbf{u}_2, \mathbf{v}_2) = (\lambda \mathbf{u}_1 + \mu \mathbf{u}_2, \lambda \mathbf{v}_1 + \mu \mathbf{v}_2). \tag{A.47}$$

The set of vectors $\{(\mathbf{u}, 0)\} \subset U \oplus V$ forms a copy of U, and $\{(0, \mathbf{v})\} \subset U \oplus V$ a copy of V. Thus U and V may be regarded as subspaces of $U \oplus V$.

If U and V are any pair of subspaces of W, we can form the space $U+V$ consisting of all elements of W that can be written as $\mathbf{u} + \mathbf{v}$ with $\mathbf{u} \in U$ and $\mathbf{v} \in V$. The decomposition $\mathbf{x} = \mathbf{u} + \mathbf{v}$ of an element $\mathbf{x} \in U + V$ into parts in U and V will be unique (in that $\mathbf{u}_1 + \mathbf{v}_1 = \mathbf{u}_2 + \mathbf{v}_2$ implies that $\mathbf{u}_1 = \mathbf{u}_2$ and $\mathbf{v}_1 = \mathbf{v}_2$) if and only if $U \cap V = \{\mathbf{0}\}$ where

$\{\mathbf{0}\}$ is the subspace containing only the zero vector. In this case $U + V$ can be identified with $U \oplus V$.

If U is a subspace of W then we can seek a *complementary space* V such that $W = U \oplus V$, or, equivalently, $W = U + V$ with $U \cap V = \{\mathbf{0}\}$. Such complementary spaces are *not* unique. Consider $\mathbb{R}^3$, for example, with U being the vectors in the xy-plane. If $\mathbf{e}$ is any vector that does not lie in this plane then the one-dimensional space spanned by $\mathbf{e}$ is a complementary space for U.

A.4.2 Quotient spaces

We have seen that if U is a subspace of W there are many complementary subspaces V such that $W = U \oplus V$. We can, however, define a *unique* space that we might denote by $W - U$ and refer to as the difference of the two spaces. It is more common, however, to see this space written as W/U and referred to as the *quotient* of W *modulo* U. This quotient space is the vector space of *equivalence classes* of vectors, where we do not distinguish between two vectors in W if their difference lies in U. In other words

$$\mathbf{x} = \mathbf{y} \pmod{U} \quad \Leftrightarrow \quad \mathbf{x} - \mathbf{y} \in U. \tag{A.48}$$

The collection of elements in W that are equivalent to $\mathbf{x}\,(\text{mod } U)$ composes a *coset*, written $\mathbf{x} + U$, a set whose elements are $\mathbf{x} + \mathbf{u}$ where $\mathbf{u}$ is any vector in U. These cosets are the elements of W/U.

When we have a linear map $A : U \to V$, the quotient space $V/\text{Im}\,A$ is often called the *co-kernel* of A.

Given a positive-definite inner product, we can define a unique *orthogonal complement* of $U \subset W$. We define $U^\perp$ to be the set

$$U^\perp = \{\mathbf{x} \in W : \langle \mathbf{x}, \mathbf{y} \rangle = 0, \ \forall \mathbf{y} \in U\}. \tag{A.49}$$

It is easy to see that this is a linear subspace and that $U \oplus U^\perp = W$. For finite-dimensional spaces

$$\boxed{\dim W/U = \dim U^\perp = \dim W - \dim U}$$

and $(U^\perp)^\perp = U$. For infinite-dimensional spaces we only have $(U^\perp)^\perp \supseteq U$. (Be careful, however. If the inner product is *not* positive definite, U and $U^\perp$ may have non-zero vectors in common.)

Although they have the same dimensions, do not confuse W/U with $U^\perp$, and in particular do not use the phrase *orthogonal complement* without specifying an inner product.

A practical example of a quotient space occurs in digital imaging. A colour camera reduces the infinite-dimensional space $\mathcal{L}$ of coloured light incident on each pixel to three numbers, R, G and B, these obtained by pairing the spectral intensity with the frequency response (an element of $\mathcal{L}^*$) of the red, green and blue detectors at that point. The space of distinguishable colours is therefore only three dimensional. Many different incident spectra will give the same output *RGB* signal, and are therefore equivalent as far as the camera is concerned. In the colour industry these equivalent colours are called *metamers*. Equivalent colours differ by spectral intensities that lie in the space $\mathcal{B}$ of *metameric black*. There is no inner product here, so it is meaningless to think of the space of distinguishable colours as being $\mathcal{B}^\perp$. It is, however, precisely what we mean by $\mathcal{L}/\mathcal{B}$.

A.4.3 Projection-operator decompositions

An operator $P : V \to V$ that obeys $P^2 = P$ is called a *projection operator*. It projects a vector $\mathbf{x} \in V$ to $P\mathbf{x} \in \operatorname{Im} P$ *along* $\operatorname{Ker} P$ – in the sense of casting a shadow onto $\operatorname{Im} P$ with the light coming from the direction $\operatorname{Ker} P$. In other words all vectors lying in $\operatorname{Ker} P$ are killed, whilst any vector already in $\operatorname{Im} P$ is left alone by P. (If $\mathbf{x} \in \operatorname{Im} P$ then $\mathbf{x} = P\mathbf{y}$ for some $\mathbf{y} \in V$, and $P\mathbf{x} = P^2\mathbf{y} = P\mathbf{y} = \mathbf{x}$.) The only vector common to both $\operatorname{Ker} P$ and $\operatorname{Im} P$ is $\mathbf{0}$, and so

$$V = \operatorname{Ker} P \oplus \operatorname{Im} P. \tag{A.50}$$

A set of projection operators P_i that are "orthogonal"

$$P_i P_j = \delta_{ij} P_i, \tag{A.51}$$

and sum to the identity operator

$$\sum_i P_i = I, \tag{A.52}$$

is called a *resolution of the identity*. The resulting equation

$$\mathbf{x} = \sum_i P_i \mathbf{x} \tag{A.53}$$

decomposes $\mathbf{x}$ uniquely into a sum of terms $P_i\mathbf{x} \in \operatorname{Im} P_i$ and so decomposes V into a direct sum of subspaces $V_i \equiv \operatorname{Im} P_i$:

$$V = \bigoplus_i V_i. \tag{A.54}$$

Exercise A.8: Let P_1 be a projection operator. Show that $P_2 = I - P_1$ is also a projection operator and $P_1P_2 = 0$. Show also that $\operatorname{Im} P_2 = \operatorname{Ker} P_1$ and $\operatorname{Ker}\ P_2 = \operatorname{Im} P_1$.

A.5 Inhomogeneous linear equations

Suppose we wish to solve the system of linear equations

$$\begin{aligned} a_{11}y_1 + a_{12}y_2 + \cdots + a_{1n}y_n &= b_1 \\ a_{21}y_1 + a_{22}y_2 + \cdots + a_{2n}y_n &= b_2 \\ \vdots \quad & \quad \vdots \\ a_{m1}y_1 + a_{m2}y_2 + \cdots + a_{mn}y_n &= b_m \end{aligned}$$

or, in matrix notation,

$$\mathbf{A}\mathbf{y} = \mathbf{b}, \tag{A.55}$$

where $\mathbf{A}$ is the m-by-n matrix with entries a_{ij}. Faced with such a problem, we should start by asking ourselves the questions:

(i) Does a solution exist?
(ii) If a solution does exist, is it unique?

These issues are best addressed by considering the matrix $\mathbf{A}$ as a linear operator $A : V \to W$, where V is n-dimensional and W is m-dimensional. The natural language is then that of the range and null-spaces of A. There is no solution to the equation $\mathbf{A}\mathbf{y} = \mathbf{b}$ when $\operatorname{Im} A$ is not the whole of W and $\mathbf{b}$ does not lie in $\operatorname{Im} A$. Similarly, the solution will not be unique if there are distinct vectors $\mathbf{x}_1$, $\mathbf{x}_2$ such that $A\mathbf{x}_1 = A\mathbf{x}_2$. This means that $A(\mathbf{x}_1 - \mathbf{x}_2) = \mathbf{0}$, or $(\mathbf{x}_1 - \mathbf{x}_2) \in \operatorname{Ker} A$. These situations are linked, as we have seen, by the range–null-space theorem:

$$\dim(\operatorname{Ker} A) + \dim(\operatorname{Im} A) = \dim V. \tag{A.56}$$

Thus, if $m > n$ there are bound to be some vectors $\mathbf{b}$ for which no solution exists. When $m < n$ the solution cannot be unique.

A.5.1 Rank and index

Suppose $V \equiv W$ (so $m = n$ and the matrix is square) and we choose an inner product, $\langle \mathbf{x}, \mathbf{y} \rangle$, on V. Then $\mathbf{x} \in \operatorname{Ker} A$ implies that, for all $\mathbf{y}$

$$0 = \langle \mathbf{y}, A\mathbf{x} \rangle = \langle A^\dagger \mathbf{y}, \mathbf{x} \rangle, \tag{A.57}$$

or that $\mathbf{x}$ is perpendicular to the range of $A^\dagger$. Conversely, let $\mathbf{x}$ be perpendicular to the range of $A^\dagger$. Then

$$\langle \mathbf{x}, A^\dagger \mathbf{y} \rangle = 0, \quad \forall \mathbf{y} \in V, \tag{A.58}$$

which means that

$$\langle A\mathbf{x}, \mathbf{y} \rangle = 0, \quad \forall \mathbf{y} \in V, \tag{A.59}$$

and, by the non-degeneracy of the inner product, this means that $A\mathbf{x} = \mathbf{0}$. The net result is that

$$\operatorname{Ker} A = (\operatorname{Im} A^\dagger)^\perp. \tag{A.60}$$

Similarly

$$\operatorname{Ker} A^\dagger = (\operatorname{Im} A)^\perp. \tag{A.61}$$

Now

$$\begin{aligned} \dim(\operatorname{Ker} A) + \dim(\operatorname{Im} A) &= \dim V, \\ \dim(\operatorname{Ker} A^\dagger) + \dim(\operatorname{Im} A^\dagger) &= \dim V, \end{aligned} \tag{A.62}$$

but

$$\begin{aligned} \dim(\operatorname{Ker} A) &= \dim(\operatorname{Im} A^\dagger)^\perp \\ &= \dim V - \dim(\operatorname{Im} A^\dagger) \\ &= \dim(\operatorname{Ker} A^\dagger). \end{aligned}$$

Thus, for finite-dimensional square matrices, we have

$$\boxed{\dim(\operatorname{Ker} A) = \dim(\operatorname{Ker} A^\dagger).}$$

In particular, the row and column rank of a square matrix coincide.

Example: Consider the matrix

$$\mathbf{A} = \begin{pmatrix} 1 & 2 & 3 \\ 1 & 1 & 1 \\ 2 & 3 & 4 \end{pmatrix}.$$

Clearly, the number of linearly independent rows is two, since the third row is the sum of the other two. The number of linearly independent columns is also two – although less obviously so – because

$$-\begin{pmatrix} 1 \\ 1 \\ 2 \end{pmatrix} + 2\begin{pmatrix} 2 \\ 1 \\ 3 \end{pmatrix} = \begin{pmatrix} 3 \\ 1 \\ 4 \end{pmatrix}.$$

Warning: The equality $\dim(\operatorname{Ker} A) = \dim(\operatorname{Ker} A^\dagger)$ need not hold in infinite-dimensional spaces. Consider the space with basis $\mathbf{e}_1, \mathbf{e}_2, \mathbf{e}_3, \ldots$ indexed by the positive integers. Define $A\mathbf{e}_1 = \mathbf{e}_2, A\mathbf{e}_2 = \mathbf{e}_3$, and so on. This operator has $\dim(\operatorname{Ker} A) = 0$. The adjoint with respect to the natural inner product has $A^\dagger\mathbf{e}_1 = \mathbf{0}$, $A^\dagger\mathbf{e}_2 = \mathbf{e}_1$, $A^\dagger\mathbf{e}_3 = \mathbf{e}_2$. Thus $\operatorname{Ker} A^\dagger = \{\mathbf{e}_1\}$, and $\dim(\operatorname{Ker} A^\dagger) = 1$. The difference $\dim(\operatorname{Ker} A) - \dim(\operatorname{Ker} A^\dagger)$ is called the *index* of the operator. The index of an operator is often related to topological properties of the space on which it acts, and in this way appears in physics as the origin of *anomalies* in quantum field theory.

A.5.2 Fredholm alternative

The results of the previous section can be summarized as saying that the *Fredholm alternative* holds for finite square matrices. The Fredholm Alternative is the set of statements

I. **Either**
   (i) $A\mathbf{x} = \mathbf{b}$ has a *unique* solution,
   or
   (ii) $A\mathbf{x} = \mathbf{0}$ has a solution.

II. If $A\mathbf{x} = \mathbf{0}$ has n linearly independent solutions, then so does $A^\dagger\mathbf{x} = \mathbf{0}$.

III. If alternative (ii) holds, then $A\mathbf{x} = \mathbf{b}$ has *no* solution unless b is orthogonal to all solutions of $A^\dagger\mathbf{x} = \mathbf{0}$.

It should be obvious that this is a recasting of the statements that

$$\dim(\operatorname{Ker} A) = \dim(\operatorname{Ker} A^\dagger),$$

and

$$(\operatorname{Ker} A^\dagger)^\perp = \operatorname{Im} A. \tag{A.63}$$

Notice that finite-dimensionality is essential here. Neither of these statements is guaranteed to be true in infinite-dimensional spaces.

A.6 Determinants

A.6.1 Skew-symmetric *n*-linear forms

You will be familiar with the elementary definition of the determinant of an n-by-n matrix $\mathbf{A}$ having entries a_{ij}:

$$\det \mathbf{A} \equiv \begin{vmatrix} a_{11} & a_{12} & \ldots & a_{1n} \\ a_{21} & a_{22} & \ldots & a_{2n} \\ \vdots & \vdots & \ddots & \vdots \\ a_{n1} & a_{n2} & \ldots & a_{nn} \end{vmatrix} \stackrel{\text{def}}{=} \epsilon_{i_1 i_2 \ldots i_n} a_{1i_1} a_{2i_2} \ldots a_{ni_n}. \tag{A.64}$$

Here, $\epsilon_{i_1 i_2 \ldots i_n}$ is the *Levi-Civita* symbol, which is skew-symmetric in all its indices and $\epsilon_{12\ldots n} = 1$. From this definition we see that the determinant changes sign if any pair of its rows are interchanged, and that it is linear in each row. In other words

$$\begin{vmatrix} \lambda a_{11} + \mu b_{11} & \lambda a_{12} + \mu b_{12} & \ldots & \lambda a_{1n} + \mu b_{1n} \\ c_{21} & c_{22} & \ldots & c_{2n} \\ \vdots & \vdots & \ddots & \vdots \\ c_{n1} & c_{n2} & \ldots & c_{nn} \end{vmatrix}$$

$$= \lambda \begin{vmatrix} a_{11} & a_{12} & \ldots & a_{1n} \\ c_{21} & c_{22} & \ldots & c_{2n} \\ \vdots & \vdots & \ddots & \vdots \\ c_{n1} & c_{n2} & \ldots & c_{nn} \end{vmatrix} + \mu \begin{vmatrix} b_{11} & b_{12} & \ldots & b_{1n} \\ c_{21} & c_{22} & \ldots & c_{2n} \\ \vdots & \vdots & \ddots & \vdots \\ c_{n1} & c_{n2} & \ldots & c_{nn} \end{vmatrix}.$$

If we consider each row as being the components of a vector in an n-dimensional vector space V, we may regard the determinant as being a skew-symmetric n-linear form, i.e. a map

$$\omega : \overbrace{V \times V \times \ldots V}^{n \text{ factors}} \to \mathbb{F} \tag{A.65}$$

which is linear in each slot,

$$\omega(\lambda \mathbf{a} + \mu \mathbf{b}, \mathbf{c}_2, \ldots, \mathbf{c}_n) = \lambda\, \omega(\mathbf{a}, \mathbf{c}_2, \ldots, \mathbf{c}_n) + \mu\, \omega(\mathbf{b}, \mathbf{c}_2, \ldots, \mathbf{c}_n), \tag{A.66}$$

and changes sign when any two arguments are interchanged,

$$\omega(\ldots, \mathbf{a}_i, \ldots, \mathbf{a}_j, \ldots) = -\omega(\ldots, \mathbf{a}_j, \ldots, \mathbf{a}_i, \ldots). \tag{A.67}$$

We will denote the space of skew-symmetric n-linear forms on V by the symbol $\bigwedge^n(V^*)$. Let ω be an arbitrary skew-symmetric n-linear form in $\bigwedge^n(V^*)$, and let $\{\mathbf{e}_1, \mathbf{e}_2, \ldots, \mathbf{e}_n\}$ be a basis for V. If $\mathbf{a}_i = a_{ij}\mathbf{e}_j$ $(i = 1, \ldots, n)$ is a collection of n vectors,[5] we compute

$$\begin{aligned} \omega(\mathbf{a}_1, \mathbf{a}_2, \ldots, \mathbf{a}_n) &= a_{1i_1} a_{2i_2} \ldots a_{ni_n} \omega(\mathbf{e}_{i_1}, \mathbf{e}_{i_2}, \ldots, \mathbf{e}_{i_n}) \\ &= a_{1i_1} a_{2i_2} \ldots a_{ni_n} \epsilon_{i_1 i_2 \ldots, i_n} \omega(\mathbf{e}_1, \mathbf{e}_2, \ldots, \mathbf{e}_n). \end{aligned} \tag{A.68}$$

In the first line we have exploited the linearity of ω in each slot, and in going from the first to the second line we have used skew-symmetry to rearrange the basis vectors in

[5] The index j on a_{ij} should really be a superscript since a_{ij} is the j-th contravariant component of the vector $\mathbf{a}_i$. We are writing it as a subscript only for compatibility with other equations in this section.

their canonical order. We deduce that all skew-symmetric n-forms are proportional to the determinant

$$\omega(\mathbf{a}_1, \mathbf{a}_2, \dots, \mathbf{a}_n) \propto \begin{vmatrix} a_{11} & a_{12} & \dots & a_{1n} \\ a_{21} & a_{22} & \dots & a_{2n} \\ \vdots & \vdots & \ddots & \vdots \\ a_{n1} & a_{n2} & \dots & a_{nn} \end{vmatrix},$$

and that the proportionality factor is the number $\omega(\mathbf{e}_1, \mathbf{e}_2, \dots, \mathbf{e}_n)$. When the number of its slots is equal to the dimension of the vector space, there is therefore essentially only *one* skew-symmetric multilinear form and $\bigwedge^n(V^*)$ is a one-dimensional vector space.

Now we use the notion of skew-symmetric n-linear forms to give a powerful definition of the determinant of an *endomorphism*, i.e. a linear map $A : V \to V$. Let ω be a non-zero skew-symmetric n-linear form. The object

$$\omega_A(\mathbf{x}_1, \mathbf{x}_2, \dots, \mathbf{x}_n) \stackrel{\text{def}}{=} \omega(A\mathbf{x}_1, A\mathbf{x}_2, \dots, A\mathbf{x}_n) \tag{A.69}$$

is also a skew-symmetric n-linear form. Since there is only one such object up to multiplicative constants, we must have

$$\omega(A\mathbf{x}_1, A\mathbf{x}_2, \dots, A\mathbf{x}_n) \propto \omega(\mathbf{x}_1, \mathbf{x}_2, \dots, \mathbf{x}_n). \tag{A.70}$$

We define "$\det A$" to be the constant of proportionality. Thus

$$\omega(A\mathbf{x}_1, A\mathbf{x}_2, \dots, A\mathbf{x}_n) = \det(A)\omega(\mathbf{x}_1, \mathbf{x}_2, \dots, \mathbf{x}_n). \tag{A.71}$$

By writing this out in a basis where the linear map A is represented by the matrix $\mathbf{A}$, we easily see that

$$\det \mathbf{A} = \det A. \tag{A.72}$$

The new definition is therefore compatible with the old one. The advantage of this more sophisticated definition is that it makes no appeal to a basis, and so shows that the determinant of an endomorphism is a basis-independent concept. A byproduct is an easy proof that $\det(AB) = \det(A)\det(B)$, a result that is not so easy to establish with the elementary definition. We write

$$\begin{aligned} \det(AB)\omega(\mathbf{x}_1, \mathbf{x}_2, \dots, \mathbf{x}_n) &= \omega(AB\mathbf{x}_1, AB\mathbf{x}_2, \dots, AB\mathbf{x}_n) \\ &= \omega(A(B\mathbf{x}_1), A(B\mathbf{x}_2), \dots, A(B\mathbf{x}_n)) \\ &= \det(A)\omega(B\mathbf{x}_1, B\mathbf{x}_2, \dots, B\mathbf{x}_n) \\ &= \det(A)\det(B)\omega(\mathbf{x}_1, \mathbf{x}_2, \dots, \mathbf{x}_n). \end{aligned} \tag{A.73}$$

Cancelling the common factor of $\omega(\mathbf{x}_1, \mathbf{x}_2, \dots, \mathbf{x}_n)$ completes the proof.

Exercise A.9: Let ω be a skew-symmetric n-linear form on an n-dimensional vector space. Assuming that ω does not vanish identically, show that a set of n vectors $\mathbf{x}_1, \mathbf{x}_2, \ldots, \mathbf{x}_n$ is linearly independent, and hence forms a basis, if, and only if, $\omega(\mathbf{x}_1, \mathbf{x}_2, \ldots, \mathbf{x}_n) \neq 0$.

Exercise A.10: Extend the pairing between V and its dual space V^* to a pairing between the one-dimensional $\bigwedge^n(V^*)$ and *its* dual space. Use this pairing, together with the result of Exercise A.5, to show that

$$\det \mathbf{A}^{\mathrm{T}} = \det A^* = [\det A]^* = [\det A]^{\mathrm{T}} = \det A = \det \mathbf{A},$$

where the "$*$" denotes the conjugate operator (and not complex conjugation) and the penultimate equality holds because transposition has no effect on a one-by-one matrix. Conclude that $\det \mathbf{A} = \det \mathbf{A}^{\mathrm{T}}$. A determinant is therefore unaffected by the interchange of its rows with its columns.

Exercise A.11: *Cauchy–Binet formula.* Let $\mathbf{A}$ be an m-by-n matrix and $\mathbf{B}$ be an n-by-m matrix. The matrix product $\mathbf{AB}$ is therefore defined, and is an m-by-m matrix. Let S be a subset of $\{1, \ldots, n\}$ with m elements, and let $\mathbf{A}_S$ be the m-by-m matrix whose columns are the columns of $\mathbf{A}$ corresponding to indices in S. Similarly let $\mathbf{B}_S$ be the m-by-m matrix whose *rows* are the rows of $\mathbf{B}$ with indices in S. Show that

$$\det \mathbf{AB} = \sum_S \det \mathbf{A}_S \det \mathbf{B}_S$$

where the sum is over all $n!/m!(n-m)!$ subsets S. If $m > n$ there are no such subsets. Show that in this case $\det \mathbf{AB} = 0$.

Exercise A.12: Let

$$\mathbf{A} = \begin{pmatrix} \mathbf{a} & \mathbf{b} \\ \mathbf{c} & \mathbf{d} \end{pmatrix}$$

be a partitioned matrix where $\mathbf{a}$ is m-by-m, $\mathbf{b}$ is m-by-n, $\mathbf{c}$ is n-by-m and $\mathbf{d}$ is n-by-n. By making a Gaussian decomposition

$$\mathbf{A} = \begin{pmatrix} \mathbf{I}_m & \mathbf{x} \\ \mathbf{0} & \mathbf{I}_n \end{pmatrix} \begin{pmatrix} \Lambda_1 & \mathbf{0} \\ \mathbf{0} & \Lambda_2 \end{pmatrix} \begin{pmatrix} \mathbf{I}_m & \mathbf{0} \\ \mathbf{y} & \mathbf{I}_n \end{pmatrix},$$

show that, for invertible $\mathbf{d}$, we have *Schur's determinant formula*[6]

$$\det \mathbf{A} = \det(\mathbf{d}) \det(\mathbf{a} - \mathbf{b}\mathbf{d}^{-1}\mathbf{c}).$$

[6] I. Schur, *J. für reine und angewandte Math.*, **147** (1917) 205.

A.6.2 The adjugate matrix

Given a square matrix

$$\mathbf{A} = \begin{pmatrix} a_{11} & a_{12} & \dots & a_{1n} \\ a_{21} & a_{22} & \dots & a_{2n} \\ \vdots & \vdots & \ddots & \vdots \\ a_{n1} & a_{n2} & \dots & a_{nn} \end{pmatrix} \tag{A.74}$$

and an element a_{ij}, we define the corresponding *minor* M_{ij} to be the determinant of the $(n-1)$-by-$(n-1)$ matrix constructed by deleting from $\mathbf{A}$ the row and column containing a_{ij}. The number

$$A_{ij} = (-1)^{i+j} M_{ij} \tag{A.75}$$

is then called the *cofactor* of the element a_{ij}. (It is traditional to use uppercase letters to denote cofactors.) The basic result involving cofactors is that

$$\sum_j a_{ij} A_{i'j} = \delta_{ii'} \det \mathbf{A}. \tag{A.76}$$

When $i = i'$, this is known as the *Laplace development* of the determinant about row i. We get zero when $i \neq i'$ because we are effectively developing a determinant with two equal rows. We now define the *adjugate matrix*,[7] $\operatorname{Adj} \mathbf{A}$, to be the transposed matrix of the cofactors:

$$(\operatorname{Adj} \mathbf{A})_{ij} = A_{ji}. \tag{A.77}$$

In terms of this we have

$$\mathbf{A}(\operatorname{Adj} \mathbf{A}) = (\det \mathbf{A})\mathbf{I}. \tag{A.78}$$

In other words

$$\mathbf{A}^{-1} = \frac{1}{\det \mathbf{A}} \operatorname{Adj} \mathbf{A}. \tag{A.79}$$

Each entry in the adjugate matrix is a polynomial of degree $n - 1$ in the entries of the original matrix. Thus, no division is required to form it, and the adjugate matrix exists even if the inverse matrix does not.

[7] Some authors rather confusingly call this the *adjoint matrix*.

Exercise A.13: It is possible to Laplace-develop a determinant about a *set* of rows. For example, the development of a 4-by-4 determinant about its first *two* rows is given by:

$$\begin{vmatrix} a_1 & b_1 & c_1 & d_1 \\ a_2 & b_2 & c_2 & d_2 \\ a_3 & b_3 & c_3 & d_3 \\ a_4 & b_4 & c_4 & d_4 \end{vmatrix} = \begin{vmatrix} a_1 & b_1 \\ a_2 & b_2 \end{vmatrix}\begin{vmatrix} c_3 & d_3 \\ c_4 & d_4 \end{vmatrix} - \begin{vmatrix} a_1 & c_1 \\ a_2 & c_2 \end{vmatrix}\begin{vmatrix} b_3 & d_3 \\ b_4 & d_4 \end{vmatrix} + \begin{vmatrix} a_1 & d_1 \\ a_2 & d_2 \end{vmatrix}\begin{vmatrix} b_3 & c_3 \\ b_4 & c_4 \end{vmatrix}$$

$$+ \begin{vmatrix} b_1 & c_1 \\ b_2 & c_2 \end{vmatrix}\begin{vmatrix} a_3 & d_3 \\ a_4 & d_4 \end{vmatrix} - \begin{vmatrix} b_1 & d_1 \\ b_2 & d_2 \end{vmatrix}\begin{vmatrix} a_3 & c_3 \\ a_4 & c_4 \end{vmatrix} + \begin{vmatrix} c_1 & d_1 \\ c_2 & d_2 \end{vmatrix}\begin{vmatrix} a_3 & b_3 \\ a_4 & b_4 \end{vmatrix}.$$

Understand why this formula is correct and, using that insight, describe the general rule.

Exercise A.14: *Sylvester's lemma.*[8] Let **A** and **B** be two n-by-n matrices. Show that

$$\det \mathbf{A} \det \mathbf{B} = \sum \det \mathbf{A}' \det \mathbf{B}',$$

where $\mathbf{A}'$ and $\mathbf{B}'$ are constructed by selecting a fixed set of $k < n$ columns of **B** (which we can, without loss of generality, take to be the first k columns) and interchanging them with k columns of **A**, preserving the order of the columns. The sum is over all $n!/k!(n-k)!$ ways of choosing columns of **A**. (Hint: show that, without loss of generality, we can take the columns of **A** to be a set of basis vectors, and that, in this case, the lemma becomes a restatement of your "general rule" from the previous problem.)

Cayley's theorem

You will know that the possible eigenvalues of the n-by-n matrix **A** are given by the roots of its *characteristic equation*

$$0 = \det(\mathbf{A} - \lambda \mathbf{I}) = (-1)^n \left(\lambda^n - \operatorname{tr}(\mathbf{A})\lambda^{n-1} + \cdots + (-1)^n \det(\mathbf{A})\right), \tag{A.80}$$

and have probably met with *Cayley's theorem* that asserts that every matrix obeys its own characteristic equation.

$$\mathbf{A}^n - \operatorname{tr}(\mathbf{A})\mathbf{A}^{n-1} + \cdots + (-1)^n \det(\mathbf{A})\mathbf{I} = \mathbf{0}. \tag{A.81}$$

The proof of Cayley's theorem involves the adjugate matrix. We write

$$\det(\mathbf{A} - \lambda \mathbf{I}) = (-1)^n \left(\lambda^n + \alpha_1 \lambda^{n-1} + \cdots + \alpha_n\right) \tag{A.82}$$

and observe that

$$\det(\mathbf{A} - \lambda \mathbf{I})\mathbf{I} = (\mathbf{A} - \lambda \mathbf{I}) \operatorname{Adj}(\mathbf{A} - \lambda \mathbf{I}). \tag{A.83}$$

[8] J. J. Sylvester, *Phil. Mag.*, **1** (1851) 295.

Now Adj $(\mathbf{A} - \lambda\mathbf{I})$ is a matrix-valued polynomial in λ of degree $n - 1$, and it can be written

$$\mathrm{Adj}\,(\mathbf{A} - \lambda\mathbf{I}) = \mathbf{C}_0\lambda^{n-1} + \mathbf{C}_1\lambda^{n-2} + \cdots + \mathbf{C}_{n-1}, \tag{A.84}$$

for some matrix coefficients $\mathbf{C}_i$. On multiplying out the equation

$$(-1)^n \left(\lambda^n + \alpha_1\lambda^{n-1} + \cdots + \alpha_n\right)\mathbf{I} = (\mathbf{A} - \lambda\mathbf{I})(\mathbf{C}_0\lambda^{n-1} + \mathbf{C}_1\lambda^{n-2} + \cdots + \mathbf{C}_{n-1}) \tag{A.85}$$

and comparing like powers of λ, we find the relations

$$\begin{aligned} (-1)^n\mathbf{I} &= -\mathbf{C}_0, \\ (-1)^n\alpha_1\mathbf{I} &= -\mathbf{C}_1 + \mathbf{A}\mathbf{C}_0, \\ (-1)^n\alpha_2\mathbf{I} &= -\mathbf{C}_2 + \mathbf{A}\mathbf{C}_1, \\ &\vdots \\ (-1)^n\alpha_{n-1}\mathbf{I} &= -\mathbf{C}_{n-1} + \mathbf{A}\mathbf{C}_{n-2}, \\ (-1)^n\alpha_n\mathbf{I} &= \quad \mathbf{A}\mathbf{C}_{n-1}. \end{aligned}$$

Multiply the first equation on the left by $\mathbf{A}^n$, the second by $\mathbf{A}^{n-1}$, and so on down to the last equation which we multiply by $\mathbf{A}^0 \equiv \mathbf{I}$. Now add. We find that the sum telescopes to give Cayley's theorem,

$$\mathbf{A}^n + \alpha_1\mathbf{A}^{n-1} + \cdots + \alpha_n\mathbf{I} = \mathbf{0},$$

as advertised.

A.6.3 Differentiating determinants

Suppose that the elements of $\mathbf{A}$ depend on some parameter x. From the elementary definition

$$\det\mathbf{A} = \epsilon_{i_1 i_2 \ldots i_n} a_{1i_1} a_{2i_2} \ldots a_{ni_n},$$

we find

$$\frac{d}{dx}\det\mathbf{A} = \epsilon_{i_1 i_2 \ldots i_n}\left(a'_{1i_1} a_{2i_2} \ldots a_{ni_n} + a_{1i_1} a'_{2i_2} \ldots a_{ni_n} + \cdots + a_{1i_1} a_{2i_2} \ldots a'_{ni_n}\right). \tag{A.86}$$

In other words,

$$\frac{d}{dx}\det\mathbf{A} = \begin{vmatrix} a'_{11} & a'_{12} & \dots & a'_{1n} \\ a_{21} & a_{22} & \dots & a_{2n} \\ \vdots & \vdots & \ddots & \vdots \\ a_{n1} & a_{n2} & \dots & a_{nn} \end{vmatrix} + \begin{vmatrix} a_{11} & a_{12} & \dots & a_{1n} \\ a'_{21} & a'_{22} & \dots & a'_{2n} \\ \vdots & \vdots & \ddots & \vdots \\ a_{n1} & a_{n2} & \dots & a_{nn} \end{vmatrix} + \cdots$$

$$+ \begin{vmatrix} a_{11} & a_{12} & \dots & a_{1n} \\ a_{21} & a_{22} & \dots & a_{2n} \\ \vdots & \vdots & \ddots & \vdots \\ a'_{n1} & a'_{n2} & \dots & a'_{nn} \end{vmatrix}.$$

The same result can also be written more compactly as

$$\frac{d}{dx}\det\mathbf{A} = \sum_{ij}\frac{da_{ij}}{dx}A_{ij}, \tag{A.87}$$

where A_{ij} is cofactor of a_{ij}. Using the connection between the adjugate matrix and the inverse, this is equivalent to

$$\frac{1}{\det\mathbf{A}}\frac{d}{dx}\det\mathbf{A} = \operatorname{tr}\left\{\frac{d\mathbf{A}}{dx}\mathbf{A}^{-1}\right\}, \tag{A.88}$$

or

$$\frac{d}{dx}\ln(\det\mathbf{A}) = \operatorname{tr}\left\{\frac{d\mathbf{A}}{dx}\mathbf{A}^{-1}\right\}. \tag{A.89}$$

A special case of this formula is the result

$$\frac{\partial}{\partial a_{ij}}\ln(\det\mathbf{A}) = \left(\mathbf{A}^{-1}\right)_{ji}. \tag{A.90}$$

A.7 Diagonalization and canonical forms

An essential part of the linear algebra tool-kit is the set of techniques for the reduction of a matrix to its simplest, *canonical form*. This is often a diagonal matrix.

A.7.1 Diagonalizing linear maps

A common task is the diagonalization of a matrix $\mathbf{A}$ representing a linear map A. Let us recall some standard material relating to this:

(i) If $A\mathbf{x} = \lambda\mathbf{x}$ for a non-zero vector $\mathbf{x}$, then $\mathbf{x}$ is said to be an *eigenvector* of A with *eigenvalue* λ.

(ii) A linear operator A on a finite-dimensional vector space is said to be *self-adjoint*, or *hermitian*, with respect to the inner product $\langle\ ,\ \rangle$ if $A = A^\dagger$, or equivalently if $\langle \mathbf{x}, A\mathbf{y}\rangle = \langle A\mathbf{x}, \mathbf{y}\rangle$ for all $\mathbf{x}$ and $\mathbf{y}$.

(iii) If A is hermitian with respect to a positive-definite inner product $\langle\ ,\ \rangle$ then all the eigenvalues λ are real. To see that this is so, we write

$$\lambda\langle \mathbf{x}, \mathbf{x}\rangle = \langle \mathbf{x}, \lambda\mathbf{x}\rangle = \langle \mathbf{x}, A\mathbf{x}\rangle = \langle A\mathbf{x}, \mathbf{x}\rangle = \langle \lambda\mathbf{x}, \mathbf{x}\rangle = \lambda^*\langle \mathbf{x}, \mathbf{x}\rangle. \tag{A.91}$$

Because the inner product is positive definite and $\mathbf{x}$ is not zero, the factor $\langle \mathbf{x}, \mathbf{x}\rangle$ cannot be zero. We conclude that $\lambda = \lambda^*$.

(iv) If A is hermitian and λ_i and λ_j are two distinct eigenvalues with eigenvectors $\mathbf{x}_i$ and $\mathbf{x}_j$, respectively, then $\langle \mathbf{x}_i, \mathbf{x}_j\rangle = 0$. To prove this, we write

$$\lambda_j\langle \mathbf{x}_i, \mathbf{x}_j\rangle = \langle \mathbf{x}_i, A\mathbf{x}_j\rangle = \langle A\mathbf{x}_i, \mathbf{x}_j\rangle = \langle \lambda_i\mathbf{x}_i, \mathbf{x}_j\rangle = \lambda_i^*\langle \mathbf{x}_i, \mathbf{x}_j\rangle. \tag{A.92}$$

But $\lambda_i^* = \lambda_i$, and so $(\lambda_i - \lambda_j)\langle \mathbf{x}_i, \mathbf{x}_j\rangle = 0$. Since, by assumption, $(\lambda_i - \lambda_j) \neq 0$ we must have $\langle \mathbf{x}_i, \mathbf{x}_j\rangle = 0$.

(v) An operator A is said to be *diagonalizable* if we can find a basis for V that consists of eigenvectors of A. In this basis, A is represented by the matrix $\mathbf{A} = \operatorname{diag}(\lambda_1, \lambda_2, \ldots, \lambda_n)$, where the λ_i are the eigenvalues.

Not all linear operators can be diagonalized. The key element determining the diagonalizability of a matrix is the *minimal polynomial equation* obeyed by the matrix representing the operator. As mentioned in the previous section, the possible eigenvalues of an N-by-N matrix $\mathbf{A}$ are given by the roots of the *characteristic equation*

$$0 = \det(\mathbf{A} - \lambda\mathbf{I}) = (-1)^N \left(\lambda^N - \operatorname{tr}(\mathbf{A})\lambda^{N-1} + \cdots + (-1)^N \det(\mathbf{A})\right).$$

This is because a non-trivial solution to the equation

$$\mathbf{A}\mathbf{x} = \lambda\mathbf{x} \tag{A.93}$$

requires the matrix $\mathbf{A} - \lambda\mathbf{I}$ to have a non-trivial null-space, and so $\det(\mathbf{A} - \lambda\mathbf{I})$ must vanish. Cayley's theorem, which we proved in the previous section, asserts that every matrix obeys its own characteristic equation:

$$\mathbf{A}^N - \operatorname{tr}(\mathbf{A})\mathbf{A}^{N-1} + \cdots + (-1)^N \det(\mathbf{A})\mathbf{I} = \mathbf{0}.$$

The matrix $\mathbf{A}$ may, however, satisfy an equation of lower degree. For example, the characteristic equation of the matrix

$$\mathbf{A} = \begin{pmatrix} \lambda_1 & 0 \\ 0 & \lambda_1 \end{pmatrix} \tag{A.94}$$

is $(\lambda - \lambda_1)^2$. Cayley therefore asserts that $(\mathbf{A} - \lambda_1\mathbf{I})^2 = \mathbf{0}$. This is clearly true, but $\mathbf{A}$ also satisfies the equation of first degree $(\mathbf{A} - \lambda_1\mathbf{I}) = \mathbf{0}$.

The equation of lowest degree satisfied by $\mathbf{A}$ is said to be the minimal polynomial equation. It is unique up to an overall numerical factor: if two distinct minimal equations of degree n were to exist, and if we normalize them so that the coefficients of $\mathbf{A}^n$ coincide, then their difference, if non-zero, would be an equation of degree $\le (n-1)$ obeyed by $\mathbf{A}$ – and a contradiction to the minimal equation having degree n.

If

$$P(\mathbf{A}) \equiv (\mathbf{A} - \lambda_1\mathbf{I})^{\alpha_1}(\mathbf{A} - \lambda_2\mathbf{I})^{\alpha_2} \cdots (\mathbf{A} - \lambda_n\mathbf{I})^{\alpha_n} = \mathbf{0} \tag{A.95}$$

is the minimal equation then each root λ_i is an eigenvalue of $\mathbf{A}$. To prove this, we select one factor of $(A - \lambda_i\mathbf{I})$ and write

$$P(\mathbf{A}) = (A - \lambda_i\mathbf{I})Q(\mathbf{A}), \tag{A.96}$$

where $Q(\mathbf{A})$ contains all the remaining factors in $P(\mathbf{A})$. We now observe that there must be some vector $\mathbf{y}$ such that $\mathbf{x} = Q(\mathbf{A})\mathbf{y}$ is not zero. If there were no such $\mathbf{y}$ then $Q(\mathbf{A}) = \mathbf{0}$ would be an equation of lower degree obeyed by $\mathbf{A}$ in contradiction to the assumed minimality of $P(\mathbf{A})$. Since

$$\mathbf{0} = P(\mathbf{A})\mathbf{y} = (\mathbf{A} - \lambda_i\mathbf{I})\mathbf{x} \tag{A.97}$$

we see that $\mathbf{x}$ is an eigenvector of $\mathbf{A}$ with eignvalue λ_i.

Because all possible eigenvalues appear as roots of the characteristic equation, the minimal equation must have the same roots as the characteristic equation, but with equal or lower multiplicities α_i.

In the special case that $\mathbf{A}$ is self-adjoint, or hermitian, with respect to a positive definite inner product $\langle\ ,\ \rangle$ the minimal equation has no repeated roots. Suppose that this were not so, and that $\mathbf{A}$ has minimal equation $(\mathbf{A} - \lambda\mathbf{I})^2 R(\mathbf{A}) = \mathbf{0}$ where $R(\mathbf{A})$ is a polynomial in $\mathbf{A}$. Then, for all vectors $\mathbf{x}$ we have

$$0 = \langle R\mathbf{x}, (\mathbf{A} - \lambda\mathbf{I})^2 R\mathbf{x}\rangle = \langle(\mathbf{A} - \lambda\mathbf{I})R\mathbf{x}, (\mathbf{A} - \lambda\mathbf{I})R\mathbf{x}\rangle. \tag{A.98}$$

Now the vanishing of the rightmost expression shows that $(\mathbf{A} - \lambda\mathbf{I})R(\mathbf{A})\mathbf{x} = \mathbf{0}$ for all $\mathbf{x}$. In other words

$$(\mathbf{A} - \lambda\mathbf{I})R(\mathbf{A}) = \mathbf{0}. \tag{A.99}$$

The equation with the repeated factor was not minimal therefore, and we have a contradiction.

If the equation of lowest degree satisfied by the matrix has no repeated roots, the matrix is diagonalizable; if there are repeated roots, it is not. The last statement should be obvious, because a diagonalized matrix satisfies an equation with no repeated roots, and this equation will hold in all bases, including the original one. The first statement, in combination with the observation that the minimal equation for a hermitian matrix has no repeated roots, shows that a hermitian (with respect to a positive definite inner product) matrix can be diagonalized.

To establish the first statement, suppose that $\mathbf{A}$ obeys the equation

$$\mathbf{0} = P(\mathbf{A}) \equiv (\mathbf{A} - \lambda_1 \mathbf{I})(\mathbf{A} - \lambda_2 \mathbf{I}) \cdots (\mathbf{A} - \lambda_n \mathbf{I}), \tag{A.100}$$

where the λ_i are all distinct. Then, setting $x \to \mathbf{A}$ in the identity[9]

$$\begin{aligned} 1 = &\frac{(x-\lambda_2)(x-\lambda_3)\cdots(x-\lambda_n)}{(\lambda_1-\lambda_2)(\lambda_1-\lambda_3)\cdots(\lambda_1-\lambda_n)} + \frac{(x-\lambda_1)(x-\lambda_3)\cdots(x-\lambda_n)}{(\lambda_2-\lambda_1)(\lambda_2-\lambda_3)\cdots(\lambda_2-\lambda_n)} + \cdots \\ &+ \frac{(x-\lambda_1)(x-\lambda_2)\cdots(x-\lambda_{n-1})}{(\lambda_n-\lambda_1)(\lambda_n-\lambda_2)\cdots(\lambda_n-\lambda_{n-1})}, \end{aligned} \tag{A.101}$$

where in each term one of the factors of the polynomial is omitted in both numerator and denominator, we may write

$$\mathbf{I} = \mathbf{P}_1 + \mathbf{P}_2 + \cdots + \mathbf{P}_n, \tag{A.102}$$

where

$$\mathbf{P}_1 = \frac{(\mathbf{A} - \lambda_2 \mathbf{I})(\mathbf{A} - \lambda_3 \mathbf{I}) \cdots (\mathbf{A} - \lambda_n \mathbf{I})}{(\lambda_1-\lambda_2)(\lambda_1-\lambda_3)\cdots(\lambda_1-\lambda_n)}, \tag{A.103}$$

etc. Clearly $\mathbf{P}_i\mathbf{P}_j = \mathbf{0}$ if $i \neq j$, because the product contains the minimal equation as a factor. Multiplying (A.102) by $\mathbf{P}_i$ therefore gives $\mathbf{P}_i^2 = \mathbf{P}_i$, showing that the $\mathbf{P}_i$ are projection operators. Further $(\mathbf{A} - \lambda_i \mathbf{I})(\mathbf{P}_i) = \mathbf{0}$, so

$$(\mathbf{A} - \lambda_i \mathbf{I})(\mathbf{P}_i \mathbf{x}) = \mathbf{0} \tag{A.104}$$

for any vector $\mathbf{x}$, and we see that $\mathbf{P}_i\mathbf{x}$, if not zero, is an eigenvector with eigenvalue λ_i. Thus $\mathbf{P}_i$ projects into the i-th eigenspace. Applying the resolution of the identity (A.102) to a vector $\mathbf{x}$ shows that it can be decomposed

$$\begin{aligned} \mathbf{x} &= \mathbf{P}_1\mathbf{x} + \mathbf{P}_2\mathbf{x} + \cdots + \mathbf{P}_n\mathbf{x} \\ &= \mathbf{x}_1 + \mathbf{x}_2 + \cdots + \mathbf{x}_n, \end{aligned} \tag{A.105}$$

where $\mathbf{x}_i$, if not zero, is an eigenvector with eigenvalue λ_i. Since any $\mathbf{x}$ can be written as a sum of eigenvectors, the eigenvectors span the space.

[9] The identity may be verified by observing that the difference of the left- and right-hand sides is a polynomial of degree $n-1$, which, by inspection, vanishes at the n points $x = \lambda_i$. But a polynomial that has more zeros than its degree must be identically zero.

Jordan decomposition

If the minimal polynomial has repeated roots, the matrix can still be reduced to the *Jordan canonical form*, which is diagonal except for some 1's immediately above the diagonal.

For example, suppose the characteristic equation for a 6-by-6 matrix $\mathbf{A}$ is

$$0 = \det(\mathbf{A} - \lambda\mathbf{I}) = (\lambda_1 - \lambda)^3(\lambda_2 - \lambda)^3, \tag{A.106}$$

but the minimal equation is

$$0 = (\lambda_1 - \lambda)^3(\lambda_2 - \lambda)^2. \tag{A.107}$$

Then the Jordan form of $\mathbf{A}$ might be

$$\mathbf{T}^{-1}\mathbf{AT} = \begin{pmatrix} \lambda_1 & 1 & 0 & 0 & 0 & 0 \\ 0 & \lambda_1 & 1 & 0 & 0 & 0 \\ 0 & 0 & \lambda_1 & 0 & 0 & 0 \\ 0 & 0 & 0 & \lambda_2 & 1 & 0 \\ 0 & 0 & 0 & 0 & \lambda_2 & 0 \\ 0 & 0 & 0 & 0 & 0 & \lambda_2 \end{pmatrix}. \tag{A.108}$$

One may easily see that (A.107) is the minimal equation for this matrix. The minimal equation alone does not uniquely specify the pattern of λ_i's and 1's in the Jordan form, though.

It is rather tedious, but quite straightforward, to show that any linear map can be reduced to a Jordan form. The proof is sketched in the following exercises:

Exercise A.15: Suppose that the linear operator T is represented by an $N \times N$ matrix, where $N > 1$. T obeys the equation

$$(T - \lambda I)^p = 0,$$

with $p = N$, but does not obey this equation for any $p < N$. Here λ is a number and I is the identity operator.

(i) Show that if T has an eigenvector, the corresponding eigenvalue must be λ. Deduce that T *cannot* be diagonalized.
(ii) Show that there exists a vector $\mathbf{e}_1$ such that $(T - \lambda I)^N \mathbf{e}_1 = 0$, but no lesser power of $(T - \lambda I)$ kills $\mathbf{e}_1$.
(iii) Define $\mathbf{e}_2 = (T - \lambda I)\mathbf{e}_1$, $\mathbf{e}_3 = (T - \lambda I)^2\mathbf{e}_1$, etc. up to $\mathbf{e}_N$. Show that the vectors $\mathbf{e}_1, \ldots, \mathbf{e}_N$ are linearly independent.

(iv) Use $\mathbf{e}_1, \dots, \mathbf{e}_N$ as a basis for your vector space. Taking

$$\mathbf{e}_1 = \begin{pmatrix} 0 \\ \vdots \\ 0 \\ 1 \end{pmatrix}, \quad \mathbf{e}_2 = \begin{pmatrix} 0 \\ \vdots \\ 1 \\ 0 \end{pmatrix}, \quad \dots, \quad \mathbf{e}_N = \begin{pmatrix} 1 \\ 0 \\ \vdots \\ 0 \end{pmatrix},$$

write out the matrix representing T in the $\mathbf{e}_i$ basis.

Exercise A.16: Let $T : V \to V$ be a linear map, and suppose that the minimal polynomial equation satisfied by T is

$$Q(T) = (T - \lambda_1 I)^{r_1}(T - \lambda_2 I)^{r_2} \dots (T - \lambda_n I)^{r_n} = 0.$$

Let V_{λ_i} denote the space of *generalized eigenvectors* for the eigenvalue λ_i. This is the set of $\mathbf{x}$ such that $(T - \lambda_i I)^{r_i}\mathbf{x} = 0$. You will show that

$$V = \bigoplus_i V_{\lambda_i}.$$

(i) Consider the set of polynomials $Q_{\lambda_i,j}(t) = (t-\lambda_i)^{-(r_i-j+1)}Q(t)$ where $j = 1, \dots, r_i$. Show that this set of $N \equiv \sum_i r_i$ polynomials forms a basis for the vector space $\mathcal{F}_{N-1}(t)$ of polynomials in t of degree no more than $N - 1$. (Since the number of $Q_{\lambda_i,j}$ is N, and this is equal to the dimension of $\mathcal{F}_{N-1}(t)$, the claim will be established if you can show that the polynomials are linearly independent. This is easy to do: suppose that

$$\sum_{\lambda_i,j} \alpha_{\lambda_i,j} Q_{\lambda_i,j}(t) = 0.$$

Set $t = \lambda_i$ and deduce that $\alpha_{\lambda_i,1} = 0$. Knowing this, differentiate with respect to t and again set $t = \lambda_i$ and deduce that $\alpha_{\lambda_i,2} = 0$, and so on.)

(ii) Since the $Q_{\lambda_i,j}$ form a basis, and since $1 \in \mathcal{F}_{N-1}$, argue that we can find $\beta_{\lambda_i,j}$ such that

$$1 = \sum_{\lambda_i,j} \beta_{\lambda_i,j} Q_{\lambda_i,j}(t).$$

Now define

$$P_i = \sum_{j=1}^{r_i} \beta_{\lambda_i,j} Q_{\lambda_i,j}(T),$$

and so

$$I = \sum_{\lambda_i} P_i, (\star).$$

Use the minimal polynomial equation to deduce that $P_i P_j = 0$ if $i \neq j$. Multiplication of $(\star)$ by P_i then shows that $P_i P_j = \delta_{ij} P_j$. Deduce from this that $(\star)$ is a resolution of the identity into a sum of mutually orthogonal projection operators P_i that project onto the spaces V_{λ_i}. Conclude that any $\mathbf{x}$ can be expanded as $\mathbf{x} = \sum_i \mathbf{x}_i$ with $\mathbf{x}_i \equiv P_i \mathbf{x} \in V_{\lambda_i}$.

(iii) Show that the decomposition also implies that $V_{\lambda_i} \cap V_{\lambda_j} = \{\mathbf{0}\}$ if $i \neq j$. (Hint: a vector in V_{λ_i} is killed by all projectors with the possible exception of P_i and a vector in V_{λ_j} will be killed by all the projectors with the possible exception of P_j.)

(iv) Put these results together to deduce that V is a direct sum of the V_{λ_i}.

(v) Combine the result of part (iv) with the ideas behind Exercise A.15 to complete the proof of the Jordan decomposition theorem.

A.7.2 Diagonalizing quadratic forms

Do not confuse the notion of diagonalizing the matrix representing a *linear map* $A : V \to V$ with that of diagonalizing the matrix representing a *quadratic form*. A (real) quadratic form is a map $Q : V \to \mathbb{R}$, which is obtained from a symmetric bilinear form $B : V \times V \to \mathbb{R}$ by setting the two arguments, $\mathbf{x}$ and $\mathbf{y}$, in $B(\mathbf{x}, \mathbf{y})$ equal:

$$Q(\mathbf{x}) = B(\mathbf{x}, \mathbf{x}). \tag{A.109}$$

No information is lost by this specialization. We can recover the non-diagonal ($\mathbf{x} \neq \mathbf{y}$) values of B from the diagonal values, $Q(\mathbf{x})$, by using the *polarization trick*

$$B(\mathbf{x}, \mathbf{y}) = \frac{1}{2}[Q(\mathbf{x} + \mathbf{y}) - Q(\mathbf{x}) - Q(\mathbf{y})]. \tag{A.110}$$

An example of a real quadratic form is the kinetic energy term

$$T(\dot{\mathbf{x}}) = \frac{1}{2} m_{ij} \dot{x}^i \dot{x}^j = \frac{1}{2} \dot{\mathbf{x}}^T \mathbf{M} \dot{\mathbf{x}} \tag{A.111}$$

in a "small vibrations" Lagrangian. Here, $\mathbf{M}$, with entries m_{ij}, is the mass matrix.

Whilst one can diagonalize such forms by the tedious procedure of finding the eigenvalues and eigenvectors of the associated matrix, it is simpler to use Lagrange's method, which is based on repeatedly completing squares.

Consider, for example, the quadratic form

$$Q = x^2 - y^2 - z^2 + 2xy - 4xz + 6yz = (x, y, z) \begin{pmatrix} 1 & 1 & -2 \\ 1 & -1 & 3 \\ -2 & 3 & -1 \end{pmatrix} \begin{pmatrix} x \\ y \\ z \end{pmatrix}. \tag{A.112}$$

We complete the square involving x:

$$Q = (x + y - 2z)^2 - 2y^2 + 10yz - 5z^2, \tag{A.113}$$

where the terms outside the squared group no longer involve x. We now complete the square in y:

$$Q = (x + y - 2z)^2 - (\sqrt{2}y - \frac{5}{\sqrt{2}}z)^2 + \frac{15}{2}z^2, \tag{A.114}$$

so that the remaining term no longer contains y. Thus, on setting

$$\begin{aligned}
\xi &= x + y - 2z, \\
\eta &= \sqrt{2}y - \frac{5}{\sqrt{2}}z, \\
\zeta &= \sqrt{\frac{15}{2}}z,
\end{aligned}$$

we have

$$Q = \xi^2 - \eta^2 + \zeta^2 = (\xi, \eta, \zeta) \begin{pmatrix} 1 & 0 & 0 \\ 0 & -1 & 0 \\ 0 & 0 & 1 \end{pmatrix} \begin{pmatrix} \xi \\ \eta \\ \zeta \end{pmatrix}. \tag{A.115}$$

If there are no x^2, y^2 or z^2 terms to get us started, then we can proceed by using $(x+y)^2$ and $(x-y)^2$. For example, consider

$$\begin{aligned}
Q &= 2xy + 2yz + 2zy \\
&= \frac{1}{2}(x+y)^2 - \frac{1}{2}(x-y)^2 + 2xz + 2yz \\
&= \frac{1}{2}(x+y)^2 + 2(x+y)z - \frac{1}{2}(x-y)^2 \\
&= \frac{1}{2}(x+y+2z)^2 - \frac{1}{2}(x-y)^2 - 4z^2 \\
&= \xi^2 - \eta^2 - \zeta^2,
\end{aligned}$$

where

$$\begin{aligned}
\xi &= \frac{1}{\sqrt{2}}(x + y + 2z), \\
\eta &= \frac{1}{\sqrt{2}}(x - y), \\
\zeta &= \sqrt{2}z.
\end{aligned}$$

A judicious combination of these two tactics will reduce the matrix representing any real quadratic form to a matrix with ± 1's and 0's on the diagonal, and zeros elsewhere. As the egregiously asymmetric treatment of x, y, z in the last example indicates, this can be done in many ways, but *Cayley's law of inertia* asserts that the *signature* – the number of $+1$'s, -1's and 0's – will always be the same. Naturally, if we allow complex numbers in the redefinitions of the variables, we can always reduce the form to one with only $+1$'s and 0's.

The essential difference between diagonalizing linear maps and diagonalizing quadratic forms is that in the former case we seek matrices $\mathbf{A}$ such that $\mathbf{A}^{-1}\mathbf{M}\mathbf{A}$ is diagonal, whereas in the latter case we seek matrices $\mathbf{A}$ such that $\mathbf{A}^{\mathrm{T}}\mathbf{M}\mathbf{A}$ is diagonal. Here, the superscript T denotes transposition.

Exercise A.17: Show that the matrix

$$\mathbf{Q} = \begin{pmatrix} a & b \\ b & c \end{pmatrix}$$

representing the quadratic form

$$Q(x, y) = ax^2 + 2bxy + cy^2$$

may be reduced to

$$\begin{pmatrix} 1 & 0 \\ 0 & 1 \end{pmatrix}, \quad \begin{pmatrix} 1 & 0 \\ 0 & -1 \end{pmatrix} \quad \text{or} \quad \begin{pmatrix} 1 & 0 \\ 0 & 0 \end{pmatrix},$$

depending on whether the *discriminant*, $ac - b^2$, is respectively greater than zero, less than zero or equal to zero.

Warning: You might be tempted to refer to the discriminant $ac - b^2$ as being the determinant of Q. It is indeed the determinant of the matrix $\mathbf{Q}$, but there is no such thing as the "determinant" of the quadratic form itself. You may compute the determinant of the matrix representing Q in some basis, but if you change basis and repeat the calculation you will get a different answer. For *real* quadratic forms, however, the *sign* of the determinant stays the same, and this is all that the discriminant cares about.

A.7.3 Block-diagonalizing symplectic forms

A skew-symmetric bilinear form $\omega : V \times V \to \mathbb{R}$ is often called a *symplectic form*. Such forms play an important role in Hamiltonian dynamics and in optics. Let $\{\mathbf{e}_i\}$ be a basis for V, and set

$$\omega(\mathbf{e}_i, \mathbf{e}_j) = \omega_{ij}. \tag{A.116}$$

If $\mathbf{x} = x^i \mathbf{e}_i$ and $\mathbf{y} = y^i \mathbf{e}_i$, we therefore have

$$\omega(\mathbf{x}, \mathbf{y}) = \omega(\mathbf{e}_i, \mathbf{e}_j) x^i y^j = \omega_{ij} x^i y^j. \tag{A.117}$$

The numbers ω_{ij} can be thought of as the entries in a real skew-symmetric matrix $\mathbf{\Omega}$, in terms of which $\omega(\mathbf{x},\mathbf{y}) = \mathbf{x}^{\mathrm{T}}\mathbf{\Omega}\mathbf{y}$. We cannot exactly "diagonalize" such a skew-symmetric matrix because a matrix with non-zero entries only on its principal diagonal is necessarily symmetric. We can do the next best thing, however, and reduce $\mathbf{\Omega}$ to *block diagonal* form with simple 2-by-2 skew matrices along the diagonal.

We begin by expanding ω as

$$\omega = \frac{1}{2}\omega_{ij}\mathbf{e}^{*i}\wedge,\mathbf{e}^{*j} \tag{A.118}$$

where the *wedge* (or *exterior*) product $\mathbf{e}^{*j}\wedge\mathbf{e}^{*j}$ of a pair of basis vectors in V^* denotes the particular skew-symmetric bilinear form

$$\mathbf{e}^{*i}\wedge\mathbf{e}^{*j}(\mathbf{e}_\alpha,\mathbf{e}_\beta) = \delta^i_\alpha\delta^j_\beta - \delta^i_\beta\delta^j_\alpha. \tag{A.119}$$

Again, if $\mathbf{x} = x^i\mathbf{e}_i$ and $\mathbf{y} = y^i\mathbf{e}_i$, we have

$$\begin{aligned}\mathbf{e}^{*i}\wedge\mathbf{e}^{*j}(\mathbf{x},\mathbf{y}) &= \mathbf{e}^{*i}\wedge\mathbf{e}^{*j}(x^\alpha\mathbf{e}_\alpha, y^\beta\mathbf{e}_\beta)\\ &= (\delta^i_\alpha\delta^j_\beta - \delta^i_\beta\delta^j_\alpha)x^\alpha y^\beta\\ &= x^iy^j - y^ix^j.\end{aligned} \tag{A.120}$$

Consequently

$$\omega(\mathbf{x},\mathbf{y}) = \frac{1}{2}\omega_{ij}(x^iy^j - y^ix^j) = \omega_{ij}x^iy^j, \tag{A.121}$$

as before. We extend the definition of the wedge product to other elements of V^* by requiring "$\wedge$" to be associative and distributive, taking note that

$$\mathbf{e}^{*i}\wedge\mathbf{e}^{*j} = -\mathbf{e}^{*j}\wedge\mathbf{e}^{*i}, \tag{A.122}$$

and so $0 = \mathbf{e}^{*1}\wedge\mathbf{e}^{*1} = \mathbf{e}^{*2}\wedge\mathbf{e}^{*2}$, etc.

We next show that there exists a basis $\{\mathbf{f}^{*i}\}$ of V^* such that

$$\omega = \mathbf{f}^{*1}\wedge\mathbf{f}^{*2} + \mathbf{f}^{*3}\wedge\mathbf{f}^{*4} + \cdots + \mathbf{f}^{*(p-1)}\wedge\mathbf{f}^{*p}. \tag{A.123}$$

Here, the integer $p \le n$ is the *rank* of ω. It is necessarily an even number.

The new basis is constructed by a skew-analogue of Lagrange's method of completing the square. If

$$\omega = \frac{1}{2}\omega_{ij}\mathbf{e}^{*i}\wedge\mathbf{e}^{*j} \tag{A.124}$$

is not identically zero, we can, after re-ordering the basis if necessary, assume that $\omega_{12} \neq 0$. Then

$$\omega = \left(\mathbf{e}^{*1} - \frac{1}{\omega_{12}}(\omega_{23}\mathbf{e}^{*3} + \cdots + \omega_{2n}\mathbf{e}^{*n})\right) \wedge (\omega_{12}\mathbf{e}^{*2} + \omega_{13}\mathbf{e}^{*3} + \cdots \omega_{1n}\mathbf{e}^{*n}) + \omega^{\{3\}} \tag{A.125}$$

where $\omega^{\{3\}} \in \bigwedge^2(V^*)$ does not contain $\mathbf{e}^{*1}$ or $\mathbf{e}^{*2}$. We set

$$\mathbf{f}^{*1} = \mathbf{e}^{*1} - \frac{1}{\omega_{12}}(\omega_{23}\mathbf{e}^{*3} + \cdots + \omega_{2n}\mathbf{e}^{*n}) \tag{A.126}$$

and

$$\mathbf{f}^{*2} = \omega_{12}\mathbf{e}^{*2} + \omega_{13}\mathbf{e}^{*3} + \cdots \omega_{1n}\mathbf{e}^{*n}. \tag{A.127}$$

Thus,

$$\omega = \mathbf{f}^{*1} \wedge \mathbf{f}^{*2} + \omega^{\{3\}}. \tag{A.128}$$

If the remainder $\omega^{\{3\}}$ is identically zero, we are done. Otherwise, we apply the same process to $\omega^{\{3\}}$ so as to construct $\mathbf{f}^{*3}$, $\mathbf{f}^{*4}$ and $\omega^{\{5\}}$; we continue in this manner until we find a remainder, $\omega^{\{p+1\}}$, that vanishes.

Let $\{\mathbf{f}_i\}$ be the basis for V dual to the basis $\{\mathbf{f}^{*i}\}$. Then $\omega(\mathbf{f}_1, \mathbf{f}_2) = -\omega(\mathbf{f}_2, \mathbf{f}_1) = \omega(\mathbf{f}_3, \mathbf{f}_4) = -\omega(\mathbf{f}_4, \mathbf{f}_3) = 1$, and so on, all other values being zero. This shows that if we define the coefficients $a^i{}_j$ by expressing $\mathbf{f}^{*i} = a^i{}_j\mathbf{e}^{*j}$, and hence $\mathbf{e}_i = \mathbf{f}_j a^j{}_i$, then the matrix Ω has been expressed as

$$\Omega = \mathbf{A}^{\mathrm{T}}\widetilde{\Omega}\mathbf{A}, \tag{A.129}$$

where $\mathbf{A}$ is the matrix with entries $a^i{}_j$, and $\widetilde{\Omega}$ is the matrix

$$\widetilde{\Omega} = \begin{pmatrix} 0 & 1 & & & \\ -1 & 0 & & & \\ & & 0 & 1 & \\ & & -1 & 0 & \\ & & & & \ddots \end{pmatrix}, \tag{A.130}$$

which contains $p/2$ diagonal blocks of

$$\begin{pmatrix} 0 & 1 \\ -1 & 0 \end{pmatrix}, \tag{A.131}$$

and all other entries are zero.

Example: Consider the skew bilinear form

$$\omega(\mathbf{x},\mathbf{y}) = \mathbf{x}^{\mathrm{T}}\boldsymbol{\Omega}\mathbf{y} = \left(x^1,x^2,x^3,x^4\right)\begin{pmatrix} 0 & 1 & 3 & 0 \\ -1 & 0 & 1 & 5 \\ -3 & -1 & 0 & 0 \\ 0 & -5 & 0 & 0 \end{pmatrix}\begin{pmatrix} y^1 \\ y^2 \\ y^3 \\ y^4 \end{pmatrix}. \tag{A.132}$$

This corresponds to

$$\omega = \mathbf{e}^{*1} \wedge \mathbf{e}^{*2} + 3\mathbf{e}^{*1} \wedge \mathbf{e}^{*3} + \mathbf{e}^{*2} \wedge \mathbf{e}^{*3} + 5\mathbf{e}^{*2} \wedge \mathbf{e}^{*4}. \tag{A.133}$$

Following our algorithm, we write ω as

$$\omega = (\mathbf{e}^{*1} - \mathbf{e}^{*3} - 5\mathbf{e}^{*4}) \wedge (\mathbf{e}^{*2} + 3\mathbf{e}^{*3}) - 15\mathbf{e}^{*3} \wedge \mathbf{e}^{*4}. \tag{A.134}$$

If we now set

$$\begin{aligned} \mathbf{f}^{*1} &= \mathbf{e}^{*1} - \mathbf{e}^{*3} - 5\mathbf{e}^{*4}, \\ \mathbf{f}^{*2} &= \mathbf{e}^{*2} + 3\mathbf{e}^{*3}, \\ \mathbf{f}^{*3} &= -15\mathbf{e}^{*3}, \\ \mathbf{f}^{*4} &= \mathbf{e}^{*4}, \end{aligned} \tag{A.135}$$

we have

$$\omega = \mathbf{f}^{*1} \wedge \mathbf{f}^{*2} + \mathbf{f}^{*3} \wedge \mathbf{f}^{*4}. \tag{A.136}$$

We have correspondingly expressed the matrix $\boldsymbol{\Omega}$ as

$$\begin{aligned} &\begin{pmatrix} 0 & 1 & 3 & 0 \\ -1 & 0 & 1 & 5 \\ -3 & -1 & 0 & 0 \\ 0 & -5 & 0 & 0 \end{pmatrix} \\ &= \begin{pmatrix} 1 & 0 & 0 & 0 \\ 0 & 1 & 0 & 0 \\ -1 & 3 & -15 & 0 \\ -5 & 0 & 0 & 1 \end{pmatrix}\begin{pmatrix} 0 & 1 & & \\ -1 & 0 & & \\ & & 0 & 1 \\ & & -1 & 0 \end{pmatrix}\begin{pmatrix} 1 & 0 & -1 & -5 \\ 0 & 1 & 3 & 0 \\ 0 & 0 & -15 & 0 \\ 0 & 0 & 0 & 1 \end{pmatrix}. \end{aligned} \tag{A.137}$$

Exercise A.18: Let Ω be a skew-symmetric $2n$-by-$2n$ matrix with entries $\omega_{ij} = -\omega_{ji}$. Define the *Pfaffian* of Ω by

$$\operatorname{Pf}\Omega = \frac{1}{2^n n!}\sum \epsilon_{i_1 i_2 \ldots i_{2n}}\omega_{i_1 i_2}\omega_{i_3 i_4}\ldots\omega_{i_{2n-1}i_{2n}}.$$

Show that $\mathrm{Pf}\,(\mathbf{M}^{\mathrm{T}}\Omega\mathbf{M}) = \det(\mathbf{M})\,\mathrm{Pf}\,(\Omega)$. By reducing Ω to a suitable canonical form, show that $(\mathrm{Pf}\,\Omega)^2 = \det\Omega$.

Exercise A.19: Let $\omega(\mathbf{x},\mathbf{y})$ be a non-degenerate skew-symmetric bilinear form on $\mathbb{R}^{2n}$, and $\mathbf{x}_1,\ldots\mathbf{x}_{2n}$ a set of vectors. Prove *Weyl's identity*

$$\mathrm{Pf}\,(\Omega)\det|\mathbf{x}_1,\ldots\mathbf{x}_{2n}| = \frac{1}{2^n n!}\sum \epsilon_{i_1,\ldots,i_{2n}}\omega(\mathbf{x}_{i_1},\mathbf{x}_{i_2})\cdots\omega(\mathbf{x}_{i_{2n-1}},\mathbf{x}_{i_{2n}}).$$

Here $\det|\mathbf{x}_1,\ldots\mathbf{x}_{2n}|$ is the determinant of the matrix whose rows are the $\mathbf{x}_i$ and Ω is the matrix corresponding to the form ω.

Now let $M:\mathbb{R}^{2n}\to\mathbb{R}^{2n}$ be a linear map. Show that

$$\begin{aligned}&\mathrm{Pf}\,(\Omega)\,(\det M)\det|\mathbf{x}_1,\ldots\mathbf{x}_{2n}|\\ &\quad= \frac{1}{2^n n!}\sum \epsilon_{i_1,\ldots,i_{2n}}\omega(M\mathbf{x}_{i_1},M\mathbf{x}_{i_2})\cdots\omega(M\mathbf{x}_{i_{2n-1}},M\mathbf{x}_{i_{2n}}).\end{aligned}$$

Deduce that if $\omega(M\mathbf{x},M\mathbf{y}) = \omega(\mathbf{x},\mathbf{y})$ for all vectors $\mathbf{x},\mathbf{y}$, then $\det M = 1$. The set of such matrices M that preserve ω compose the *symplectic group* $\mathrm{Sp}(2n,\mathbb{R})$.

Appendix B
Fourier series and integrals

Fourier series and Fourier integral representations are the most important examples of the expansion of a function in terms of a complete orthonormal set. The material in this appendix reviews features peculiar to these special cases, and is intended to complement the general discussion of orthogonal series in Chapter 2.

B.1 Fourier series

A function defined on a finite interval may be expanded as a Fourier *series*.

B.1.1 Finite Fourier series

Suppose we have measured $f(x)$ in the interval $[0, L]$, but only at the discrete set of points $x = na$, where a is the sampling interval and $n = 0, 1, \ldots, N-1$, with $Na = L$. We can then represent our data $f(na)$ by a *finite* Fourier series. This representation is based on the geometric sum

$$\sum_{m=0}^{N-1} e^{ik_m(n'-n)a} = \frac{e^{2\pi i(n-n')a} - 1}{e^{2\pi i(n'-n)a/N} - 1}, \tag{B.1}$$

where $k_m \equiv 2\pi m/Na$. For integer n, and n', the expression on the right-hand side of (B.1) is zero unless $n' - n'$ is an integer multiple of N, when it becomes indeterminate. In this case, however, each term on the left-hand side is equal to unity, and so their sum is equal to N. If we restrict n and n' to lie between 0 and $N-1$, we have

$$\sum_{m=0}^{N-1} e^{ik_m(n'-n)a} = N\delta_{n'n}. \tag{B.2}$$

Inserting (B.2) into the formula

$$f(na) = \sum_{n'=0}^{N-1} f(n'a)\,\delta_{n'n} \tag{B.3}$$

shows that

$$f(na) = \sum_{m=0}^{N-1} a_m e^{-ik_m na}, \quad \text{where} \quad a_m \equiv \frac{1}{N}\sum_{n=0}^{N-1} f(na) e^{ik_m na}. \tag{B.4}$$

This is the finite Fourier representation.

When $f(na)$ is real, it is convenient to make the k_m sum symmetric about $k_m = 0$ by taking $N = 2M + 1$ and setting the summation limits to be $\pm M$. The finite geometric sum then becomes

$$\sum_{m=-M}^{M} e^{im\theta} = \frac{\sin(2M+1)\theta/2}{\sin\theta/2}. \tag{B.5}$$

We set $\theta = 2\pi(n' - n)/N$ and use the same tactics as before to deduce that

$$f(na) = \sum_{m=-M}^{M} a_m \, e^{-ik_m na}, \tag{B.6}$$

where again $k_m = 2\pi m/L$, with $L = Na$, and

$$a_m = \frac{1}{N} \sum_{n=0}^{2M} f(na) \, e^{ik_m na}. \tag{B.7}$$

In this form it is manifest that f being real both implies and is implied by $a_{-m} = a_m^*$.

These finite Fourier expansions are algebraic identities. No limits have to be taken, and so no restrictions need be placed on $f(na)$ for them to be valid. They are all that is needed for processing experimental data.

Although the initial $f(na)$ was defined only for the finite range $0 \leq n \leq N - 1$, the Fourier sum (B.4) or (B.7) is defined for any n, and so extends f to a periodic function of n with period N.

B.1.2 Continuum limit

Now we wish to derive a Fourier representation for functions defined everywhere on the interval $[0, L]$, rather than just at the sampling points. The natural way to proceed is to build on the results from the previous section by replacing the interval $[0, L]$ with a discrete lattice of $N = 2M + 1$ points at $x = na$, where a is a small lattice spacing which we ultimately take to zero. For any non-zero a the continuum function $f(x)$ is thus replaced by the finite set of numbers $f(na)$. If we stand back and blur our vision so that we can no longer perceive the individual lattice points, a plot of this discrete function will look little different from the original continuum $f(x)$. In other words, provided that f is slowly varying on the scale of the lattice spacing, $f(an)$ can be regarded as a smooth function of $x = an$.

The basic "integration rule" for such smooth functions is that

$$a \sum_n f(an) \to \int f(an) \, a \, dn \to \int f(x) \, dx, \tag{B.8}$$

as a becomes small. A sum involving a Kronecker δ will become an integral containing a Dirac delta function:

$$a \sum_n f(na) \frac{1}{a} \delta_{nm} = f(ma) \to \int f(x)\, \delta(x-y)\, dx = f(y). \tag{B.9}$$

We can therefore think of the delta function as arising from

$$\frac{\delta_{nn'}}{a} \to \delta(x-x'). \tag{B.10}$$

In particular, the divergent quantity $\delta(0)$ (in x space) is obtained by setting $n = n'$, and can therefore be understood to be the reciprocal of the lattice spacing, or, equivalently, the number of lattice points per unit volume.

Now we take the formal continuum limit of (B.7) by letting $a \to 0$ and $N \to \infty$ while keeping their product $Na = L$ fixed. The *finite* Fourier representation

$$f(na) = \sum_{m=-M}^{M} a_m e^{-\frac{2\pi i m}{Na} na} \tag{B.11}$$

now becomes an *infinite* series

$$f(x) = \sum_{m=-\infty}^{\infty} a_m\, e^{-2\pi i m x/L}, \tag{B.12}$$

whereas

$$a_m = \frac{a}{Na} \sum_{n=0}^{N-1} f(na) e^{\frac{2\pi i m}{Na} na} \to \frac{1}{L} \int_0^L f(x) e^{2\pi i m x/L}\, dx. \tag{B.13}$$

The series (B.12) is the Fourier expansion for a function on a finite interval. The sum is equal to $f(x)$ in the interval $[0, L]$. Outside, it produces L-periodic translates of the original f.

This Fourier expansion (B.12), (B.13) is the same series that we would obtain by using the $L^2[0, L]$ orthonormality

$$\frac{1}{L} \int_0^L e^{2\pi i m x/L}\, e^{-2\pi i n x/L}\, dx = \delta_{nm}, \tag{B.14}$$

and using the methods of Chapter 2. The arguments adduced there, however, guarantee convergence only in the L^2 sense. While our present "continuum limit" derivation is only heuristic, it does suggest that for reasonably behaved periodic functions f the Fourier series (B.12) converges *pointwise* to $f(x)$. A continuous periodic function possessing a continuous first derivative is sufficiently "well-behaved" for pointwise convergence.

Furthermore, if the function f is smooth then the convergence is uniform. This is useful to know, but we often desire a Fourier representation for a function with discontinuities. A stronger result is that if f is *piecewise continuous* in $[0, L]$ – i.e., continuous with the exception of at most a finite number of discontinuities – and its first derivative is *also* piecewise continuous, then the Fourier series will converge pointwise (but not uniformly[1]) to $f(x)$ at points where $f(x)$ is continuous, and to its average

$$F(x) = \frac{1}{2} \lim_{\epsilon \to 0} \{f(x+\epsilon) + f(x-\epsilon)\} \tag{B.15}$$

at those points where $f(x)$ has jumps. In Section B.3.2 we shall explain why the series converges to this average, and examine the nature of this convergence.

Most functions of interest to engineers are piecewise continuous, and this result is then all that they require. In physics, however, we often have to work with a broader class of functions, and so other forms of convergence become relevant. In quantum mechanics, in particular, the probability interpretation of the wavefunction requires only convergence in the L^2 sense, and this demands no smoothness properties at all – the Fourier representation converging to f whenever the L^2 norm $\|f\|^2$ is finite.

Half-range Fourier series

The exponential series

$$f(x) = \sum_{m=-\infty}^{\infty} a_m \, e^{-2\pi i m x/L} \tag{B.16}$$

can be re-expressed as the trigonometric sum

$$f(x) = \frac{1}{2} A_0 + \sum_{m=1}^{\infty} \{A_m \cos(2\pi m x/L) + B_m \sin(2\pi m x/L)\}, \tag{B.17}$$

where

$$\begin{aligned} A_m &= \begin{cases} 2a_0 & m = 0, \\ a_m + a_{-m}, & m > 0, \end{cases} \\ B_m &= i(a_{-m} - a_m). \end{aligned} \tag{B.18}$$

This is called a *full-range* trigonometric Fourier series for functions defined on $[0, L]$. In Chapter 2 we expanded functions in series containing only sines. We can expand any function $f(x)$ defined on a finite interval as such a *half-range* Fourier series. To do this, we regard the given domain of $f(x)$ as being the half interval $[0, L/2]$ (hence the name).

[1] If a sequence of continuous functions converges uniformly, then its limit function is continuous.

We then extend $f(x)$ to a function on the whole of $[0, L]$ and expand as usual. If we extend $f(x)$ by setting $f(x + L/2) = -f(x)$ then the A_m are all zero and we have

$$f(x) = \sum_{m=1}^{\infty} B_m \sin(2\pi mx/L), \quad x \in [0, L/2], \tag{B.19}$$

where

$$B_m = \frac{4}{L} \int_0^{L/2} f(x) \sin(2\pi mx/L)\, dx. \tag{B.20}$$

Alternatively, we may extend the range of definition by setting $f(x+L/2) = f(L/2-x)$. In this case it is the B_m that become zero and we have

$$f(x) = \frac{1}{2}A_0 + \sum_{m=1}^{\infty} A_m \cos(2\pi mx/L), \quad x \in [0, L/2], \tag{B.21}$$

with

$$A_m = \frac{4}{L} \int_0^{L/2} f(x) \cos(2\pi mx/L)\, dx. \tag{B.22}$$

The difference between a full-range and a half-range series is therefore seen principally in the continuation of the function outside its initial interval of definition. A full range series repeats the function periodically. A half-range sine series changes the sign of the continued function each time we pass to an adjacent interval, whilst the half-range cosine series reflects the function as if each interval endpoint were a mirror.

B.2 Fourier integral transforms

When the function we wish to represent is defined on the entirety of $\mathbb{R}$ then we must use the Fourier *integral* representation.

B.2.1 Inversion formula

We formally obtain the Fourier integral representation from the Fourier series for a function defined on $[-L/2, L/2]$. Start from

$$f(x) = \sum_{m=-\infty}^{\infty} a_m\, e^{-\frac{2\pi im}{L}x}, \tag{B.23}$$

$$a_m = \frac{1}{L} \int_{-L/2}^{L/2} f(x)\, e^{\frac{2\pi im}{L}x}\, dx, \tag{B.24}$$

and let L become large. The discrete $k_m = 2\pi m/L$ then merge into the continuous variable k and

$$\sum_{m=-\infty}^{\infty} \to \int_{-\infty}^{\infty} dm = L \int_{-\infty}^{\infty} \frac{dk}{2\pi}. \tag{B.25}$$

The product La_m remains finite, and becomes a function $\tilde{f}(k)$. Thus

$$f(x) = \int_{-\infty}^{\infty} \tilde{f}(k)\, e^{-ikx} \frac{dk}{2\pi}, \tag{B.26}$$

$$\tilde{f}(k) = \int_{-\infty}^{\infty} f(x)\, e^{ikx}\, dx. \tag{B.27}$$

This is the Fourier integral transform and its inverse.

It is good practice when doing Fourier transforms in physics to treat x and k asymmetrically: always put the 2π's with the dk's. This is because, as (B.25) shows, $dk/2\pi$ has the physical meaning of the number of Fourier modes per unit (spatial) volume with wavenumber between k and $k + dk$.

The Fourier representation of the Dirac delta function is

$$\delta(x - x') = \int_{-\infty}^{\infty} \frac{dk}{2\pi} e^{ik(x-x')}. \tag{B.28}$$

Suppose we put $x = x'$. Then "$\delta(0)$", which we earlier saw can be interpreted as the inverse lattice spacing, and hence the density of lattice points, is equal to $\int_{-\infty}^{\infty} \frac{dk}{2\pi}$. This is the total number of Fourier modes per unit length.

Exchanging x and k in the integral representation of $\delta(x - x')$ gives us the Fourier representation for $\delta(k - k')$:

$$\int_{-\infty}^{\infty} e^{i(k-k')x}\, dx = 2\pi\, \delta(k - k'). \tag{B.29}$$

Thus $2\pi\delta(0)$ (in k space), although mathematically divergent, has the physical meaning $\int dx$, the volume of the system. It is good practice to put a 2π with each $\delta(k)$ because this combination has a direct physical interpretation.

Take care to note that the symbol $\delta(0)$ has a very different physical interpretation depending on whether δ is a delta function in x or in k space.

Parseval's identity

Note that with the Fourier transform pair defined as

$$\tilde{f}(k) = \int_{-\infty}^{\infty} e^{ikx} f(x)\, dx \tag{B.30}$$

$$f(x) = \int_{-\infty}^{\infty} e^{-ikx} \tilde{f}(k) \frac{dk}{2\pi}, \tag{B.31}$$

Parseval's theorem takes the form

$$\int_{-\infty}^{\infty} |f(x)|^2 \, dx = \int_{-\infty}^{\infty} |\tilde{f}(k)|^2 \frac{dk}{2\pi}. \tag{B.32}$$

Parseval's theorem tells us that the Fourier transform is a unitary map from $L^2(\mathbb{R}) \to L^2(\mathbb{R})$.

B.2.2 The Riemann–Lebesgue lemma

There is a reciprocal relationship between the rates at which a function and its Fourier transform decay at infinity. The more rapidly the function decays, the more high-frequency modes it must contain – and hence the slower the decay of its Fourier transform. Conversely, the smoother a function the fewer high-frequency modes it contains and the faster the decay of its transform. Quantitative estimates of this version of Heisenberg's uncertainty principle are based on the *Riemann–Lebesgue lemma*.

Recall that a function f is in $L^1(\mathbb{R})$ if it is integrable (this condition excludes the delta function) and goes to zero at infinity sufficiently rapidly that

$$\|f\|_1 \equiv \int_{-\infty}^{\infty} |f| \, dx < \infty. \tag{B.33}$$

If $f \in L^1(\mathbb{R})$ then its Fourier transform

$$\tilde{f}(k) = \int_{-\infty}^{\infty} f(x) e^{ikx} \, dx \tag{B.34}$$

exists, is a continuous function of k and

$$|\tilde{f}(k)| \le \|f\|_1. \tag{B.35}$$

The Riemann–Lebesgue lemma asserts that if $f \in L^1(\mathbb{R})$ then

$$\lim_{k \to \infty} \tilde{f}(k) = 0. \tag{B.36}$$

We will not give the proof. For f integrable in the Riemann sense, it is not difficult, being almost a corollary of the definition of the Riemann integral. We must point out, however, that the "$|\ldots|$" modulus sign is essential in the $L^1(\mathbb{R})$ condition. For example, the integral

$$I = \int_{-\infty}^{\infty} \sin(x^2) \, dx \tag{B.37}$$

is convergent, but only because of extensive cancellations. The $L^1(\mathbb{R})$ norm

$$\int_{-\infty}^{\infty} |\sin(x^2)| \, dx \tag{B.38}$$

is *not* finite, and whereas the Fourier transform of $\sin(x^2)$, i.e.

$$\int_{-\infty}^{\infty} \sin(x^2)\, e^{ikx}\, dx = \sqrt{\pi}\, \cos\left(\frac{k^2+\pi}{4}\right), \tag{B.39}$$

is also convergent, it does not decay to zero as k grows large.

The Riemann–Lebesgue lemma tells us that the Fourier transform maps $L^1(\mathbb{R})$ into $C_\infty(\mathbb{R})$, the latter being the space of continuous functions vanishing at infinity. Be careful: this map is only *into* and not *onto*. The inverse Fourier transform of a function vanishing at infinity does not necessarily lie in $L^1(\mathbb{R})$.

We link the smoothness of $f(x)$ to the rapid decay of $\widetilde{f}(k)$, by combining Riemann–Lebesgue with integration by parts. Suppose that both f and f' are in $L^1(\mathbb{R})$. Then

$$\widetilde{[f']}(k) \equiv \int_{-\infty}^{\infty} f'(x)\, e^{ikx}\, dx = -ik \int_{-\infty}^{\infty} f(x)\, e^{ikx}\, dx = -ik\widetilde{f}(k) \tag{B.40}$$

tends to zero. (No boundary terms arise from the integration by parts because for both f and f' to be in $L^1(\mathbb{R})$ the function f must tend to zero at infinity.) Since $k\widetilde{f}(k)$ tends to zero, $\widetilde{f}(k)$ itself must go to zero faster than $1/k$. We can continue in this manner and see that each additional derivative of f that lies in $L^1(\mathbb{R})$ buys us an extra power of $1/k$ in the decay rate of $\widetilde{f}$ at infinity. If any derivative possesses a jump discontinuity, however, *its* derivative will contain a delta function, and a delta function is not in $L^1(\mathbb{R})$. Thus, if n is the largest integer for which $k^n\widetilde{f}(k) \to 0$ we may expect $f^{(n)}(x)$ to be somewhere discontinuous. For example, the function $f(x) = e^{-|x|}$ has a first derivative that lies in $L^1(\mathbb{R})$, but this derivative is discontinuous. The Fourier transform $\widetilde{f}(k) = 2/(1+k^2)$ therefore decays as $1/k^2$, but no faster.

B.3 Convolution

Suppose that $f(x)$ and $g(x)$ are functions on the real line $\mathbb{R}$. We define their *convolution* $f * g$, when it exists, by

$$[f * g](x) \equiv \int_{-\infty}^{\infty} f(x-\xi)\, g(\xi)\, d\xi\,. \tag{B.41}$$

A change of variable $\xi \to x - \xi$ shows that, despite the apparently asymmetric treatment of f and g in the definition, the $*$ product obeys $f * g = g * f$.

B.3.1 The convolution theorem

Now, let $\widetilde{f}(k)$ denote the Fourier transforms of f, i.e.

$$\widetilde{f}(k) = \int_{-\infty}^{\infty} e^{ikx} f(x)\, dx. \tag{B.42}$$

We claim that

$$\widetilde{[f * g]} = \tilde{f}\,\tilde{g}. \tag{B.43}$$

The following computation shows that this claim is correct:

$$\begin{aligned}
\widetilde{[f * g]}(k) &= \int_{-\infty}^{\infty} e^{ikx} \left(\int_{-\infty}^{\infty} f(x-\xi)\, g(\xi)\, d\xi \right) dx \\
&= \int_{-\infty}^{\infty} \int_{-\infty}^{\infty} e^{ikx} f(x-\xi)\, g(\xi)\, d\xi\, dx \\
&= \int_{-\infty}^{\infty} \int_{-\infty}^{\infty} e^{ik(x-\xi)}\, e^{ik\xi}\, f(x-\xi)\, g(\xi)\, d\xi\, dx \\
&= \int_{-\infty}^{\infty} \int_{-\infty}^{\infty} e^{ikx'} e^{ik\xi}\, f(x')\, g(\xi)\, d\xi\, dx' \\
&= \left(\int_{-\infty}^{\infty} e^{ikx'} f(x')\, dx' \right) \left(\int_{-\infty}^{\infty} e^{ik\xi} g(\xi)\, d\xi \right) \\
&= \tilde{f}(k)\tilde{g}(k).
\end{aligned} \tag{B.44}$$

Note that we have freely interchanged the order of integrations. This is not always permissible, but it is allowed if $f, g \in L^1(\mathbb{R})$, in which case $f * g$ is also in $L^1(\mathbb{R})$.

B.3.2 Apodization and Gibbs' phenomenon

The convolution theorem is useful for understanding what happens when we truncate a Fourier series at a finite number of terms, or cut off a Fourier integral at a finite frequency or wavenumber.

Consider, for example, the cut-off Fourier integral representation

$$f_\Lambda(x) \equiv \frac{1}{2\pi} \int_{-\Lambda}^{\Lambda} \tilde{f}(k) e^{-ikx}\, dk, \tag{B.45}$$

where $\tilde{f}(k) = \int_{-\infty}^{\infty} f(x)\, e^{ikx}\, dx$ is the Fourier transform of f. We can write this as

$$f_\Lambda(x) = \frac{1}{2\pi} \int_{-\infty}^{\infty} \theta_\Lambda(k) \tilde{f}(k)\, e^{-ikx}\, dk \tag{B.46}$$

where $\theta_\Lambda(k)$ is unity if $|k| < \Lambda$ and zero otherwise. Written this way, the Fourier transform of f_Λ becomes the product of the Fourier transform of the original f with θ_Λ. The function f_Λ itself is therefore the convolution

$$f_\Lambda(x) = \int_{-\infty}^{\infty} \delta_\Lambda^{\mathrm{D}}(x-\xi) f(\xi)\, d\xi \tag{B.47}$$

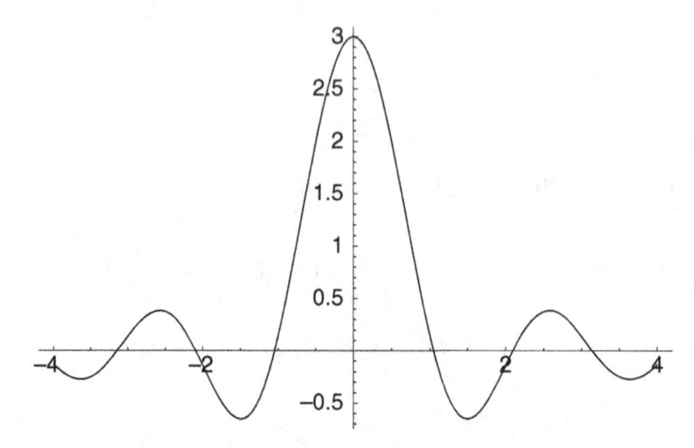

Figure B.1 A plot of $\pi\delta^{\mathrm{D}}_{\Lambda}(x)$ for $\Lambda = 3$.

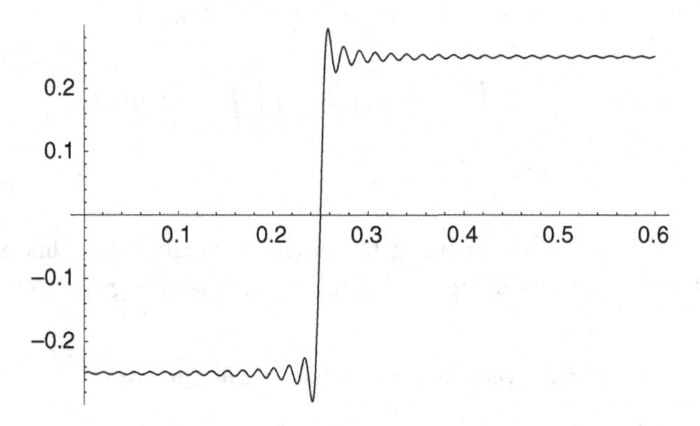

Figure B.2 The Gibbs phenomenon: a Fourier reconstruction of a piecewise constant function that jumps discontinuously from $y = -0.25$ to $+0.25$ at $x = 0.25$.

of f with

$$\delta^{\mathrm{D}}_{\Lambda}(x) \equiv \frac{1}{\pi}\frac{\sin(\Lambda x)}{x} = \frac{1}{2\pi}\int_{-\infty}^{\infty} \theta_{\Lambda}(k)e^{-ikx}\,dk, \tag{B.48}$$

which is the inverse Fourier transform of $\theta_{\Lambda}(x)$. We see that $f_{\Lambda}(x)$ is a kind of local average of the values of $f(x)$ smeared by the approximate delta function $\delta^{\mathrm{D}}_{\Lambda}(x)$; see Figure B.1. The superscript D stands for "Dirichlet", and $\delta^{\mathrm{D}}_{\Lambda}(x)$ is known as the *Dirichlet kernel*.

When $f(x)$ can be treated as a constant on the scale ($\approx 2\pi/\Lambda$) of the oscillation in $\delta^{\mathrm{D}}_{\Lambda}(x)$, all that matters is that $\int_{-\infty}^{\infty}\delta^{\mathrm{D}}_{\Lambda}(x)\,dx = 1$, and so $f_{\Lambda}(x) \approx f(x)$. This is the case if $f(x)$ is smooth and Λ is sufficiently large. However, if $f(x)$ possesses a discontinuity at x_0, say, then we can never treat it as a constant and the rapid oscillations in $\delta^{\mathrm{D}}_{\Lambda}(x)$ cause a "ringing" in $f_{\Lambda}(x)$ whose amplitude does not decrease (although the *width* of the region surrounding x_0 in which the effect is noticeable will decrease) as Λ grows. This ringing is known as *Gibbs' phenomenon* (Figure B.2).

The amplitude of the ringing is largest immediately on either side of the point of discontinuity, where it is about 9% of the jump in f. This magnitude is determined by the area under the central spike in $\delta^{\mathrm{D}}_{\Lambda}(x)$, which is

$$\frac{1}{\pi}\int_{-\pi/\Lambda}^{\pi/\Lambda}\frac{\sin(\Lambda x)}{x}\,dx = 1.18\ldots, \tag{B.49}$$

independent of Λ. For x *exactly* at the point of discontinuity, $f_\Lambda(x)$ receives equal contributions from both sides of the jump and hence converges to the average

$$\lim_{\Lambda\to\infty} f_\Lambda(x) = \frac{1}{2}\Big\{f(x_+) + f(x_-)\Big\}, \tag{B.50}$$

where $f(x_\pm)$ are the limits of f taken from the right and left, respectively. When $x = x_0 - \pi/\Lambda$, however, the central spike lies entirely to the left of the point of discontinuity and

$$\begin{aligned} f_\Lambda(x) &\approx \frac{1}{2}\{(1+1.18)f(x_-) + (1-1.18)f(x_+)\} \\ &\approx f(x_-) + 0.09\{f(x_-) - f(x_+)\}. \end{aligned} \tag{B.51}$$

Consequently, $f_\Lambda(x)$ overshoots its target $f(x_-)$ by approximately 9% of the discontinuity. Similarly when $x = x_0 + \pi/\Lambda$

$$f_\Lambda(x) \approx f(x_+) + 0.09\{f(x_+) - f(x_-)\}. \tag{B.52}$$

The ringing is a consequence of the abrupt truncation of the Fourier sum. If, instead of a sharp cutoff, we gradually de-emphasize the higher frequencies by the replacement

$$\widetilde{f}(k) \to \widetilde{f}(k)\, e^{-\alpha k^2/2} \tag{B.53}$$

then

$$\begin{aligned} f_\alpha(x) &= \frac{1}{2\pi}\int_{-\infty}^{\infty}\widetilde{f}(k)e^{-\alpha k^2}e^{-ikx}\,dk \\ &= \int_{-\infty}^{\infty}\delta^{\mathrm{G}}_{\alpha}(x-\xi)f(y)\,d\xi \end{aligned} \tag{B.54}$$

where

$$\delta^{\mathrm{G}}_{\alpha}(x) = \frac{1}{\sqrt{2\pi\alpha}}e^{-x^2/2\alpha}, \tag{B.55}$$

is a non-oscillating Gaussian approximation to a delta function. The effect of this convolution is to smooth out, or *mollify*, the original f, resulting in a C^∞ function. As

α becomes small, the limit of $f_\alpha(x)$ will again be the local average of $f(x)$, so at a discontinuity f_α will converge to the mean $\frac{1}{2}\{f(x_+)+f(x_-)\}$.

When reconstructing a signal from a finite range of its Fourier components – for example from the output of an aperture-synthesis radio-telescope – it is good practice to smoothly suppress the higher frequencies in such a manner. This process is called *apodizing* (i.e. cutting off the feet of) the data. If we fail to apodize then any interesting sharp feature in the signal will be surrounded by "diffraction ring" artefacts.

Exercise B.1: Suppose that we exponentially suppress the higher frequencies by multiplying the Fourier amplitude $\tilde{f}(k)$ by $e^{-\epsilon|k|}$. Show that the original signal is smoothed by convolution with a *Lorentzian* approximation to a delta function

$$\delta_\epsilon^{\mathrm{L}}(x-\xi) = \frac{1}{\pi}\frac{\epsilon}{\epsilon^2+(x-\xi)^2}.$$

Observe that

$$\lim_{\epsilon\to 0}\delta_\epsilon^{\mathrm{L}}(x) = \delta(x).$$

Exercise B.2: Consider the apodized Fourier series

$$f_r(\theta) = \sum_{n=-\infty}^{\infty} a_n r^{|n|} e^{in\theta},$$

where the parameter r lies in the range $0 < r < 1$, and the coefficients are

$$a_n \equiv \frac{1}{2\pi}\int_0^{2\pi} e^{-in\theta} f(\theta)\, d\theta.$$

Assuming that it is legitimate to interchange the order of the sum and integral, show that

$$\begin{aligned} f_r(\theta) &= \int_0^{2\pi} \delta_r^{\mathrm{P}}(\theta-\theta') f(\theta') d\theta' \\ &\equiv \frac{1}{2\pi}\int_0^{2\pi}\left(\frac{1-r^2}{1-2r\cos(\theta-\theta')+r^2}\right) f(\theta') d\theta'. \end{aligned}$$

Here the superscript P stands for Poisson because $\delta_r^{\mathrm{P}}(\theta)$ is the *Poisson kernel* that solves the Dirichlet problem in the unit disc. Show that $\delta_r^{\mathrm{P}}(\theta)$ tends to a delta function as $r \to 1$ from below.

Exercise B.3: *The periodic Hilbert transform.* Show that in the limit $r \to 1$ the sum

$$\sum_{n=-\infty}^{\infty} \operatorname{sgn}(n) e^{in\theta} r^{|n|} = \frac{re^{i\theta}}{1-re^{i\theta}} - \frac{re^{-i\theta}}{1-re^{-i\theta}}, \quad 0 < r < 1$$

becomes the principal-part distribution

$$P\left(i\cot\left(\frac{\theta}{2}\right)\right).$$

Let $f(\theta)$ be a smooth function on the unit circle, and define its *Hilbert transform* $\mathcal{H}f$ to be

$$(\mathcal{H}f)(\theta) = \frac{1}{2\pi}P\int_0^{2\pi} f(\theta')\cot\left(\frac{\theta-\theta'}{2}\right)d\theta'.$$

Show that the original function can be recovered from $(\mathcal{H}f)(\theta)$, together with knowledge of the angular average $\bar{f} = \int_0^{2\pi} f(\theta)\,d\theta/2\pi$, as

$$\begin{aligned} f(\theta) &= -\frac{1}{2\pi}P\int_0^{2\pi}(\mathcal{H}f)(\theta')\cot\left(\frac{\theta-\theta'}{2}\right)d\theta' + \frac{1}{2\pi}\int_0^{2\pi} f(\theta')\,d\theta' \\ &= -(\mathcal{H}^2 f))(\theta) + \bar{f}. \end{aligned}$$

Exercise B.4: Find a closed-form expression for the sum

$$\sum_{n=-\infty}^{\infty} |n|\, e^{in\theta} r^{2|n|}, \quad 0 < r < 1.$$

Now let $f(\theta)$ be a smooth function defined on the unit circle and

$$a_n = \frac{1}{2\pi}\int_0^{2\pi} f(\theta)e^{-in\theta}\,d\theta$$

its n-th Fourier coefficient. By taking a limit $r \to 1$, show that

$$\pi\sum_{n=-\infty}^{\infty}|n|\,a_n a_{-n} = \frac{\pi}{4}\int_0^{2\pi}\int_0^{2\pi}[f(\theta)-f(\theta')]^2\operatorname{cosec}^2\left(\frac{\theta-\theta'}{2}\right)\frac{d\theta}{2\pi}\frac{d\theta'}{2\pi},$$

both the sum and integral being convergent. Show that these last two expressions are equal to

$$\frac{1}{2}\iint_{r<1}|\nabla\varphi|^2\, r dr d\theta,$$

where $\varphi(r,\theta)$ is the function harmonic in the unit disc, whose boundary value is $f(\theta)$.

Exercise B.5: Let $\widetilde{f}(k)$ be the Fourier transform of the smooth real function $f(x)$. Take a suitable limit in the previous problem to show that

$$S[f] \equiv \frac{1}{4\pi}\int_{-\infty}^{\infty}\int_{-\infty}^{\infty}\left\{\frac{f(x)-f(x')}{x-x'}\right\}^2 dx dx' = \frac{1}{2}\int_{-\infty}^{\infty}|k|\left|\widetilde{f}(k)\right|^2\frac{dk}{2\pi}.$$

Exercise B.6: By taking a suitable limit in Exercise B.3 show that, when acting on smooth functions f such that $\int_{-\infty}^{\infty}|f|\,dx$ is finite, we have $\mathcal{H}(\mathcal{H}f) = -f$, where

$$(\mathcal{H}f)(x) = \frac{P}{\pi}\int_{-\infty}^{\infty}\frac{f(x')}{x-x'}\,dx'$$

defines the Hilbert transform of a function on the real line. (Because $\mathcal{H}$ gives zero when acting on a constant, some condition, such as $\int_{-\infty}^{\infty}|f|\,dx$ being finite, is necessary if $\mathcal{H}$ is to be invertible.)

B.4 The Poisson summation formula

Suppose that $f(x)$ is a smooth function that tends rapidly to zero at infinity. Then the series

$$F(x) = \sum_{n=-\infty}^{\infty} f(x+nL) \tag{B.56}$$

converges to a smooth function of period L. It therefore has a Fourier expansion

$$F(x) = \sum_{m=-\infty}^{\infty} a_m\, e^{-2\pi imx/L}. \tag{B.57}$$

We can compute the Fourier coefficients a_m by integrating term-by-term

$$\begin{aligned} a_m &= \frac{1}{L}\int_0^L F(x)\, e^{2\pi imx/L}\,dx \\ &= \frac{1}{L}\sum_{n=-\infty}^{\infty}\int_0^L f(x+nL)\, e^{2\pi imx/L}\,dx \\ &= \frac{1}{L}\int_{-\infty}^{\infty} f(x)\, e^{2\pi imx/L}\,dx \\ &= \frac{1}{L}\widetilde{f}(2\pi m/L). \end{aligned} \tag{B.58}$$

Thus

$$\sum_{n=-\infty}^{\infty} f(x+nL) = \frac{1}{L}\sum_{m=-\infty}^{\infty}\widetilde{f}(2\pi m/L)e^{-2\pi imx/L}. \tag{B.59}$$

When we set $x=0$, this last equation becomes

$$\sum_{n=-\infty}^{\infty} f(nL) = \frac{1}{L}\sum_{m=-\infty}^{\infty}\widetilde{f}(2\pi m/L). \tag{B.60}$$

The equality of this pair of doubly infinite sums is known as the *Poisson summation formula*.

Example: As the Fourier transform of a Gaussian is another Gaussian, the Poisson formula with $L = 1$ applied to $f(x) = \exp(-\kappa x^2)$ gives

$$\sum_{m=-\infty}^{\infty} e^{-\kappa m^2} = \sqrt{\frac{\pi}{\kappa}} \sum_{m=-\infty}^{\infty} e^{-m^2\pi^2/\kappa}, \tag{B.61}$$

and (rather more usefully) applied to $\exp(-\frac{1}{2}tx^2 + ix\theta)$ gives

$$\sum_{n=-\infty}^{\infty} e^{-\frac{1}{2}tn^2+in\theta} = \sqrt{\frac{2\pi}{t}} \sum_{n=-\infty}^{\infty} e^{-\frac{1}{2t}(\theta+2\pi n)^2}. \tag{B.62}$$

The last identity is known as *Jacobi's imaginary transformation*. It reflects the equivalence of the eigenmode expansion and the method-of-images solution of the diffusion equation

$$\frac{1}{2}\frac{\partial^2\varphi}{\partial x^2} = \frac{\partial\varphi}{\partial t} \tag{B.63}$$

on the unit circle. Notice that when t is small the sum on the right-hand side converges very slowly, whereas the sum on the left converges very rapidly. The opposite is true for large t. The conversion of a slowly converging series into a rapidly converging one is a standard application of the Poisson summation formula. It is the prototype of many *duality maps* that exchange a physical model with a large coupling constant for one with weak coupling.

If we take the limit $t \to 0$ in (B.62), the right-hand side approaches a sum of delta functions, and so gives us the useful identity

$$\frac{1}{2\pi}\sum_{n=-\infty}^{\infty} e^{inx} = \sum_{n=-\infty}^{\infty} \delta(x + 2\pi n). \tag{B.64}$$

The right-hand side of (B.64) is sometimes called the "Dirac comb".

Gauss sums

The Poisson sum formula

$$\sum_{m=-\infty}^{\infty} e^{-\kappa m^2} = \sqrt{\frac{\pi}{\kappa}} \sum_{m=-\infty}^{\infty} e^{-m^2\pi^2/\kappa} \tag{B.65}$$

remains valid for complex κ, provided that $\mathrm{Re}\,\kappa > 0$. We can therefore consider the special case

$$\kappa = i\pi\frac{p}{q} + \epsilon, \tag{B.66}$$

where ϵ is a positive real number and p and q are positive integers whose product pq we assume to be *even*. We investigate what happens to (B.65) as $\epsilon \to 0$.

The left-hand side of (B.65) can be decomposed into the double sum

$$\sum_{m=-\infty}^{\infty} \sum_{r=0}^{q-1} e^{-i\pi(p/q)(r+mq)^2} e^{-\epsilon(r+mq)^2}. \tag{B.67}$$

Because pq is even, each term in $e^{-i\pi(p/q)(r+mq)^2}$ is independent of m. At the same time, the small ϵ limit of the infinite sum

$$\sum_{m=-\infty}^{\infty} e^{-\epsilon(r+mq)^2}, \tag{B.68}$$

being a Riemann sum for the integral

$$\int_{-\infty}^{\infty} e^{-\epsilon q^2 m^2} dm = \frac{1}{q}\sqrt{\frac{\pi}{\epsilon}}, \tag{B.69}$$

becomes independent of r, and so a common factor of all terms in the finite sum over r.

If ϵ is small, we can make the replacement,

$$\kappa^{-1} = \frac{\epsilon - i\pi p/q}{\epsilon^2 + \pi^2 p^2/q^2} \to \frac{\epsilon - i\pi p/q}{\pi^2 p^2/q^2}, \tag{B.70}$$

after which, the right-hand side contains the double sum

$$\sum_{m=-\infty}^{\infty} \sum_{r=0}^{p-1} e^{i\pi(q/p)(r+mp)^2} e^{-\epsilon(q^2/p^2)(r+mp)^2}. \tag{B.71}$$

Again each term in $e^{i\pi(q/p)(r+mp)^2}$ is independent of m, and

$$\sum_{m=-\infty}^{\infty} e^{-\epsilon(q^2/p^2)(r+mp)^2} \to \int_{-\infty}^{\infty} e^{-\epsilon q^2 m^2} dm = \frac{1}{q}\sqrt{\frac{\pi}{\epsilon}} \tag{B.72}$$

becomes independent of r. Also

$$\lim_{\epsilon \to 0} \left\{ \sqrt{\frac{\pi}{\kappa}} \right\} = e^{-i\pi/4} \sqrt{\frac{q}{p}}. \tag{B.73}$$

Thus, after cancelling the common factor of $(1/q)\sqrt{\pi/\epsilon}$, we find that

$$\frac{1}{\sqrt{q}} \sum_{r=0}^{q-1} e^{-i\pi(p/q)r^2} = e^{-i\pi/4} \frac{1}{\sqrt{p}} \sum_{r=0}^{p-1} e^{i\pi(q/p)r^2}, \quad pq \text{ even}. \tag{B.74}$$

This Poisson-summation-like equality of *finite* sums is known as the *Landsberg–Schaar identity*. No purely algebraic proof is known.

Gauss considered the special case $p = 2$, in which case we get

$$\frac{1}{\sqrt{q}}\sum_{r=0}^{q-1} e^{-2\pi i r^2/q} = e^{-i\pi/4}\frac{1}{\sqrt{2}}(1 + e^{i\pi q/2}) \tag{B.75}$$

or, more explicitly

$$\sum_{r=0}^{q-1} e^{-2\pi i r^2/q} = \begin{cases} (1-i)\sqrt{q}, & q = 0 \pmod 4, \\ \sqrt{q}, & q = 1 \pmod 4, \\ 0, & q = 2 \pmod 4, \\ -i\sqrt{q}, & q = 3 \pmod 4. \end{cases} \tag{B.76}$$

The complex conjugate result is perhaps slightly prettier:

$$\sum_{r=0}^{q-1} e^{2\pi i r^2/q} = \begin{cases} (1+i)\sqrt{q}, & q = 0 \pmod 4, \\ \sqrt{q}, & q = 1 \pmod 4, \\ 0, & q = 2 \pmod 4, \\ i\sqrt{q}, & q = 3 \pmod 4. \end{cases} \tag{B.77}$$

Gauss used these sums to prove the law of quadratic reciprocity.

Exercise B.7: By applying the Poisson summation formula to the Fourier transform pair

$$f(x) = e^{-\epsilon|x|}e^{-ix\theta}, \quad \text{and} \quad \tilde{f}(k) = \frac{2\epsilon}{\epsilon^2 + (k-\theta)^2},$$

where $\epsilon > 0$, deduce that

$$\frac{\sinh \epsilon}{\cosh \epsilon - \cos(\theta - \theta')} = \sum_{n=-\infty}^{\infty} \frac{2\epsilon}{\epsilon^2 + (\theta - \theta' + 2\pi n)^2}. \tag{B.78}$$

Hence show that the Poisson kernel is equivalent to an infinite periodic sum of Lorentzians

$$\frac{1}{2\pi}\left(\frac{1 - r^2}{1 - 2r\cos(\theta - \theta') + r^2}\right) = -\frac{1}{\pi}\sum_{n=-\infty}^{\infty} \frac{\ln r}{(\ln r)^2 + (\theta - \theta' + 2\pi n)^2}.$$

References

This is a list of books mentioned by name in the text. They are not all in print, but should be found in any substantial physics library. The publication data refer to the most recently available printings.

R. Abrahams, J. E. Marsden, T. Ratiu, *Foundations of Mechanics* (Benjamin Cummings, 2nd edition, 1978).

V. I. Arnold, *Mathematical Methods of Classical Mechanics* (Springer Verlag, 2nd edition, 1997).

J. A. de Azcárraga, J. M. Izquierdo, *Lie groups, Lie Algebras, Cohomology and some Applications in Physics* (Cambridge University Press, 1995).

G. Baym, *Lectures on Quantum Mechanics* (Addison Wesley, 1969).

C. M. Bender, S. A. Orszag, *Advanced Mathematical Methods for Scientists and Engineers* (Springer Verlag, 1999).

G. Birkhoff, S. MacLane, *Survey of Modern Algebra* (A. K. Peters, 1997).

M. Born, E. Wolf, *Principles of Optics* (Cambridge University Press, 7th edition, 1999).

A. Erdélyi, W. Magnus, F. Oberhettinger, F. G. Tricomi, *Higher Transcendental Functions* (in 3 vols), and *Tables of Integral Transforms* (2 vols), known collectively as "The Bateman Manuscript Project" (McGraw-Hill, 1953).

F. G. Friedlander, M. Joshi, *Introduction to the Theory of Distributions* (Cambridge University Press, 2nd edition, 1999).

P. R. Halmos, *Finite Dimensional Vector Spaces* (Springer Verlag, 1993).

J. L. Lagrange, *Analytic Mechanics* (English translation, Springer Verlag, 2001).

L. D. Landau, E. M. Lifshitz, *Quantum Mechanics (Non-relativistic Theory)* (Pergamon Press, 1981).

M. J. Lighthill, *Generalized Functions* (Cambridge University Press, 1960).

M. J. Lighthill, *Waves in Fluids* (Cambridge University Press, 2002).

J. C. Maxwell, *A Treatise on Electricity and Magnetism* (Dover reprint edition, 1954).

C. W. Misner, K. S. Thorne, J. A. Wheeler, *Gravitation* (W. H. Freeman, 1973).

P. M. Morse, H. Feshbach, *Methods of Theoretical Physics* (McGraw-Hill, 1953).

N. I. Muskhelishvili, *Singular Integral Equations* (Dover, reprint of 2nd edition, 1992).

R. G. Newton, *Scattering Theory of Waves and Particles* (Dover, reprint of 2nd edition, 2002).

A. Perelomov, *Generalized Coherent States and their Applications* (Springer-Verlag, 1986).

M. Reed, B. Simon, *Methods of Modern Mathematical Physics* (4 vols.) (Academic Press, 1978–80).

L. I. Schiff, *Quantum Mechanics* (McGraw-Hill, 3rd edition, 1968).

M. Spivak, *A Comprehensive Introduction to Differential Geometry* (5 vols.) (Publish or Perish, 2nd edition, 1979).

I. Stackgold, *Boundary Value Problems of Mathematical Physics*, Vols. I and II (SIAM, 2000).

G. N. Watson, *A Treatise on the Theory of Bessel Functions* (Cambridge University Press, 2nd edition, 1955).

E. T. Whittaker, G. N. Watson, *A Course of Modern Analysis* (Cambridge University Press, 4th edition, 1996).

H. Weyl, *The Classical Groups* (Princeton University Press, reprint edition, 1997).

Index

Printed in the United States
By Bookmasters